Basic College Mathematics

SECOND EDITION

K. Elayn Martin-Gay

University of New Orleans

PRENTICE HALL
Upper Saddle River, New Jersey 07458

Library of Congress Cataloging-in-Publication Data

Martin-Gay, K. Elayn
    Basic college mathematics/K. Elayn Martin-Gay.–2nd ed.
      p. cm.
    Includes index.
    ISBN 0-13-067699-3 (pbk.: alk. paper)–ISBN 0-13-008747-5 (alk. paper)
    1. Mathematics.   I. Title.

QA39.3.M37 2003
510—dc21   2001058807

Executive Acquisition Editor: Karin E. Wagner
Editor in Chief: Christine Hoag
Project Manager: Mary Beckwith
Vice President/Director of Production and Manufacturing: David W. Riccardi
Executive Managing Editor: Kathleen Schiaparelli
Senior Managing Editor: Linda Mihatov Behrens
Production Management: Elm Street Publishing Services, Inc.
Production Assistant: Nancy Bauer
Manufacturing Buyer: Alan Fischer
Manufacturing Manager: Trudy Pisciotti
Executive Marketing Manager: Eilish Collins Main
Marketing Assistant: Annett Uebel
Development Editor: Kathy Sessa-Federico
Editor in Chief, Development: Carol Trueheart
Media Project Manager, Developmental Math: Audra J. Walsh
Editorial Assistant: Heather Balderson
Art Director: Maureen Eide
Assistant to the Art Director: John Christiana
Interior Designer: Circa 86
Cover Designer: Jack Robol
Art Editor: Thomas Benfatti
Managing Editor, Audio/Video Assets: Grace Hazeldine
Creative Director: Carole Anson
Director of Creative Services: Paul Belfanti
Photo Researcher: Diane Austin
Photo Editor: Beth Boyd
Cover Art: Seaform detail. Handblown glass, by Dale Chihuly. Photo by Terry Rishel.
Art Studio: Artworks:
        Senior Manager: Patricia Burns
        Production Manager: Ronda Whitson
        Production Technologies Manager: Matthew Haas
        Project Coordinator: Jessica Einsig
        Illustrators: Kathyrn Anderson, Dan Knopsnyder, Mark Landis
        Quality Assurance: Stacy Smith, Pamela Taylor, Timothy Nguyen
Compositor: Preparé/Emilcomp

© 2003, 1999 by Prentice-Hall, Inc.
Upper Saddle River, New Jersey 07458

Photo Credits appear on page I8, which constitutes a continuation of the copyright page.

Printed in the United States of America
10 9 8 7 6 5

ISBN: 0-13-067699-3 (paperback) 0-13-008747-5 (case bound)

Pearson Education Ltd.
Pearson Education Australia Pty., Limited
Pearson Education Singapore, Pte. Ltd.
Pearson Education North Asia Ltd.
Pearson Education Canada, Ltd.
Pearson Educacíon de Mexico, S.A. de C.V.
Pearson Education—Japan
Pearson Education Malaysia, Pte. Ltd.

CONTENTS

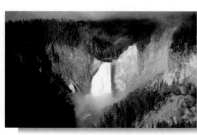

To my mother, Barbara M. Miller,
and her husband, Leo Miller,
and to the memory of my father,
Robert J. Martin

PREFACE

About This Book

Basic College Mathematics, Second Edition was written to provide a **solid foundation** in the basics of college mathematics, including the topics of whole numbers, fractions, decimals, ratio, and proportion, percent, and measurement as well as introductions to geometry, statistics and probability, and algebra topics. Specific care was taken to ensure that students have the most **up-to-date relevant** text preparation for their next mathematics course or for nonmathematical courses that require an understanding of basic mathematical concepts. I have tried to achieve this by writing a user-friendly text that is keyed to objectives and contains many worked-out examples. As suggested by the AMATYC Crossroads Document and the NCTM Standards (plus Addenda), real-life and real-data applications, data interpretation, conceptual understanding, problem solving, writing, cooperative learning, appropriate use of technology, mental mathematics, number sense, estimation, critical thinking, and geometric concepts are emphasized and integrated throughout the book.

The many factors that contributed to the success of the first edition have been retained. In preparing the Second Edition, I considered comments and suggestions of colleagues, students, and many users of the prior edition throughout the country.

Basic College Mathematics, Second Edition is part of a series of texts that can include *Introductory Algebra,* Second Edition; *Prealgebra,* Third Edition; *Intermediate Algebra,* Second Edition; and a combined text, *Algebra A Combined Approach*, Second Edition. Throughout the series pedagogical features are designed to develop student proficiency in algebra and problem solving, and to prepare students for future courses.

Key Pedagogical Features and Changes in the Second Edition

Readability and Connections   I have tried to make the writing style as clear as possible while still retaining the mathematical integrity of the content. When a new topic is presented, an effort has been made to relate the new ideas to those that students may already know. Constant reinforcement and connections within problem-solving strategies, data interpretation, geometry, patterns, graphs, and situations from everyday life can help students gradually master both new and old information. In addition, each section begins with a list of objectives covered in the section. Clear organization of section material based on objectives further enhances readability.

Problem-Solving Process   This is formally introduced in Chapter 1 with a four-step process that is integrated throughout the text. The four steps are **Understand**, **Translate**, **Solve**, and **Interpret**. The repeated use of these steps in a variety of examples shows their wide applicability. Reinforcing the steps can increase students' comfort level and confidence in tackling problems.

Applications and Connections   Every effort was made to include as many interesting and relevant real-life applications as possible throughout the text in both worked-out examples and exercise sets. In the Second Edition, the applications have been thoroughly revised and updated, and the number of applications has increased. The applications help to motivate students and strengthen their understanding of mathematics in the real world. They show connections to a wide range of fields including agriculture, allied health, anthropology, art, astronomy, biology, business, chemistry, construction, consumer affairs, earth science, education, entertainment, environmental issues, finance, geography, government, history, medicine, music, nutrition, physics, sports, travel, and weather. Many of the applications are based on recent real data. Sources for data include newspapers, magazines, publicly held companies, government agencies, special-interest groups, research organizations, and reference books. Opportunities for obtaining your own real data are also included. See the Applications Index on page xvi.

Practice Problems   Throughout the text, each worked-out example has a parallel Practice Problem placed next to the example in the margin. These invite students to be actively involved in the learning process before beginning the end-of-section exercise set. Practice Problems immediately reinforce a skill after it is developed. Answers appear at the bottom of the page for quick reference.

Concept Checks   These margin exercises are appropriately placed throughout the text. They allow students to gauge their grasp of an idea as it is being explained in the text. Concept Checks stress conceptual understanding at the point of use and help suppress misconceived notions before they start. Answers appear at the bottom of the page.

Increased Integration of Geometry Concepts   In addition to the traditional topics in basic math courses, this text contains a strong emphasis on problem solving and geometric concepts, which are integrated throughout. The geometry concepts presented are those most important to a student's understanding, and I have included many applications and exercises devoted to this topic. These are marked with the the geometry icon. Also, Chapter 8, Geometry, provides a focused treatment of the topics.

Helpful Hints   Helpful Hints contain practical advice on applying mathematical concepts. These are found throughout the text and strategically placed where students are most likely to need immediate reinforcement. Helpful Hints are highlighted for quick reference.

Visual Reinforcement of Concepts   The Second Edition contains a wealth of graphics, models, photographs, and illustrations to visually clarify and reinforce concepts. These include new and updated bar graphs, line graphs, calculator screens, application illustrations, and geometric figures.

Calculator Explorations   These optional explorations offer point-of-use intruction, through examples and exercises, on the proper use of scientific calculators as tools in the mathematical problem-solving process. Placed appropriately throughout the text, Calculator Explorations also reinforce concepts learned in the corresponding section and motivate discovery-based learning.

Additional exercises building on the skill developed in the Explorations may be found in exercise sets throughout the text. Exercise requiring a calculator are marked with the ▦ icon.

Study Skills Reminders   New Study Skills Reminder boxes are integrated throughout the text. They are strategically placed to constantly remind and encourage students as they hone their study skills. A new **Section 1.1**, Tips on Success in Mathematics, provides an overview of the Study Skills needed to succeed in math. These are reinforced by the Study Skills Reminder boxes throughout the text.

Focus On   Appropriately placed throughout each chapter, these are divided into Focus on Mathematical Connections, Focus on Business and Career, Focus on the Real World, and Focus on History. They are written to help students develop effective habits for engaging in investigations of other branches of mathematics, understanding the importance of mathematics in various careers and in the world of business, and seeing the relevance of mathematics in both the present and past through critical thinking exercises and group activities.

Chapter Highlights   Found at the end of each chapter, these contain key definitions, concepts, and examples to help students understand and retain what they have learned and help them organize their notes and study for tests.

Chapter Activity   These features occur once per chapter at the end of the chapter, often serving as a chapter wrap-up. For individual or group completion, the Chapter Activity, usually hands-on or data-based, complements and extends to concepts of the chapter, allowing students to make decisions and interpretations and to think and write about algebra.

Integrated Reviews   These "mid-chapter reviews" are appropriately placed once per chapter. Integrated Reviews allow students to review and assimilate the many different skills learned separately over several sections before moving on to related material in the chapter.

Pretests   Each chapter begins with a pretest that is designed to help students identify areas where they need to pay special attention in the upcoming chapter.

Chapter Review and Test   The end of each chapter contains a review of topics introduced in the chapter. The Chapter Review offers exercises that are keyed to sections of the chapter. The Chapter Test is a practice test and is not keyed to sections of the chapter.

Cumulative Review   These features are found at the end of each chapter (except Chapter 1). Each problem contained in the Cumulative Review is an earlier worked example in the text that is referenced in the back of the book along with the answer. Students who need to see a complete worked-out solution, with explanation, can do so by turning to the appropriate example in the text.

Student Resource Icons   At the beginning of each section, videotape and CD, tutorial software, Prentice Hall Tutor Center, and solutions manual icons are displayed. These icons help reinforce that these learning aids are available should students wish to use them to help them review concepts and skills at their own pace. These items have direct correlation to the text and emphasize the text's methods of solution.

Functional Use of Color and New Design   Elements of this text are highlighted with color or design to make it easier for students to read and study. Special care has been taken to use color within solutions to examples or in the art to **help clarify, distinguish, or connect concepts**.

Exercise Sets   Each text section ends with an Exercise Set. Each exercise in the set, except those found in parts labeled Review and Preview or Combining Concepts, is keyed to one of the objectives of the section. Wherever possible, a specific example is also referenced. In addition to the approximately 4400 exercises in end-of-section exercise sets, exercises may also be found in the Pretests, Integrated Reviews, Chapter Reviews, Chapter Tests, and Cumulative Reviews.

   Exercises and examples marked with a video icon have been worked out step-by-step by the author in the videos that accompany this text.

Throughout the exercises in the text there is an emphasis on data and graphical interpretation via tables, charts, and graphs. The ability to interpret data and read and create a variety of types of graphs is developed gradually so students become comfortable with it. Chapter 9 gives a concise introduction to statistical graphs and descriptive statistics. Similarly, geometric concepts—such as perimeter and area—are integrated throughout the text. Exercises and examples marked with a geometry icon △ have been identified for convenience. In addition, Chapter 8 is a focused geometry chapter.

Each exercise set contains one or more of the following features.

Mental Math   Found at the beginning of an exercise set, these mental warmups reinforce concepts found in the accompanying section and increase students' confidence before they tackle an exercise set. By relying on their own mental skills, students increase not only their confidence in themselves but also their number sense and estimation ability.

Review and Preview   These exercises occur in each exercise set (except for those in Chapter 1) after the exercises keyed to the objectives of the section. Review and Preview problems are keyed to earlier sections and review concepts learned earlier in the text that are needed in the next section or in the next chapter. These exercises show the links between earlier topics and later material.

Combining Concepts   These exercises are found at the end of each exercise set after the Review and Preview exercises. Combining Concepts exercises require students to combine several concepts from that section or to take the concepts of the section a step further by combining them with concepts learned in previous sections. For instance, sometimes students are required to combine the concepts of the section with the problem-solving process they learned in Chapter 1 to try their hand at solving an application problem.

Writing Exercises   These exercises occur in almost every exercise set and are marked with an icon. They require students to assimilate information and provide a written response to explain concepts or justify their thinking. Guidelines recommended by the American Mathematical Association of Two Year Colleges (AMATYC) and other professional groups recommend incorporating writing in mathematics courses to reinforce concepts.

Vocabulary Checks   Vocabulary Checks, **new to this edition**, provide an opportunity for students to become more familiar with the use of mathematical terms as they strengthen their verbal skills. These appear at the end of the chapter before the Chapter Highlights.

Data and Graphical Interpretation   There is an emphasis on data interpretation in exercises via tables and graphs. The ability to interpret data and read and create a variety of types of graphs is developed gradually so students become comfortable with it. In addition, Chapter 9 gives a concise introduction to statistical graphs and descriptive statistics.

Internet Excursions   These exercises occur once per chapter. Internet Excursions require students to use the Internet as a data-collection tool to complete the exercises, allowing students first-hand experience with manipulating and working with real data.

Key Content Features in the Second Edition

Overview   This new edition retains many of the factors that have contributed to its success. Even so, **every section of the text was carefully re-examined**. Throughout the new edition you will find numerous new applications, examples, and many real-life applications and exercises. For example, look at the exercise sets of Sections 2.3, 3.6, or 4.1. Some sections have internal re-organization to better clarify and enhance the presentation.

Chapter 1 now begins with Tips for Success in Mathematics (Section 1.1). **New Study Skills Reminder** boxes have been inserted throughout the text. These boxes reinforce the tips from Section 1.1. They are placed strategically to encourage students to hone their study skills.

△ **Increased Integration of Geometry Concepts**   In addition to the traditional topics in basic mathematics courses, this text contains a strong emphasis on problem solving, and geometric concepts are integrated throughout. The geometry concepts presented are those most important to a student's understanding, and I have included many **applications and exercises** devoted to this topic. These are marked with a geometry icon. Also, Chapter 8 focuses on geometry.

New Examples   Detailed step-by-step examples were added, deleted, replaced, or updated as needed. Many of these reflect real life.

Exercise Sets Revised and Updated   The exercise sets have been carefully examined and extensively revised. The **real-world and real-data applications** have been thoroughly updated and many new applications are included. In addition, an **increased number of challenging problems** have been included in the new edition. **Writing exercises** are now included in most exercise sets and new **Vocabulary Checks** have been added to the end of the chapter to help students become proficient in the language of mathematics.

Chapters 2 and 3 Revised   Chapter 2, Multiplying Fractions, and Chapter 3, Adding and Subtracting Fractions, have been revised as follows:

- Increased emphasis was given throughout the chapters on the role of 0 in the numerator and denominator of a fraction.

- Many new exercises were added. Many operations on fractions with larger numerators or denominators were included.

- A significant number of new fraction applications have been added. In addition, a new Section 3.6, Fractions and Problem Solving, has been included. This new section includes applications solved by all four arithmetic operations on fractions as well as applications requiring more than one operation to solve.

- The Integrated Review in Chapter 3 now includes adding, subtracting, multiplying, and dividing fractions.

Enhanced Supplements Package   The Second Edition is supported by a wealth of supplements designed for **added effectiveness and efficiency**. New items include the MathPro 5 on-line tutorial with diagnostic and unique video clip feature, a new computerized testing system (TestGen-EQ with Quiz-Master), Prentice Hall Tutor Center, digitized videos on CD, and Instructor's CD series. Please see the list of supplements for descriptions.

Options for On-line and Distance Learning

For maximum convenience, Prentice Hall offers on-line interactivity and delivery options for a variety of distance learning needs. Instructors may access or adopt these in conjunction with this text.

http://www.prenhall.com/martin-gay_basic

The **Companion Web** site includes basic distance learning access to provide links to the text's Internet Excursions and a selection of on-line self quizzes. Email is available.

WebCT WebCT includes distance learning access to content found in the Martin-Gay companion Web site plus more. WebCT provides tools to create, manage, and use on-line course materials. Save time and take advantage of items such as on-line help, communication tools, and access to instructor and student manuals. Your college may already have WebCT's software installed on their server or you may choose to download it. Contact your local Prentice Hall sales representative for details.

BlackBoard Visit http://www.prenhall.com/demo. For distance learning access to content and features from the Martin-Gay companion Web site plus more, Blackboard provides simple templates and tools.

Course Compass™ Powered by BlackBoard. Visit http://www.prenhall.com/demo.

Supplements for the Instructor

Printed Supplements

Annotated Instructor's Edition (0-13-075223-1)

- Answers to all exercises printed on the same text page.
- Teaching Tips throughout the text placed at key points in the margin.

Instructor's Solution Manual (0-13-067690-X)

- Solutions to even-numbered section exercises.
- Solutions to every (even and odd) Mental Math exercise.
- Solutions to every (even and odd) Practice Problem (margin exercise).
- Solutions to every (even and odd) exercise found in the Pretests, Integrated Reviews (mid-chapter reviews), Chapter Reviews, Chapter Tests, and Cumulative Reviews.

Instructor's Resource Manual with Tests (0-13-075221-5)

- Notes to the Instructor that include an introduction to Interactive Learning, Interpreting Graphs and Data, Alternative Assessment, Using Technology, and Helping Students Succeed.
- Two free-response Pretests per chapter.
- Eight Chapter Tests per chapter (3 multiple-choice, 5 free-response).
- Two Cumulative Review Tests (one multiple-choice, one free-response) every two chapters (after chapters 2, 4, 6, 9).
- Eight Final Exams (3 multiple-choice, 5 free-response).
- Twenty additional exercises per section for added test exercises if needed.
- Group Activities (an average of two per chapter; providing short group activities in a convenient, ready-to-use format).
- Answers to all items.

Media Supplements

TestGen-EQ with QuizMaster CD-ROM (Windows/Macintosh) (0-13-075220-7)

- Algorithmically driven, text-specific testing program.
- Networkable for administering tests and capturing grades on-line.
- Edit and add your own questions to create a nearly unlimited number of tests and worksheets.
- Use the new "Function Plotter" to create graphs.

- Tests can be easily exported to HTML so they can be posted to the Web for student practice.
- Includes an email function for network users, enabling instructors to send a message to a specific student or an entire group.
- Network-based reports and summaries for a class or student and for cumulative or selected scores are available.

MathPro 5 Instructor Version

- On-line, customizable tutorial, diagnostic, and assessment program for anytime, anywhere tutorial support.
- Text specific at the learning objective level.
- Diagnostic option identifies student skills, provides individual learning plan, and tutorial reinforcement.
- Integration of TestGen-EQ allows for testing to operate within the tutorial environment.
- Course management tracking of tutorial and testing activity.

MathPro Explorer 4.0

- Network Version IBM/Mac 0-13-075229-0.
- Enables instructors to create either customized or algorithmically generated practice quizzes from any section of a chapter.
- Includes email function for network users, enabling instructors to send a message to a specific student or to an entire group.
- Network-based reports and summaries for a class or student and for cumulative or selected scores.

Instructor's CD Series (0-13-047357-X)

- Written and presented by Elayn Martin-Gay.
- Contains suggestions for presenting course material, utilizing the integrated resource package, time-saving tips, and much more.

Companion Web Site http://www.prenhall.com/martin-gay_basic

- Create a customized on-line syllabus with Syllabus Manager.
- Links related to the Internet Excursions in each chapter allow students to find and retrieve real data for use in guided problem solving.
- Assign quizzes or monitor student self quizzes by having students email results, such as true/false reading quizzes or vocabulary check quizzes.
- Destination links provide additional opportunities to explore related sites.

Supplements for the Student

Printed Supplements

Student's Solution Manual (0-13-075222-3)

- Solutions to odd-numbered section exercises.
- Solutions to every (even and odd) Mental Math exercise.
- Solutions to every (even and odd) Practice Problem (margin exercise).
- Solutions to every (even and odd) exercise found in the Pretests, Integrated Reviews (mid-chapter reviews), Chapter Reviews, Chapter Tests, and Cumulative Reviews.

Study Skills Notebook (0-13-047356-1)

Media Supplements

MathPro 5 (Student Version)

- Online, customizable tutorial, diagnostic, and assessment software.
- Text specific to the learning objective level, providing anytime, anywhere tutorial support.

- Algorithmically driven for virtually unlimited practice problems with immediate feedback.
- "Watch" screen videoclips by K. Elayn Martin-Gay.
- Step-by Step solutions.
- Summary of Progress.

MathPro 4.0 Explorer Student Version (0-13-075232-0)

- Available on CD-ROM for stand alone use or can be networked in the school laboratory.
- Text specific tutorial exercises and instructions at the objective level.
- Algorithmically generated Practice Problems.
- "Watch" screen videoclips by K. Elayn Martin-Gay.

Videotape Series (0-13-075228-2)

- Written and presented by Elayn Martin-Gay.
- Keyed to each section of the text.
- Step-by-step solutions to exercises from each section of the text. Exercises that are worked in the videos are marked with a video icon.

New Digitized Lecture Videos on CD-ROM (0-13-075218-5)

- The entire set of *Basic College Mathematics,* Second Edition lecture videotapes in digital form.
- Convenient access anytime to video tutorial support from a computer at home or on campus.
- Available shrinkwrapped with the text or stand-alone.

New Prentice Hall Tutor Center

- Staffed with developmental math instructors and open 5 days a week, 7 hours per day.
- Obtain help for examples and exercises in Martin-Gay, *Basic College Mathematics,* Second Edition via toll-free telephone, fax, or email.
- The Prentice Hall Tutor Center is accessed through a registration number that may be bundled with a new text or purchased separately with a used book. Visit http://www.prenhall.com/tutorcenter to learn more.

Companion Web Site www.prenhall.com/martin-gay_basic

- Links related to the Internet Excursions in each chapter allow you to collect data to solve specific internet exercises.

Acknowledgments

First, as usual, I would like to thank my husband, Clayton, for his constant encouragement. I would also like to thank my children, Eric and Bryan, for their sense of humor and especially for asking Dad to cook the bacon that I always used to burn.

I would also like to thank my extended family for their invaluable help and also their sense of humor. Their contributions are too numerous to list. They are Rod, Karen, and Adam Pasch; Michael, Christopher, Matthew, and Jessica Callac; Stuart, Earline, Melissa, Mandy, Bailey, and Ethan Martin; Mark, Sabrina, and Madison Martin; Leo and Barbara Miller; and Jewett Gay.

I would like to thank the following reviewers for their input and suggestions:

Maria Cristina Berisso, *Shasta College*
Ellen O'Connell, *Triton College*
Charles Odion, *Houston Community College*
Neal Rogers, *Santa Ana College*
Togba Sapolucia, *Northeast College*
Jimmie A. Van Alphen, *Ozarks Technical Community College*

There were many people who helped me develop this text and I will attempt to thank some of them here. John Crumley was invaluable for contributing to the overall accuracy of this text. Emily Keaton and Kathy Sessa-Federico were also invaluable for their many suggestions and contributions during the development and writing of this first edition. Ingrid Mount at Elm Street Publishing Services provided guidance throughout the production process. I thank Richard Semmler and Jenny Crawford for all their work on the solutions, text, and accuracy. Lastly, a special thank you to my project manager Mary Beckwith and executive editor Karin Wagner, for their support and assistance throughout the development and production of this text and to all the staff at Prentice Hall: Linda Behrens, Alan Fischer, Maureen Eide, Grace Hazeldine, Tom Benfatti, Eilish Main, John Tweeddale, Chris Hoag, Paul Corey, and Tim Bozik.

K. Elayn Martin-Gay

About the Author

K. Elayn Martin-Gay has taught mathematics at the University of New Orleans for more than 20 years and has received numerous teaching awards including the local University Alumni Association's Award for Excellence in Teaching.

Over the years, Elayn has developed a videotaped lecture series to help her students understand algebra better. This highly successful video material is the basis for her books: *Basic College Mathematics*, Second Edition; *Prealgebra*, Third Edition; *Introductory Algebra*, Second Edition; *Intermediate Algebra*, Second Edition; *Algebra A Combined Approach*, Second Edition; and her hardback series: *Beginning Algebra*, Third Edition; *Intermediate Algebra*, Third Edition; *Beginning and Intermediate Algebra*, Third Edition; and *Intermediate Algebra: A Graphing Approach*, Second Edition.

APPLICATIONS INDEX

HIGHLIGHTS OF BASIC COLLEGE MATHEMATICS, SECOND EDITION

Basic College Mathematics, Second Edition is the primary learning tool in a fully integrated learning package to help you succeed in this course. Author K. Elayn Martin-Gay focuses on enhancing the traditional emphasis of mastering the basics with innovative pedagogy and a meaningful learning program. There are three goals that drive her authorship:

▲ **Master and apply skills and concepts**

▲ **Build confidence**

▲ **Increase motivation**

Take a few moments now to examine some of the features that have been incorporated into *Basic College Mathematics, Second Edition* to help students excel.

Multiplying and Dividing Fractions
CHAPTER 2

Fractions are numbers, and like whole numbers, they can be added, subtracted, multiplied, and divided. Fractions are very useful and appear frequently in everyday language, in common phrases like "half an hour," "quarter of a pound," and "third of a cup." This chapter introduces the concept of fractions, presents some basic vocabulary, and demonstrates how to multiply and divide fractions.

2.1 Introduction to Fractions and Mixed Numbers

2.2 Factors and Prime Factorization

2.3 Simplest Form of a Fraction

Integrated Review— Summary on Fractions, Mixed Numbers, and Factors

2.4 Multiplying Fractions

2.5 Dividing Fractions

Established in 1872, Yellowstone National Park was the first national park in the United States. Today, the National Park Service operates a total of 384 areas ranging from 55 national parks such as Yellowstone and the Grand Canyon, to 75 national monuments such as Devil's Tower in Wyoming, to four national parkways such as the Blue Ridge Parkway in North Carolina and Virginia. The largest park in the system is Wrangell-St. Elias National Park and Preserve in Alaska, with 13,200,000 acres of land. The smallest park in the system is Thaddeus Kosciuszko National Memorial in Pennsylvania, covering only $\frac{1}{50}$ of an acre of land. In Example 10 and Exercise 60 in Section 2.3 on pages 124 and 126, we find the fraction of national parks that can be found in Utah and Alaska or Hawaii.

101

◀ **REAL WORLD APPLICATIONS**

Chapter-opening real-world applications introduce you to everyday situations that are applicable to the mathematics you will learn in the upcoming chapter, showing the relevance of mathematics in daily life.

Become a Confident Problem Solver

A goal of this text is to help you develop problem-solving abilities.

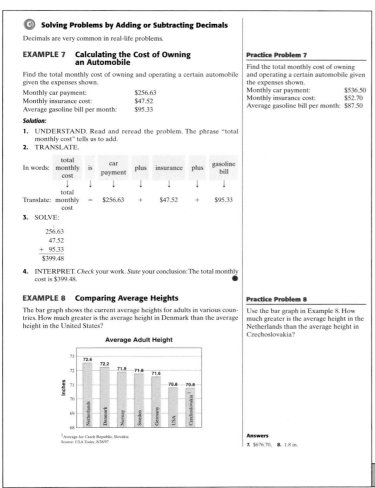

C **Solving Problems by Adding or Subtracting Decimals**

Decimals are very common in real-life problems.

EXAMPLE 7   Calculating the Cost of Owning an Automobile

Find the total monthly cost of owning and operating a certain automobile given the expenses shown.

Monthly car payment:	$256.63
Monthly insurance cost:	$47.52
Average gasoline bill per month:	$95.33

Solution:

1. UNDERSTAND. Read and reread the problem. The phrase "total monthly cost" tells us to add.
2. TRANSLATE.

In words:	total monthly cost	is	car payment	plus	insurance	plus	gasoline bill
	↓	↓	↓	↓	↓	↓	↓
Translate:	total monthly cost	=	$256.63	+	$47.52	+	$95.33

3. SOLVE:

```
   256.63
    47.52
 +  95.33
  $399.48
```

4. INTERPRET. *Check* your work. *State* your conclusion: The total monthly cost is $399.48. ●

EXAMPLE 8   Comparing Average Heights

The bar graph shows the current average heights for adults in various countries. How much greater is the average height in Denmark than the average height in the United States?

Average Adult Height

Inches — 73, 72, 71, 70, 69, 68

Netherlands 72.6, Denmark 72.2, Norway 71.9, Sweden 71.8, Germany 71.6, USA 70.8, Czechoslovakia [1] 70.8

[1] Average for Czech Republic, Slovakia
Source: *USA Today*, 8/28/97

Practice Problem 7

Find the total monthly cost of owning and operating a certain automobile given the expenses shown.

Monthly car payment:	$536.50
Monthly insurance cost:	$52.70
Average gasoline bill per month:	$87.50

Practice Problem 8

Use the bar graph in Example 8. How much greater is the average height in the Netherlands than the average height in Czechoslovakia?

Answers

7. $676.70, **8.** 1.8 in.

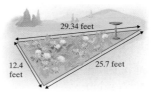

Page 243

◄ **GENERAL STRATEGY FOR PROBLEM-SOLVING**

Save time by having a plan. This text's organization can help you. Note the outlined problem-solving steps, *Understand, Translate, Solve,* and *Interpret.*

Problem solving is introduced early, emphasized and integrated throughout the book. The author provides patient explanations and illustrates how to apply the problem-solving procedure to the in-text examples.

GEOMETRY ►

Geometry concepts are integrated throughout the text. Examples and exercises involving geometric concepts are now identified with a triangle icon. △ This text includes chapter 8 on Geometry as well.

△ **53.** A landscape architect is planning a border for a flower garden that's shaped like a triangle. The sides of the garden measure 12.4 feet, 29.34 feet, and 25.7 feet. Find the amount of border material needed.

29.34 feet
12.4 feet
25.7 feet

Page 247

xxiii

Master and Apply Basic Skills and Concepts

K. Elayn Martin-Gay provides thorough explanations of key concepts and enlivens the content by integrating successful and innovative pedagogy. *Basic College Mathematics, Second Edition* integrates skill building throughout the text and provides problem-solving strategies and hints along the way. These features have been included to enhance your understanding of mathematical concepts.

◄ CONCEPT CHECKS

Concept Checks are special margin exercises found in most sections. Work these to help gauge your grasp of the concept being developed in the text.

Concept Check

Explain how you could estimate the sum:
$5\frac{1}{9} + 14\frac{10}{11}$

Page 188

Page 68

COMBINING CONCEPTS ►

Combining Concepts exercises are found at the end of each exercise set. Solving these exercises will expose you to the way mathematical ideas build upon each other.

Combining Concepts

The following table shows the top five leading U.S. advertisers in 2000 and the amount of money spent in that year on ads. Use this table to answer Exercises 57 and 58.

Company	Year 2000 Ad Expenditures
General Motors Corp.	$2,883,215,000
DaimlerChrysler AG	$1,671,764,000
Procter & Gamble Co.	$1,542,847,000
Philip Morris Cos. Inc.	$1,538,040,000
Time Warner Inc.	$1,318,082,000

(*Source:* CMR, a Taylor Nelson Sofres company)

57. Find the average amount of money spent on ads for the year by the top two companies.

58. Find the average amount of money spent on ads by DaimlerChrysler, Procter & Gamble, Philip Morris, and Time Warner.

PRACTICE PROBLEMS ►

Practice Problems occur in the margins next to every Example. Work these problems after an example to immediately reinforce your understanding.

Practice Problem 5 △

Find the perimeter of the polygon shown. (A centimeter is a unit of length in the metric system.)

5 centimeters
4 centimeters
8 centimeters
10 centimeters

Practice Problem 6 △

A new shopping mall has a floor plan in the shape of a triangle. Each of the mall's ...perimeter

EXAMPLE 5   Find the perimeter of the polygon shown.

3 inches
2 inches
1 inch
4 inches
3 inches

Solution: To find the perimeter (distance around), we add the lengths of the sides.

$$2\,\text{in.} + 3\,\text{in.} + 1\,\text{in.} + 3\,\text{in.} + 4\,\text{in.} = 13\,\text{in.}$$

The perimeter is 13 inches.   ●

EXAMPLE 6   Calculating the Perimeter of a Building

The largest commercial building in the world under one roof is the flower auction building of the cooperative VBA in Aalsmeer, Netherlands. The floor plan is a rectangle that measures 776 meters by 639 meters. Find the perimeter of this building. (A meter is a unit of length in the metric system.) (*Source: The Handy Science Answer Book*, Visible Ink Press)

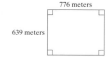

776 meters
639 meters

Page 120

85. The number 2 is a prime number. All other even natural numbers are composite numbers. Explain why.

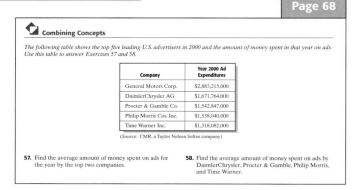

Page 20

▲ WRITING EXERCISES

New Writing Exercises, marked by an icon, ✎ are now found in most practice sets.

Test Yourself and Check Your Understanding

Good exercise sets and an abundance of worked-out examples are essential for building student confidence. The exercises you will find in this worktext are intended to help you build skills and understand concepts as well as motivate and challenge you. In addition, features like Chapter Highlights, Chapter Reviews, Chapter Tests, and Cumulative Reviews are found at the end of each chapter to help you study and organize your notes.

Chapter 2 Pretest

1. Use a fraction to represent the shaded area of the figure.

2. Write each mixed number as an improper fraction.
   a. $3\frac{4}{7}$ **b.** $9\frac{5}{8}$

Page 102

◀ PRETESTS

Pretests open each chapter. Take a Pretest to evaluate where you need the most help before beginning a new chapter.

Page 129

Name _____ Section _____ Date _____

Integrated Review—Summary on Fractions, Mixed Numbers, and Factors

Use a fraction to represent the shaded area of each figure or figure group.

1.

2.

INTEGRATED REVIEWS ▶

Integrated Reviews serve as mid-chapter reviews and help you to assimilate the new skills you have learned separately over several sections.

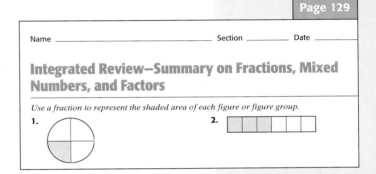

Review and Preview

Simplify each fraction. See Section 2.3.

67. $\frac{10}{12}$ **68.** $\frac{8}{12}$ **69.** $\frac{20}{24}$ **70.** $\frac{22}{24}$ **71.** $\frac{50}{75}$ **72.** $\frac{30}{65}$

◆ Combining Concepts

Solve.

73. In 2000, about $\frac{11}{67}$ of the total weight of mail delivered by the United States Postal Service was first-class mail. That same year, about $\frac{75}{134}$ of the total weight of mail delivered by the United States Postal Service was standard mail. Which of these two categories account for a greater portion of the mail handled by weight? (*Source*: U.S. Postal Service)

74. The National Park System (NPS) in the United States includes a wide variety of park types. National military parks account for $\frac{3}{128}$ of all NPS

Page 202

◀ REVIEW AND PREVIEW

Review and Preview exercises review concepts learned earlier in the text that are needed in the next section or chapters.

Page 213

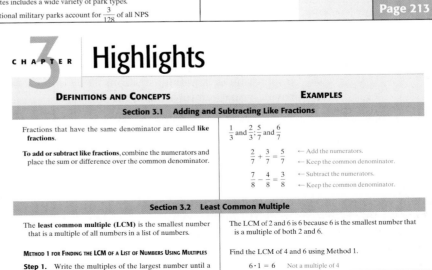

CHAPTER 3

Highlights

DEFINITIONS AND CONCEPTS	EXAMPLES
Section 3.1 Adding and Subtracting Like Fractions	
Fractions that have the same denominator are called **like fractions**.	$\frac{1}{3}$ and $\frac{2}{3}$; $\frac{5}{7}$ and $\frac{6}{7}$
To add or subtract like fractions, combine the numerators and place the sum or difference over the common denominator.	$\frac{2}{7} + \frac{3}{7} = \frac{5}{7}$ ← Add the numerators. ← Keep the common denominator. $\frac{7}{8} - \frac{4}{8} = \frac{3}{8}$ ← Subtract the numerators. ← Keep the common denominator.
Section 3.2 Least Common Multiple	
The **least common multiple (LCM)** is the smallest number that is a multiple of all numbers in a list of numbers.	The LCM of 2 and 6 is 6 because 6 is the smallest number that is a multiple of both 2 and 6.
METHOD 1 FOR FINDING THE LCM OF A LIST OF NUMBERS USING MULTIPLES	Find the LCM of 4 and 6 using Method 1.
Step 1. Write the multiples of the largest number until a	$6 \cdot 1 = 6$ Not a multiple of 4

CHAPTER HIGHLIGHTS ▶

Found at the end of every chapter, the Chapter Highlights contain key definitions, concepts, and examples to help students understand and retain what they have learned.

Increase Motivation

Throughout *Basic College Mathematics, Second Edition*, K. Elayn Martin-Gay provides interesting real-world applications to strengthen your understanding of the relevance of math in everyday life. When a new topic is presented, an effort has been made to relate the new ideas to those that students may already know. The Second Edition increases emphasis on visualization to clarify and reinforce key concepts.

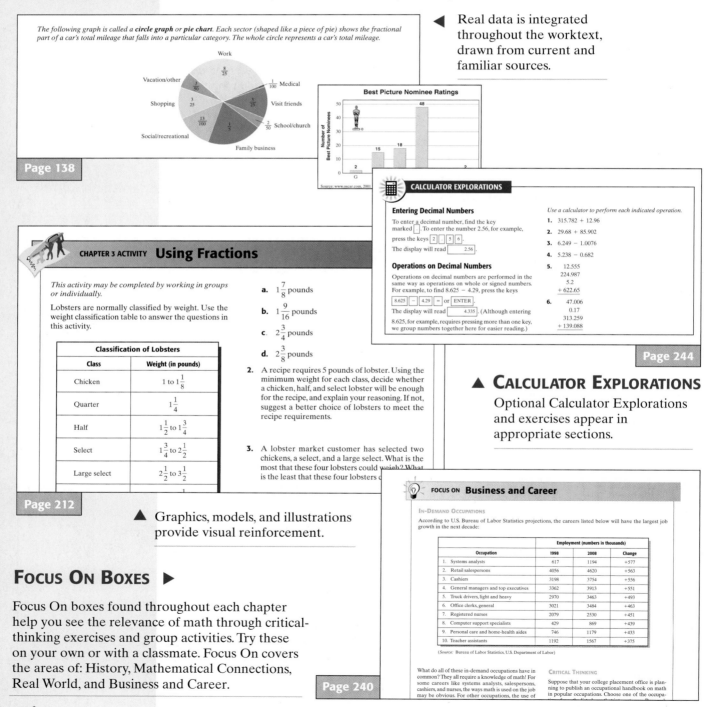

◀ Real data is integrated throughout the worktext, drawn from current and familiar sources.

The following graph is called a **circle graph** or **pie chart**. Each sector (shaped like a piece of pie) shows the fractional part of a car's total mileage that falls into a particular category. The whole circle represents a car's total mileage.

Page 138

Page 244

▲ **CALCULATOR EXPLORATIONS**
Optional Calculator Explorations and exercises appear in appropriate sections.

CHAPTER 3 ACTIVITY Using Fractions

This activity may be completed by working in groups or individually.

Lobsters are normally classified by weight. Use the weight classification table to answer the questions in this activity.

Classification of Lobsters

Class	Weight (in pounds)
Chicken	1 to $1\frac{1}{8}$
Quarter	$1\frac{1}{4}$
Half	$1\frac{1}{2}$ to $1\frac{3}{4}$
Select	$1\frac{3}{4}$ to $2\frac{1}{2}$
Large select	$2\frac{1}{2}$ to $3\frac{1}{2}$

a. $1\frac{7}{8}$ pounds

b. $1\frac{9}{16}$ pounds

c. $2\frac{3}{4}$ pounds

d. $2\frac{3}{8}$ pounds

2. A recipe requires 5 pounds of lobster. Using the minimum weight for each class, decide whether a chicken, half, and select lobster will be enough for the recipe, and explain your reasoning. If not, suggest a better choice of lobsters to meet the recipe requirements.

3. A lobster market customer has selected two chickens, a select, and a large select. What is the most that these four lobsters could weigh? What is the least that these four lobsters c...

Page 212

▲ Graphics, models, and illustrations provide visual reinforcement.

FOCUS ON BOXES ▶

Focus On boxes found throughout each chapter help you see the relevance of math through critical-thinking exercises and group activities. Try these on your own or with a classmate. Focus On covers the areas of: History, Mathematical Connections, Real World, and Business and Career.

Page 240

Build Confidence

Several features of this text can be helpful in building your confidence and mathematical competence. As you study, also notice the connections the author makes to relate new material to ideas that you may already know.

◀ **TIPS FOR SUCCESS**

New coverage of study skills in Section 1.1 reinforces this important component to success in this course.

1.1 Tips for Success in Mathematics

Before reading this section, remember that your instructor is your best source for information. Please see your instructor for any additional help or information.

Ⓐ Getting Ready for This Course

Now that you have decided to take this course, remember that a *positive attitude* will make all the difference in the world. Your belief that you can succeed is just as important as your commitment to this course. that you are ready for this course by having the time and positive at

Page 3

Page 134

STUDY SKILLS REMINDER

Are you organized?

Have you ever had trouble finding a completed assignment? When it's time to study for a test, are your notes neat and organized? Have you ever had trouble reading your own mathematics handwriting? (Be honest—I have.)

When any of these things happen, it's time to get organized. Here are a few suggestions:

Write your notes and complete your homework assignment in a notebook with pockets (spiral or ring binder.) Take class notes in this notebook, and then follow the notes with your completed homework assignment. When you receive graded papers or handouts, place them in the notebook pocket so that you will not lose them.

Remember to mark (possibly with an exclamation point) any

STUDY SKILLS REMINDERS ▶

New Study Skills Reminders are integrated throughout the book to reinforce section 1.1 and encourage the development of strong study skills.

Mental Math

Ⓐ *Identify the numerator and the denominator of each fraction. See Examples 1 and 2.*

1. $\frac{1}{2}$ 2. $\frac{1}{4}$ 3. $\frac{10}{3}$

4. $\frac{53}{21}$ 5. $\frac{3}{7}$ 6. $\frac{11}{15}$

Page 109

◀ **MENTAL MATH**

Mental Math warm-up exercises reinforce concepts found in the accompanying section and can increase your confidence before beginning an exercise set.

Page 60

HELPFUL HINTS ▶

Found throughout the text, these contain practical advice on applying mathematical concepts. They are strategically placed where you are most likely to need immediate reinforcement.

Helpful Hint

Since division and multiplication are reverse operations, don't forget that a division problem can be checked by multiplying.

Chapter 3 VOCABULARY CHECK

Fill in each blank with one of the words or phrases listed below.

   equivalent   least common multiple   like   mixed number

1. Fractions that have the same denominator are called _____ fractions.
2. The _____ is the smallest number that is a multiple of all numbers in a list of numbers.
3. _____ fractions represent the same portion of a whole.
4. A _____ has a whole number part and a fraction part.

Page 213

◀ **VOCABULARY CHECKS**

New Vocabulary Checks allow you to write your answers to questions about chapter content and strengthen verbal skills.

Enrich Your Learning

Seek out these additional Student Resources to match your personal learning style.

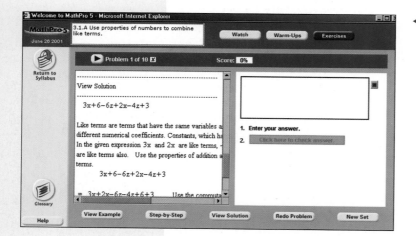

◄ MathPro 5 is the online customizable tutorial, diagnostic and assessment software. It is text-specific to the learning objective level and provides anytime, anywhere tutorial support. It provides:
- Diagnostic review of student skills
- Virtually unlimited practice problems with immediate feedback
- Video clips by K. Elayn Martin-Gay
- Step-by-step solutions
- Summary of progress

MathPro 4 is available on CD-ROM for standalone use or can be networked in the school laboratory.

Text-specific videos, available on CD or VHS, are hosted by the award-winning teacher and author of *Basic College Mathematics*. They cover each objective in every chapter section as a supplementary review. ►

◄ Prentice Hall Tutor Center provides text-specific tutoring via phone, fax, and e-mail. Visit http://prenhall.com/tutorcenter for details.

ALSO AVAILABLE:

▲ Student Solutions Manual

▲ Study Skills Notebook

▲ How to Study Math

▲ Math on the Internet

▲ *The New York Times/ Themes of the Times*

Ask your instructor or bookstore about these additional study aids.

The Whole Numbers

Mathematics is an important tool for everyday life. Knowing basic mathematical skills can help simplify tasks such as creating a monthly budget. Whole numbers are the basic building blocks of mathematics. The whole numbers answer the question "How many?"

This chapter covers basic operations on whole numbers. Knowledge of these operations provides a good foundation on which to build further mathematical skills.

The American Kennel Club (AKC) was founded in 1884. This nonprofit organization is devoted to the raising and welfare of purebred dogs. The AKC recognizes 148 different dog breeds, ranging in size from the Chihuahua to the Great Dane. In addition to sanctioning dog shows, the AKC maintains a registry of individual purebred dogs. AKC registration means that a dog's parents and ancestors were purebreds. In Exercises 53–56 on page 16, we will see how whole numbers can be used to keep track of and compare American Kennel Club registrations.

Name _____ Section _____ Date _____

Chapter 1 Pretest

1. Determine the place value of the digit 7 in the whole number 5732.

2. Write the whole number 23,490 in words.

3. Add: 58 + 29

4. Multiply: 413
$$\times \quad 9$$

Subtract. Check by adding.

5.   857
     $-\ 231$

6.   51
     $-\ 19$

Solve.

7. Karen Lewis is reading a 329-page novel. If she has just finished reading page 193, how many more pages must she read to finish the novel?

8. Round 9045 to the nearest ten.

9. Round each number to the nearest hundred to find an estimated sum.

     382
     436
     2084
     $+\ \ 176$

10. Use the distributive property to rewrite the following expression: 9(3 + 11)

△ **11.** Find the perimeter of the following figure:

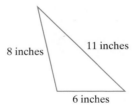

8 inches   11 inches

6 inches

△ **12.** Find the area of the following rectangle:

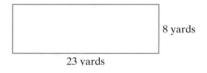

8 yards

23 yards

13. The seats in the history lecture hall are arranged in 32 rows with 18 seats in each row. Find how many seats are in this room.

Divide and then check by multiplying.

14. 2187 ÷ 9

15. $\dfrac{5361}{12}$

Solve.

16. Find the average of the following list of numbers: 29, 36, 84, 41, 6, 12, 65

17. Paul Crandall left $29,640 in his will to go to his three favorite nephews. If each boy was to receive the same amount of money, how much did each boy receive?

18. Write 9·9·9·9·9·9·9 using exponential notation.

19. Evaluate: 7^4

20. Simplify: 36 + 18 ÷ 6

Answers

1. _____
2. _____
3. _____
4. _____
5. _____
6. _____
7. _____
8. _____
9. _____
10. _____
11. _____
12. _____
13. _____
14. _____
15. _____
16. _____
17. _____
18. _____
19. _____
20. _____

1.1 Tips for Success in Mathematics

Before reading this section, remember that your instructor is your best source for information. Please see your instructor for any additional help or information.

A Get ready for this course.

B Understand some general tips for success.

C Understand how to use this text.

D Get help as soon as you need it.

E Learn how to prepare for and take an exam.

F Develop good time management.

SSM   SG   CD & VIDEO   MATH PRO   WEB
TUTOR CENTER

A Getting Ready for This Course

Now that you have decided to take this course, remember that a *positive attitude* will make all the difference in the world. Your belief that you can succeed is just as important as your commitment to this course. Make sure that you are ready for this course by having the time and positive attitude that it takes to succeed.

Next, make sure that you have scheduled your math course at a time that will give you the best chance for success. For example, if you are also working, you may want to check with your employer to make sure that your work hours will not conflict with your course schedule.

On the day of your first class period, double-check your schedule and allow yourself extra time to arrive in case of traffic problems or difficulty locating your classroom. Make sure that you bring at least your textbook, paper, and a writing instrument. Are you required to have a lab manual, graph paper, calculator, or some other supply besides this text? If so, also bring this material with you.

B General Tips for Success

Below are some general tips that will increase your chance for success in a mathematics class. Many of these tips will also help you in other courses you may be taking.

Exchange names and phone numbers with at least one other person in class. This contact person can be a great help if you miss an assignment or want to discuss math concepts or exercises that you find difficult.

Choose to attend all class periods. If possible, sit near the front of the classroom. This way, you will see and hear the presentation better. It may also be easier for you to participate in classroom activities.

Do your homework. You've probably heard the phrase "practice makes perfect" in relation to music and sports. It also applies to mathematics. You will find that the more time you spend solving mathematics problems, the easier the process becomes. Be sure to schedule enough time to complete your assignments before the next class period.

Check your work. Review the steps you made while working a problem. Learn to check your answers in the original problems. You may also compare your answers with the answers to selected exercises section in the back of the book. If you have made a mistake, try to figure out what went wrong. Then correct your mistake. If you can't find what went wrong don't erase your work or throw it away. Bring your work to your instructor, a tutor in a math lab, or a classmate. It is easier for someone to find where you had trouble if they look at your original work.

Learn from your mistakes. Everyone, even your instructor, makes mistakes. Use your errors to learn and to become a better math student. The key is finding and understanding your errors. Was your mistake a careless one, or did you make it because you can't read your own math writing? If so, try to work more slowly or write more neatly and make a conscious effort to carefully check your work. Did you make a mistake because you don't understand a concept? Take the time to review the concept or ask questions to better understand it.

Know how to get help if you need it. It's all right to ask for help. In fact, it's a good idea to ask for help whenever there is something that you don't understand. Make sure you know when your instructor has office hours and how to find his or her office. Find out whether math tutoring services are available on your campus. Check out the hours, location, and requirements of the tutoring service. Know whether videotapes or software are available and how to access these resources.

Organize your class materials, including homework assignments, graded quizzes and tests, and notes from your class or lab. All of these items will make valuable references throughout your course and when studying for upcoming tests and the final exam. Make sure that you can locate these materials when you need them.

Read your textbook before class. Reading a mathematics textbook is unlike reading for fun, such as reading a newspaper. Your pace will be much slower. It is helpful to have paper and a pencil with you when you read. Try to work out examples on your own as you encounter them in your text. You should also write down any questions that you want to ask in class. When you read a mathematics textbook, sometimes some of the information in a section will be unclear. But after you hear a lecture or watch a videotape on that section, you will understand it much more easily than if you had not read your text beforehand.

Don't be afraid to ask questions. You are not the only person in class with questions. Other students are normally grateful that someone has spoken up.

Hand in assignments on time. This way you can be sure that you will not lose points for being late. Show every step of a problem and be neat and organized. Also be sure that you understand which problems are assigned for homework. You can always double-check the assignment with another student in your class.

Ⓒ Using This Text

There are many helpful resources that are available to you in this text. It is important that you become familiar with and use these resources. They should increase your chances for success in this course.

- Each example in every section has a parallel Practice Problem. As you read a section, try each Practice Problem after you've finished the corresponding example. This "learn-by-doing" approach will help you grasp ideas before you move on to other concepts.

- The main section of exercises in each exercise set is referenced by an objective, such as Ⓐ or Ⓑ, and also an example(s). Use this referencing if you have trouble completing an assignment from the exercise set.

- If you need extra help in a particular section, look at the beginning of the section to see what videotapes and software are available.

- Make sure that you understand the meaning of the icons that are beside many exercises. The video icon 📷 tells you that the corresponding exercise may be viewed on the videotape that corresponds to that section. The pencil icon ✎ tells you that this exercise is a writing exercise in which you should answer in complete sentences. The △ icon tells you that the exercise involves geometry.

- Integrated Reviews in each chapter offer you a chance to practice—in one place—the many concepts that you have learned separately over several sections.

- There are many opportunities at the end of each chapter to help you understand the concepts of the chapter.

  Chapter Highlights contain chapter summaries and examples.
  Chapter Reviews contain review problems organized by section.
  Chapter Tests are sample tests to help you prepare for an exam.
  Cumulative Reviews are reviews consisting of material from the beginning of the book to the end of that particular chapter.

See the preface at the beginning of this text for a more thorough explanation of the features of this text.

Ⓓ Getting Help

If you have trouble completing assignments or understanding the mathematics, get help as soon as you need it! This tip is presented as an objective on its own because it is so important. In mathematics, usually the material presented in one section builds on your understanding of the previous section. This means that if you don't understand the concepts covered during a class period, there is a good chance that you will not understand the concepts covered during the next class period. If this happens to you, get help as soon as you can.

Where can you get help? Many suggestions have been made in the section on where to get help, and now it is up to you to do it. Try your instructor, a tutoring center, or a math lab, or you may want to form a study group with fellow classmates. If you do decide to see your instructor or go to a tutoring center, make sure that you have a neat notebook and are ready with your questions.

Ⓔ Preparing for and Taking an Exam

Make sure that you allow yourself plenty of time to prepare for a test. If you think that you are a little "math anxious," it may be that you are not preparing for a test in a way that will ensure success. The way that you prepare for a test in mathematics is important. To prepare for a test:

1. Review your previous homework assignments.
2. Review any notes from class and section-level quizzes you have taken. (If this is a final exam, also review chapter tests you have taken.)
3. Review concepts and definitions by reading the Highlights at the end of each chapter.
4. Practice working out exercises by completing the Chapter Review found at the end of each chapter. (If this is a final exam, go through a Cumulative Review. There is one found at the end of each chapter except Chapter 1. Choose the review found at the end of the latest chapter that you have covered in your course.) *Don't stop here!*
5. It is important that you place yourself in conditions similar to test conditions to find out how you will perform. In other words, as soon as you feel that you know the material, get a few blank sheets of paper and take a sample test. There is a Chapter Test available at the end of each chapter, or you can work selected problems from the Chapter Review. Your instructor may also provide you with a review sheet. During this sample test, do not use your notes or your textbook. Then check your sample test. If you are not satisfied with the results, study the areas that you are weak in and try again.
6. On the day of the test, allow yourself plenty of time to arrive at where you will be taking your exam.

When taking your test:

1. Read the directions on the test carefully.
2. Read each problem carefully as you take the test. Make sure that you answer the question asked.
3. Watch your time and pace yourself so that you can attempt each problem on your test.
4. If you have time, check your work and answers.
5. Do not turn your test in early. If you have extra time, spend it double-checking your work.

Ⓕ Managing Your Time

As a college student, you know the demands that classes, homework, work, and family place on your time. Some days you probably wonder how you'll ever get everything done. One key to managing your time is developing a schedule. Here are some hints for making a schedule:

1. Make a list of all of your weekly commitments for the term. Include classes, work, regular meetings, extracurricular activities, etc. You may also find it helpful to list such things as laundry, regular workouts, grocery shopping, etc.
2. Next, estimate the time needed for each item on the list. Also make a note of how often you will need to do each item. Don't forget to include time estimates for the reading, studying, and homework you do outside of your classes. You may want to ask your instructor for help estimating the time needed.
3. In the exercise set below, you are asked to block out a typical week on the schedule grid given. Start with items with fixed time slots like classes and work.
4. Next, include the items on your list with flexible time slots. Think carefully about how best to schedule some items such as study time.
5. Don't fill up every time slot on the schedule. Remember that you need to allow time for eating, sleeping, and relaxing! You should also allow a little extra time in case some items take longer than planned.
6. If you find that your weekly schedule is too full for you to handle, you may need to make some changes in your workload, classload, or in other areas of your life. You may want to talk to your advisor, manager or supervisor at work, or someone in your college's academic counseling center for help with such decisions.

Note: In this chapter, we begin a feature called Study Skills Reminder. The purpose of this feature is to remind you of some of the information given in this section and to further expand on some topics in this section.

EXERCISE SET 1.1

1. What is your instructor's name?

2. What are your instructor's office location and office hours?

3. What is the best way to contact your instructor?

4. What does the ✎ icon mean?

5. What does the 📷 icon mean?

6. What does the △ icon mean?

7. Do you have the name and contact information of at least one other student in class?

8. Will your instructor allow you to use a calculator in this class?

9. Are videotapes and/or tutorial software available to you? If so, where?

10. Is there a tutoring service available? If so, what are its hours?

11. Have you attempted this course before? If so, write down ways that you might improve your chances of success during this second attempt.

12. List some steps that you can take if you begin having trouble understanding the material or completing an assignment.

13. Read or reread objective **F** and fill out the schedule grid below.

	Monday	Tuesday	Wednesday	Thursday	Friday	Saturday	Sunday
7:00 a.m.							
8:00 a.m.							
9:00 a.m.							
10:00 a.m.							
11:00 a.m.							
12:00 a.m.							
1:00 p.m.							
2:00 p.m.							
3:00 p.m.							
4:00 p.m.							
5:00 p.m.							
6:00 p.m.							
7:00 p.m.							
8:00 p.m.							
9:00 p.m.							

1.2 Place Value and Names for Numbers

The **digits** 0, 1, 2, 3, 4, 5, 6, 7, 8, and 9 can be used to write numbers. For example, the **whole numbers** are

0, 1, 2, 3, 4, 5, 6, 7, 8, 9, 10, 11, ...

The three dots (...) after the 11 mean that this list continues indefinitely. That is, there is no largest whole number. The smallest whole number is 0.

OBJECTIVES

A Find the place value of a digit in a whole number.

B Write a whole number in words and in standard form.

C Write a whole number in expanded form.

D Read tables.

SSM
TUTOR CENTER   SG   CD & VIDEO   MATH PRO   WEB

A Finding the Place Value of a Digit in a Whole Number

The position of each digit in a number determines its **place value**. A place-value chart is shown next for the whole number 48,337,000.

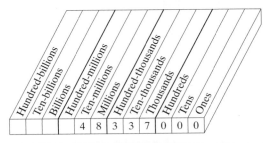

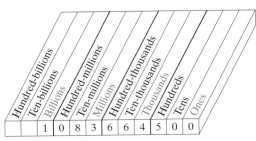

The two 3s in 48,337,000 represent different amounts because of their different placements. The place value of the 3 on the left is hundred-thousands. The place value of the 3 on the right is ten-thousands.

EXAMPLES   Find the place value of the digit 4 in each whole number.

1. 48,761
      ↑
   ten-thousands

2. 249
      ↑
     tens

3. 524,007,656
              ↑
          millions

●

B Writing a Whole Number in Words and in Standard Form

A whole number such as 1,083,664,500 is written in **standard form**. Notice that commas separate the digits into groups of three, starting from the right. Each group of three digits is called a **period**. The names of the first four periods are shown in red.

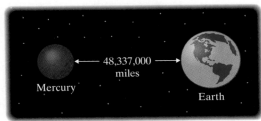

Practice Problems 1–3

Find the place value of the digit 7 in each whole number.

1. 72,589,620
2. 67,890
3. 50,722

Answers

1. ten-millions, **2.** thousands, **3.** hundreds

Writing a Whole Number in Words

To write a whole number in words, write the number in each period followed by the name of the period. (The ones period is usually not written.) This same procedure can be used to read a whole number.

For example, we write 1,083,664,500 as

one **billion**,

eighty-three **million**,

six hundred sixty-four **thousand**,

five **hundred**

> **Helpful Hint**
>
> The name of the ones period is not used when reading and writing whole numbers. For example,
>
> 9,265
>
> is read as
>
> "nine **thousand**, two **hundred** sixty-five."

Practice Problems 4–5

Write each number in words.

4. 67
5. 395

EXAMPLES   Write each number in words.

4. 85   eighty-five
5. 126   one hundred twenty-six   ●

> **Helpful Hint**
>
> The word "and" is *not* used when reading and writing whole numbers. It is used when reading and writing mixed numbers and some decimal values, as shown later in this text.

Practice Problem 6

Write 321,670,200 in words.

EXAMPLE 6   Write 106,052,447 in words.

Solution: 106,052,447 is written as

one hundred six **million**, fifty-two **thousand**, four **hundred** forty-seven   ●

Try the Concept Check in the margin.

Concept Check

True or false? When writing a check for $2600, the word name we write for the dollar amount of the check would be "two thousand sixty." Explain your answer.

Writing a Whole Number in Standard Form

To write a whole number in standard form, write the number in each period, followed by a comma.

Answers

4. sixty-seven,   **5.** three hundred ninety-five,
6. three hundred twenty-one million, six hundred seventy thousand, two hundred

Concept Check: false

EXAMPLES   Write each number in standard form.

7. sixty-one   61

8. eight hundred five   805

9. two million, five hundred sixty-four thousand, three hundred fifty

2,564,350

10. nine thousand, three hundred eighty-six

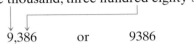

9,386   or   9386

> **Helpful Hint**
>
> A comma may or may not be inserted in a four-digit number. For example, both
>
>   9,386   and   9386
>
> are acceptable ways of writing nine thousand, three hundred eighty-six.

Practice Problems 7–10

Write each number in standard form.

7.   twenty-nine
8.   seven hundred ten
9.   twenty-six thousand seventy-one
10. six thousand, five hundred seven

C **Writing a Whole Number in Expanded Form**

The place value of a digit can be used to write a number in expanded form. The **expanded form** of a number shows each digit of the number with its place value. For example, 5672 is written in expanded form as

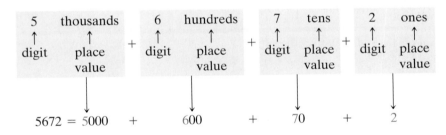

$$5672 = 5000 \;+\; 600 \;+\; 70 \;+\; 2$$

EXAMPLE 11   Write 706,449 in expanded form.

Solution: $700{,}000 + 6000 + 400 + 40 + 9$

Practice Problem 11

Write 1,047,608 in expanded form.

D **Reading Tables**

Now that we know about place value and names for whole numbers, we introduce one way that whole number data may be presented. **Tables** are often used to organize and display facts that contain numbers. The table below shows the countries that have won the most medals during the 2000 Olympic summer games. (Although the medals are truly won by athletes from the various countries, for simplicity we will state that countries have won the medals.)

Answers

7. 29, **8.** 710, **9.** 26,071, **10.** 6507
11. 1,000,000 + 40,000 + 7000 + 600 + 8

Most Medals 2000 Olympic Summer Games				
	Gold	**Silver**	**Bronze**	**Total**
United States	39	25	33	97
Russia	32	28	28	88
China	28	16	15	59
Australia	16	25	17	58
Germany	14	17	26	57
France	13	14	11	38
Italy	13	8	13	34
Cuba	11	11	7	29
Great Britain	11	10	7	28
South Korea	8	9	11	28
Romania	11	6	9	26
Netherlands	12	9	4	25

(*Source:* The Sydney Morning Herald)

For example, by reading from left to right along the row marked "U.S." we find that the United States won 39 gold, 25 silver, and 33 bronze medals during the 2000 Summer Games.

Practice Problem 12

Use the Summer Games table to answer the following questions:

a. How many bronze medals did Australia win during the Summer Games of the 2000 Olympics?

b. Which countries shown won more than 30 gold medals?

Answers

12. a. 17, **b.** United States and Russia

EXAMPLE 12

Use the Summer Games table to answer each question.

a. How many silver medals did Italy win during the Summer Games of the Olympics?

b. Which country shown has won fewer gold medals than Great Britain?

Solution:

a. Find "Italy" in the left column. Then read from left to right until the "Silver" column is reached. We find that Italy won 8 silver medals.

b. Great Britain won 11 gold medals while South Korea won 8, so South Korea won fewer gold medals than Great Britain.

EXERCISE SET 1.2

(A) *Determine the place value of the digit 5 in each whole number. See Examples 1 through 3.*

1. 352

2. 905

3. 5890

4. 6527

5. 62,500,000

6. 79,050,000

7. 5,070,099

8. 51,682,700

(B) *Write each whole number in words. See Examples 4 through 6.*

9. 5420

10. 3165

11. 26,990

12. 42,009

13. 1,620,000

14. 3,204,000

15. 53,520,170

16. 47,033,107

Write each number in the sentence in words. See Examples 4 through 6.

17. At this writing, the population of Bermuda is 62,997. (*Source:* CIA's *The World Factbook 2000*)

18. The Goodyear blimp *Eagle* holds 202,700 cubic feet of helium. (*Source:* The Goodyear Tire & Rubber Company)

Bermuda

Hamilton

19. The world's tallest building, the PETRONAS Twin Towers in Kuala Lumpur, Malaysia, is 1483 feet tall. (*Source:* Council on Tall Buildings and Urban Habitat)

20. Liz Harold has the number 16,820,409 showing on her calculator display.

21. Each day, UPS delivers 13,600,000 packages and documents worldwide. (*Source:* United Parcel Service of America, Inc.)

22. The highest point in New Mexico is Wheeler Peak, at an elevation of 13,161 feet. (*Source:* U.S. Geological Survey)

Write each whole number in standard form. See Examples 7 through 10.

23. Six thousand, five hundred eight

24. Three thousand, three hundred seventy

25. Twenty-nine thousand, nine hundred

26. Forty-two thousand, six

27. Six million, five hundred four thousand, nineteen

28. Ten million, thirty-seven thousand, sixteen

29. Three million, fourteen

30. Seven million, twelve

Write the whole number in each sentence in standard form. See Examples 7 through 10.

31. Hank Aaron holds the career record for home runs in Major League baseball, with a total of seven hundred fifty-five home runs. (*Source:* Major League Baseball)

32. The average distance between Earth and the Sun is more than 93 million miles.

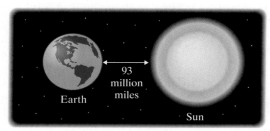

33. The average annual salary for an NHL hockey player for the 2000–2001 season was one million, four hundred thousand dollars. (*Source:* NHL Players Association)

34. In 2000, there were one hundred million, eight hundred thousand U.S. households that owned at least one television set. (*Source:* Nielsen Media Research)

35. As of 2001, there were one thousand, two hundred forty-four species classified as either threatened or endangered in the United States. (*Source:* U.S. Fish & Wildlife Service)

36. The list price for a 2001 Lamborghini Diablo is two hundred seventy-four thousand, nine hundred dollars. (*Source:* Automotive Information Center)

Write each whole number in expanded form. See Example 11.

37. 406　　　　　**38.** 789　　　　　**39.** 5290　　　　　**40.** 6040

41. 62,407　　　　　**42.** 20,215　　　　　**43.** 30,680　　　　　**44.** 99,032

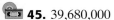 **45.** 39,680,000　　　　　　　　　**46.** 47,703,029

The table shows the six tallest mountains in New England and their elevations. Use this table to answer Exercises 47 through 52. See Example 12.

Mountain	Elevation (in feet)
Boott Spur (NH)	5492
Mt. Adams (NH)	5774
Mt. Clay (NH)	5532
Mt. Jefferson (NH)	5712
Mt. Sam Adams (NH)	5584
Mt. Washington (NH)	6288

(*Source:* U.S. Geological Survey)

47. Write the elevation of Mt. Clay in words.

48. Write the elevation of Mt. Washington in words.

49. Write the height of Boott Spur in expanded form.

50. Write the height of Mt. Jefferson in expanded form.

51. Which mountain is the tallest in New England?

52. Which mountain is the second tallest in New England?

...ble shows the top ten breeds of dogs in 2000 according to the American Kennel Club. Use this table to answer Exer-
, 53 through 56. See Example 12.

Top Ten American Kennel Club Registrations in 2000	
Breed	**Number of Registered Dogs**
Beagle	52,026
Boxer	38,803
Chihuahua	43,096
Dachshund	54,773
German shepherd dog	57,660
Golden retriever	66,300
Labrador retriever	172,841
Poodle	45,868
Shih Tzu	37,599
Yorkshire terrier	43,574

(*Source:* American Kennel Club)

53. Which breed has the most American Kennel Club registrations? Write the number of registrations for this breed in words.

54. Which of the listed breeds has the fewest registrations? Write the number of registered dogs for this breed in words.

55. Which breed has more dogs registered, Chihuahua or Golden retriever?

56. Which breed has fewer dogs registered, Beagle or Yorkshire terrier?

 Combining Concepts

57. Write the largest four-digit number that can be made from the digits 3, 6, 7, and 2 if each digit must be used once. ____ ____ ____ ____

58. Write the largest five-digit number that can be made using the digits 4, 5, and 3 if each digit must be used at least once. ____ ____ , ____ ____

59. If a number is given in words, describe the process used to write this number in standard form.

60. If a number is written in standard form, describe the process used to write this number in expanded form.

61. The Pro-Football Hall of Fame was established on September 7, 1963, in this town. Use the information and the diagram below to find the name of the town.

- Alliance is east of Massillon.
- Dover is between Canton and New Philadelphia.
- Massillon is not next to Alliance.
- Canton is north of Dover.

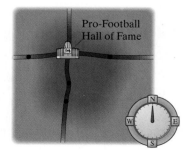

1.3   Adding Whole Numbers

 Adding Whole Numbers

OBJECTIVES

Ⓐ Add whole numbers.
Ⓑ Find the perimeter of a polygon.
Ⓒ Solve problems by adding whole numbers.

SSM   TUTOR CENTER   SG   CD & VIDEO   MATH PRO   WEB

If one computer in an office has a 2-megabyte memory and a second computer has a 4-megabyte memory, the total memory in the two computers can be found by adding 2 and 4.

$$2 \text{ megabytes} + 4 \text{ megabytes} = 6 \text{ megabytes}$$

The **sum** is 6 megabytes of memory. Each of the numbers 2 and 4 is called an **addend**.

$$\underset{\text{addend}}{2} \;+\; \underset{\text{addend}}{4} \;=\; \underset{\text{sum}}{6}$$

To add whole numbers, we add the digits in the ones place, then the tens place, then the hundreds place, and so on. For example, let's add 2236 + 160.

```
  2236
+  160
  2396
```

Line up numbers vertically so that the place values correspond. Then add digits in corresponding place values, starting with the ones place.

sum of ones
sum of tens
sum of hundreds
sum of thousands

EXAMPLE 1   Add: 23 + 136

Solution:
```
   23
+ 136
  159
```

●

When the sum of digits in corresponding place values is more than 9, "carrying" is necessary. For example, to add 365 + 89, add the ones-place digits first.

```
  ¹
  365
+  89
    4
```

5 ones + 9 ones = **14 ones** or **1 ten** + **4 ones**
Write the 4 ones in the ones place and carry the 1 ten to the tens place.

Next, add the tens-place digits.

Practice Problem 1

Add: 7235 + 542

Answer

1. 7777

$$\begin{array}{r} \overset{1\ 1}{3\,6\,5} \\ +\ 8\,9 \\ \hline 5\,4 \end{array}$$

1 ten + 6 tens + 8 tens = **15 tens** or **1 hundred + 5 tens**

Write the 5 tens in the tens place and carry the 1 hundred to the hundreds place.

Next, add the hundreds-place digits.

$$\begin{array}{r} \overset{1\ 1}{365} \\ +\ \ 89 \\ \hline 454 \end{array}$$

1 hundred + 3 hundreds = 4 hundreds

Write the 4 hundreds in the hundreds place.

Practice Problem 2

Add: 27,364 + 92,977

EXAMPLE 2   Add: 34,285 + 149,761

Solution:

$$\begin{array}{r} \overset{1\ 1\ 1}{34,285} \\ +\ 149,761 \\ \hline 184,046 \end{array}$$

●

Try the Concept Check in the margin.

Concept Check

What is wrong with the following computation?

$$\begin{array}{r} 394 \\ +\ 283 \\ \hline 577 \end{array}$$

Before we continue adding whole numbers, let's review some properties of addition that you may have already discovered. The first property that we will review is the **addition property of 0**. This property reminds us that the sum of 0 and any number is that same number.

> ### Addition Property of 0
>
> The sum of 0 and any number is that number. For example,
>
> $$7 + 0 = 7$$
> $$0 + 7 = 7$$

Next, notice that we can add any two whole numbers in any order and the sum is the same. For example,

$$4 + 5 = 9 \qquad \text{and} \qquad 5 + 4 = 9$$

We call this special property of addition the **commutative property of addition**.

> ### Commutative Property of Addition
>
> Changing the **order** of two addends does not change their sum. For example,
>
> $$2 + 3 = 5 \qquad \text{and} \qquad 3 + 2 = 5$$

Another property that can help us when adding numbers is the **associative property of addition**. This property states that when adding numbers, the grouping of the numbers can be changed without changing the sum. We use parentheses to group numbers. They indicate what numbers to add first. For example, let's use two different groupings to find the sum of $2 + 1 + 5$.

$$2 + \underline{(1 + 5)} = 2 + 6 = 8$$

Answers

2. 120,341

Concept Check: forgot to carry 1 hundred to the hundreds place

Also,

$$(2 + 1) + 5 = 3 + 5 = 8$$

Both groupings give a sum of 8.

Associative Property of Addition

Changing the grouping of addends does not change their sum. For example,

$$3 + (5 + 7) = 3 + 12 = 15 \quad \text{and} \quad (3 + 5) + 7 = 8 + 7 = 15$$

The commutative and associative properties tell us that we can add whole numbers using any order and grouping that we want.

When adding several numbers, it is often helpful to look for two or three numbers whose sum is 10, 20, and so on.

EXAMPLE 3   Add: $13 + 2 + 7 + 8 + 9$

Solution: $13 + 2 + 7 + 8 + 9 = 39$

$$20 + 10 + 9$$

$$39$$

EXAMPLE 4   Add: $1647 + 246 + 32 + 85$

Solution:
$$
\begin{array}{r}
{\scriptstyle 1\,2\,2} \\
1647 \\
246 \\
32 \\
+\quad 85 \\
\hline
2010
\end{array}
$$

B Finding the Perimeter of a Polygon

A special application of addition is finding the perimeter of a polygon. A **polygon** can be described as a flat figure formed by line segments connected at their ends. (For more review, see Appendix C.) Geometric figures such as triangles, squares, and rectangles are called polygons.

Triangle   Square   Rectangle

The **perimeter** of a polygon is the *distance around* the polygon. This means that the perimeter of a polygon is the sum of the lengths of its sides.

Practice Problem 3

Add: $11 + 7 + 8 + 9 + 13$

Practice Problem 4

Add: $19 + 5042 + 638 + 526$

Answers

3. 48,   **4.** 6225

Practice Problem 5

Find the perimeter of the polygon shown. (A centimeter is a unit of length in the metric system.)

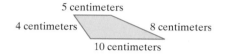

5 centimeters
4 centimeters
8 centimeters
10 centimeters

Practice Problem 6

A new shopping mall has a floor plan in the shape of a triangle. Each of the mall's three sides is 532 feet. Find the perimeter of the building.

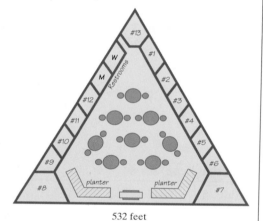

532 feet

EXAMPLE 5

Find the perimeter of the polygon shown.

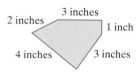

2 inches
3 inches
1 inch
4 inches
3 inches

Solution: To find the perimeter (distance around), we add the lengths of the sides.

$$2 \text{ in.} + 3 \text{ in.} + 1 \text{ in.} + 3 \text{ in.} + 4 \text{ in.} = 13 \text{ in.}$$

The perimeter is 13 inches.

EXAMPLE 6   Calculating the Perimeter of a Building

The largest commercial building in the world under one roof is the flower auction building of the cooperative VBA in Aalsmeer, Netherlands. The floor plan is a rectangle that measures 776 meters by 639 meters. Find the perimeter of this building. (A meter is a unit of length in the metric system.) (*Source: The Handy Science Answer Book*, Visible Ink Press)

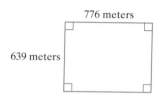

776 meters
639 meters

Solution: Recall that opposite sides of a rectangle have the same lengths. To find the perimeter of this building, we add the lengths of the sides. The sum of the lengths of its sides is

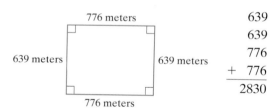

776 meters
639 meters   639 meters
776 meters

$$
\begin{array}{r}
639 \\
639 \\
776 \\
+ \ 776 \\
\hline
2830
\end{array}
$$

The perimeter of the building is 2830 meters.

© Solving Problems by Adding

Often, real-life problems occur that can be solved by writing an addition statement. The first step in solving any word problem is to *understand* the problem by reading it carefully. Descriptions of problems solved through addition *may* include any of these key words or phrases:

Key Words or Phrases	Example	Symbols
added to	5 added to 7	7 + 5
plus	0 plus 78	0 + 78
increased by	12 increased by 6	12 + 6
more than	11 more than 25	25 + 11
total	the total of 8 and 1	8 + 1
sum	the sum of 4 and 133	4 + 133

Answers

5. 27 cm,   **6.** 1596 ft

To solve a word problem that involves addition, we first use the facts described to write an addition statement. Then we write the corresponding solution of the real-life problem. It is sometimes helpful to write the statement in words (brief phrases) and then translate to numbers.

EXAMPLE 7   Finding a Salary

The governor's salary in the state of Florida was recently increased by $7867. If the old salary was $112,304, find the new salary. (*Source: The World Almanac and Book of Facts*, 2000)

Solution: The key phrase here is "increased by," which suggests that we add. To find the new salary, we add the increase, $7867, to the old salary.

In Words		Translate to Numbers
old salary	→	112,304
+ increase	→	+ 7867
new salary	→	120,171

The Florida governor's salary is now $120,171.

EXAMPLE 8   Determining the Number of Baseball Cards in a Collection

Alan Mayfield collects baseball cards. He has 109 cards for the New York Yankees, 96 for the Chicago White Sox, 79 for the Kansas City Royals, 42 for the Seattle Mariners, 67 for the Oakland Athletics, and 52 for the California Angels. How many cards does he have in total?

Solution: The key word here is "total." To find the total number of Alan's baseball cards, we find the sum of the quantities for each team.

In Words		Translate to Numbers
New York Yankees cards	→	109
Chicago White Sox cards	→	96
Kansas City Royals cards	→	79
Seattle Mariners cards	→	42
Oakland Athletics cards	→	67
+ California Angels cards	→	+ 52
Total cards	→	445

Alan has a total of 445 baseball cards..

Practice Problem 7

Texas produces 90 million pounds of pecans per year. Georgia is the world's top pecan producer and produces 15 million pounds more pecans than Texas. How much does Georgia produce? (*Source:* Absolute Trivia.com)

Practice Problem 8

Elham Abo-Zahrah collects thimbles. She has 42 glass thimbles, 17 steel thimbles, 37 porcelain thimbles, 9 silver thimbles, and 15 plastic thimbles. How many thimbles are in her collection?

Answers

7. 105 million lb,   **8.** 120 thimbles

CALCULATOR EXPLORATIONS

Adding Numbers

To add numbers on a calculator, find the keys marked $\boxed{+}$ and $\boxed{=}$ or $\boxed{\text{ENTER}}$.

For example, to add 5 and 7 on a calculator, press the keys $\boxed{5}$ $\boxed{+}$ $\boxed{7}$ $\boxed{=}$ or $\boxed{\text{ENTER}}$.

The display will read $\boxed{\qquad 12}$.
Thus, $5 + 7 = 12$.

To add 687 and 981 on a calculator, press the keys $\boxed{687}$ $\boxed{+}$ $\boxed{981}$ $\boxed{=}$ or $\boxed{\text{ENTER}}$.

The display will read $\boxed{\qquad 1668}$.

Thus, $687 + 981 = 1668$. (Although entering 687, for example, requires pressing more than one key, here numbers are grouped together for easier reading.)

Use a calculator to add.

1. $89 + 45$ **2.** $76 + 97$

3. $285 + 55$ **4.** $8773 + 652$

5.   985   **6.**   465

      1210   9888

       562   620

  $+$ 77 $+$ 1550

Mental Math

Find each sum.

1. 5 + 7   **2.** 20 + 30   **3.** 5000 + 4000   **4.** 4300 + 26   **5.** 1620 + 0   **6.** 6 + 126 + 4

EXERCISE SET 1.3

A *Add. See Examples 1 through 4.*

1. 14
      + 22

2. 27
      + 31

3. 62
      + 30

4. 37
      + 42

5. 12
       13
      + 24

6. 23
       45
      + 30

 7. 5267
      + 132

8. 236
      + 6243

9. 53 + 64

10. 41 + 74

11. 22 + 49

12. 35 + 47

13. 38 + 79   **14.** 92 + 37   **15.** 8
                                                      9
                                                      2
                                                      5
                                                    + 1

16. 3
        5
        8
        5
      + 7

17. 6
        21
        14
        9
      + 12

18. 12
        4
        8
        26
      + 10

19. 81
        17
        23
        79
      + 12

20. 64
        28
        56
        25
      + 32

21. 62 + 18 + 14

22. 23 + 49 + 18

23. 40 + 800 + 70   **24.** 30 + 900 + 20   **25.** 7542 + 49 + 682   **26.** 1624 + 1832 + 1976

 27. 24 + 9006 + 489 + 2407   **28.** 16 + 748 + 1056 + 770

29. 627
        628
      + 629

30. 427
        383
      + 229

31. 6820
        4271
      + 5626

32. 6789
        4321
      + 5555

33. 507
        593
      + 10

34. 864
        733
      + 356

35. 4200
        2107
      + 2692

36. 5000	**37.** 49	**38.** 26	**39.** 121,742	**40.** 504,218
400	628	582	57,279	321,920
+ 3021	5762	4763	6586	38,507
	+ 29,462	+ 62,511	+ 426,782	+ 594,687

B *Find the perimeter of each figure. See Examples 5 and 6.*

△ **41.**

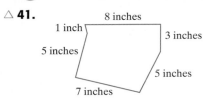

△ **42.**

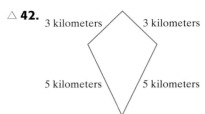

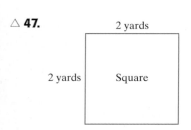 **43.**

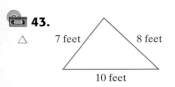

△ **44.**

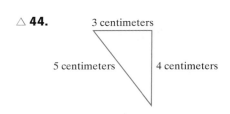

△ **45.**

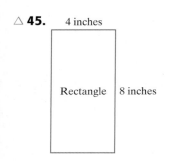

△ **46.**

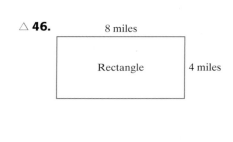

△ **47.**

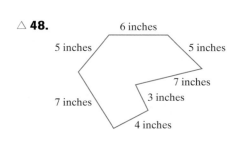

△ **48.**

24

Solve. See Examples 7 and 8.

49. The highest point in South Carolina is Sassafras Mountain at 3560 feet above sea level. The highest point in North Carolina is Mt. Mitchell, whose peak is 3124 feet higher than Sassafras Mountain. Find how high Mt. Mitchell is. (*Source:* U.S. Geological Survey)

50. The distance from Kansas City, Kansas, to Hays, Kansas, is 285 miles. Colby, Kansas, is 98 miles farther from Kansas City than Hays. Find how far it is from Kansas City to Colby.

△ **51.** Leo Callier is installing an invisible fence in his backyard. How many feet of wiring are needed to enclose the yard below?

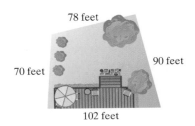

78 feet
90 feet
70 feet
102 feet

△ **52.** A homeowner is considering adding gutters around her home. Find the perimeter of her rectangular home.

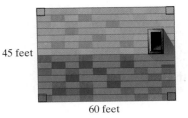

45 feet
60 feet

53. Dan Marino holds the NFL career record for most passes completed. He completed 2305 passes from the beginning of his NFL career in 1983 through 1989. He completed another 2662 passes from 1990 through 1999, his last season before retiring from professional football. How many passes did he complete during his NFL career? (*Source:* National Football League)

54. In 2000, Harley-Davidson sold 158,817 of its motorcycles domestically. In addition, 45,775 Harley-Davidson motorcycles were sold internationally. What was the total number of Harley-Davidson motorcycles sold in 2000? (*Source:* Harley-Davidson, Inc.)

55. During the spring of 2001, Kellogg Company acquired Keebler Foods Company. Before the merger, Kellogg employed 15,000 people and Keebler employed 13,000 people. Assuming there were no layoffs after the merger, how many employees did the newly combined company have? (*Source:* Kellogg Company)

56. The DVD video format was introduced in March 1997. From 1997 through 1999, a total of 5,423,786 DVD players were sold in the United States. A total of 8,498,545 DVD players were sold in the United States in 2000. How many DVD players in all were sold from 1997 through the end of 2000? (*Source:* Consumer Electronics Association)

57. Wilma Rudolph, who won three gold medals in track and field events in the 1960 Summer Olympics, was born in 1940. Marion Jones, who also won three gold medals in track and field events but in the 2000 Summer Olympics, was born 35 years later. In what year was Marion Jones born?

58. In 1999, there were 7153 Blockbuster video rental stores worldwide. In 2000, the Blockbuster store tally had increased by 524 stores. How many Blockbuster stores were there in 2000? (*Source:* Blockbuster Inc.)

 Combining Concepts

The table shows the number of Target stores in ten states. Use this table to answer Exercises 59 through 64.

The Top States for Target Stores in 2001

State	Number of Stores
Arizona	28
California	153
Florida	68
Georgia	31
Illinois	52
Indiana	30
Michigan	49
Minnesota	56
Ohio	37
Texas	91

(*Source:* Target Corporation)

59. Which state has the most Target stores?

60. Which of the states listed in the table has the fewest number of Target stores?

61. What is the total number of Target stores located in the three states with the most Target stores?

62. Which pair of neighboring states have more Target stores, Indiana and Illinois or Michigan and Ohio?

63. How many Target stores are located in the ten states listed in the table? Use a calculator to check your total.

64. Target operates stores in a total of 46 states. There are 396 Target stores located in the states not listed in the table. What is the total number of Target stores in the United States?

65. In your own words, explain the commutative property of addition.

66. In your own words, explain the associative property of addition.

67. Add: $78,962 + 129,968,350 + 36,462,880$

68. Add: $56,468,980 + 1,236,785 + 986,768,000$

1.4 Subtracting Whole Numbers

Ⓐ **Subtracting Whole Numbers**

If you have $5 and someone gives you $3, you have a total of $8, since $5 + 3 = 8$. Similarly, if you have $8 and then someone borrows $3, you have $5 left. **Subtraction** is finding the **difference** of two numbers.

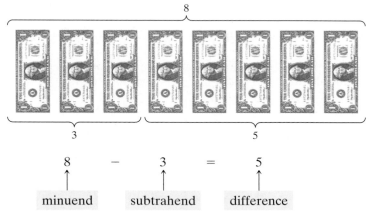

$$8 \quad - \quad 3 \quad = \quad 5$$

minuend   subtrahend   difference

Notice that addition and subtraction are very closely related. In fact, subtraction is defined in terms of addition.

$$8 - 3 = 5 \text{ because } 5 + 3 = 8$$

This means that subtraction can be *checked* by addition, and we say that addition and subtraction are reverse operations.

EXAMPLE 1   Subtract. Check each answer by adding.

a. $12 - 9$ **b.** $11 - 6$ **c.** $5 - 5$ **d.** $7 - 0$

Solution:

a. $12 - 9 = 3$ because $3 + 9 = 12$
b. $11 - 6 = 5$ because $5 + 6 = 11$
c. $5 - 5 = 0$ because $0 + 5 = 5$
d. $7 - 0 = 7$ because $7 + 0 = 7$

Look again at Examples 1(c) and 1(d).

1(c) $5 - 5 = 0$

same number   difference is 0

1(d) $7 - 0 = 7$

a number minus 0   difference is the same number

These two examples illustrate the subtraction properties of 0.

Subtraction Properties of 0

The difference of any number and that same number is 0. For example,

$$11 - 11 = 0$$

The difference of any number and 0 is that same number. For example,

$$45 - 0 = 45$$

Practice Problem 1

Subtract. Check each answer by adding.

a. $14 - 9$
b. $9 - 9$
c. $4 - 0$

When subtraction involves numbers of two or more digits, it is more convenient to subtract vertically. For example, to subtract $893 - 52$,

$$
\begin{array}{r}
8\ 9\ 3 \quad \leftarrow \text{minuend} \\
-\ \ 5\ 2 \quad \leftarrow \text{subtrahend} \\
\hline
8\ 4\ 1 \quad \leftarrow \text{difference}
\end{array}
$$

Line up the numbers vertically so that the minuend is on top and the place values correspond. Subtract in corresponding places, starting with the ones place.

$$
\begin{array}{l}
3 - 2 \\
9 - 5 \\
8 - 0
\end{array}
$$

To check, add.

$$
\begin{array}{r}
\text{difference} \\
+\ \text{subtrahend} \\
\hline
\text{minuend}
\end{array}
\quad \text{or} \quad
\begin{array}{r}
841 \\
+\ 52 \\
\hline
893
\end{array}
$$

Since this is the original minuend, the problem checks.

Practice Problem 2

Subtract. Check by adding.

a. $4689 - 253$

b. $981 - 630$

EXAMPLE 2   Subtract: $7826 - 505$. Check by adding.

Solution:
$$
\begin{array}{r}
7826 \\
-\ 505 \\
\hline
7321
\end{array}
$$

Check:
$$
\begin{array}{r}
7321 \\
+\ 505 \\
\hline
7826
\end{array}
$$

B Subtracting with Borrowing

When a digit in the second number (subtrahend) is larger than the corresponding digit in the first number (minuend), **borrowing** is necessary. For example, consider

$$
\begin{array}{r}
81 \\
-\ 63
\end{array}
$$

Since the 3 in the ones place of 63 is larger than the 1 in the ones place of 81, borrowing is necessary. We borrow 1 ten from the tens place and add it to the ones place.

Borrowing

$$
\underset{\text{tens}}{8} - \underset{\text{ten}}{1} = \underset{\text{tens}}{7} \quad \rightarrow \quad
\begin{array}{r}
7\ 11 \leftarrow 1 \text{ ten} + 1 \text{ one} = 11 \text{ ones} \\
8\!\!\!/\ 1\!\!\!/ \\
-\ 6\ 3
\end{array}
$$

Now we subtract the ones-place digits and then the tens-place digits.

$$
\begin{array}{r}
\overset{7\ \ 11}{8\!\!\!/\ 1\!\!\!/} \\
-\ 6\ 3 \\
\hline
1\ 8 \leftarrow 11 - 3 = 8 \\
 \ \ 7 - 6 = 1
\end{array}
$$

Check:

$$
\begin{array}{r}
18 \\
+\ 63 \\
\hline
81
\end{array}
$$
The original minuend

EXAMPLE 3   Subtract: 43 − 29. Check by adding.

Solution:
$$\begin{array}{r} \overset{3\ 13}{\cancel{4}\ \cancel{3}} \\ -\ 2\ 9 \\ \hline 1\ 4 \end{array}$$

Check:
$$\begin{array}{r} 14 \\ +\ 29 \\ \hline 43 \end{array}$$

Sometimes we may have to borrow from more than one place. For example, to subtract 7631 − 152, we first borrow from the tens place.

$$\begin{array}{r} 76\overset{2\ 11}{\cancel{3}\ \cancel{1}} \\ -\ 1\ 5\ 2 \\ \hline 9 \end{array} \leftarrow 11-2=9$$

In the tens place, 5 is greater than 2, so we borrow again. This time we borrow from the hundreds place.

6 hundreds − **1 hundred** = 5 hundreds

1 hundred + 2 tens or
10 tens + 2 tens = 12 tens

$$\begin{array}{r} 7\overset{5\ \cancel{6}\ 11}{\cancel{6}\ \cancel{3}\ \cancel{1}} \\ -\ 1\ 5\ 2 \\ \hline 7\ 4\ 7\ 9 \end{array}$$

Check:
$$\begin{array}{r} 7479 \\ +\ 152 \\ \hline 7631 \end{array}$$ The original minuend

EXAMPLE 4   Subtract: 900 − 174. Check by adding.

Solution: In the ones place, 4 is larger than 0, so we borrow from the tens place. But the tens place of 900 is 0, so to borrow from the tens place we must first borrow from the hundreds place.

$$\begin{array}{r} \overset{8\ 10}{\cancel{9}\ \cancel{0}}\ 0 \\ -\ 1\ 7\ 4 \end{array}$$

Now borrow from the tens place.

$$\begin{array}{r} \overset{8\ \overset{9}{\cancel{10}}\ 10}{\cancel{9}\ \cancel{0}\ \cancel{0}} \\ -\ 1\ 7\ 4 \\ \hline 7\ 2\ 6 \end{array}$$

Check:
$$\begin{array}{r} \overset{1\ 1}{726} \\ +\ 174 \\ \hline 900 \end{array}$$

C   Solving Problems by Subtracting

Descriptions of real-life problems that suggest solving by subtraction include these key words or phrases:

Practice Problem 3

Subtract. Check by adding.

a. $\begin{array}{r} 227 \\ -\ 175 \end{array}$

b. $\begin{array}{r} 1136 \\ -\ 914 \end{array}$

c. $\begin{array}{r} 8627 \\ -\ 4119 \end{array}$

Practice Problem 4

Subtract. Check by adding.

a. $\begin{array}{r} 400 \\ -\ 164 \end{array}$

b. $\begin{array}{r} 200 \\ -\ 45 \end{array}$

c. $\begin{array}{r} 1000 \\ -\ 762 \end{array}$

Answers
3. a. 52,   **b.** 222,   **c.** 4508
4. a. 236,   **b.** 155,   **c.** 238

Concept Check

In each of the following problems, identify which number is the minuend and which number is the subtrahend.

a. What is the result when 9 is subtracted from 20?

b. What is the difference of 15 and 8?

c. Find a number that is 15 fewer than 23.

Practice Problem 5

The radius of Earth is 6378 kilometers. The radius of Mars is 2981 kilometers less than the radius of Earth. What is the radius of Mars? (*Source:* National Space Science Data Center)

Key Words or Phrases	Examples	Symbols
subtract	subtract 5 from 8	$8 - 5$
difference	the difference of 10 and 2	$10 - 2$
less	17 less 3	$17 - 3$
take away	14 take away 9	$14 - 9$
decreased by	7 decreased by 5	$7 - 5$
subtracted from	9 subtracted from 12	$12 - 9$

Try the Concept Check in the margin.

EXAMPLE 5   Finding the Radius of a Planet

The radius of Venus is 6052 kilometers. The radius of Mercury is 3612 kilometers less than the radius of Venus. Find the radius of Mercury. (*Source:* National Space Science Data Center)

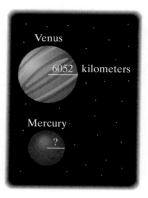

Solution:

In Words		Translate to Numbers
radius of Venus	$\longrightarrow$	6052
less 3612	$\longrightarrow$	$-\ 3612$
radius of Mercury	$\longrightarrow$	2440

The radius of Mercury is 2440 kilometers.   ●

Practice Problem 6

A new suit originally priced at $92 is now on sale for $47. How much money was taken off the original price?

EXAMPLE 6   Calculating Miles Per Gallon

A subcompact car gets 42 miles per gallon of gas. A full-size car gets 17 miles per gallon of gas. How many more miles per gallon does the subcompact car get than the full-size car?

Solution:

	In Words		Translate to Numbers
	subcompact miles per gallon	$\longrightarrow$	$\overset{3\ \ 12}{\cancel{4}\,\cancel{2}}$
$-$	full-size miles per gallon	$\longrightarrow$	$-\ 1\ 7$
	more miles per gallon		$2\ 5$

The subcompact car gets 25 more miles per gallon than the full-size car.   ●

Since subtraction and addition are reverse operations, don't forget that a subtraction problem can be checked by adding.

Graphs can be used to visualize data. The graph shown next is called a **bar graph**.

EXAMPLE 7   Reading a Bar Graph

The graph below shows the ratings of Best Picture nominees since PG-13 was introduced in 1984. In this graph, each bar represents a different rating, and the height of each bar represents the number of Best Picture nominees for that rating.

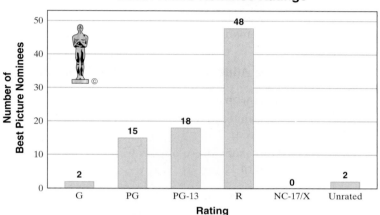

Best Picture Nominee Ratings

Source: www.oscar.com, 2001

a. Which rating did most Best Picture nominees have?

b. How many more Best Picture nominees were rated PG-13 than PG?

Solution:

a. The rating for most Best Picture nominees is the one corresponding to the highest bar, which is an R rating.

b. The number of Best Picture nominees rated PG-13 is 18. The number of Best Picture nominees rated PG is 15. To find how many more pictures were rated PG-13, we subtract.

$$18 - 15 = 3$$

Three more Best Picture nominees were rated PG-13 than PG.

Practice Problem 7

Use the graph in Example 7 to answer the following:

a. Which rating had the least number of Best Picture nominees?

b. How many more Best Picture nominees were rated R than G?

Answers

7. a. NC-17/X,   **b.** 46

CALCULATOR EXPLORATIONS

Subtracting Numbers

To subtract numbers on a calculator, find the keys marked $\boxed{-}$ and $\boxed{=}$ or $\boxed{\text{ENTER}}$.

For example, to find $83 - 49$ on a calculator, press the keys $\boxed{83}$ $\boxed{-}$ $\boxed{49}$ $\boxed{=}$ or $\boxed{\text{ENTER}}$.

The display will read $\boxed{34}$. Thus, $83 - 49 = 34$.

Use a calculator to subtract.

1. $865 - 95$ **2.** $76 - 27$

3. $147 - 38$ **4.** $366 - 87$

5. $9625 - 647$ **6.** $10,711 - 8925$

FOCUS ON **Mathematical Connections**

MODELING SUBTRACTION OF WHOLE NUMBERS

A mathematical concept can be represented or modeled in many different ways. For instance, subtraction can be represented by the following symbolic model:

$$11 - 4$$

The following verbal models can also represent subtraction of these same quantities:

"Four subtracted from eleven" or
"Eleven take away four"

Physical models can also represent mathematical concepts. In these models, a number is represented by that many objects. For example, the number 5 can be represented by five pennies, squares, paper clips, tiles, or bottle caps.

A physical representation of the number 5

Take-Away Model for Subtraction: 11 − 4

- Start with 11 objects.
- Take 4 objects away.
- How many objects remain?

Comparison Model for Subtraction: 11 − 4

- Start with a set of 11 of one type of object and a set of 4 of another type of object.

- Make as many pairs that include one object of each type as possible.

- How many more objects are in the larger set?

Missing Addend Model for Subtraction: 11 − 4

- Start with 4 objects.
- Continue adding objects until a total of 11 is reached.
- How many more objects were needed to give a total of 11?

CRITICAL THINKING

Use an appropriate physical model for subtraction to solve each of the following problems. Explain your reasoning for choosing each model.

1. Sneha has assembled 12 computer components so far this shift. If his quota is 20 components, how many more components must he assemble to reach his quota?
2. Yuko wants to plant 14 daffodil bulbs in her yard. She planted 5 bulbs in the front yard. How many bulbs does she have left for planting in the backyard?
3. Todd is 19 years old and his sister Tanya is 13 years old. How much older is Todd than Tanya?

Name _____ Section _____ Date _____

Mental Math

Find each difference.

1. 9 − 2

2. 6 − 6

3. 5 − 0

4. 44 − 22

5. 93 − 93

6. 700 − 400

7. 700 − 300

8. 700 − 700

9. 600 − 100

10. 600 − 0

EXERCISE SET 1.4

 Subtract. Check by adding. See Examples 1 and 2.

1. 67
  − 23

2. 72
  − 41

3. 82
  − 22

4. 27
  − 10

5. 389
  − 124

6. 572
  − 321

7. 677
  − 423

8. 766
  − 324

9. 998
  − 453

10. 912
  − 610

11. 749
  − 149

12. 257
  − 257

 Subtract. Check by adding. See Examples 1 through 4.

13. 62
  − 37

14. 55
  − 29

15. 70
  − 25

16. 80
  − 37

17. 938
  − 792

18. 436
  − 275

19. 922
  − 634

20. 674
  − 299

21. 600
  − 432

22. 300
  − 149

23. 42
  − 36

24. 73
  − 29

25.   923
        − 476

26.   813
        − 227

27.   6283
        − 560

28.   5349
        − 720

29.   533
        − 29

30.   724
        − 16

31.   200
        − 111

32.   300
        − 211

33.   1983
        − 1904

34.   1983
        − 1914

35.   56,422
        − 16,508

36.   76,652
        − 29,498

37. 50,000 − 17,289

38. 40,000 − 23,582

39. 7020 − 1979

40. 6050 − 1878

41. 51,111 − 19,898

42. 62,222 − 39,898

43. Subtract 5 from 9.

44. Subtract 9 from 21.

45. Find the difference of 41 and 21.

46. Find the difference of 16 and 5.

47. Subtract 56 from 63.

48. Subtract 41 from 59.

Solve. See Examples 5 through 7.

49. Dyllis King is reading a 503-page book. If she has just finished reading page 239, how many more pages must she read to finish the book?

50. When Lou and Judy Zawislak began a trip, the odometer read 55,492. When the trip was over, the odometer read 59,320. How many miles did they drive on their trip?

51. During the 2000–2001 regular season, Jerry Stackhouse of the Detroit Pistons led the NBA in total points scored with 2380. The Philadelphia 76ers' Allen Iverson placed second for total points scored with 2207. How many more points did Stackhouse score than Iverson during the 2000–2001 regular season? (*Source:* National Basketball Association)

52. In 1999, Americans bought 233,125 Ford Expeditions. In 2000, 19,642 fewer Expeditions were sold in the United States. How many Expeditions were sold in the United States in 2000? (*Source:* Ford Motor Company)

53. The peak of Mt. McKinley in Alaska is 20,320 feet above sea level. The peak of Long's Peak in Colorado is 14,255 feet above sea level. How much higher is the peak of Mt. McKinley than Long's Peak? (*Source:* U.S. Geological Survey)

Mt. McKinley, Alaska   Long's Peak, Colorado

54. On one day in May the temperature in Paddin, Indiana, dropped 27 degrees from 2 p.m. to 4 p.m. If the temperature at 2 p.m. was 73° Fahrenheit, what was the temperature at 4 p.m.?

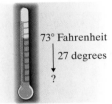

73° Fahrenheit

27 degrees

?

55. Buhler Gomez has a total of $539 in his checking account. If he writes a check for each of the items below, how much money will be left in his account?

South Central Bell	$27
Central LA Electric Co.	$101
Mellon Finance	$236

56. Pat Salanki's blood cholesterol level is 243. The doctor tells him it should be decreased to 185. How much of a decrease is this?

57. The distance from Kansas City to Denver is 645 miles. Hays, Kansas, lies on the road between the two and is 287 miles from Kansas City. What is the distance between Hays and Denver?

58. Alan Little is trading his car in on a new car. The new car costs $15,425. His car is worth $7998. How much more money does he need to buy the new car?

59. A new VCR with remote control costs $525. Prunella Pasch has $914 in her savings account. How much will she have left in her savings account after she buys the VCR?

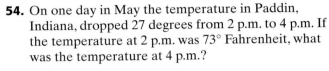 **60.** A stereo that regularly sells for $547 is discounted by $99 in a sale. What is the sale price?

61. The population of Florida grew from 12,937,926 in 1990 to 15,982,378 in 2000. What was Florida's population increase over this time period? (*Source:* U.S. Census Bureau)

62. The population of El Paso, Texas, was 515,342 in 1990 and 563,662 in 2000. By how much did the population of El Paso grow from 1990 to 2000? (*Source:* U.S. Census Bureau)

63. In 1996, there were 45,305 cocker spaniels registered with the American Kennel Club. In 2000, there were 15,912 fewer cocker spaniels registered. How many cocker spaniels were registered with the AKC in 2000? (*Source:* American Kennel Club)

64. In the United States, there were 41,589 tornadoes from 1950 through 2000. In all, 13,205 of these tornadoes occurred from 1990 through 2000. How many tornadoes occurred during the period prior to 1990? (*Source:* Storm Prediction Center, National Weather Service)

65. Jo Keen and Trudy Waterbury were candidates for student government president. Who won the election if the votes were cast as follows? By how many votes did the winner win?

Class	Candidate	
	Jo	Trudy
Freshman	276	295
Sophomore	362	122
Junior	201	312
Senior	179	18

66. Two students submitted advertising budgets for a student government fund-raiser.

	Student A	Student B
Radio ads	$600	$300
Newspaper ads	$200	$400
Posters	$150	$240
Handbills	$120	$170

If $1200 is available for advertising, how much excess would each budget have?

67. Until recently, the world's largest permanent maze was located in Ruurlo, Netherlands. This maze of beech hedges covers 94,080 square feet. A new hedge maze using hibiscus bushes at the Dole Plantation in Wahiawa, Hawaii, covers 100,000 square feet. How much larger is the Dole Plantation maze than the Ruurlo maze? (*Source: The Guinness Book of Records*)

68. There were only 27 California condors in the entire world in 1987. By 2001, the number of California condors had increased to 184. How much of an increase was this? (*Source:* California Department of Fish and Game)

69. Papa John's is the third largest pizza chain in the United States. In 1999, there were 2486 Papa John's restaurants worldwide. By 2000, the number of Papa John's restaurants had grown to 2817. How many new Papa John's restaurants were added during 2000? (*Source:* Papa John's International Inc.)

70. The Mackinac Bridge is a suspension bridge that connects the lower and upper peninsulas of Michigan across the Straits of Mackinac. Its total length is 26,372 feet. The Lake Pontchartrain Bridge is a twin concrete trestle bridge in Slidell, Louisiana. Its total length is 28,547 feet. Which bridge is longer and by how much? (*Sources:* Mackinac Bridge Authority and Federal Highway Administration, Bridge Division)

The bar graph shows the number of airplane movements (landings and takeoffs of an aircraft) in 2000 for the top five airports in the world. Use this graph to answer Exercises 71 through 74. See Example 7.

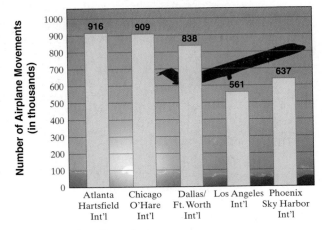

Top Five Airports in 2000

Number of Airplane Movements (in thousands)

Atlanta Hartsfield Int'l	916
Chicago O'Hare Int'l	909
Dallas/ Ft. Worth Int'l	838
Los Angeles Int'l	561
Phoenix Sky Harbor Int'l	637

Source: Airports Council International

71. Which airport is the busiest?

72. Which airports have fewer than 700 thousand airplane movements per year?

73. How many more airplane movements per year does the Chicago O'Hare International Airport have than the Phoenix Sky Harbor International Airport?

74. How many more airplane movements per year does the Atlanta Hartsfield International Airport have than the Dallas/Ft. Worth International Airport?

Combining Concepts

The table shows the top ten leading television advertisers in the United States in 2000 and the amount of money spent in that year on television ads. Use this table to answer Exercises 75 through 79.

Advertiser	Amount Spent on TV Ads in 2000
DaimlerChrysler AG	$826,194,200
General Motors Corp.	$582,423,300
Ford Motor Co. Dealers Association	$410,105,200
Ford Motor Co.	$305,798,500
Honda Motor Co. Ltd.	$305,089,900
Verizon Communications	$287,416,500
Nissan Motor Co. Ltd.	$245,449,700
Toyota Motor Corp.	$225,498,300
McDonald's Corp.	$223,790,000
General Mills Inc.	$199,736,600

(*Source:* Television Bureau of Advertising, Inc.)

75. Which companies spent more than $500 million on television ads?

76. Which companies spent fewer than $300 million on television ads?

77. How much more money did DaimlerChrysler AG spend on television ads than General Motors Corp.?

78. How much more money did Verizon Communications spend on television ads than McDonald's Corp.?

79. Find the total amount of money spent by these ten companies on television ads.

80. The local college library is having a Million Pages of Reading promotion. The freshmen have read a total of 289,462 pages; the sophomores have read a total of 369,477 pages; the juniors have read a total of 218,287 pages; and the seniors have read a total of 121,685 pages. Have they reached a goal of one million pages? If not, how many more pages need to be read?

Fill in the missing digits in each problem.

81.
```
   526_
 - 2_85
 ------
   28_4
```

82.
```
   10,_4_
  - 8 5_4
  -------
    _710
```

83. Is there a commutative property of subtraction? In other words, does order matter when subtracting? Why or why not?

1.5 Rounding and Estimating

Ⓐ Round whole numbers.

Ⓑ Use rounding to estimate sums and differences.

Ⓒ Solve problems by estimating.

SSM
TUTOR CENTER   SG   CD & VIDEO   MATH PRO   WEB

Ⓐ Rounding Whole Numbers

Rounding a whole number means approximating it. A rounded whole number is often easier to use, understand, and remember than the precise whole number. For example, instead of trying to remember the Iowa state population as 2,851,792, it is much easier to remember it rounded to the nearest million: 3 million people.

To understand rounding, let's look at the following illustrations. The whole number 36 is closer to 40 than 30, so 36 rounded to the nearest ten is 40.

The whole number 52 rounded to the nearest ten is 50 because 52 is closer to 50 than to 60.

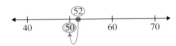

In trying to round 25 to the nearest ten, we see that 25 is halfway between 20 and 30. It is not closer to either number. In such a case, we round to the larger ten, that is, to 30.

To round a whole number without using a number line, follow these steps:

Rounding Whole Numbers to a Given Place Value

Step 1. Locate the digit to the right of the given place value.

Step 2. If this digit is 5 or greater, add 1 to the digit in the given place value and replace each digit to its right by 0.

Step 3. If this digit is less than 5, replace it and each digit to its right by 0.

EXAMPLE 1   Round 568 to the nearest ten.

Solution: 5 6 ⑧   The digit to the right of the tens place is the ones place, which is circled.
↑
tens place

5 6 ⑧   Since the circled digit is 5 or greater, add 1 to the 6 in the tens place and replace the digit to the right by 0.
↑
Add 1.   Replace with 0.

We find that 568 rounded to the nearest ten is 570.

Practice Problem 1

Round to the nearest ten.

a. 46
b. 731
c. 125

Answers

 1. a. 50,   **b.** 730,   **c.** 130

Practice Problem 2

Round to the nearest thousand.

a. 56,702

b. 7444

c. 291,500

Practice Problem 3

Round to the nearest hundred.

a. 2777

b. 38,152

c. 762,955

Concept Check

Round each of the following numbers to the nearest *hundred*. Explain your reasoning.

a. 79

b. 33

Practice Problem 4

Round each number to the nearest ten to find an estimated sum.

$$
\begin{array}{r}
79 \\
35 \\
42 \\
21 \\
+\ 98 \\
\hline
\end{array}
$$

Answers

2. a. 57,000, **b.** 7000, **c.** 292,000, **3. a.** 2800,
b. 38,200, **c.** 763,000, **4.** 280

Concept Check: **a.** 100, **b.** 0

EXAMPLE 2   Round 278,362 to the nearest thousand.

Solution: Thousands place

278, ③ 62

Do not add 1.   Replace with zeros.

3 is less than 5.

The number 278,362 rounded to the nearest thousand is 278,000.

EXAMPLE 3   Round 248,982 to the nearest hundred.

Solution: Hundreds place

248,9 ⑧ 2

Add 1.

8 is greater than or equal to 5.

9 + 1 = 10, so replace the digit 9 by 0 and carry 1 to the place value to the left.

$$2\ \ \overset{8+\ 1}{4}\ \overset{}{8},\ \overset{0}{9}\ \ 8\ \ 2$$

Add 1.   Replace with zeros.

The number 248,982 rounded to the nearest hundred is 249,000.

Try the Concept Check in the margin.

B   Estimating Sums and Differences

By rounding addends, we can estimate sums. An estimated sum is appropriate when an exact sum is not necessary. To estimate the sum shown, round each number to the nearest hundred and then add.

$$
\begin{array}{rll}
768 & \text{rounds to} & 800 \\
1952 & \text{rounds to} & 2000 \\
225 & \text{rounds to} & 200 \\
+\ 149 & \text{rounds to} & +\ 100 \\
\hline
& & 3100
\end{array}
$$

The estimated sum is 3100, which is close to the exact sum of 3094.

EXAMPLE 4

Round each number to the nearest hundred to find an estimated sum.

$$
\begin{array}{r}
294 \\
625 \\
1071 \\
+\ 349 \\
\hline
\end{array}
$$

Solution:

$$
\begin{array}{rll}
294 & \text{rounds to} & 300 \\
625 & \text{rounds to} & 600 \\
1071 & \text{rounds to} & 1100 \\
+\ 349 & \text{rounds to} & +\ 300 \\
\hline
& & 2300
\end{array}
$$

The estimated sum is 2300. (The exact sum is 2339.)

EXAMPLE 5

Round each number to the nearest hundred to find an estimated difference.

$$
\begin{array}{r}
4725 \\
- 2879 \\
\end{array}
$$

Solution:

$$
\begin{array}{lll}
4725 & \text{rounds to} & 4700 \\
- 2879 & \text{rounds to} & - 2900 \\
\hline
& & 1800 \\
\end{array}
$$

The estimated difference is 1800. (The exact difference is 1846.) ●

 Solving Problems by Estimating

Making estimates is often the quickest way to solve real-life problems when their solutions do not need to be exact.

EXAMPLE 6   Estimating Distances

Jose Guillermo is trying to estimate quickly the distance from Temple, Texas, to Brenham, Texas. Round each distance given on the map to the nearest ten to estimate the total distance.

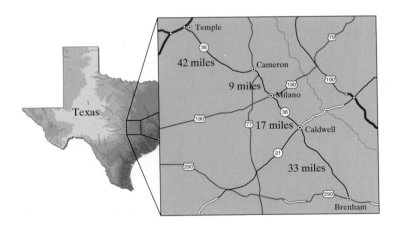

Solution:

Distance		Estimation
42	rounds to	40
9	rounds to	10
17	rounds to	20
+ 33	rounds to	+ 30
		100

It is approximately 100 miles from Temple to Brenham. (The exact distance is 101 miles.) ●

Practice Problem 5

Round each number to the nearest thousand to find an estimated difference.

$$
\begin{array}{r}
4725 \\
- 2879 \\
\end{array}
$$

Practice Problem 6

Tasha Kilbey is trying to estimate how far it is from Grove, Kansas, to Hays, Kansas. Round each given distance on the map to the nearest ten to estimate the total distance.

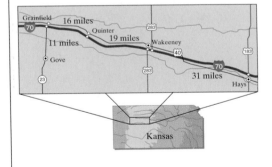

Answers

5. 2000, **6.** 80 mi

Practice Problem 7

In a recent year, there were 120,624 reported cases of chicken pox, 22,866 reported cases of tuberculosis, and 45,970 reported cases of salmonellosis in the United States. Round each number to the nearest ten-thousand to estimate the total number of cases reported for these diseases. (*Source:* Centers for Disease Control and Prevention)

EXAMPLE 7 Estimating Data

In three recent years the numbers of reported cases of mumps in the United States were 906, 1537, and 1692. Round each number to the nearest hundred to estimate the total number of cases reported over this period. (*Source:* Centers for Disease Control and Prevention)

Solution:

Number of Cases		Estimation
906	rounds to	900
1537	rounds to	1500
+ 1692	rounds to	+ 1700
		4100

The approximate number of cases reported over this period is 4100. ●

EXERCISE SET 1.5

In class → evens only
Homework → odds only

A *Round each whole number to the given place. See Examples 1 through 3.*

1. 632 to the nearest ten

630

2. 273 to the nearest ten

270

3. 635 to the nearest ten

640

4. 275 to the nearest ten

280

5. 792 to the nearest ten

790

6. 394 to the nearest ten

390

7. 395 to the nearest ten

400

8. 582 to the nearest ten

580

9. 1096 to the nearest ten

1,100

10. 2198 to the nearest ten

2200

11. 42,682 to the nearest thousand

4,300

12. 42,682 to the nearest ten-thousand

40,000

13. 248,695 to the nearest hundred

248,700

14. 179,406 to the nearest hundred

179,300

15. 36,499 to the nearest thousand

36,000

16. 96,501 to the nearest thousand

17. 99,995 to the nearest ten

100,000

18. 39,994 to the nearest ten

19. 59,725,642 to the nearest ten-million

60,000,000

20. 39,523,698 to the nearest million

40,000,000

Complete the table by estimating the given number to the given place value.

		Tens	Hundreds	Thousands
21.	5281	5280	5300	5,000
22.	7619	7,620	7,600	8,000
23.	9444	9430	9300	8,000
24.	7777	7,780	7,800	8,000
25.	14,876	14880	14 900	15,000
26.	85,049	85,050	85,000	85,000

27. Estimate to the nearest thousand the active duty U.S. Air Force personnel level in 2000, 355,654. (*Source:* U.S. Department of Defense)

356,000

28. Estimate to the nearest hundred-thousand the number of passengers handled in 2000 by the Atlanta Hartsfield International Airport, 80,171,036. (*Source:* Airports Council International)

80,200,000

Round each number to the indicated place.

29. Kareem Abdul-Jabbar holds the NBA record for points scored, a total of 38,387 over his NBA career. Round this number to the nearest thousand. (*Source:* National Basketball Association)

38,000

30. It takes 10,759 days for Saturn to make a complete orbit around the Sun. Round this number to the nearest hundred. (*Source:* National Space Science Data Center)

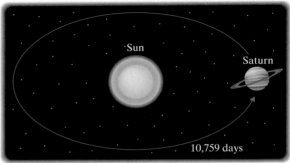

Sun

Saturn

10,759 days

31. According to the 2000 U.S. Census, the population of the United States is 281,421,906. Round this population figure to the nearest million. (*Source:* U.S. Census Bureau)

281,000,000

32. The top U.S. advertiser in 2000 was General Motors Corp., which spent a total of $2,883,214,638 on advertising in all types of media. Round this advertising expense to the nearest hundred-million. (*Source:* CMR, a Taylor Nelson Sofres company)

33. The average salary for a Major League baseball player during the 2001 season was $2,264,403. Round this average salary to the nearest hundred-thousand. (*Source:* Major League Baseball Players Association)

$2,300,000

34. In 2000, the Procter & Gamble Company had $39,951,000,000 in sales. Round this sales figure to the nearest billion. (*Source:* The Procter & Gamble Company)

B *Estimate the sum or difference by rounding each number to the nearest ten. See Examples 4 and 5.*

35.
```
  29
  35
  42
+ 16
────
 130
```

36.
```
  62   60
  72   70
  15   20
+ 19   20
────  ────
      170
```

37.
```
  649   650
- 272   270
─────  ────
        380
```

38.
```
  555   500
- 235   240
─────  ────
        320
```

Estimate the sum or difference by rounding each number to the nearest hundred. See Examples 4 and 5.

39.
```
  1812   1800
  1776   1800
+ 1945   1900
──────  ─────
        5500
```

40.
```
  2010   2000
  2001   2000
+ 1984   2000
──────  ─────
        6000
```

41.
```
  1774   1800
- 1492   1500
──────  ─────
         300
```

42.
```
  1989   2000
- 1870   1900
──────  ─────
         100
```

43.
```
  2995   3000
  1649   1600
+ 3940   3900
──────  ─────
        8500
```

44.
```
  799   800
  1655  1700
+ 271   300
─────  ─────
       2800
```

Estimation is useful to check for incorrect answers when using a calculator. For example, pressing a key too hard may result in a double digit, while pressing a key too softly may result in the number not appearing in the display.

Two of the given calculator answers below are incorrect. Find them by estimating each sum.

45. 362 + 419   781   correct

46. 522 + 785   1307   correct

47. 432 + 679 + 198   1139   incorrect

48. 229 + 443 + 606   1278   correct

49. 7806 + 5150   12,956   correct

50. 5233 + 4988   9011   incorrect

51. 31,439 + 18,781   50,220   correct

52. 68,721 + 52,335   121,056   correct

C *Solve each problem by estimating. See Examples 6 and 7.*

53. Campo Appliance Store advertises three refrigerators on sale at $799, $1299, and $999. Round each cost to the nearest hundred to estimate the total cost.

$3\,1,000$

54. Jared Nuss scored 89, 92, 100, 67, 75, and 79 on his calculus tests. Round each score to the nearest ten to estimate his total score.

$90, 90, 100, 70, 80, 80$

510

55. Arlene Neville wants to estimate quickly the distance from Stockton to LaCrosse. Round each distance given on the map to the nearest ten miles to estimate the total distance.

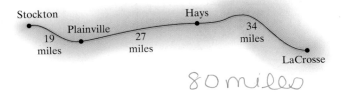

Stockton
Plainville
19 miles
27 miles
Hays
34 miles
LaCrosse

$80\ miles$

56. Carmelita Watkins is pricing new stereo systems. One system sells for $1895 and another system sells for $1524. Round each price to the nearest hundred dollars to estimate the difference in price of these systems.

1900
1500
400

57. The peak of Mt. McKinley, in Alaska, is 20,320 feet above sea level. The top of Mt. Rainier, in Washington, is 14,410 feet above sea level. Round each height to the nearest thousand to estimate the difference in elevation of these two peaks. (*Source:* U.S. Geological Survey)

$6000\,ft$

58. The Gonzales family took a trip and traveled 458, 489, 377, 243, 69, and 702 miles on six consecutive days. Round each distance to the nearest hundred to estimate the distance they traveled.

59. In 2000 the population of Chicago was 2,896,016, and the population of Philadelphia was 1,517,550. Round each population to the nearest hundred-thousand to estimate how much larger Chicago was than Philadelphia. (*Source:* U.S. Census Bureau, 2000 census)

$1,400,000$

60. The distance from Kansas City to Boston is 1429 miles and from Kansas City to Chicago, 530 miles. Round each distance to the nearest hundred to estimate how much farther Boston is from Kansas City than Chicago is.

61. In the 1964 presidential election, Lyndon Johnson received 41,126,233 votes and Barry Goldwater received 27,174,898 votes. Round each number of votes to the nearest million to estimate the number of votes by which Johnson won the election.

$14,000,000\ votes$

62. Enrollment figures at Normal State University showed an increase from 49,713 credit hours in 1988 to 51,746 credit hours in 1989. Round each number to the nearest thousand to estimate the increase.

63. Head Start is a national program that provides developmental and social services for America's low-income preschool children ages three to five. Enrollment figures in Head Start programs showed an increase from 750,696 children in 1995 to 857,664 children in 2000. Round each number of children to the nearest thousand to estimate this increase. (*Source:* Head Start Bureau)

$107,000\ children$

64. In 2000, General Motors produced 271,800 Saturn cars. Similarly, in 1999 only 232,570 Saturns were produced. Round each number of cars to the nearest thousand to estimate the increase in Saturn production from 1999 to 2000. (*Source:* General Motors Corporation)

$2000 \rightarrow 272,000$
$1999 \rightarrow 233,000$
$39,000$

The following table (from Section 1.4) shows the top ten leading television advertisers in the United States for 2000 and the amount of money spent in that year on television ads. Use this table to answer Exercises 65 through 68.

Advertiser	Amount Spent on TV Ads in 2000
DaimlerChrysler AG	$826,194,200
General Motors Corp.	$582,423,300
Ford Motor Co. Dealers Association	$410,105,200
Ford Motor Co.	$305,798,500
Honda Motor Co. Ltd.	$305,089,900
Verizon Communications	$287,416,500
Nissan Motor Co. Ltd.	$245,449,700
Toyota Motor Corp.	$225,498,300
McDonald's Corp.	$223,790,000
General Mills Inc.	$199,736,600

(*Source:* Television Bureau of Advertising, Inc.)

65. Approximate the amount of money spent on television advertising by General Motors Corp. to the nearest hundred-million.

$600,000,000

66. Approximate the amount of money spent on television advertising by Nissan Motor Co. Ltd. to the nearest hundred-million.

67. Approximate the amount of money spent on television advertising by McDonald's Corp. to the nearest million.

$224,000,000

68. Approximate the amount of money spent on television advertising by Ford Motor Co. to the nearest million.

 Combining Concepts

69. A number rounded to the nearest hundred is 8600. Determine the smallest possible number.

70. Determine the largest possible number.

71. On August 23, 1989, it was estimated that 1,500,000 people joined hands in a human chain stretching 370 miles to protest the fiftieth anniversary of the pact that allowed what was then the Soviet Union to annex the Baltic nations in 1939. If the estimate of the number of people is to the nearest hundred-thousand, determine the largest possible number of people in the chain.

72. In your own words, explain how to round a number to the nearest thousand.

△ **73.** Estimate the perimeter by first rounding each length to the nearest hundred.

5950 miles   7693 miles

8203 miles

46

1.6 Multiplying Whole Numbers

Suppose that we wish to count the number of desks in a classroom. The desks are arranged in 5 rows, and each row has 6 desks.

OBJECTIVES

Ⓐ Use the properties of multiplication.

Ⓑ Multiply whole numbers.

Ⓒ Find the area of a rectangle.

Ⓓ Solve problems by multiplying whole numbers.

SSM
TUTOR CENTER   SG   CD & VIDEO   MATH PRO   WEB

6 desks in each row

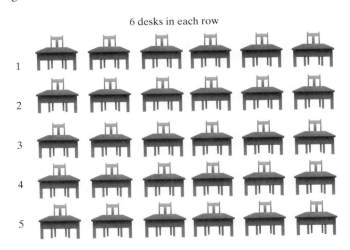

Adding 5 sixes gives the total number of desks: $6 + 6 + 6 + 6 + 6 = 30$ desks. When each addend is the same, we refer to this as **repeated addition**.
   Multiplication is repeated addition but with different notation.

$$\underbrace{6 + 6 + 6 + 6 + 6}_{\text{5 sixes}} = \underset{\text{factor}}{5} \quad \underset{}{\times} \quad \underset{\text{factor}}{6} \quad = \quad \underset{\text{product}}{30}$$

The $\times$ is called a **multiplication sign**. The numbers 5 and 6 are called **factors**. The number 30 is called the **product**. The notation 5×6 is read as "five times six." The symbols $\cdot$ and ( ) can also be used to indicate multiplication.

$$5 \times 6 = 30, \quad 5 \cdot 6 = 30, \quad (5)(6) = 30, \quad \text{and} \quad 5(6) = 30$$

Try the Concept Check in the margin.

Ⓐ Using the Properties of Multiplication

As with addition, we memorize products of one-digit whole numbers and then use certain properties of multiplication to multiply larger numbers. (If necessary, review the multiplication of one-digit numbers in Appendix B.) Notice in the appendix that when any number is multiplied by 0, the result is always 0. This is called the **multiplication property of 0**.

Multiplication Property of 0

The product of 0 and any number is 0. For example,

$$5 \cdot 0 = 0$$
$$0 \cdot 8 = 0$$

   Also notice in the appendix that when any number is multiplied by 1, the result is always the original number. We call this result the **multiplication property of 1**.

Concept Check

a. Rewrite $4 + 4 + 4 + 4 + 4 + 4 + 4$ using multiplication.

b. Rewrite 3×16 as repeated addition. Is there more than one way to do this? If so, show all ways.

Answers

Concept Check: **a.** $7 \times 4 = 28$
b. $16 + 16 + 16 = 48$; yes,
   $3 + 3 + 3 + 3 + 3 + 3 + 3 + 3 + 3 + 3 +$
   $3 + 3 + 3 + 3 + 3 + 3 = 48$

Multiplication Property of 1

The product of 1 and any number is that same number. For example,

$1 \cdot 9 = 9$

$6 \cdot 1 = 6$

Practice Problem 1

Multiply.

a. 3×0

b. $4(1)$

c. $(0)(34)$

d. $1 \cdot 76$

EXAMPLE 1 Multiply.

a. 6×1 **b.** $0(8)$ **c.** $1 \cdot 45$ **d.** $(75)(0)$

Solution:

a. $6 \times 1 = 6$ **b.** $0(8) = 0$

c. $1 \cdot 45 = 45$ **d.** $(75)(0) = 0$ ●

Like addition, multiplication is commutative and associative. Notice that when multiplying two numbers, the order of these numbers can be changed without changing the product. For example,

$3 \cdot 5 = 15$ and $5 \cdot 3 = 15$

This property is the **commutative property of multiplication**.

Commutative Property of Multiplication

Changing the **order** of two factors does not change their product. For example,

$9 \cdot 2 = 18$ and $2 \cdot 9 = 18$

Another property that can help us when multiplying is the **associative property of multiplication**. This property states that when multiplying numbers, the grouping of the numbers can be changed without changing the product. For example,

$2 \cdot (3 \cdot 4) = 2 \cdot 12 = 24$

Also,

$(2 \cdot 3) \cdot 4 = 6 \cdot 4 = 24$

Both groupings give a product of 24.

Associative Property of Multiplication

Changing the **grouping** of factors does not change their product. For example,

$5 \cdot (3 \cdot 2) = (5 \cdot 3) \cdot 2$

$5 \cdot (3 \cdot 2) = 5 \cdot 6 = 30$ and $(5 \cdot 3) \cdot 2 = 15 \cdot 2 = 30$

With these properties, along with the **distributive property**, we can find the product of any whole numbers. The distributive property says that multiplication **distributes** over addition. For example, notice that $3(2 + 5)$ is the same as $3 \cdot 2 + 3 \cdot 5$.

Answers

1. **a.** 0, **b.** 4, **c.** 0, **d.** 76

$$3\underbrace{(2 + 5)} = 3(7) = 21$$

$$\underbrace{3 \cdot 2} + 3 \cdot 5 = 6 + 15 = 21$$

Notice in $3(2 + 5) = 3 \cdot 2 + 3 \cdot 5$ that each number inside the parentheses is multiplied by 3.

Distributive Property

Multiplication distributes over addition. For example,

$$2(3 + 4) = 2 \cdot 3 + 2 \cdot 4$$

EXAMPLE 2   Rewrite each using the distributive property.

a. $3(4 + 5)$ **b.** $10(6 + 8)$ **c.** $2(7 + 3)$

Solution: Using the distributive property, we have

a. $3(4 + 5) = 3 \cdot 4 + 3 \cdot 5$
b. $10(6 + 8) = 10 \cdot 6 + 10 \cdot 8$
c. $2(7 + 3) = 2 \cdot 7 + 2 \cdot 3$

B   **Multiplying Whole Numbers**

Let's use the distributive property to multiply 7(48). To do so, we begin by writing the expanded form of 48 and then applying the distributive property.

$$7(48) = 7(40 + 8)$$
$$= 7 \cdot 40 + 7 \cdot 8 \quad \text{Apply the distributive property.}$$
$$= 280 + 56 \quad \text{Multiply.}$$
$$= 336 \quad \text{Add.}$$

This is how we multiply whole numbers. When multiplying whole numbers, we will use the following notation.

$$\begin{array}{r} \overset{5}{48} \\ \times\ 7 \\ \hline 336 \end{array} \leftarrow 7 \cdot 8 = 56$$

Write 6 in the ones place and carry 5 to the tens place.

$7 \cdot 4 = 28$ and $28 + 5 = 33$

EXAMPLE 3   Multiply: $\begin{array}{r} 25 \\ \times\ 8 \end{array}$

Solution: $\begin{array}{r} \overset{4}{25} \\ \times\ 8 \\ \hline 200 \end{array}$

To multiply larger whole numbers, use the following similar notation. Multiply 89×52.

Practice Problem 2

Rewrite each using the distributive property.

a. $5(2 + 3)$
b. $9(8 + 7)$
c. $3(6 + 1)$

Practice Problem 3

Multiply.
a. $\begin{array}{r} 36 \\ \times\ 4 \end{array}$ b. $\begin{array}{r} 92 \\ \times\ 9 \end{array}$

Answers
2. a. $5(2 + 3) = 5 \cdot 2 + 5 \cdot 3$,
b. $9(8 + 7) = 9 \cdot 8 + 9 \cdot 7$,
c. $3(6 + 1) = 3 \cdot 6 + 3 \cdot 1$,
3. a. 144,   **b.** 828

Step 1

$$\begin{array}{r} \overset{1}{89} \\ \times\ 52 \\ \hline 178 \end{array}$$ ← Multiply 89 × 2.

Step 2

$$\begin{array}{r} \overset{4}{89} \\ \times\ 52 \\ \hline 178 \\ 4450 \end{array}$$ ← Multiply 89 × 50.

Step 3

$$\begin{array}{r} 89 \\ \times\ 52 \\ \hline 178 \\ 4450 \\ \hline 4628 \end{array}$$ Add.

The numbers 178 and 4450 are called **partial products**. The sum of the partial products, 4628, is the product of 89 and 52.

Practice Problem 4

Multiply.

a. $\begin{array}{r} 594 \\ \times\ 72 \\ \hline \end{array}$ b. $\begin{array}{r} 306 \\ \times\ 81 \\ \hline \end{array}$

EXAMPLE 4   Multiply: 236 × 86

Solution:

$$\begin{array}{r} 236 \\ \times\ 86 \\ \hline 1416 \\ 18,880 \\ \hline 20,296 \end{array}$$

← 6(236)
← 80(236)
Add.

Practice Problem 5

Multiply.

a. $\begin{array}{r} 726 \\ \times\ 142 \\ \hline \end{array}$ b. $\begin{array}{r} 4 \\ \times\ 288 \\ \hline \end{array}$

EXAMPLE 5   Multiply: 631 × 125

Solution:

$$\begin{array}{r} 631 \\ \times\ 125 \\ \hline 3155 \\ 12,620 \\ 63,100 \\ \hline 78,875 \end{array}$$

← 5(631)
← 20(631)
← 100(631)
Add.

Try the Concept Check in the margin.

Concept Check

Find and explain the error in the following multiplication problem:

$$\begin{array}{r} 102 \\ \times\ 33 \\ \hline 306 \\ 306 \\ \hline 612 \end{array}$$

C   Finding the Area of a Rectangle

A special application of multiplication is finding the area of a region. Area measures the amount of surface of a region. For example, we measure a plot of land or the living space of a home by area. The figures show two examples of units of area measure. (A centimeter is a unit of length in the metric system.)

Actual size

1 inch | 1 square inch |

1 inch

Actual size

| 1 square cm | 1 centimeter

1 centimeter

To measure the area of a geometric figure such as the rectangle shown, count the number of square units that cover the region.

How many of these are in this?

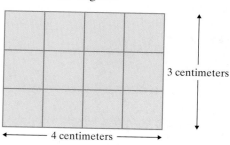

3 centimeters

4 centimeters

This rectangular region contains 12 square units, each 1 square centimeter. Thus, the area is 12 square centimeters. This total number of squares can be found by counting or by multiplying **4 · 3**(length · width).

$$\text{Area of a rectangle} = \text{length} \cdot \text{width}$$
$$= (4\,\text{centimeters})(3\,\text{centimeters})$$
$$= 12\,\text{square centimeters}$$

In this section, we find the areas of rectangles only. In later sections, we find the areas of other geometric regions.

△ EXAMPLE 6   Finding the Area of a State

The state of Colorado is in the shape of a rectangle whose length is 380 miles and whose width is 280 miles. Find its area.

Solution: The area of a rectangle is the product of its length and its width.

$$\text{Area} = \text{length} \cdot \text{width}$$
$$= (380\,\text{miles})(280\,\text{miles})$$
$$= 106,400\,\text{square miles}$$

The area of Colorado is 106,400 square miles.

Solving Problems by Multiplying

There are several words or phrases that indicate the operation of multiplication. Some of these are as follows:

Key Words or Phrases	Example	Symbols
multiply	multiply 5 by 7	$5 \cdot 7$
product	the product of 3 and 2	$3 \cdot 2$
times	10 times 13	$10 \cdot 13$

Many key words or phrases describing real-life problems that suggest addition might be better solved by multiplication instead. For example, to find the **total** cost of 8 shirts, each selling for $27, we can either add $27 + 27 + 27 + 27 + 27 + 27 + 27 + 27$, or we can multiply $8(27)$.

Practice Problem 6   △

The state of Wyoming is in the shape of a rectangle whose length is 360 miles and whose width is 280 miles. Find its area.

Answer

6. 100,800 sq mi

Practice Problem 7

A computer printer can print 240 characters per second in draft mode. How many total characters can it print in 15 seconds?

Practice Problem 8

Softball T-shirts come in two styles: plain at $6 each and striped at $7 each. The team orders 4 plain shirts and 5 striped shirts. Find the total cost of the order.

Practice Problem 9

If an average page in a book contains 259 words, estimate, rounding to the nearest hundred, the total number of words contained on 195 pages.

Answers

7. 3600 characters,   **8.** $59,   **9.** 60,000 words

EXAMPLE 7   Finding Disk Space

A single-density computer disk can hold 370 thousand bytes of information. How many total bytes can 42 such disks hold?

Solution: Forty-two disks will hold 42×370 thousand bytes.

In Words		Translate to Numbers
bytes per disk	$\rightarrow$	370
$\times$ disks	$\rightarrow$	$\times$ 42
		740
		14,800
total bytes		15,540

Forty-two disks will hold 15,540 thousand bytes.

EXAMPLE 8   Budgeting Money

Earline Martin agrees to take her children and their cousins to the San Antonio Zoo. The ticket price for each child is $4 and for each adult, $6. If 8 children and 1 adult plan to go, how much money is needed for admission?

Solution: If the price of one child's ticket is $4, the price for 8 children is $8 \cdot 4 = \$32$. The price of one adult ticket is $6, so the total cost is

In Words		Translate to Numbers
price of 8 children	$\rightarrow$	32
$+$ price of adult	$\rightarrow$	$+$ 6
total cost		38

The total cost is $38.

EXAMPLE 9   Estimating Word Count

The average page of a book contains 259 words. Estimate, rounding to the nearest ten, the total number of words contained on 22 pages.

Solution: The exact number of words is 259×22. Estimate this product by rounding each factor to the nearest ten.

$$
\begin{array}{rcr}
259 & \text{rounds to} & \overset{1}{260} \\
\times \ 22 & \text{rounds to} & \times \ 20 \\
\hline
& & 5200
\end{array}
$$

There are approximately 5200 words contained on 22 pages.

CALCULATOR EXPLORATIONS

Multiplying Numbers

To multiply numbers on a calculator, find the keys marked $\boxed{\times}$ and $\boxed{=}$ or $\boxed{\text{ENTER}}$. For example, to find $31 \cdot 66$ on a calculator, press the keys $\boxed{31}$ $\boxed{\times}$ $\boxed{66}$ $\boxed{=}$ or $\boxed{\text{ENTER}}$. The display will read $\boxed{\qquad 2046}$. Thus, $31 \cdot 66 = 2046$.

Use a calculator to multiply.

1. 72×48

2. 81×92

3. $163 \cdot 94$

4. $285 \cdot 144$

5. $983(277)$

6. $1562(843)$

Name _____ Section _____ Date _____

Mental Math

(A) *Multiply. See Example 1.*

1. $1 \cdot 24$ **2.** $55 \cdot 1$ **3.** $0 \cdot 19$ **4.** $27 \cdot 0$

5. $8 \cdot 0 \cdot 9$ **6.** $7 \cdot 6 \cdot 0$ **7.** $87 \cdot 1$ **8.** $1 \cdot 41$

EXERCISE SET 1.6

(A) *Use the distributive property to rewrite each expression. See Example 2.*

1. $4(3 + 9)$ **2.** $5(8 + 2)$ **3.** $2(4 + 6)$ **4.** $6(1 + 4)$ **5.** $10(11 + 7)$ **6.** $12(12 + 3)$

(B) *Multiply. See Example 3.*

7. $\begin{array}{r} 42 \\ \times\ 6 \\ \hline \end{array}$ **8.** $\begin{array}{r} 79 \\ \times\ 3 \\ \hline \end{array}$ **9.** $\begin{array}{r} 624 \\ \times\ 3 \\ \hline \end{array}$ **10.** $\begin{array}{r} 638 \\ \times\ 5 \\ \hline \end{array}$

11. $\begin{array}{r} 277 \\ \times\ 6 \\ \hline \end{array}$ **12.** $\begin{array}{r} 882 \\ \times\ 2 \\ \hline \end{array}$ **13.** $\begin{array}{r} 1062 \\ \times\ 5 \\ \hline \end{array}$ **14.** $\begin{array}{r} 9021 \\ \times\ 3 \\ \hline \end{array}$

Multiply. See Examples 4 and 5.

15. $\begin{array}{r} 298 \\ \times\ 14 \\ \hline \end{array}$ **16.** $\begin{array}{r} 591 \\ \times\ 72 \\ \hline \end{array}$ **17.** $\begin{array}{r} 231 \\ \times\ 47 \\ \hline \end{array}$ **18.** $\begin{array}{r} 526 \\ \times\ 23 \\ \hline \end{array}$ **19.** $\begin{array}{r} 809 \\ \times\ 14 \\ \hline \end{array}$ **20.** $\begin{array}{r} 307 \\ \times\ 16 \\ \hline \end{array}$

21. $(620)(40)$ **22.** $(720)(80)$ **23.** $(998)(12)(0)$ **24.** $(593)(47)(0)$ **25.** $(590)(1)(10)$

26. $(240)(1)(20)$ **27.** $\begin{array}{r} 1234 \\ \times\ 48 \\ \hline \end{array}$ **28.** $\begin{array}{r} 1357 \\ \times\ 79 \\ \hline \end{array}$ **29.** $\begin{array}{r} 609 \\ \times\ 234 \\ \hline \end{array}$ **30.** $\begin{array}{r} 505 \\ \times\ 127 \\ \hline \end{array}$

31. $\begin{array}{r} 5621 \\ \times\ 324 \\ \hline \end{array}$ **32.** $\begin{array}{r} 1234 \\ \times\ 567 \\ \hline \end{array}$ **33.** $\begin{array}{r} 1941 \\ \times\ 235 \\ \hline \end{array}$ **34.** $\begin{array}{r} 1876 \\ \times\ 437 \\ \hline \end{array}$ **35.** $\begin{array}{r} 589 \\ \times\ 110 \\ \hline \end{array}$ **36.** $\begin{array}{r} 426 \\ \times\ 110 \\ \hline \end{array}$

Estimate the products by rounding each factor to the nearest hundred. See Example 9.

37. 576×354 **38.** 982×650 **39.** 604×451 **40.** 111×999

(C) *Find the area of each rectangle. See Example 6.*

41.

9 meters

7 meters

△ **42.** 4 inches

12 inches

53

△ **43.**

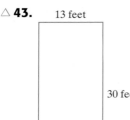

13 feet

30 feet

△ **44.**

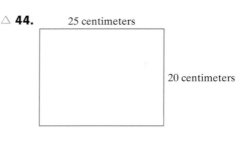

25 centimeters

20 centimeters

Ⓒ Ⓓ *Solve. See Examples 6 through 9.*

45. One tablespoon of olive oil contains 125 calories. How many calories are in 3 tablespoons of olive oil? (*Source: Home and Garden Bulletin No. 72*, U.S. Department of Agriculture).

46. One ounce of hulled sunflower seeds contains 14 grams of fat. How many grams of fat are in 6 ounces of hulled sunflower seeds? (*Source: Home and Garden Bulletin No. 72*, U.S. Department of Agriculture).

47. The textbook for a course in Civil War history costs $54. There are 35 students in the class. Find the total cost of the history books for the class.

48. The seats in the mathematics lecture hall are arranged in 12 rows with 6 seats in each row. Find how many seats are in this room.

49. A case of canned peas has *two layers* of cans. In each layer are 8 rows with 12 cans in each row. Find how many cans are in a case.

50. An apartment building has *three floors*. Each floor has five rows of apartments with four apartments in each row. Find how many apartments there are.

△ **51.** A plot of land measures 90 feet by 110 feet. Find its area.

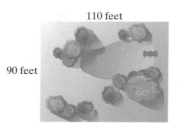

110 feet

90 feet

△ **52.** A house measures 45 feet by 60 feet. Find the floor area of the house.

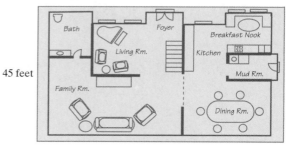

45 feet

60 feet

△ **53.** Recall from an earlier section that the largest commercial building in the world under one roof is the flower auction building of the cooperative VBA in Aalsmeer, Netherlands. The floor plan is a rectangle that measures 776 meters by 639 meters. Find the area of this building. (A meter is a unit of length in the metric system.) (*Source: The Handy Science Answer Book*, Visible Ink Press)

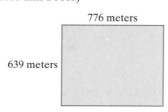

776 meters

639 meters

△ **54.** The largest lobby can be found at the Hyatt Regency in San Francisco, CA. It is in the shape of a rectangle that measures 350 feet by 160 feet. Find its area.

55. A pixel is a rectangular dot on a graphing calculator screen. If a graphing calculator screen contains 62 pixels in a row and 94 pixels in a column, find the total number of pixels on a screen.

56. A high-performance recordable compact disc can hold 700 megabytes (MB) of information. How many MBs can 17 discs hold?

57. A line of print on a computer contains 60 characters (letters, spaces, punctuation marks). Find how many characters there are in 25 lines.

58. An average cow eats 3 pounds of grain per day. Find how much grain a cow eats in a year. (Assume 365 days in 1 year.)

59. One ounce of Planters® Dry Roasted Peanuts has 160 calories. How many calories are in 8 ounces? (*Source:* RJR Nabisco, Inc.)

60. One ounce of Planters® Dry Roasted Peanuts has 13 grams of fat. How many grams of fat are in 8 ounces? (*Source:* RJR Nabisco, Inc.)

61. The diameter of the planet Saturn is 9 times as great as the diameter of Earth. The diameter of Earth is 7927 miles. Find the diameter of Saturn.

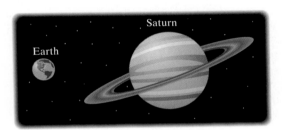

62. The planet Uranus orbits the Sun every 84 Earth years. Find how many Earth days two orbits take. (Assume 365 days in 1 year.)

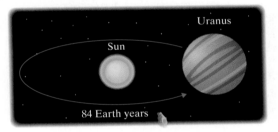

63. A window washer in New York City is bidding for a contract to wash the windows of a 23-story building. To write a bid, the number of windows in the building is needed. If there are 7 windows in each row of windows on 2 sides of the building and 4 windows per row on the other 2 sides of the building, find the total number of windows.

64. In North America, the average toy expenditure per child is $328 per year. On average, how much is spent on toys for a child by the time he or she reaches age 18? (*Source:* The NPD Group Worldwide)

65. Hershey's main chocolate factory in Hershey, Pennsylvania, uses 700,000 quarts of milk each day. How many quarts of milk would be used during the month of March, assuming that chocolate is made at the factory every day of the month? (*Source:* Hershey Foods Corp.)

66. Among older Americans (age 65 years and older), there are 4 times as many widows as widowers. There were 1,994,000 widowers in 2000. How many widows were there in 2000? (*Sources:* Administration on Aging, U.S. Census Bureau)

67. In 2000, the average cost of the Head Start program was $5951 per child. That year, 857,664 children participated in the program. Round each number to the nearest thousand and estimate the total cost of the Head Start program in 2000. (*Source:* Head Start Bureau)

68. American households currently spend an average of $4810 on food each year. A small community encompasses 1643 households. Round each number to the nearest hundred and estimate the total annual food expenditure for the residents of this community. (*Source:* U.S. Bureau of Labor Statistics)

 Combining Concepts

In a survey, college students were asked to name their favorite fruit. The results are shown in the picture graph. Use this graph to answer Exercises 69 through 72.

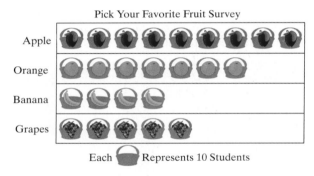

69. How many students chose grapes as their favorite fruit?

70. How many students chose bananas as their favorite fruit?

71. Which two fruits were the most popular?

72. Which two fruits were chosen by a total of 110 students?

Fill in the missing digits in each problem.

73.
```
      4_
   ×  _3
     126
    3780
    3906
```

74.
```
      _7
   ×  6_
     171
    3420
    3591
```

75. Explain how to multiply two 2-digit numbers using partial products.

76. A slice of enriched white bread has 65 calories. One tablespoon of jam has 55 calories, and one tablespoon of peanut butter has 95 calories. Suppose a peanut butter and jelly sandwich is made with two slices of white bread, one tablespoon of jam, and one tablespoon of peanut butter. How many calories are in two such peanut butter and jelly sandwiches? (*Source: Home and Garden Bulletin No. 72*, U.S. Department of Agriculture)

77. During the NBA's 2000–2001 season, Kobe Bryant of the Los Angeles Lakers scored 61 three-point field goals, 640 two-point field goals, and 475 free throws (worth one point each). How many points did Kobe Bryant score during the 2000–2001 season? (*Source:* National Basketball Association)

1.7 Dividing Whole Numbers

Suppose three people pooled their money and bought a raffle ticket at a local fund-raiser. Their ticket was the winner and they won a $60 cash prize. They then divided the prize into three equal parts so that each person received $20.

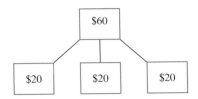

OBJECTIVES

Ⓐ Divide whole numbers.

Ⓑ Perform long division.

Ⓒ Solve problems that require dividing by whole numbers.

Ⓓ Find the average of a list of numbers.

SSM
TUTOR CENTER   SG   CD & VIDEO   MATH PRO   WEB

Ⓐ Dividing Whole Numbers

The process of separating a quantity into equal parts is called **division**. Division can be symbolized by several notations.

$$\overset{\text{quotient}}{\underset{\text{divisor}}{3\overline{)60}}}\;\overset{20}{} \leftarrow \text{dividend}$$

$$\underset{\text{divisor}}{\frac{60}{3}} = 20 \leftarrow \text{quotient} \quad(\text{dividend})$$

$$\underset{\text{dividend}\;\;\text{divisor}}{60 \div 3} = \overset{\text{quotient}}{20}$$

(In the notation $\frac{60}{3}$, the bar separating 60 and 3 is called a **fraction bar**.) Just as subtraction is the reverse of addition, division is the reverse of multiplication. This means that division can be checked by multiplication.

$$3\overline{)60}^{\,20} \quad \text{because} \quad 20 \cdot 3 = 60$$

Since multiplication and division are related in this way, you can use the multiplication table in Appendix B to review quotients of one-digit divisors if necessary.

EXAMPLE 1   Find each quotient. Check by multiplying.

a. $42 \div 7$ **b.** $\dfrac{81}{9}$ **c.** $4\overline{)24}$

Solution:

a. $42 \div 7 = 6$ because $6 \cdot 7 = 42$

b. $\dfrac{81}{9} = 9$ because $9 \cdot 9 = 81$

c. $4\overline{)24}^{\,6}$ because $6 \cdot 4 = 24$

EXAMPLE 2   Find each quotient. Check by multiplying.

a. $1\overline{)8}$ **b.** $11 \div 1$ **c.** $\dfrac{9}{9}$ **d.** $7 \div 7$ **e.** $\dfrac{10}{1}$ **f.** $6\overline{)6}$

Solution:

a. $1\overline{)8}^{\,8}$ because $8 \cdot 1 = 8$

b. $11 \div 1 = 11$ because $11 \cdot 1 = 11$

Practice Problem 1

Find each quotient. Check by multiplying.

a. $8\overline{)48}$

b. $35 \div 5$

c. $\dfrac{49}{7}$

Practice Problem 2

Find each quotient. Check by multiplying.

a. $\dfrac{8}{8}$ b. $3 \div 1$ c. $1\overline{)12}$

d. $2 \div 1$ e. $\dfrac{5}{1}$

Answers

1. a. 6, **b.** 7, **c.** 7, **2. a.** 1, **b.** 3, **c.** 12,
d. 2, **e.** 5

c. $\dfrac{9}{9} = 1$ because $1 \cdot 9 = 9$

d. $7 \div 7 = 1$ because $1 \cdot 7 = 7$

e. $\dfrac{10}{1} = 10$ because $10 \cdot 1 = 10$

f. $6\overline{)6}^{\,1}$ because $1 \cdot 6 = 6$ ●

Example 2 illustrates important properties of division as described next:

Division Properties of 1

The quotient of any number and that same number is 1. For example,

$$8 \div 8 = 1 \qquad \dfrac{7}{7} = 1 \qquad 4\overline{)4}^{\,1}$$

The quotient of any number and 1 is that same number. For example,

$$9 \div 1 = 9 \qquad \dfrac{6}{1} = 6 \qquad 1\overline{)3}^{\,3} \qquad \dfrac{0}{1} = 0$$

Practice Problem 3

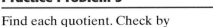

Find each quotient. Check by multiplying.

a. $\dfrac{0}{7}$ b. $5\overline{)0}$ c. $9 \div 0$

d. $0 \div 6$

EXAMPLE 3   Find each quotient. Check by multiplying.

a. $9\overline{)0}$ **b.** $0 \div 12$ **c.** $\dfrac{0}{5}$ **d.** $\dfrac{3}{0}$

Solution:

a. $9\overline{)0}^{\,0}$ because $0 \cdot 9 = 0$

b. $0 \div 12 = 0$ because $0 \cdot 12 = 0$

c. $\dfrac{0}{5} = 0$ because $0 \cdot 5 = 0$

d. If $\dfrac{3}{0} = $ a **number**, then the **number** times $0 = 3$. Recall that any number multiplied by 0 is 0 and not 3. We say, then, that $\dfrac{3}{0}$ is **undefined**.   ●

Example 3 illustrates important division properties of 0.

Division Properties of 0

The quotient of 0 and any number (except 0) is 0. For example,

$$0 \div 9 = 0 \qquad \dfrac{0}{5} = 0 \qquad 14\overline{)0}^{\,0}$$

The quotient of any number and 0 is not a number. We say that

$$\dfrac{3}{0}, \quad 0\overline{)3}, \quad \text{and} \quad 3 \div 0$$

are **undefined**.

Answers

3. a. 0, **b.** 0, **c.** undefined, **d.** 0

Performing Long Division

When dividends are larger, the quotient can be found by a process called **long division**. For example, let's divide 2541 by 3.

$$3\overline{)2541}$$

We can't divide 3 into 2, so we try dividing 3 into the first two digits.

$$\begin{array}{r} 8 \\ 3\overline{)2541} \end{array}$$ $25 \div 3 = 8$ with 1 left, so our best estimate is 8. We place 8 over the 5 in 25.

Next, multiply 8 and 3 and subtract this product from 25. Make sure that this difference is less than the divisor.

$$\begin{array}{r} 8 \\ 3\overline{)2541} \\ -24 \\ \hline 1 \end{array}$$ $8(3) = 24$
$25 - 24 = 1$, and 1 is less than the divisor 3.

Bring down the next digit and go through the process again.

$$\begin{array}{r} 84 \\ 3\overline{)2541} \\ -24\downarrow \\ \hline 14 \\ -12 \\ \hline 2 \end{array}$$ $14 \div 3 = 4$ with 2 left

$4(3) = 12$
$14 - 12 = 2$

Once more, bring down the next digit and go through the process.

$$\begin{array}{r} 847 \\ 3\overline{)2541} \\ -24 \\ \hline 14 \\ -12 \\ \hline 21 \\ -21 \\ \hline 0 \end{array}$$ $21 \div 3 = 7$

$7(3) = 21$
$21 - 21 = 0$

The quotient is 847. To check, see that $847 \times 3 = 2541$.

EXAMPLE 4   Divide: $3705 \div 5$. Check by multiplying.

Solution:

$$\begin{array}{r} 7 \\ 5\overline{)3705} \\ -35\downarrow \\ \hline 20 \end{array}$$ $37 \div 5 = 7$ with 2 left. Place this estimate, 7, over the 7 in 37.

$7(5) = 35$
$37 - 35 = 2$, and 2 is less than the divisor 5.
↑── Bring down the 0.

$$\begin{array}{r} 74 \\ 5\overline{)3705} \\ -35 \\ \hline 20 \\ -20\downarrow \\ \hline 05 \end{array}$$ $20 \div 5 = 4$

$4(5) = 20$
$20 - 20 = 0$, and 0 is less than the divisor 5.
↑── Bring down the 5.

Practice Problem 4

Divide. Check by multiplying.

a. $6\overline{)5382}$

b. $4\overline{)2212}$

Answers

4. a. 897,   **b.** 553

$$
\begin{array}{r}
741 \\
5\overline{)3705} \\
-35 \\
\hline
20 \\
-20 \\
\hline
5 \\
-5 \\
\hline
0
\end{array}
$$

$5 \div 5 = 1$

$1(5) = 5$

$5 - 5 = 0$

Check:

$$
\begin{array}{r}
741 \\
\times \quad 5 \\
\hline
3705
\end{array}
$$

> **Helpful Hint**
>
> Since division and multiplication are reverse operations, don't forget that a division problem can be checked by multiplying.

Practice Problem 5

Divide and check.

a. $3\overline{)2397}$

b. $7\overline{)2520}$

EXAMPLE 5   Divide and check: $1872 \div 9$

Solution:

$$
\begin{array}{r}
208 \\
9\overline{)1872} \\
-18 \\
\hline
07 \\
-0 \\
\hline
72 \\
-72 \\
\hline
0
\end{array}
$$

$2(9) = 18$

$18 - 18 = 0$; bring down the 7.

$0(9) = 0$

$7 - 0 = 7$; bring down the 2.

$8(9) = 72$

$72 - 72 = 0$

Check: $208 \cdot 9 = 1872$

Naturally, quotients don't always "come out even." Making 4 rows out of 26 chairs, for example, isn't possible if each row is supposed to have exactly the same number of chairs. Each of 4 rows can have 6 chairs, but 2 chairs are still left over.

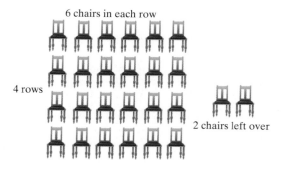

6 chairs in each row

4 rows

2 chairs left over

We signify "leftovers" or **remainders** in this way:

$$
\begin{array}{r}
6 \quad \text{R} 2 \\
4\overline{)26}
\end{array}
$$

The **whole number part of the quotient** is 6; the **remainder part of the quotient** is 2. Checking by multiplying,

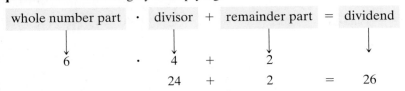

$$24 + 2 = 26$$

EXAMPLE 6 Divide and check: $2557 \div 7$

Solution:

```
      365   R 2
  7)2557
   − 21↓|         3(7) = 21
     45|          25 − 21 = 4; bring down the 5.
   − 42↓          6(7) = 42
     37           45 − 42 = 3; bring down the 7.
   − 35           5(7) = 35
      2           37 − 35 = 2; the remainder is 2.
```

Check: $365 \cdot 7 + 2 = 2557$

whole number part · divisor + remainder part = dividend

Practice Problem 6

Divide and check.

a. $5\overline{)949}$

b. $6\overline{)4399}$

EXAMPLE 7 Divide and check: $56{,}717 \div 8$

Solution:

```
     7089    R 5
  8)56717
   − 56↓||        7(8) = 56
     07||         Subtract and bring down the 7.
   −  0↓|         0(8) = 0
     71|          Subtract and bring down the 1.
   − 64↓          8(8) = 64
     77           Subtract and bring down the 7.
   − 72           9(8) = 72
      5           Subtract. The remainder is 5.
```

Check: $7089 \cdot 8 + 5 = 56{,}717$

whole number part · divisor + remainder part = dividend

Practice Problem 7

Divide and check.

a. $5\overline{)40{,}841}$

b. $7\overline{)22{,}430}$

When the divisor has more than one digit, the same pattern applies. For example, let's find $1358 \div 23$.

```
      5
  23)1358          135 ÷ 23 = 5 with 20 left over. Our estimate is 5.
     115↓          5(23) = 115
     208           135 − 115 = 20. Bring down the 8.
```

Now we continue estimating.

Answers

6. a. 189 R 4, **b.** 733 R 1, **7. a.** 8168 R 1,
b. 3204 R 2

$$\begin{array}{r} 59 \quad \text{R1} \\ 23\overline{)1358} \\ -115 \\ \hline 208 \\ -207 \\ \hline 1 \end{array}$$

208 ÷ 23 = 9 with 1 left over.

9(23) = 207

208 − 207 = 1. The remainder is 1.

To check, see that $59 \cdot 23 + 1 = 1358$.

EXAMPLE 8   Divide: $6819 \div 17$

Solution:

$$\begin{array}{r} 401 \quad \text{R 2} \\ 17\overline{)6819} \\ -68 \\ \hline 01 \\ -0 \\ \hline 19 \\ -17 \\ \hline 2 \end{array}$$

4(17) = 68

Subtract and bring down the 1.

0(17) = 0

Subtract and bring down the 9.

1(17) = 17

Subtract. The remainder is 2.

To check, see that $401 \cdot 17 + 2 = 6819$.

EXAMPLE 9   Divide: $51,600 \div 403$

Solution:

$$\begin{array}{r} 128 \quad \text{R 16} \\ 403\overline{)51600} \\ -403 \\ \hline 1130 \\ -806 \\ \hline 3240 \\ -3224 \\ \hline 16 \end{array}$$

1(403) = 403

Subtract and bring down the 0.

2(403) = 806

Subtract and bring down the 0.

8(403) = 3224

Subtract. The remainder is 16.

To check, see that $128 \cdot 403 + 16 = 51,600$.

(A) Solving Problems by Dividing

Below are some key words and phrases that indicate the operation of multiplication:

Key Words or Phrases	Examples	Symbols
divide	divide 10 by 5	$10 \div 5$ or $\dfrac{10}{5}$
quotient	the quotient of 64 and 4	$64 \div 4$ or $\dfrac{64}{4}$
divided by	9 divided by 3	$9 \div 3$ or $\dfrac{9}{3}$
divided or **shared equally among**	$100 divided equally among five people	$100 \div 5$ or $\dfrac{100}{5}$

Try the Concept Check in the margin.

Practice Problem 8

Divide: $5740 \div 19$

Practice Problem 9

Divide: $16,589 \div 247$

Concept Check

Which of the following is the correct way to represent "the quotient of 20 and 5"? Or are both correct? Explain your answer.

a. $5 \div 20$

b. $20 \div 5$

Answers

8. 302 R 2,   **9.** 67 R 40

Concept Check: **a.** incorrect,   **b.** correct

EXAMPLE 10   Finding Shared Earnings

Zachary, Tyler, and Stephanie McMillan share a paper route to earn money for college expenses. The total in their fund after expenses was $2895. How much is each person's equal share?

Solution: Each person's equal share is (total) ÷ (number of people) or

$$2895 \div 3$$

Then

$$
\begin{array}{r}
965 \\
3\overline{)2895} \\
-27 \\
\hline
19 \\
-18 \\
\hline
15 \\
-15 \\
\hline
0
\end{array}
$$

Each person's share is $965.

EXAMPLE 11   Calculating Shipping Needs

How many boxes are needed to ship 56 pairs of Nikes to a shoe store in Texarkana if 9 pairs of shoes will fit in each shipping box?

Solution:

$$
\begin{array}{ccccc}
\text{number} & = & \text{total pairs} & \div & \text{how many} \\
\text{of boxes} & & \text{of shoes} & & \text{pairs in a box} \\
\downarrow & & \downarrow & & \downarrow \\
\text{number} & = & 56 & \div & 9 \\
\text{of boxes} & & & &
\end{array}
$$

$$
\begin{array}{r}
6 \quad \text{R } 2 \\
9\overline{)56} \\
-54 \\
\hline
2
\end{array}
$$

There are 6 full boxes with 2 pairs of shoes left over, so 7 boxes will be needed.

EXAMPLE 12   Dividing Holiday Favors Among Students

Mary Schultz has 48 kindergarten students. She buys 260 stickers as Thanksgiving Day favors for her students. Can she divide the stickers up equally among her students? If not, how many stickers will be left over?

Solution:

$$
\begin{array}{ccc}
\text{number of} & \div & \text{number of} \\
\text{stickers} & & \text{students} \\
\downarrow & & \downarrow \\
260 & \div & 48
\end{array}
$$

$$
\begin{array}{r}
5 \quad \text{R } 20 \\
48\overline{)260} \\
-240 \\
\hline
20
\end{array}
$$

No, the stickers cannot be divided equally among her students since there is a nonzero remainder. There will be 20 stickers left over.

Practice Problem 10

Marina, Manual, and Min bought 10 dozen high-density computer diskettes to share equally. How many diskettes did each person get?

Practice Problem 11

Peanut butter and cheese cracker sandwiches come in 6 sandwiches to a package. How many full packages are formed with 195 sandwiches?

Practice Problem 12

Calculators can be packed 24 to a box. If 497 calculators are to be packed but only full boxes are shipped, how many full boxes will be shipped? How many calculators are left over and not shipped?

Answers

10. 40 diskettes,   **11.** 32 full packages,
12. 20 full boxes; 17 calculators left over

D Finding Averages

A special application of division (and addition) is finding the average of a list of numbers. The **average** of a list of numbers is the sum of the numbers divided by the *number* of numbers.

$$\text{average} = \frac{\text{sum of numbers}}{\textit{number} \text{ of numbers}}$$

Practice Problem 13

To compute a safe time to wait for reactions to occur after allergy shots are administered, a lab technician is given a list of elapsed times between administered shots and reactions. Find the average of the times 5 minutes, 7 minutes, 20 minutes, 6 minutes, 9 minutes, 3 minutes, and 48 minutes.

Answer

13. 14 min

EXAMPLE 13 Averaging Scores

Liam Reilly's scores in his mathematics class so far are 93, 86, 71, and 82. Find his average score.

Solution: To find his average score, we find the sum of his scores and divide by 4, the number of scores.

$$
\begin{array}{r}
93 \\
86 \\
71 \\
+\,82 \\
\hline
332 \;\text{sum}
\end{array}
\qquad
\text{average} = \frac{332}{4} = 83
\qquad
\begin{array}{r}
83 \\
4\overline{)332} \\
-32 \\
\hline
12 \\
-12 \\
\hline
0
\end{array}
$$

His average score is 83.

CALCULATOR EXPLORATIONS

Dividing Numbers

To divide numbers on a calculator, find the keys marked $\div$ and $=$ or ENTER . For example, to find $435 \div 5$ on a calculator, press the keys $\boxed{435}$ $\boxed{\div}$ $\boxed{5}$ $\boxed{=}$ or ENTER . The display will read $\boxed{87}$. Thus, $435 \div 5 = 87$.

Use a calculator to divide.

1. $848 \div 16$

2. $564 \div 12$

3. $95\overline{)5890}$

4. $27\overline{)1053}$

5. $\dfrac{32{,}886}{126}$

6. $\dfrac{143{,}088}{264}$

7. $0 \div 315$

8. $315 \div 0$

Name _____ Section _____ Date _____

Mental Math

 Find each quotient. See Examples 1 through 3.

1. $40 \div 8$ **2.** $72 \div 9$ **3.** $45 \div 5$ **4.** $24 \div 3$ **5.** $0 \div 5$

6. $0 \div 8$ **7.** $9 \div 1$ **8.** $12 \div 1$ **9.** $\dfrac{16}{16}$ **10.** $\dfrac{49}{49}$

11. $\dfrac{25}{5}$ **12.** $\dfrac{45}{9}$ **13.** $6 \div 0$ **14.** $\dfrac{12}{0}$ **15.** $7 \div 1$

16. $6 \div 6$ **17.** $0 \div 4$ **18.** $7 \div 0$ **19.** $16 \div 2$ **20.** $18 \div 3$

EXERCISE SET 1.7

 Divide and then check by multiplying. See Examples 4 and 5.

1. $9\overline{)108}$ **2.** $5\overline{)85}$ **3.** $6\overline{)222}$ **4.** $8\overline{)640}$ **5.** $3\overline{)1014}$ **6.** $4\overline{)504}$

Divide and then check by multiplying. See Examples 6 and 7.

7. $6\overline{)98}$ **8.** $7\overline{)422}$ **9.** $2\overline{)1127}$ **10.** $3\overline{)1240}$

11. $186 \div 5$ **12.** $167 \div 3$ **13.** $2121 \div 8$ **14.** $333 \div 4$

Divide and then check by multiplying. See Examples 8 and 9.

15. $23\overline{)1127}$ **16.** $42\overline{)2016}$ **17.** $55\overline{)715}$ **18.** $32\overline{)1856}$ **19.** $97\overline{)9449}$

20. 1938 ÷ 44

21. 3708 ÷ 18

22. 7224 ÷ 12

23. 6578 ÷ 13

24. 5670 ÷ 14

25. 9299 ÷ 46

26. 2539 ÷ 64

27. $\dfrac{10,620}{236}$

28. $\dfrac{5781}{123}$

29. $\dfrac{10,194}{103}$

30. $\dfrac{23,048}{240}$

31. 20,619 ÷ 102

32. 40,803 ÷ 203

33. 45,046 ÷ 223

34. 164,592 ÷ 543

C *Solve. See Examples 10 through 12.*

35. Kathy Gomez teaches Spanish lessons for $85 per student for a 5-week session. From one group of students, she collects $4930. Find how many students are in the group.

36. Martin Thieme teaches American Sign Language classes for $55 per student for a 7-week session. He collects $1430 from the group of students. Find how many students are in the group.

37. Twenty-one people pooled their money and bought lottery tickets. One ticket won a prize of $5,292,000. Find how many dollars each person received.

38. The gravity of Jupiter is 318 times as strong as the gravity of Earth, so objects on Jupiter weigh 318 times as much as they weigh on Earth. If a person would weigh 52,470 pounds on Jupiter, find how much the person weighs on Earth.

39. A truck hauls wheat to a storage granary. It carries a total of 5810 bushels of wheat in 14 trips. How much does the truck haul each trip if each trip it hauls the same amount?

5810 bushels

40. The white stripes dividing the lanes on a highway are 25 feet long, and the spaces between them are 25 feet long. Find how many whole stripes there are in 1 mile of highway. (A mile is 5280 feet.)

66

41. There is a bridge over highway I-35 every three miles. The first bridge is at the beginning of the 265-mile stretch of highway. Find how many bridges there are over 265 miles of I-35.

42. An 18-hole golf course is 5580 yards long. If the distance to each hole is the same, find the distance between holes.

43. Wendy Holladay has a piece of rope 185 feet long that she wants to cut into pieces for an experiment in her second-grade class. Each piece of rope is to be 8 feet long. Determine whether she has enough rope for her 22-student class. Determine the amount extra or the amount short.

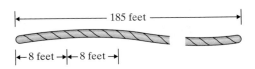

44. Jesse White is in the requisitions department of Central Electric Lighting Company. Light poles along a highway are placed 492 feet apart. The first light pole is at the beginning of the 1-mile strip. Find how many poles he should order for a 1-mile strip of highway. (A mile is 5280 feet.)

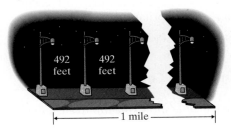

45. Marshall Faulk of the St. Louis Rams led the NFL in touchdowns during the 2000 football season, scoring a total of 156 points as touchdowns. If a touchdown is worth 6 points, how many touchdowns did Faulk make during 2000? (*Source:* National Football League)

46. Broad Peak in Pakistan is the twelfth-tallest mountain in the world. Its elevation is 26,400 feet. A mile is 5280 feet. How many miles high is Broad Peak? (*Source:* National Geographic Society)

47. Find how many yards are in 1 mile. (A mile is 5280 feet; a yard is 3 feet.)

48. Find how many whole feet are in 1 rod. (A mile is 5280 feet; 1 mile is 320 rods.)

Find the average of each list of numbers. See Example 13.

49. 14, 22, 45, 18, 30, 27

50. 37, 26, 15, 29, 29, 51, 22

51. 204, 968, 552, 268

52. 121, 200, 185, 176, 163

53. 86, 79, 81, 69, 80

54. 92, 96, 90, 85, 92, 79

The normal monthly temperature in degrees Fahrenheit for Minneapolis, Minnesota, is given in the table. Use this table to answer Exercises 55 and 56. (Source: National Climatic Data Center)

January	12°	July	74°
February	18°	August	71°
March	31°	September	61°
April	46°	October	49°
May	59°	November	33°
June	68°	December	18°

55. Find the average temperature for December, January, and February.

56. Find the average temperature for the entire year.

Combining Concepts

The following table shows the top five leading U.S. advertisers in 2000 and the amount of money spent in that year on ads. Use this table to answer Exercises 57 and 58.

Company	Year 2000 Ad Expenditures
General Motors Corp.	$2,883,215,000
DaimlerChrysler AG	$1,671,764,000
Procter & Gamble Co.	$1,542,847,000
Philip Morris Cos. Inc.	$1,538,040,000
Time Warner Inc.	$1,318,082,000

(*Source:* CMR, a Taylor Nelson Sofres company)

57. Find the average amount of money spent on ads for the year by the top two companies.

58. Find the average amount of money spent on ads by DaimlerChrysler, Procter & Gamble, Philip Morris, and Time Warner.

In Example 13 in this section, we found that the average of 93, 86, 71, and 82 is 83. Use this information to answer Exercises 59 and 60.

59. If the number 71 is removed from the list of numbers, does the average increase or decrease? Explain why.

60. If the number 93 is removed from the list of numbers, does the average increase or decrease? Explain why.

61. Without computing it, tell whether the average of 126, 135, 198, 113 is 86. Explain why it is or why it is not.

△ **62.** If the area of a rectangle is 30 square feet and its width is 3 feet, what is its length?

30 square feet	3 feet
?	

Integrated Review—Operations on Whole Numbers

Perform each indicated operation.

1.   23
   46
$+\,79$

2.   7006
$-\,451$

3.   36
$\times\,45$

4. $8\overline{)4496}$

5. $1\cdot 79$

6. $\dfrac{36}{0}$

7. $9 \div 1$

8. $9 \div 9$

9. $0 \cdot 13$

10. $7 \cdot 0 \cdot 8$

11. $0 \div 2$

12. $12 \div 4$

13.   4219
$-\,1786$

14.   1861
$+\,7965$

15. $5\overline{)1068}$

16.   1259
$\times\,\;\;63$

17. $3 \cdot 9$

18. $45 \div 5$

19.   207
$-\,\;\;69$

20.   207
$+\,\;\;69$

21. $7\overline{)7695}$

22. $9\overline{)1000}$

23. $32\overline{)21,222}$

24. $65\overline{)70,000}$

25. $4000 - 2976$

26. $10,000 - 101$

27.   303
$\times\,101$

28. $(475)(100)$

Answers

1. _____
2. _____
3. _____
4. _____
5. _____
6. _____
7. _____
8. _____
9. _____
10. _____
11. _____
12. _____
13. _____
14. _____
15. _____
16. _____
17. _____
18. _____
19. _____
20. _____
21. _____
22. _____
23. _____
24. _____
25. _____
26. _____
27. _____
28. _____

29. _____

30. _____

31. _____

32. _____

33. _____

34. _____

39. _____

40. _____

41. _____

42. _____

29. 7)0

30. $\dfrac{14}{0}$

31. $\dfrac{0}{6}$

32. $0 \div 105$

33. Subtract 14 from 100.

34. Find the difference of 43 and 21.

Complete the table by rounding the given number to the given place value.

		Tens	Hundreds	Thousands
35.	8625			
36.	1553			
37.	10,901			
38.	432,198			

Find the perimeter and area of each figure.

△ **39.**

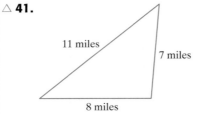

Square   5 feet

△ **40.**

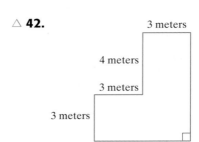

14 inches

Rectangle   7 inches

Find the perimeter of each figure.

△ **41.**

11 miles

7 miles

8 miles

△ **42.**

3 meters

4 meters

3 meters

3 meters

1.8 An Introduction to Problem Solving

A ## Solving Problems Involving Addition, Subtraction, Multiplication, or Division

In this section, we decide which operation to perform in order to solve a problem. Don't forget the key words and phrases that help indicate which operation to use. Some of these are listed below and were introduced earlier in the chapter. Also included are several words and phrases that translate to the symbol "=".

OBJECTIVES

A Solve problems by adding, subtracting, multiplying, or dividing whole numbers.

B Solve problems that require more than one operation.

SSM
TUTOR CENTER   SG   CD & VIDEO   MATH PRO   WEB

Addition (+)	Subtraction (−)	Multiplication (·)	Division (÷)	Equality (=)
sum	difference	product	quotient	equals
plus	minus	times	divide	is equal to
added to	subtract	multiply	shared equally	is/was
more than	less than		among	
increased by	decreased by		divided by	
total	less			

The following problem-solving steps may be helpful to you:

Problem-Solving Steps

1. UNDERSTAND the problem. Some ways of doing this are to read and reread the problem, construct a drawing, look for key words to identify an operation, and estimate a solution.
2. TRANSLATE the problem. That is, write the problem in short form using words, and then translate to numbers and symbols.
3. SOLVE the problem. Carry out the indicated operation from step 2.
4. INTERPRET the results. *Check* the proposed solution in the stated problem and *state* your conclusions. Write your results with the correct units attached.

EXAMPLE 1   Calculating the Length of a River

The Hudson River in New York State is 306 miles long. The Snake River in the northwestern United States is 732 miles longer than the Hudson River. How long is the Snake River? (*Source:* U.S. Department of the Interior)

Solution:

1. UNDERSTAND. Read and reread the problem, and then draw a picture. Notice that we are told that Snake River is 732 miles longer than the Hudson River. The phrase "longer than" means that we add.

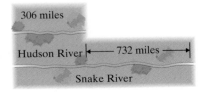

306 miles

Hudson River |◄——— 732 miles ———►|

Snake River

Practice Problem 1

The Bank of America Building is the second-tallest building in San Francisco, California, at 779 feet. The tallest building in San Francisco is the Transamerica Pyramid, which is 74 feet taller than the Bank of America Building. How tall is the Transamerica Pyramid? (*Source: The World Almanac, 2001*)

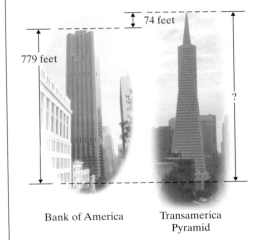

Bank of America   Transamerica Pyramid

Answer

1. 853 ft

2. TRANSLATE.

In words: Snake River │ is │ 732 miles │ longer than │ the Hudson River
                ↓   ↓   ↓   ↓   ↓
Translate: Snake River =   732   +   306

3. SOLVE:
$$
\begin{array}{r}
732 \\
+\ 306 \\
\hline
1038
\end{array}
$$

4. INTERPRET. *Check* your work. *State* your conclusion: The Snake River is 1038 miles long. ●

Practice Problem 2

Four friends bought a lottery ticket and won $65,000. If each person is to receive the same amount of money, how much does each person receive?

EXAMPLE 2   Filling a Shipping Order

How many cases can be filled with 9900 cans of jalapeños if each case holds 48 cans? How many cans will be left over? Will there be enough cases to fill an order for 200 cases?

Solution:

1. UNDERSTAND. Read and reread the problem. Draw a picture to help visualize the situation.

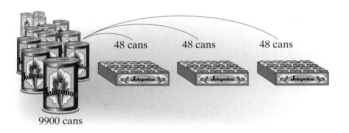

Since each case holds 48 cans, we want to know how many 48s there are in 9900. We find this by dividing.

2. TRANSLATE.

In words: Number of cases │ is │ 9900 │ divided by │ 48
                  ↓   ↓   ↓   ↓   ↓
Translate: Number of cases =   9900   ÷   48

3. SOLVE:

$$
\begin{array}{r}
206\ \text{R}\ 12 \\
48\overline{)9900} \\
-96 \\
\hline
300 \\
-288 \\
\hline
12
\end{array}
$$

4. INTERPRET. *Check* your work. *State* your conclusion: 206 cases will be filled, with 12 cans left over. There will be enough cases to fill an order for 200 cases. ●

EXAMPLE 3   Calculating Budget Costs

The director of a learning lab at a local community college is working on next year's budget. Thirty-three new video players are needed at a cost of $540 each. What is the total cost of these video players?

Solution:

1. UNDERSTAND. Read and reread the problem, and then draw a diagram.

33 Video Players

$540 $540   ... $540

From the phrase "total cost," we might decide to solve this problem by adding. This would work, but repeated addition, or multiplication, would save time.

2. TRANSLATE.

In words:	Total cost	is	number of video players	times	cost of a video player
	↓	↓	↓	↓	↓
Translate:	Total cost	=	33	×	$540

3. SOLVE:

$$\begin{array}{r} \overset{1}{5}40 \\ \times\ 33 \\ \hline 1620 \\ 1620\ \ \\ \hline 17{,}820 \end{array}$$

4. INTERPRET. *Check* your work. *State* your conclusion: The total cost of the video players is $17,820. ●

EXAMPLE 4   Calculating a Public School Teacher's Salary

In 2000, the average salary of a public school teacher in California was $47,680. For the same year, the average salary for a public school teacher in Louisiana was $14,571 less than this. What was the average public school teacher's salary in Louisiana? (*Source:* National Education Association)

Solution:

1. UNDERSTAND. Read and reread the problem. Notice that we are told that the Louisiana salary is $14,571 less than the California salary. The phrase "less than" indicates subtraction.

2. TRANSLATE. Remember that order matters when subtracting, so be careful when translating.

In words:	Louisiana salary	is	California salary	minus	$14,571
	↓	↓	↓	↓	↓
Translate:	Louisiana salary	=	47,680	−	14,571

3. SOLVE:

$$
\begin{array}{r}
47{,}680 \\
-\ 14{,}571 \\
\hline
33{,}109
\end{array}
$$

4. INTERPRET. *Check* your work. *State* your conclusion: The average Louisiana teacher's salary in 2000 was $33,109.   ●

B **Solving Problems That Require More Than One Operation**

We must sometimes use more than one operation to solve a problem.

Practice Problem 5   △

A gardener is trying to decide how much fertilizer to buy for his yard. He knows that his lot is in the shape of a rectangle that measures 90 feet by 120 feet. He also knows that the floor of his house is in the shape of a rectangle that measures 45 feet by 65 feet. How much area of the lot is not covered by the house?

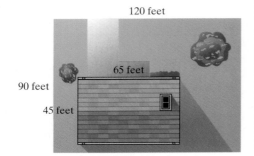

EXAMPLE 5   Planting a New Garden

A gardener bought enough plants to fill a rectangular garden with length 30 feet and width 20 feet. Because of shading problems from a nearby tree, the gardener changed the width of the garden to 15 feet. If the area is to remain the same, what is the new length of the garden?

Solution:

1. UNDERSTAND. Read and reread the problem. Then draw a picture to help visualize the problem.

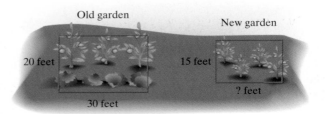

2. TRANSLATE. First we find the area of the old garden. Recall from Section 1.6 that the area of a rectangle is its length times its width.

In words: Area   =   length   times   width
Translate: Area   =   30   ×   20

3. SOLVE. 30(20) = 600, so the area is 600 square feet. Since the area of the new garden is to be 600 square feet also, we need to see how many 15s there are in 600. This means division.

$$
\begin{array}{r}
40 \\
15\overline{)600} \\
-60 \\
\hline
00
\end{array}
$$

4. INTERPRET. *Check* your work. *State* your conclusion: The length of the new garden is 40 feet.   ●

Answer

5. 7875 sq ft

Name _____ Section _____ Date _____

EXERCISE SET 1.8

A *Solve. See Examples 1 through 4.*

1. 41 increased by 8 is what number? 49

2. What is the product of 12 and 9?
$12 \cdot 9 = 108$

3. What is the quotient of 1185 and 5? 237

4. 78 decreased by 12 is what number?
$78 - 12 = 66$

5. What is the total of 35 and 7? 42

6. What is the difference of 48 and 8?
$48 - 8 = 40$

7. 60 times 10 is what number? 600

8. 60 divided by 10 is what number?
$60 \div 10 = 6$

△ **9.** A vacant lot in the shape of a rectangle measures 120 feet by 80 feet. What is the area of the lot?

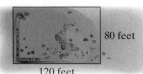
80 feet
120 feet

$9600 ft^2$

△ **10.** A parking lot in the shape of a rectangle measures 100 feet by 150 feet. What is the area of the parking lot?

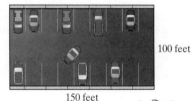
100 feet
150 feet

$100 \cdot 150 = 15000$

11. The Henrick family bought a house for $85,700 and later sold the house for $101,200. How much money did they make by selling the house?

$ 15,500

12. Three people dream of equally sharing a $147 million lottery. How much would each person receive if they have the winning ticket?

$147 \div 3 = 49$ million

13. There are 24 hours in a day. How many hours are in a week?

168 hr

14. There are 60 minutes in an hour. How many minutes are in a day?

60 × 24 = 1440

15. The Goodyear Tire & Rubber Company maintains a fleet of five blimps. The *Spirit of Goodyear* can hold 202,700 cubic feet of helium. Its smaller sister, the *Spirit of Europe*, can hold 132,700 fewer cubic feet of helium than *Spirit of Goodyear*. How much helium can *Spirit of Europe* hold? (*Source:* Goodyear Tire & Rubber Company)

70,000 cu ft

16. In the game of Monopoly, a player must own all properties in a color group before building houses. The yellow color-group properties are Atlantic Avenue, Ventnor Avenue, and Marvin Gardens. These cost $260, $260, and $280, respectively, when purchased from the bank. What total amount must a player pay to the bank before houses can be built on the yellow properties? (*Source:* Hasbro, Inc.)

260 + 260 + 280 = 800

17. Yellowstone National Park in Wyoming was the first national park in the United States. It was created in 1872. One of the more recent additions to the National Park System is Governors Island National Monument in New York. It was established in 2001. How much older is Yellowstone than Governors Island? (*Source:* National Park Service)

129 yr

18. Razor scooters were introduced in 2000. Radio Flyer Wagons were first introduced 83 years earlier. In what year were Radio Flyer Wagons introduced? (*Source:* Toy Industry Association, Inc.)

2000 − 83 = 1917

19. Since their introduction, the number of LEGO building bricks that have been sold is equivalent to the world's current population of approximately 6 billion people owning 52 LEGO bricks each. About how many LEGO bricks have been sold since their introduction? (*Source:* LEGO Company)

64

20. In April 2001, the average weekly pay for a production worker in the United States was $486 per week. At that rate, how much would a production worker have earned working a 52-week year? (*Source:* U.S. Bureau of Labor Statistics)

```
    486
  x  52
    972
  24300
  25272
```

21. The May Department Stores Company operates Lord & Taylor, Foley's, Filene's, Kaufmann's, and other department stores around the country. It also operates 55 Robinsons-May stores in California, Nevada, and Arizona. In 2000, Robinsons-May had sales of $2,200,000,000. What is the average amount of sales made by each Robinsons-May store? (*Source:* The May Department Stores Company)

$40,000,000

22. In 2000, the United States Postal Service delivered approximately 1,200,000,000 pieces of Priority Mail. The total weight of all items sent Priority Mail that year was approximately 2,400,000,000 pounds. What was the average weight of an item sent Priority Mail during 2000? (*Source:* United States Postal Service)

1 200 000 000
2 400 000 000
_____
        2

3,600,000,000 ÷ 2 =
                    1,800,000,000

23. In 2000, the average weekly pay for a financial records processing supervisor in the United States was $640. If such a supervisor works 40 hours in one week, what is his or her hourly pay? (*Source:* U.S. Bureau of Labor Statistics) $16

24. In 2000, the average weekly pay for a locomotive engineer in the United States was $920. If a locomotive engineer works 40 hours in one week, what is his or her hourly pay? (*Source:* U.S. Bureau of Labor Statistics)

        23   920 ÷ 40 = 23
    40)920
      -80
      120

25. Three ounces of canned tuna in oil has 165 calories. How many calories does 1 ounce have? (*Source:* Home and Garden Bulletin No. 72, U.S. Department of Agriculture) 55 calories

26. A whole cheesecake has 3360 calories. If the cheesecake is cut into 12 equal pieces, how many calories will each piece have? (*Source:* Home and Garden Bulletin No. 72, U.S. Department of Agriculture)

         280
    12)3360
       24   280 calories
       96
       96
       00

27. An average of 13,756 tickets per Major League Soccer (MLS) game were sold in 2000. A total of 192 games were played during the 2000 season. What was the total number of MLS tickets sold during 2000? (*Source:* Major League Soccer)

2,641,152 tickets

28. Merlakia Jones of the WNBA's Cleveland Rockers scored an average of 11 points per basketball game during the 2000 regular season. She played a total of 32 games during the season. What was the total number of points she scored during 2000? (*Source:* Women's National Basketball Association)

     32
   × 11
   ____
     32
    32
   ____
    352

29. The enrollment of all students in elementary and secondary schools in the United States in 2008 is projected to be 54,268,000. Of these students, 16,234,000 are expected to be enrolled in secondary schools. How many students are expected to be enrolled in elementary schools in 2008? (*Source:* National Center for Education Statistics)

38,034,000 students

30. Kroger is one of the largest grocery retailers in the United States. In 2001, Kroger operated a total of 3143 supermarkets and convenience stores. Of this total, 789 were convenience stores. How many super markets did Kroger operate in 2001? (*Source:* The Kroger Company)

3143
789
2,354

2,354 Groc. Stores

31. The length of the southern boundary of the conterminous United States is 1933 miles. The length of the northern boundary of the conterminous United States is 2054 miles longer than this. What is the length of the northern boundary? (*Source:* U.S. Geological Survey)

3987 miles

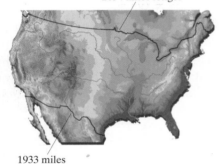

2054 miles longer

1933 miles

32. In humans, 14 muscles are required to smile. It takes 29 more muscles to frown. How many muscles does it take to frown?

14
29
55 muscles

33. Marcel Rockett receives a paycheck every four weeks. Find how many paychecks he receives in a year. (A year has 52 weeks.)

13 pay checks

34. A loan of $6240 is to be paid in 48 equal payments. How much is each payment?

in class

B *Solve. See Example 5.*

35. Find the total cost of 3 sweaters at $38 each and 5 shirts at $25 each.

$239

36. Find the total cost of 10 computers at $2100 each and 7 boxes of diskettes at $12 each.

in class

37. A college student has $950 in an account. She spends $205 from the account on books and then deposits $300 in the account. How much money is now in the account?

Handwritten:
Has 950
spends 205
deposits 300

950
205
745
300
1045

△ 38. The Meish's yard is in the shape of a rectangle and measures 50 feet by 75 feet. In their yard, they have a rectangular swimming pool that measures 15 feet by 25 feet. How much of their yard is not part of the swimming pool?

Handwritten:
50 | 15 [ ] 25 | 75

50
75
650
3500
3750

25
15
125
250
375

3750
375
3375

The table shows the menu from Corky's, a concession stand at the county fair. Use this menu to answer Exercises 39 through 41.

39. A couple orders the following items: 1 hot dog, 1 hamburger, 1 order of onion rings, and 2 sodas. How much will their order cost?

Handwritten: $12

3
4
3
2
12

Corky's Concession Stand Menu	
Item	**Price**
Hot dog	$3
Hamburger	$4
Soda	$1
Onion rings	$3
French fries	$2
Apple	$1
Candy bar	$2

40. A hungry college student is debating between the following two orders:
   a. a hamburger, an order of onion rings, a candy bar, and a soda.
   b. a hot dog, an apple, an order of french fries, and a soda.
   Which order will be cheaper? By how much?

Handwritten: In class

41. A family of four is debating between the following two orders:
   a. 6 hot dogs, 4 orders of onion rings, and 4 sodas.
   b. 4 hamburgers, 4 orders of french fries, 2 apples, and 4 sodas.
   Will the family save any money by ordering (b) instead of (a)? If so, how much?

Handwritten:
a. 6 × 3 = 18   4 × 3 = 12 + 4 = 34
b. 4 × 4 = 18   4 × 2 = 8   + 2 + 4 = 32
They will save $4 by choosing B.

Combining Concepts

42. In 2000, the United States Postal Service issued approximately 230,000,000 money orders worth $29,945,200,000. Round the value of the money orders issued to the nearest hundred-million to estimate the average value of each money order. (*Source:* United States Postal Service)

Handwritten: in class

43. In 2001, there were about 500 Hilton Hotels worldwide with a total of 147,667 guestrooms. Round the number of guestrooms to the nearest thousand to estimate the average number of guestrooms per hotel. (*Source:* Hilton Hotels Corporation)

Handwritten: 296 guest rooms

Use the table to answer Exercises 44 through 50.

Top Corporations Receiving U.S. Patents in 2000		
Company	Country	Number of Patents
Canon Kabushiki Kaisha	Japan	1890
Fujitsu Limited	Japan	1147
International Business Machines Corporation	U.S.	2886
Lucent Technologies Inc.	U.S.	1411
Micron Technology, Inc.	U.S.	1304
Motorola Inc.	U.S.	1196
NEC Corporation	Japan	2020
Samsung Electronics Co., Ltd.	South Korea	1441
Sony Corporation	Japan	1385
Toshiba Corporation	Japan	1232

(*Source:* United States Patent and Trademark Office)

44. Which company listed received the most patents in 2000?

45. Which company listed received the least patents in 2000? Fujitsu Limited

46. How many more patents did the company with the most patents receive than the company with the least patents?

47. How many more patents did Samsung receive than Sony? 56 patents

48. How many more patents did Lucent Technologies receive than Toshiba?

49. Which company received more patents, Motorola or Fujitsu? How many more patents did it receive? Motorola 1196 1147 49 more 49

50. Did the U.S. or Japanese companies listed receive more patents overall? How many more patents did they receive?

Internet Excursions

  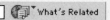

Go To: http://www.prenhall.com/martin-gay_basic   What's Related

This World Wide Web site provides access to a site that allows the user to find the distance, as the crow flies, between two cities. It also gives population and elevation figures for each city. Visit this site and answer the following questions:

51. Choose three cities in your state: (A) _____, (B) _____, and (C) _____.

52. Find the distance between city A and city B. Then find the distance from city B to city C. If a crow were to fly directly from city A to city B and then continue on to city C, how far would it fly?

53. List the population for each city. Which city has the greatest population? How many more people does that city have than the city with the least population?

1.9 Exponents and Order of Operations

A   Using Exponential Notation

OBJECTIVES

Ⓐ Write repeated factors using exponential notation.

Ⓑ Evaluate expressions containing exponents.

Ⓒ Use the order of operations.

Ⓓ Find the area of a square.

SSM   TUTOR CENTER   SG   CD & VIDEO   MATH PRO   WEB

An **exponent** is a shorthand notation for repeated multiplication. When the same number is a factor several times, an exponent may be used. In the product

$$\underbrace{2 \cdot 2 \cdot 2 \cdot 2 \cdot 2}_{\text{2 is a factor 5 times.}}$$

Using an exponent, this product can be written as

$$2^5 \qquad \text{Read as "two to the fifth power."}$$

(exponent points to 5; base points to 2)

Thus,

$$2 \cdot 2 \cdot 2 \cdot 2 \cdot 2 = 2^5$$

This is called **exponential notation**. The **exponent**, 5, indicates how many times the **base**, 2, is a factor.

Certain expressions are used when reading exponential notation.

$$5 = 5^1 \text{ is read as "five to the first power."}$$

$$5 \cdot 5 = 5^2 \text{ is read as "five to the second power" or "five squared."}$$

$$5 \cdot 5 \cdot 5 = 5^3 \text{ is read as "five to the third power" or "five cubed."}$$

$$5 \cdot 5 \cdot 5 \cdot 5 = 5^4 \text{ is read as "five to the fourth power."}$$

Usually, an exponent of 1 is not written, so when no exponent appears, we assume that the exponent is 1. For example, $2 = 2^1$ and $7 = 7^1$.

EXAMPLES   Write using exponential notation.

1. $4 \cdot 4 \cdot 4 = 4^3$
2. $7 \cdot 7 = 7^2$
3. $5 \cdot 5 \cdot 5 \cdot 5 = 5^4$
4. $6 \cdot 6 \cdot 6 \cdot 8 \cdot 8 \cdot 8 \cdot 8 \cdot 8 = 6^3 \cdot 8^5$

B   Evaluating Exponential Expressions

To **evaluate** an exponential expression, we write the expression as a product and then find the value of the product.

EXAMPLES   Evaluate.

5. $8^2 = 8 \cdot 8 = 64$
6. $7^1 = 7$
7. $2^5 = 2 \cdot 2 \cdot 2 \cdot 2 \cdot 2 = 32$
8. $5 \cdot 6^2 = 5 \cdot 6 \cdot 6 = 180$

Example 8 illustrates an important property: An exponent applies only to its base. The exponent 2, in $5 \cdot 6^2$, applies only to its base, 6.

Helpful Hint

An exponent applies only to its base. For example, $4 \cdot 2^3$ means $4 \cdot 2 \cdot 2 \cdot 2$.

Practice Problems 1–4

Write using exponential notation.

1. $2 \cdot 2 \cdot 2$
2. $3 \cdot 3$
3. $10 \cdot 10 \cdot 10 \cdot 10 \cdot 10 \cdot 10$
4. $5 \cdot 5 \cdot 4 \cdot 4 \cdot 4$

Practice Problems 5–8

Evaluate.

5. 2^3
6. 5^2
7. 10^1
8. $4 \cdot 5^2$

Answers

1. 2^3, **2.** 3^2, **3.** 10^6, **4.** $5^2 \cdot 4^3$, **5.** 8, **6.** 25, **7.** 10, **8.** 100

Copyright 2003 Prentice-Hall Inc

Concept Check

Which of the following statements is correct?

a. 3^6 is the same as $6 \cdot 6 \cdot 6$.

b. "Eight to the fourth power" is the same as 8^4.

c. "Ten squared" is the same as 10^3.

d. 11^2 is the same as $11 \cdot 2$.

Answer

Concept Check: b

Helpful Hint

Don't forget that 2^4, for example, is *not* $2 \cdot 4$. 2^4 means repeated multiplication of the same factor.

$$2^4 = 2 \cdot 2 \cdot 2 \cdot 2 = 16, \quad \text{whereas } 2 \cdot 4 = 8$$

Try the Concept Check in the margin.

C Using the Order of Operations

Suppose that you are in charge of taking inventory at a local bookstore. An employee has given you the number of a certain book in stock as the expression

$$3 + 2 \cdot 10$$

To calculate the value of this expression, do you add first or multiply first? If you add first, the answer is 50. If you multiply first, the answer is 23.

Contents: 10 books
Contents: 10 books

Mathematical symbols wouldn't be very useful if two values were possible for one expression. Thus, mathematicians have agreed that, given a choice, we multiply first.

$$3 + 2 \cdot 10 = 3 + 20 \qquad \text{Multiply.}$$
$$= 23 \qquad \text{Add.}$$

This agreement is one of several **order of operations** agreements.

Order of Operations

1. Perform all operations within grouping symbols such as parentheses or brackets.

2. Evaluate any expressions with exponents.

3. Multiply or divide in order from left to right.

4. Add or subtract in order from left to right.

For example, using the order of operations, let's evaluate $2^3 \cdot 4 - (10 \div 5)$.

$$2^3 \cdot 4 - (10 \div 5) = 2^3 \cdot 4 - 2 \qquad \text{Simplify inside parentheses.}$$

$$= 8 \cdot 4 - 2 \qquad \text{Write } 2^3 \text{ as 8.}$$

$$= 32 - 2 \qquad \text{Multiply } 8 \cdot 4.$$

$$= 30 \qquad \text{Subtract.}$$

EXAMPLE 9   Simplify: $2 \cdot 4 - 3 \div 3$

Solution: There are no parentheses and no exponents, so we start by multiplying and dividing, from left to right.

$$2 \cdot 4 - 3 \div 3 = 8 - 3 \div 3 \qquad \text{Multiply.}$$
$$= 8 - 1 \qquad \text{Divide.}$$
$$= 7 \qquad \text{Subtract.}$$

EXAMPLE 10   Simplify: $4^2 \div 2 \cdot 4$

Solution: We start by evaluating 4^2.

$$4^2 \div 2 \cdot 4 = 16 \div 2 \cdot 4 \qquad \text{Write } 4^2 \text{ as 16.}$$

Next we multiply or divide *in order* from left to right. Since division appears before multiplication from left to right, we divide first, then multiply.

$$16 \div 2 \cdot 4 = 8 \cdot 4 \qquad \text{Divide.}$$
$$= 32 \qquad \text{Multiply.}$$

EXAMPLE 11   Simplify: $(8 - 6)^2 + 2^3 \cdot 3$

Solution: $(8 - 6)^2 + 2^3 \cdot 3 = 2^2 + 2^3 \cdot 3 \qquad \text{Simplify inside parentheses.}$

$$= 4 + 8 \cdot 3 \qquad \text{Write } 2^2 \text{ as 4 and } 2^3 \text{ as 8.}$$
$$= 4 + 24 \qquad \text{Multiply.}$$
$$= 28 \qquad \text{Add.}$$

EXAMPLE 12   Simplify: $4^3 + [3^2 - (10 \div 2)] - 7 \cdot 3$

Solution: Here we begin with the innermost set of parentheses.

$$4^3 + [3^2 - (10 \div 2)] - 7 \cdot 3 = 4^3 + [3^2 - 5] - 7 \cdot 3 \qquad \text{Simplify inside parentheses.}$$
$$= 4^3 + [9 - 5] - 7 \cdot 3 \qquad \text{Write } 3^2 \text{ as 9.}$$
$$= 4^3 + 4 - 7 \cdot 3 \qquad \text{Simplify inside brackets.}$$
$$= 64 + 4 - 7 \cdot 3 \qquad \text{Write } 4^3 \text{ as 64.}$$
$$= 64 + 4 - 21 \qquad \text{Multiply.}$$
$$= 47 \qquad \text{Add and subtract from left to right.}$$

EXAMPLE 13   Simplify: $\dfrac{7 - 2 \cdot 3 + 3^2}{2^2 + 1}$

Solution: Here, the fraction bar is like a grouping symbol. We simplify above and below the fraction bar separately.

$$\frac{7 - 2 \cdot 3 + 3^2}{2^2 + 1} = \frac{7 - 2 \cdot 3 + 9}{4 + 1} \qquad \text{Evaluate } 3^2 \text{ and } 2^2.$$
$$= \frac{7 - 6 + 9}{5} \qquad \text{Multiply } 2 \cdot 3 \text{ in the numerator and add 4 and 1 in the denominator.}$$
$$= \frac{10}{5} \qquad \text{Add and subtract from left to right.}$$
$$= 2 \qquad \text{Divide.}$$

Practice Problem 9

Simplify: $16 \div 4 - 2$

Practice Problem 10

Simplify: $18 \div 3^2 \cdot 2^2$

Practice Problem 11

Simplify: $(9 - 8)^3 + 3 \cdot 2^4$

Practice Problem 12

Simplify: $24 \div [20 - (3 \cdot 4)] + 2^3 - 5$

Practice Problem 13

Simplify: $\dfrac{60 - 5^2 + 1}{3(1 + 1)}$

Answers

9. 2,   **10.** 8,   **11.** 49,   **12.** 6,   **13.** 6

D Finding the Area of a Square

Since a square is a special rectangle, we can find its area by finding the product of its length and its width.

Area of a rectangle = length · width

By recalling that each side of a square has the same measurement, we can use the following procedure to find its area:

$$\text{Area of a square} = \text{length} \cdot \text{width}$$
$$= \text{side} \cdot \text{side}$$
$$= (\text{side})^2$$

Practice Problem 14

Find the area of a square whose side measures 11 centimeters.

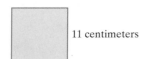

11 centimeters

Answer

14. 121 sq cm

EXAMPLE 14   Find the area of a square whose side measures 5 inches.

Solution: Area of a square = $(\text{side})^2$

$$= (5 \text{ inches})^2$$
$$= 25 \text{ square inches}$$

5 inches

The area of the square is 25 square inches.

CALCULATOR EXPLORATIONS

Exponents

To evaluate an exponent such as 4^7 on a calculator, find the keys marked $\boxed{y^x}$ and $\boxed{=}$ or $\boxed{\text{ENTER}}$. To evaluate 4^7, press the keys $\boxed{4}$ $\boxed{y^x}$ $\boxed{7}$ $\boxed{=}$ or $\boxed{\text{ENTER}}$. The display will read $\boxed{16384}$. Thus, $4^7 = 16{,}384$.

Use a calculator to evaluate.

1. 3^6 **2.** 5^6
3. 4^5 **4.** 7^6
5. 2^{11} **6.** 6^8

Order of Operations

To see whether your calculator has the order of operations built in, evaluate $5 + 2 \cdot 3$ by pressing the keys $\boxed{5}$ $\boxed{+}$ $\boxed{2}$ $\boxed{\times}$ $\boxed{3}$ $\boxed{=}$ or $\boxed{\text{ENTER}}$. If the display reads $\boxed{11}$, your calculator does have the order of operations built in.

This means that most of the time you can key in a problem exactly as it is written and the calculator will perform operations in the proper order. When evaluating an expression containing parentheses, key in the parentheses. (If an expression contains brackets, key in parentheses.) For example, to evaluate $2[25 - (8 + 4)] - 11$, press the keys $\boxed{2}$ $\boxed{\times}$ $\boxed{(}$ $\boxed{25}$ $\boxed{-}$ $\boxed{(}$ $\boxed{8}$ $\boxed{+}$ $\boxed{4}$ $\boxed{)}$ $\boxed{)}$ $\boxed{-}$ $\boxed{11}$ $\boxed{=}$ or $\boxed{\text{ENTER}}$.

The display will read $\boxed{15}$.

Use a calculator to evaluate.

7. $7^4 + 5^3$
8. $12^4 - 8^4$
9. $63 \cdot 75 - 43 \cdot 10$
10. $8 \cdot 22 + 7 \cdot 16$
11. $4(15 \div 3 + 2) - 10 \cdot 2$
12. $155 - 2(17 + 3) + 185$

EXERCISE SET 1.9

A *Write using exponential notation. See Examples 1 through 4.*

1. $3 \cdot 3 \cdot 3 \cdot 3$
3^4
81

2. $5 \cdot 5 \cdot 5$
5^3

3. $7 \cdot 7 \cdot 7 \cdot 7 \cdot 7 \cdot 7 \cdot 7 \cdot 7$
7^8

4. $6 \cdot 6 \cdot 6 \cdot 6 \cdot 6$

5. $12 \cdot 12 \cdot 12$
12^3

6. $10 \cdot 10$
10^2

7. $6 \cdot 6 \cdot 5 \cdot 5 \cdot 5$
$6^2 5^3$

8. $4 \cdot 4 \cdot 4 \cdot 3 \cdot 3$
$4^3 3^2$

9. $9 \cdot 9 \cdot 9 \cdot 8$
$9^3 \cdot 8$

10. $7 \cdot 7 \cdot 7 \cdot 4$
$7^3 \cdot 4$

11. $3 \cdot 2 \cdot 2 \cdot 2 \cdot 2 \cdot 2$
$3 \cdot 2^5$

12. $4 \cdot 6 \cdot 6 \cdot 6 \cdot 6$
$4 \cdot 6^4$

13. $3 \cdot 2 \cdot 2 \cdot 5 \cdot 5 \cdot 5$
$3 \cdot 2^2 \cdot 5^3$

14. $6 \cdot 6 \cdot 2 \cdot 9 \cdot 9 \cdot 9 \cdot 9$
$6^2 \cdot 2 \cdot 9^4$

B *Evaluate. See Examples 5 through 8.*

15. 5^2
$\ast 25$

16. 6^2
$\ast 36$

17. 5^3
$25 \cdot 5$
$\ast 272$

18. 6^3
$36 \cdot 6$
$\ast 396$

19. 2^6
$2 \cdot 2 = 4$
$2 \cdot 2 = 4 \quad \ast 6$
$2 \cdot 2 = 4 \quad \sqrt{4}$
$\ast 64 \quad \overline{64}$

20. 2^7
64
2
$\ast 128$

21. 2^{10}
$\ast 1024$

22. 1^{12}
$\ast 12$

23. 7^1
$\ast 7$

24. 8^1
$\ast 8$

25. 3^5
$\ast 243$

26. 5^4

27. 2^8
256

28. 3^3

29. 4^3
64

30. 4^4

31. 9^2
81

32. 8^2

33. 9^3
729

34. 8^3

35. 10^2
100

36. 10^3

37. 10^4
$10,000$

38. 10^5

39. 10^1
10

40. 14^1

41. 1920^1
1920

42. 6849^1

43. 3^6
729

44. 4^5

C *Simplify. See Examples 9 through 13.*

45. $15 + 3 \cdot 2$
21

46. $24 + 6 \cdot 3$
8

47. $20 - 4 \cdot 3$
8

48. $17 - 2 \cdot 6$

49. $5 \cdot 9 - 16$
29

50. $8 \cdot 4 - 10$

51. $28 \div 4 - 3$
4

52. $42 \div 7 - 6$

53. $14 + \dfrac{24}{8}$
17

54. $32 + \dfrac{8}{2}$

 **55.** $6 \cdot 5 + 8 \cdot 2$ **56.** $3 \cdot 4 + 9 \cdot 1$ **57.** $0 \div 6 + 4 \cdot 7$ **58.** $0 \div 8 + 7 \cdot 6$ **59.** $6 + 8 \div 2$

46   28   10

60. $6 + 9 \div 3$ **61.** $(6 + 8) \div 2$ **62.** $(6 + 9) \div 3$ **63.** $(6^2 - 4) \div 8$ **64.** $(7^2 - 7) \div 7$

7.   4

65. $(3 + 5^2) \div 2$ **66.** $(13 + 6^2) \div 7$  **67.** $6^2 \cdot (10 - 8)$ **68.** $5^3 \div (10 + 15)$ **69.** $\dfrac{18 + 6}{2^4 - 4}$

14   72   2

70. $\dfrac{15 + 17}{5^2 - 3^2}$ **71.** $(2 + 5) \cdot (8 - 3)$ **72.** $(9 - 7) \cdot (12 + 18)$  **73.** $\dfrac{7(9 - 6) + 3}{4^{3^2} - 3}$

35

74. $\dfrac{5(12 - 7) - 4}{5^2 - 2^3 - 10}$ **75.** $5 \div 0 + 24$ **76.** $18 - 7 \div 0$ **77.** $3^4 - [35 - (12 - 6)]$

undefined   52

78. $[40 - (8 - 2)] - 2^5$  **79.** $(7 \cdot 5) + [9 \div (3 \div 3)]$ **80.** $(18 \div 6) + [(3 + 5) \cdot 2]$

44

81. $8 \cdot [4 + (6 - 1) \cdot 2] - 50 \cdot 2$ **82.** $35 \div [3^2 + (9 - 7) - 2^2] + 10 \cdot 3$

12

83. $7^2 - \{18 - [40 \div (4 \cdot 2) + 2] + 5^2\}$ **84.** $29 - \{5 + 3[8 \cdot (10 - 8)] - 50\}$

13

D *Find the area of each square. See Example 14.*

△ **85.**
20 miles   400 mi2

△ **86.**
4 meters

△ **87.**
8 centimeters   64 cm2

△ **88.**
31 feet

△ **89.** The Eiffel Tower stands on a square base measuring 100 meters on each side. Find the area of the base.

10,000mi2

△ **90.** A square lawn that measures 72 feet on each side is to be fertilized. If 5 bags of fertilizer are available and each bag can fertilize 1000 square feet, is there enough fertilizer to cover the lawn?

 Combining Concepts

Insert grouping symbols (parentheses) so that each given expression evaluates to the given number.

91. $2 + 3 \cdot 6 - 2$; evaluate to 28

$(2+3) \cdot 6 \cdot 2$

93. $24 \div 3 \cdot 2 + 2 \cdot 5$; evaluate to 14

$24 \div (3 \cdot 2) + 2 \cdot 5$

92. $2 + 3 \cdot 6 - 2$; evaluate to 20

94. $24 \div 3 \cdot 2 + 2 \cdot 5$; evaluate to 15

△ **95.** A building contractor is bidding on a contract to install gutters on seven homes in a retirement community, all in the shape shown. To estimate her cost of materials, she needs to know the total perimeter of all seven homes. Find the total perimeter.

14,000ft2

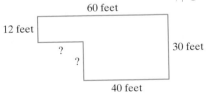

Simplify.

96. $25^3 \cdot (45 - 7 \cdot 5) \cdot 5$

97. $(7 + 2^4)^5 - (3^5 - 2^4)^2$

98. Explain why $2 \cdot 3^2$ is not the same as $(2 \cdot 3)^2$.

CHAPTER 1 ACTIVITY Investigating Endangered and Threatened Species

This activity may be completed by working in groups or individually.

An **endangered** species is one that is thought to be in danger of becoming extinct throughout all or a major part of its habitat. A **threatened** species is one that may become endangered. The Division of Endangered Species at the U.S. Fish and Wildlife Service keeps close tabs on the state of threatened and endangered wildlife in the United States and around the world. The table below was compiled from data in the Division of Endangered Species' box score published on June 30, 2001. The "Total Species" column gives the total number of endangered and threatened species for each group.

1. Round each number of *endangered animal species* to the nearest ten to estimate the Animal Total.
2. Round each number of *endangered plant species* to the nearest ten to estimate the Plant Total.
3. Add the exact numbers of endangered animal species to find the exact Animal Total and record it in the table in the Endangered Species column. Add the exact numbers of endangered plant species to find the Plant Total and record it in the table in the Endangered Species column. Then find the total number of endangered species (animals and plants combined) and record this number in the table as the Grand Total in the Endangered Species column.
4. Find the Animal Total, Plant Total, and Grand Total for the Total Species column. Record these values in the table.
5. Use the data in the table to complete the Threatened Species column.
6. Write a paragraph discussing the conclusions that can be drawn from the table.
7. (Optional) The Division of Endangered Species updates its endangered/threatened species box score monthly. Visit the current box score of endangered and threatened species at http://ecos.fws.gov/tess/html/boxscore.html on the World Wide Web. How have the figures changed since June 30, 2001? Write a paragraph summarizing the changes in the numbers of endangered and threatened animals and plants.

Endangered and Threatened Species Worldwide

	Group	Endangered Species	Threatened Species	Total Species
Animals	Mammals	314		340
	Birds	253		273
	Reptiles	78		115
	Amphibians	18		27
	Fishes	81		125
	Clams	63		71
	Snails	21		32
	Insects	37		46
	Arachnids	12		12
	Crustaceans	18		21
	Animal Total			
Plants	Flowering Plants	566		707
	Conifers	2		5
	Ferns and Allies	24		26
	Lichens	2		2
	Plant Total			
	Grand Total			

(*Source:* U.S. Fish and Wildlife Service, Division of Endangered Species)

Chapter 1 VOCABULARY CHECK

Fill in each blank with one of the words or phrases listed below.

place value   whole numbers   perimeter   exponent
area

1. The _____ are 0, 1, 2, 3, . . .
2. The _____ of a polygon is its distance around or the sum of the lengths of its sides.
3. The position of each digit in a number determines its _____.
4. An _____ is a shorthand notation for repeated multiplication of the same factor.
5. To find the _____ of a rectangle, multiply length times width.

CHAPTER 1

Highlights

DEFINITIONS AND CONCEPTS	EXAMPLES

Section 1.2 Place Value and Names for Numbers

The **whole numbers** are 0, 1, 2, 3, 4, 5, . . .
The position of each digit in a number determines its **place value**. A place-value chart is shown next with the names of the periods given.

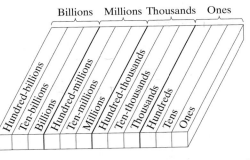

Billions   Millions   Thousands   Ones

Hundred-billions, Ten-billions, Billions, Hundred-millions, Ten-millions, Millions, Hundred-thousands, Ten-thousands, Thousands, Hundreds, Tens, Ones

To write a whole number in words, write the number in the period followed by the name of the period. The name of the ones period is not included.

0, 14, 968, 5,268,619

9,078,651,002 is written as nine billion, seventy-eight million, six hundred fifty-one thousand, two.

Section 1.3 Adding Whole Numbers

To add whole numbers, add the digits in the ones place, then the tens place, then the hundreds place, and so on, carrying when necessary.

Find the sum:

$$
\begin{array}{r}
\overset{2\,1\,1}{2689} \quad \leftarrow \quad \text{addend} \\
1735 \quad \leftarrow \quad \text{addend} \\
+\ \ 662 \quad \leftarrow \quad \text{addend} \\
\hline
5086 \quad \leftarrow \quad \text{sum}
\end{array}
$$

The **perimeter** of a polygon is its distance around or the sum of the lengths of its sides.

△ Find the perimeter of the polygon shown.

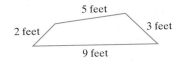

5 feet
2 feet   3 feet
9 feet

The perimeter is 5 feet + 3 feet + 9 feet + 2 feet = 19 feet.

DEFINITIONS AND CONCEPTS	EXAMPLES

Section 1.4   Subtracting Whole Numbers

To subtract whole numbers, subtract the digits in the ones place, then the tens place, then the hundreds place, and so on, borrowing when necessary.

Subtract:

$$
\begin{array}{r}
{\overset{8\ 15}{7\cancel{9}\cancel{5}4}} \leftarrow \quad \text{minuend} \\
-\ 5673 \leftarrow \quad \text{subtrahend} \\
\hline
2281 \leftarrow \quad \text{difference}
\end{array}
$$

Section 1.5   Rounding and Estimating

ROUNDING WHOLE NUMBERS TO A GIVEN PLACE VALUE

Step 1. Locate the digit to the right of the given place value.

Step 2. If this digit is 5 or greater, add 1 to the digit in the given place value and replace each digit to its right with 0.

Step 3. If this digit is less than 5, replace it and each digit to its right with 0.

Round 15,721 to the nearest thousand.

$$
15,\overset{\uparrow}{\textcircled{7}}\underbrace{21}
$$

Add 1 ⌐   Replace with zeros.

Since the circled digit is 5 or greater, add 1 to the given place value and replace digits to its right with zeros.

15,721 rounded to the nearest thousand is 16,000.

Section 1.6   Multiplying Whole Numbers

To multiply 73 and 58, for example, multiply 73 and 8, then 73 and 50. The sum of these partial products is the product of 73 and 58. Use the notation to the right.

$$
\begin{array}{r}
73 \leftarrow \quad \text{factor} \\
\times\ 58 \leftarrow \quad \text{factor} \\
\hline
584 \leftarrow \quad 73 \times 8 \\
3650 \leftarrow \quad 73 \times 50 \\
\hline
4234 \leftarrow \quad \text{product}
\end{array}
$$

To find the **area** of a rectangle, multiply length times width.

Find the area of the rectangle shown.

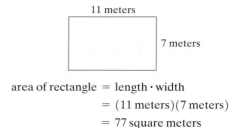

11 meters

7 meters

$$
\begin{aligned}
\text{area of rectangle} &= \text{length} \cdot \text{width} \\
&= (11 \text{ meters})(7 \text{ meters}) \\
&= 77 \text{ square meters}
\end{aligned}
$$

Section 1.7   Dividing Whole Numbers

To divide larger whole numbers, use the process called **long division** as shown to the right.

$$
\begin{array}{r}
507 \text{ R } 2 \leftarrow \quad \text{quotient} \\
\text{divisor} \rightarrow \quad 14\overline{)7100} \qquad \leftarrow \quad \text{dividend} \\
-70\downarrow \qquad 5(14) = 70 \\
\hline
10\big| \qquad \text{Subtract and bring down the 0.} \\
-\ 0\downarrow \qquad 0(14) = 0 \\
\hline
100 \qquad \text{Subtract and bring down the 0.} \\
98 \qquad 7(14) = 98 \\
\hline
2 \qquad \text{Subtract. The remainder is 2.}
\end{array}
$$

To check, see that $507 \cdot 14 + 2 = 7100$.

The **average** of a list of numbers is

$$
\text{average} = \frac{\text{sum of numbers}}{\text{number of numbers}}
$$

Find the average of 23, 35, and 38.

$$
\text{average} = \frac{23 + 35 + 38}{3} = \frac{96}{3} = 32
$$

DEFINITIONS AND CONCEPTS	**EXAMPLES**

Section 1.8   An Introduction to Problem Solving

PROBLEM-SOLVING STEPS

1. UNDERSTAND the problem.

2. TRANSLATE the problem.

3. SOLVE the problem.

4. INTERPRET the results.

Suppose that 225 tickets are sold for each performance of a play. How many tickets are sold for 5 performances?

1. UNDERSTAND. Read and reread the problem. Since we want the number of tickets for 5 performances, we multiply.

2. TRANSLATE.

number of tickets	is	number of performances	times	tickets per performance
↓	↓	↓	↓	↓
Number of tickets	=	5	·	225

3. SOLVE:
$$\begin{array}{r} \overset{1\,2}{225} \\ \times\ \ 5 \\ \hline 1125 \end{array}$$

4. INTERPRET. **Check** your work and **state** your conclusion: There are 1125 tickets sold for 5 performances.

Section 1.9   Exponents and Order of Operations

An **exponent** is a shorthand notation for repeated multiplication of the same factor.

ORDER OF OPERATIONS

Simplify expressions using the following order: If grouping symbols such as parentheses ( ) or brackets [ ] are present, simplify expressions within those first, starting with the innermost set. If fraction bars are present, simplify above and below the fraction bar separately.

1. Simplify any expressions with exponents.
2. Perform multiplications or divisions in order from left to right.
3. Perform additions or subtractions in order from left to right.

The **area of a square** is $(\text{side})^2$.

$$3^4 = \underbrace{3 \cdot 3 \cdot 3 \cdot 3}_{4 \text{ factors of } 3} = 81$$

base

Simplify: $\dfrac{5 + 3^2}{2(7 - 6)}$

Simplify above and below the fraction bar separately.

$$\frac{5 + 3^2}{2(7 - 6)} = \frac{5 + 9}{2(1)} \qquad \begin{array}{l}\text{Evaluate } 3^2 \\ \text{Subtract: } 7 - 6.\end{array}$$

$$= \frac{14}{2} \qquad\qquad \begin{array}{l}\text{Add.} \\ \text{Multiply.}\end{array}$$

$$= 7 \qquad\qquad \text{Divide.}$$

Find the area of the square shown.

9 inches

$$\text{Area of the square} = (\text{side})^2$$
$$= (9 \text{ inches})^2$$
$$= 81 \text{ square inches}$$

JUDGING DISTANCES

Do you know how to estimate a distance without using a tape measure? One easy way to do this is to use the length of your own stride. First, measure the length of your stride in inches. You can do this with the following steps:

- Lay a yardstick on the floor.
- Stand next to the yardstick with feet together so that both heels line up with the 0-mark on the yardstick.
- Take a normal-sized step forward.
- Find the whole-inch mark nearest the toe of the foot that is farthest from the 0-mark. This is roughly the length of your stride.

To judge a distance, simply pace it off using normal-sized strides. Multiply the number of strides by the length of your stride to get a rough estimate of the distance.

Suppose you need to measure a distance that can't easily be paced, such as a pond or a busy street. You can easily "transfer" the distance to a more easily paced area by using a baseball cap.

- Stand at the edge of the pond or street while wearing a baseball cap.
- Bend your head until your chin rests on your chest.
- Pull the bill of the cap up or down until it appears to touch the other side of the pond or street.
- Without moving your head or the cap, pivot your body to the right until you have a clear path straight ahead for walking.
- Notice where the bill seems to be touching the ground now. The distance to this point is the same as the distance across the pond or street you are measuring.
- Pace off the distance to this point and find an estimate of the distance as before.

Pace off this distance

Suppose a distance estimate using this method is 1302 inches. To write the distance in terms of feet, divide the estimate by 12. The quotient is the number of whole feet, and the remainder is the number of inches. A distance of 1302 inches is the same as 108 feet 6 inches.

$$
\begin{array}{r}
108\ \text{R}\ 6 \\
12\overline{)1302} \\
\underline{12} \\
10 \\
\underline{0} \\
102 \\
\underline{96} \\
6
\end{array}
$$

GROUP ACTIVITY

Materials: yardstick, baseball cap

Use the baseball cap procedure to estimate the distance across a river, stream, pond, or busy road on or near your campus. Have each person in your group estimate the same distance using the length of his or her own stride. Compare your results. Write a brief report summarizing your findings. Be sure to include what distance your group estimated, the length of each member's stride, the number of strides each member paced off, and each member's distance estimate. Conclude by discussing reasons for any differences in estimates.

Chapter 1 Review

(1.2) *Determine the place value of the digit 4 in each whole number.*

1. 5480

2. 46,200,120

Write each whole number in words.

3. 5480

4. 46,200,120

Write each whole number in expanded form.

5. 6279

6. 403,225,000

Write each whole number in standard form.

7. Fifty-nine thousand, eight hundred

8. Six billion, three hundred four million

The following table shows the populations of the ten largest cities in the United States. Use this table to answer Exercises 9 through 12.

Rank	City	2000	1990	1980
1	New York, NY	8,008,278	7,322,564	7,071,639
2	Los Angeles, CA	3,694,820	3,485,398	2,968,528
3	Chicago, IL	2,896,016	2,783,726	3,005,072
4	Houston, TX	1,953,631	1,630,553	1,595,138
5	Philadelphia, PA	1,517,550	1,585,577	1,688,210
6	Phoenix, AZ	1,321,045	983,403	789,704
7	San Diego, CA	1,223,400	1,110,549	875,538
8	Dallas, TX	1,188,580	1,006,877	904,599
9	San Antonio, TX	1,144,646	935,933	785,940
10	Detroit, MI	951,270	1,027,974	1,203,368

(*Source:* U.S. Census Bureau)

9. Find the population of Houston, Texas, in 1990.

10. Find the population of Los Angeles, California, in 1980.

11. Find the increase in population for Phoenix, Arizona, from 1980 to 2000.

12. Find the decrease in population for Detroit, Michigan, from 1990 to 2000.

(1.3) *Add.*

13. 7 + 6

14. 8 + 9

15. 3 + 0

16. 0 + 10

17. 25 + 8 + 5

18. 27 + 41

19. 32 + 24

20. 19 + 21

21. 47 + 63

22. 77 + 43

23. 567 + 383

24. 463 + 787

25. 591 + 623 + 497

26. 5982 + 1647 + 2238

27. The distance from Chicago to New York City is 714 miles. The distance from New York City to New Delhi, India, is 7318 miles. Find the total distance from Chicago to New Delhi if traveling through New York City.

28. Susan Summerline earned salaries of $62,589, $65,340, and $69,770 during the years 1998, 1999, and 2000, respectively. Find her total earnings during those three years.

Find the perimeter of each figure.

△ **29.**
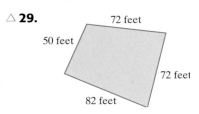
72 feet

50 feet

72 feet

82 feet

△ **30.** 11 kilometers   20 kilometers

35 kilometers

(1.4)   *Subtract and then check.*

31. 42 − 9

32. 67 − 24

33. 93 − 79

34. 60 − 27

35. 599 − 237

36. 462 − 397

37. 583 − 279

38. 600 − 124

39. 4000 − 1886

40. 4268 − 3947

41. Bob Roma is proofreading the Yellow Pages for his county. If he has finished 315 pages of the total 712 pages, how many pages does he have left to proofread?

42. Shelly Winters bought a new car listed at $28,425. She received a discount of $1599 and a factory rebate of $1200. Find how much she paid for the car.

The following bar graph shows the monthly savings account balance for a freshman attending a local community college. Use this graph to answer Exercises 43 through 46.

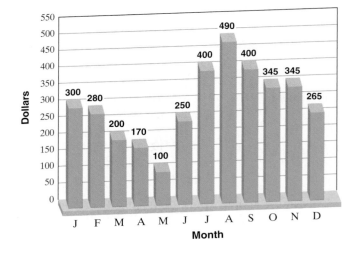

43. During what month was the balance the least?

44. During what month was the balance the greatest?

45. For what months was the balance greater than $350?

46. For what months was the balance less than $200?

(1.5) *Round to the given place.*

47. 93 to the nearest ten

48. 45 to the nearest ten

49. 467 to the nearest ten

50. 493 to the nearest hundred

51. 4832 to the nearest hundred

52. 57,534 to the nearest thousand

53. 49,683,712 to the nearest million

54. 768,542 to the nearest hundred-thousand

Estimate the sum or difference by rounding each number to the nearest hundred.

55. 4892 + 647 + 1876

56. 5925 − 1787

57. In 2001, there were 68,490,000 households in the United States subscribing to cable television services. Round this number to the nearest million. (*Source:* Nielsen Media Research-NTI)

58. In 2000, the total number of employees working for U.S. airlines was 679,967. Round this number to the nearest thousand. (*Source:* The Air Transport Association of America, Inc.)

(1.6) *Multiply.*

59. $6 \cdot 7$

60. $8 \cdot 3$

61. $5(0)$

62. $0(9)$

63. $\begin{array}{r} 47 \\ \times 30 \\ \hline \end{array}$

64. $\begin{array}{r} 69 \\ \times 42 \\ \hline \end{array}$

65. $20(8)(5)$

66. $25(9 \times 4)$

67. $\begin{array}{r} 48 \\ \times 77 \\ \hline \end{array}$

68. $\begin{array}{r} 77 \\ \times 22 \\ \hline \end{array}$

69. $49 \cdot 49 \cdot 0$

70. $62 \cdot 88 \cdot 0$

71. $\begin{array}{r} 586 \\ \times 29 \\ \hline \end{array}$

72. $\begin{array}{r} 242 \\ \times 37 \\ \hline \end{array}$

73. $\begin{array}{r} 642 \\ \times 177 \\ \hline \end{array}$

74. $\begin{array}{r} 347 \\ \times 129 \\ \hline \end{array}$

75. $\begin{array}{r} 1026 \\ \times 401 \\ \hline \end{array}$

76. $\begin{array}{r} 2107 \\ \times 302 \\ \hline \end{array}$

Estimate each product by rounding each factor to the given place.

77. 49 · 32; tens

78. 586 · 357; hundreds

79. 5231 · 243; hundreds

80. 7836 · 912; hundreds

81. One ounce of Swiss cheese contains 8 grams of fat. How many grams of fat are in 3 ounces of Swiss cheese? (*Source: Home and Garden Bulletin No. 72,* U.S. Department of Agriculture)

82. There were 5283 students enrolled at Weskan State University in the fall semester. Each paid $927 in tuition. Find the total tuition collected.

Find the area of each rectangle.

△ **83.**

12 miles

5 miles

△ **84.** 20 centimeters

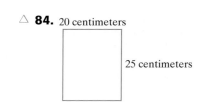

25 centimeters

(1.7) *Divide and then check.*

85. 18 ÷ 6

86. 36 ÷ 9

87. 42 ÷ 7

88. 25 ÷ 5

89. 27 ÷ 5

90. 18 ÷ 4

91. 16 ÷ 0

92. 0 ÷ 8

93. 9 ÷ 9

94. 10 ÷ 1

95. 918 ÷ 0

96. 0 ÷ 668

97. $5\overline{)75}$

98. $8\overline{)159}$

99. $26\overline{)626}$

100. $6\overline{)336}$

101. $32\overline{)49}$

102. $19\overline{)680}$

103. $20\overline{)10,000}$

104. $43\overline{)909}$

105. $47\overline{)23,782}$

106. $30\overline{)480}$

107. $16\overline{)3192}$

108. $25\overline{)5000}$

109. One foot is 12 inches. Find how many feet there are in 5496 inches.

110. Find the average of the numbers 76, 49, 32, and 47.

(1.8) *Solve.*

111. 20 increased by 5 is what number?

112. The difference of 20 and 5 is what number?

113. The product of 20 and 5 is what number?

114. The quotient of 20 and 5 is what number?

115. A box can hold 24 cans of corn. How many boxes can be filled with 648 cans of corn?

116. If a ticket to a movie costs $6, how much do 32 tickets cost?

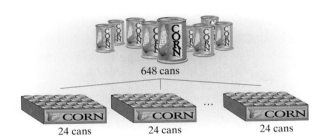

648 cans

CORN   CORN   ...   CORN

24 cans   24 cans   24 cans

117. Aspirin was 100 years old in 1997 and was the first U.S. drug made in tablet form. Today, people take 11 billion tablets a year for heart disease prevention and 4 billion tablets a year for headaches. How many more tablets are taken a year for heart disease prevention? (*Source:* Bayer Market Research)

118. The cost to banks when a person uses an ATM (Automatic Teller Machine) is 27¢. The cost to banks when a person deposits a check with a teller is 48¢ more. How much is this cost?

(1.9) *Write using exponential notation.*

119. $7 \cdot 7 \cdot 7 \cdot 7$

120. $6 \cdot 6 \cdot 3 \cdot 3 \cdot 3$

121. $4 \cdot 2 \cdot 2 \cdot 2 \cdot 3 \cdot 3$

122. $5 \cdot 5 \cdot 7 \cdot 7 \cdot 7 \cdot 2 \cdot 2$

Simplify.

123. 7^2

124. 2^6

125. $5^3 \cdot 3^2$

126. $4^1 \cdot 10^2 \cdot 7^2$

127. $18 \div 3 + 7$

128. $12 - 8 \div 4$

129. $\dfrac{(6^2 - 3)}{3^2 + 2}$

130. $\dfrac{16 - 8}{2^3}$

131. $2 + 3[1 + (20 - 17) \cdot 3]$

132. $21 - [2^4 - (7 - 5) - 10] + 8 \cdot 2$

Find the area of each square.

△ **133.**

7 meters

△ **134.**

3 inches

Chapter 1 Test

Evaluate.

1. $59 + 82$

2. $600 - 487$

3. $\begin{array}{r} 496 \\ \times\ 30 \\ \hline \end{array}$

4. $52,896 \div 69$

5. $2^3 \cdot 5^2$

6. $6^1 \cdot 2^3$

7. $98 \div 1$

8. $0 \div 49$

9. $62 \div 0$

10. $(2^4 - 5) \cdot 3$

11. $16 + 9 \div 3 \cdot 4 - 7$

12. $2[(6 - 4)^2 + (22 - 19)^2] + 10$

13. Round 52,369 to the nearest thousand.

Estimate each sum or difference by rounding each number to the nearest hundred.

14. $6289 + 5403 + 1957$

15. $4267 - 2738$

Solve.

16. Twenty-nine cans of Sherwin-Williams paint cost $493. How much was each can?

17. Admission to a movie costs $7 per ticket. The Math Club has 17 members who are going to a movie together. What is the total cost of their tickets?

18. Jo McElory is looking at two new refrigerators for her apartment. One costs $599 and the other costs $725. How much more expensive is the higher-priced one?

19. In 2000, California was the state with the most National Football League players, with a total of 208. California had 41 more NFL players than Texas. How many NFL players called Texas home? (*Source:* National Football League)

20. One tablespoon of white granulated sugar contains 45 calories. How many calories are in 8 tablespoons of white granulated sugar? (*Source: Home and Garden Bulletin No. 72*, U.S. Department of Agriculture)

1. _____

2. _____

3. _____

4. _____

5. _____

6. _____

7. _____

8. _____

9. _____

10. _____

11. _____

12. _____

13. _____

14. _____

15. _____

16. _____

17. _____

18. _____

19. _____

20. _____

Find the perimeter and the area of each figure.

△ **21.**

Square	5 centimeters

△ **22.**

20 yards

Rectangle	10 yards

The following table shows the top grossing movies for 1999 and 2000. Use this table to answer Exercises 23 through 25.

1999

Movie	Gross
Star Wars: Episode 1—The Phantom Menace	$431,065,444
The Sixth Sense	$293,501,675
Toy Story 2	$245,823,397
Austin Powers: The Spy Who Shagged Me	$205,399,422
The Matrix	$171,383,253

2000

Movie	Gross
How the Grinch Stole Christmas	$260,031,035
Cast Away	$233,630,478
Mission: Impossible 2	$215,397,307
Gladiator	$187,670,866
What Women Want	$182,805,123

(*Source:* The Internet Movie Database Ltd)

23. Find the amount earned by the movie *Cast Away* in 2000.

24. How much more did *Mission: Impossible 2* in 2000 earn than *The Matrix* in 1999?

25. How much more did the top-grossing movie of 1999 earn than the top-grossing movie in 2000?

Multiplying and Dividing Fractions

Fractions are numbers, and like whole numbers, they can be added, subtracted, multiplied, and divided. Fractions are very useful and appear frequently in everyday language, in common phrases like "half an hour," "quarter of a pound," and "third of a cup." This chapter introduces the concept of fractions, presents some basic vocabulary, and demonstrates how to multiply and divide fractions.

Established in 1872, Yellowstone National Park was the first national park in the United States. Today, the National Park Service operates a total of 384 areas ranging from 55 national parks such as Yellowstone and the Grand Canyon, to 75 national monuments such as Devil's Tower in Wyoming, to four national parkways such as the Blue Ridge Parkway in North Carolina and Virginia. The largest park in the system is Wrangell-St. Elias National Park and Preserve in Alaska, with 13,200,000 acres of land. The smallest park in the system is Thaddeus Kosciuszko National Memorial in Pennsylvania, covering only $\frac{1}{50}$ of an acre of land. In Example 10 and Exercise 60 in Section 2.3 on pages 124 and 126, we find the fraction of national parks that can be found in Utah and Alaska or Hawaii.

Name _____ Section _____ Date _____

Chapter 2 Pretest

1. _____

2. a. _____

b. _____

3. a. _____

b. _____

4. _____

5. _____

6. _____

7. _____

8. _____

9. _____

10. _____

11. _____

12. _____

13. _____

14. _____

15. _____

16. _____

17. _____

18. _____

19. _____

20. _____

1. Use a fraction to represent the shaded area of the figure.

2. Write each mixed number as an improper fraction.

  a. $3\frac{4}{7}$ **b.** $9\frac{5}{8}$

3. Write each improper fraction as a mixed number or a whole number.

  a. $\frac{52}{13}$ **b.** $\frac{71}{6}$

4. List all the factors of 36.

Identify each number as prime or composite.

5. 19

6. 91

7. Find the prime factorization of 80.

Write each fraction in simplest form.

8. $\frac{42}{72}$

9. $\frac{13}{91}$

Determine whether each pair of fractions are equal.

10. $\frac{6}{108}$ and $\frac{1}{18}$

11. $\frac{30}{134}$ and $\frac{5}{22}$

12. There are 12 inches in 1 foot. What fraction of a foot is 8 inches?

Multiply. Write each answer in simplest form.

13. $\frac{3}{5} \cdot \frac{2}{7}$

14. $\frac{4}{9} \cdot \frac{3}{28}$

15. $1\frac{3}{5} \cdot 1\frac{1}{4}$

16. Pam Wiseman has a take-home pay of $800 a month. $\frac{2}{5}$ of her monthly income is spent on her car payment. How much is her monthly car payment?

17. Find the reciprocal of $\frac{12}{19}$.

Divide. Write all answers in simplest form.

18. $\frac{4}{5} \div \frac{2}{15}$

19. $5\frac{5}{8} \div 3$

20. Jillian Grant used $1\frac{1}{2}$ gallons of gas to drive $34\frac{1}{2}$ miles. How many miles could she drive using 1 gallon of gas?

2.1 Introduction to Fractions and Mixed Numbers

OBJECTIVES

Ⓐ Identify the numerator and the denominator of a fraction.

Ⓑ Simplify fractions that simplify to 0 or 1.

Ⓒ Write a fraction to represent the shaded part of a figure.

Ⓓ Identify proper fractions, improper fractions, and mixed numbers.

Ⓔ Write mixed numbers as improper fractions.

Ⓕ Write improper fractions as mixed numbers or whole numbers.

SSM TUTOR CENTER   SG   CD & VIDEO   MATH PRO   WEB

Ⓐ Identifying Numerators and Denominators

Whole numbers are used to count whole things or units, such as cars, ball games, horses, dollars, and people. To refer to a part of a whole, fractions are used. For example, a whole baseball game is divided into nine parts called innings. If a player pitched 5 complete innings, the fraction $\frac{5}{9}$ can be used to show the part of a whole game that he or she pitched. The 9 in the fraction $\frac{5}{9}$ is called the **denominator**, and it refers to the total number of equal parts (innings) in the whole game. The 5 in the fraction $\frac{5}{9}$ is called the **numerator**, and it tells how many of those equal parts (innings) the pitcher pitched.

$$\frac{5}{9} \quad \begin{array}{l} \leftarrow \text{ how many of the parts being considered} \\ \leftarrow \text{ number of equal parts in the whole} \end{array}$$

EXAMPLES   Identify the numerator and the denominator of each fraction.

1. $\frac{3}{7} \quad \begin{array}{l} \leftarrow \text{ numerator} \\ \leftarrow \text{ denominator} \end{array}$

2. $\frac{13}{5} \quad \begin{array}{l} \leftarrow \text{ numerator} \\ \leftarrow \text{ denominator} \end{array}$

⬤

Ⓑ Fractions that Simplify to 0 or 1

Before we continue further, don't forget from Section 1.7 that the fraction bar indicates division. Let's review some division properties for 1 and 0.

$$\frac{9}{9} = 1 \text{ because } 1 \cdot 9 = 9 \qquad \frac{11}{1} = 11 \text{ because } 11 \cdot 1 = 11$$

$$\frac{0}{6} = 0 \text{ because } 0 \cdot 6 = 0$$

$$\frac{6}{0} \text{ is undefined because there is no number that when multiplied by } 0 \text{ gives 6.}$$

In general, we can say the following.

Let n be any whole number.

$$\frac{n}{n} = 1 \text{ as long as } n \text{ is not 0.} \qquad \frac{0}{n} = 0 \text{ as long as } n \text{ is not 0.}$$

$$\frac{n}{1} = n \qquad \frac{n}{0} \text{ is undefined.}$$

EXAMPLES   Simplify.

3. $\frac{5}{5} = 1$ **4.** $\frac{0}{7} = 0$ **5.** $\frac{10}{1} = 10$ **6.** $\frac{3}{0}$ is undefined

Ⓒ Writing Fractions to Represent Shaded Areas of Figures

One way to become familiar with the concept of fractions is to visualize fractions with shaded figures. We can then write a fraction to represent the shaded area of the figure.

Practice Problems 1–2

Identify the numerator and the denominator of each fraction.

1. $\frac{9}{2}$ 2. $\frac{10}{17}$

Practice Problems 3–6

Simplify.

3. $\frac{0}{2}$ 4. $\frac{8}{8}$

5. $\frac{4}{0}$ 6. $\frac{20}{1}$

Answers

1. numerator = 9, denominator = 2,
2. numerator = 10, denominator = 17,
3. 0, **4.** 1, **5.** undefined, **6.** 20

Practice Problems 7–8

Write a fraction to represent the shaded area of each figure.

7.

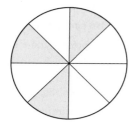

8.

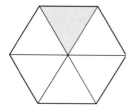

Practice Problems 9–10

Draw and shade a part of a diagram to represent each fraction.

9. $\frac{2}{3}$ of a diagram

10. $\frac{7}{11}$ of a diagram

Answers

7. $\frac{3}{8}$, **8.** $\frac{1}{6}$,

9. answers may vary; for example,

10. answers may vary; for example,

EXAMPLES   Write a fraction to represent the shaded area of each figure.

7. The figure is divided into 5 equal parts, and 2 of them are shaded. Thus $\frac{2}{5}$ of the figure is shaded.

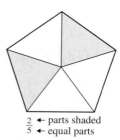

$\frac{2}{5}$ ← parts shaded
← equal parts

8. The figure is divided into 4 equal parts, and 3 of them are shaded. Thus, $\frac{3}{4}$ of the figure is shaded.

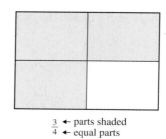

$\frac{3}{4}$ ← parts shaded
← equal parts

EXAMPLES   Draw and shade a part of a diagram to represent each fraction.

9. $\frac{5}{6}$ of a diagram

We can use a geometric figure such as a rectangle and divide it into 6 equal parts. Then we will shade 5 of the equal parts.

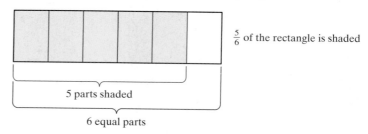

$\frac{5}{6}$ of the rectangle is shaded

5 parts shaded

6 equal parts

10. $\frac{3}{8}$ of a diagram

If you'd like, our diagram can consist of 8 triangles of the same size. We will shade 3 of the triangles.

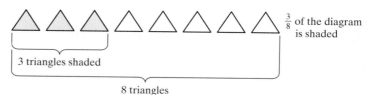

$\frac{3}{8}$ of the diagram is shaded

3 triangles shaded

8 triangles

EXAMPLE 11   Writing Fractions from Real-Life Data

Of the nine planets in our solar system, two are closer to the Sun than the Earth is. What fraction of the planets are closer to the Sun than the Earth is?

Solution: The fraction closer to the Sun than the Earth is:

$\dfrac{2}{9}$ ← number of planets closer
      ← number of planets in our solar system

Thus, $\dfrac{2}{9}$ of the planets in our solar system are closer to the Sun than the Earth is.

●

(D) Identifying Proper Fractions, Improper Fractions, and Mixed Numbers

A **proper fraction** is a fraction whose numerator is less than its denominator. Proper fractions are less than 1. The shaded portion of the triangle's area is represented by $\dfrac{2}{3}$.

$\dfrac{2}{3}$

An **improper fraction** is a fraction whose numerator is greater than or equal to its denominator. Improper fractions are greater than or equal to 1. The shaded part of the group of circles' area is $\dfrac{9}{4}$. The shaded part of the rectangle's area is $\dfrac{6}{6}$. (Recall from earlier that $\dfrac{6}{6}$ simplifies to 1 and notice that 1 whole figure or rectangle was shaded.)

Whole circle

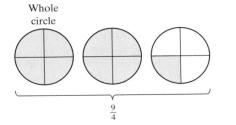

$\dfrac{9}{4}$

$\dfrac{6}{6}$

A **mixed number** contains a whole number and a fraction. Mixed numbers are greater than 1. Earlier, we wrote the shaded part of the group of circles below as the improper fraction 9/4. Now let's write the shaded part as a

Practice Problem 11

Of the nine planets in our solar system, seven are farther from the Sun than Venus is. What fraction of the planets are farther from the Sun than Venus is?

Answer

11. $\dfrac{7}{9}$

mixed number. The shaded part of the group of circles' area is $2\frac{1}{4}$. (Read "two and one-fourth.")

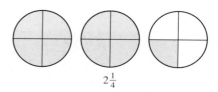

$$2\frac{1}{4}$$

Try the Concept Check in the margin.

Helpful Hint

The mixed number $2\frac{1}{4}$ represents $2 + \frac{1}{4}$.

EXAMPLES

Represent the shaded part of each figure group's area as both an improper fraction and a mixed number.

12.

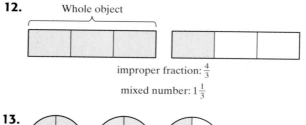

improper fraction: $\frac{4}{3}$

mixed number: $1\frac{1}{3}$

13.

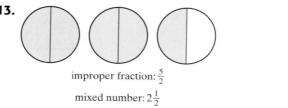

improper fraction: $\frac{5}{2}$

mixed number: $2\frac{1}{2}$

Try the Concept Check in the margin.

E   Writing Mixed Numbers as Improper Fractions

Notice from Examples 12 and 13 that mixed numbers and improper fractions were both used to represent the shaded area of the figure groups. For example,

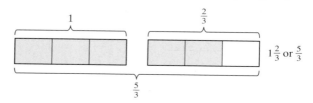

$1\frac{2}{3}$ or $\frac{5}{3}$

The following steps may be used to write a mixed number as an improper fraction:

Concept Check

Identify each as a proper fraction, improper fraction, or mixed number.

a. $\frac{6}{7}$ b. $\frac{13}{12}$

c. $\frac{2}{2}$ d. $\frac{99}{101}$

e. $1\frac{7}{8}$ f. $\frac{93}{74}$

Practice Problems 12–13

Represent the shaded part of each figure group as both an improper fraction and a mixed number.

12.

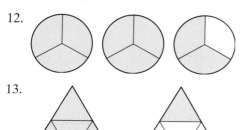

13.

Concept Check

Is $2\frac{1}{8}$ closer to the number 2 or the number 3? If you were to estimate $2\frac{1}{8}$ by a whole number, which number would you choose? Why?

Writing a Mixed Number as an Improper Fraction

To write a mixed number as an improper fraction:

Step 1. Multiply the whole number by the denominator of the fraction.

Step 2. Add the numerator of the fraction to the product from Step 1.

Step 3. Write the sum from Step 2 as the numerator of the improper fraction over the original denominator.

For example,

$$1\frac{2}{3} = \frac{3 \cdot 1 + 2}{3} = \frac{3 + 2}{3} = \frac{5}{3}$$

EXAMPLE 14   Write each as an improper fraction.

a. $4\frac{2}{9}$ **b.** $1\frac{8}{11}$

Solution:

a. $4\frac{2}{9} = \frac{9 \cdot 4 + 2}{9} = \frac{36 + 2}{9} = \frac{38}{9}$

b. $1\frac{8}{11} = \frac{11 \cdot 1 + 8}{11} = \frac{11 + 8}{11} = \frac{19}{11}$ ●

(F) Writing Improper Fractions as Mixed Numbers or Whole Numbers

Just as there are times when an improper fraction is preferred, sometimes a mixed or a whole number better suits a situation. To write improper fractions as mixed or whole numbers, just remember that the fraction bar means division.

Writing an Improper Fraction as a Mixed Number or a Whole Number

To write an improper fraction as a mixed number or a whole number;

Step 1. Divide the denominator into the numerator.

Step 2. The whole number part of the mixed number is the quotient. The fraction part of the mixed number is the remainder over the original denominator.

$$\text{quotient} \frac{\text{remainder}}{\text{original denominator}}$$

For example,

$$\frac{5}{3} = 3\overline{)5} = 1\frac{2}{3} \quad \leftarrow \text{remainder}$$
$$\phantom{\frac{5}{3} = 3\overline{)5} = 1\frac{2}{3}} \quad \leftarrow \text{original denominator}$$
$$\underset{\uparrow}{\underline{3}}$$
$$2 \quad \text{quotient}$$

Practice Problem 14

Write each as an improper fraction.

a. $2\frac{5}{7}$ b. $5\frac{1}{3}$

c. $9\frac{3}{10}$ d. $1\frac{1}{5}$

Answers

14. a. $\frac{19}{7}$, **b.** $\frac{16}{3}$, **c.** $\frac{93}{10}$, **d.** $\frac{6}{5}$

Practice Problem 15

Write each as a mixed number or a whole number.

a. $\dfrac{8}{5}$ b. $\dfrac{17}{6}$

c. $\dfrac{48}{4}$ d. $\dfrac{35}{4}$

e. $\dfrac{51}{7}$ f. $\dfrac{21}{20}$

EXAMPLE 15   Write each as a mixed number or a whole number.

a. $\dfrac{30}{7}$ b. $\dfrac{16}{15}$ c. $\dfrac{84}{6}$

Solution:

a.
$$7)\overline{30} \quad \begin{array}{r} 4 \\ 28 \\ \hline 2 \end{array} \qquad \dfrac{30}{7} = 4\dfrac{2}{7}$$

b.
$$15)\overline{16} \quad \begin{array}{r} 1 \\ 15 \\ \hline 1 \end{array} \qquad \dfrac{16}{15} = 1\dfrac{1}{15}$$

c.
$$6)\overline{84} \quad \begin{array}{r} 14 \\ 6 \\ \hline 24 \\ 24 \\ \hline 0 \end{array} \qquad \dfrac{84}{6} = 14 \qquad \text{Since the remainder is 0, the result is the whole number 14.} \quad \bullet$$

> **Helpful Hint**
>
> When the remainder is 0, the improper fraction is a whole number. For example, $\dfrac{92}{4} = 23$.
>
> $$4)\overline{92} \quad \begin{array}{r} 23 \\ 8 \\ \hline 12 \\ 12 \\ \hline 0 \end{array}$$

Answers

15. a. $1\dfrac{3}{5}$,   **b.** $2\dfrac{5}{6}$,   **c.** 12,   **d.** $8\dfrac{3}{4}$,   **e.** $7\dfrac{2}{7}$,

f. $1\dfrac{1}{20}$

Name _____ Section _____ Date _____

Mental Math

A *Identify the numerator and the denominator of each fraction. See Examples 1 and 2.*

1. $\dfrac{1}{2}$

2. $\dfrac{1}{4}$

3. $\dfrac{10}{3}$

4. $\dfrac{53}{21}$

5. $\dfrac{3}{7}$

6. $\dfrac{11}{15}$

EXERCISE SET 2.1

B *Simplify. See Examples 3 through 6.*

1. $\dfrac{21}{21}$ **2.** $\dfrac{1}{0}$ **3.** $\dfrac{0}{2}$ **4.** $\dfrac{8}{8}$ **5.** $\dfrac{13}{1}$ **6.** $\dfrac{8}{0}$

7. $\dfrac{0}{5}$ **8.** $\dfrac{14}{1}$ **9.** $\dfrac{0}{10}$ **10.** $\dfrac{16}{1}$ **11.** $\dfrac{18}{18}$ **12.** $\dfrac{0}{17}$

13. $\dfrac{9}{0}$ **14.** $\dfrac{5}{0}$ **15.** $\dfrac{9}{1}$ **16.** $\dfrac{12}{12}$

C *Write a fraction to represent the shaded area of each figure. See Examples 7 and 8.*

17.

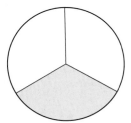

18.

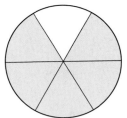

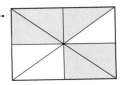 **19.**

20.

21.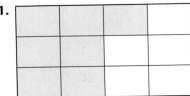

22.

109

23.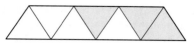

24.

25.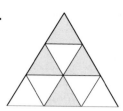

26.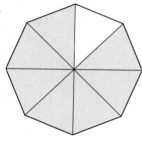

Draw and shade a part of a diagram to represent each fraction. See Examples 9 and 10.

27. $\frac{5}{8}$ of a diagram

28. $\frac{1}{4}$ of a diagram

29. $\frac{1}{5}$ of a diagram

30. $\frac{3}{5}$ of a diagram

31. $\frac{6}{7}$ of a diagram

32. $\frac{7}{9}$ of a diagram

33. $\frac{4}{4}$ of a diagram

34. $\frac{6}{6}$ of a diagram

Write each fraction. See Example 11.

35. Of the 131 students at a small private school, 42 are freshmen. What fraction of the students are freshmen?

36. Of the 78 executives at a private accounting firm, 61 are women. What fraction of the executives are women?

37. From Exercise 35, how many students are *not* freshmen? What fraction of the students are *not* freshmen?

38. From Exercise 36, how many of the executives are men? What fraction of the executives are men?

39. According to a recent study, four out of ten visits to U.S. hospital emergency rooms are for an injury. What fraction of emergency room visits are injury-related? (*Source:* National Center for Health Statistics)

40. The average American driver spends about 1 hour each day in a car or truck. What fraction of a day does an average American driver spend in a car or truck? (*Hint:* How many hours are in a day?) (*Source:* U.S. Department of Transportation—Federal Highway Administration)

41. As of 2001, the United States has had 43 different presidents. A total of eight U.S. presidents were born in the state of Virginia, more than any other state. What fraction of U.S. presidents were born in Virginia? (*Source: 1998 World Almanac and Book of Facts*)

Eight U.S. Presidents

42. Of the nine planets in our solar system, four have days that are longer than the 24-hour Earth day. What fraction of the planets have longer days than Earth has? (*Source:* National Space Science Data Center)

D *Write the shaded area in each figure group as (a) a mixed number and (b) an improper fraction. See Examples 12 and 13.*

43.
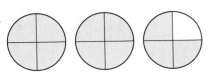

44.

45.
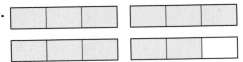

46.

47.
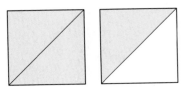

48.
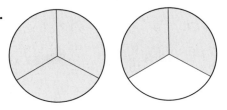

49.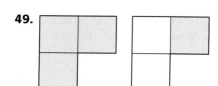

50.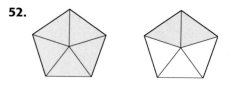

51.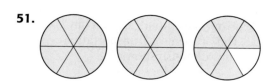

52.

53.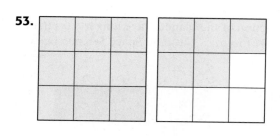

54.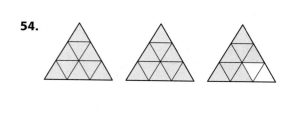

 Write each mixed number as an improper fraction. See Example 14.

55. $2\dfrac{1}{3}$ **56.** $6\dfrac{3}{4}$ **57.** $3\dfrac{2}{3}$ **58.** $3\dfrac{3}{8}$ **59.** $2\dfrac{5}{8}$

60. $12\dfrac{3}{5}$ **61.** $2\dfrac{11}{15}$ **62.** $4\dfrac{7}{8}$ **63.** $11\dfrac{6}{7}$ **64.** $12\dfrac{2}{3}$

65. $6\dfrac{5}{8}$ **66.** $8\dfrac{9}{10}$ **67.** $3\dfrac{3}{5}$ **68.** $17\dfrac{7}{12}$ **69.** $4\dfrac{13}{24}$

112

70. $10\frac{14}{27}$ **71.** $6\frac{6}{13}$ **72.** $5\frac{17}{25}$ **73.** $9\frac{7}{20}$ **74.** $12\frac{7}{15}$

F *Write each improper fraction as a mixed number or a whole number. See Example 15.*

75. $\frac{17}{5}$ **76.** $\frac{13}{7}$ **77.** $\frac{42}{13}$ **78.** $\frac{39}{3}$ **79.** $\frac{47}{15}$

80. $\frac{65}{12}$ **81.** $\frac{198}{6}$ **82.** $\frac{112}{7}$ **83.** $\frac{225}{15}$ **84.** $\frac{23}{5}$

85. $\frac{37}{8}$ **86.** $\frac{46}{21}$ **87.** $\frac{18}{17}$ **88.** $\frac{149}{143}$ **89.** $\frac{247}{23}$

90. $\frac{437}{53}$ **91.** $\frac{46}{11}$ **92.** $\frac{67}{17}$ **93.** $\frac{200}{3}$ **94.** $\frac{300}{7}$

Review and Preview

Simplify. See Section 1.9.

95. 3^2 **96.** 4^3 **97.** 5^3 **98.** 3^4

99. 7^2 **100.** 5^4 **101.** $2^3 \cdot 3$ **102.** $4^2 \cdot 5$

103. Write $38\frac{41}{79}$ as an improper fraction.

104. Write $\frac{2178}{31}$ as a mixed number.

105. The United States Marine Corps (USMC) has five principal training centers in California, three in North Carolina, two in South Carolina, one in Arizona, one in Hawaii, and one in Virginia. What fraction of the total USMC principal training centers are located in California? (*Source:* U.S. Department of Defense)

106. Habitat for Humanity is a nonprofit organization that helps provide affordable housing to families in need. Habitat for Humanity does its work of building and renovating houses through 1500 local affiliates in the United States and 300 international affiliates. What fraction of the total Habitat for Humanity affiliates are located in the United States? (*Source:* Habitat for Humanity International)

107. The Public Broadcasting Service (PBS) provides programming to the noncommercial public TV stations of the United States. The table shows a breakdown of the public television licensees by type. Each licensee operates one or more PBS member TV stations. What fraction of the public television licensees are universities or colleges?

108. The table shows the number of Wendy's restaurants operated in various regions. What fraction of Wendy's restaurants are located in the United States?

Wendy's Restaurants

Region	Number
United States	5095
Canada	324
Outside North America	373

(*Source:* Wendy's International, Inc.)

Public Television Licensees

Type	Number
Community organizations	88
Universities/colleges	55
State authorities	21
Local education/municipal authorities	7

(*Source:* The Public Broadcasting Service)

109. In your own words, explain how to write an improper fraction as a mixed number.

110. In your own words, explain how to write a mixed number as an improper fraction.

2.2 Factors and Prime Factorization

To perform many operations with fractions, it is necessary to be able to factor a number. In this section, only the **natural numbers**—1, 2, 3, 4, 5, and so on—will be considered.

Try the Concept Check in the margin.

Ⓐ Finding Factors of Numbers

Recall that when numbers are multiplied to form a product, each number is called a factor. Since $3 \cdot 4 = 12$, both 3 and 4 are **factors** of 12, and $3 \cdot 4$ is called a **factorization** of 12.

The two-number factorizations of 12 are

$$1 \cdot 12 \quad 2 \cdot 6 \quad 3 \cdot 4$$

Thus, we say that the factors of 12 are 1, 2, 3, 4, 6, and 12.

> **Helpful Hint**
>
> A **factor** of a number divides the number evenly (with a remainder of 0). For example,
>
> $$\frac{12}{1\overline{)12}} \quad \frac{6}{2\overline{)12}} \quad \frac{4}{3\overline{)12}} \quad \frac{3}{4\overline{)12}} \quad \frac{2}{6\overline{)12}} \quad \frac{1}{12\overline{)12}}$$

EXAMPLE 1   Find all the factors of 20.

Solution: First we write all the two-number factorizations of 20.

$$1 \cdot 20 = 20$$
$$2 \cdot 10 = 20$$
$$4 \cdot 5 = 20$$

The factors of 20 are 1, 2, 4, 5, 10, and 20.

Ⓑ Identifying Prime and Composite Numbers

Of all the ways to factor a number, one special way is called the **prime factorization**. To help us write prime factorizations, we first review prime and composite numbers.

> **Prime Numbers**
>
> A **prime number** is a natural number that has exactly two different factors, 1 and itself.

EXAMPLE 2   Determine whether each number is prime. Explain your answers.

$$3, \quad 9, \quad 11, \quad 17, \quad 26$$

Solution: The number 3 is prime. Its only factors are 1 and 3.
The number 9 is not prime. It has more than two factors: 1, 3, and 9.
The number 11 is prime. Its only factors are 1 and 11.
The number 17 is prime. Its only factors are 1 and 17.
The number 26 is not prime. Its factors are 1, 2, 13, and 26.

OBJECTIVES

Ⓐ Find the factors of a number.

Ⓑ Identify prime and composite numbers.

Ⓒ Find the prime factorization of a number.

SSM   SG   CD & VIDEO   MATH PRO   WEB
TUTOR CENTER

Concept Check

How are the natural numbers and the whole numbers alike? How are they different?

Practice Problem 1

Find all the factors of each number.
a. 15   b. 7

Practice Problem 2

Determine whether each number is prime. Explain your answers.
15, 11, 24, 29, 39

Answers

1. a. 1, 3, 5, 15, **b.** 1, 7, **2.** 11, 29 are prime. 15, 24, and 39 are not prime.

Concept Check: answers may vary

The first ten prime numbers are

$$2, 3, 5, 7, 11, 13, 17, 19, 23, 29$$

It would be helpful to memorize these.

If a natural number other than 1 is not a prime number, it is called a **composite number**.

Composite Numbers

A **composite number** is any natural number, other than 1, that is not prime.

 Helpful Hint

The natural number 1 is neither prime nor composite.

C Finding Prime Factorizations

Now we are ready to find **prime factorizations** of numbers.

Prime Factorization

The **prime factorization** of a number is a factorization in which all the factors are prime numbers.

For example, the prime factorization of 12 is $2 \cdot 2 \cdot 3$ because

$$12 = \underbrace{2 \cdot 2 \cdot 3}$$

This product is 12, and each number is a prime number.

There is only one prime factorization for any given number. In other words, the prime factorization of a number is unique.

Practice Problem 3

Find the prime factorization of 28.

EXAMPLE 3 Find the prime factorization of 45.

Solution: The first prime number, 2, does not divide 45 evenly (with a remainder of 0). The second prime number, 3, does, so we divide 45 by 3.

$$3\overline{)45} \quad \overset{15}{}$$

Because 15 is not prime and 3 also divides 15 evenly, we divide by 3 again.

$$
\begin{array}{r}
5 \\
3\overline{)15} \\
3\overline{)45}
\end{array}
$$

The quotient, 5, is a prime number, so we are finished. The prime factorization of 45 is

$$45 = 3 \cdot 3 \cdot 5 \quad \text{or} \quad 45 = 3^2 \cdot 5,$$

using exponents. ●

There are a few quick **divisibility tests** to determine whether a number is divisible by the primes 2, 3, or 5. (A number is divisible by 2, for example, if 2 divides it evenly.)

Answer

3. $2^2 \cdot 7$

Divisibility Tests

A whole number is divisible by:

- **2** if the last digit is 0, 2, 4, 6, or 8.

    ↓

    132 is divisible by 2 since the last digit is a 2.
- **3** if the sum of the digits is divisible by 3.

    144 is divisible by 3 since $1 + 4 + 4 = 9$ is divisible by 3.
- **5** if the last digit is 0 or 5.

    ↓

    1115 is divisible by 5 since the last digit is a 5.

Helpful Hint

Here are a few other divisibility tests you may find interesting. A whole number is divisible by:

- **4** if its last two digits are divisible by 4.

    1712 is divisible by 4.
- **6** if it's divisible by 2 and 3.

    9858 is divisible by 6.
- **9** if the sum of its digits is divisible by 9.

    5238 is divisible by 9 since $5 + 2 + 3 + 8 = 18$ is divisible by 9.

EXAMPLE 4   Find the prime factorization of 180.

Solution: We divide 180 by 2 and continue dividing until the quotient is no longer divisible by 2. We then divide by the next largest prime number, 3, until the quotient is no longer divisible by 3. We continue this process until the quotient is a prime number.

$$
\begin{array}{r}
5 \\
3\overline{)\ 15} \\
3\overline{)\ 45} \\
2\overline{)\ 90} \\
2\overline{)180}
\end{array}
$$

Thus, the prime factorization of 180 is

$$180 = 2 \cdot 2 \cdot 3 \cdot 3 \cdot 5 \quad \text{or} \quad 180 = 2^2 \cdot 3^2 \cdot 5,$$

using exponents.

EXAMPLE 5   Find the prime factorization of 945.

Solution: This number is not divisible by 2 but is divisible by 3. We will begin then by dividing 945 by 3.

$$
\begin{array}{r}
7 \\
5\overline{)\ 35} \\
3\overline{)105} \\
3\overline{)315} \\
3\overline{)945}
\end{array}
$$

Thus, the prime factorization of 945 is

$$945 = 3 \cdot 3 \cdot 3 \cdot 5 \cdot 7 \quad \text{or} \quad 945 = 3^3 \cdot 5 \cdot 7$$

Practice Problem 4

Find the prime factorization of 75.

Practice Problem 5

Find the prime factorization of 756.

Answers

4. $3 \cdot 5^2$,   **5.** $2^2 \cdot 3^3 \cdot 7$

Another way to find the prime factorization is to use a factor tree, as shown in the next example.

Practice Problem 6

Use a factor tree to find the prime factorization of 70.

EXAMPLE 6 Use a factor tree to find the prime factorization of 18.

Solution: We begin by writing 18 as a product of two numbers, say, $2 \cdot 9$.

$$
\begin{array}{c}
18 \\
\diagup \diagdown \\
2 \cdot 9
\end{array}
$$

The number 2 is prime, but 9 is not. So we write 9 as $3 \cdot 3$.

$$
\begin{array}{c}
18 \\
\diagup \diagdown \\
2 \cdot 9 \\
\downarrow \quad \downarrow \diagdown \\
2 \cdot 3 \cdot 3
\end{array}
$$

Each factor is now prime, so the prime factorization is

$$18 = 2 \cdot 3 \cdot 3 \quad \text{or} \quad 18 = 2 \cdot 3^2,$$

using exponents.

Don't forget that multiplication is commutative, so $2 \cdot 3 \cdot 3$ can also be written as $3 \cdot 3 \cdot 2$ or $3 \cdot 2 \cdot 3$. Any of these can be called the one prime factorization of 18.

Practice Problem 7

Use a factor tree to find the prime factorization of each number.

a. 10

b. 30

c. 72

EXAMPLE 7 Use a factor tree to find the prime factorization of 24.

Solution:
$$
\begin{array}{c}
24 \\
\diagup \diagdown \\
4 \cdot 6 \\
\diagup \diagdown \quad \downarrow \diagdown \\
2 \cdot 2 \cdot 2 \cdot 3
\end{array}
$$

The prime factorization of 24 is

$$24 = 2 \cdot 2 \cdot 2 \cdot 3 \quad \text{or} \quad 2^3 \cdot 3,$$

using exponents.

Try the Concept Check in the margin.

Concept Check

True or false? Two different numbers can have exactly the same prime factorization. Explain your answer.

> **Helpful Hint**
>
> When using a factor tree, we will arrive at the same prime factorization of a number no matter what original factors we use. For example, let's factor 24 again from Example 7.

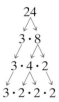

$$
\begin{array}{c}
24 \\
\diagup \diagdown \\
3 \cdot 8 \\
\downarrow \quad \diagup \diagdown \\
3 \cdot 4 \cdot 2 \\
\downarrow \diagup \diagdown \downarrow \\
3 \cdot 2 \cdot 2 \cdot 2
\end{array}
$$

Answers

6. $2 \cdot 5 \cdot 7$, **7. a.** $2 \cdot 5$, **b.** $2 \cdot 3 \cdot 5$, **c.** $2^3 \cdot 3^2$

Concept Check: false

EXERCISE SET 2.2

A *List all the factors of each number. See Example 1.*

1. 8   **2.** 6   **3.** 25   **4.** 30   **5.** 4   **6.** 9

7. 18   **8.** 24   **9.** 7   **10.** 5   **11.** 80   **12.** 100

13. 12   **14.** 20   **15.** 34   **16.** 26

B *Identify each number as prime or composite. See Example 2.*

17. 7   **18.** 5   **19.** 4   **20.** 49   **21.** 10   **22.** 13

23. 29   **24.** 45   **25.** 6   **26.** 2   **27.** 15   **28.** 21

29. 31   **30.** 27   **31.** 33   **32.** 51

C *Find the prime factorization of each number. Write any repeated factors using exponents. See Examples 3 through 7.*

33. 12   **34.** 20   **35.** 15   **36.** 21   **37.** 40   **38.** 63

39. 36   **40.** 64   **41.** 39   **42.** 33   **43.** 48   **44.** 28

45. 54   **46.** 56   **47.** 60   **48.** 100   **49.** 110   **50.** 140

51. 88   **52.** 93   **53.** 128   **54.** 81   **55.** 150   **56.** 175

57. 300   **58.** 360   **59.** 240   **60.** 400   **61.** 945   **62.** 504

63. 700   **64.** 1000   **65.** 882   **66.** 405   **67.** 637   **68.** 539

Review and Preview

Round each whole number to the indicated place value. See Section 1.5.

69. 4267 hundreds

70. 7,658,240 ten-thousands

71. 4,286,340 tens

72. 19,764 thousands

73. 55,342 hundreds

74. 10,292,876 millions

75. 3499 tens

76. 2,437,831 hundred-thousands

77. 1247 thousands

78. 9485 tens

Combining Concepts

Find the prime factorization of each number.

79. 3600   **80.** 1350   **81.** 34,020   **82.** 2464   **83.** 131,625

84. In your own words, define a prime number.

85. The number 2 is a prime number. All other even natural numbers are composite numbers. Explain why.

2.3 Simplest Form of a Fraction

A   Writing Fractions in Simplest Form

Fractions that represent the same portion of a whole are called **equivalent fractions**.

$$\frac{1}{3} \qquad \frac{2}{6} \qquad \frac{4}{12}$$

For example, $\frac{1}{3}, \frac{2}{6}$, and $\frac{4}{12}$ all represent the same shaded portion of the rectangle's area, so they are equivalent fractions.

$$\frac{1}{3} = \frac{2}{6} = \frac{4}{12}$$

A special form of a fraction is called **simplest form**.

Simplest Form of a Fraction

A fraction is written in **simplest form** or **lowest terms** when the numerator and the denominator have no common factors other than 1.

For example, the fraction $\frac{2}{6}$ *is not* in simplest form because 2 and 6 both have a factor of 2. That is, 2 is a common factor of 2 and 6. The fraction $\frac{1}{3}$ *is* in simplest form because 1 and 3 have no common factor other than 1. The process of writing a fraction in simplest form is called **simplifying** the fraction.

We can use the prime factorization of a number to help us write a fraction in simplest form.

Writing a Fraction in Simplest Form

To write a fraction in simplest form, write the prime factorization of the numerator and the denominator and then divide both by all common factors.

For example,

$$\frac{2}{6} = \frac{2}{2 \cdot 3} = \frac{2 \div 2}{2 \cdot 3 \div 2} = \frac{1}{3} \qquad \text{Divide the numerator and the denominator by 2.}$$

In the future, we will use the following notation to show dividing the numerator and denominator by common factors:

$$\frac{2}{6} = \frac{\overset{1}{\cancel{2}}}{\underset{1}{\cancel{2}} \cdot 3} = \frac{1}{1 \cdot 3} = \frac{1}{3}$$

Practice Problem 1

Write in simplest form: $\dfrac{30}{45}$

Practice Problem 2

Write in simplest form: $\dfrac{39}{51}$

Practice Problem 3

Write in simplest form: $\dfrac{9}{50}$

Practice Problem 4

Write in simplest form: $\dfrac{49}{63}$

Practice Problem 5

Write in simplest form: $\dfrac{24}{20}$

Concept Check

Which is the correct way to simplify the fraction $\dfrac{15}{25}$? Or are both correct? Explain.

a. $\dfrac{15}{25} = \dfrac{3 \cdot \overset{1}{\cancel{5}}}{5 \cdot \cancel{5}} = \dfrac{3}{5}$ b. $\dfrac{\overset{1}{\cancel{15}}}{\cancel{25}} = \dfrac{11}{21}$

Practice Problem 6

Write in simplest form: $\dfrac{8}{56}$

Answers

1. $\dfrac{2}{3}$, **2.** $\dfrac{13}{17}$, **3.** $\dfrac{9}{50}$, **4.** $\dfrac{7}{9}$, **5.** $\dfrac{6}{5}$ or $1\dfrac{1}{5}$, **6.** $\dfrac{1}{7}$

Concept Check: **a.** correct, **b.** incorrect

EXAMPLE 1 Write in simplest form: $\dfrac{12}{20}$

Solution: First, we write the prime factorization of the numerator and the denominator.

$$\frac{12}{20} = \frac{2 \cdot 2 \cdot 3}{2 \cdot 2 \cdot 5}$$

Next, we divide the numerator and the denominator by all common factors.

$$\frac{12}{20} = \frac{\overset{1}{\cancel{2}} \cdot \overset{1}{\cancel{2}} \cdot 3}{\underset{1}{\cancel{2}} \cdot \underset{1}{\cancel{2}} \cdot 5} = \frac{3}{5}$$

EXAMPLE 2 Write in simplest form: $\dfrac{42}{66}$

Solution:

$$\frac{42}{66} = \frac{\overset{1}{\cancel{2}} \cdot \overset{1}{\cancel{3}} \cdot 7}{\underset{1}{\cancel{2}} \cdot \underset{1}{\cancel{3}} \cdot 11} = \frac{7}{11}$$

EXAMPLE 3 Write in simplest form: $\dfrac{10}{27}$

Solution:

$$\frac{10}{27} = \frac{2 \cdot 5}{3 \cdot 3 \cdot 3}$$

Since 10 and 27 have no common factors, $\dfrac{10}{27}$ is already in simplest form.

EXAMPLE 4 Write in simplest form: $\dfrac{30}{108}$

Solution:

$$\frac{30}{108} = \frac{\overset{1}{\cancel{2}} \cdot \overset{1}{\cancel{3}} \cdot 5}{2 \cdot \underset{1}{\cancel{2}} \cdot \underset{1}{\cancel{3}} \cdot 3 \cdot 3} = \frac{5}{18}$$

EXAMPLE 5 Write in simplest form: $\dfrac{72}{26}$

Solution:

$$\frac{72}{26} = \frac{\overset{1}{\cancel{2}} \cdot 2 \cdot 2 \cdot 3 \cdot 3}{\underset{1}{\cancel{2}} \cdot 13} = \frac{36}{13},$$

which can also be written as

$$2\frac{10}{13}$$

Try the Concept Check in the margin.

EXAMPLE 6 Write in simplest form: $\dfrac{6}{60}$

Solution:

$$\frac{6}{60} = \frac{\overset{1}{\cancel{2}} \cdot \overset{1}{\cancel{3}}}{2 \cdot \underset{1}{\cancel{2}} \cdot \underset{1}{\cancel{3}} \cdot 5} = \frac{1}{10}$$

Helpful Hint

Be careful when all factors of the numerator or denominator are divided out. In Example 6, the numerator was $1 \cdot 1 = 1$, so the final result was $\frac{1}{10}$.

In the fraction of Example 6, $\frac{6}{60}$, you may have immediately noticed that the largest common factor of 6 and 60 is 6. If so, you may simply divide out that common factor.

$$\frac{6}{60} = \frac{\overset{1}{\cancel{6}}}{\underset{1}{\cancel{6} \cdot 10}} = \frac{1}{10} \qquad \text{Divide out a common factor of 6.}$$

Notice that the result, $\frac{1}{10}$, is in simplest form. If it were not, we would repeat the same procedure until the result was in simplest form.

EXAMPLE 7   Write in simplest form: $\frac{45}{75}$

Solution: You may write the prime factorizations of 45 and 75 or you may notice that these two numbers have a common factor of 15.

$$\frac{45}{75} = \frac{3 \cdot \overset{1}{\cancel{15}}}{5 \cdot \underset{1}{\cancel{15}}} = \frac{3}{5}$$

The numerator and denominator of $\frac{3}{5}$ have no common factors other than 1, so $\frac{3}{5}$ is in simplest form.

B   **Determining Whether Two Fractions Are Equivalent**

How can we check to see whether a simplified fraction is equivalent to an original fraction? Any two fractions are equivalent if their **cross products** are equal. This test for equality is shown in the next example.

EXAMPLE 8   Determine whether $\frac{6}{60}$ and $\frac{1}{10}$ are equivalent.

Solution:

$$\frac{6}{60} \quad \frac{1}{10}$$

$60 \cdot 1 = 60$ These cross products are
$6 \cdot 10 = 60$ equal, so the fractions are equivalent.

$$\frac{6}{60} = \frac{1}{10}$$

Practice Problem 7

Write in simplest form: $\frac{42}{48}$.

Practice Problem 8

Determine whether $\frac{7}{9}$ and $\frac{21}{27}$ are equivalent.

Answers

7. $\frac{7}{8}$,   **8.** are equivalent

Practice Problem 9

Determine whether $\frac{4}{13}$ and $\frac{5}{18}$ are equivalent.

EXAMPLE 9   Determine whether $\frac{8}{11}$ and $\frac{19}{26}$ are equivalent.

Solution:

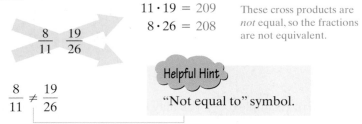

$$11 \cdot 19 = 209$$
$$8 \cdot 26 = 208$$

These cross products are *not* equal, so the fractions are not equivalent.

$$\frac{8}{11} \neq \frac{19}{26}$$

Helpful Hint

"Not equal to" symbol.

C **Solving Problems by Writing Fractions in Simplest Form**

Many real-life problems can be solved by writing fractions. To make the answers clearer, these fractions should be written in simplest form.

EXAMPLE 10   **Calculating the Fraction of Parks in Utha**

As of 2001, there were 55 national parks in the United States. Five of these parks are located in the state of Utah. What fraction of the national parks in the United States can be found in Utah? Write the fraction in simplest form. (*Source:* National Park Service)

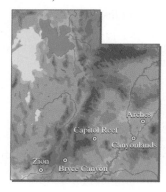

Practice Problem 10

Eighty pigs were used in a recent study of olestra, a calorie-free fat substitute. A group of 12 of these pigs were fed a diet high in fat. What fraction of the pigs were fed the high-fat diet in this study? Write your answer in simplest form. (*Source:* from a study conducted by the Procter & Gamble Company)

Solution: First we determine the fraction of parks found in Utah.

$$\frac{5}{55} \quad \begin{array}{l} \leftarrow \text{ national parks in Utah} \\ \leftarrow \text{ total national parks} \end{array}$$

Next we simplify the fraction.

$$\frac{5}{55} = \frac{\overset{1}{\cancel{5}}}{\underset{1}{\cancel{5}} \cdot 11} = \frac{1}{11}$$

Thus, $\frac{1}{11}$ of the United States' national parks are in Utah.

Answers

9. are not equivalent,   **10.** $\frac{3}{20}$

EXERCISE SET 2.3

A *Write each fraction in simplest form. See Examples 1 through 7.*

1. $\dfrac{3}{12}$

2. $\dfrac{5}{20}$

3. $\dfrac{7}{35}$

4. $\dfrac{9}{48}$

 5. $\dfrac{14}{16}$

6. $\dfrac{18}{34}$

7. $\dfrac{24}{30}$

8. $\dfrac{70}{80}$

9. $\dfrac{35}{42}$

10. $\dfrac{25}{55}$

11. $\dfrac{63}{72}$

12. $\dfrac{56}{64}$

13. $\dfrac{21}{49}$

14. $\dfrac{14}{35}$

15. $\dfrac{24}{40}$

16. $\dfrac{36}{54}$

17. $\dfrac{36}{63}$

18. $\dfrac{39}{42}$

19. $\dfrac{12}{15}$

20. $\dfrac{18}{24}$

21. $\dfrac{25}{40}$

22. $\dfrac{36}{42}$

23. $\dfrac{27}{90}$

24. $\dfrac{60}{150}$

25. $\dfrac{36}{24}$

26. $\dfrac{60}{36}$

27. $\dfrac{40}{64}$

28. $\dfrac{28}{60}$

29. $\dfrac{70}{196}$

30. $\dfrac{98}{126}$

31. $\dfrac{66}{308}$

32. $\dfrac{65}{234}$

33. $\dfrac{55}{85}$

34. $\dfrac{78}{90}$

35. $\dfrac{189}{216}$

36. $\dfrac{84}{189}$

37. $\dfrac{75}{350}$

38. $\dfrac{72}{420}$

39. $\dfrac{288}{480}$

40. $\dfrac{135}{585}$

B *Determine whether each pair of fractions is equivalent. See Examples 8 and 9.*

41. $\dfrac{10}{15}$ and $\dfrac{6}{9}$

42. $\dfrac{9}{12}$ and $\dfrac{15}{20}$

43. $\dfrac{7}{11}$ and $\dfrac{5}{8}$

44. $\dfrac{6}{7}$ and $\dfrac{7}{8}$

45. $\dfrac{10}{13}$ and $\dfrac{12}{15}$

46. $\dfrac{2}{3}$ and $\dfrac{29}{43}$

47. $\dfrac{3}{9}$ and $\dfrac{6}{18}$

48. $\dfrac{3}{11}$ and $\dfrac{33}{121}$

49. $\dfrac{4}{10}$ and $\dfrac{6}{15}$

50. $\dfrac{2}{5}$ and $\dfrac{4}{11}$

51. $\dfrac{1}{7}$ and $\dfrac{2}{8}$

52. $\dfrac{2}{8}$ and $\dfrac{5}{20}$

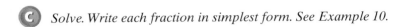

C *Solve. Write each fraction in simplest form. See Example 10.*

53. A work shift for an employee at McDonald's consists of 8 hours. What fraction of the employee's work shift is represented by 2 hours?

54. Two thousand baseball caps were sold one year at the U.S. Open Golf Tournament. What fractional part of this total does 200 caps represent?

55. There are 5280 feet in a mile. What fraction of a mile is represented by 2640 feet?

56. There are 100 centimeters in 1 meter. What fraction of a meter is 20 centimeters?

57. As of 2000, a total of 414 individuals from around the world had flown in space. Of these, 261 were citizens of the United States. What fraction of individuals who have flown in space were Americans? (*Source: Congressional Research Service*)

58. In 2001, Hallmark Cards employed 25,000 full-time employees worldwide. About 5600 employees worked at the Hallmark headquarters in Kansas City, Missouri. What fraction of Hallmark employees worked in Kansas City? (*Source:* Hallmark Cards, Inc.)

59. Fifteen states in the United States have Ritz-Carlton hotels. (*Source:* Marriott International)
  a. What fraction of states can claim at least one Ritz-Carlton hotel?
  b. How many states do not have a Ritz-Carlton hotel?
  c. Write the fraction of states without a Ritz-Carlton hotel.

60. As of 2001, there were 55 national parks in the United States. Ten of these parks are located in Alaska or Hawaii. (*Source:* National Park Service)
  a. What fraction of the national parks in the United States can be found in Alaska or Hawaii?
  b. How many of the national parks in the United States are found outside Alaska and Hawaii?
  c. Write the fraction of national parks found in states other than Alaska and Hawaii.

Review and Preview

Multiply. See Section 1.6.

61. $\begin{array}{r} 91 \\ \times\ 4 \end{array}$

62. $\begin{array}{r} 73 \\ \times\ 8 \end{array}$

63. $\begin{array}{r} 387 \\ \times\ 6 \end{array}$

64. $\begin{array}{r} 562 \\ \times\ 9 \end{array}$

65. $\begin{array}{r} 72 \\ \times\ 35 \end{array}$

66. $\begin{array}{r} 238 \\ \times\ 26 \end{array}$

Combining Concepts

Determine whether each is true or false. If false, explain why.

67. $\dfrac{14}{42} = \dfrac{\overset{1}{\cancel{2}} \cdot \overset{1}{\cancel{7}}}{\underset{1}{\cancel{2}} \cdot 3 \cdot \underset{1}{\cancel{7}}} = \dfrac{0}{3}$

68. A proper fraction cannot be equivalent to an improper fraction.

Write each fraction in simplest form.

69. $\dfrac{3975}{6625}$

70. $\dfrac{9506}{12,222}$

There are generally considered to be eight basic blood types. The table shows the number of people with the various blood types in a typical group of 100 blood donors. Use the table to answer Exercises 71 through 75. Write each answer in simplest form.

Distribution of Blood Types in Blood Donors	
Blood Type	**Number of People**
O Rh-positive	37
O Rh-negative	7
A Rh-positive	36
A Rh-negative	6
B Rh-positive	9
B Rh-negative	1
AB Rh-positive	3
AB Rh-negative	1

(*Source:* American Red Cross Biomedical Services)

71. What fraction of blood donors have blood type A Rh-positive?

72. What fraction of blood donors have an O blood type?

73. What fraction of blood donors have an AB blood type?

74. What fraction of blood donors have a B blood type?

75. What fraction of blood donors have the negative Rh-factor?

FOCUS ON the Real World

BLOOD AND BLOOD DONATION

Blood is the workhorse of the body. It carries to the body's tissues everything they need, from nutrients to antibodies to heat. Blood also carries away waste products like carbon dioxide. Blood contains three types of cells—red blood cells, white blood cells, and platelets—suspended in clear, watery fluid called plasma. Blood is $\frac{11}{20}$ plasma, and plasma itself is $\frac{9}{10}$ water. In the average healthy adult human, blood accounts for $\frac{1}{11}$ of a person's body weight.

Roughly every 2 seconds someone in the United States needs blood. Although only $\frac{1}{20}$ of eligible donors donate blood, the American Red Cross is still able to collect nearly 6 million volunteer donations of blood each year. This volume makes Red Cross Biomedical Services the largest blood supplier for blood transfusions in the United States.

The modern Red Cross blood donation program has its roots in World War II. In 1940, Dr. Charles Drew headed the Red Cross-sponsored American blood collection efforts for bombing victims in Great Britain. During that time, Dr. Drew developed techniques for separating plasma from blood cells that allowed mass production of plasma. Dr. Drew had discovered that plasma could be preserved longer than whole blood. He also found that dried plasma could be stored longer than its liquid form. By 1941, Dr. Drew had become the first medical director of the first American Red Cross Blood Bank in the United States. Plasma collected through this program saved the lives of many wounded civilians and Allied soldiers by reducing the high rate of death from shock.

GROUP ACTIVITY

Contact your local Red Cross Blood Service office. Find out how many people donated blood in your area in the past two months. Ask whether it is possible to get a breakdown of the blood donations by blood type. (For more on blood type, see Exercises 71 through 75 in Section 2.3.)

1. Research the population of the area served by your local Red Cross Blood Service office. Write the fraction of the local population who gave blood in the past two months.

2. Use the breakdown by blood type to write the fraction of donors giving each type of blood.

Name _____ Section _____ Date _____

Integrated Review—Summary on Fractions, Mixed Numbers, and Factors

Use a fraction to represent the shaded area of each figure or figure group.

1.

2.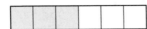

3.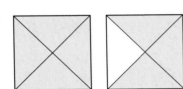

4. In a survey, 73 people out of 85 get fewer than 8 hours of sleep each night. What fraction of people in the survey get fewer than 8 hours of sleep?

Simplify.

5. $\dfrac{11}{11}$

6. $\dfrac{17}{1}$

7. $\dfrac{0}{3}$

8. $\dfrac{7}{0}$

9. $\dfrac{25}{0}$

10. $\dfrac{23}{23}$

11. $\dfrac{25}{1}$

12. $\dfrac{0}{19}$

Write each mixed number as an improper fraction.

13. $3\dfrac{1}{8}$

14. $5\dfrac{3}{5}$

15. $9\dfrac{6}{7}$

16. $20\dfrac{1}{7}$

Write each improper fraction as a mixed number or a whole number.

17. $\dfrac{20}{7}$

18. $\dfrac{55}{11}$

19. $\dfrac{39}{8}$

20. $\dfrac{98}{11}$

Answers

1. _____
2. _____
3. _____
4. _____
5. _____
6. _____
7. _____
8. _____
9. _____
10. _____
11. _____
12. _____
13. _____
14. _____
15. _____
16. _____
17. _____
18. _____
19. _____
20. _____

List the factors of each number.

21. 35   **22.** 40   **23.** 72   **24.** 13

Write the prime factorization of each number. Write any repeated factors using exponents.

25. 65   **26.** 70   **27.** 252

28. 315   **29.** 441   **30.** 286

Write each fraction in simplest form.

31. $\dfrac{2}{14}$ **32.** $\dfrac{20}{24}$ **33.** $\dfrac{18}{38}$ **34.** $\dfrac{42}{110}$ **35.** $\dfrac{32}{64}$

36. $\dfrac{72}{80}$ **37.** $\dfrac{54}{135}$ **38.** $\dfrac{90}{240}$ **39.** $\dfrac{165}{210}$ **40.** $\dfrac{245}{385}$

Determine whether each pair of fractions is equivalent.

41. $\dfrac{7}{8}$ and $\dfrac{9}{10}$ **42.** $\dfrac{10}{12}$ and $\dfrac{15}{18}$

43. Of the 50 states, 2 states are not adjacent to any other states. What fraction of the states are not adjacent to other states? Write the fraction in simplest form.

44. In 2000, 761 films were released and rated. Of these, 146 were rated PG-13. What fraction were rated PG-13?

2.4 Multiplying Fractions

Ⓐ Multiplying Fractions

OBJECTIVES

Ⓐ Multiply fractions.

Ⓑ Multiply fractions and mixed numbers or whole numbers.

Ⓒ Solve problems by multiplying fractions.

SSM
TUTOR CENTER   SG   CD & VIDEO   MATH PRO   WEB

Let's use a diagram to discover how fractions are multiplied. For example, to multiply $\frac{1}{2}$ and $\frac{3}{4}$, we find $\frac{1}{2}$ of $\frac{3}{4}$. To do this, we begin with a diagram showing $\frac{3}{4}$ of a rectangle's area shaded.

 $\frac{3}{4}$ of the rectangle's area is shaded.

To find $\frac{1}{2}$ of $\frac{3}{4}$, we heavily shade $\frac{1}{2}$ of the part that is already shaded.

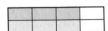

By counting smaller rectangles, we see that $\frac{3}{8}$ of the larger rectangle is now heavily shaded, so that $\frac{1}{2}$ of $\frac{3}{4}$ is $\frac{3}{8}$. This means that

$$\frac{1}{2} \cdot \frac{3}{4} = \frac{3}{8}$$ Notice that $\frac{1}{2} \cdot \frac{3}{4} = \frac{1 \cdot 3}{2 \cdot 4} = \frac{3}{8}$.

Multiplying Fractions

To multiply two fractions, multiply the numerators and multiply the denominators.

If a, b, c, and d represent positive whole numbers, we have

$$\frac{a}{b} \cdot \frac{c}{d} = \frac{a \cdot c}{b \cdot d}$$

EXAMPLES   Multiply.

1. $\dfrac{2}{3} \cdot \dfrac{5}{11} = \dfrac{2 \cdot 5}{3 \cdot 11} = \dfrac{10}{33}$

2. $\dfrac{1}{4} \cdot \dfrac{1}{2} = \dfrac{1 \cdot 1}{4 \cdot 2} = \dfrac{1}{8}$

EXAMPLE 3   Multiply and simplify: $\dfrac{6}{7} \cdot \dfrac{14}{27}$

Solution:

$$\frac{6}{7} \cdot \frac{14}{27} = \frac{6 \cdot 14}{7 \cdot 27}$$

We can simplify by finding the prime factorization and dividing out common factors.

$$\frac{6 \cdot 14}{7 \cdot 27} = \frac{2 \cdot \overset{1}{\cancel{3}} \cdot 2 \cdot \overset{1}{\cancel{7}}}{\underset{1}{\cancel{7}} \cdot \underset{1}{\cancel{3}} \cdot 3 \cdot 3}$$

$$= \frac{4}{9}$$

Practice Problems 1–2

Multiply.

1. $\dfrac{3}{8} \cdot \dfrac{5}{7}$ 2. $\dfrac{1}{3} \cdot \dfrac{1}{6}$

Practice Problem 3

Multiply and simplify: $\dfrac{6}{11} \cdot \dfrac{5}{8}$

Answers

1. $\dfrac{15}{56}$,   2. $\dfrac{1}{18}$,   3. $\dfrac{15}{44}$

In simplifying a product, it may be possible to identify common factors without actually writing the prime factorization. For example,

$$\frac{10}{11} \cdot \frac{1}{20} = \frac{10 \cdot 1}{11 \cdot 20} = \frac{\overset{1}{\cancel{10}} \cdot 1}{11 \cdot \cancel{10} \cdot 2} = \frac{1}{22}$$

Practice Problem 4

Multiply and simplify: $\dfrac{4}{15} \cdot \dfrac{3}{8}$

EXAMPLE 4 Multiply and simplify: $\dfrac{23}{32} \cdot \dfrac{4}{7}$

Solution: Notice that 4 and 32 have a common factor of 4.

$$\frac{23}{32} \cdot \frac{4}{7} = \frac{23 \cdot 4}{32 \cdot 7} = \frac{23 \cdot \overset{1}{\cancel{4}}}{\cancel{4} \cdot 8 \cdot 7} = \frac{23}{56}$$

After multiplying two fractions, always check to see whether the product can be simplified.

Practice Problems 5–7

Multiply.

5. $\dfrac{2}{5} \cdot \dfrac{15}{17}$

6. $\dfrac{4}{11} \cdot \dfrac{33}{16}$

7. $\dfrac{8}{15} \cdot \dfrac{3}{11} \cdot \dfrac{25}{16}$

EXAMPLES Multiply.

5. $\dfrac{3}{4} \cdot \dfrac{8}{5} = \dfrac{3 \cdot 8}{4 \cdot 5} = \dfrac{3 \cdot \overset{1}{\cancel{4}} \cdot 2}{\cancel{4} \cdot 5} = \dfrac{6}{5}$

6. $\dfrac{6}{13} \cdot \dfrac{26}{30} = \dfrac{6 \cdot 26}{13 \cdot 30} = \dfrac{\overset{1}{\cancel{6}} \cdot \overset{1}{\cancel{13}} \cdot 2}{\cancel{13} \cdot \cancel{6} \cdot 5} = \dfrac{2}{5}$

7. $\dfrac{1}{3} \cdot \dfrac{2}{5} \cdot \dfrac{9}{16} = \dfrac{1 \cdot 2 \cdot 9}{3 \cdot 5 \cdot 16} = \dfrac{1 \cdot \overset{1}{\cancel{2}} \cdot \overset{1}{\cancel{3}} \cdot 3}{\cancel{3} \cdot 5 \cdot \cancel{2} \cdot 8} = \dfrac{3}{40}$

B Multiplying Fractions and Mixed Numbers or Whole Numbers

When multiplying a fraction and a mixed or a whole number, remember that mixed and whole numbers can be written as fractions.

Multiplying Fractions and Mixed Numbers or Whole Numbers

To multiply a fraction and a mixed number or a whole number, first write the mixed or whole number as a fraction and then multiply as usual.

Practice Problem 8

Multiply and simplify: $2\dfrac{1}{2} \cdot \dfrac{8}{15}$

EXAMPLE 8 Multiply: $3\dfrac{1}{3} \cdot \dfrac{7}{8}$

Solution: The mixed number $3\dfrac{1}{3}$ can be written as the fraction $\dfrac{10}{3}$. Then,

$$3\frac{1}{3} \cdot \frac{7}{8} = \frac{10}{3} \cdot \frac{7}{8} = \frac{\overset{1}{\cancel{2}} \cdot 5 \cdot 7}{3 \cdot \cancel{2} \cdot 4} = \frac{35}{12} \quad \text{or} \quad 2\frac{11}{12}$$

Don't forget that a whole number can be written as a fraction by writing the whole number over 1. For example,

$$20 = \frac{20}{1} \quad \text{and} \quad 7 = \frac{7}{1}$$

Answers

4. $\dfrac{1}{10}$, 5. $\dfrac{6}{17}$, 6. $\dfrac{3}{4}$, 7. $\dfrac{5}{22}$, 8. $\dfrac{4}{3}$ or $1\dfrac{1}{3}$

EXAMPLES   Multiply.

9. $1\frac{2}{3} \cdot 2\frac{1}{4} = \frac{5}{3} \cdot \frac{9}{4} = \frac{5 \cdot 9}{3 \cdot 4} = \frac{5 \cdot \overset{1}{\cancel{3}} \cdot 3}{\underset{1}{\cancel{3}} \cdot 4} = \frac{15}{4}$ or $3\frac{3}{4}$

10. $\frac{3}{4} \cdot 20 = \frac{3}{4} \cdot \frac{20}{1} = \frac{3 \cdot 20}{4 \cdot 1} = \frac{3 \cdot \overset{1}{\cancel{4}} \cdot 5}{\underset{1}{\cancel{4}} \cdot 1} = \frac{15}{1}$ or 15

Recall from Section 1.6 that zero multiplied by any number is zero. This is true of fractions and mixed numbers also.

EXAMPLES   Multiply.

11. $0 \cdot \frac{3}{5} = 0$

12. $2\frac{3}{8} \cdot 0 = 0$

Try the Concept Check in the margin.

ⓒ Solving Problems by Multiplying Fractions

To solve real-life problems that involve multiplying fractions, we use our four problem-solving steps from Chapter 1. In Example 11, a new key word that implies multiplication is used. That key word is "of."

EXAMPLE 13   Finding the Number of Roller Coasters in an Amusement Park

Cedar Point is an amusement park located in Sandusky, Ohio. Its collection of 68 rides is the largest in the world. Of the rides, $\frac{7}{34}$ are roller coasters. How many roller coasters are in Cedar Point's collection of rides? (*Source:* Cedar Fair, L.P.)

Solution:

1. UNDERSTAND the problem. To do so, read and reread the problem. We are told that $\frac{7}{34}$ of Cedar Point's rides are roller coasters. The word "of" here means multiplication.

2. TRANSLATE.

In words:	Number of roller coasters	is	$\frac{7}{34}$	of	total rides at Cedar Point
	↓	↓	↓	↓	↓
Translate:	Number of roller coasters	=	$\frac{7}{34}$	·	68

3. SOLVE:

$$\frac{7}{34} \cdot 68 = \frac{7}{34} \cdot \frac{68}{1} = \frac{7 \cdot 68}{34 \cdot 1} = \frac{7 \cdot \overset{1}{\cancel{34}} \cdot 2}{\underset{1}{\cancel{34}} \cdot 1} = \frac{14}{1} \quad \text{or} \quad 14$$

4. INTERPRET. *Check* your work. *State* your conclusion: The number of roller coasters at Cedar Point is 14.

Helpful Hint

To help visualize a fractional part of a whole number, look at the diagram below.

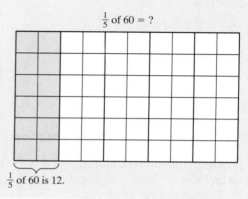

$\frac{1}{5}$ of 60 = ?

$\frac{1}{5}$ of 60 is 12.

STUDY SKILLS REMINDER

Are you organized?

Have you ever had trouble finding a completed assignment? When it's time to study for a test, are your notes neat and organized? Have you ever had trouble reading your own mathematics handwriting? (Be honest—I have.)

When any of these things happen, it's time to get organized. Here are a few suggestions:

Write your notes and complete your homework assignment in a notebook with pockets (spiral or ring binder.) Take class notes in this notebook, and then follow the notes with your completed homework assignment. When you receive graded papers or handouts, place them in the notebook pocket so that you will not lose them.

Remember to mark (possibly with an exclamation point) any note(s) that seem extra important to you. Also remember to mark (possibly with a question mark) any notes or homework that you are having trouble with. Don't forget to see your instructor or a math tutor to help you with the concepts or exercises that you are having trouble understanding.

Also, if you are having trouble reading your own handwriting, *slow down* and write your mathematics work clearly!

EXERCISE SET 2.4

A *Multiply. Write each answer in simplest form. See Examples 1 through 7 and 11.*

1. $\dfrac{1}{3} \cdot \dfrac{2}{5}$

2. $\dfrac{2}{3} \cdot \dfrac{4}{7}$

3. $\dfrac{6}{5} \cdot \dfrac{1}{7}$

4. $\dfrac{7}{3} \cdot \dfrac{2}{3}$

5. $\dfrac{3}{10} \cdot \dfrac{3}{8}$

6. $\dfrac{2}{5} \cdot \dfrac{7}{11}$

7. $\dfrac{7}{8} \cdot \dfrac{2}{3}$

8. $\dfrac{5}{9} \cdot \dfrac{7}{4}$

9. $\dfrac{2}{7} \cdot \dfrac{5}{8}$

10. $\dfrac{5}{8} \cdot \dfrac{1}{3}$

11. $\dfrac{1}{2} \cdot \dfrac{2}{15}$

12. $\dfrac{3}{8} \cdot \dfrac{5}{12}$

13. $\dfrac{5}{8} \cdot \dfrac{9}{4}$

14. $\dfrac{8}{15} \cdot \dfrac{5}{32}$

15. $\dfrac{5}{28} \cdot \dfrac{2}{25}$

16. $\dfrac{4}{25} \cdot \dfrac{5}{8}$

17. $0 \cdot \dfrac{8}{9}$

18. $\dfrac{11}{12} \cdot 0$

19. $\dfrac{18}{20} \cdot \dfrac{36}{99}$

20. $\dfrac{5}{32} \cdot \dfrac{64}{100}$

21. $\dfrac{19}{37} \cdot 0$

22. $0 \cdot \dfrac{1}{10}$

23. $\dfrac{3}{2} \cdot \dfrac{7}{3}$

24. $\dfrac{15}{2} \cdot \dfrac{3}{5}$

25. $\dfrac{14}{21} \cdot \dfrac{15}{16} \cdot \dfrac{1}{2}$

26. $\dfrac{15}{20} \cdot \dfrac{2}{7} \cdot \dfrac{21}{5}$

27. $\dfrac{1}{10} \cdot \dfrac{1}{11}$

28. $\dfrac{1}{9} \cdot \dfrac{1}{4}$

29. $\dfrac{3}{8} \cdot \dfrac{9}{10}$

30. $\dfrac{4}{5} \cdot \dfrac{8}{25}$

31. $\dfrac{6}{15} \cdot \dfrac{5}{16}$

32. $\dfrac{9}{20} \cdot \dfrac{10}{90}$

33. $\dfrac{7}{72} \cdot \dfrac{9}{49}$

34. $\dfrac{3}{80} \cdot \dfrac{2}{27}$

35. $\dfrac{11}{20} \cdot \dfrac{1}{7} \cdot \dfrac{5}{22}$

36. $\dfrac{27}{32} \cdot \dfrac{10}{13} \cdot \dfrac{16}{30}$

37. $\dfrac{1}{3} \cdot \dfrac{2}{7} \cdot \dfrac{1}{5}$

38. $\dfrac{3}{5} \cdot \dfrac{1}{2} \cdot \dfrac{3}{7}$

39. $\dfrac{9}{20} \cdot 0 \cdot \dfrac{4}{19}$

40. $\dfrac{8}{11} \cdot \dfrac{4}{7} \cdot 0$

41. $\dfrac{3}{14} \cdot \dfrac{6}{25} \cdot \dfrac{5}{27} \cdot \dfrac{7}{6}$

42. $\dfrac{7}{8} \cdot \dfrac{9}{20} \cdot \dfrac{12}{22} \cdot \dfrac{11}{14}$

B *Multiply. Write each answer in simplest form. See Examples 8 through 10.*

43. $3 \cdot \dfrac{1}{4}$

44. $\dfrac{2}{3} \cdot 6$

45. $\dfrac{5}{8} \cdot 4$

46. $3 \cdot \dfrac{7}{8}$

47. $1\dfrac{1}{4} \cdot \dfrac{4}{25}$

48. $2\dfrac{1}{5} \cdot \dfrac{6}{7}$

49. $\dfrac{2}{5} \cdot 4\dfrac{1}{6}$

50. $\dfrac{3}{22} \cdot 3\dfrac{2}{3}$

51. $\dfrac{2}{3} \cdot 1$

52. $4 \cdot \dfrac{5}{9}$

53. $2\dfrac{1}{5} \cdot 3\dfrac{1}{2}$ **54.** $2\dfrac{1}{4} \cdot 7\dfrac{1}{8}$ **55.** $3\dfrac{4}{5} \cdot 6\dfrac{2}{7}$ **56.** $5\dfrac{5}{6} \cdot 7\dfrac{1}{5}$ **57.** $\dfrac{3}{4} \cdot 16 \cdot \dfrac{1}{2}$

58. $\dfrac{7}{8} \cdot 24 \cdot \dfrac{1}{3}$ **59.** $5 \cdot 2\dfrac{1}{2}$ **60.** $6 \cdot 3\dfrac{1}{3}$ **61.** $1\dfrac{1}{5} \cdot 12\dfrac{1}{2}$ **62.** $1\dfrac{1}{6} \cdot 7\dfrac{1}{5}$

63. $\dfrac{3}{10} \cdot 15 \cdot 2\dfrac{1}{2}$ **64.** $\dfrac{11}{20} \cdot 12 \cdot 3\dfrac{1}{3}$ **65.** $3\dfrac{1}{2} \cdot 1\dfrac{3}{4} \cdot 2\dfrac{2}{3}$ **66.** $4\dfrac{1}{2} \cdot 2\dfrac{1}{9} \cdot 1\dfrac{1}{5}$

67. $9\dfrac{5}{7} \cdot 8\dfrac{1}{5} \cdot 0$ **68.** $4\dfrac{11}{13} \cdot 0 \cdot 12\dfrac{1}{13}$ **69.** $12\dfrac{4}{5} \cdot 6\dfrac{7}{8} \cdot \dfrac{26}{77}$ **70.** $14\dfrac{2}{5} \cdot 8\dfrac{1}{3} \cdot \dfrac{11}{16}$

C   *Solve. Write each answer in simplest form. See Example 11.*

71. Each turn of a screw sinks it $\dfrac{3}{16}$ of an inch deeper into a piece of wood. Find how deep the screw is after 8 turns.

72. A veterinarian's dipping vat holds 36 gallons of liquid. She normally fills it $\dfrac{5}{6}$ full of a medicated flea dip solution. Find how many gallons of solution are normally in the vat.

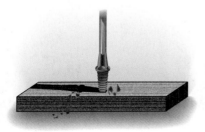

36 gallons

$\frac{5}{6}$ full

73. Julie Froelich drives $5\dfrac{1}{2}$ miles a day to and from the Star Five television station, where she is the anchor woman for the evening news 5 days a week. How far does she drive in a week?

74. The Braybendre family has a take-home pay of $1400 a month. They spend $\dfrac{2}{7}$ of their monthly income on their house payment. How much is their house payment?

75. An estimate for the measure of an adult's wrist is $\dfrac{1}{4}$ of the waist size. If Jorge has a 34-inch waist, estimate the size of his wrist.

76. The market value of a house is $\dfrac{3}{4}$ of its appraised value. If the appraised value of Pauline's house is $60,000, find its market value.

77. The plans for a deck call for $\frac{2}{5}$ of a 4-foot post to be underground. Find the length of the post that is to be buried.

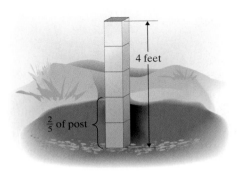

4 feet

$\frac{2}{5}$ of post

△ **78.** The radius of a circle is one-half of its diameter as shown. If the diameter of a circle is $\frac{3}{8}$ of an inch, what is its radius?

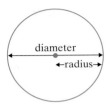

diameter

←radius→

79. A patient was told that no more than $\frac{1}{5}$ of his calories should come from fat. If his diet consists of 3000 calories a day, how many of these calories can come from fat?

80. A recipe calls for $\frac{1}{3}$ of a cup of flour. How much flour should be used if only $\frac{1}{2}$ of the recipe is being made?

81. A special on a cruise to the Bahamas is advertised to be $\frac{2}{3}$ of the regular price. If the regular price is $2757, what is the sale price?

82. The Gonzales recently sold their house for $102,000, but $\frac{3}{50}$ of this amount goes to the real estate companies that helped them sell their house. How much money do the Gonzales pay to the real estate companies?

83. A sidewalk is built 6 bricks wide by laying each brick side by side. How many inches wide is the sidewalk if each brick measures $3\frac{1}{4}$ inches wide?

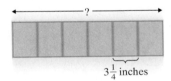

?

$3\frac{1}{4}$ inches

84. The nutrition label on a can of crushed pineapple shows 9 grams of carbohydrates for each cup of pineapple. How many grams of carbohydrates are in a $2\frac{1}{2}$-cup can?

Find the area of each rectangle. Recall that area = (length)(width).

△ **85.**

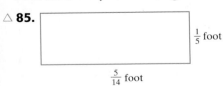

$\frac{1}{5}$ foot

$\frac{5}{14}$ foot

△ **86.**

$\frac{1}{2}$ mile

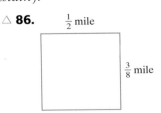

$\frac{3}{8}$ mile

△ **87.**

$1\frac{3}{4}$ yards

2 yards

△ **88.**

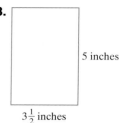

5 inches

$3\frac{1}{2}$ inches

△ **89.** A model for a proposed computer chip measures $5\frac{1}{2}$ inches by $5\frac{1}{2}$ inches. Find its area.

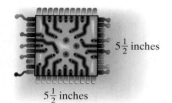

$5\frac{1}{2}$ inches

$5\frac{1}{2}$ inches

△ **90.** The Saltalamachio's are planning to build a deck that measures $4\frac{1}{2}$ yards by $6\frac{1}{3}$ yards. Find the area of their proposed deck.

The following graph is called a **circle graph** *or* **pie chart***. Each sector (shaped like a piece of pie) shows the fractional part of a car's total mileage that falls into a particular category. The whole circle represents a car's total mileage.*

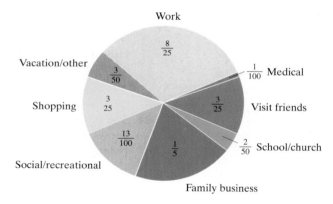

Work

$\frac{8}{25}$

Vacation/other

$\frac{3}{50}$

$\frac{1}{100}$ Medical

$\frac{3}{25}$ Visit friends

Shopping

$\frac{3}{25}$

$\frac{13}{100}$

$\frac{1}{5}$

$\frac{2}{50}$ School/church

Social/recreational

Family business

Source: The American Automobile Manufacturers Association and
The National Automobile Dealers Association

In one year, the Rodriguez family drove 12,000 miles in the family car. Use the circle graph to determine how many of these miles might be expected to fall in the categories shown in Exercises 91 through 94.

91. Work   **92.** Shopping   **93.** Family business   **94.** Medical

Review and Preview

Divide. See Section 1.7

95. $8\overline{)1648}$ **96.** $7\overline{)3920}$ **97.** $23\overline{)1300}$ **98.** $31\overline{)2500}$

Combining Concepts

99. In 2000, there were 34,800,000 Americans age 65 or older. About $\frac{11}{20}$ of these older Americans had annual incomes under $15,000. How many older Americans had incomes of less than $15,000? (*Source:* U.S. Census Bureau)

100. In 2000, there were 669 public radio stations broadcasting in the United States. Of these stations, $\frac{79}{223}$ were operated by a nonprofit community organization. How many public radio stations were operated by a nonprofit community organization in 2000? (*Source:* Corporation for Public Broadcasting)

Internet Excursions

Go To: http://www.prenhall.com/martin-gay_basic   What's Related

This World Wide Web site will provide access to a site called the Low Fat Vegetarian Archive, or a related site. It contains more than 2500 recipes for healthy low-fat and fat-free vegetarian dishes. Users can search the recipes by category, such as Mexican foods or desserts, or by ingredient.

101. Visit this site and find a recipe that is interesting to you. Be sure that the ingredient list contains at least two measures that are fractions (for example, $\frac{1}{3}$ cup). Print or copy the recipe. Then rewrite the list of ingredients to show the amounts that would be needed to make $\frac{1}{2}$ of the recipe.

102. Find a different recipe at this site that is interesting to you. Make sure that the recipe lists at least two fractional measures. Print or copy the recipe. Then rewrite the list of ingredients to show the amounts that would be needed to triple the recipe.

MODELING FRACTIONS

There are several different physical models for representing fractions.

SET MODEL

In this model, a fraction represents the portion of a set of objects that has a certain characteristic. For example, in the set of 10 shapes, 3 are hearts. That is, $\frac{3}{10}$ of the shapes are hearts.

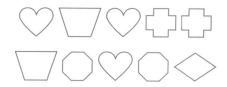

AREA MODEL

In this model, a shape is divided into a number of equal-sized regions. A fraction can be represented by shading some of the regions. For example, both of the following models represent the fraction $\frac{7}{8}$.

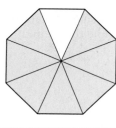

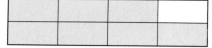

NUMBER LINE MODEL

For this model, draw a line and label a point 0 and a point to its right, 1. Now subdivide this distance from 0 to 1 depending on the denominator of the fraction that is to be represented. For example, the fraction $\frac{3}{4}$ is graphed by subdividing the portion of the number line between 0 and 1 into four equal lengths and placing a dot at the mark that represents $\frac{3}{4}$ of the distance between 0 and 1.

CRITICAL THINKING

1. **a.** Represent the fraction $\frac{5}{6}$ in two different ways using the set model.

    b. Represent the fraction $\frac{5}{6}$ in two different ways using the area model.

    c. Represent the fraction $\frac{5}{6}$ using the number line model.

2. Which model do you prefer? Why?

3. Which model do you think would be most useful for representing multiplication of fractions? Explain how this could be done.

2.5 Dividing Fractions

Ⓐ Finding Reciprocals of Fractions

Before we can divide fractions, we need to know how to find the **reciprocal** of a fraction.

Reciprocal of a Fraction

Two numbers are **reciprocals** of each other if their product is 1.

For example,

$$\frac{2}{3} \cdot \frac{3}{2} = \frac{2 \cdot 3}{3 \cdot 2} = \frac{6}{6} = 1 \qquad \text{so } \frac{2}{3} \text{ and } \frac{3}{2} \text{ are reciprocals.}$$

$$4 \cdot \frac{1}{4} = \frac{4}{1} \cdot \frac{1}{4} = \frac{4 \cdot 1}{1 \cdot 4} = \frac{4}{4} = 1 \qquad \text{so } 4 \text{ and } \frac{1}{4} \text{ are reciprocals.}$$

Finding the Reciprocal of a Fraction

To find the reciprocal of a fraction, interchange its numerator and denominator.

For example, the reciprocal of $\dfrac{6}{11}$ is $\dfrac{11}{6}$.

EXAMPLES   Find the reciprocal of each fraction.

1. The reciprocal of $\dfrac{5}{6}$ is $\dfrac{6}{5}$.

$$\frac{5}{6} \cdot \frac{6}{5} = \frac{5 \cdot 6}{6 \cdot 5} = \frac{30}{30} = 1$$

2. The reciprocal of $\dfrac{11}{8}$ is $\dfrac{8}{11}$.

$$\frac{11}{8} \cdot \frac{8}{11} = \frac{11 \cdot 8}{8 \cdot 11} = \frac{88}{88} = 1$$

3. The reciprocal of $\dfrac{1}{3}$ is $\dfrac{3}{1}$ or 3.

$$\frac{1}{3} \cdot \frac{3}{1} = \frac{1 \cdot 3}{3 \cdot 1} = \frac{3}{3} = 1$$

4. The reciprocal of 5, or $\dfrac{5}{1}$, is $\dfrac{1}{5}$.

$$\frac{5}{1} \cdot \frac{1}{5} = \frac{5 \cdot 1}{1 \cdot 5} = \frac{5}{5} = 1$$

Helpful Hint

Every number except 0 has a reciprocal. The number 0 has no reciprocal because there is no number that when multiplied by 0 gives a result of 1.

Ⓑ Dividing Fractions

Division of fractions has the same meaning as division of whole numbers. For example,

10 ÷ 5 means: How many 5s are there in 10?

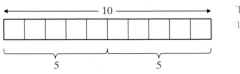

There are two 5s in 10, so
10 ÷ 5 = 2

OBJECTIVES

Ⓐ Find the reciprocal of a fraction.

Ⓑ Divide fractions.

Ⓒ Divide fractions and mixed numbers or whole numbers.

Ⓓ Solve problems by dividing fractions.

SSM
TUTOR CENTER   SG   CD & VIDEO   MATH PRO   WEB

Practice Problems 1–4

Find the reciprocal of each number.

1. $\dfrac{4}{9}$ 2. $\dfrac{15}{7}$

3. 7   4. $\dfrac{1}{8}$

Answers

1. $\dfrac{9}{4}$, **2.** $\dfrac{7}{15}$, **3.** $\dfrac{1}{7}$, **4.** 8

$\frac{3}{4} \div \frac{1}{8}$ means: How many $\frac{1}{8}$s are there in $\frac{3}{4}$?

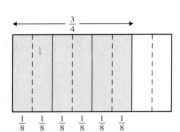

There are six $\frac{1}{8}$s in $\frac{3}{4}$,

so $\frac{3}{4} \div \frac{1}{8} = 6$

We can use reciprocals to divide fractions.

Dividing Fractions

To divide two fractions, multiply the first fraction by the reciprocal of the second fraction.

If a, b, c, and d represent positive whole numbers, then

$$\frac{a}{b} \div \frac{c}{d} = \frac{a}{b} \cdot \frac{d}{c} = \frac{a \cdot d}{b \cdot c}$$

$\lfloor$reciprocal$\rfloor$

For example,

multiply by reciprocal

$$\frac{3}{4} \div \frac{1}{8} = \frac{3}{4} \cdot \frac{8}{1} = \frac{3 \cdot 8}{4 \cdot 1} = \frac{3 \cdot 2 \cdot \overset{1}{\cancel{4}}}{\underset{1}{\cancel{4}} \cdot 1} = \frac{6}{1} \text{ or } 6$$

Just as when you are multiplying fractions, always check to see whether your answer can be simplified when you divide fractions.

EXAMPLES   Divide and simplify.

5. $\frac{7}{8} \div \frac{2}{9} = \frac{7}{8} \cdot \frac{9}{2} = \frac{7 \cdot 9}{8 \cdot 2} = \frac{63}{16}$

6. $\frac{5}{16} \div \frac{3}{4} = \frac{5}{16} \cdot \frac{4}{3} = \frac{5 \cdot 4}{16 \cdot 3} = \frac{5 \cdot \overset{1}{\cancel{4}}}{\underset{1}{\cancel{4}} \cdot 4 \cdot 3} = \frac{5}{12}$

7. $\frac{2}{5} \div \frac{1}{2} = \frac{2}{5} \cdot \frac{2}{1} = \frac{2 \cdot 2}{5 \cdot 1} = \frac{4}{5}$

After dividing two fractions, always check to see whether the product can be simplified.

Helpful Hint

When dividing fractions, do *not* look for common factors to divide out until you rewrite the division as multiplication.

Do not try to divide out these two 2s.

$$\frac{1}{2} \div \frac{2}{3} = \frac{1}{2} \cdot \frac{3}{2} = \frac{3}{4}$$

Practice Problems 5–7

Divide and simplify.

5. $\frac{3}{2} \div \frac{14}{5}$

6. $\frac{8}{7} \div \frac{2}{9}$

7. $\frac{4}{9} \div \frac{1}{2}$

Answers

5. $\frac{15}{28}$, **6.** $\frac{36}{7}$ or $5\frac{1}{7}$, **7.** $\frac{8}{9}$

Dividing Fractions and Mixed Numbers or Whole Numbers

Just as with multiplying, remember that mixed or whole numbers should be written as fractions before you divide them.

> **Dividing Fractions and Mixed Numbers or Whole Numbers**
>
> To divide with a mixed number or a whole number, first write the mixed or whole number as a fraction and then divide as usual.

EXAMPLES   Divide.

8. $\dfrac{3}{4} \div 5 = \dfrac{3}{4} \cdot \dfrac{1}{5} = \dfrac{3 \cdot 1}{4 \cdot 5} = \dfrac{3}{20}$

9. $\dfrac{11}{18} \div 2\dfrac{5}{6} = \dfrac{11}{18} \div \dfrac{17}{6} = \dfrac{11}{18} \cdot \dfrac{6}{17} = \dfrac{11 \cdot 6}{18 \cdot 17} = \dfrac{11 \cdot \overset{1}{\cancel{6}}}{\underset{1}{\cancel{6}} \cdot 3 \cdot 17} = \dfrac{11}{51}$

10. $5\dfrac{2}{3} \div 2\dfrac{5}{9} = \dfrac{17}{3} \div \dfrac{23}{9} = \dfrac{17}{3} \cdot \dfrac{9}{23} = \dfrac{17 \cdot 9}{3 \cdot 23} = \dfrac{17 \cdot \overset{1}{\cancel{3}} \cdot 3}{\underset{1}{\cancel{3}} \cdot 23} = \dfrac{51}{23}$

or $2\dfrac{5}{23}$

Recall from Section 1.7 that the quotient of 0 and any number (except 0) is 0. This is true of fractions and mixed numbers also. For example,

$$0 \div \dfrac{7}{8} = 0 \cdot \dfrac{8}{7} = 0 \qquad \text{Recall that 0 multiplied by any number is 0.}$$

Also recall from Section 1.7 that the quotient of any number and 0 is not a number. This is also true of fractions and mixed numbers. For example, to find $\dfrac{7}{8} \div 0$, or $\dfrac{7}{8} \div \dfrac{0}{1}$, we would need to find the reciprocal of $0\left(\text{or } \dfrac{0}{1}\right)$. As we mentioned in the helpful hint at the beginning of this section, 0 has no reciprocal because there is no number that when multiplied by 0 gives a result of 1. Thus,

$$\dfrac{7}{8} \div 0 \text{ is undefined.}$$

EXAMPLES   Divide.

11. $0 \div \dfrac{2}{21} = 0 \cdot \dfrac{21}{2} = 0$

12. $1\dfrac{3}{4} \div 0 \text{ is undefined.}$

Try the Concept Check in the margin.

D ## Solving Problems by Dividing Fractions

To solve real-life problems that involve dividing fractions, we continue to use our four problem-solving steps.

Practice Problems 8–10

Divide.

8. $\dfrac{4}{9} \div 5$ **9.** $\dfrac{8}{15} \div 3\dfrac{4}{5}$

10. $3\dfrac{2}{5} \div 2\dfrac{2}{15}$

Practice Problems 11–12

Divide.

11. $\dfrac{14}{17} \div 0$

12. $0 \div 2\dfrac{1}{8}$

Concept Check

a. Which of the following is the correct way to divide $\dfrac{2}{5}$ by $\dfrac{3}{4}$? Or are both correct? Explain.

1. $\dfrac{5}{2} \cdot \dfrac{3}{4}$ **2.** $\dfrac{2}{5} \cdot \dfrac{4}{3}$

b. How could you estimate the quotient $8\dfrac{6}{7} \div 2\dfrac{4}{5}$?

Answers
8. $\dfrac{4}{45}$, **9.** $\dfrac{8}{57}$, **10.** $\dfrac{51}{32}$ or $1\dfrac{19}{32}$, **11.** undefined, **12.** 0
Concept Check: **a.** 2 is correct, **b.** Round the mixed numbers to the nearest whole numbers and divide.

Practice Problem 13

A designer of women's clothing designs a woman's dress that requires $2\frac{1}{7}$ yards of material. How many dresses can be made from a 30-yard bolt of material?

EXAMPLE 13   Calculating Manufacturing Materials Needed

In a manufacturing process, a metal-cutting machine cuts strips $1\frac{3}{5}$ inches long from a piece of metal stock. How many such strips can be cut from a 48-inch piece of stock?

Solution:

1. UNDERSTAND the problem. To do so, read and reread the problem. Then draw a diagram:

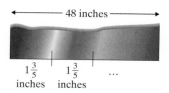

We want to know how many $1\frac{3}{5}$s there are in 48.

2. TRANSLATE.

In words:	Number of strips	is	48	divided by	$1\frac{3}{5}$
	↓	↓	↓	↓	↓
Translate:	Number of strips	=	48	÷	$1\frac{3}{5}$

3. SOLVE:

$$48 \div 1\frac{3}{5} = 48 \div \frac{8}{5} = \frac{48}{1}\cdot\frac{5}{8} = \frac{48\cdot 5}{1\cdot 8} = \frac{\overset{1}{\cancel{8}}\cdot 6\cdot 5}{1\cdot\underset{1}{\cancel{8}}} = \frac{30}{1} \text{ or } 30$$

4. INTERPRET. *Check* your work. *State* your conclusion: Thirty strips can be cut from the 48-inch piece of stock. ●

Answer

13. 14 dresses

EXERCISE SET 2.5

A *A Find the reciprocal of each fraction. See Examples 1 through 4.*

1. $\dfrac{4}{7}$

2. $\dfrac{9}{10}$

3. $\dfrac{1}{11}$

4. $\dfrac{1}{20}$

5. 15

6. 13

7. $\dfrac{12}{7}$

8. $\dfrac{10}{3}$

B *Divide. Write each answer in simplest form. See Examples 5 through 7.*

9. $\dfrac{2}{3} \div \dfrac{5}{6}$

10. $\dfrac{5}{8} \div \dfrac{2}{3}$

11. $\dfrac{6}{15} \div \dfrac{12}{5}$

12. $\dfrac{4}{15} \div \dfrac{8}{3}$

13. $\dfrac{8}{9} \div \dfrac{1}{2}$

14. $\dfrac{10}{11} \div \dfrac{4}{5}$

15. $\dfrac{3}{7} \div \dfrac{5}{6}$

16. $\dfrac{16}{27} \div \dfrac{8}{15}$

17. $\dfrac{3}{5} \div \dfrac{4}{5}$

18. $\dfrac{11}{16} \div \dfrac{13}{16}$

19. $\dfrac{11}{20} \div \dfrac{3}{11}$

20. $\dfrac{9}{20} \div \dfrac{2}{9}$

21. $\dfrac{1}{10} \div \dfrac{10}{1}$

22. $\dfrac{3}{13} \div \dfrac{13}{3}$

23. $\dfrac{7}{9} \div \dfrac{7}{3}$

24. $\dfrac{6}{11} \div \dfrac{6}{5}$

25. $\dfrac{3}{7} \div \dfrac{4}{7}$

26. $\dfrac{5}{8} \div \dfrac{3}{8}$

27. $\dfrac{7}{8} \div \dfrac{5}{6}$

28. $\dfrac{3}{8} \div \dfrac{5}{8}$

29. $\dfrac{7}{45} \div \dfrac{4}{25}$

30. $\dfrac{14}{52} \div \dfrac{1}{13}$

31. $\dfrac{2}{37} \div \dfrac{1}{7}$

32. $\dfrac{100}{158} \div \dfrac{10}{79}$

33. $\dfrac{3}{25} \div \dfrac{27}{40}$

34. $\dfrac{6}{15} \div \dfrac{7}{10}$

35. $\dfrac{11}{12} \div \dfrac{11}{12}$

36. $\dfrac{7}{13} \div \dfrac{7}{13}$

37. $\dfrac{11}{85} \div \dfrac{7}{5}$

38. $\dfrac{13}{84} \div \dfrac{3}{16}$

39. $\dfrac{8}{13} \div 0$

40. $0 \div \dfrac{4}{11}$

41. $\dfrac{27}{100} \div \dfrac{3}{20}$

42. $\dfrac{25}{128} \div \dfrac{5}{32}$

43. $0 \div \dfrac{7}{8}$

44. $\dfrac{2}{3} \div 0$

45. $\dfrac{25}{126} \div \dfrac{125}{441}$

46. $\dfrac{65}{495} \div \dfrac{26}{231}$

C *Divide. Write each answer in simplest form. See Examples 8 through 10.*

47. $\dfrac{2}{3} \div 4$

48. $\dfrac{5}{6} \div 10$

49. $8 \div \dfrac{3}{5}$

50. $7 \div \dfrac{2}{11}$

51. $2\dfrac{1}{2} \div \dfrac{1}{2}$

52. $4\dfrac{2}{3} \div \dfrac{2}{5}$

53. $\dfrac{5}{12} \div 2\dfrac{1}{3}$

54. $\dfrac{4}{15} \div 2\dfrac{1}{2}$

55. $3\dfrac{3}{7} \div 3\dfrac{1}{3}$

56. $2\dfrac{5}{6} \div 4\dfrac{6}{7}$

57. $12 \div \dfrac{1}{8}$

58. $9 \div \dfrac{1}{6}$

59. $4\dfrac{5}{11} \div 1\dfrac{2}{5}$

60. $8\dfrac{2}{7} \div 3\dfrac{1}{7}$

61. $1\dfrac{4}{9} \div 2\dfrac{5}{6}$

62. $3\dfrac{1}{10} \div 2\dfrac{1}{5}$

63. $2\dfrac{3}{8} \div 0$

64. $0 \div 15\dfrac{4}{7}$

65. $\dfrac{33}{50} \div 1$

66. $1 \div \dfrac{13}{17}$

67. $0 \div 7\dfrac{9}{10}$

68. $20\dfrac{1}{5} \div 0$

69. $1 \div \dfrac{18}{35}$

70. $\dfrac{17}{75} \div 1$

71. $10\dfrac{5}{9} \div 16\dfrac{2}{3}$

72. $20\dfrac{5}{6} \div 137\dfrac{1}{2}$

D *Solve. Write each answer in simplest form. (Some applications require division and some require multiplication to solve.) See Example 11.*

73. Sharon Bollen is to take $3\dfrac{1}{3}$ tablespoons of medicine per day in 4 equally-divided doses. How much medicine is to be taken in each dose?

74. Vonshay Bartlet wants to build a $5\dfrac{1}{2}$-foot-tall bookcase with 6 equally-spaced shelves. How far apart should he space the shelves?

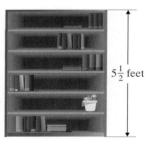

$5\frac{1}{2}$ feet

75. A small airplane used $58\dfrac{3}{4}$ gallons of fuel to fly an $8\dfrac{1}{2}$ hour trip. How many gallons of fuel were used for each hour?

76. Kesha Jonston is planning a Memorial Day barbecue. She has $27\dfrac{3}{4}$ pounds of hamburger. How many quarter-pound hamburgers can she make?

77. Movie theater owners received a total of $7660 million in movie admission tickets, about $\frac{7}{10}$ of this amount was for R-rated movies. Final the amount of money received from R-rated movies. (*Source: Motion Picture Association of America*)

78. The Oregon National Historic Trail is 2,170 miles long. It begins in Independence, Missouri, and ends in Oregon City, Oregon. Manfred Coulon has hiked $\frac{2}{5}$ of the trail before. How many miles has he hiked? (*Source:* National Park Service)

79. In July 2001, the average price of aluminum was $64\frac{1}{2}$ ¢ per pound. During that time, Severo Gutierrez received 903¢ for aluminum cans that he sold for recycling at a scrap metal center. Assuming that he received the average price, how many pounds of aluminum cans did Severo recycle? (*Source:* London Metal Exchange)

80. The record for rainfall during a 24-hour period in Alaska is $15\frac{1}{5}$ inches. This record was set in Angoon, Alaska, in October 1982. How much rain fell per hour on average? (*Source:* National Climatic Data Center)

81. An order for 125 custom-made candle stands was placed with Mr. Levi, the manager of Just For You, Inc. The worker assigned to the job can produce $2\frac{3}{5}$ candle stands per hour. Using this worker, how many work hours will be required to complete the order?

82. Yoko's Fine Jewelry sells a $\frac{3}{4}$-carat gem for $450. At this price, what is the cost of one carat?

△ **83.** The area of the rectangle below is 12 square meters. If its width is $2\frac{4}{7}$ meters, find its length.

Rectangle	$2\frac{4}{7}$ meters

△ **84.** The perimeter of the square below is $23\frac{1}{2}$ feet. Find the length of each side.

Square

Review and Preview

Perform each indicated operation. See Sections 1.3 and 1.4.

85.
```
   27
   76
+  98
```

86.
```
  811
   42
+  69
```

87.
```
  968
- 772
```

88.
```
  882
- 773
```

89.
```
  2000
-  431
```

90.
```
  500
-  92
```

91. The FedEx Express air fleet includes 258 Cessnas. These Cessnas make up $\frac{129}{320}$ of the FedEx fleet. How many aircraft make up the entire FedEx Express air fleet? (*Source:* FedEx Corporation)

92. One-third of all native flowering plant species in the United States are at risk of becoming extinct. That translates into 5144 at-risk flowering plant species. Based on this data, how many flowering plant species are native to the United States overall? (*Source:* The Nature Conservancy) (*Hint:* How many $\frac{1}{3}$s are in 5144?)

93. During the 2000 NFL regular season, the Baltimore Ravens lost four games. These losses represented $\frac{1}{4}$ of the regular-season games they played. How many regular-season games did the Ravens play? (*Source:* National Football League)

94. In your own words, explain how to divide fractions.

 CHAPTER 2 ACTIVITY # Comparing Fractions

This activity may be completed by working in groups or individually.

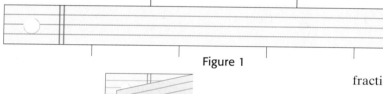

Figure 1

Figure 2

Make seven identical copies of the strip shown in Figure 1. Cut out each strip.

1. Fold and crease one of the strips (as shown in Figure 2) to represent two equal halves. Now unfold the strip so that it lies flat again. You have just created a fraction strip for the denominator 2.

2. Make similar fraction strips for the denominators 3, 4, 5, 6, 8, and 12 by folding. (Note: The markings along the top of Figure 1 will help you gauge equal thirds. The markings along the bottom of Figure 1 will help you gauge equal fifths.)

3. Shade one of the fraction strips to represent $\frac{1}{2}$ and another to represent $\frac{2}{5}$. Compare the fraction strips to decide which fraction is larger or whether the fractions are equal.

4. Shade one of the fraction strips to represent $\frac{3}{8}$ and another to represent $\frac{2}{3}$. Compare the fraction strips to decide which fraction is larger or whether the fractions are equal.

5. Shade one of the fraction strips to represent $\frac{3}{4}$ and another to represent $\frac{5}{6}$. Compare the fraction strips to decide which fraction is larger or whether the fractions are equal.

6. Shade one of the fraction strips to represent $\frac{2}{8}$ and another to represent $\frac{3}{12}$. (To reuse a fraction strip, just flip it over and use the blank side.) Compare the fraction strips to decide which fraction is larger or whether the fractions are equal.

7. Shade one of the fraction strips to represent $\frac{4}{5}$ and another to represent $\frac{7}{12}$. (To reuse a fraction strip just flip it over and use the blank side.) Compare the fraction strips to decide which fraction is larger or whether the fractions are equal.

Chapter 2 VOCABULARY CHECK

Fill in each blank with one of the words or phrases listed below.

mixed number	reciprocals	equivalent
composite number	improper fraction	simplest form
prime number	proper fraction	

1. Two numbers are _____ of each other if their product is 1.
2. A _____ is a natural number greater than 1 that is not prime.
3. Fractions that represent the same portion of a whole are called _____ fractions.
4. An _____ is a fraction whose numerator is greater than or equal to its denominator.
5. A _____ is a natural number greater than 1 whose only factors are 1 and itself.
6. A fraction is in _____ when the numerator and the denominator have no factors in common other than 1.
7. A _____ is one whose numerator is less than its denominator.
8. A _____ contains a whole number part and a fraction part.

CHAPTER

Highlights

DEFINITIONS AND CONCEPTS	EXAMPLES

Section 2.1 Introduction to Fractions and Mixed Numbers

A **fraction** is of the form

$$\frac{\text{numerator}}{\text{denominator}}$$
← number of parts considered
← number of equal parts in the whole

Write a fraction to represent the shaded part of the figure.

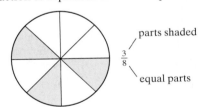

parts shaded

$\dfrac{3}{8}$

equal parts

A fraction is called a **proper fraction** if its numerator is less than its denominator.

$$\frac{1}{3}, \frac{2}{5}, \frac{7}{8}, \frac{100}{101}$$

A fraction is called an **improper fraction** if its numerator is greater than or equal to its denominator.

$$\frac{5}{4}, \frac{2}{2}, \frac{9}{7}, \frac{101}{100}$$

A **mixed number** contains a whole number and a fraction.

$$1\frac{1}{2}, 5\frac{7}{8}, 25\frac{9}{10}$$

TO WRITE A MIXED NUMBER AS AN IMPROPER FRACTION

1. Multiply the whole number part of the mixed number by the denominator of the fraction.
2. Add the numerator of the fraction to the product from step 1.
3. Write this sum as the numerator of the improper fraction over the original denominator.

$$5\frac{2}{7} = \frac{5 \cdot 7 + 2}{7} = \frac{35 + 2}{7} = \frac{37}{7}$$

DEFINITIONS AND CONCEPTS	EXAMPLES

Section 2.1   Introduction to Fractions and Mixed Numbers *(continued)*

To Write an Improper Fraction as a Mixed Number or a Whole Number

1. Divide the denominator into the numerator.
2. The whole number part of the mixed number is the quotient. The fraction is the remainder over the original denominator.

$$\text{quotient} \frac{\text{remainder}}{\text{original denominator}}$$

$$\frac{17}{3} = 5\frac{2}{3}$$

$$\begin{array}{r} 5 \\ 3\overline{)17} \\ \underline{15} \\ 2 \end{array}$$

Section 2.2   Factors and Prime Factorization

A **prime number** is a natural number that has exactly two different factors, 1 and itself.

$$2, 3, 5, 7, 11, 13, 17, \ldots$$

A **composite number** is any natural number other than 1 that is not prime.

$$4, 6, 8, 9, 10, 12, 14, 15, 16, \ldots$$

The prime factorization of a number is unique.

Write the prime factorization of 60.

$$60 = 6 \cdot 10$$
$$= 2 \cdot 3 \cdot 2 \cdot 5$$

Section 2.3   Simplest Form of a Fraction

Fractions that represent the same portion of a whole are called **equivalent fractions**.

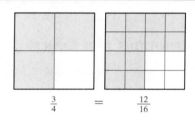

$$\frac{3}{4} \quad = \quad \frac{12}{16}$$

A fraction is in **simplest form** or **lowest terms** when the numerator and the denominator have no common factors other than 1.

Write in simplest form: $\dfrac{30}{36}$

$$\frac{30}{36} = \frac{\overset{1}{\cancel{2}} \cdot \overset{1}{\cancel{3}} \cdot 5}{\underset{1}{\cancel{2}} \cdot 2 \cdot \underset{1}{\cancel{3}} \cdot 3} = \frac{5}{6}$$

Two fractions are equivalent if their cross products are equal.

Determine whether $\dfrac{7}{8}$ and $\dfrac{21}{24}$ are equivalent.

$$\frac{7}{8} = \frac{21}{24}$$

$$\left. \begin{array}{l} 8 \cdot 21 = 168 \\ 7 \cdot 24 = 168 \end{array} \right\}$$

Since the cross products are equal, the fractions are equivalent.

$$\frac{7}{8} = \frac{21}{24}$$

DEFINITIONS AND CONCEPTS	EXAMPLES

To multiply two fractions, multiply the numerators and multiply the denominators.

Multiply.

$$\frac{7}{8} \cdot \frac{3}{5} = \frac{7 \cdot 3}{8 \cdot 5} = \frac{21}{40}$$

$$\frac{3}{4} \cdot \frac{1}{6} = \frac{3 \cdot 1}{4 \cdot 6} = \frac{\overset{1}{\cancel{3}} \cdot 1}{4 \cdot \underset{1}{\cancel{6}}} = \frac{1}{8}$$

$$2\frac{1}{3} \cdot \frac{1}{9} = \frac{7}{3} \cdot \frac{1}{9} = \frac{7 \cdot 1}{3 \cdot 9} = \frac{7}{27}$$

Two numbers are **reciprocals** of each other if their product is 1.

To divide two fractions, multiply the first fraction by the reciprocal of the second fraction.

The reciprocal of $\frac{3}{5}$ is $\frac{5}{3}$.

Divide.

$$\frac{3}{10} \div \frac{7}{9} = \frac{3}{10} \cdot \frac{9}{7} = \frac{3 \cdot 9}{10 \cdot 7} = \frac{27}{70}$$

Have you decided to successfully complete this course?

Ask yourself if one of your current goals is to successfully complete this course.

If it is not a goal of yours, ask yourself why? One common reason is fear of failure. Amazingly enough, fear of failure alone can be strong enough to keep many of us from doing our best in any endeavor. Another common reason is that you simply haven't taken the time to make successfully completing this course one of your goals.

If you are taking this mathematics course, then successfully completing this course probably should be one of your goals. To make it a goal, start by writing this goal in your mathematics notebook. Then read or reread Section 1.1 and make a commitment to try the suggestions in this section.

If successfully completing this course is already a goal of yours, also read or reread Section 1.1 and try some suggestions in this section so that you are actively working toward your goal.

Good luck and don't forget that a positive attitude will make a big difference.

Name _____ Section _____ Date _____

Chapter 2 Review

(2.1) *Determine whether each number is an improper fraction, a proper fraction, or a mixed number.*

1. $\dfrac{11}{23}$ **2.** $\dfrac{9}{8}$ **3.** $\dfrac{1}{2}$ **4.** $2\dfrac{1}{4}$ **5.** $\dfrac{0}{3}$ **6.** $5\dfrac{6}{7}$

Write a fraction to represent the shaded area.

7.

8.

9.

10.

11. A basketball player made 11 free throws out of 12 during a game. What fraction of free throws did he make?

12. A new car lot contained 23 blue cars out of a total of 131 cars. What fraction of cars on the lot are blue?

Write each improper fraction as a mixed number or a whole number.

13. $\dfrac{15}{4}$ **14.** $\dfrac{39}{13}$ **15.** $\dfrac{275}{6}$ **16.** $\dfrac{125}{4}$

Write each mixed number as an improper fraction.

17. $1\frac{1}{5}$

18. $2\frac{8}{9}$

19. $3\frac{11}{12}$

20. $5\frac{10}{17}$

(2.2) Identify each number as prime or composite.

21. 51

22. 17

23. 27

24. 21

List all factors of each number.

25. 42

26. 30

Find the prime factorization of each number.

27. 68

28. 90

29. 785

30. 255

(2.3) Write each fraction in simplest form.

31. $\frac{12}{28}$

32. $\frac{15}{27}$

33. $\frac{25}{75}$

34. $\frac{36}{72}$

35. $\frac{29}{32}$

36. $\frac{18}{23}$

37. $\frac{45}{27}$

38. $\frac{42}{30}$

39. $\dfrac{48}{6}$
40. $\dfrac{54}{9}$
41. $\dfrac{140}{150}$
42. $\dfrac{84}{140}$

(2.4) *Multiply. Write each answer in simplest form.*

43. $\dfrac{3}{5} \cdot \dfrac{1}{2}$
44. $\dfrac{6}{7} \cdot \dfrac{5}{12}$
45. $\dfrac{7}{8} \cdot \dfrac{2}{3}$
46. $\dfrac{6}{15} \cdot \dfrac{5}{8}$

47. $\dfrac{24}{5} \cdot \dfrac{15}{8}$
48. $\dfrac{27}{21} \cdot \dfrac{7}{18}$
49. $5 \cdot \dfrac{7}{8}$
50. $6 \cdot \dfrac{5}{12}$

51. $\dfrac{39}{3} \cdot \dfrac{7}{13} \cdot \dfrac{5}{21}$
52. $\dfrac{42}{5} \cdot \dfrac{15}{6} \cdot \dfrac{7}{9}$
53. $1\dfrac{5}{8} \cdot \dfrac{2}{3}$
54. $3\dfrac{6}{11} \cdot \dfrac{5}{13}$

55. $4\dfrac{1}{6} \cdot 2\dfrac{2}{5}$
56. $5\dfrac{2}{3} \cdot 2\dfrac{1}{4}$
57. $\dfrac{3}{4} \cdot 8 \cdot 4\dfrac{1}{8}$
58. $2\dfrac{1}{9} \cdot 3 \cdot \dfrac{1}{38}$

59. There are 58 calories in 1 ounce of turkey. How many calories are there in a $3\dfrac{1}{2}$-ounce serving of turkey?

60. There are $3\dfrac{1}{3}$ grams of fat in each ounce of lean hamburger. How many grams of fat are in a 4-ounce hamburger patty?

△ **61.** Find the area of each rectangle.

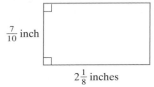

$\frac{7}{10}$ inch

$2\frac{1}{8}$ inches

△ **62.**

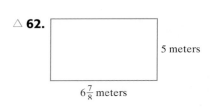

5 meters

$6\frac{7}{8}$ meters

(2.5) *Find the reciprocal of each fraction.*

63. 7

64. $\frac{1}{8}$

65. $\frac{14}{23}$

66. $\frac{17}{5}$

Divide. Write each answer in simplest form.

67. $\frac{3}{4} \div \frac{3}{8}$

68. $\frac{21}{4} \div \frac{7}{5}$

69. $\frac{18}{5} \div \frac{2}{5}$

70. $\frac{9}{2} \div \frac{1}{3}$

71. $\frac{5}{3} \div 2$

72. $5 \div \frac{15}{8}$

73. $6\frac{3}{4} \div 1\frac{2}{7}$

74. $5\frac{1}{2} \div 2\frac{1}{11}$

75. $\frac{7}{2} \div 1\frac{1}{2}$

76. $1\frac{3}{5} \div \frac{1}{4}$

77. A suit is on sale for $120, which is $\frac{3}{5}$ of the regular price. What is the regular price of the suit?

78. Herman Heltznutt walks 5 days a week for a total distance of $5\frac{1}{4}$ miles per week. If he walks the same distance each day, find the distance he walks each day.

Name _____ Section _____ Date _____

Chapter 2 Test

Perform each indicated operation. Write each answer in simplest form.

1. $\dfrac{4}{4} \div \dfrac{3}{4}$ **2.** $\dfrac{4}{3} \cdot \dfrac{4}{4}$ **3.** $2 \cdot \dfrac{1}{8}$ **4.** $\dfrac{2}{3} \cdot \dfrac{8}{15}$

5. $8 \div \dfrac{1}{2}$ **6.** $13\dfrac{1}{2} \div 3$ **7.** $\dfrac{3}{8} \cdot \dfrac{16}{6} \cdot \dfrac{4}{11}$ **8.** $5\dfrac{1}{4} \div \dfrac{7}{12}$

9. $\dfrac{16}{3} \div \dfrac{3}{12}$ **10.** $3\dfrac{1}{3} \cdot 6\dfrac{3}{4}$ **11.** $12 \div 3\dfrac{1}{3}$ **12.** $\dfrac{14}{5} \cdot \dfrac{25}{21} \cdot 2$

Write each mixed number as an improper fraction.

13. $7\dfrac{2}{3}$ **14.** $3\dfrac{6}{11}$

Write each improper fraction as a mixed number or a whole number.

15. $\dfrac{23}{5}$ **16.** $\dfrac{75}{4}$

△ **17.** Find the area of the figure.

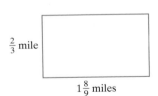

$\frac{2}{3}$ mile

$1\frac{8}{9}$ miles

18. During a 258-mile trip, a car used $10\dfrac{3}{4}$ gallons of gas. How many miles would we expect the car to travel on 1 gallon of gas?

Write each fraction in simplest form.

19. $\dfrac{24}{210}$ **20.** $\dfrac{42}{70}$

Answers

1. _____

2. _____

3. _____

4. _____

5. _____

6. _____

7. _____

8. _____

9. _____

10. _____

11. _____

12. _____

13. _____

14. _____

15. _____

16. _____

17. _____

18. _____

19. _____

20. _____

21. How many square yards of artificial turf are necessary to cover a football field, including the end zones and 10 yards beyond the sidelines? (*Hint:* A football field measures $100 \times 53\frac{1}{3}$ yards, and the end zones are 10 yards deep.)

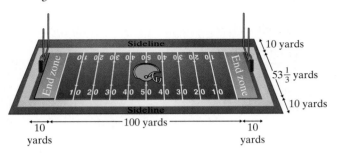

Find the prime factorization of each number.

22. 280

23. 84

24. Prior to an oil spill, the stock in an oil company sold for $120 per share. As a result of the liability that the company incurred from the spill, the price per share fell to $\frac{3}{4}$ of the price before the spill. What did the stock sell for after the spill?

Name _____ Section _____ Date _____

Cumulative Review

1. Find the place value of the digit 4 in the whole number 48,761.

2. Write the number eight hundred five in standard form.

3. Add: 34,285 + 149,761

△ **4.** Find the perimeter of the polygon shown.

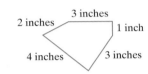

2 inches   3 inches
1 inch
4 inches   3 inches

5. The governor's salary in the state of Florida was recently increased by $7867. If the old salary was $112,304, find the new salary. (*Source: The World Almanac and Book of Facts*, 2000)

6. Subtract: 7826 − 505. Check by adding.

7. The graph below shows the ratings of Best Picture nominees since PG-13 was introduced in 1984. On this graph, each bar represents a different rating, and the height of each bar represents the number of Best Picture nominees for that rating. (*Source:* www.oscar.com, 2001)

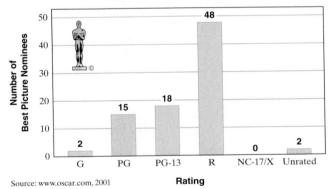

Best Picture Nominee Ratings

Number of Best Picture Nominees

Source: www.oscar.com, 2001

Rating

a. Which rating did most Best Picture nominees have?
b. How many more Best Picture nominees were rated PG-13 than PG?

8. Round 568 to the nearest ten.

9. Round each number to the nearest hundred to find an estimated difference.
       4725
     −2879

Answers

1. _____

2. _____

3. _____

4. _____

5. _____

6. _____

7. a. _____

   b. _____

8. _____

9. _____

10. a. _____

b. _____

c. _____

d. _____

11. a. _____

b. _____

c. _____

12. a. _____

b. _____

c. _____

d. _____

13. _____

14. _____

15. _____

16. _____

17. _____

18. _____

19. _____

20. a. _____

b. _____

21. _____

22. _____

23. _____

24. _____

25. _____

10. Multiply.
   a. 6×1
   b. $0(8)$
   c. $1 \cdot 45$
   d. $(75)(0)$

11. Rewrite each using the distributive property.
   a. $3(4 + 5)$
   b. $10(6 + 8)$
   c. $2(7 + 3)$

12. Find each quotient. Check by multiplying.
   a. $9\overline{)0}$
   b. $0 \div 12$
   c. $\dfrac{0}{5}$
   d. $\dfrac{3}{0}$

13. Divide and check: $1872 \div 9$

14. How many boxes are needed to ship 56 pairs of Nikes to a shoe store in Texarkana if 9 pairs of shoes will fit in each shipping box?

△ 15. A gardener bought enough plants to fill a rectangular garden with length 30 feet and width 20 feet. Because of shading problems from a nearby tree, the gardener changed the width of the garden to 15 feet. If the area is to remain the same, what is the new length of the garden?

Write using exponential notation.

16. $4 \cdot 4 \cdot 4$

17. $6 \cdot 6 \cdot 6 \cdot 8 \cdot 8 \cdot 8 \cdot 8 \cdot 8$

18. Simplify: $2 \cdot 4 - 3 \div 3$

19. Use a fraction to represent the shaded part of the figure.

20. Write as improper fractions.
   a. $4\dfrac{2}{9}$
   b. $1\dfrac{8}{11}$

21. Find all the factors of 20.

22. Write in simplest form: $\dfrac{42}{66}$

23. Multiply: $3\dfrac{1}{3} \cdot \dfrac{7}{8}$

24. Find the reciprocal of $\dfrac{1}{3}$.

25. Divide and simplify: $\dfrac{5}{16} \div \dfrac{3}{4}$

Adding and Subtracting Fractions

Having learned what fractions are and how to multiply and divide them in Chapter 2, we are ready to continue our study of fractions. In this chapter, we learn how to add and subtract fractions and mixed numbers. We then conclude this chapter with solving problems using fractions.

The American lobster, *Homarus americanus*, is big business in New England. Lobster fishers catch more than 80 million pounds of lobsters annually. Typically, more than $\frac{4}{5}$ of the lobster catch is made off the coasts of Maine, Massachusetts, and Rhode Island. Most American lobsters are a dark blue-green or gray color when alive. There are rare instances of blue, yellow, and white lobsters. However, the only red lobsters are ones that have been cooked. In Example 8 on page 189 and in the Chapter Activity, we see how fractions are used in the lobster fishing industry.

Name _____ Section _____ Date _____

Chapter **3** Pretest

Add and simplify.

1. $\dfrac{1}{9} + \dfrac{7}{9}$ **2.** $\dfrac{5}{12} + \dfrac{7}{12}$ **3.** $\dfrac{1}{6} + \dfrac{3}{4}$ **4.** $\dfrac{3}{20} + \dfrac{4}{15}$

5. $\begin{array}{r} 5\dfrac{4}{9} \\ + 3\dfrac{3}{4} \\ \hline \end{array}$

Subtract and simplify.

6. $\dfrac{8}{17} - \dfrac{3}{17}$ **7.** $\dfrac{7}{18} - \dfrac{5}{18}$ **8.** $\dfrac{11}{14} - \dfrac{2}{21}$ **9.** $\dfrac{4}{5} - \dfrac{2}{3}$

10. $\begin{array}{r} 10\dfrac{2}{5} \\ - 4\dfrac{5}{6} \\ \hline \end{array}$

Find the LCM of the following lists of numbers:

11. 6, 15 **12.** 20, 24, 45

Write each fraction as an equivalent fraction using the given denominator.

13. $\dfrac{2}{5} = \dfrac{}{35}$ **14.** $\dfrac{13}{9} = \dfrac{}{27}$

15. Insert $<$ or $>$ to make a true statement. **16.** Evaluate: $\left(\dfrac{3}{5}\right)^3$
$\dfrac{4}{7} \quad \dfrac{3}{5}$

17. Use the order of operations to simplify the following expression:
$\dfrac{2}{9} + \dfrac{1}{4} \cdot \dfrac{2}{3}$

18. Find the average of the following list of numbers: $\dfrac{1}{2}, \dfrac{2}{3}$, and $\dfrac{1}{30}$

19. Peggy Brown mixes $\dfrac{1}{4}$ pound of cashews with $\dfrac{5}{8}$ pound of peanuts. How much does the mixture weigh?

20. A plumber cuts a pipe $2\dfrac{5}{8}$ feet long from a pipe 8 feet long. How long is the remaining piece?

21. A recipe calls for $2\dfrac{3}{4}$ cups of sugar. If $2\dfrac{1}{2}$ recipes are being made, how much sugar will be needed?

Answers list:
1.
2.
3.
4.
5.
6.
7.
8.
9.
10.
11.
12.
13.
14.
15.
16.
17.
18.
19.
20.
21.

3.1 Adding and Subtracting Like Fraction

Fractions with the same denominators are called **like fractions**. Fractions that have different denominators are called **unlike fractions**.

Like Fractions	Unlike Fractions
$\dfrac{2}{5}$ and $\dfrac{3}{5}$	$\dfrac{2}{5}$ and $\dfrac{3}{4}$
$\dfrac{5}{21}, \dfrac{16}{21}$, and $\dfrac{7}{21}$	$\dfrac{5}{7}$ and $\dfrac{5}{9}$

O B J E C T I V E S

Ⓐ Add like fractions.

Ⓑ Subtract like fractions.

Ⓒ Solve problems by adding or subtracting like fractions.

SSM
TUTOR CENTER   SG   CD & VIDEO   MATH PRO   WEB

Ⓐ Adding Like Fractions

To see how we add like fractions (fractions with the same denominator), study the figures below:

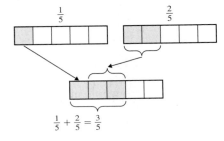

$$\frac{1}{5} + \frac{2}{5} = \frac{3}{5}$$

Adding Like Fractions

To add like fractions, add the numerators and write the sum over the common denominator. If a, b, and c represent nonzero whole numbers, we have

$$\frac{a}{c} + \frac{b}{c} = \frac{a+b}{c}$$

For example,

$$\frac{1}{4} + \frac{2}{4} = \frac{1+2}{4} = \frac{3}{4} \qquad \text{Add the numerators.}$$
$$\text{Keep the denominator.}$$

Helpful Hint

As usual, don't forget to write all answers in simplest form.

EXAMPLES   Add and simplify.

1. $\dfrac{2}{7} + \dfrac{3}{7} = \dfrac{2+3}{7} = \dfrac{5}{7}$ ⟵ Add the numerators.
        ⟵ Keep the common denominator.

2. $\dfrac{3}{16} + \dfrac{7}{16} = \dfrac{3+7}{16} = \dfrac{10}{16} = \dfrac{\overset{1}{\cancel{2}} \cdot 5}{\underset{1}{\cancel{2}} \cdot 8} = \dfrac{5}{8}$

3. $\dfrac{7}{8} + \dfrac{6}{8} + \dfrac{3}{8} = \dfrac{7+6+3}{8} = \dfrac{16}{8} = 2$

Try the Concept Check in the margin.

Practice Problems 1–3

Add and simplify.

1. $\dfrac{5}{9} + \dfrac{2}{9}$ 2. $\dfrac{5}{8} + \dfrac{1}{8}$

3. $\dfrac{10}{11} + \dfrac{1}{11} + \dfrac{7}{11}$

Concept Check

Find and correct the error in the following:

$$\frac{1}{5} + \frac{1}{5} = \frac{2}{10}$$

Answers

1. $\dfrac{7}{9}$, **2.** $\dfrac{3}{4}$, **3.** $\dfrac{18}{11} = 1\dfrac{7}{11}$

Concept Check: We don't add denominators together; correct solution: $\dfrac{1}{5} + \dfrac{1}{5} = \dfrac{2}{5}$.

B Subtracting Like Fractions

To see how we subtract like fractions (fractions with the same denominator), study the following figure:

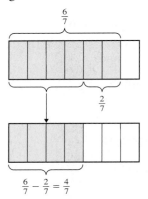

$$\frac{6}{7} - \frac{2}{7} = \frac{4}{7}$$

Subtracting Like Fractions

To subtract like fractions, subtract the numerators and write the difference over the common denominator.

If a, b, and c represent nonzero whole numbers, then

$$\frac{a}{c} - \frac{b}{c} = \frac{a - b}{c}$$

For example,

$$\frac{4}{5} - \frac{2}{5} = \frac{4 - 2}{5} = \frac{2}{5}$$ Subtract the numerators.
Keep the denominator.

EXAMPLES   Subtract and simplify.

4. $\dfrac{8}{9} - \dfrac{1}{9} = \dfrac{8 - 1}{9} = \dfrac{7}{9}$ ⟵ Subtract the numerators.
⟵ Keep the common denominator.

5. $\dfrac{7}{8} - \dfrac{5}{8} = \dfrac{7 - 5}{8} = \dfrac{2}{8} = \dfrac{\overset{1}{\cancel{2}}}{\underset{1}{\cancel{2}} \cdot 4} = \dfrac{1}{4}$

●

Practice Problems 4–5

Subtract and simplify.

4. $\dfrac{7}{12} - \dfrac{2}{12}$

5. $\dfrac{9}{10} - \dfrac{1}{10}$

C Solving Problems by Adding or Subtracting Like Fractions

Many real-life problems involve finding the perimeters of square or rectangular areas such as pastures, swimming pools, and so on. We can use our knowledge of adding fractions to find perimeters.

△ **EXAMPLE 6**   Find the perimeter of the rectangle.

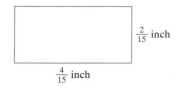

$\frac{2}{15}$ inch

$\frac{4}{15}$ inch

Practice Problem 6   △

Find the perimeter of the square.

$\frac{1}{7}$ mile

Answers

4. $\dfrac{5}{12}$,   **5.** $\dfrac{4}{5}$,   **6.** $\dfrac{4}{7}$ mi

Solution: Recall that perimeter means distance around and that opposite sides of a rectangle are the same length.

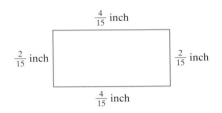

$$\text{Perimeter} = \frac{2}{15} + \frac{4}{15} + \frac{2}{15} + \frac{4}{15} = \frac{2 + 4 + 2 + 4}{15}$$

$$= \frac{12}{15} = \frac{\overset{1}{\cancel{3}} \cdot 4}{\underset{1}{\cancel{3}} \cdot 5} = \frac{4}{5}$$

The perimeter of the rectangle is $\frac{4}{5}$ inch.

We can combine our skills in adding and subtracting fractions with our four problem-solving steps from Chapter 1 to solve many kinds of real-life problems.

EXAMPLE 7   Total Amount of an Ingredient in a Recipe

A recipe calls for $\frac{1}{3}$ of a cup of flour at the beginning and $\frac{2}{3}$ of a cup of flour later. How much total flour is needed to make the recipe?

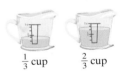

$\frac{1}{3}$ cup $\frac{2}{3}$ cup

Solution:
1. UNDERSTAND the problem. To do so, read and reread the problem. Since we are finding total flour, we add.

2. TRANSLATE.

In words:	total flour	is	flour at the beginning	added to	flour later
	↓	↓	↓	↓	↓
Translate:	total flour	=	$\frac{1}{3}$	+	$\frac{2}{3}$

3. SOLVE: $\dfrac{1}{3} + \dfrac{2}{3} = \dfrac{1+2}{3} = \dfrac{\overset{1}{\cancel{3}}}{\underset{1}{\cancel{3}}} = 1$

4. INTERPRET. *Check* your work. *State* your conclusion: The total flour needed for the recipe is 1 cup.

Practice Problem 7

If a piano student practices the piano $\frac{3}{8}$ of an hour in the morning and $\frac{1}{8}$ of an hour in the evening, how long did she practice that day?

Answer

7. $\frac{1}{2}$ h

Practice Problem 8

A jogger ran $\frac{13}{4}$ miles on Monday and $\frac{7}{4}$ miles on Wednesday. How much farther did he run on Monday than on Wednesday?

EXAMPLE 8 Calculating Distance

The distance from home to the World Gym is $\frac{7}{8}$ of a mile and from home to the post office is $\frac{5}{8}$ of a mile. How much farther is it from home to the World Gym than from home to the post office?

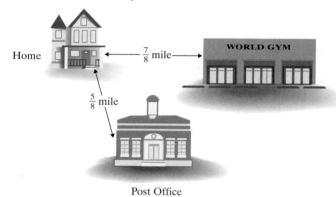

Post Office

Solution:

1. UNDERSTAND. Read and reread the problem. The phrase "How much farther" tells us to subtract distances.

2. TRANSLATE.

In words:	distance farther	is	home to World Gym distance	minus	home to post office distance
	↓	↓	↓	↓	↓
Translate:	distance farther	=	$\frac{7}{8}$	−	$\frac{5}{8}$

3. SOLVE: $\frac{7}{8} - \frac{5}{8} = \frac{7-5}{8} = \frac{2}{8} = \frac{\overset{1}{\cancel{2}}}{\underset{1}{\cancel{2} \cdot 4}} = \frac{1}{4}$

4. INTERPRET. *Check* your work. *State* your conclusion: The distance from home to the World Gym is $\frac{1}{4}$ mile farther than from home to the post office.

●

Answer

8. $1\frac{1}{2}$ mi

Mental Math

State whether the fractions in each list are like or unlike fractions.

1. $\dfrac{7}{8}, \dfrac{7}{10}$

2. $\dfrac{2}{3}, \dfrac{4}{9}$

3. $\dfrac{9}{10}, \dfrac{1}{10}$

4. $\dfrac{8}{11}, \dfrac{2}{11}$

5. $\dfrac{2}{31}, \dfrac{30}{31}, \dfrac{19}{31}$

6. $\dfrac{3}{10}, \dfrac{3}{11}, \dfrac{3}{13}$

7. $\dfrac{5}{12}, \dfrac{7}{12}, \dfrac{12}{11}$

8. $\dfrac{1}{5}, \dfrac{2}{5}, \dfrac{4}{5}$

EXERCISE SET 3.1

A *Add and simplify. See Examples 1 through 3.*

1. $\dfrac{1}{7} + \dfrac{2}{7}$

2. $\dfrac{5}{9} + \dfrac{2}{9}$

3. $\dfrac{1}{10} + \dfrac{1}{10}$

4. $\dfrac{1}{4} + \dfrac{1}{4}$

5. $\dfrac{2}{9} + \dfrac{4}{9}$

6. $\dfrac{3}{10} + \dfrac{2}{10}$

7. $\dfrac{6}{20} + \dfrac{1}{20}$

8. $\dfrac{5}{24} + \dfrac{7}{24}$

9. $\dfrac{3}{14} + \dfrac{4}{14}$

10. $\dfrac{2}{8} + \dfrac{3}{8}$

11. $\dfrac{10}{11} + \dfrac{3}{11}$

12. $\dfrac{4}{17} + \dfrac{10}{17}$

13. $\dfrac{4}{13} + \dfrac{2}{13} + \dfrac{1}{13}$

14. $\dfrac{5}{11} + \dfrac{1}{11} + \dfrac{2}{11}$

15. $\dfrac{7}{18} + \dfrac{3}{18} + \dfrac{2}{18}$

16. $\dfrac{7}{15} + \dfrac{4}{15} + \dfrac{1}{15}$

B *Subtract and simplify. See Examples 4 and 5.*

17. $\dfrac{10}{11} - \dfrac{4}{11}$

18. $\dfrac{9}{13} - \dfrac{5}{13}$

19. $\dfrac{4}{5} - \dfrac{1}{5}$

20. $\dfrac{7}{8} - \dfrac{4}{8}$

21. $\dfrac{7}{4} - \dfrac{3}{4}$

22. $\dfrac{18}{5} - \dfrac{3}{5}$

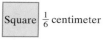 **23.** $\dfrac{7}{8} - \dfrac{1}{8}$

24. $\dfrac{5}{6} - \dfrac{1}{6}$

25. $\dfrac{25}{12} - \dfrac{15}{12}$

26. $\dfrac{30}{20} - \dfrac{15}{20}$

27. $\dfrac{11}{10} - \dfrac{3}{10}$

28. $\dfrac{14}{15} - \dfrac{4}{15}$

29. $\dfrac{27}{33} - \dfrac{8}{33}$

30. $\dfrac{37}{45} - \dfrac{18}{45}$

C *Find the perimeter of each figure. (Hint: Recall that perimeter means distance around.) See Example 6.*

 31.

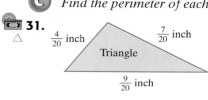

$\dfrac{4}{20}$ inch $\dfrac{7}{20}$ inch

Triangle

$\dfrac{9}{20}$ inch

△ **32.**

Square $\dfrac{1}{6}$ centimeter

△ **33.**

$\dfrac{5}{12}$ meter   Rectangle

$\dfrac{7}{12}$ meter

△ **34.**

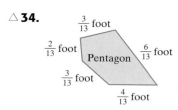

$\dfrac{3}{13}$ foot

$\dfrac{2}{13}$ foot $\dfrac{6}{13}$ foot

Pentagon

$\dfrac{3}{13}$ foot

$\dfrac{4}{13}$ foot

Solve. Write each answer in simplest form. See Examples 7 and 8.

 35. Emil Vasquez, a bodybuilder, worked out $\dfrac{7}{8}$ of an hour one morning before school and $\dfrac{5}{8}$ of an hour that evening. How long did he work out that day?

36. A recipe for Heavenly Hash cake calls for $\dfrac{3}{4}$ cup of sugar and later $\dfrac{1}{4}$ cup of sugar. How much sugar is needed to make the recipe?

37. Jenny Joyce, a railroad inspector, must inspect $\frac{19}{20}$ of a mile of railroad track. If she has already inspected $\frac{5}{20}$ of a mile, how much more does she need to inspect?

38. Scott Davis has run $\frac{11}{8}$ miles already and plans to complete $\frac{16}{8}$ miles. To do this, how much farther must he run?

39. When people take aspirin, $\frac{31}{50}$ of the time it is used to treat some type of pain. Approximately $\frac{7}{50}$ of all aspirin use is for treating headaches. What fraction of aspirin use is for treating pain other than headaches? (*Source*: Bayer Market Research)

40. In the United States, about $\frac{27}{98}$ of all households owning at least one television set use broadcast TV services only. Approximately $\frac{65}{98}$ of all households owning at least one television set subscribe to cable TV services. What fraction of television-owning households use either broadcast or cable services? (*Source*: Carmel Group)

41. According to a recent survey, $\frac{23}{100}$ of American adults who purchase books online spend less than $50 per order, and $\frac{22}{100}$ spend between $50 and $99 per order. What fraction of American adults who purchase books online spend less than $100 per online book order? (*Source*: American Booksellers Association)

42. The commonwealth of Massachusetts gets $\frac{48}{100}$ of its operating revenue from taxes and $\frac{13}{100}$ of its revenue from lottery sales. What fraction of Massachusetts' operating revenue comes from lottery sales and taxes? (*Source*: Massachusetts State Lottery Commission)

43. In 2000, the fraction of states in the United States with maximum interstate highway speed limits up to and including 70 mph was $\frac{39}{50}$. The fraction of states with 70 mph speed limits was $\frac{18}{50}$. What fraction of states had speed limits that were less than 70 mph? (*Source*: Insurance Institute for Highway Safety)

44. In the United States in 2000, about $\frac{59}{100}$ of all network advertisements were 30-second ads. About $\frac{32}{100}$ of all network advertisements were 15-second ads. What fraction of network advertisements were either 15 seconds or 30 seconds long in 2000? (*Source*: CMR-MediaWatch)

Review and Preview

Write the prime factorization of each number. See Section 2.2.

45. 10   **46.** 12   **47.** 8   **48.** 20   **49.** 55   **50.** 28

 Combining Concepts

Perform each indicated operation.

51. $\dfrac{3}{8} + \dfrac{7}{8} - \dfrac{5}{8}$

52. $\dfrac{12}{20} - \dfrac{1}{20} - \dfrac{3}{20}$

53. $\dfrac{4}{11} + \dfrac{5}{11} - \dfrac{3}{11} + \dfrac{2}{11}$

54. $\dfrac{9}{12} + \dfrac{1}{12} - \dfrac{3}{12} - \dfrac{5}{12}$

Solve. Write each answer in simplest form.

55. Of all the television sets owned by American households, $\dfrac{23}{100}$ are located in a master bedroom, $\dfrac{10}{100}$ are located in a child's bedroom, and $\dfrac{9}{100}$ are located in another bedroom. What fraction of television sets owned by American households are *not* located in a bedroom? (*Source:* Television Bureau of Advertising, Inc.)

56. Mike Cannon jogged $\dfrac{3}{8}$ of a mile from home and then rested. Then he continued jogging for another $\dfrac{3}{8}$ of a mile until he discovered his watch had fallen off. He walked back along the same path for $\dfrac{4}{8}$ of a mile until he found his watch. Find how far he was from his *starting point*.

57. In your own words, explain how to add like fractions.

58. In your own words, explain how to subtract like fractions.

 STUDY SKILLS REMINDER

What should you do the day of an exam?

On the day of an exam, try the following:

- Allow yourself plenty of time to arrive.
- Read the directions on the test carefully.
- Read each problem carefully as you take your test. Make sure that you answer the question asked.
- Watch your time and pace yourself so that you may attempt each problem on your test.
- If you have time, check your work and answers.
- Do not turn your test in early. If you have extra time, spend it double-checking your work.

Good luck!

3.2 Least Common Multiple

A Finding the Least Common Multiple Using Multiples

A multiple of a number is the product of that number and a natural number. For example,
multiples of 5 are

$$\begin{array}{cccccccc}5\cdot1 & 5\cdot2 & 5\cdot3 & 5\cdot4 & 5\cdot5 & 5\cdot6 & 5\cdot7 & 5\cdot8\\ \downarrow & \downarrow & \downarrow & \downarrow & \downarrow & \downarrow & \downarrow & \downarrow\\ 5, & 10, & 15, & 20, & 25, & 30, & 35, & 40, \ldots\end{array}$$

Multiples of 4 are

4, 8, 12, 16, 20, 24, 28, 32, 36, 40, 44, ...

Common multiples of both 4 and 5 are numbers that are found in both lists above. If we study the lists of multiples and extend them we have

Common multiples of 4 and 5: 20, 40, 60, 80, ...

We call the smallest number in the list of common multiples the **least common multiple (LCM)**. From the list of common multiples of 4 and 5, we see that the LCM of 4 and 5 is 20.

EXAMPLE 1   Find the LCM of 6 and 8.

Solution: Multiples of 6: 6, 12, 18, ⟨24⟩, 30, 36, 42, ⟨48⟩, ...

Multiples of 8: 8, 16, ⟨24⟩, 32, 40, ⟨48⟩, 56, ...

The common multiples are 24, 48, ... The least common multiple (LCM) is 24. ●

Listing all the multiples of every number in a list can be cumbersome and tedious. We can condense the procedure shown in Example 1 with the following steps:

> **Method 1: Finding the LCM of a List of Numbers Using Multiples of the Largest Number**
>
> **Step 1.** Write the multiples of the largest number (starting with the number itself) until you find one that is a multiple of the other numbers.
> **Step 2.** The multiple found in step 1 is the LCM.

EXAMPLE 2   Find the LCM of 6 and 9.

Solution: We write the multiples of 9 until we find a number that is also a multiple of 6.

$$9\cdot1 = 9 \qquad \text{Not a multiple of 6}$$
$$9\cdot2 = 18 \qquad \text{A multiple of 6}$$

The LCM of 6 and 9 is 18. ●

EXAMPLE 3   Find the LCM of 7 and 14.

Solution: We write the multiples of 14 until we find one that is also a multiple of 7.

$$14\cdot1 = 14 \qquad \text{A multiple of 7}$$

The LCM of 7 and 14 is 14. ●

OBJECTIVES

- Ⓐ Find the least common multiple (LCM) using multiples.
- Ⓑ Find the LCM using prime factorization.
- Ⓒ Write equivalent fractions.

SSM
TUTOR CENTER   SG   CD & VIDEO   MATH PRO   WEB

Practice Problem 1

Find the LCM of 15 and 25.

Practice Problem 2

Find the LCM of 8 and 10.

Practice Problem 3

Find the LCM of 8 and 16.

Answers

1. 75, **2.** 40, **3.** 16

Practice Problem 4

Find the LCM of 25 and 30.

EXAMPLE 4   Find the LCM of 12 and 20.

Solution: We write the multiples of 20 until we find one that is also a multiple of 12.

$$20 \cdot 1 = 20 \quad \text{Not a multiple of 12}$$
$$20 \cdot 2 = 40 \quad \text{Not a multiple of 12}$$
$$20 \cdot 3 = 60 \quad \text{A multiple of 12}$$

The LCM of 12 and 20 is 60.   ●

B **Finding the LCM Using Prime Factorization**

Method 1 for finding multiples works fine for smaller numbers, but may get tedious for larger numbers. A second method that uses prime factorization may be easier to use for larger numbers.

For example, to find the LCM of 270 and 84, let's look at the prime factorization of each.

$$270 = 2 \cdot 3 \cdot 3 \cdot 3 \cdot 5$$
$$84 = 2 \cdot 2 \cdot 3 \cdot 7$$

Recall that the LCM must be a multiple of both 270 and 84. Thus, to build the LCM, we start with all the factors of one number, say 270. Since the LCM must also be a multiple of 84, we look at the factors of 84, or $2 \cdot 2 \cdot 3 \cdot 7$. Of these factors, 2 and 3 are already in the factorization of 270, but 7 and an additional factor of 2 are not. Thus, the factors 2 and 7 are also needed in the LCM.

Factors of 270
$(2 \cdot 3 \cdot 3 \cdot 3 \cdot 5) \cdot 2 \cdot 7$ Factors of 84
LCM of 270 and 84

This method 2 is summarized below:

Method 2: Finding the LCM of a List of Numbers Using Prime Factorization

Step 1.   Write the prime factorization of each number.

Step 2.   For each factor that's different from the prime factors in step 1, circle the largest number of factors found in any one factorization.

Step 3.   The LCM is the product of the circled factors.

Practice Problem 5

Find the LCM of 40 and 108.

EXAMPLE 5   Find the LCM of 72 and 60.

Solution: First we write the prime factorization of each number.

$$72 = 2 \cdot 2 \cdot 2 \cdot 3 \cdot 3$$
$$60 = 2 \cdot 2 \cdot 3 \cdot 5$$

For the prime factors shown, we circle the largest number of factors found in either factorization.

$$72 = (2 \cdot 2 \cdot 2)(3 \cdot 3)$$
$$60 = 2 \cdot 2 \cdot 3 \cdot (5)$$

Answers

4. 150,   **5.** 1080

The LCM is the product of the circled factors.

$$\text{LCM} = 2 \cdot 2 \cdot 2 \cdot 3 \cdot 3 \cdot 5 = 360$$

The LCM is 360.

> **Helpful Hint**
>
> If the number of factors of a prime number are equal, circle either one, but not both. For example,
>
> $12 = ⓶ \cdot ② \cdot ③$
> $15 = 3 \cdot ⑤$ Circle either 3, but not both.
>
> The LCM is $2 \cdot 2 \cdot 3 \cdot 5 = 60$.

EXAMPLE 6   Find the LCM of 15, 18, and 54.

Solution: $15 = 3 \cdot ⑤$
$18 = ② \cdot 3 \cdot 3$
$54 = 2 \cdot ③ \cdot ③ \cdot ③$

The LCM is $2 \cdot 3 \cdot 3 \cdot 3 \cdot 5$ or 270.

EXAMPLE 7   Find the LCM of 11 and 33.

Solution: $11 = ⑪$
$33 = ③ \cdot 11$

The LCM is $3 \cdot 11$ or 33.

Ⓒ Writing Equivalent Fractions

To add or subtract unlike fractions in the next section, we first write equivalent fractions with the LCM as the denominator. Recall that fractions that represent the same portion of a whole are called "equivalent fractions."

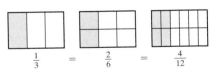

$$\frac{1}{3} = \frac{2}{6} = \frac{4}{12}$$

To write $\dfrac{1}{3}$ as an equivalent fraction with a denominator of 12, we multiply by 1.

$$\frac{1}{3} = \frac{1}{3} \cdot 1 = \frac{1}{3} \cdot \frac{4}{4} = \frac{1 \cdot 4}{3 \cdot 4} = \frac{4}{12}$$

Recall that
$\frac{4}{4} = 1$

So $\dfrac{1}{3} = \dfrac{4}{12}$. We use the following notation to write equivalent fractions:

$$\frac{1}{3} = \frac{1 \cdot 4}{3 \cdot 4} = \frac{4}{12}$$

Practice Problem 6

Find the LCM of 20, 24, and 120.

Practice Problem 7

Find the LCM of 7 and 21.

Answers

6. 120, **7.** 21

Concept Check

Which of the following is not equivalent to $\frac{3}{4}$?

a. $\frac{6}{8}$ b. $\frac{18}{24}$ c. $\frac{9}{14}$ d. $\frac{30}{40}$

Practice Problem 8

Write an equivalent fraction with the indicated denominator: $\frac{7}{8} = \frac{}{56}$

Practice Problem 9

Write an equivalent fraction with the indicated denominator.

$$\frac{3}{5} = \frac{}{15}$$

Concept Check

True or false? When the fraction $\frac{2}{9}$ is rewritten as an equivalent fraction with 27 as the denominator, the result is $\frac{2}{27}$.

Answers

8. $\frac{49}{56}$, **9.** $\frac{9}{15}$

Concept Check: **c.**

Concept Check: false; the correct result would be $\frac{6}{27}$

Try the Concept Check in the margin.

EXAMPLE 8 Write an equivalent fraction with the indicated denominator.

$$\frac{3}{4} = \frac{}{20}$$

Solution: $\frac{3}{4} = \frac{}{20}$ $4 \cdot 5 = 20$

Since $4 \cdot 5 = 20$, we multiply the numerator and the denominator of $\frac{3}{4}$ by 5.

$$\frac{3}{4} = \frac{3 \cdot 5}{4 \cdot 5} = \frac{15}{20}$$ Multiply the numerator and the denominator by 5.

Helpful Hint

To check Example 8, write $\frac{15}{20}$ in simplest form.

$$\frac{15}{20} = \frac{3 \cdot \overset{1}{5}}{4 \cdot \underset{1}{5}} = \frac{3}{4}, \text{ the original fraction.}$$

If the original fraction is in lowest terms, we can check our work by writing the new equivalent fraction in simplest form. This form should be the original fraction.

EXAMPLE 9 Write an equivalent fraction with the indicated denominator.

$$\frac{1}{2} = \frac{}{14}$$

Solution: Since $2 \cdot 7 = 14$, we multiply the numerator and the denominator by 7.

$$\frac{1}{2} = \frac{1 \cdot 7}{2 \cdot 7} = \frac{7}{14}$$

Thus, $\frac{1}{2} = \frac{7}{14}$.

Try the Concept Check in the margin.

EXERCISE SET 3.2

A **B** *Find the LCM of each list of numbers. See Examples 1 through 7.*

1. 3, 4

2. 4, 6

 3. 9, 15

4. 12, 20

5. 12, 18

6. 12, 15

7. 24, 36

8. 42, 70

9. 18, 21

10. 24, 45

11. 15, 25

12. 21, 14

13. 8, 24

14. 15, 90

15. 6, 7

16. 15, 8

17. 25, 15, 6

18. 4, 6, 18

19. 34, 68

20. 25, 175

21. 84, 294

22. 48, 54

23. 30, 36, 50

24. 5, 10, 25

25. 3, 21, 51

26. 70, 98, 100

27. 11, 33, 121

28. 10, 15, 100

29. 8, 6, 27

30. 6, 25, 10

31. 4, 6, 10, 15

32. 25, 3, 15, 10

C *Write each fraction as an equivalent fraction with the given denominator. See Examples 8 and 9.*

33. $\dfrac{4}{7} = \dfrac{}{35}$

34. $\dfrac{3}{5} = \dfrac{}{20}$

35. $\dfrac{2}{3} = \dfrac{}{21}$

36. $\dfrac{1}{6} = \dfrac{}{24}$

37. $\dfrac{2}{5} = \dfrac{}{25}$

38. $\dfrac{9}{10} = \dfrac{}{70}$ **39.** $\dfrac{1}{2} = \dfrac{}{30}$ **40.** $\dfrac{1}{3} = \dfrac{}{30}$ **41.** $\dfrac{10}{7} = \dfrac{}{21}$ **42.** $\dfrac{5}{3} = \dfrac{}{21}$

43. $\dfrac{3}{4} = \dfrac{}{28}$ **44.** $\dfrac{4}{5} = \dfrac{}{45}$ **45.** $\dfrac{2}{3} = \dfrac{}{45}$ **46.** $\dfrac{2}{3} = \dfrac{}{75}$ **47.** $\dfrac{4}{9} = \dfrac{}{81}$

48. $\dfrac{5}{11} = \dfrac{}{88}$ **49.** $\dfrac{4}{3} = \dfrac{}{9}$ **50.** $\dfrac{3}{2} = \dfrac{}{4}$ **51.** $\dfrac{9}{5} = \dfrac{}{10}$ **52.** $\dfrac{7}{4} = \dfrac{}{12}$

Review and Preview

Add or subtract as indicated. See Section 3.1.

53. $\dfrac{7}{10} - \dfrac{2}{10}$ **54.** $\dfrac{8}{13} - \dfrac{3}{13}$ **55.** $\dfrac{1}{5} + \dfrac{1}{5}$ **56.** $\dfrac{1}{8} + \dfrac{3}{8}$

57. $\dfrac{23}{18} - \dfrac{15}{18}$ **58.** $\dfrac{36}{30} - \dfrac{12}{30}$ **59.** $\dfrac{2}{9} + \dfrac{1}{9} + \dfrac{6}{9}$ **60.** $\dfrac{2}{12} + \dfrac{7}{12} + \dfrac{3}{12}$

Combining Concepts

Write each fraction as an equivalent fraction with the indicated denominator.

61. $\dfrac{37}{165} = \dfrac{}{3630}$ **62.** $\dfrac{108}{215} = \dfrac{}{4085}$

63. In your own words, explain how to find the LCM of two numbers.

64. In your own words, explain how to write a fraction as an equivalent fraction with a given denominator.

3.3 Adding and Subtracting Unlike Fractions

Ⓐ Adding Unlike Fractions

In this section we add and subtract fractions with unlike denominators. To add or subtract these unlike fractions, we first write the fractions as equivalent fractions with a common denominator and then add or subtract the like fractions. The common denominator that we use is the LCM of the denominators. This denominator is called the **least common denominator (LCD)**.

To begin, let's add the unlike fractions $\frac{3}{4} + \frac{1}{6}$. The LCM of denominators 4 and 6 is 12. This means that the LCD of denominators 4 and 6 is 12. So we write each fraction as an equivalent fraction with a denominator of 12.

$$\frac{3}{4} = \frac{3 \cdot 3}{4 \cdot 3} = \frac{9}{12} \text{ and } \frac{1}{6} = \frac{1 \cdot 2}{6 \cdot 2} = \frac{2}{12}$$

Then,

$$\frac{3}{4} + \frac{1}{6} = \frac{9}{12} + \frac{2}{12} = \frac{11}{12}$$

Adding or Subtracting Unlike Fractions

Step 1. Find the LCD of the denominators of the fractions.

Step 2. Write each fraction as an equivalent fraction whose denominator is the LCD.

Step 3. Add or subtract the like fractions.

Step 4. Write the sum or difference in simplest form.

EXAMPLE 1   Add: $\frac{2}{5} + \frac{4}{15}$

Solution: **Step 1.**   The LCD of the denominators 5 and 15 is 15.

Step 2.

$$\frac{2}{5} = \frac{2 \cdot 3}{5 \cdot 3} = \frac{6}{15}, \quad \frac{4}{15} = \frac{4}{15} \qquad \leftarrow \text{This fraction already has a denominator of 15.}$$

Step 3.

$$\frac{2}{5} + \frac{4}{15} = \frac{6}{15} + \frac{4}{15} = \frac{10}{15}$$

Step 4.   Write in simplest form.

$$\frac{10}{15} = \frac{2 \cdot \overset{1}{\cancel{5}}}{3 \cdot \underset{1}{\cancel{5}}} = \frac{2}{3}$$

EXAMPLE 2   Add: $\frac{2}{15} + \frac{3}{10}$

Solution:

Step 1.   The LCD of 15 and 10 is 30.

Step 2. $\frac{2}{15} = \frac{2 \cdot 2}{15 \cdot 2} = \frac{4}{30} \qquad \frac{3}{10} = \frac{3 \cdot 3}{10 \cdot 3} = \frac{9}{30}$

OBJECTIVES

Ⓐ Add unlike fractions.

Ⓑ Subtract unlike fractions.

Ⓒ Solve problems by adding or subtracting unlike fractions.

SSM
TUTOR CENTER   SG   CD & VIDEO   MATH PRO   WEB

Practice Problem 1

Add: $\frac{4}{7} + \frac{3}{14}$

Practice Problem 2

Add: $\frac{5}{6} + \frac{2}{9}$

Answers

1. $\frac{11}{14}$, **2.** $\frac{19}{18} = 1\frac{1}{18}$

Step 3. $\dfrac{2}{15} + \dfrac{3}{10} = \dfrac{4}{30} + \dfrac{9}{30} = \dfrac{13}{30}$

Step 4. $\dfrac{13}{30}$ is in simplest form.

Practice Problem 3

Add: $\dfrac{2}{5} + \dfrac{4}{9}$

EXAMPLE 3 Add: $\dfrac{2}{3} + \dfrac{1}{7}$

Solution: The LCD of 3 and 7 is 21.

$$\dfrac{2}{3} + \dfrac{1}{7} = \dfrac{2\cdot 7}{3\cdot 7} + \dfrac{1\cdot 3}{7\cdot 3}$$
$$= \dfrac{14}{21} + \dfrac{3}{21}$$
$$= \dfrac{17}{21}$$

Practice Problem 4

Add: $\dfrac{1}{4} + \dfrac{4}{5} + \dfrac{9}{10}$

EXAMPLE 4 Add: $\dfrac{1}{2} + \dfrac{2}{3} + \dfrac{5}{6}$

Solution: The LCD of 2, 3, and 6 is 6.

$$\dfrac{1}{2} + \dfrac{1}{3} + \dfrac{1}{6} = \dfrac{1\cdot 3}{2\cdot 3} + \dfrac{2\cdot 2}{3\cdot 2} + \dfrac{5}{6}$$
$$= \dfrac{3}{6} + \dfrac{4}{6} + \dfrac{5}{6}$$
$$= \dfrac{12}{6} = 2$$

Try the Concept Check in the margin.

Concept Check

Find and correct the error in the following: $\dfrac{2}{9} + \dfrac{4}{11} = \dfrac{6}{20} = \dfrac{3}{10}$

B ## Subtracting Unlike Fractions

As indicated in the box on page 177, we follow the same steps when subtracting unlike fractions as when adding them.

EXAMPLE 5 Subtract: $\dfrac{2}{5} - \dfrac{3}{20}$

Solution:

Step 1. The LCD of the denominators 5 and 20 is 20.

Step 2. $\dfrac{2}{5} = \dfrac{2\cdot 4}{5\cdot 4} = \dfrac{8}{20}$ $\dfrac{3}{20} = \dfrac{3}{20}$ ← The fraction already has a denominator of 20.

Practice Problem 5

Subtract: $\dfrac{7}{12} - \dfrac{5}{24}$

Step 3. $\dfrac{2}{5} - \dfrac{3}{20} = \dfrac{8}{20} - \dfrac{3}{20} = \dfrac{5}{20}$

Step 4. Write in simplest form.

Practice Problem 6

Subtract: $\dfrac{9}{10} - \dfrac{3}{7}$

$$\dfrac{5}{20} = \dfrac{\cancel{5}}{\cancel{5}\cdot 4} = \dfrac{1}{4}$$

Answers

3. $\dfrac{38}{45}$, **4.** $\dfrac{39}{20}$ or $1\dfrac{19}{20}$, **5.** $\dfrac{3}{8}$, **6.** $\dfrac{33}{70}$

Concept Check: When adding unlike fractions, we don't add the denominators. Correct solution:
$\dfrac{2}{9} + \dfrac{4}{11} = \dfrac{22}{99} + \dfrac{36}{99} = \dfrac{58}{99}$

EXAMPLE 6 Subtract: $\dfrac{10}{11} - \dfrac{2}{3}$

Solution:

Step 1. The LCD of 11 and 3 is 33.

Step 2. $\dfrac{10}{11} = \dfrac{10 \cdot 3}{11 \cdot 3} = \dfrac{30}{33}$ $\dfrac{2}{3} = \dfrac{2 \cdot 11}{3 \cdot 11} = \dfrac{22}{33}$

Step 3. $\dfrac{10}{11} - \dfrac{2}{3} = \dfrac{30}{33} - \dfrac{22}{33} = \dfrac{8}{33}$

Step 4. $\dfrac{8}{33}$ is in simplest form.

EXAMPLE 7   Subtract: $\dfrac{11}{12} - \dfrac{2}{9}$

Solution: The LCD of 12 and 9 is 36.

$$\dfrac{11}{12} - \dfrac{2}{9} = \dfrac{11 \cdot 3}{12 \cdot 3} - \dfrac{2 \cdot 4}{9 \cdot 4}$$

$$= \dfrac{33}{36} - \dfrac{8}{36}$$

$$= \dfrac{25}{36}$$

Practice Problem 7

Subtract: $\dfrac{7}{8} - \dfrac{5}{6}$

(C) Solving Problems by Adding or Subtracting Unlike Fractions

Very often, real-world problems involve adding or subtracting unlike fractions.

EXAMPLE 8   Finding Total Weight

A freight truck has $\dfrac{1}{4}$ ton of computers, $\dfrac{1}{3}$ ton of televisions, and $\dfrac{3}{8}$ ton of small appliances. Find the total weight of its load.

Practice Problem 8

To repair her sidewalk, a homeowner must pour small amounts of cement in three different locations. She needs $\dfrac{3}{5}$ of a cubic yard, $\dfrac{2}{10}$ of a cubic yard, and $\dfrac{2}{15}$ of a cubic yard for these locations. Find the total amount of concrete the homeowner needs. If she bought enough cement mix for 1 cubic yard, did she buy enough?

Solution:

1. UNDERSTAND. Read and reread the problem. The phrase "total weight" tells us to add.
2. TRANSLATE.

In words:

total weight	is	weight of computers	plus	weight of televisions	plus	weight of appliances
↓	↓	↓	↓	↓	↓	↓

Translate:

total weight	=	$\dfrac{1}{4}$	+	$\dfrac{1}{3}$	+	$\dfrac{3}{8}$

Answers

7. $\dfrac{1}{24}$,   **8.** $\dfrac{14}{15}$ cu yd; yes

3. SOLVE: The LCD is 24.

$$\frac{1}{4} + \frac{1}{3} + \frac{3}{8} = \frac{1 \cdot 6}{4 \cdot 6} + \frac{1 \cdot 8}{3 \cdot 8} + \frac{3 \cdot 3}{8 \cdot 3}$$

$$= \frac{6}{24} + \frac{8}{24} + \frac{9}{24}$$

$$= \frac{23}{24}$$

4. INTERPRET. *Check* the solution. *State* your conclusion: The total weight of the truck's load is $\frac{23}{24}$ ton. ●

Practice Problem 9

Find the difference in length of two boards if one board is $\frac{4}{5}$ of a foot long and the other is $\frac{2}{3}$ of a foot long.

$\frac{4}{5}$ of a foot

$\frac{2}{3}$ of a foot

EXAMPLE 9   Calculating Flight Time

A flight from Tucson to Phoenix, Arizona, requires $\frac{5}{12}$ of an hour. If the plane has been flying $\frac{1}{4}$ of an hour, find how much time remains before landing.

Arizona

Phoenix

$\frac{5}{12}$ hour

Tucson

Solution:

1. UNDERSTAND. Read and reread the problem. The phrase "how much time remains" tells us to subtract.

2. TRANSLATE.

In words:	time remaining	is	flight time from Tucson to Phoenix	minus	flight time already passed
	↓	↓	↓	↓	↓
Translate:	time remaining	$=$	$\frac{5}{12}$	$-$	$\frac{1}{4}$

3. SOLVE: The LCD is 12.

$$\frac{5}{12} - \frac{1}{4} = \frac{5}{12} - \frac{1 \cdot 3}{4 \cdot 3}$$

$$= \frac{5}{12} - \frac{3}{12}$$

$$= \frac{2}{12} = \frac{\overset{1}{\cancel{2}}}{\underset{1}{\cancel{2} \cdot 6}} = \frac{1}{6}$$

4. INTERPRET. *Check* the solution. *State* your conclusion: The flight time remaining is $\frac{1}{6}$ of an hour. ●

EXERCISE SET 3.3

A *Add and simplify. See Examples 1 through 4.*

1. $\frac{2}{3} + \frac{1}{6}$

2. $\frac{5}{6} + \frac{1}{12}$

3. $\frac{1}{2} + \frac{1}{3}$

4. $\frac{2}{3} + \frac{1}{4}$

5. $\frac{2}{11} + \frac{2}{33}$

6. $\frac{5}{9} + \frac{1}{3}$

7. $\frac{3}{14} + \frac{3}{7}$

8. $\frac{2}{5} + \frac{2}{15}$

9. $\frac{11}{35} + \frac{2}{7}$

10. $\frac{2}{5} + \frac{3}{25}$

11. $\frac{5}{12} + \frac{1}{9}$

12. $\frac{7}{12} + \frac{5}{18}$

13. $\frac{7}{15} + \frac{5}{12}$

14. $\frac{5}{8} + \frac{3}{20}$

15. $\frac{2}{28} + \frac{2}{21}$

16. $\frac{6}{25} + \frac{7}{35}$

17. $\frac{19}{40} + \frac{3}{14}$

18. $\frac{11}{20} + \frac{8}{22}$

19. $\frac{5}{11} + \frac{3}{13}$

20. $\frac{3}{7} + \frac{9}{17}$

21. $\frac{5}{7} + \frac{1}{8} + \frac{1}{2}$

22. $\frac{10}{13} + \frac{7}{10} + \frac{1}{5}$

23. $\frac{5}{11} + \frac{3}{9} + \frac{2}{3}$

24. $\frac{7}{18} + \frac{2}{9} + \frac{5}{6}$

25. $\frac{1}{3} + \frac{1}{9} + \frac{1}{27}$

26. $\frac{1}{4} + \frac{1}{16} + \frac{1}{64}$

27. $\frac{11}{20} + \frac{3}{5} + \frac{1}{3}$

28. $\frac{2}{7} + \frac{13}{28} + \frac{2}{5}$

B *Subtract and simplify. See Examples 5 through 7.*

29. $\frac{7}{8} - \frac{3}{16}$

30. $\frac{5}{13} - \frac{3}{26}$

31. $\frac{5}{6} - \frac{3}{7}$

32. $\frac{3}{4} - \frac{1}{7}$

33. $\frac{5}{7} - \frac{1}{8}$

34. $\dfrac{10}{13} - \dfrac{7}{10}$ **35.** $\dfrac{5}{11} - \dfrac{3}{9}$ **36.** $\dfrac{7}{18} - \dfrac{2}{9}$ ⬛ **37.** $\dfrac{11}{35} - \dfrac{2}{7}$ **38.** $\dfrac{2}{5} - \dfrac{3}{25}$

39. $\dfrac{5}{12} - \dfrac{1}{9}$ **40.** $\dfrac{7}{12} - \dfrac{5}{18}$ **41.** $\dfrac{7}{15} - \dfrac{5}{12}$ **42.** $\dfrac{5}{8} - \dfrac{3}{20}$ **43.** $\dfrac{3}{28} - \dfrac{2}{21}$

44. $\dfrac{6}{25} - \dfrac{7}{35}$ **45.** $\dfrac{1}{100} - \dfrac{1}{1000}$ **46.** $\dfrac{9}{28} - \dfrac{3}{40}$ **47.** $\dfrac{10}{26} - \dfrac{3}{8}$ **48.** $\dfrac{21}{44} - \dfrac{11}{36}$

C *Find the perimeter of each geometric figure. (Hint: Recall that perimeter means distance around.)*

⬛ **49.**
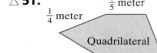
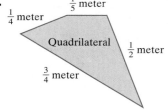
$\frac{1}{3}$ centimeter   Parallelogram

$\frac{4}{5}$ centimeter

△ **50.**
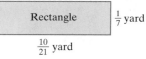
$\frac{3}{8}$ mile $\frac{5}{8}$ mile   Triangle

$\frac{1}{2}$ mile

△ **51.**
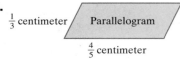
$\frac{1}{5}$ meter

$\frac{1}{4}$ meter

Quadrilateral $\frac{1}{2}$ meter

$\frac{3}{4}$ meter

△ **52.**
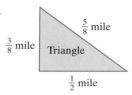
Rectangle $\frac{1}{7}$ yard

$\frac{10}{21}$ yard

Solve. See Examples 8 and 9.

53. Jillian Frize has $\dfrac{5}{8}$ of a ream of paper.

  a. If she places $\dfrac{3}{8}$ of the paper in her printer, what

  portion of the ream remains?

  b. If a ream of paper contains 500 sheets, how many
  sheets of paper remain?

54. Martell Wilson, a sales clerk at a Godiva candy shop, mixes $\dfrac{3}{4}$ pound of chocolate-covered almonds with $\dfrac{5}{8}$ pound of coconut creams in a gift container. How much does the mixture weigh?

55. About $\frac{13}{20}$ of American students ages 10 to 17 name math, science, or art as their favorite subject in school. Art is the favorite subject for about $\frac{4}{25}$ of these students. For what fraction of students this age is math or science their favorite subject? (*Source*: Peter D. Hart Research Associates for the National Science Foundation)

56. Together, the United States' and Japan's postal services handle $\frac{49}{100}$ of the world's mail volume. Japan's postal service alone handles $\frac{3}{50}$ of the world's mail. What fraction of the world's mail is handled by the postal service of the United States? (*Source*: U.S. Postal Service)

The table gives the fraction of Americans who eat pasta at various intervals. Use this table for Exercises 57 and 58.

How Often Americans Eat Pasta	
Frequency	**Fraction**
3 times per week	$\frac{31}{100}$
1 or 2 times per week	$\frac{23}{50}$
1 or 2 times per month	$\frac{17}{100}$
Less often	$\frac{3}{50}$

(*Source*: Princeton Survey Research)

57. What fraction of Americans eat pasta 1, 2, or 3 times a week?

58. What fraction of Americans eat pasta 1 or 2 times a month or less often?

Review and Preview

Multiply or divide as indicated. See Sections 2.4 and 2.5.

59. $1\frac{1}{2} \cdot 3\frac{1}{3}$

60. $2\frac{5}{6} \div 5$

61. $4 \div 7\frac{1}{4}$

62. $4\frac{3}{4} \cdot 5\frac{1}{5}$

63. $3 \cdot 2\frac{1}{9}$

64. $6\frac{2}{7} \cdot 14$

▲ Combining Concepts

Perform each indicated operation.

▦ **65.** $\dfrac{30}{55} + \dfrac{1000}{1760}$; the LCD of 55 and 1760 is 1760.

▦ **66.** $\dfrac{19}{26} - \dfrac{968}{1352}$; the LCD of 26 and 1352 is 1352.

✎ **67.** In your own words, describe how to add two fractions with different denominators.

The table shows the fraction of the world's land area taken up by each continent. Use the table to answer Exercises 68 through 72.

Continent	Fraction of World's Land Area
Africa	$\dfrac{4}{19}$
Antarctica	$\dfrac{5}{57}$
Asia	$\dfrac{17}{57}$
Australia	$\dfrac{1}{19}$
Europe	$\dfrac{4}{57}$
North America	$\dfrac{3}{19}$
South America	$\dfrac{7}{57}$

(*Source:* Based on data from the National Geographic Society)

68. What fraction of the world's land area is accounted for by North and South America?

69. What fraction of the world's land area is accounted for by Asia and Europe?

▦ **70.** If the total land area of Earth's surface is approximately 57,000,000 square miles, estimate the combined land area of the North and South American continents.

▦ **71.** If the total land area of Earth's surface is approximately 57,000,000 square miles, estimate the combined land area of the European and Asian continents.

▦ **72.** Antarctica is generally considered to be uninhabited. What fraction of the world's land area is accounted for by inhabited continents?

Subtract from left to right.

73. $\dfrac{2}{3} - \dfrac{1}{4} - \dfrac{2}{5}$

74. $\dfrac{9}{10} - \dfrac{7}{200} - \dfrac{1}{3}$

Name _____ Section _____ Date _____

Integrated Review—Operations on Fractions

Find the LCM of each list of numbers.

1. 5, 6   **2.** 3, 7   **3.** 2, 14

4. 5, 25   **5.** 4, 20, 25   **6.** 6, 18, 30

Write each fraction as an equivalent fraction with the indicated denominator.

7. $\dfrac{3}{8} = \dfrac{}{24}$ **8.** $\dfrac{7}{9} = \dfrac{}{36}$ **9.** $\dfrac{1}{4} = \dfrac{}{40}$

10. $\dfrac{2}{5} = \dfrac{}{30}$ **11.** $\dfrac{11}{15} = \dfrac{}{75}$ **12.** $\dfrac{5}{6} = \dfrac{}{48}$

Add or subtract as indicated. Simplify if necessary.

13. $\dfrac{3}{8} + \dfrac{1}{8}$ **14.** $\dfrac{7}{10} - \dfrac{3}{10}$ **15.** $\dfrac{4}{15} + \dfrac{9}{15}$

16. $\dfrac{17}{24} - \dfrac{3}{24}$ **17.** $\dfrac{1}{4} + \dfrac{1}{2}$ **18.** $\dfrac{1}{3} - \dfrac{1}{5}$

19. $\dfrac{7}{9} - \dfrac{2}{5}$ **20.** $\dfrac{3}{10} + \dfrac{2}{25}$ **21.** $\dfrac{7}{8} + \dfrac{1}{20}$

22. $\dfrac{5}{12} - \dfrac{2}{18}$ **23.** $\dfrac{1}{11} - \dfrac{1}{11}$ **24.** $\dfrac{3}{17} - \dfrac{2}{17}$

25. $\dfrac{9}{11} - \dfrac{2}{3}$ **26.** $\dfrac{1}{6} - \dfrac{1}{7}$ **27.** $\dfrac{2}{9} + \dfrac{1}{18}$

1. _____
2. _____
3. _____
4. _____
5. _____
6. _____
7. _____
8. _____
9. _____
10. _____
11. _____
12. _____
13. _____
14. _____
15. _____
16. _____
17. _____
18. _____
19. _____
20. _____
21. _____
22. _____
23. _____
24. _____
25. _____
26. _____
27. _____

28. $\dfrac{4}{13} + \dfrac{2}{26}$ **29.** $\dfrac{2}{9} + \dfrac{1}{18} + \dfrac{1}{3}$ **30.** $\dfrac{3}{10} + \dfrac{1}{5} + \dfrac{6}{25}$

Perform the indicated operation.

31. $\dfrac{9}{10} + \dfrac{2}{3}$ **32.** $\dfrac{9}{10} - \dfrac{2}{3}$ **33.** $\dfrac{9}{10} \cdot \dfrac{2}{3}$

34. $\dfrac{9}{10} \div \dfrac{2}{3}$ **35.** $\dfrac{21}{25} - \dfrac{3}{70}$ **36.** $\dfrac{21}{25} + \dfrac{3}{70}$

37. $\dfrac{21}{25} \div \dfrac{3}{70}$ **38.** $\dfrac{21}{25} \cdot \dfrac{3}{70}$ **39.** $3\dfrac{7}{8} \cdot 2\dfrac{2}{3}$

40. $3\dfrac{7}{8} \div 2\dfrac{2}{3}$ **41.** $\dfrac{2}{9} + \dfrac{5}{27} + \dfrac{1}{2}$ **42.** $\dfrac{3}{8} + \dfrac{11}{16} + \dfrac{2}{3}$

43. $11\dfrac{7}{10} \div 3\dfrac{3}{100}$ **44.** $7\dfrac{1}{4} \div 3\dfrac{3}{5}$ **45.** $\dfrac{14}{15} - \dfrac{4}{27}$

46. $\dfrac{9}{14} - \dfrac{11}{32}$

3.4 Adding and Subtracting Mixed Numbers

A Adding Mixed Numbers

Recall that a mixed number has a whole number part and a fraction part.

$$2\frac{3}{8} \text{ means } 2 + \frac{3}{8}$$

whole number, fraction

Try the Concept Check in the margin.

Adding or Subtracting Mixed Numbers

To add or subtract mixed numbers, add or subtract the fractions and then add or subtract the whole numbers.

For example,

$$\begin{array}{r} 2\frac{2}{7} \\ + 6\frac{3}{7} \\ \hline 8\frac{5}{7} \end{array}$$
← Add the fractions;
then add the whole numbers.

EXAMPLE 1   Add: $2\frac{1}{3} + 5\frac{3}{8}$

Solution: The LCD is 24.

$$\begin{array}{r} 2\frac{1\cdot 8}{3\cdot 8} = 2\frac{8}{24} \\ + 5\frac{3\cdot 3}{8\cdot 3} = 5\frac{9}{24} \\ \hline 7\frac{17}{24} \end{array}$$
← Add the fractions.
Add the whole numbers.

EXAMPLE 2   Add: $3\frac{4}{5} + 1\frac{4}{15}$

Solution: The LCD of 5 and 15 is 15.

$$\begin{array}{r} 3\frac{4}{5} = 3\frac{12}{15} \\ + 1\frac{4}{15} = 1\frac{4}{15} \\ \hline 4\frac{16}{15} \end{array}$$
Add the fractions; then add the whole numbers.
Notice that the fractional part is improper.

Since $\frac{16}{15}$ is $1\frac{1}{15}$ we can write the sum as

$$4\frac{16}{15} = 4 + 1\frac{1}{15} = 5\frac{1}{15}$$

OBJECTIVES

- A Add mixed numbers.
- B Subtract mixed numbers.
- C Solve problems by adding or subtracting mixed numbers.

SSM TUTOR CENTER   SG   CD & VIDEO   MATH PRO   WEB

Concept Check

Which of the following are equivalent to 7?

a. $6\frac{5}{5}$ b. $6\frac{7}{7}$

c. $5\frac{8}{4}$ d. $6\frac{17}{17}$

e. all of these

Practice Problem 1

Add: $4\frac{2}{5} + 5\frac{3}{10}$

Practice Problem 2

Add: $2\frac{5}{14} + 5\frac{6}{7}$

Answers

1. $9\frac{7}{10}$, **2.** $8\frac{3}{14}$

Concept Check: e

Practice Problem 3

Add: $10 + 2\frac{1}{7} + 3\frac{1}{5}$

Concept Check

Explain how you could estimate the sum:
$5\frac{1}{9} + 14\frac{10}{11}$

Practice Problem 4

Subtract: $29\frac{7}{8} - 13\frac{3}{16}$

Practice Problem 5

Subtract: $9\frac{7}{15} - 5\frac{4}{5}$

Answers

3. $15\frac{12}{35}$, **4.** $16\frac{11}{16}$, **5.** $3\frac{2}{3}$

Concept Check: Round each mixed number to the nearest whole number and add. $5\frac{1}{9}$ rounds to 5 and $14\frac{10}{11}$ rounds to 15, and their sum is $5 + 15 = 20$.

EXAMPLE 3 Add: $1\frac{4}{5} + 4 + 2\frac{1}{2}$

Solution: The LCD of 5 and 2 is 10.

$$1\frac{4}{5} = 1\frac{8}{10}$$
$$4 = 4$$
$$+2\frac{1}{2} = 2\frac{5}{10}$$
$$7\frac{13}{10} = 7 + 1\frac{3}{10} = 8\frac{3}{10}$$

Try the Concept Check in the margin.

B **Subtracting Mixed Numbers**

EXAMPLE 4 Subtract: $9\frac{3}{7} - 5\frac{2}{21}$

Solution: The LCD of 7 and 21 is 21.

$$9\frac{3}{7} = 9\frac{9}{21} \leftarrow \text{The LCD of 7 and 21 is 21.}$$
$$-5\frac{2}{21} = -5\frac{2}{21}$$
$$4\frac{7}{21} \leftarrow \text{Subtract the fractions.}$$
↑ Subtract the whole numbers.

Then $4\frac{7}{21}$ simplifies to $4\frac{1}{3}$. The difference is $4\frac{1}{3}$.

When subtracting mixed numbers, borrowing may be needed, as shown in the next example.

EXAMPLE 5 Subtract: $7\frac{3}{14} - 3\frac{6}{7}$

Solution: The LCD of 7 and 14 is 14.

$$7\frac{3}{14} = 7\frac{3}{14} \quad \text{Notice that we cannot subtract } \frac{12}{14} \text{ from } \frac{3}{14}, \text{ so we borrow from the whole number 7.}$$
$$-3\frac{6}{7} = -3\frac{12}{14}$$

borrow 1 from 7

$$7\frac{3}{14} = 6 + 1\frac{3}{14} = 6 + \frac{17}{14} \text{ or } 6\frac{17}{14}$$
Now subtract.

$$7\frac{3}{14} = 7\frac{3}{14} = 6\frac{17}{14}$$
$$-3\frac{6}{7} = -3\frac{12}{14} = -3\frac{12}{14}$$
$$3\frac{5}{14} \leftarrow \text{Subtract the fractions.}$$
↑ Subtract the whole numbers.

Try the Concept Check in the margin.

EXAMPLE 6   Subtract: $12 - 8\frac{3}{7}$

Solution: $12 = 11\frac{7}{7}$ Borrow 1 from 12 and write it as $\frac{7}{7}$.

$$-8\frac{3}{7} = -8\frac{3}{7}$$

$$3\frac{4}{7}$$ ← Subtract the fractions.

Subtract the whole numbers.

C   Solving Problems by Adding or Subtracting Mixed Numbers

Now that we know how to add and subtract mixed numbers, we can solve real-life problems.

EXAMPLE 7   Calculating Total Weight

Sarah Grahamm purchases two packages of ground round. One package weighs $2\frac{3}{8}$ pounds and the other $1\frac{4}{5}$ pounds. What is the combined weight of the ground round?

Solution:

1. UNDERSTAND. Read and reread the problem. The phrase "combined weight" tells us to add.

2. TRANSLATE.

In words:	combined weight	is	weight of one package	plus	weight of second package
	↓	↓	↓	↓	↓
Translate:	combined weight	=	$2\frac{3}{8}$	+	$1\frac{4}{5}$

3. Solve: $2\frac{3}{8} = 2\frac{15}{40}$

$$+1\frac{4}{5} = 1\frac{32}{40}$$

$$3\frac{47}{40} = 4\frac{7}{40}$$

4. INTERPRET. *Check* your work. *State* your conclusion: The combined weight of the ground round is $4\frac{7}{40}$ pounds.

Concept Check

In the subtraction problem $5\frac{1}{4} - 3\frac{3}{4}$, $5\frac{1}{4}$ must be rewritten because $\frac{3}{4}$ cannot be subtracted from $\frac{1}{4}$. Why is it incorrect to rewrite $5\frac{1}{4}$ as $5\frac{5}{4}$?

Practice Problem 6

Subtract: $25 - 10\frac{2}{9}$

Practice Problem 7

Two rainbow trout weigh $2\frac{1}{2}$ pounds and $3\frac{2}{3}$ pounds. What is the total weight of the two trout?

Answers

6. $14\frac{7}{9}$,   **7.** $6\frac{1}{6}$ lb

Concept Check: Rewrite $5\frac{1}{4}$ as $4\frac{5}{4}$ by borrowing from the 5.

Practice Problem 8

The measurement around the trunk of a tree just below shoulder height is called its girth. The largest known American beech tree in the United States has a girth of $23\frac{1}{4}$ feet. The largest known sugar maple tree in the United States has a girth of $19\frac{5}{12}$ feet. How much larger is the girth of the largest known American beech tree than the girth of the largest known sugar maple tree? (*Source: American Forests*)

girth

EXAMPLE 8 Finding Legal Lobster Size

Lobster fisherman must measure the upper body shells of the lobsters they catch. Lobsters that are too small are thrown back into the ocean. Each state has its own size standard for lobsters to help control the breeding stock. In 1988, Massachusetts increased its legal lobster size from $3\frac{3}{16}$ inches to $3\frac{7}{32}$ inches. How much of an increase was this? (*Source*: Peabody Essex Museum, Salem, Massachusetts)

Solution:

1. UNDERSTAND. Read and reread the problem carefully. The word "increase" found in the problem might make you think that we add to solve the problem. But the phrase "how much of an increase" tells us to subtract to find the increase.

2. TRANSLATE.

In words:	increase	is	new lobster size	minus	old lobster size
	↓	↓	↓	↓	↓
Translate:	increase	=	$3\frac{7}{32}$	−	$3\frac{3}{16}$

3. Solve: $3\dfrac{7}{32} = 3\dfrac{7}{32}$

$$-3\frac{3}{16} = 3\frac{6}{32}$$
$$\overline{\phantom{-3\frac{3}{16} = }\;\frac{1}{32}}$$

4. INTERPRET. *Check* your work. *State* your conclusion: The increase in lobster size is $\frac{1}{32}$ of an inch. ●

Answer

8. $3\frac{5}{6}$ ft

EXERCISE SET 3.4

Add. See Examples 1 through 3.

1. $4\frac{7}{10}$
$+ 2\frac{1}{10}$

2. $7\frac{4}{9}$
$+ 3\frac{2}{9}$

3. $10\frac{3}{14}$
$+ 3\frac{4}{7}$

4. $12\frac{5}{12}$
$+ 4\frac{1}{6}$

5. $9\frac{1}{5}$
$+ 8\frac{2}{25}$

6. $6\frac{2}{13}$
$+ 8\frac{7}{26}$

7. $1\frac{5}{6}$
$+ 5\frac{3}{8}$

8. $2\frac{5}{12}$
$+ 1\frac{5}{8}$

9. $3\frac{1}{2}$
$+ 4\frac{1}{8}$

10. $9\frac{3}{4}$
$+ 2\frac{1}{8}$

11. $8\frac{2}{5}$
$+ 11\frac{2}{3}$

12. $7\frac{3}{7}$
$+ 3\frac{1}{5}$

13. $15\frac{1}{6}$
$+ 13\frac{5}{12}$

14. $21\frac{3}{10}$
$+ 11\frac{3}{5}$

15. $40\frac{9}{10}$
$+ 15\frac{8}{27}$

16. $102\frac{5}{8}$
$+ 96\frac{21}{25}$

17. $3\frac{5}{8}$
$2\frac{1}{6}$
$+ 7\frac{3}{4}$

18. $4\frac{1}{3}$
$9\frac{2}{5}$
$+ 3\frac{1}{6}$

19. $12\frac{3}{14}$
10
$+ 25\frac{5}{12}$

20. $8\frac{2}{9}$
32
$+ 9\frac{10}{21}$

B *Subtract. See Examples 4 through 6.*

21. $4\frac{7}{10}$
$- 2\frac{1}{10}$

22. $7\frac{4}{9}$
$- 3\frac{2}{9}$

23. $10\frac{13}{14}$
$- 3\frac{4}{7}$

24. $12\frac{5}{12}$
$- 4\frac{1}{6}$

25. $9\frac{1}{5}$
$- 8\frac{6}{25}$

26. $6\frac{2}{13}$
$- 4\frac{7}{26}$

27. $15\frac{4}{7}$
$- 9\frac{11}{14}$

28. $23\frac{3}{5}$
$- 8\frac{8}{15}$

29. $5\frac{2}{3} - 3\frac{1}{5}$

30. $5\frac{3}{8} - 2\frac{13}{20}$

31. $47\frac{4}{18} - 23\frac{19}{24}$

32. $6\frac{1}{6} - 5\frac{11}{14}$

33. 10
$- 8\frac{1}{5}$

34. 23
$- 17\frac{3}{4}$

35. $22\frac{7}{8}$
$- 7$

36. $27\dfrac{3}{21}$
-9

37. $7\dfrac{3}{16}$
$-5\dfrac{3}{18}$

38. $8\dfrac{5}{12}$
$-7\dfrac{5}{14}$

39. $11\dfrac{3}{5}$
$-9\dfrac{11}{15}$

40. $9\dfrac{2}{5}$
$-7\dfrac{1}{10}$

41. 6
$-2\dfrac{4}{9}$

42. 8
$-1\dfrac{7}{10}$

43. $63\dfrac{1}{6}$
$-47\dfrac{5}{12}$

44. $86\dfrac{2}{15}$
$-27\dfrac{3}{10}$

45. $29\dfrac{9}{11}$
-12

46. $48\dfrac{4}{5}$
-24

47. $33\dfrac{11}{20}$
$-15\dfrac{19}{30}$

48. $54\dfrac{7}{30}$
$-38\dfrac{29}{50}$

C *Solve. See Examples 7 and 8.*

49. Jerald Divis, a tax consultant, takes $3\dfrac{1}{2}$ hours to prepare a personal tax return and $5\dfrac{7}{8}$ hours to prepare a business return. How much longer does it take him to prepare the business return?

50. Shamalika Corning, a trim carpenter, cuts a board $3\dfrac{3}{8}$ feet long from one that's 6 feet long. How long is the remaining piece?

51. On four consecutive days, Jose Gonzalez, a concert pianist, practiced for $2\dfrac{1}{2}$ hours, $1\dfrac{2}{3}$ hours, $2\dfrac{1}{4}$ hours, and $3\dfrac{5}{6}$ hours. Find his total practice time.

52. Coach Marlene Lawfield was preparing her team for a tennis tournament and enforced this practice schedule: Monday, $2\dfrac{1}{2}$ hours; Tuesday, $2\dfrac{2}{3}$ hours; Wednesday, $1\dfrac{3}{4}$ hours; and Thursday, $1\dfrac{9}{16}$ hours. How long did the team practice that week before Friday's tournament?

53. If Tucson's average annual rainfall is $11\dfrac{1}{4}$ inches and Yuma's is $3\dfrac{3}{5}$ inches, how much more rain, on average, does Tucson get than Yuma?

54. A pair of crutches needs adjustment. One crutch is 43 inches and the other is $41\dfrac{5}{8}$ inches. Find how much the short crutch should be lengthened to make both crutches the same length.

55. Charlotte Dowlin has $15\frac{2}{3}$ feet of plastic pipe. She cuts off a $2\frac{1}{2}$-foot length and then a $3\frac{1}{4}$-foot length. If she now needs a 10-foot piece of pipe, will the remaining piece do?

56. Jessica Callac takes $2\frac{3}{4}$ hours to clean her room. Her brother Matthew takes $1\frac{1}{3}$ hours to clean his room. If they start at the same time, how long does Matthew have to wait for Jessica to finish?

57. If the total weight allowable without overweight charges is 50 pounds and a traveler's luggage weighs $60\frac{5}{8}$ pounds, on how many pounds will the traveler's overweight charges be based?

58. What is the difference between interest rates of $11\frac{1}{2}$% and $9\frac{3}{4}$%?

59. The longest floating pontoon bridge in the United States is the Evergreen Point Bridge in Seattle, Washington. It is 2526 yards long. The second-longest pontoon bridge in the United States is the Hood Canal Bridge in Point Gamble, Washington which is $2173\frac{2}{3}$ yards long. How much longer is the Evergreen Point Bridge than the Hood Canal Bridge? (*Source*: Federal Highway Administration)

60. The record for largest rainbow trout ever caught is $42\frac{1}{8}$ pounds and was set in Alaska in 1970. The record for largest tiger trout ever caught is $20\frac{13}{16}$ pounds and was set in Michigan in 1978. How much more did the record-setting rainbow trout weigh than the record-setting tiger trout? (*Source*: International Game Fish Association)

61. Located on an island in New York City's harbor, the Statue of Liberty is one of the largest statues in the world. The copper figure is $46\frac{1}{20}$ meters tall from feet to tip of torch. The figure stands on a pedestal that is $46\frac{47}{50}$ meters feet tall. What is the overall height of the Statue of Liberty from the base of the pedestal to the tip of the torch? (*Source*: National Park Service)

The following table lists some upcoming total eclipses of the Sun that will be visible in North America. The duration of each eclipse is listed in the table. Use the table to answer Exercises 62 through 64.

Total Solar Eclipses Visible from North America	
Date of Eclipse	Duration (in minutes)
August 1, 2008	$2\frac{9}{20}$
August 21, 2017	$2\frac{2}{3}$
April 8, 2024	$4\frac{7}{15}$

(*Source:* NASA/Goddard Space Flight Center)

62. What is the total duration for the three eclipses?

63. How much longer will the April 8, 2024, eclipse be than the August 21, 2017, eclipse?

64. How much longer will the August 21, 2017, eclipse be than the August 1, 2008, eclipse?

Find the perimeter of each figure.

△ **65.**

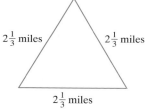

$2\frac{1}{3}$ miles $2\frac{1}{3}$ miles

$2\frac{1}{3}$ miles

△ **66.**

$3\frac{1}{4}$ yards $3\frac{1}{4}$ yards

$3\frac{1}{4}$ yards $3\frac{1}{4}$ yards

$3\frac{1}{4}$ yards

△ **67.**

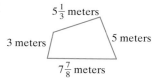

$5\frac{1}{3}$ meters

3 meters   5 meters

$7\frac{7}{8}$ meters

△ **68.**

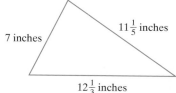

7 inches $11\frac{1}{5}$ inches

$12\frac{1}{3}$ inches

Review and Preview

Evaluate each expression. See Section 1.9.

69. 2^3

70. 3^2

71. 5^2

72. 2^5

73. 3^4

74. 1^{10}

75. 4^3

76. 9^2

Solve.

77. Carmen's Candy Clutch is famous for its "Nutstuff," a special blend of nuts and candy. A Supreme box of Nutstuff has $2\frac{1}{4}$ pounds of nuts and $3\frac{1}{2}$ pounds of candy. A Deluxe box has $1\frac{3}{8}$ pounds of nuts and $4\frac{1}{4}$ pounds of candy. Which box is heavier and by how much?

78. Willie Cassidie purchased three Supreme boxes and two Deluxe boxes of Nutstuff from Carmen's Candy Clutch. What is the total weight of his purchase?

79. Explain in your own words why $9\frac{13}{9}$ is equal to $10\frac{4}{9}$.

80. In your own words, explain how to borrow when subtracting mixed numbers.

Internet Excursions

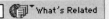
http://www.prenhall.com/martin-gay_basic

Did you know that a 2- × -4-inch board doesn't actually measure 2 inches by 4 inches? Visit this Web site to find the actual sizes of all kinds of pieces of lumber.

81. Visit this Web site to find the actual sizes of a nominal (or so-called) 2- × -4-inch board, a 6- × -8-inch timber, and a 1- × -12-inch common board. How much smaller is the actual width and thickness compared to the nominal width and thickness of each piece of lumber?

82. Visit this Web site to find the actual sizes of a nominal 1- × -6-inch common board and a nominal 6- × -10-inch timber. Compare the actual widths and thicknesses of these two pieces of lumber. How much wider and thicker is the timber than the board?

Are you prepared for a test on Chapter 3?

Below I have listed some *common trouble areas* for students in Chapter 3. After studying for your test—but before taking your test—read these.

Make sure you remember how to perform different operations on fractions !!! Try to add, subtract, multiply, then divide $\frac{3}{5}$ and $\frac{7}{15}$. Check your results below.

$$\frac{3}{5} + \frac{7}{15} = \frac{3 \cdot 3}{5 \cdot 3} + \frac{7}{15} = \frac{9}{15} + \frac{7}{15} = \frac{16}{15} = 1\frac{1}{15}$$

To add or subtract, need common denominators.

$$\frac{3}{5} - \frac{7}{15} = \frac{3 \cdot 3}{5 \cdot 3} - \frac{7}{15} = \frac{9}{15} - \frac{7}{15} = \frac{2}{15}$$

$$\frac{3}{5} \cdot \frac{7}{15} = \frac{3 \cdot 7}{5 \cdot 15} = \frac{\overset{1}{\cancel{3}} \cdot 7}{5 \cdot \underset{1}{\cancel{3}} \cdot 5} = \frac{7}{25}$$

$$\frac{3}{5} \div \frac{7}{15} = \frac{3}{5} \cdot \frac{15}{7} = \frac{\overset{1}{\cancel{3}} \cdot 15}{\underset{1}{\cancel{3}} \cdot 7} = \frac{15}{7} = 2\frac{1}{7}$$

To divide, multiply by the reciprocal.

3.5 Order, Exponents, and the Order of Operations

OBJECTIVES

A Compare fractions.
B Evaluate fractions raised to powers.
C Review operations on fractions.
D Use the order of operations.

SSM
TUTOR CENTER   SG   CD & VIDEO   MATH PRO   WEB

A Comparing Fractions

Recall that whole numbers can be shown on a number line using equally spaced distances.

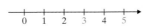

From the number line, we can see the order of numbers. For example, we can see that 3 is less than 5 because 3 is to the left of 5.

For any two numbers on a number line, the number to the left is always the smaller number, and the number to the right is always the larger number.

We use the **inequality symbols** $<$ or $>$ to write the order of numbers.

Inequality Symbols

$<$ means *is less than.*

$>$ means *is greater than.*

For example,

$\underbrace{3 \text{ is less than } 5}$ or $\underbrace{5 \text{ is greater than } 3}$

$\qquad 3 < 5 \qquad\qquad\qquad\qquad 5 > 3$

We can compare fractions the same way. To see fractions on a number line, divide the spaces between whole numbers into equal parts.

For example, let's compare $\dfrac{2}{5}$ and $\dfrac{4}{5}$.

$$\frac{5}{5}=1$$

Since $\dfrac{4}{5}$ is to the right of $\dfrac{2}{5}$,

$$\frac{2}{5} < \frac{4}{5} \qquad \text{Notice that } 2 < 4 \text{ also.}$$

Comparing Fractions

To determine which of two fractions is greater,

Step 1. Write the fractions as like fractions.

Step 2. The fraction with the greater numerator is the greater fraction.

EXAMPLE 1   Insert $<$ or $>$ to form a true statement.

$$\frac{3}{10} \qquad \frac{2}{7}$$

Solution: The LCD is 70.

$$\frac{3}{10} = \frac{3 \cdot 7}{10 \cdot 7} = \frac{21}{70} \qquad \frac{2}{7} = \frac{2 \cdot 10}{7 \cdot 10} = \frac{20}{70}$$

Practice Problem 1

Insert $<$ or $>$ to form a true statement.

$$\frac{8}{9} \qquad \frac{9}{10}$$

Answer

1. $<$

Since $21 > 20$, then $\dfrac{21}{70} > \dfrac{20}{70}$ or

$$\dfrac{3}{10} > \dfrac{2}{7}$$

Practice Problem 2

Insert $<$ or $>$ to form a true statement.

$$\dfrac{4}{7} \qquad \dfrac{3}{5}$$

EXAMPLE 2   Insert $<$ or $>$ to form a true statement.

$$\dfrac{9}{10} \qquad \dfrac{11}{12}$$

Solution: The LCD is 60.

$$\dfrac{9}{10} = \dfrac{9\cdot 6}{10\cdot 6} = \dfrac{54}{60} \qquad \dfrac{11}{12} = \dfrac{11\cdot 5}{12\cdot 5} = \dfrac{55}{60}$$

Since $54 < 55$, then $\dfrac{54}{60} < \dfrac{55}{60}$ or

$$\dfrac{9}{10} < \dfrac{11}{12}$$

Helpful Hint

If we think of $<$ and $>$ as arrowheads, a true statement is always formed when the arrow points to the smaller number.

$$\dfrac{2}{3} > \dfrac{1}{3} \qquad\qquad \dfrac{5}{6} < \dfrac{7}{6}$$

points to smaller number   points to smaller number

B Evaluating Fractions Raised to Powers

Recall from Section 1.9 that exponents indicate repeated multiplication.

exponent
↓
$$5^3 = 5\cdot 5\cdot 5 = 125$$
↑
base   3 factors of 5

Exponents mean the same when the base is a fraction. For example,

$$\left(\dfrac{1}{3}\right)^4 = \dfrac{1}{3}\cdot\dfrac{1}{3}\cdot\dfrac{1}{3}\cdot\dfrac{1}{3} = \dfrac{1}{81}$$

Practice Problems 3–5

Evaluate each expression.

3. $\left(\dfrac{1}{5}\right)^2$ 4. $\left(\dfrac{2}{3}\right)^3$ 5. $\left(\dfrac{1}{4}\right)^2\left(\dfrac{2}{3}\right)^3$

Answers

2. $<$,   3. $\dfrac{1}{25}$,   4. $\dfrac{8}{27}$,   5. $\dfrac{1}{54}$

EXAMPLES   Evaluate each expression.

3. $\left(\dfrac{1}{4}\right)^2 = \dfrac{1}{4}\cdot\dfrac{1}{4} = \dfrac{1}{16}$

4. $\left(\dfrac{3}{5}\right)^3 = \dfrac{3}{5}\cdot\dfrac{3}{5}\cdot\dfrac{3}{5} = \dfrac{27}{125}$

5. $\left(\dfrac{1}{6}\right)^2\cdot\left(\dfrac{3}{4}\right)^3 = \left(\dfrac{1}{6}\cdot\dfrac{1}{6}\right)\cdot\left(\dfrac{3}{4}\cdot\dfrac{3}{4}\cdot\dfrac{3}{4}\right) = \dfrac{1\cdot 1\cdot \overset{1}{\cancel{3}}\cdot \overset{1}{\cancel{3}}\cdot 3}{2\cdot \underset{1}{\cancel{3}}\cdot 2\cdot \underset{1}{\cancel{3}}\cdot 4\cdot 4\cdot 4} = \dfrac{3}{256}$

C Reviewing Operations on Fractions

To get ready to use the order of operations with fractions, let's first review operations on fractions that we have learned.

EXAMPLES   Perform each indicated operation.

6. $\dfrac{1}{2} \div \dfrac{8}{7} = \dfrac{1}{2} \cdot \dfrac{7}{8} = \dfrac{1 \cdot 7}{2 \cdot 8} = \dfrac{7}{16}$

7. $\dfrac{3}{5} + \dfrac{7}{10} = \dfrac{3 \cdot 2}{5 \cdot 2} + \dfrac{7}{10} = \dfrac{6}{10} + \dfrac{7}{10} = \dfrac{13}{10} = 1\dfrac{3}{10}$ The LCD is 10.

8. $\dfrac{2}{9} \cdot \dfrac{3}{11} = \dfrac{2 \cdot 3}{9 \cdot 11} = \dfrac{2 \cdot \overset{1}{\cancel{3}}}{\underset{1}{\cancel{3}} \cdot 3 \cdot 11} = \dfrac{2}{33}$

9. $\dfrac{6}{7} - \dfrac{1}{3} = \dfrac{6 \cdot 3}{7 \cdot 3} - \dfrac{1 \cdot 7}{3 \cdot 7} = \dfrac{18}{21} - \dfrac{7}{21} = \dfrac{11}{21}$ The LCD is 21.   ●

Practice Problems 6–9

Perform each indicated operation.

6. $\dfrac{3}{7} \div \dfrac{2}{11}$

7. $\dfrac{5}{12} + \dfrac{1}{6}$

8. $\dfrac{2}{3} \cdot \dfrac{9}{10}$

9. $\dfrac{11}{12} - \dfrac{2}{5}$

D Using the Order of Operations

The order of operations that we use on whole numbers applies to expressions containing fractions also.

Order of Operations

1. Perform all operations within parentheses or brackets.
2. Evaluate any expressions with exponents.
3. Multiply or divide in order from left to right.
4. Add or subtract in order from left to right.

EXAMPLE 10   Simplify: $\dfrac{1}{5} + \dfrac{2}{3} \cdot \dfrac{4}{5}$

Solution:

$$\dfrac{1}{5} + \dfrac{2}{3} \cdot \dfrac{4}{5} = \dfrac{1}{5} + \dfrac{8}{15}$$ Multiply first.

$$= \dfrac{1 \cdot 3}{5 \cdot 3} + \dfrac{8}{15}$$ The LCD is 15.

$$= \dfrac{3}{15} + \dfrac{8}{15}$$

$$= \dfrac{11}{15}$$ Add.

Practice Problem 10

Simplify: $\dfrac{2}{9} + \dfrac{3}{8} \cdot \dfrac{4}{9}$

Answers

● **6.** $\dfrac{33}{14} = 2\dfrac{5}{14}$,   **7.** $\dfrac{7}{12}$,   **8.** $\dfrac{3}{5}$,   **9.** $\dfrac{31}{60}$,   **10.** $\dfrac{7}{18}$

Practice Problem 11

Simplify: $\left(\dfrac{2}{5}\right)^2 \div \left(\dfrac{3}{5} - \dfrac{11}{25}\right)$

EXAMPLE 11 Simplify: $\left(\dfrac{2}{3}\right)^2 \div \left(\dfrac{8}{27} + \dfrac{2}{3}\right)$

Solution:

$$\left(\dfrac{2}{3}\right)^2 \div \left(\dfrac{8}{27} + \dfrac{2}{3}\right) = \left(\dfrac{2}{3}\right)^2 \div \left(\dfrac{8}{27} + \dfrac{18}{27}\right) \quad \text{Write } \dfrac{2}{3} \text{ as } \dfrac{18}{27}.$$

$$= \left(\dfrac{2}{3}\right)^2 \div \dfrac{26}{27} \quad \text{Simplify inside the parentheses.}$$

$$= \dfrac{4}{9} \div \dfrac{26}{27} \quad \text{Write } \left(\dfrac{2}{3}\right)^2 \text{ as } \dfrac{4}{9}.$$

$$= \dfrac{4}{9} \cdot \dfrac{27}{26}$$

$$= \dfrac{2 \cdot 2 \cdot 3 \cdot \cancel{9}}{\cancel{9} \cdot \cancel{2} \cdot 13}$$

$$= \dfrac{6}{13}$$

Recall that the average of a list of numbers is their sum divided by the number of numbers in the list.

Practice Problem 12

Find the average of $\dfrac{1}{2}, \dfrac{3}{8}$, and $\dfrac{7}{24}$.

EXAMPLE 12 Find the average of $\dfrac{1}{3}, \dfrac{2}{5}$, and $\dfrac{2}{9}$.

Solution: The average is their sum, divided by 3.

$$\left(\dfrac{1}{3} + \dfrac{2}{5} + \dfrac{2}{9}\right) \div 3 = \left(\dfrac{15}{45} + \dfrac{18}{45} + \dfrac{10}{45}\right) \div 3 \quad \text{The LCD is 45.}$$

$$= \dfrac{43}{45} \div 3 \quad \text{Add.}$$

$$= \dfrac{43}{45} \cdot \dfrac{1}{3}$$

$$= \dfrac{43}{135} \quad \text{Multiply.}$$

Try the Concept Check in the margin.

Concept Check

What should be done first to simplify

$$3\left[\left(\dfrac{1}{4}\right)^2 + \dfrac{3}{2}\left(\dfrac{6}{7} - \dfrac{1}{3}\right)\right]?$$

Answers

11. 1, **12.** $\dfrac{7}{18}$

Concept Check: $\dfrac{6}{7} - \dfrac{1}{3}$

EXERCISE SET 3.5

A *Insert < or > to form a true statement. See Examples 1 and 2.*

1. $\dfrac{7}{9}$ $\quad$ $\dfrac{6}{9}$

2. $\dfrac{12}{17}$ $\quad$ $\dfrac{13}{17}$

3. $\dfrac{3}{3}$ $\quad$ $\dfrac{5}{3}$

4. $\dfrac{3}{23}$ $\quad$ $\dfrac{4}{23}$

5. $\dfrac{9}{42}$ $\quad$ $\dfrac{5}{21}$

6. $\dfrac{17}{20}$ $\quad$ $\dfrac{5}{6}$

7. $\dfrac{9}{8}$ $\quad$ $\dfrac{17}{16}$

8. $\dfrac{3}{8}$ $\quad$ $\dfrac{14}{40}$

9. $\dfrac{3}{4}$ $\quad$ $\dfrac{2}{3}$

10. $\dfrac{5}{7}$ $\quad$ $\dfrac{16}{21}$

11. $\dfrac{3}{5}$ $\quad$ $\dfrac{9}{14}$

12. $\dfrac{3}{10}$ $\quad$ $\dfrac{7}{25}$

13. $\dfrac{27}{100}$ $\quad$ $\dfrac{7}{25}$

14. $\dfrac{10}{17}$ $\quad$ $\dfrac{19}{34}$

15. $\dfrac{1}{10}$ $\quad$ $\dfrac{1}{11}$

16. $\dfrac{2}{5}$ $\quad$ $\dfrac{1}{3}$

B *Evaluate each expression. See Examples 3 through 5.*

17. $\left(\dfrac{1}{2}\right)^4$

18. $\left(\dfrac{1}{7}\right)^2$

19. $\left(\dfrac{2}{5}\right)^3$

20. $\left(\dfrac{3}{4}\right)^3$

21. $\left(\dfrac{4}{7}\right)^3$

22. $\left(\dfrac{2}{3}\right)^4$

23. $\left(\dfrac{2}{9}\right)^2$

24. $\left(\dfrac{7}{11}\right)^2$

25. $\left(\dfrac{3}{4}\right)^2 \cdot \left(\dfrac{2}{3}\right)^3$

26. $\left(\dfrac{1}{6}\right)^2 \cdot \left(\dfrac{9}{10}\right)^2$

27. $\dfrac{9}{10}\left(\dfrac{2}{5}\right)^2$

28. $\dfrac{7}{11}\left(\dfrac{3}{10}\right)^2$

C *Perform each indicated operation. See Examples 6 through 9.*

29. $\dfrac{2}{15} + \dfrac{3}{5}$

30. $\dfrac{9}{10} \div \dfrac{2}{3}$

31. $\dfrac{3}{7} \cdot \dfrac{1}{5}$

32. $\dfrac{5}{12} + \dfrac{5}{6}$

33. $1 - \dfrac{4}{9}$

34. $5 - \dfrac{2}{3}$

35. $4\dfrac{2}{9} + 5\dfrac{9}{11}$

36. $7\dfrac{3}{7} + 6\dfrac{3}{5}$

37. $\dfrac{5}{6} - \dfrac{3}{4}$

38. $\dfrac{3}{8} \cdot \dfrac{1}{11}$

39. $\dfrac{6}{11} \div \dfrac{2}{3}$

40. $\dfrac{7}{10} - \dfrac{3}{25}$

41. $0 \cdot \dfrac{9}{10}$

42. $\dfrac{5}{6} \cdot 0$

43. $0 \div \dfrac{9}{10}$

44. $\dfrac{5}{6} \div 0$

45. $\dfrac{20}{35} \cdot \dfrac{7}{10}$

46. $\dfrac{11}{20} + \dfrac{7}{15}$

47. $\dfrac{4}{7} - \dfrac{6}{11}$

48. $\dfrac{18}{25} \div \dfrac{3}{5}$

D *Use the order of operations to simplify each expression. See Examples 10 and 11.*

49. $\dfrac{1}{5} + \dfrac{1}{3} \cdot \dfrac{1}{4}$

50. $\dfrac{1}{2} + \dfrac{1}{6} \cdot \dfrac{1}{3}$

51. $\dfrac{5}{6} \div \dfrac{1}{3} \cdot \dfrac{1}{4}$

52. $\dfrac{7}{8} \div \dfrac{1}{4} \cdot \dfrac{1}{7}$

53. $\frac{1}{5} \cdot \left(2\frac{5}{6} - \frac{1}{3}\right)$

54. $\frac{4}{7} \cdot \left(6 - 2\frac{1}{2}\right)$

55. $2 \cdot \left(\frac{1}{4} + \frac{1}{5}\right) + 2$

56. $\frac{2}{5} \cdot \left(5 - \frac{1}{2}\right) - 1$

57. $\left(\frac{3}{4}\right)^2 \div \left(\frac{3}{4} - \frac{1}{12}\right)$

59. $\left(\frac{8}{9}\right)^2 \div \left(2 - \frac{2}{3}\right)$

59. $\left(\frac{2}{3} - \frac{5}{9}\right)^2$

60. $\left(1 - \frac{2}{5}\right)^3$

61. $\left(\frac{3}{4} + \frac{1}{8}\right)^2 - \left(\frac{1}{2} + \frac{1}{8}\right)$

62. $\left(\frac{1}{6} + \frac{1}{3}\right)^3 + \left(\frac{2}{5} \cdot \frac{3}{4}\right)^2$

Find the average of each list of numbers. See Example 12.

63. $\frac{5}{6}$ and $\frac{2}{3}$

64. $\frac{1}{2}$ and $\frac{4}{7}$

65. $\frac{1}{3}, \frac{1}{4},$ and $\frac{1}{6}$

66. $\frac{1}{5}, \frac{3}{10},$ and $\frac{3}{20}$

Review and Preview

Simplify each fraction. See Section 2.3.

67. $\frac{10}{12}$

68. $\frac{8}{12}$

69. $\frac{20}{24}$

70. $\frac{22}{24}$

71. $\frac{50}{75}$

72. $\frac{30}{65}$

 Combining Concepts

Solve.

73. In 2000, about $\frac{11}{67}$ of the total weight of mail delivered by the United States Postal Service was first-class mail. That same year, about $\frac{75}{134}$ of the total weight of mail delivered by the United States Postal Service was standard mail. Which of these two categories account for a greater portion of the mail handled by weight? (*Source*: U.S. Postal Service)

74. The National Park System (NPS) in the United States includes a wide variety of park types. National military parks account for $\frac{3}{128}$ of all NPS parks, and $\frac{1}{24}$ of NPS parks are classified as national preserves. Which category, national military park or national preserve, is bigger? (*Source*: National Park Service)

75. Approximately $\frac{7}{10}$ of U.S. adults have a savings account. About $\frac{11}{25}$ of U.S. adults have a non-interest bearing checking account. Which type of banking service, savings account or non-interest checking account, do adults in the United States use more? (*Source*: Scarborough Research/USData.com, Inc.)

76. About $\frac{127}{500}$ of U.S. adults rent one or two videos per month. Approximately $\frac{31}{200}$ of U.S. adults rent three or four videos per month. Which video rental category, 1–2 videos or 3–4 videos per month, is bigger? (*Source*: Telenation/Market Facts, Inc.)

77. Calculate $\frac{2^3}{3}$ and $\left(\frac{2}{3}\right)^3$. Do both of these expressions simplify to the same number? Explain why or why not.

78. Calculate $\left(\frac{1}{2}\right)^2 \cdot \left(\frac{3}{4}\right)^2$ and $\left(\frac{1}{2} \cdot \frac{3}{4}\right)^2$. Do both of these expressions simplify to the same number? Explain why or why not.

3.6 Fractions and Problem Solving

Ⓐ **Solving Problems Containing Fractions or Mixed Numbers**

Now that we know how to add, subtract, multiply, and divide fractions and mixed numbers, we can solve problems containing these numbers.

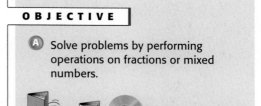
OBJECTIVE

Ⓐ Solve problems by performing operations on fractions or mixed numbers.

SSM
TUTOR CENTER   SG   CD & VIDEO   MATH PRO   WEB

△ **EXAMPLE 1**

In 2001, Sony produced a camcorder that was the smallest. It measures 5 inches by $2\frac{1}{2}$ inches by $1\frac{3}{4}$ inches and can store 30 minutes of moving images. Find the volume of a box with these dimensions. (*Source: Guinness World Records*, 2001)

Solution:

1. UNDERSTAND. Read and reread the problem. The phrase "volume of a box" tells us what to do. Recall that volume of a box is the product of its length, width, and height. Since we are multiplying, it makes no difference which measurement we call length, width, or height.

2. TRANSLATE.

In words:

volume of a box	is	length	·	width	·	height
↓	↓	↓		↓		↓

Translate: volume of a box $=$ 5 in. · $2\frac{1}{2}$ in. · $1\frac{3}{4}$ in.

3. SOLVE:

$$5 \text{ in.} \cdot 2\frac{1}{2}\text{ in.} \cdot 1\frac{3}{4}\text{ in.} = \frac{5}{1}\cdot\frac{5}{2}\cdot\frac{7}{4} \quad \text{cubic inches}$$

$$= \frac{5\cdot5\cdot7}{1\cdot2\cdot4} \quad \text{cubic inches}$$

$$= \frac{175}{8} \text{ or } 21\frac{7}{8} \quad \text{cubic inches}$$

4. INTERPRET. *Check* your work. *State* your conclusion: The volume of a box that measures 5 inches by $2\frac{1}{2}$ inches by $1\frac{3}{4}$ inches is $21\frac{7}{8}$ cubic inches. ●

EXAMPLE 2   Given the following diagram, find L, its total length.

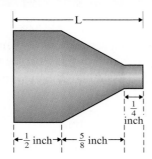

Solution:

1. UNDERSTAND. Read and reread the problem. Then study the diagram. The phrase "total length" tells us what to do.

Practice Problem 1   △

Find the volume of a box that measures $4\frac{1}{3}$ feet by $1\frac{1}{2}$ feet by $3\frac{1}{3}$ feet.

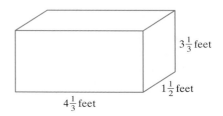

Practice Problem 2

Given the following diagram, find W, its total width.

Answers

1. $21\frac{2}{3}$ cu ft,   **2.** $\frac{5}{8}$ in.

2. TRANSLATE. It makes no difference which length we call first, second, or third length.

In words:	total length	is	first length	+	second length	+	third length
	↓	↓	↓		↓		↓
Translate:	total length	=	$\frac{1}{2}$ in.	+	$\frac{5}{8}$ in.	+	$\frac{1}{4}$ in.

3. SOLVE:

$$\frac{1}{2} + \frac{5}{8} + \frac{1}{4} = \frac{1 \cdot 4}{2 \cdot 4} + \frac{5}{8} + \frac{1 \cdot 2}{4 \cdot 2}$$
$$= \frac{4}{8} + \frac{5}{8} + \frac{2}{8}$$
$$= \frac{11}{8} \text{ or } 1\frac{3}{8}$$

4. INTERPRET. *Check* your work. *State* your conclusion: The total length, L, is $1\frac{3}{8}$ inches.

Many problems require more than one operation to solve. For example, see the application below: ●

EXAMPLE 3

A contractor is considering buying land to develop a subdivision for single-family homes. Suppose she buys 44 acres and calculates that $4\frac{1}{4}$ acres of this land will be used for roads and a retention pond. How many $\frac{3}{4}$-acre lots can she sell using the rest of the acreage?

Solution:

1. UNDERSTAND. Read and reread the problem. The phrase "using the rest of the acreage" tells is that initially we are to subtract.

2a. TRANSLATE. First, let's calculate the amount of acreage that can be used for lots.

In words:	acreage for lots	is	total acreage	minus	acreage for roads and a pond
	↓	↓	↓	↓	↓
Translate:	acreage for lots	=	44	−	$4\frac{1}{4}$

Suppose that 25 acres of land are purchased, but because of roads and wetlands concerns, $6\frac{2}{3}$ acres cannot be developed into lots. How many $\frac{5}{6}$-acre lots can the rest of the land be divided into?

Answer

3. 22 lots

3a. SOLVE:

$$44 \quad = \quad 43\frac{4}{4}$$

$$\underline{-\ 4\frac{1}{4} \quad = \quad -\ 4\frac{1}{4}}$$

$$39\frac{3}{4}$$

2b. TRANSLATE. Now that we know $39\frac{3}{4}$ acres can be used for lots, we calculate how many $\frac{3}{4}$ acres are in $39\frac{3}{4}$. This means that we divide.

In words:	number of $\frac{3}{4}$-acre lots	is	acreage for lots	divided by	size of each lot
	↓	↓	↓	↓	↓
Translate:	number of $\frac{3}{4}$-acre lots	$=$	$39\frac{3}{4}$	$\div$	$\frac{3}{4}$

3b. SOLVE:

$$39\frac{3}{4} \div \frac{3}{4} = \frac{159}{4} \cdot \frac{4}{3} = \frac{\overset{53}{\cancel{159}} \cdot \overset{1}{\cancel{4}}}{\underset{1}{\cancel{4}} \cdot \underset{1}{\cancel{3}}} = \frac{53}{1} \text{ or } 53$$

4.   INTERPRET. *Check* your work. *State* your conclusion: The contractor can sell $53\frac{3}{4}$-acre lots.   ●

FOCUS ON **Mathematical Connections**

INDUCTIVE REASONING

Inductive reasoning is the process of drawing a general conclusion from just a few observations. Many times in mathematics, we observe similarities among numbers or calculations and notice a pattern emerging. Identifying a pattern in this way uses inductive reasoning.

For example, look at the following list of numbers. The dots at the end of the list indicate that the list continues indefinitely. What do you notice?

$$2, 4, 6, 8, 10, 12, \ldots$$

You probably noticed the pattern that each number in the list is 2 greater than the previous number and that all of the numbers are even numbers. If we were asked to guess the next number in the list, we could be confident that "14" would be a good response.
Let's try finding another pattern in a list of numbers.

$$10, 13, 18, 25, 34, 45, \ldots$$

What do you notice? It might be useful to find the difference between successive numbers in the list. Using the differences found below, we can see that each successive difference is 2 greater than the previous difference. We can guess that the next difference will be 13, so the next number in the list is probably $45 + 13 = 58$.

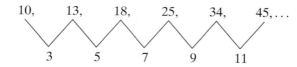

CRITICAL THINKING

Give the next two numbers in each list.

1. $5, 8, 11, 14, 17, 20, \ldots$

2. $100, 95, 90, 85, 80, \ldots$

3. $\dfrac{1}{2}, \dfrac{1}{3}, \dfrac{1}{4}, \dfrac{1}{5}, \ldots$

4. $5, 1, 6, 1, 1, 7, 1, 1, 1, 8, 1, 1, 1, \ldots$

5. $3, 4, 6, 9, 13, 18, \ldots$

6. $2, 4, 8, 16, 32, \ldots$

Name _____ Section _____ Date _____

EXERCISE SET 3.6

Solve. See Examples 1 through 3.

1. A nacho recipe calls for $\frac{1}{3}$ cup chedder cheese and $\frac{1}{2}$ cup jalapeño cheese. Find the total amount of cheese in the recipe.

2. A recipe for brownies calls for $1\frac{2}{3}$ cups of sugar. If you are doubling the recipe, how much sugar do you need?

3. A decorative wall in Ben and Joy Lander's garden is to be built using brick that is $2\frac{3}{4}$ inches wide and a mortar joint that is $\frac{1}{2}$ inch wide. Use the diagram to find the height of the wall.

4. Suppose that Ben and Joy Lander (from Exercise 3) decide that they want one more layer of bricks with a mortar joint below and above that layer. Find the new height of the wall.

height   mortar joint

5. Doug and Claudia Scaggs recently drove $290\frac{1}{4}$ miles on $13\frac{1}{2}$ gallons of gas. Calculate how many miles per gallon they get in their vehicle.

6. A contractor is using 18 acres of his land to sell $\frac{3}{4}$-acre lots. How many lots can he sell?

7. The life expectancy of a circulating coin is 30 years. The life expectancy of a circulating dollar bill is only $\frac{1}{20}$ as long. Find the life expectancy of circulating paper money. (*Source:* The U.S. Mint)

8. The Indian Head one-cent coin of 1859–1864 was made of copper and nickel only. If $\frac{3}{25}$ of the coin was nickel, what part of the coin was copper? (*Source:* The U.S. Mint)

9. The Gauge Act of 1846 set the standard gauge for U.S. railroads at $56\frac{1}{2}$ inches. (See figure.) If the standard gauge in Spain is $65\frac{9}{10}$ inches, how much wider is Spain's standard gauge than the U.S. standard gauge? (*Source:* San Diego Railroad Museum)

10. The standard railroad track gauge (see figure) in Spain is $65\frac{9}{10}$ inches, while in neighboring Portugal it is $65\frac{11}{20}$ inches. Which gauge is wider and by how much? (*Source:* San Diego Museum)

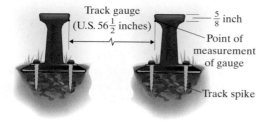

Track gauge (U.S. $56\frac{1}{2}$ inches)
$\frac{5}{8}$ inch
Point of measurement of gauge
Track spike

11. Mark Nguyen is a tailor making costumes for a play. He needs enough material for 1 large shirt that requires $1\frac{1}{2}$ yards of material and 5 small shirts that each require $\frac{3}{4}$ yard of material. He finds a 5-yard remnant of material on sale. Is 5 yards of material enough to make all 6 shirts? If not, how much more material does he need?

12. A carpenter has a 12-foot board to be used to make window sills. If each sill requires $2\frac{5}{16}$ feet, how many sills can be made from the 12-foot board?

13. A plumber has a 10-foot piece of PVC pipe. How many $\frac{9}{5}$-foot pieces can be cut from the 10-foot piece?

14. A beanbag manufacturer makes a large beanbag requiring $4\frac{1}{3}$ yards of vinyl fabric and a smaller size requiring $3\frac{1}{4}$ yards. A 100-yard roll of fabric is to be used to make 12 large beanbags. How many smaller beanbags can be made from the remaining piece?

15. Suppose that the cross section of a piece of pipe looks like the diagram shown. What is the outer diameter?

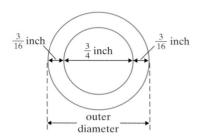

$\frac{3}{16}$ inch $\frac{3}{4}$ inch $\frac{3}{16}$ inch

outer diameter

16. Suppose that the cross section of a piece of pipe looks like the diagram shown. What is the inner diameter?

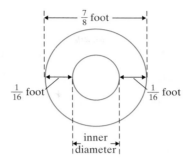

$\frac{7}{8}$ foot

$\frac{1}{16}$ foot $\frac{1}{16}$ foot

inner diameter

17. A recipe for chocolate chip cookies calls for $2\frac{1}{2}$ cups of flour. If you are making $1\frac{1}{2}$ recipes, how many cups of flour are needed?

18. A recipe for a homemade cleaning solution calls for $1\frac{3}{4}$ cups of vinegar. If you are tripling the recipe, how much vinegar is needed?

19. A total solar eclipse on November 23, 2003, will last $1\frac{19}{20}$ minutes and can be viewed from Antarctica only. The next total solar eclipse on April 8, 2005, will last $\frac{7}{10}$ of a minute and can be viewed in parts of the Pacific Ocean and northwestern South America. How much longer is the 2003 solar eclipse? (*Source: 2001 World America*)

△ **20.** The world's smallest cell phone measures $3\frac{1}{5}$ inches by $1\frac{7}{10}$ inches by 1 inch. Find the volume of a box with those dimensions. (*Source: Guinness World Records*, 2001)

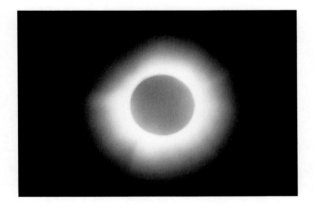

△ **21.** The Polaroid Pop Shot, the world's first disposable instant camera, can take color photographs measuring $4\frac{1}{2}$ inches by $2\frac{1}{2}$ inches. Find the area of a photograph. (*Source: Guinness World Records*, 2001)

22. The pole vault record for the 1986 Summer Olympics was $10\frac{5}{6}$ feet. The record for the 2000 Summer Olympics was a little over $19\frac{1}{3}$ feet. Find the difference of these heights. (*Source: 2001 World Almanac*)

23. A stack of $\frac{5}{8}$-inch-wide sheetrock has a height of $41\frac{7}{8}$ inches. How many sheets of sheetrock are in the stack?

24. A stack of $\frac{5}{4}$-inch-wide books has a height of $28\frac{3}{4}$ inches. How many books are in the stack?

25. William Arcencio is remodeling his home. In order to save money, he is upgrading the plumbing himself. He needs 12 pieces of copper tubing, each $\frac{3}{4}$ of a foot long. If he has a 10-foot piece of tubing, will that be enough? How much more does he need or how much tubing will he have left over?

26. Trishelle Dallam is building a bookcase. Each shelf will be $2\frac{3}{8}$ feet long, and she needs wood for 7 shelves.
   a. How many shelves can she cut from an 8-foot board?
   b. Based on your answer for part a, how many 8-foot boards will she need?

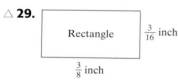

Recall that the average of a list of numbers is their sum divided by the number of numbers in the list. Use this procedure for Exercises 27 and 28.

27. A female lion had 4 cubs. They weighed $2\frac{1}{8}$, $2\frac{7}{8}$, $3\frac{1}{4}$, and $3\frac{1}{2}$ pounds. What is the average cub weight?

28. Three brook trout were caught, tagged, and then released. They weighed $1\frac{1}{2}$, $1\frac{3}{8}$, and $1\frac{7}{8}$ pounds. Find their average weight.

Find the area and perimeter of each figure.

△ **29.**
Rectangle $\frac{3}{16}$ inch
$\frac{3}{8}$ inch

△ **30.**
Square $1\frac{7}{10}$ mile

For Exercises 31 through 34, see the diagram. (Source: www.usflag.org)

31. The length of the U.S. flag is $1\frac{9}{10}$ its width. If a flag is being designed with a width of $2\frac{1}{2}$ feet, find its length.

32. The width of the Union portion the U.S. flag is $\frac{7}{13}$ of the width of the flag. If a flag is being designed with a width of $2\frac{1}{2}$ feet, find the width of the Union portion.

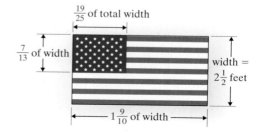

33. There are 13 stripes of equal width in the flag. If the width of a flag is $2\frac{1}{2}$ feet, find the width of each stripe.

34. The length of the Union portion of the flag is $\frac{19}{25}$ of the total width. If the width of a flag is $2\frac{1}{2}$ feet, find the length of the Union portion.

 35. Suppose that you are finding the average of $1\frac{3}{4}$, $1\frac{1}{8}$, and $1\frac{9}{10}$. Can the average be $2\frac{1}{4}$? Can the average be $\frac{15}{16}$? Why or why not?

Combining Concepts

36. Coins were practically made by hand in the late 1700s. Back then, it took 3 years to produce our nation's first million coins. Today, it takes only $\frac{11}{13,140}$ as long to produce the same amount. Calculate how long it takes today in hours to produce one million coins. (*Hint:* First convert 3 years to equivalent hours.) (*Source:* The U.S. Mint)

37. The largest suitcase measures $13\frac{1}{3}$ feet by $8\frac{3}{4}$ feet by $4\frac{4}{25}$ feet. Find its volume. (*Source: Guinness World Records*, 2001)

The figure shown is for Exercises 38 and 39.

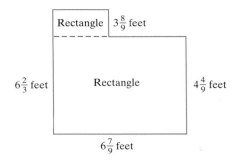

38. Find the area of the figure. (Hint: The area of the figure can be found by finding the sum of the areas of the rectangles shown in the figure.)

39. Find the perimeter of the figure.

This activity may be completed by working in groups or individually.

Lobsters are normally classified by weight. Use the weight classification table to answer the questions in this activity.

Classification of Lobsters	
Class	**Weight (in pounds)**
Chicken	1 to $1\frac{1}{8}$
Quarter	$1\frac{1}{4}$
Half	$1\frac{1}{2}$ to $1\frac{3}{4}$
Select	$1\frac{3}{4}$ to $2\frac{1}{2}$
Large select	$2\frac{1}{2}$ to $3\frac{1}{2}$
Jumbo	Over $3\frac{1}{2}$

(*Source:* The Maine Lobster Promotion Council)

1. A lobster fisher has kept four lobsters from a lobster trap. Classify each lobster if they have the following weights:

a. $1\frac{7}{8}$ pounds

b. $1\frac{9}{16}$ pounds

c. $2\frac{3}{4}$ pounds

d. $2\frac{3}{8}$ pounds

2. A recipe requires 5 pounds of lobster. Using the minimum weight for each class, decide whether a chicken, half, and select lobster will be enough for the recipe, and explain your reasoning. If not, suggest a better choice of lobsters to meet the recipe requirements.

3. A lobster market customer has selected two chickens, a select, and a large select. What is the most that these four lobsters could weigh? What is the least that these four lobsters could weigh?

4. A lobster market customer wishes to buy three quarters. If lobsters sell for $7 per pound, how much will the customer owe for her purchase?

5. Why do you think there is no classification for lobsters weighing under 1 pound?

Chapter 3   VOCABULARY CHECK

Fill in each blank with one of the words or phrases listed below.

equivalent   least common multiple   like   mixed number

1. Fractions that have the same denominator are called _____ fractions.
2. The _____ is the smallest number that is a multiple of all numbers in a list of numbers.
3. _____ fractions represent the same portion of a whole.
4. A _____ has a whole number part and a fraction part.

CHAPTER 3 | Highlights

DEFINITIONS AND CONCEPTS	EXAMPLES

Section 3.1   Adding and Subtracting Like Fractions

Fractions that have the same denominator are called **like fractions**.

$\frac{1}{3}$ and $\frac{2}{3}$; $\frac{5}{7}$ and $\frac{6}{7}$

To add or subtract like fractions, combine the numerators and place the sum or difference over the common denominator.

$\frac{2}{7} + \frac{3}{7} = \frac{5}{7}$ ← Add the numerators.
   ← Keep the common denominator.

$\frac{7}{8} - \frac{4}{8} = \frac{3}{8}$ ← Subtract the numerators.
   ← Keep the common denominator.

Section 3.2   Least Common Multiple

The **least common multiple (LCM)** is the smallest number that is a multiple of all numbers in a list of numbers.

The LCM of 2 and 6 is 6 because 6 is the smallest number that is a multiple of both 2 and 6.

METHOD 1 FOR FINDING THE LCM OF A LIST OF NUMBERS USING MULTIPLES

Step 1. Write the multiples of the largest number until a multiple common to all numbers in the list is found.

Step 2. The multiple found in step 1 is the LCM.

Find the LCM of 4 and 6 using Method 1.

$6 \cdot 1 = 6$ Not a multiple of 4
$6 \cdot 2 = 12$ A multiple of 4

The LCM is 12.

METHOD 2 FOR FINDING THE LCM OF A LIST OF NUMBERS USING PRIME FACTORIZATION

Step 1. Write the prime factorization of each number.

Step 2. For each factor that's different from the prime factors in step 1, circle the largest number of factors found in any one factorization.

Step 3. The LCM is the product of the circled factors.

Equivalent fractions represent the same portion of a whole.

Find the LCM of 6 and 20 using Method 2.

$6 = 2 \cdot \boxed{3}$
$20 = \boxed{2 \cdot 2} \cdot \boxed{5}$

The LCM is

$2 \cdot 2 \cdot 3 \cdot 5 = 60$

Write an equivalent fraction with the indicated denominator.

$\frac{2}{8} = \frac{}{16}$

$\frac{2 \cdot 2}{8 \cdot 2} = \frac{4}{16}$

DEFINITIONS AND CONCEPTS	EXAMPLES

Section 3.3   Adding and Subtracting Unlike Fractions

TO ADD OR SUBTRACT FRACTIONS WITH UNLIKE DENOMINATORS

Step 1. Find the LCD.

Step 2. Write equivalent fractions with the LCD as the denominator.

Step 3. Add or subtract the like fractions.

Step 4. Write the result in simplest form.

Add: $\dfrac{3}{20} + \dfrac{2}{5}$

Step 1. The LCD is 20.

Step 2. $\dfrac{2}{5} = \dfrac{2 \cdot 4}{5 \cdot 4} = \dfrac{8}{20}$

Step 3. $\dfrac{3}{20} + \dfrac{2}{5} = \dfrac{3}{20} + \dfrac{8}{20} = \dfrac{11}{20}$

Step 4. $\dfrac{11}{20}$ is in simplest form.

Section 3.4   Adding and Subtracting Mixed Numbers

To add or subtract mixed numbers, add or subtract the fractions and then add or subtract the whole numbers.

Add: $2\dfrac{1}{2} + 5\dfrac{7}{8}$

$$2\dfrac{1}{2} = 2\dfrac{4}{8}$$
$$+ 5\dfrac{7}{8} = 5\dfrac{7}{8}$$
$$\overline{\phantom{+ 5\dfrac{7}{8} = }\, 7\dfrac{11}{8} = 7 + 1\dfrac{3}{8} = 8\dfrac{3}{8}}$$

Section 3.5   Order, Exponents, and the Order of Operations

To compare like fractions, compare the numerators. The order of the fractions is the same as the order of the numerators.

To compare unlike fractions, first write them with a common denominator and then compare the resulting like fractions.

Exponents mean repeated multiplication when the base is a whole number or a fraction.

ORDER OF OPERATIONS

1. Simplify inside parentheses first.
2. Simplify any expressions with exponents.
3. Multiply or divide in order from left to right.
4. Add or subtract in order from left to right.

Compare $\dfrac{3}{10}$ and $\dfrac{4}{10}$.

$$\dfrac{3}{10} < \dfrac{4}{10} \text{ since } 3 < 4$$

Compare $\dfrac{2}{5}$ and $\dfrac{3}{7}$.

$$\dfrac{2}{5} = \dfrac{2 \cdot 7}{5 \cdot 7} = \dfrac{14}{35} \qquad \dfrac{3}{7} = \dfrac{3 \cdot 5}{7 \cdot 5} = \dfrac{15}{35}$$

Since $14 < 15$, then

$$\dfrac{14}{35} < \dfrac{15}{35} \quad \text{or} \quad \dfrac{2}{5} < \dfrac{3}{7}$$

$$\left(\dfrac{1}{2}\right)^3 = \dfrac{1}{2} \cdot \dfrac{1}{2} \cdot \dfrac{1}{2} = \dfrac{1}{8}$$

Perform each indicated operation.

$$\dfrac{1}{2} + \dfrac{2}{3} \cdot \dfrac{1}{5} = \dfrac{1}{2} + \dfrac{2}{15} \qquad \text{Multiply.}$$

$$= \dfrac{1 \cdot 15}{2 \cdot 15} + \dfrac{2 \cdot 2}{15 \cdot 2} \qquad \text{The LCD is 30.}$$

$$= \dfrac{15}{30} + \dfrac{4}{30}$$

$$= \dfrac{19}{30} \qquad \text{Add.}$$

| Section 3.6 | Fractions and Problem Solving |

PROBLEM-SOLVING STEPS

A stack of $\frac{3}{4}$-inch plywood has a height of $50\frac{1}{4}$ inches. How many sheets of plywood are in the stack?

1. UNDERSTAND the problem.

1. UNDERSTAND. Read and reread the problem. We want to know how many $\frac{3}{4}$'s are in $50\frac{1}{4}$, so we divide.

2. TRANSLATE the problem.

2. TRANSLATE.

number of sheets in stack	is	height of a stack	÷	height of a sheet
number of sheets in stack	=	$50\frac{1}{4}$	÷	$\frac{3}{4}$

3. SOLVE the problem.

3. SOLVE. $50\frac{1}{4} \div \frac{3}{4} = \frac{201}{4} \cdot \frac{4}{3}$

$$= \frac{\overset{67}{\cancel{201}} \cdot \overset{1}{\cancel{4}}}{\underset{1}{\cancel{4}} \cdot \underset{1}{\cancel{3}}}$$

$$= 67$$

4. INTERPRET the results.

4. INTERPRET. *Check* your work and *state* your conclusion. There are 67 sheets of plywood in the stack.

215

STUDY SKILLS REMINDER

How are your homework assignments going?

It is so important in mathematics to keep up with homework. Why? Many concepts build on each other. Often, your understanding of a day's lecture in mathematics depends on an understanding of the previous day's material.

Remember that completing your homework assignment involves a lot more than attempting a few of the problems assigned.

To complete a homework assignment, remember these four things:

1. Attempt all of it.
2. Check it.
3. Correct it.
4. If needed, ask questions about it.

Chapter 3 Review

(3.1) *Add or subtract as indicated. Simplify your answers.*

1. $\dfrac{7}{11} + \dfrac{3}{11}$

2. $\dfrac{4}{9} + \dfrac{2}{9}$

3. $\dfrac{5}{12} - \dfrac{3}{12}$

4. $\dfrac{3}{10} - \dfrac{1}{10}$

5. $\dfrac{11}{15} - \dfrac{1}{15}$

6. $\dfrac{4}{21} - \dfrac{1}{21}$

7. $\dfrac{4}{15} + \dfrac{3}{15} + \dfrac{2}{15}$

8. $\dfrac{3}{20} + \dfrac{7}{20} + \dfrac{2}{20}$

9. $\dfrac{1}{12} + \dfrac{11}{12}$

10. $\dfrac{3}{4} + \dfrac{1}{4}$

11. $\dfrac{11}{25} + \dfrac{6}{25} + \dfrac{2}{25}$

12. $\dfrac{4}{21} + \dfrac{1}{21} + \dfrac{11}{21}$

Solve.

13. Gregor Krowsky studied math for $\dfrac{3}{8}$ of an hour and geography for $\dfrac{1}{8}$ of an hour. How long did he study?

14. Beryl Goldstein mixed $\dfrac{5}{8}$ of a gallon of water with $\dfrac{1}{8}$ of a gallon of punch concentrate. Then she and her friends drank $\dfrac{3}{8}$ of a gallon of the punch. How much was left?

15. One evening Mark Alorenzo did $\dfrac{3}{8}$ of his homework before supper, another $\dfrac{2}{8}$ of it while his children did their homework, and $\dfrac{1}{8}$ after his children went to bed. What part of his homework did he do that evening?

△ **16.** The Simpson's will be fencing in their land. In order to do this, they need to find its perimeter. Find the perimeter of their land.

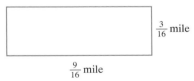

$\frac{3}{16}$ mile

$\frac{9}{16}$ mile

(3.2) *Find the LCM of each list of numbers.*

17. 5, 11

18. 20, 30

19. 20, 24

20. 12, 21

21. 12, 21, 63

22. 6, 8, 18

Write each fraction as an equivalent fraction with the given denominator.

23. $\dfrac{7}{8} = \dfrac{}{64}$

24. $\dfrac{2}{3} = \dfrac{}{30}$

25. $\dfrac{7}{11} = \dfrac{}{33}$

26. $\dfrac{10}{13} = \dfrac{}{26}$

27. $\dfrac{4}{15} = \dfrac{}{60}$

28. $\dfrac{5}{12} = \dfrac{}{60}$

(3.3) *Add or subtract as indicated. Simplify your answers.*

29. $\dfrac{7}{18} + \dfrac{2}{9}$ **30.** $\dfrac{4}{13} - \dfrac{1}{26}$ **31.** $\dfrac{1}{3} + \dfrac{1}{4}$ **32.** $\dfrac{2}{3} + \dfrac{1}{4}$ **33.** $\dfrac{5}{11} + \dfrac{2}{55}$

34. $\dfrac{4}{15} + \dfrac{1}{5}$ **35.** $\dfrac{7}{12} - \dfrac{1}{9}$ **36.** $\dfrac{7}{18} - \dfrac{2}{9}$ **37.** $\dfrac{4}{9} + \dfrac{5}{6}$ **38.** $\dfrac{9}{14} - \dfrac{3}{7}$

Find the perimeter of each figure.

△ **39.**

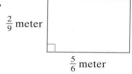

$\dfrac{2}{9}$ meter

$\dfrac{5}{6}$ meter

△ **40.** $\dfrac{1}{5}$ foot $\qquad$ $\dfrac{3}{5}$ foot

$\dfrac{7}{10}$ foot

41. Find the difference in length of two scarves if one scarf is $\dfrac{5}{12}$ of a yard long and the other is $\dfrac{2}{3}$ of a yard long.

$\dfrac{5}{12}$ of a yard

$\dfrac{2}{3}$ of a yard

42. Truman Kalzote cleaned $\dfrac{3}{5}$ of his house yesterday and $\dfrac{1}{10}$ of it today. How much of the house has been cleaned?

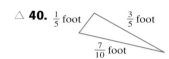

(3.4) *Add or subtract as indicated. Simplify your answers.*

43. $\begin{array}{r} 31\dfrac{2}{7} \\ + 14\dfrac{10}{21} \\ \hline \end{array}$

44. $\begin{array}{r} 24\dfrac{4}{5} \\ + 35\dfrac{1}{5} \\ \hline \end{array}$

45. $\begin{array}{r} 69\dfrac{5}{22} \\ - 36\dfrac{7}{11} \\ \hline \end{array}$

46. $\begin{array}{r} 36\dfrac{3}{20} \\ - 32\dfrac{5}{6} \\ \hline \end{array}$

47. $\begin{array}{r} 29\dfrac{2}{9} \\ 27\dfrac{7}{18} \\ + 54\dfrac{2}{3} \\ \hline \end{array}$

48. $\begin{array}{r} 7\dfrac{3}{8} \\ 9\dfrac{5}{6} \\ + 3\dfrac{1}{12} \\ \hline \end{array}$

49. $\begin{array}{r} 9\dfrac{3}{5} \\ - 4\dfrac{1}{7} \\ \hline \end{array}$

50. $\begin{array}{r} 8\dfrac{3}{11} \\ - 5\dfrac{1}{5} \\ \hline \end{array}$

Solve.

51. Two packages of soup bones weigh $3\dfrac{3}{4}$ pounds and $2\dfrac{3}{5}$ pounds. Find their combined weight.

52. A ribbon $5\dfrac{1}{2}$ yards long is cut from a reel of ribbon with 50 yards on it. Find the length of the piece remaining on the reel.

53. The average annual snowfall at a certain ski resort is $62\frac{3}{10}$ inches. Last year it had $54\frac{1}{2}$ inches. How many inches below average was last year's snowfall?

△**54.** Find the perimeter of a rectangular sheet of gift wrap that is $2\frac{1}{4}$ feet by $3\frac{1}{3}$ feet.

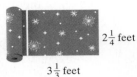

$2\frac{1}{4}$ feet

$3\frac{1}{3}$ feet

△ **55.** Find the perimeter of a sheet of shelf paper needed to fit exactly a square drawer $1\frac{1}{4}$ feet long on each side.

$1\frac{1}{4}$ feet

56. Dinah's homemade canned peaches contain $15\frac{3}{5}$ ounces per can. A can of Amy's brand contains $15\frac{5}{8}$ ounces per can. Amy's brand weighs how much more than Dinah's?

(3.5) *Insert $<$ or $>$ to form a true statement.*

57. $\dfrac{5}{11}$ $\dfrac{6}{11}$

58. $\dfrac{4}{35}$ $\dfrac{3}{35}$

59. $\dfrac{5}{14}$ $\dfrac{16}{42}$

60. $\dfrac{6}{35}$ $\dfrac{17}{105}$

61. $\dfrac{7}{8}$ $\dfrac{6}{7}$

62. $\dfrac{7}{10}$ $\dfrac{2}{3}$

Evaluate each expression. Use the order of operations to simplify.

63. $\left(\dfrac{3}{7}\right)^2$

64. $\left(\dfrac{4}{5}\right)^3$

65. $\left(\dfrac{1}{2}\right)^4 \cdot \left(\dfrac{3}{5}\right)^2$

66. $\left(\dfrac{1}{3}\right)^2 \cdot \left(\dfrac{9}{10}\right)^2$

67. $\left(\dfrac{6}{7} - \dfrac{3}{14}\right)^2$

68. $\dfrac{2}{7} \cdot \left(\dfrac{1}{5} + \dfrac{3}{10}\right)$

69. $\dfrac{2}{5} + \left(\dfrac{2}{5}\right)^2 - \dfrac{3}{25}$

70. $\dfrac{1}{4} + \left(\dfrac{1}{2}\right)^2 - \dfrac{3}{8}$

71. $\left(\dfrac{1}{3}\right)^2 - \dfrac{2}{27}$

72. $\dfrac{9}{10} \div \left(\dfrac{1}{5} + \dfrac{1}{20}\right)$

73. $\left(\dfrac{3}{4} + \dfrac{1}{2}\right) \div \left(\dfrac{4}{9} + \dfrac{1}{3}\right)$

74. $\left(\dfrac{3}{8} - \dfrac{1}{16}\right) \div \left(\dfrac{1}{2} - \dfrac{1}{8}\right)$

75. Saturn has 28 moons. The planet Uranus has only $\frac{3}{4}$ as many. Find the number of moons for Uranus. (*Source:* NASA)

Saturn   Uranus

76. James Hardaway just bought $5\frac{7}{8}$ acres of land adjacent to the $9\frac{3}{4}$ acres he already owned. How much land does he now own?

77. Linda Taneff has a board that is $10\frac{2}{3}$ feet in length. She plans to cut it into 5 equal lengths to use for a bookshelf. Find the length of each piece.

78. A recipe for pico de gallo calls for $1\frac{1}{2}$ tablespoons of cilantro. Milton McGowen has $8\frac{3}{4}$ tablespoons available. How many whole recipes of pico de gallo can he make?

Find the unknown measurements.

△ **79.**

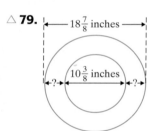

$18\frac{7}{8}$ inches

$10\frac{3}{8}$ inches

?   ?

△ **80.**

$\frac{1}{2}$ yard

$\frac{3}{5}$ yard

$1\frac{1}{30}$ yard

$1\frac{3}{10}$ yard

?

$1\frac{8}{15}$ yard

Find the perimeter and area of each rectangle. Attach the proper units to each. Remember that perimeter is measured in units and area is measured in square units.

△ **81.**

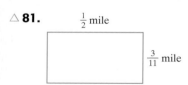

$\frac{1}{2}$ mile

$\frac{3}{11}$ mile

△ **82.**

$\frac{5}{12}$ meter

$\frac{3}{4}$ meter

Chapter 3 Test

1. Find the LCM of 4 and 15.

2. Find the LCM of 8, 9, and 12.

Insert < or > to form a true statement.

3. $\dfrac{5}{6} \quad \dfrac{26}{30}$

4. $\dfrac{7}{8} \quad \dfrac{8}{9}$

Perform each indicated operation. Simplify your answers.

5. $\dfrac{7}{9} + \dfrac{1}{9}$

6. $\dfrac{8}{15} - \dfrac{2}{15}$

7. $\dfrac{9}{10} + \dfrac{2}{5}$

8. $\dfrac{1}{6} + \dfrac{3}{14}$

9. $\dfrac{7}{8} - \dfrac{1}{3}$

10. $\dfrac{6}{21} - \dfrac{1}{7}$

11. $\dfrac{9}{20} + \dfrac{2}{3}$

12. $\dfrac{16}{25} - \dfrac{1}{2}$

13. $\dfrac{11}{12} + \dfrac{3}{8} + \dfrac{5}{24}$

14. $\begin{array}{r} 3\frac{7}{8} \\ 7\frac{2}{5} \\ +\,2\frac{3}{4} \\ \hline \end{array}$

15. $\begin{array}{r} 8\frac{2}{9} \\ 12 \\ +\,10\frac{1}{15} \\ \hline \end{array}$

16. $\begin{array}{r} 5\frac{1}{6} \\ -\,3\frac{7}{8} \\ \hline \end{array}$

17. $\begin{array}{r} 19 \\ -\,2\frac{3}{11} \\ \hline \end{array}$

18. $\dfrac{2}{7} \cdot \left(6 - \dfrac{1}{6}\right)$

19. $\left(\dfrac{2}{3}\right)^4$

20. $\left(\dfrac{1}{2} + \dfrac{1}{3}\right) \div \left(\dfrac{1}{2}\right)^2$

21. $\left(\dfrac{4}{5}\right)^2 + \left(\dfrac{1}{2}\right)^3$

22. $\left(\dfrac{3}{4}\right)^2 \div \left(\dfrac{2}{3} + \dfrac{5}{6}\right)$

Answers

1. _____
2. _____
3. _____
4. _____
5. _____
6. _____
7. _____
8. _____
9. _____
10. _____
11. _____
12. _____
13. _____
14. _____
15. _____
16. _____
17. _____
18. _____
19. _____
20. _____
21. _____
22. _____

23. A carpenter cuts a piece $2\frac{3}{4}$ feet long from a cedar plank that is $6\frac{1}{2}$ feet long. How long is the remaining piece?

24. In 2000, there were about 7400 theaters in the U.S. Single-screen theaters account for $\frac{8}{25}$ of these. Find the number of single-screen theaters. (*Source:* National Screen Service)

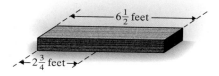

As shown in the circle graph, the market for backpacks is divided among five companies. For instance, Wilderness, Inc.'s, backpack accounts for $\frac{1}{4}$ of all backpack sales. Use the graph to answer Questions 25 and 26.

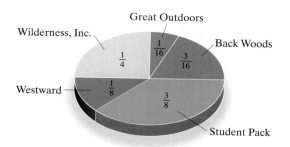

25. What fraction of backpack sales goes to Back Woods and Westward combined?

26. If a total of 500,000 backpacks are sold each year, how many backpacks does Wilderness, Inc., sell?

Find the perimeter of each figure. For Exercise 28, find the area also.

△ **27.**

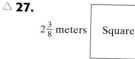

$2\frac{3}{8}$ meters | Square

△ **28.**

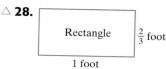

Rectangle | $\frac{2}{3}$ foot

1 foot

△ **29.**

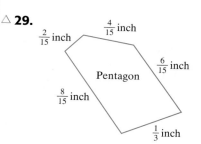

$\frac{4}{15}$ inch

$\frac{2}{15}$ inch

$\frac{6}{15}$ inch

Pentagon

$\frac{8}{15}$ inch

$\frac{1}{3}$ inch

Cumulative Review

Write each number in words.

1. 85

2. 126

3. Add: 23 + 136

4. Subtract: 43 − 29. Then check by adding.

5. Round 278,362 to the nearest thousand.

6. Multiply: 236 × 86

7. Find each quotient and then check the answer by multiplying.

  a. $1\overline{)8}$

  b. 11 ÷ 1

  c. $\dfrac{9}{9}$

  d. 7 ÷ 7

  e. $\dfrac{10}{1}$

  f. $6\overline{)6}$

8. The Hudson River in New York State is 306 miles long. The Snake River, in the northwestern United States, is 732 miles longer than the Hudson River. How long is the Snake River? (*Source*: U.S. Department of the Interior)

Evaluate.

9. 8^2

10. 2^5

Write the shaded part as an improper fraction and a mixed number.

11.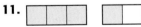

12.

13. Of the numbers 3, 9, 11, 17, 26, which are prime?

14. Find the prime factorization of 180.

1. _____

2. _____

3. _____

4. _____

5. _____

6. _____

7. a. _____

  b. _____

  c. _____

  d. _____

  e. _____

  f. _____

8. _____

9. _____

10. _____

11. _____

12. _____

13. _____

14. _____

15. _____

16. _____

17. _____

18. _____

19. _____

20. _____

21. _____

22. _____

23. _____

24. _____

25. _____

15. Write $\dfrac{72}{26}$ in simplest form.

16. Determine whether $\dfrac{8}{11}$ and $\dfrac{19}{26}$ are equivalent.

Multiply.

17. $\dfrac{2}{3} \cdot \dfrac{5}{11}$

18. $\dfrac{1}{4} \cdot \dfrac{1}{2}$

Divide.

19. $\dfrac{11}{18} \div 2\dfrac{5}{6}$

20. $5\dfrac{2}{3} \div 2\dfrac{5}{9}$

21. Add and simplify: $\dfrac{3}{16} + \dfrac{7}{16}$

22. Find the LCM of 6 and 8.

23. Add: $\dfrac{1}{2} + \dfrac{1}{3} + \dfrac{1}{6}$

24. Subtract: $9\dfrac{3}{7} - 5\dfrac{4}{21}$

25. Simplify: $\left(\dfrac{2}{3}\right)^2 \div \left(\dfrac{8}{27} + \dfrac{2}{3}\right)$

Decimals

Decimal numbers represent parts of a whole, just like fractions. In this chapter, we learn to perform arithmetic operations using decimals and to analyze the relationship between factions and decimals. We also learn how decimals are used in the real world. For instance, we use decimal numbers in our money system. One penny is 0.01 dollar and one dime is 0.10 dollar. Among many other uses, decimals also express averages, such as a batting average. A baseball player with a 0.333 batting average is a pretty good batter.

D id you know that eating chocolate dates as far back as 200 B.C.? The ancient Mayans and Aztecs of Central America ate mashed cacao beans for centuries before cocoa, the powdered product of the cacao bean, was introduced to Europe by Spanish explorers. Europeans then developed cocoa into the chocolate confection we know today. However, American Milton S. Hershey became the first to mass-produce milk chocolate at an affordable price. He founded the Hershey Chocolate Company in 1894 and introduced the Hershey's milk chocolate bar in 1900. In 1903 he established a community (Hershey, Pennsylvania) around the building of what is today the largest chocolate factory in the world. Soon after, Hershey's Kisses and Hershey's milk chocolate bars with almonds were introduced. Today, Hershey plants in the United States can produce 12 billion Kisses per year and are the largest single users of almonds in the country. In Exercise 29 on page 253, we will see how a decimal number can be used to describe Hershey's milk chocolate bar production.

1. _____

2. _____

3. _____

4. _____

5. _____

6. _____

7. _____

8. _____

9. _____

10. _____

11. _____

12. _____

13. _____

14. _____

15. _____

16. _____

17. _____

18. _____

19. _____

20. _____

Chapter 4 Pretest

1. Write the decimal five and four hundredths in standard form.

2. Write the following decimal as a fraction or a mixed number in lowest terms: 0.68

3. Write the following fraction as a decimal: $\dfrac{81}{1000}$

4. Insert $<$, $>$, or $=$ to form a true statement.
0.213   0.2421

5. Round 364.6551 to the nearest tenth.

6. Add: $13.165 + 59.08$

7. Subtract: $91.307 - 43.59$

Multiply.

8. 5.4×0.26

9. 27.01×0.001

△ **10.** Find the circumference of a circle whose radius is 11 inches. Then use the approximation 3.14 for π to approximate the circumference.

Divide.

11. $15 \div 0.003$

12. $1.26 \div 0.28$

13. $\dfrac{17.23}{1000}$

Perform the indicated operation. Estimate to see whether each proposed result is reasonable.

14. $5.8 + 3.4 + 7.9$

15. 4×129.21

Perform the indicated operation.

16. $3.2(7 - 5.6)$

17. $(1.4)^2 - 0.82$

18. $\dfrac{3}{5} - (0.3)(0.15)$

19. Write the following numbers in order from smallest to largest:
$\dfrac{7}{18}$, $\dfrac{1}{3}$, 0.37

20. Jennifer Kline was hired to baby-sit for 7.6 hours. Her hourly rate is $3.25. How much money did Jennifer make on this job?

4.1   Introduction to Decimals

 **Decimal Notation and Writing Decimals in Words**

Like fractional notation, decimal notation is used to denote a part of a whole. Numbers written in decimal notation are called **decimal numbers,** or simply **decimals.** The decimal 17.758 has three parts.

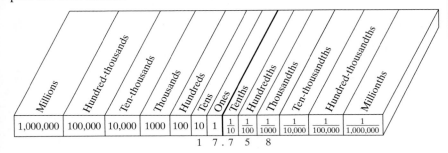

   In Section 1.2, we introduced place value for whole numbers. Place names and place values for the whole number part of a decimal number are exactly the same, as shown next. Place names and place values for the decimal part are also shown.

Millions	Hundred-thousands	Ten-thousands	Thousands	Hundreds	Tens	Ones	Tenths	Hundredths	Thousandths	Ten-thousandths	Hundred-thousandths	Millionths
1,000,000	100,000	10,000	1000	100	10	1	$\frac{1}{10}$	$\frac{1}{100}$	$\frac{1}{1000}$	$\frac{1}{10,000}$	$\frac{1}{100,000}$	$\frac{1}{1,000,000}$

$$1 \quad 7 \; . \; 7 \quad 5 \quad 8$$

Notice that the value of each place is $\frac{1}{10}$ of the value of the place to its left.

For example,

$$1 \cdot \frac{1}{10} = \frac{1}{10}$$

$\underset{\text{ones}}{\uparrow} \qquad \underset{\text{tenths}}{\uparrow}$

$$\frac{1}{10} \cdot \frac{1}{10} = \frac{1}{100}$$

$\underset{\text{tenths}}{\uparrow} \qquad \underset{\text{hundredths}}{\uparrow}$

The decimal number 17.758 means

1 ten	+	7 ones	+	7 tenths	+	5 hundredths	+	8 thousandths
or $1 \cdot 10$ +		$7 \cdot 1$ +		$7 \cdot \frac{1}{10}$ +		$5 \cdot \frac{1}{100}$ +		$8 \cdot \frac{1}{1000}$
or 10 +		7 +		$\frac{7}{10}$ +		$\frac{5}{100}$ +		$\frac{8}{1000}$

Writing (or Reading) a Decimal in Words

Step 1. Write the whole number part in words.

Step 2. Write "and" for the decimal point.

Step 3. Write the decimal part in words as though it were a whole number, followed by the place value of the last digit.

Practice Problem 1

Write the decimal 8.7 in words.

Practice Problem 2

Write the decimal 97.28 in words.

Practice Problem 3

Write the decimal 302.1056 in words.

Practice Problem 4

Write the decimal 72.1085 in words.

Practice Problems 5–6

Write each decimal in standard form.

5. Three hundred and ninety-six hundredths

6. Thirty-nine and forty-two thousandths

Answers

1. eight and seven tenths,   **2.** ninety-seven and twenty-eight hundredths,   **3.** three hundred two and one thousand fifty-six ten-thousandths, **4.** seventy-two and one thousand eighty-five ten-thousandths,   **5.** 300.96,   **6.** 39.042

EXAMPLE 1   Write the decimal 1.3 in words.

Solution: one and three tenths   ●

EXAMPLE 2

Write the decimal in the following sentence in words: The Golden Jubilee Diamond is a 545.67 carat cut diamond. (*Source: The Guinness Book of Records*)

Solution: five hundred forty-five and sixty-seven hundredths   ●

EXAMPLE 3   Write the decimal 19.5023 in words.

Solution: nineteen and five thousand twenty-three ten-thousandths   ●

EXAMPLE 4

Write the decimal in the following sentence in words: The oldest known fragments of the Earth's crust are Zircon crystals; they were discovered in Australia and are thought to be 4.276 billion years old. (*Source: The Guinness Book of Records*)

Solution: four and two hundred seventy-six thousandths   ●

B Writing Decimals in Standard Form

A decimal written in words can be written in standard form by reversing the above procedure.

EXAMPLES   Write each decimal in standard form.

5. Forty-eight and twenty-six hundredths is

    48.26
    └── hundredths place

6. Six and ninety-five thousandths is

    6.095
    └── thousandths place   ●

Helpful Hint

When converting a decimal from words to decimal notation, make sure the last digit is in the correct place by inserting 0s if necessary. For example,

Two and thirty-eight thousandths is 2.038

thousandths place

Writing Decimals as Fractions

Once you master reading and writing decimals, writing a decimal as a fraction follows naturally.

Decimal	In Words	Fraction
0.7	seven tenths	$\dfrac{7}{10}$
0.51	fifty-one hundredths	$\dfrac{51}{100}$
0.009	nine thousandths	$\dfrac{9}{1000}$
0.05	five hundredths	$\dfrac{5}{100} = \dfrac{1}{20}$

Notice that the number of decimal places in a decimal number is the same as the number of zeros in the denominator of the equivalent fraction. We can use this fact to write decimals as fractions.

$$0.\underline{51} = \frac{51}{\underline{100}}$$

2 decimal places   2 zeros

$$0.\underline{009} = \frac{9}{\underline{1000}}$$

3 decimal places   3 zeros

EXAMPLE 7   Write 0.43 as a fraction.

Solution: $0.43 = \dfrac{43}{100}$

2 decimal places   2 zeros

EXAMPLE 8   Write 5.6 as a mixed number.

Solution: $5.6 = 5\dfrac{6}{10} = 5\dfrac{3}{5}$ In simplest form

1 decimal place   1 zero

Practice Problem 7

Write 0.037 as a fraction.

Practice Problem 8

Write 14.97 as a mixed number.

Answers

7. $\dfrac{37}{1000}$,   **8.** $14\dfrac{97}{100}$

Practice Problems 9–11

Write each decimal as a fraction or mixed number. Write your answer in simplest form.

9. 0.12
10. 57.8
11. 209.986

EXAMPLES Write each decimal as a fraction or a mixed number. Write your answer in simplest form.

9. $0.125 = \dfrac{125}{1000} = \dfrac{1}{8}$

10. $23.5 = 23\dfrac{5}{10} = 23\dfrac{1}{2}$

11. $105.083 = 105\dfrac{83}{1000}$

●

D **Writing Fractions as Decimals**

If the denominator of a fraction is a power of 10, we can write it as a decimal by reversing the procedure above.

Practice Problems 12–15

Write each fraction as a decimal.

12. $\dfrac{12}{100}$

13. $\dfrac{59}{100}$

14. $\dfrac{9}{1000}$

15. $\dfrac{172}{10}$

EXAMPLES Write each fraction as a decimal.

12. $\dfrac{8}{10} = 0.8$
    ↑   ↑
    1 zero   1 decimal place

13. $\dfrac{87}{10} = 8.7$
    ↑   ↑
    1 zero   1 decimal place

14. $\dfrac{18}{1000} = 0.018$
    ↑   ↑
    3 zeros   3 decimal places

15. $\dfrac{507}{100} = 5.07$
    ↑   ↑
    2 zeros   2 decimal places

●

Answers

9. $\dfrac{3}{25}$,   10. $57\dfrac{4}{5}$,   11. $209\dfrac{493}{500}$,   12. 0.12,

13. 0.59,   14. 0.009,   15. 17.2

Mental Math

Determine the place value for the digit 7 in each number.

1. 70

2. 700

3. 0.7

4. 0.07

EXERCISE SET 4.1

A *Write each decimal number in words. See Examples 1 through 4.*

1. 6.52

2. 7.59

3. 16.23

4. 47.65

5. 0.205

6. 0.495

7. 167.009

8. 233.056

9. 200.005

10. 5000.02

11. The English Channel Tunnel is 31.04 miles long. (*Source: Railway Directory & Year Book*)

12. Saturn makes a complete orbit of the Sun every 29.48 years. (*Source:* National Space Science Data Center)

13. The recommended daily allowance of riboflavin for teenage boys between the ages of 15 and 18 is 1.8 milligrams. (*Source:* Food and Nutrition Board of the Institute of Medicine, National Academy of Sciences)

14. The top-rated television series for the 2000–2001 viewing season was Survivor II, which received a rating of 17.4. (*Source:* Nielsen Media Research)

B *Write each decimal number in standard form. See Examples 5 and 6.*

15. Six and five tenths

16. Three and nine tenths

17. Nine and eight hundredths

18. Twelve and six hundredths

19. Five and six hundred twenty-five thousandths

20. Four and three hundred ninety-nine thousandths

21. Sixty-four ten-thousandths

22. Thirty-eight ten-thousandths

23. The record rainfall amount for a 24-hour period in Alabama is thirty-two and fifty-two hundredths inches. This record was set at Dauphin Island Sea Lab in 1997. (*Source:* National Climatic Data Center)

24. The United States Postal Service vehicle fleet averages nine and sixty-two hundredths miles per gallon of fuel. (*Source:* United States Postal Service)

25. Americans consume an average of fifteen and eight-tenths pounds of watermelon annually. (*Source:* Agricultural Marketing Service, U.S. Department of Agriculture)

26. Shaquille O'Neal of the NBA's Los Angeles Lakers scored an average of twenty-eight and seven tenths points per basketball game during the 2000–2001 regular season. (*Source:* National Basketball Association)

Write each decimal as a fraction or a mixed number. Write your answer in simplest form. See Examples 7 through 11.

27. 0.3

28. 0.7

29. 0.27

30. 0.39

31. 0.8

32. 0.4

33. 0.15

34. 0.64

35. 5.47

36. 6.3

37. 0.048

38. 0.082

39. 7.008

40. 9.005

41. 15.802

42. 11.406

43. 0.3005

44. 0.2006

45. 487.32

46. 298.62

D *Write each fraction as a decimal. See Examples 12 through 15.*

47. $\dfrac{6}{10}$

48. $\dfrac{3}{10}$

49. $\dfrac{45}{100}$

50. $\dfrac{75}{100}$

51. $\dfrac{37}{10}$

52. $\dfrac{28}{10}$

53. $\dfrac{268}{1000}$

54. $\dfrac{709}{1000}$

55. $\dfrac{9}{100}$

56. $\dfrac{7}{100}$

57. $\dfrac{4026}{1000}$

58. $\dfrac{3601}{1000}$

59. $\dfrac{28}{1000}$

60. $\dfrac{63}{1000}$

61. $\dfrac{563}{10}$

62. $\dfrac{206}{10}$

Review and Preview

Round 47,261 to the indicated place value. See Section 1.2.

63. tens

64. hundreds

65. thousands

66. ten-thousands

Combining Concepts

67. In your own words, describe how to write a decimal as a fraction or a mixed number.

68. In your own words, describe how to write a fraction as a decimal.

69. Write 0.00026849576 in words.

70. Write $7\dfrac{12}{100}$ as a decimal.

71. Write $17\dfrac{268}{1000}$ as a decimal.

4.2 Order and Rounding

Ⓐ **Comparing Decimals**

One way to compare decimals is to compare their graphs on a number line. Recall that for any two numbers on a number line, the number to the left is smaller and the number to the right is larger. The decimals 0.5 and 0.8 are graphed as follows:

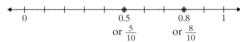

Comparing decimals by comparing their graphs on a number line can be time consuming. Another way to compare the size of decimals is to compare digits in corresponding places.

Comparing Two Decimals

Compare digits in the same places from left to right. When two digits are not equal, the number with the larger digit is the larger decimal. If necessary, insert 0s after the last digit to the right of the decimal point to continue comparing.

Compare hundredths-place digits

$$28.253 \qquad\qquad 28.263$$
$$\uparrow \qquad\qquad\qquad \uparrow$$
$$5 \quad < \quad 6$$
$$\text{so } 28.253 \quad < \quad 28.263$$

Helpful Hint

For any decimal, inserting 0s after the last digit to the right of the decimal point does not change the value of the number.

$$7.6 = 7.60 = 7.600, \text{ and so on}$$

When a whole number is written as a decimal, the decimal point is placed to the right of the ones digit.

$$25 = 25.0 = 25.00, \text{ and so on}$$

EXAMPLE 1   Insert $<$, $>$, or $=$ to form a true statement.

0.378   0.368

Solution:

0. 3 78   0. 3 68   The tenths places are the same.

0.3 7 8   0.3 6 8   The hundredths places are different.

Since $7 > 6$, then $0.378 > 0.368$.

EXAMPLE 2   Insert $<$, $>$, or $=$ to form a true statement.

0.052   0.236

Solution: 0. 0 52 $<$ 0. 2 36   0 is smaller than 2 in the tenths place.
                    $\uparrow$ $\uparrow$

Practice Problem 1

Insert $<$, $>$, or $=$ to form a true statement.

13.208   13.28

Practice Problem 2

Insert $<$, $>$, or $=$ to form a true statement.

0.12   0.086

Answers

1. $<$, **2.** $>$

Practice Problem 3

Insert $<$, $>$, or $=$ to form a true statement.

0.00985   0.076

Practice Problem 4

Write the decimals in order from smallest to largest.

14.605, 14.65, 13.9, 14.006

Practice Problem 5

Round 123.7817 to the nearest thousandth.

Practice Problem 6

Round 123.7817 to the nearest tenth.

EXAMPLE 3   Insert $<$, $>$, or $=$ to form a true statement.

0.52   0.063

Solution: $0.\,5\,2 > 0.\,0\,63$ 0 is smaller than 5 in the tenths place.

EXAMPLE 4   Write the decimals in order from smallest to largest.

7.035, 8.12, 7.03, 7.1

Solution: By comparing the ones digits, the decimal 8.12 is the largest number. To write the rest of the decimals in order, we compare digits to the right of the decimal point. We will insert zeros to help us compare.

7.035   7.030   7.100

By comparing digits to the right of the decimal point, we can now arrange the decimals from smallest to largest.

7.030, 7.035, 7.100, 8.12   or
7.03, 7.035, 7.1, 8.12

B Rounding Decimals

We **round the decimal part** of a decimal number in nearly the same way as we round whole numbers. The only difference is that we drop digits to the right of the rounding place, instead of replacing these digits by 0s. For example,

24.954   rounded to the nearest hundredth is   24.95

Rounding Decimals to a Place Value to the Right of the Decimal Point

Step 1.   Locate the digit to the right of the given place value.

Step 2.   If this digit is 5 or greater, add 1 to the digit in the given place value and drop all digits to its right. If this digit is less than 5, drop all digits to the right of the given place value.

EXAMPLE 5   Round 736.2359 to the nearest tenth.

Solution:

Step 1.   We locate the digit to the right of the tenths place.

tenths place
$736.2\,③\,59$
digit to the right

Step 2.   Since the digit to the right is less than 5, we drop it and all digits to its right.

Thus, 736.2359 rounded to the nearest tenth is 736.2.

EXAMPLE 6   Round 736.2359 to the nearest hundredth.

Solution:

Step 1.   We locate the digit to the right of the hundredths place.

hundredths place
$736.23\,⑤\,9$
digit to the right

Step 2.   Since the digit to the right is 5, we add 1 to the digit in the hundredths place and drop all digits to the right of the hundredths place.

Thus, 736.2359 rounded to the nearest hundredth is 736.24.

Rounding often occurs with money amounts. Since there are 100 cents in a dollar, each cent is $\frac{1}{100}$ of a dollar. This means that if we want to round to the nearest cent, we round to the nearest hundredth of a dollar.

EXAMPLE 7

The price of a gallon of gasoline in Aimsville is currently $1.0279. Round this to the nearest cent.

Solution:

$$\$1.02\,\textcircled{7}\,9$$

hundredths place — digit to the right

Since the digit to the right is greater than 5, we add 1 to the hundredths digit and drop all digits to the right of the hundredths digit. Thus, $1.0279 rounded to the nearest cent is $1.03. ●

EXAMPLE 8   Round $0.098 to the nearest cent.

Solution:

$$\$0.09\,\textcircled{8}$$

hundredths place — digit to the right

Since the digit to the right is greater than 5, we add 1 to the hundredths digit and drop all digits to the right of the hundredths digit.

$$\begin{array}{r} 0.09 \\ +\,0.01 \\ \hline 0.10 \end{array}$$ Add 1 hundredth to 9 hundredths.

Thus, $0.098 rounded to the nearest cent is $0.10. ●

Try the Concept Check in the margin.

EXAMPLE 9   Determining State Taxable Income

A high school teacher's taxable income is $41,567.72. The tax tables in the teacher's state use amounts to the nearest dollar. Round the teacher's income to the nearest whole dollar.

Solution: Rounding to the nearest whole dollar means rounding to the nearest ones place.

$$\$41,567.\,\underline{7}\,2$$

ones place — digit to the right

Since the digit to the right is 5 or greater, we add 1 to the ones place and drop all digits to the right of the ones place.
　　Thus, the teacher's income rounded to the nearest dollar is $41,568. ●

Practice Problem 7

In Cititown, the price of a gallon of gasoline is $1.0789. Round this to the nearest cent.

Practice Problem 8

Round $1.095 to the nearest cent.

Concept Check

1756.0894 rounded to the nearest ten is

a.　1756.1
b.　1760.0894
c.　1760

Practice Problem 9

Water bills in Gotham City are always rounded to the nearest dime. Lois's water bill was $24.43. Round her bill to the nearest dime (tenth).

Answers

7. $1.08,　**8.** $1.10,　**9.** $24.40

Concept Check:　c

FOCUS ON History

MULTICULTURAL FRACTION USE

Many ancient cultures were familiar with the use of fractions and used them in everyday life. Here are some interesting facts about the history of fractions:

■ Although the ancient Egyptians were comfortable with the idea of fractions, their system of hieroglyphic numbers allowed them to write fractions with 1 only as the numerator such as $\frac{1}{3}, \frac{1}{7}$, or $\frac{1}{10}$. To write fractions with numerators other than 1, the Egyptians had to write a sum of fractions with 1 as the numerator, and they preferred never to use the same denominator in a sum of fractions twice! For instance, rather than representing $\frac{2}{5}$ as $\frac{1}{5} + \frac{1}{5}$ (which uses the denominator 5 twice), they would they would use $\frac{1}{3} + \frac{1}{15}$. Similarly, $\frac{2}{13}$ would not be represented by $\frac{1}{13} + \frac{1}{13}$ but by $\frac{1}{8} + \frac{1}{52} + \frac{1}{104}$.

■ The ancient Babylonians also dealt with fractions but in a more direct way than the Egyptians did. For instance, they thought of the fraction $\frac{2}{7}$ as the product of 2 and the reciprocal of 7. They compiled detailed lists of reciprocals and could calculate with fractions very easily and quickly.

■ The Hindus were the first to write fractions with the numerator above the denominator. Later, Arab mathematicians copied Hindu notation and improved it by inserting a horizontal bar between the numerator and denominator to separate the two.

EXERCISE SET 4.2

A *Insert* <, >, *or* = *to form a true statement. See Examples 1 through 3.*

1. 0.15   0.16   **2.** 0.12   0.15   **3.** 0.57   0.54   **4.** 0.59   0.52

5. 0.098   0.1   **6.** 0.0756   0.2   **7.** 0.54900   0.549   **8.** 0.98400   0.984

9. 167.908   167.980   **10.** 519.3405   519.3054   **11.** 420,000   0.000042   **12.** 0.000987   987,000

Write the decimals in order from smallest to largest.

13. 0.006, 0.06, 0.0061   **14.** 0.082, 0.008, 0.080   **15.** 0.042, 0.36, 0.03

16. 0.21, 0.056, 0.065   **17.** 1.1, 1.16, 1.01, 1.09   **18.** 3.6, 3.069, 3.09, 3.06

19. 21.001, 20.905, 21.03, 21.12   **20.** 36.050, 35.72, 35.702, 35.072

B *Round each decimal to the given place value. See Examples 5 and 6.*

21. 0.57, nearest tenth   **22.** 0.54, nearest tenth   **23.** 0.234, nearest hundredth

24. 0.452, nearest hundredth   **25.** 0.5942, nearest thousandth   **26.** 63.4523, nearest thousandth

27. 98,207.23, nearest ten   **28.** 68,934.543, nearest ten   **29.** 12.342, nearest tenth

30. 42.9878, nearest thousandth   **31.** 17.667, nearest hundredth   **32.** 0.766, nearest hundredth

33. 0.501, nearest tenth   **34.** 0.602, nearest tenth   **35.** 0.1295, nearest thousandth

36. 0.8295, nearest thousandth   **37.** 3829.34, nearest ten   **38.** 4520.876, nearest hundred

Round each monetary amount to the nearest cent or dollar as indicated. See Examples 7 through 9.

39. $0.067, nearest cent   **40.** $0.025, nearest cent   **41.** $42,650.14, nearest dollar

42. $768.95, nearest dollar   **43.** $26.95, nearest dollar   **44.** $14,769.52, nearest dollar

45. $0.1992, nearest cent

46. $0.7633, nearest cent

47. Which number(s) rounds to 0.26?
0.26559   0.26499   0.25786   0.25186

48. Which number(s) rounds to 0.06?
0.0612   0.066   0.0586   0.0506

Round each number to the given place value.

49. The attendance at a Mets baseball game was reported to be 39,867. Round this number to the nearest thousand.

50. A used office desk is advertised at $19.95 by Drawley's Office Furniture. Round this price to the nearest dollar.

51. During the 2001 Boston Marathon, Catherine Ndereba of Kenya was the first woman to cross the finish line. Her time was 2.398056 hours. Round this time to the nearest hundredth. (*Source*: Boston Athletic Association)

52. The population density of the state of Louisiana is 102.5794 people per square mile. Round this population density to the nearest tenth. (*Source*: U.S. Census Bureau)

53. The length of a day on Mars is 24.6229 hours. Round this figure to the nearest thousandth. (*Source*: National Space Science Date Center)

54. Venus makes a complete orbit around the Sun every 224.695 days. Round this figure to the nearest whole day. (*Source*: National Space Science Data Center)

55. Millennium Force is a roller coaster at Cedar Point, an amusement park in Sandusky, Ohio. At the time of its debut, Millennium Force was the world's tallest and fastest roller coaster. A ride on the Millennium Force lasts about 2.75 minutes. Round this figure to the nearest tenth. (*Source*: Cedar Fair, L.P.)

56. During the 2000 NFL season, the average length of a Baltimore Ravens' punt was 40.2 yards. Round this figure to the nearest whole yard. (*Source*: National Football League)

Review and Preview

Perform each indicated operation. See Sections 1.3 and 1.4.

57. 3452 + 2314

58. 8945 + 4536

59. 94 − 23

60. 82 − 47

61. 482 − 239

62. 4002 − 3897

 Combining Concepts

The table gives the leading bowling averages for the Professional Bowlers Association for each of the years listed. Use the table to answer Exercises 63 through 65.

Year	Bowler	Average Score
1991	Norm Duke	218.208
1992	Dave Ferraro	219.702
1993	Walter Ray Williams, Jr.	222.980
1994	Norm Duke	222.830
1995	Mike Aulby	225.490
1996	Walter Ray Williams, Jr.	225.370
1997	Walter Ray Williams, Jr.	222.008
1998	Walter Ray Williams, Jr.	226.130
1999	Parker Bohn III	228.040
2001	Chris Barnes	221.000

(*Source:* Professional Bowlers Association)

63. What is the highest average score on the list? Which bowler achieved that average?

64. What is the lowest average score on the list? Which bowler had that average?

65. Make a list of the leading averages in order from greatest to least for the years shown in the table.

66. Write a 4-digit number that rounds to 26.3.

67. Write a 5-digit number that rounds to 1.7.

68. Explain how to identify the value of the 9 in the decimal 486.3297.

Internet Excursions

 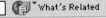

Go To: http://www.prenhall.com/martin-gay_basic

This World Wide Web site will direct you to the National Weather Service's Interactive Weather Information Network, or a related site. The table gives up-to-date weather summaries and forecasts for selected cities across the nation. The weather summary includes the previous day's high and low temperatures as well as precipitation amounts in inches, if any.

69. Record the date and time that the table was created. Scan the list of selected cities. How many cities received precipitation? Which city received the most precipitation?

70. Find the five cities that received the most precipitation. List the cities along with their precipitation amounts in order from most to least precipitation.

FOCUS ON **Business and Career**

According to U.S. Bureau of Labor Statistics projections, the careers listed below will have the largest job growth in the next decade:

Occupation	Employment (numbers in thousands)		
	1998	2008	Change
1. Systems analysts	617	1194	+577
2. Retail salespersons	4056	4620	+563
3. Cashiers	3198	3754	+556
4. General managers and top executives	3362	3913	+551
5. Truck drivers, light and heavy	2970	3463	+493
6. Office clerks, general	3021	3484	+463
7. Registered nurses	2079	2530	+451
8. Computer support specialists	429	869	+439
9. Personal care and home-health aides	746	1179	+433
10. Teacher assistants	1192	1567	+375

(*Source:* Bureau of Labor Statistics, U.S. Department of Labor)

What do all of these in-demand occupations have in common? They all require a knowledge of math! For some careers like systems analysts, salespersons, cashiers, and nurses, the ways math is used on the job may be obvious. For other occupations, the use of math may not be quite as obvious. However, tasks common to many jobs, such as filling in a time sheet or a mileage log, writing up an expense report, planning a budget, figuring a bill, ordering supplies, completing a packing list, and even making a work schedule, all require math.

CRITICAL THINKING

Suppose that your college placement office is planning to publish an occupational handbook on math in popular occupations. Choose one of the occupations from the list above that interests you. Research the occupation. Then write a brief entry for the occupational handbook that describes how a person in that career would use math in his or her job. Include an example if possible.

4.3   Adding and Subtracting Decimals

Ⓐ   Adding Decimals

Adding decimals is similar to adding whole numbers. We add digits in corresponding place values from right to left, carrying if necessary. To make sure that digits in corresponding place values are added, we line up the decimal points vertically.

OBJECTIVES

Ⓐ   Add decimals.

Ⓑ   Subtract decimals.

Ⓒ   Solve problems that involve adding or subtracting decimals.

SSM   TUTOR CENTER   SG   CD & VIDEO   MATH PRO   WEB

Adding or Subtracting Decimals

Step 1.   Write the decimals so that the decimal points line up vertically.

Step 2.   Add or subtract as with whole numbers.

Step 3.   Place the decimal point in the sum or difference so that it lines up vertically with the decimal points in the problem.

EXAMPLE 1   Add: 23.85 + 1.604

Solution: First we line up the decimal points vertically.

```
   23.850     Write one 0.
+   1.604
   ↑
line up decimal points
```

Then we add the digits from right to left as for whole numbers.

```
    ¹
   23.850
+   1.604
   25.454
   ↑ ——— Place the decimal point in the sum so that all decimal points line up.
```

Practice Problem 1

Add.

a.   15.52 + 2.371

b.   20.06 + 17.612

c.   0.125 + 122.8

> **Helpful Hint**
>
> Recall that 0's may be placed after the last digit to the right of the decimal point without changing the value of the decimal. This may be used to help line up place values when adding decimals.
>
> ```
>    3.2       becomes       3.200     Insert two 0s.
>   15.567                  15.567
> +  0.11                 +  0.110     Insert one 0.
>                           18.877     Add.
> ```

EXAMPLE 2   Add: 763.7651 + 22.001 + 43.89

Solution: First we line up the decimal points.

```
   763.7651
    22.0010     Write one 0.
+   43.8900     Write two 0s.
   829.6561     Add.
```

Practice Problem 2

Add.

a.   34.567 + 129.43 + 2.8903

b.   11.21 + 46.013 + 362.526

> **Helpful Hint**
>
> Don't forget that the decimal point in a whole number is after the last digit.

Answers

1. a. 17.891,   **b.** 37.672,   **c.** 122.925,
2. a. 166.8873,   **b.** 419.749

Practice Problem 3

Add: 26.072 + 119

Concept Check

What is wrong with the following calculation of the sum of 7.03, 2.008, 19.16, and 3.1415?

$$
\begin{array}{r}
7.03 \\
2.008 \\
19.16 \\
+\ 3.1415 \\
\hline
3.6042
\end{array}
$$

Practice Problem 4

Subtract. Check your answers.

a. 82.75 − 15.9
b. 126.032 − 95.71

Practice Problem 5

Subtract. Check your answers.

a. 5.8 − 3.92
b. 9.72 − 4.068

Practice Problem 6

Subtract. Check your answers.

a. 53 − 29.31
b. 120 − 68.22

Answers

3. 145.072, **4. a.** 66.85, **b.** 30.322, **5. a.** 1.88,
b. 5.652, **6. a.** 23.69, **b.** 51.78
Concept Check: The decimal places are not lined up properly.

EXAMPLE 3   Add: 45 + 2.06

Solution:

$$
\begin{array}{r}
45.00 \\
+\ 2.06 \\
\hline
47.06
\end{array}
$$

Write the decimal point and two 0s.
Line up decimal points.
Add.

Try the Concept Check in the margin.

B   Subtracting Decimals

Subtracting decimals is similar to subtracting whole numbers. We line up digits and subtract from right to left, borrowing when needed.

EXAMPLE 4   Subtract: 35.218 − 23.65. Check your answer.

Solution: First we line up the decimal points.

$$
\begin{array}{r}
35.2\overset{4\ \ 11\ 11}{18} \\
-\ 23.650 \\
\hline
11.568
\end{array}
$$

Write one 0.
Subtract.

Recall that we can check a subtraction problem by adding.

$$
\begin{array}{r}
11.568 \\
+\ 23.650 \\
\hline
35.218
\end{array}
$$

Difference
Subtrahend
Minuend

EXAMPLE 5   Subtract: 3.5 − 0.068. Check your answer.

Solution:

$$
\begin{array}{r}
3.500 \\
-\ 0.068 \\
\hline
3.432
\end{array}
$$

Write two 0s.
Line up decimal points.
Subtract.

Check:

$$
\begin{array}{r}
3.432 \\
+\ 0.068 \\
\hline
3.500
\end{array}
$$

Difference
Subtrahend
Minuend

EXAMPLE 6   Subtract: 85 − 17.31. Check your answer.

Solution:

$$
\begin{array}{r}
85.00 \\
-\ 17.31 \\
\hline
67.69
\end{array}
$$

Check:

$$
\begin{array}{r}
67.69 \\
+\ 17.31 \\
\hline
85.00
\end{array}
$$

Difference
Subtrahend
Minuend

Solving Problems by Adding or Subtracting Decimals

Decimals are very common in real-life problems.

EXAMPLE 7   Calculating the Cost of Owning an Automobile

Find the total monthly cost of owning and operating a certain automobile given the expenses shown.

Monthly car payment:	$256.63
Monthly insurance cost:	$47.52
Average gasoline bill per month:	$95.33

Solution:

1. UNDERSTAND. Read and reread the problem. The phrase "total monthly cost" tells us to add.

2. TRANSLATE.

In words:	total monthly cost	is	car payment	plus	insurance	plus	gasoline bill
	↓	↓	↓	↓	↓	↓	↓
Translate:	total monthly cost	=	$256.63	+	$47.52	+	$95.33

3. SOLVE:

$$\begin{array}{r} \overset{1}{2}56.63 \\ 47.52 \\ +95.33 \\ \hline \$399.48 \end{array}$$

4. INTERPRET. *Check* your work. *State* your conclusion: The total monthly cost is $399.48.   ●

EXAMPLE 8   Comparing Average Heights

The bar graph shows the current average heights for adults in various countries. How much greater is the average height in Denmark than the average height in the United States?

Average Adult Height

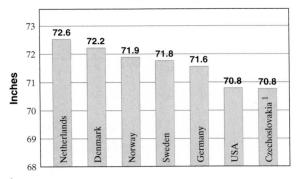

[1] Average for Czech Republic, Slovakia
Source: *USA Today*, 8/28/97

Practice Problem 7

Find the total monthly cost of owning and operating a certain automobile given the expenses shown.

Monthly car payment:	$536.50
Monthly insurance cost:	$52.70
Average gasoline bill per month:	$87.50

Practice Problem 8

Use the bar graph in Example 8. How much greater is the average height in the Netherlands than the average height in Czechoslovakia?

Answers

7. $676.70,   **8.** 1.8 in.

Solution:

1. UNDERSTAND. Read and reread the problem. Since we want to know "how much greater," we subtract.

2. TRANSLATE.

	How much greater	is	Denmark's average height	minus	U.S. average height
	↓	↓	↓	↓	↓
In words:					
Translate:	How much greater	=	72.2	−	70.8

3. SOLVE:

$$\begin{array}{r} 7\overset{1}{\cancel{2}}.\overset{12}{\cancel{2}} \\ -70.8 \\ \hline 1.4 \end{array}$$

4. INTERPRET. *Check* your work. *State* your conclusion: The average height in Denmark is 1.4 inches more than the average U.S. height. ●

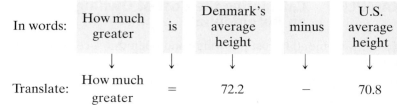

CALCULATOR EXPLORATIONS

Entering Decimal Numbers

To enter a decimal number, find the key marked $\boxed{.}$. To enter the number 2.56, for example, press the keys $\boxed{2}$ $\boxed{.}$ $\boxed{5}$ $\boxed{6}$.
The display will read $\boxed{2.56}$.

Operations on Decimal Numbers

Operations on decimal numbers are performed in the same way as operations on whole or signed numbers. For example, to find $8.625 - 4.29$, press the keys

$\boxed{8.625}$ $\boxed{-}$ $\boxed{4.29}$ $\boxed{=}$ or $\boxed{\text{ENTER}}$.

The display will read $\boxed{4.335}$. (Although entering 8.625, for example, requires pressing more than one key, we group numbers together here for easier reading.)

Use a calculator to perform each indicated operation.

1. $315.782 + 12.96$

2. $29.68 + 85.902$

3. $6.249 - 1.0076$

4. $5.238 - 0.682$

5. $\begin{array}{r} 12.555 \\ 224.987 \\ 5.2 \\ +\,622.65 \\ \hline \end{array}$

6. $\begin{array}{r} 47.006 \\ 0.17 \\ 313.259 \\ +\,139.088 \\ \hline \end{array}$

Name _____ Section _____ Date _____

Mental Math

 State the sum.

1. 0.3
   + 0.2

2. 0.4
   + 0.5

3. 1.00
   + 0.26

4. 3.00
   + 0.19

5. 7.6
   + 1.3

6. 4.5
   + 3.2

7. 0.9
   − 0.3

8. 0.6
   − 0.2

EXERCISE SET 4.3

 Add. See Examples 1 through 3.

1. 1.3 + 2.2

2. 2.5 + 4.1

3. 5.7 + 1.13

4. 2.31 + 6.4

5. 0.003 + 0.091

6. 0.004 + 0.085

7. 19.23 + 602.782

8. 47.14 + 409.567

9. 490 + 93.09

10. 600 + 83.0062

11. 234.89
   + 230.67

12. 734.89
   + 640.56

13. 100.009
     6.08
  +   9.034

14. 200.89
     7.49
  + 62.83

15. 24.6 + 2.39 + 0.0678

16. 32.4 + 1.58 + 0.0934

17. 45.023
   3.006
 + 8.403

18. 65.0028
   5.0903
 + 6.9003

B Subtract and check. See Examples 4 through 6.

19. 8.8 − 2.3

20. 7.6 − 2.1

21. 18 − 2.7

22. 28 − 3.3

23. 654.9
  − 56.67

24. 863.23
  − 39.453

25. 5.9 − 4.07

26. 6.4 − 3.04

27. 923.5 − 61.9

28. 845.93 − 45.8

29. 500.34 − 123.45

30. 600.74 − 463.98

31. 1000
  − 123.4

32. 2000
  − 327.47

33. 200 − 5.6

34. 800 − 8.9

35. 3 − 0.0012

36. 7 − 0.097

37. Subtract 6.7 from 23

38. Subtract 9.2 from 45

Solve. See Examples 7 and 8.

39. Find the total monthly cost of owning and maintaining a car given the information shown.

Monthly car payment:	$275.36
Monthly insurance cost:	$83.00
Average cost of gasoline per month:	$81.60
Average maintenance cost per month:	$14.75

40. Find the total monthly cost of owning and maintaining a car given the information shown.

Monthly car payment:	$306.42
Monthly insurance cost:	$53.50
Average cost of gasoline per month:	$123.00
Average maintenance cost per month:	$23.50

41. Gasoline was $1.339 per gallon on one day and $1.479 per gallon the next day. By how much did the price change?

42. A pair of eyeglasses costs a total of $347.89. The frames of the glasses are $97.23. How much do the lenses of the eyeglasses cost?

43. Ann-Margaret Tober bought a book for $32.48. If she paid with two $20 bills, what was her change?

44. Phillip Guillot bought a car part for $8.26. If he paid with a $10 bill, what was his change?

45. Americans' consumption of sugar is on the rise. During 1990, Americans consumed an average of 136.8 pounds of sugar in its various forms such as refined white sugar, honey, and corn sweeteners. By 2000, the average American was consuming 150.1 pounds of sugar products per year. How much more sugar was the average American consuming annually in 2000 than in 1990? (*Source*: Economic Research Service, U.S. Department of Agriculture)

46. In June 1996, the average wage for U.S. production workers was $11.81 per hour. Five years later in June 2001, this average wage had climbed to $14.29 per hour. How much of an increase was this? (*Source*: Bureau of Labor Statistics)

47. The average annual rainfall in Houston, Texas, is 46.07 inches. The average annual rainfall in New Orleans, Louisiana, is 61.88 inches. On average, how much more rain does New Orleans receive annually than Houston? (*Source*: National Climatic Data Center)

48. The average wind speed at the weather station on Mt. Washington in New Hampshire is 35.3 miles per hour. The average wind speed in Chicago, Illinois, is 10.4 miles per hour. How much faster is the average wind speed on Mt. Washington than in Chicago? (*Source*: National Climatic Data Center)

49. Brazilian Helio Castroneves won the 2001 Indianapolis 500 auto race with an average speed of 153.601 miles per hour. This was 14.006 miles per hour slower than the average speed of the winner of the 2000 Indianapolis 500. What was the average speed of the Indianapolis 500 winner in 2000? (*Source*: Indianapolis Motor Speedway)

50. It costs $3.04 to send a 2-pound package locally via parcel post at a U.S. Post Office. A 6-pound package costs $3.88 to send locally via parcel post. What is the total cost of sending a 2-pound and a 6-pound package locally via parcel post? (*Source*: United States Postal Service)

51. The snowiest city in the United States is Blue Canyon, California, which receives an average of 111.6 more inches of snow than the second-snowiest city. The second-snowiest city in the United States is Marquette, Michigan. Marquette receives an average of 129.2 inches of snow annually. How much snow does Blue Canyon receive on average each year? (*Source*: National Climatic Data Center)

52. The driest city in the world is Aswan, Egypt, which receives an average of only 0.02 inches of rain per year. Yuma, Arizona, is the driest city in the United States. Yuma receives an average of 2.63 more inches of rain each year than Aswan. What is the average annual rainfall in Yuma? (*Source*: National Climatic Data Center)

△ **53.** A landscape architect is planning a border for a flower garden that's shaped like a triangle. The sides of the garden measure 12.4 feet, 29.34 feet, and 25.7 feet. Find the amount of border material needed.

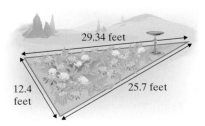

29.34 feet

12.4 feet

25.7 feet

△ **54.** A contractor needs to buy railing to completely enclose a newly built deck. Find the amount of railing needed.

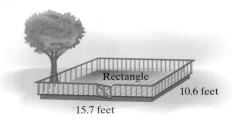

Rectangle

10.6 feet

15.7 feet

The table shows the average retail price of a gallon of gasoline (all grades and formulations) in the United States in May of each of the years shown. Use this table to answer Exercises 55 and 56.

Year	Gasoline Price (dollars per gallon)
1997	1.255
1998	1.109
1999	1.180
2000	1.579
2001	1.748

(*Source:* Energy Information Administration)

55. How much more was the average cost of a gallon of gasoline in 1997 than in 1999?

56. How much more was the average cost of a gallon of gasoline in 2001 than in 1998?

The following table shows spaceflight information for astronaut James A. Lovell. Use this table to answer Exercises 57 and 58.

Spaceflights of James A. Lovell		
Year	Mission	Duration (in hours)
1965	Gemini 6	330.583
1966	Gemini 12	94.567
1968	Apollo8	147.0
1970	Apollo 13	142.9

(*Source:* NASA)

57. Find the total time spent in spaceflight by astronaut James A. Lovell.

58. Find the total time James A. Lovell spent in spaceflight on all Apollo missions.

The following table shows the top five chocolate-consuming nations in the world. Use this table to answer Exercises 59 through 63.

The World's Top Chocolate-Consuming Countries	
Country	Pounds of Chocolate per Person
Belgium	13.9
Germany	15.8
Norway	16.0
Switzerland	22.0
United Kingdom	14.5

(*Source:* Hershey Foods Corporation)

59. Which country in the table has the greatest chocolate consumption per person?

60. Which country in the table has the least chocolate consumption per person?

61. How much more is the greatest chocolate consumption than the least chocolate consumption shown in the table?

62. How much more chocolate does the average German consume than the average citizen of the United Kingdom?

63. Make a new chart listing the countries and their corresponding chocolate consumptions in order from greatest to least.

Review and Preview

Multiply. See Sections 1.6 and 2.4.

64. $23 \cdot 2$

65. $46 \cdot 3$

66. $43 \cdot 90$

67. $30 \cdot 32$

68. $\left(\dfrac{2}{3}\right)^2$

69. $\left(\dfrac{1}{5}\right)^3$

70. $\dfrac{12}{7} \cdot \dfrac{14}{3}$

71. $\dfrac{25}{36} \cdot \dfrac{24}{40}$

 ## Combining Concepts

72. Laser beams can be used to measure the distance to the Moon. One measurement showed the distance to the moon to be 256,435.235 miles. A later measurement showed that the distance is 256,436.012 miles. Find how much farther away the Moon is in the second measurement compared with the first.

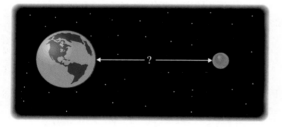

73. Explain how adding or subtracting decimals is similar to adding or subtracting whole numbers.

Find the unknown length in each figure.

△ **74.**

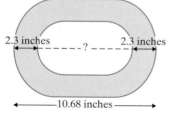

2.3 inches   ?   2.3 inches

10.68 inches

△ **75.**

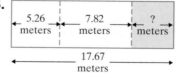

5.26 meters   7.82 meters   ? meters

17.67 meters

4.4 Multiplying Decimals

Ⓐ Multiplying Decimals

Multiplying decimals is similar to multiplying whole numbers. The only difference is that we place a decimal point in the product. To discover where a decimal point is placed in the product, let's multiply 0.6×0.03. We first write each decimal as an equivalent fraction and then multiply.

$$0.6 \qquad \times \qquad 0.03 \qquad = \frac{6}{10} \times \frac{3}{100} = \frac{18}{1000} = 0.018$$

1 decimal place ↑   2 decimal places ↑   3 decimal places ↑

Now let's multiply 0.03×0.002.

$$0.03 \qquad \times \qquad 0.002 \qquad = \frac{3}{100} \times \frac{2}{1000} = \frac{6}{100,000} = 0.00006$$

2 decimal place ↑   3 decimal places ↑   5 decimal places ↑

Instead of writing decimals as fractions each time we want to multiply, we notice a pattern from these examples and state a rule that we can use:

Multiplying Decimals

Step 1. Multiply the decimals as though they are whole numbers.

Step 2. The decimal point in the product is placed so that the number of decimal places in the product is equal to the *sum* of the number of decimal places in the factors.

EXAMPLE 1 Multiply: 23.6×0.78

Solution:

$$
\begin{array}{r}
23.6 \\
\times\ 0.78 \\
\hline
1888 \\
16520 \\
\hline
18.408
\end{array}
$$

23.6 — 1 decimal place
× 0.78 — 2 decimal places
18.408 — 3 decimal places

EXAMPLE 2 Multiply: 0.283×0.3

Solution:

$$
\begin{array}{r}
0.283 \\
\times\ \ \ 0.3 \\
\hline
0.0849
\end{array}
$$

0.283 — 3 decimal places
× 0.3 — 1 decimal place
0.0849 — 4 decimal places

└─ Insert one 0 since the product must have 4 decimal places.

EXAMPLE 3 Multiply: 0.0531×16

Solution:

$$
\begin{array}{r}
0.0531 \\
\times\ \ \ \ \ 16 \\
\hline
3186 \\
05310 \\
\hline
0.8496
\end{array}
$$

0.0531 — 4 decimal places
× 16 — 0 decimal places
0.8496 — 4 decimal places

Try the Concept Check in the margin.

Practice Problem 1

Multiply: 45.9×0.42

Practice Problem 2

Multiply: 0.112×0.6

Practice Problem 3

Multiply: 0.0721×48

Concept Check

True or false? The number of decimal places in the product of 0.261 and 0.78 is 6. Explain.

Answers

1. 19.278, **2.** 0.0672, **3.** 3.4608

Concept Check: false: 3 decimal places + 2 decimal places means 5 decimal places in the product

B Multiplying by Powers of 10

There are some patterns that occur when we multiply a number by a power of 10 such as 10, 100, 1000, 10,000, and so on.

$23.6951 \times 10 = 236.951$ Move the decimal point *1 place* to the *right*.

1 zero

$23.6951 \times 100 = 2369.51$ Move the decimal point *2 places* to the *right*.

2 zeros

$23.6951 \times 100,000 = 2,369,510.$ Move the decimal point *5 places* to the *right* (insert a 0).

5 zeros

Notice that we move the decimal point the same number of places as there are zeros in the power of 10.

Multiplying Decimals by Powers of 10 such as 10, 100, 1000, 10,000 . . .

Move the decimal point to the *right* the same number of places as there are *zeros* in the power of 10.

Practice Problems 4–6

Multiply.

4. 23.7×10
5. 203.004×100
6. 1.15×1000

EXAMPLES   Multiply.

4. $7.68 \times 10 = 76.8$ 7.68
5. $23.702 \times 100 = 2370.2$ 23.702
6. $76.3 \times 1000 = 76,300$ 76.300

There are also powers of 10 that are less than 1. The decimals 0.1, 0.01, 0.001, 0.0001, and so on are examples of powers of 10 less than 1. Notice the pattern when we multiply by these powers of 10:

$569.2 \times 0.1 = 56.92$ Move the decimal point *1 place* to the *left*.

1 decimal place

$569.2 \times 0.01 = 5.692$ Move the decimal point *2 places* to the *left*.

2 decimal places

$569.2 \times 0.0001 = 0.05692$ Move the decimal point *4 places* to the *left* (insert one 0).

4 decimal places

Multiplying Decimals by Powers of 10 such as 0.1, 0.01, 0.001, 0.0001 . . .

Move the decimal point to the *left* the same number of places as there are *decimal places* in the power of 10.

Practice Problems 7–9

Multiply.

7. 7.62×0.1
8. 1.9×0.01
9. 7682×0.001

Answers

4. 237, **5.** 20,300.4, **6.** 1150, **7.** 0.762, **8.** 0.019, **9.** 7.682

EXAMPLES   Multiply.

7. $42.1 \times 0.1 = 4.21$ 42.1
8. $76,805 \times 0.01 = 768.05$ 76,805.
9. $9.2 \times 0.001 = 0.0092$ 0009.2

Many times we see large numbers written, for example, in the form 267.1 million rather than in the longer standard notation. The next example shows how to interpret these numbers:

284.8 million

EXAMPLE 10

In 2001, the population of the United States was estimated to be 284.8 million. Write this number in standard notation. (*Source*: U.S. Census Bureau)

Solution: 284.8 million = 284.8 × 1 million
= 284.8 × 1,000,000 = 284,800,000.   ●

(c) Finding the Circumference of a Circle

Recall that the distance around a polygon is called its **perimeter**. The distance around a circle is given a special name called the **circumference**, and this distance depends on the radius or the diameter of the circle.

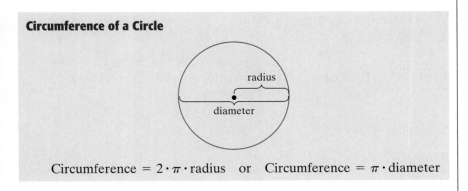

Circumference of a Circle

Circumference = 2 · π · radius   or   Circumference = π · diameter

The symbol π is the Greek letter pi, pronounced "pie." It is a number between 3 and 4. The number π rounded to two decimal places is 3.14, and a fraction approximation for π is $\frac{22}{7}$.

Practice Problem 10

In 2000, there were 115.9 million households in the United States. Write this number in standard notation. (*Source*: U.S. Census Bureau)

Answer

10. 115,900,000

Practice Problem 11

Find the circumference of a circle whose radius is 11 meters. Then use the approximation 3.14 for π to approximate this circumference.

Practice Problem 12

Elaine Rehmann is fertilizing her garden. She uses 5.6 ounces of fertilizer per square yard. The garden measures 60.5 square yards. How much fertilizer does she need?

EXAMPLE 11

Find the circumference of a circle whose radius is 5 inches. Then use the approximation 3.14 for π to approximate the circumference.

Solution:

$$\begin{aligned} \text{Circumference} &= 2 \cdot \pi \cdot \text{radius} \\ &= 2 \cdot \pi \cdot 5 \text{ inches} \\ &= 10\pi \text{ inches} \end{aligned}$$

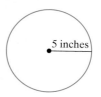

5 inches

Next, we replace π with the approximation 3.14.

$$\begin{aligned} \text{Circumference} &= 10\,\pi \text{ inches} \\ (\text{"is approximately"}) \to \quad &\approx 10(3.14) \text{ inches} \\ &= 31.4 \text{ inches} \end{aligned}$$

The *exact* circumference or distance around the circle is 10π inches, which is *approximately* 31.4 inches. ●

Ⓓ Solving Problems by Multiplying Decimals

The solutions to many real-life problems are found by multiplying decimals. We continue using our four problem-solving steps to solve such problems.

EXAMPLE 12   Finding the Total Cost of Materials for a Job

A college student is hired to paint a billboard with paint costing $2.49 per quart. If the job requires 3 quarts of paint, what is the total cost of the paint?

Solution:

1. UNDERSTAND. Read and reread the problem. The phrase "total cost" might make us think addition, but since this problem requires repeated addition, let's multiply.

2. TRANSLATE.

In words:	Total cost	is	cost per quart of paint	times	number of quarts
	↓	↓	↓	↓	↓
Translate:	Total cost	=	2.49	×	3

3. SOLVE:

$$\begin{array}{r} \overset{1\ 2}{2.49} \\ \times \quad 3 \\ \hline 7.47 \end{array}$$

4. INTERPRET. *Check* your work. *State* your conclusion: The total cost of the paint is $7.47. ●

Answers

11. 22π m; 69.08 m,   **12.** 338.8 oz

Name _____ Section _____ Date _____

A *Multiply. See Examples 1 through 3.*

1. 0.2
   × 0.6

2. 0.7
   × 0.9

 3. 1.2
   × 0.5

4. 6.8
   × 0.3

5. 0.26
   ×   5

6. 0.19
   ×   6

7. 5.3
   × 4.2

8. 6.2
   × 3.8

9. 5.62
   × 7.7

10. 8.03
   × 5.5

11. 1.0047
   ×   8.2

12. 2.0005
   ×   5.5

13. 490.2
   × 0.023

14. 300.9
   × 0.032

15. 16.003
   ×   5.31

16. 31.006
   ×   3.71

B *Multiply. See Examples 4 through 9.*

17. 6.5×10

18. 7.2×100

19. 6.5×0.1

20. 7.2×0.01

21. 7.093×100

22. 6.046×1000

23. 0.06×0.01

24. 4.7×0.1

25. 9.1×1000

26. 0.5×10

27. 37.62×0.001

28. 14.3×0.001

Write each number in standard form. See Example 10.

29. The storage silos at the main Hershey chocolate factory in Hershey, Pennsylvania, can hold enough cocoa beans to make 5.5 billion Hershey's milk chocolate bars. (*Source*: Hershey Foods Corporation)

30. There are 844 thousand places to eat out in the United States. (*Source*: National Restaurant Association)

31. About 36.4 million American households own at least one dog. (*Source*: American Pet Products Manufacturers Association)

32. The most-visited national park in the United States is the Blue Ridge Parkway in Virginia and North Carolina. An estimated 19.15 million people visited the park in 2000. (*Source*: National Park Service)

33. The Blue Streak is the oldest roller coaster at Cedar Point, an amusement park in Sandusky, Ohio. Since 1964, it has given more than 49.8 million rides. (*Source*: Cedar Fair, L.P.)

34. In 2001, sales by the American restaurant industry were expected to reach an average of $1.1 billion on a typical day. (*Source*: National Restaurant Association)

C *Find the circumference of each circle. Then use the approximation 3.14 for π and approximate each circumference. See Example 11.*

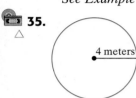
35.
4 meters

△ **36.**
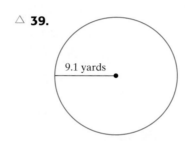
8 feet

△ **37.**
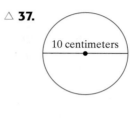
10 centimeters

△ **38.**
22 inches

△ **39.**
9.1 yards

△ **40.**
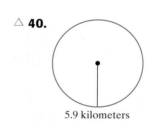
5.9 kilometers

D *Solve. See Example 12.*

41. A 1-ounce serving of cream cheese contains 6.2 grams of saturated fat. How much saturated fat is in 4 ounces of cream cheese? (*Source: Home and Garden Bulletin No. 72*; U.S. Department of Agriculture)

42. A 3.5-ounce serving of lobster meat contains 0.1 gram of saturated fat. How much saturated fat do 3 servings of lobster meat contain? (*Source*: The National Institute of Health)

43. The average cost of driving a car in 2001 was $0.51 per mile. How much would it have cost to drive a car 8750 miles in 2001? (*Source*: American Automobile Association)

44. In the first half of 2001, a U.S. airline passenger paid $0.1410, on average, to fly 1 mile. How much would it have cost to fly from Atlanta, Georgia, to Minneapolis, Minnesota, a distance of 905 miles? Round to the nearest cent. (*Source*: Air Transport Association of America, Inc.)

45. A meter is a unit of length in the metric system that is approximately equal to 39.37 inches. Sophia Wagner is 1.65 meters tall. Find her approximate height in inches.

46. The doorway to a room is 2.15 meters tall. Approximate this height in inches. (*Hint*: See Exercise 45.)

47. Jose Severos, an electrician for Central Power and Light, worked 40 hours last week. Calculate his pay before taxes for last week if his hourly wage is $13.88.

48. Maribel Chin, an assembly line worker, worked 20 hours last week. Her hourly rate is $8.52 per hour. Calculate Maribel's pay before taxes.

Review and Preview

Divide. See Sections 1.7 and 2.5.

49. $130 \div 5$

50. $495 \div 27$

51. $2016 \div 56$

52. $1863 \div 69$

53. $2920 \div 365$

54. $2916 \div 6$

55. $\dfrac{24}{7} \div \dfrac{8}{21}$

56. $\dfrac{162}{25} \div \dfrac{9}{75}$

Combining Concepts

57. Find how far radio waves travel in 20.6 seconds. (Radio waves travel at a speed of $1.86 \times 100,000$ miles per second.)

58. If it takes radio waves approximately 8.3 minutes to travel from the Sun to the Earth, find approximately how far it is from the Sun to the Earth. (*Hint*: See Exercise 57.)

59. In 1893, the first ride known as a Ferris wheel was constructed by George Washington Gale Ferris. Its diameter was 250 feet. Find its circumference. Give an exact answer and a two-decimal-place approximation, using 3.14 for π.

Source: *The Handy Science Answer Book*, Visible Ink Press, 1994

60. The radius of the Earth is approximately 3950 miles. Find the distance around the Earth at the equator. Give an exact answer and a two-decimal-place approximation, using 3.14 for π. (*Hint*: Find the circumference of a circle with radius 3950 miles.)

3950 miles

61. In your own words, explain how to find the number of decimal places in a product of decimal numbers.

62. In your own words, explain how to multiply by a power of 10.

MODELING MULTIPLICATION WITH DECIMAL NUMBERS

We can use an area model to represent multiplying decimal numbers. For instance, we can think of the 10×10 grid shown below on the left as representing 1, or a whole. We can show the product 0.5×0.2 by shading five tenths of the grid along one side of the square and shading two tenths of the grid along the other side.

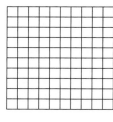 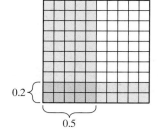

CRITICAL THINKING

1. What does the region where the shading overlaps represent? See whether you can figure it out by comparing it to the product of 0.5×0.2 found numerically. (Remember that one square within the grid represents one hundredth.)

2. How could you use this type of area model to represent the product of a decimal number (such as 0.6) and a whole number (such as 2)?

3. Use the model to show each of the following products:

   a. 0.4×0.4

   b. 0.9×0.1

   c. 0.8×2

   d. 0.7×1.5

4.5 Dividing Decimals

 Dividing Decimals

Dividing decimal numbers is similar to dividing whole numbers. The only difference is that we place a decimal point in the quotient. If the divisor is a whole number, we place the decimal point in the quotient directly above the decimal point in the dividend, and then divide as with whole numbers. Recall that division can be checked by multiplication.

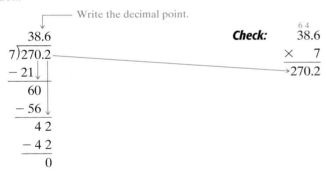

$$
\begin{array}{r}
0.26 \quad \leftarrow \text{quotient} \\
\text{divisor} \rightarrow 32\overline{)8.32} \quad \leftarrow \text{dividend} \\
-6\,4 \\
\hline
192 \\
-192 \\
\hline
0
\end{array}
$$

Check:
$$
\begin{array}{r}
0.26 \quad \text{Quotient} \\
\times \quad 32 \quad \text{Divisor} \\
\hline
52 \\
7\,8 \\
\hline
8.32 \quad \text{Dividend}
\end{array}
$$

O B J E C T I V E S

 Divide decimals.

 Divide decimals by powers of 10.

 Solve problems by dividing decimals.

SSM
TUTOR CENTER   SG   CD & VIDEO   MATH PRO   WEB

Dividing by a Whole Number

Step 1. Place the decimal point in the quotient directly above the decimal point in the dividend.

Step 2. Divide as with whole numbers.

EXAMPLE 1   Divide: $270.2 \div 7$. Check your answer.

Solution:

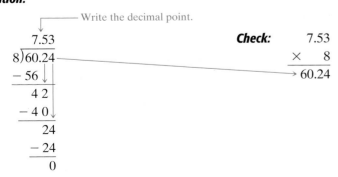

Write the decimal point.

$$
\begin{array}{r}
38.6 \\
7\overline{)270.2} \\
-21 \\
\hline
60 \\
-56 \\
\hline
4\,2 \\
-4\,2 \\
\hline
0
\end{array}
$$

Check:
$$
\begin{array}{r}
{}^{6\,4} \\
38.6 \\
\times \quad 7 \\
\hline
270.2
\end{array}
$$

The quotient is 38.6.

EXAMPLE 2   Divide: $60.24 \div 8$. Check your answer.

Solution:

Write the decimal point.

$$
\begin{array}{r}
7.53 \\
8\overline{)60.24} \\
-56 \\
\hline
4\,2 \\
-4\,0 \\
\hline
24 \\
-24 \\
\hline
0
\end{array}
$$

Check:
$$
\begin{array}{r}
7.53 \\
\times \quad 8 \\
\hline
60.24
\end{array}
$$

Sometimes to continue dividing we need to insert zeros after the last digit in the dividend.

Practice Problem 1

Divide: §517.2 ÷ 6. Check your answer.

Practice Problem 2

Divide: $26.19 \div 9$. Check your answer.

Answers

1. 86.2, **2.** 2.91

Practice Problem 3

Divide and check.

a. $0.4 \div 8$

b. $13.62 \div 12$

EXAMPLE 3   Divide: $0.5 \div 4$. Check your answer.

Solution:

$$
\begin{array}{r}
0.125 \\
4\overline{)0.500} \\
\underline{4} \\
10 \\
\underline{-8} \\
20 \\
\underline{-20} \\
0
\end{array}
$$

Insert two 0s to continue dividing.

Check:
$$
\begin{array}{r}
\overset{1\,2}{0.125} \\
\times \quad 4 \\
\hline
0.500
\end{array}
$$

If the divisor is not a whole number, before we divide we need to move the decimal point to the right until the divisor is a whole number.

$$1.5\overline{)64.85}$$
divisor $\nearrow$ $\quad$ $\nwarrow$ dividend

To understand how this works, let's rewrite

$$1.5\overline{)64.85} \quad \text{as} \quad \frac{64.85}{1.5}$$

and then multiply the numerator and the denominator by 10.

$$\frac{64.85}{1.5} = \frac{64.85 \times 10}{1.5 \times 10} = \frac{648.5}{15},$$

which can be written as $15.\overline{)648.5}$. Notice that

$$1.5\overline{)64.85} \quad \text{is equivalent to} \quad 15.\overline{)648.5}$$

The decimal points in the dividend and the divisor were both moved one place to the right, and the divisor is now a whole number. This procedure is summarized next:

Dividing by a Decimal

Step 1.   Move the decimal point in the divisor to the right until the divisor is a whole number.

Step 2.   Move the decimal point in the dividend to the right the *same number of places* as the decimal point was moved in Step 1.

Step 3.   Divide. Place the decimal point in the quotient directly over the moved decimal point in the dividend.

Practice Problem 4

Divide: $166.88 \div 5.6$

EXAMPLE 4   Divide: $10.764 \div 2.3$

Solution: We move the decimal points in the divisor and the dividend one place to the right so that the divisor is a whole number.

$$2.3\overline{)10.764} \quad \text{becomes} \quad
\begin{array}{r}
4.68 \\
23.\overline{)107.64} \\
\underline{92} \\
15\,6 \\
\underline{13\,8} \\
1\,84 \\
\underline{1\,84} \\
0
\end{array}$$

Answers

3. a. 0.05, **b.** 1.135, **4.** 29.8

EXAMPLE 5   Divide: $5.264 \div 0.32$

Solution:

$$0.32\overline{)5.264}$$ becomes
$$\begin{array}{r} 16.45 \\ 32\overline{)526.40} \\ -32 \\ \hline 206 \\ -192 \\ \hline 14\ 4 \\ -12\ 8 \\ \hline 1\ 60 \\ -1\ 60 \\ \hline 0 \end{array}$$ Insert one 0.   ●

Try the Concept Check in the margin.

EXAMPLE 6

Divide: $17.5 \div 0.48$. Round the quotient to the nearest hundredth.

Solution: First we move the decimal points in the divisor and the dividend two places. Then we divide and round the quotient to the nearest hundredth.

hundredths place

$$\begin{array}{r} 36.458 \approx 36.46 \\ 48.\overline{)1750.000} \\ -144 \\ \hline 310 \\ -288 \\ \hline 220 \\ -192 \\ \hline 280 \\ -240 \\ \hline 400 \\ -384 \\ \hline 16 \end{array}$$

"is approximately"

When rounding to the nearest hundredth, carry the division process out to one more decimal place, the thousandths place.

●

Try the Concept Check in the margin.

B   Dividing Decimals by Powers of 10

There are patterns that occur when we divide decimals by powers of 10 such as 10, 100, 1000, and so on.

$$\frac{569.2}{10} = 56.92$$ Move the decimal point *1 place* to the *left*.

1 zero

$$\frac{569.2}{10,000} = 0.05692$$ Move the decimal point *4 places* to the *left*.

4 zeros

This pattern suggests the following rule:

> **Dividing Decimals by Powers of 10 such as 10, 100, or 1000**
>
> Move the decimal point of the dividend to the *left* the same number of places as there are *zeros* in the power of 10.

Practice Problems 7–8

Divide.

7. $\dfrac{28}{1000}$

8. $\dfrac{8.56}{100}$

EXAMPLES Divide.

7. $\dfrac{786}{10,000} = 0.0786$ Move the decimal point *4 places* to the *left*.

4 zeros

8. $\dfrac{0.12}{10} = 0.012$ Move the decimal point *1 place* to the *left*.

1 zero

C **Solving Problems by Dividing Decimals**

Many real-life problems involve dividing decimals.

Practice Problem 9

A bag of fertilizer covers 1250 square feet of lawn. Tim Parker's lawn measures 14,800 square feet. How many bags of fertilizer does he need? If he can buy only whole bags of fertilizer, how many whole bags does he need?

EXAMPLE 9 **Calculating Materials Needed for a Job**

A gallon of paint covers a 250-square-foot area. If Betty Adkins wishes to paint a wall that measures 1450 square feet, how many gallons of paint does she need? If she can buy only gallon containers of paint, how many gallon containers does she need?

Solution:

1. UNDERSTAND. Read and reread the problem. To find the number of gallons, we divide 1450 by 250.

2. TRANSLATE.

In words:	number of gallons	is	square feet	divided by	square feet per gallon
Translate:	number of gallons	=	1450	÷	250

3. SOLVE:

$$
\begin{array}{r}
5.8 \\
250\overline{)1450.0} \\
-1250 \\
\hline
200\ 0 \\
-200\ 0 \\
\hline
0
\end{array}
$$

4. INTERPRET. *Check* your work. *State* your conclusion: Betty needs 5.8 gallons of paint. If she can buy only gallon containers of paint, she needs 6 gallon containers of paint to complete the job.

EXERCISE SET 4.5

 A *Divide. See Examples 1 through 5.*

1. $5\overline{)0.47}$ **2.** $2\overline{)11.8}$ **3.** $0.06\overline{)18}$ **4.** $0.04\overline{)20}$ **5.** $0.82\overline{)4.756}$

6. $0.92\overline{)3.312}$ **7.** $5.5\overline{)36.3}$ **8.** $2.2\overline{)21.78}$ **9.** $6.195 \div 15$ **10.** $0.54 \div 12$

11. Divide 4.2 by 0.6   **12.** Divide 3.6 by 0.9.   **13.** $0.27\overline{)1.296}$ **14.** $0.34\overline{)2.176}$ **15.** $0.02\overline{)42}$

16. $0.03\overline{)24}$ **17.** $0.6\overline{)18}$ **18.** $0.4\overline{)20}$ **19.** $0.005\overline{)35}$ **20.** $0.0007\overline{)35}$

21. $7.2\overline{)70.56}$ **22.** $6.3\overline{)52.92}$ **23.** $5.4\overline{)51.84}$ **24.** $7.7\overline{)33.88}$ **25.** $\dfrac{1.215}{0.027}$ **26.** $\dfrac{3.213}{0.051}$

Divide. Round the quotients as indicated. See Example 6.

27. Divide 429.34 by 2.4 and round the quotient to the nearest hundred.

28. Divide 54.8 by 2.6 and round the quotient to the nearest ten.

29. Divide 0.549 by 0.023 and round the quotient to the nearest hundredth.

30. Divide 0.0453 by 0.98 and round the quotient to the nearest thousandth.

31. Divide 45.23 by 0.4 and round the quotient to the nearest ten.

32. Divide 983.32 by 0.061 and round the quotient to the nearest thousand.

33. $\dfrac{54.982}{100}$ **34.** $\dfrac{342.54}{100}$ **35.** $\dfrac{12.9}{1000}$ **36.** $\dfrac{13.49}{10}$ **37.** $\dfrac{87}{10}$ **38.** $\dfrac{0.27}{1000}$

C *Solve. See Example 9.*

39. Dorren Schmidt pays $73.86 per month to pay back a loan of $1772.64. In how many months will the loan be paid off?

△ **40.** Josef Jones is painting the walls of a room. The walls have a total area of 546 square feet. A quart of paint covers 52 square feet. If he must buy paint in whole quarts, how many quarts does he need?

41. The leading monetary winner in men's professional golf in 2000 was Tiger Woods. He earned $9,188,321. Suppose he had earned this working 40 hours per week for a year. Determine his hourly wage to the nearest cent. (*Source:* PGA TOUR, Inc.)

42. Juanita Gomez bought unleaded gasoline for her car at $1.169 per gallon. She wanted to keep a record of how many gallons her car is using but forgot to write down how many gallons she purchased. She wrote a check for $27.71 to pay for it. How many gallons, to the nearest tenth of a gallon, did she buy?

△ **43.** A pound of fertilizer covers 39 square feet of lawn. Vivian Bulgakov's lawn measures 7883.5 square feet. How much fertilizer, to the nearest tenth of a pound, does she need to buy?

44. A page of a book contains about 1.5 kilobytes of information. If a computer disk can hold 740 kilobytes of information, how many pages of a book can be stored on one computer disk? Round to the nearest tenth of a page.

45. There are approximately 39.37 inches in 1 meter. How many meters, to the nearest tenth of a meter, are there in 200 inches?

← 1 meter →
← ≈39.37 inches →

46. There are approximately 2.54 centimeters in 1 inch. How many inches are there in 50 centimeters? Round to the nearest tenth.

← 1 inch →
← ≈2.54 cm →

47. In the United States, an average child will wear down 730 crayons by his or her tenth birthday. Find the number of boxes of 64 crayons this is equivalent to. Round to the nearest tenth. (*Source:* Binney & Smith Inc.)

48. In 2000, American farmers received an average of $243 per 100 chickens. What was the average price per chicken? (*Source:* National Agricultural Statistics Service)

49. During the 24 Hours of the Le Mans endurance auto race in 2001, the winning team of Frank Biela, Tom Kristensen, and Emanuele Pirro drove a total of 2707.44 miles in 24 hours. What was their average speed in miles per hour? Round to the nearest tenth. (*Source:* Automobile Club de l'Ouest)

50. In 2000, Kenyan runner Tegla Loroupe set a new world record for the women's 20,000-meter event. Her time for the event was 3926.6 seconds. What was her average speed in meters per second? Round to the nearest tenth. (*Source:* International Amateur Athletic Federation)

51. Katie Smith of the WNBA's Minnesota Lynx scored a total of 646 points during the 32 basketball games she played in the 2000 regular season. What was the average number of points she scored per game? Round to the nearest hundredth. (*Source:* Women's National Basketball Association)

52. During the 2000 Major League Soccer season, the Chicago Fire was the top-scoring team with a total of 67 goals throughout the season. The Chicago Fire played 32 games. What was the average number of goals the team scored per game? Round to the nearest hundredth. (*Source:* Major League Soccer)

Review and Preview

Round each decimal to the given place value. See Section 4.2.

53. 345.219, nearest hundredth

54. 902.155, nearest hundredth

55. 1000.994, nearest tenth

56. 234.1029, nearest thousandth

Use the order of operations to simplify each expression. See Section 1.9.

57. $2 + 3 \cdot 6$ **58.** $9 \cdot 2 + 8 \cdot 3$ **59.** $20 - 10 \div 5$ **60.** $(20 - 10) \div 5$

 Combining Concepts

Recall from Section 1.7 that the average of a list of numbers is their total divided by how many numbers there are in the list. Use this procedure to find the average of the test scores listed in Exercises 61 and 62. If necessary, round to the nearest tenth.

61. 86, 78, 91, 85

62. 56, 75, 80

63. In 2000, American manufacturers shipped approximately 942.5 million music CDs to retailers. How many music CDs were shipped per week on average? (*Source:* Recording Industry Association of America)

△ **64.** The area of the rectangle is 38.7 square feet. If its width is 4.5 feet, find its length.

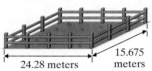

Area is 38.7 square feet. 4.5 feet
?

△ **65.** The perimeter of the square is 180.8 centimeters. Find the length of a side.

Perimeter is 180.8 centimeters.
?

△ **66.** Don Larson is building a horse corral that's shaped like a rectangle with dimensions of 24.28 meters by 15.675 meters. He plans to make a four-wire fence; that is, he will string four wires around the corral. How much wire will he need?

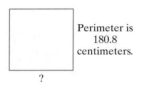

24.28 meters   15.675 meters

67. When dividing decimals, describe the process you use to place the decimal point in the quotient.

68. In your own words, describe how to quickly divide a number by a power of 10 such as 10, 100, 1000, etc.

Integrated Review—Operations on Decimals

Perform the indicated operations.

1. $1.6 + 0.97$

2. $3.2 + 0.85$

3. $9.8 - 0.9$

4. $10.2 - 6.7$

5. $\begin{array}{r} 0.8 \\ \times\, 0.2 \\ \hline \end{array}$

6. $\begin{array}{r} 0.6 \\ \times\, 0.4 \\ \hline \end{array}$

7. $8\overline{)2.16}$

8. $6\overline{)3.12}$

9. $\begin{array}{r} 9.6 \\ \times\, 0.5 \\ \hline \end{array}$

10. $\begin{array}{r} 8.7 \\ \times\, 0.7 \\ \hline \end{array}$

11. $\begin{array}{r} 123.6 \\ -\ \ 48.04 \\ \hline \end{array}$

12. $\begin{array}{r} 325.2 \\ -\ \ 36.08 \\ \hline \end{array}$

13. $25 + 0.026$

14. $0.125 + 44$

15. $3.4\overline{)29.24}$

16. $1.9\overline{)10.26}$

17. 2.8×100

18. 1.6×1000

19. $\begin{array}{r} 96.21 \\ 7.028 \\ +\ 121.7 \\ \hline \end{array}$

20. $\begin{array}{r} 0.268 \\ 1.93 \\ +\ 142.881 \\ \hline \end{array}$

21. $25.76 \div 46$

22. $27.09 \div 43$

23. $\begin{array}{r} 12.004 \\ \times\ \ \ \ 2.3 \\ \hline \end{array}$

24. $\begin{array}{r} 28.006 \\ \times\ \ \ \ 5.2 \\ \hline \end{array}$

Answers

1. _____
2. _____
3. _____
4. _____
5. _____
6. _____
7. _____
8. _____
9. _____
10. _____
11. _____
12. _____
13. _____
14. _____
15. _____
16. _____
17. _____
18. _____
19. _____
20. _____
21. _____
22. _____
23. _____
24. _____

25. _____

26. _____

27. _____

28. _____

29. _____

30. _____

31. _____

32. _____

33. _____

34. _____

35. _____

36. _____

37. _____

38. _____

39. _____

40. _____

41. _____

42. _____

43. _____

44. _____

25. Subtract 4.6 from 10.

26. Subtract 0.26 from 18.

27.   268.19
        + 146.25

28.   860.18
        + 434.85

29. $\dfrac{2.958}{0.087}$

30. $\dfrac{1.708}{0.061}$

31. $160 - 43.19$

32. $120 - 101.21$

33. 15.62×10

34. $15.62 \div 10$

35. $15.62 + 10$

36. $15.62 - 10$

37. $117.26 \div 2.6$

38. 117.26×2.6

39. $117.26 - 2.6$

40. $117.26 + 2.6$

41. 0.0072×0.06

42. 0.0025×0.03

43. $0.0072 + 0.06$

44. $0.03 - 0.0025$

4.6 Estimating and Order of Operations

A Estimating Operations on Decimals

Estimating sums, differences, products, and quotients of decimal numbers is an important skill whether you use a calculator or perform decimal operations by hand. When you can estimate results as well as calculate them, you can judge whether the calculations are reasonable. If they aren't, you know you've made an error somewhere along the way.

OBJECTIVES

A Estimate operations on decimals.

B Simplify expressions containing decimals using the order of operations.

SSM
TUTOR CENTER   SG   CD & VIDEO   MATH PRO   WEB

EXAMPLE 1

Subtract: $78.62 − $16.85. Then estimate the difference by rounding each decimal to the nearest dollar and then subtracting to see whether the proposed result is reasonable.

Solution:

Given		Estimate
$78.62	rounds to	$79
− $16.85	rounds to	− $17
$61.77		$62

The estimated difference is $62, so $61.77 is reasonable.   ●

Try the Concept Check in the margin.

Practice Problem 1

Subtract: $65.34 − $14.68. Then estimate the difference to see whether the proposed result is reasonable.

Concept Check

Why shouldn't the sum 21.98 + 42.36 be estimated as 30 + 50 = 80?

EXAMPLE 2

Multiply: 28.06 × 1.95. Then estimate the product to see whether the proposed result is reasonable.

Solution:

Given	Estimate 1	Estimate 2
28.06	28	30
× 1.95	× 2	× 2
14030	56	60
252540		
280600		
54.7170		

The answer 54.7170 is reasonable.   ●

As shown in Example 2, estimated results will vary depending on what estimates are used. Notice that estimating results is a good way to see whether the decimal point has been correctly placed.

Practice Problem 2

Multiply: 30.26 × 2.98. Then estimate the product to see whether the proposed solution is reasonable.

Answers

1. $50.66,   **2.** 90.1748

Concept Check: Each number is rounded incorrectly. The estimate is too high.

Practice Problem 3

Divide: 713.7 ÷ 91.5. Then estimate the quotient to see whether the proposed answer is reasonable.

EXAMPLE 3

Divide: 272.356 ÷ 28.4. Then estimate the quotient to see whether the proposed result is reasonable.

Solution:

Given		**Estimate**	

$$
\begin{array}{r}
9.59 \\
284.\overline{)2723.56} \\
-2556 \\
\hline
1675 \\
-1420 \\
\hline
2556 \\
-2556 \\
\hline
0
\end{array}
$$

$$
\begin{array}{r}
9 \\
30\overline{)270}
\end{array}
$$

or

$$
\begin{array}{r}
9 \\
30\overline{)270}
\end{array}
$$

The estimate is 9, so 9.59 is reasonable. ●

Practice Problem 4

Referring to the figure below, estimate how much farther it is between Dodge City and Pratt than it is between Garden City and Dodge City by rounding each given distance to the nearest mile.

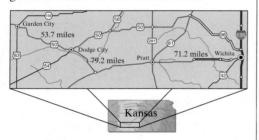

EXAMPLE 4   Estimating Distance

Refer to the figure in the margin to estimate the distance in miles between Garden City, Kansas, and Wichita, Kansas, by rounding each given distance to the nearest ten.

Solution:

Calculated Distance		**Estimate**
53.7	rounds to	50
79.2	rounds to	80
+ 71.2	rounds to	+ 70
		200

The distance between Garden City and Wichita is approximately 200 miles. (The calculated distance is 204.1 miles.) ●

B ## Simplifying Expressions Containing Decimals

In the remaining examples, we review the order of operations by simplifying expressions that contain decimals.

Order of Operations

1. Perform all operations within grouping symbols such as parentheses or brackets.
2. Evaluate any expressions with exponents.
3. Multiply or divide in order from left to right.
4. Add or subtract in order from left to right.

Answers

3. 7.8, **4.** 25 mi

EXAMPLE 5   Simplify: $0.5(8.6 - 1.2)$

Solution: According to the order of operations, we simplify inside the parentheses first.

$$0.5(8.6 - 1.2) = 0.5(7.4) \quad \text{Subtract.}$$
$$= 3.7 \quad \text{Multiply.}$$

Practice Problem 5

Simplify: $8.69(3.2 - 1.8)$

EXAMPLE 6   Simplify: $(1.3)^2$

Solution: We recall the meaning of an exponent.

$$(1.3)^2 = (1.3)(1.3) \quad \text{Use the definition of an exponent.}$$
$$= 1.69 \quad \text{Multiply.}$$

Practice Problem 6

Simplify: $(0.7)^2$

EXAMPLE 7   Simplify: $\dfrac{0.7 + 1.84}{0.4}$

Solution: First we simplify the numerator of the fraction. Then we divide.

$$\frac{0.7 + 1.84}{0.4} = \frac{2.54}{0.4} \quad \text{Simplify the numerator.}$$
$$= 6.35 \quad \text{Divide.}$$

Practice Problem 7

Simplify: $\dfrac{8.78 - 2.8}{20}$

EXAMPLE 8   Simplify: $5.68 + (0.9)^2 \div 100$

Solution:

$$5.68 + 0.9^2 \div 100 = 5.68 + 0.81 \div 100 \quad \text{Simplify } 0.9^2.$$
$$= 5.68 + 0.0081 \quad \text{Divide.}$$
$$= 5.6881 \quad \text{Add.}$$

Practice Problem 8

Simplify: $20.06 - (1.2)^2 \div 10$

Answers

5. 12.166,   **6.** 0.49,   **7.** 0.299,   **8.** 19.916

STUDY SKILLS REMINDER

Are you getting all the mathematics help that you need?

Remember that, in addition to your instructor, there are many places to get help with your mathematics course. For example, see what is available from the list below.

- This text has an accompanying video lesson for every section in this text.

- The back of this book contains answers to odd-numbered exercises and selected solutions.

- MathPro is available with this text. It is a tutorial software program with lessons corresponding to each section in the text.

- A student solutions manual is available that contains worked-out solutions to odd-numbered exercises as well as solutions to every exercise in the Chapter Pretests, Integrated Reviews, Chapter Reviews, Chapter Tests, and Cumulative Reviews.

- Don't forget to check with your instructor for other local resources available to you, such as a tutor center.

EXERCISE SET 4.6

A *Perform each indicated operation. Then estimate to see whether each proposed result is reasonable. See Examples 1 through 3.*

1. 2.1 + 5.8 + 4.1   **2.** 6.1 + 5.9 + 3.1   **3.** 4.9 − 2.1   **4.** 7.3 − 3.8   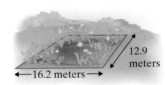**5.** 6 × 483.11

6. 98.2 × 405.8   **7.** 62.16 ÷ 14.8   **8.** 186.75 ÷ 24.9   **9.** 69.2 + 32.1 + 48.5

10. 179.52 + 221.47 + 397.23   **11.** 34.92 − 12.03   **12.** 349.87 − 251.32

Solve. See Example 4.

△ **13.** Estimate the perimeter of the rectangle by rounding each measurement to the nearest whole inch.

5.9 inches

12.2 inches

△ **14.** Joe Armani wants to put plastic edging around his rectangular flower garden to help control weeds. The length and width are 16.2 meters and 12.9 meters, respectively. Find how much edging he needs by rounding each measurement to the nearest whole meter.

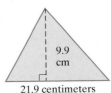

12.9 meters

←16.2 meters→

△ **15.** Estimate the perimeter of the triangle by rounding each measurement to the nearest whole foot.

11.8 feet   12.9 feet

14.2 feet

△ **16.** The area of a triangle is area $= \frac{1}{2} \cdot$ base $\cdot$ height

Approximate the area of a triangle whose base is 21.9 centimeters and whose height is 9.9 centimeters. Round each measurement to the nearest whole centimeter.

9.9 cm

21.9 centimeters

△ **17.** Use 3.14 for π to approximate the circumference of a circle whose radius is 7 meters.

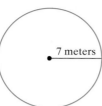

7 meters

△ **18.** Use 3.14 for π to approximate the circumference of a circle whose radius is 6 centimeters.

6 centimeters

19. Mike and Sandra Hallahan are taking a trip and plan to travel about 1550 miles. Their car gets 32.8 miles to the gallon. Approximate how many gallons of gasoline their car will use. Round the result to the nearest gallon.

20. Refer to Exercise 19. Suppose gasoline costs $1.059 per gallon. Approximate to the nearest cent how much money Mike and Sandra need for gasoline for their 1550-mile trip.

21. Ken and Wendy Holladay purchased a new car. They pay $198.79 per month on their car loan for 5 years. Approximate how much they will pay for the car altogether by rounding the monthly payment to the nearest hundred.

22. Yoshikazu Sumo is purchasing the following groceries. He has only a $10 bill in his pocket. Estimate the cost of each item to see whether he has enough money: Bread, $1.49; milk, $2.09; carrots, $0.97; corn, $0.89; salt, $0.53; and butter, $2.89.

23. Estimate the total distance to the nearest mile between Grove City and Jerome by rounding each distance to the nearest mile.

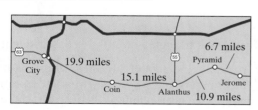

△ **24.** Estimate the perimeter of the figure shown by rounding each measurement to the nearest whole foot.

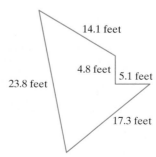

14.1 feet

4.8 feet

5.1 feet

23.8 feet

17.3 feet

25. The all-time top five movies (that is, those that have earned the most money at the box office in the United States) along with the approximate amount of money they have earned are listed in the table. Estimate the total amount of money that these movies have earned.

All-Time Top Five American Movies	
Movie	Total Box Office Receipts
Titanic (1997)	$600.7 million
Star Wars (1977)	$460.9 million
Star Wars: Episode I— The Phantom Menace (1999)	$431.1 million
E.T. the Extra-Terrestrial (1982)	$399.8 million
Jurassic Park (1993)	$356.8 million

(*Source:* The Internet Movie Database Ltd.)

26. San Marino is one of the smallest countries in the world. In 2000, it had a population density of 1171 people per square mile. The area of San Marino is about 23 square miles. Use this data to estimate the population of San Marino in 2000. (*Source: The World Factbook*, U.S. Central Intelligence Agency)

27. Cape Verde is a group of islands off the western coast of Africa. In 2000, it had a population density of 258 people per square mile. The area of Cape Verde is about 1557 square miles. Use this data to estimate the population of Cape Verde in 2000. (*Source: The World Factbook*, U.S. Central Intelligence Agency)

28. In 2000, American manufacturers shipped approximately 942.5 million music CDs to retailers. The cost of these shipments was $13,214.5 million. Use this data to estimate the cost of an individual CD. Round to the nearest cent. (*Source*: Recording Industry Association of America)

B *Simplify each expression. See Examples 5 through 8.*

29. $(0.4)^2$

30. $300 - 100 \times 2.3$

31. $\dfrac{1 + 0.8}{0.6}$

32. $(0.09)^2$

33. $1.4(2 - 1.8)$

34. $\dfrac{0.29 + 1.69}{3}$

35. $30.03 + 5.1 \times 9.9$

36. $60 - 6.02 \times 8.97$

37. $7.8 - 4.83 \div 2.1$

38. $90 - 62.1 \div 2.7 + 8.6$ **39.** $(2.3)^2$

40. $(8.2)(100) - (8.2)(10)$ **41.** $(3.1 + 0.7)(2.9 - 0.9)$

42. $\dfrac{3.19 - 0.707}{1.3}$ **43.** $\dfrac{(4.5)^2}{100}$ **44.** $0.9(6.5 - 5.6)$ **45.** $\dfrac{7 + 0.74}{0.06}$ **46.** $(1.5)^2 + 0.5$

47. $7.8 + 1.1 \times 100 - 3.6$ **48.** $9.6 - 7.8 \div 10 + 1.2$ **49.** $(0.3)^2 + 2(14.6 - 0.03)$

50. $5(20.6 - 2.06) - (0.8)^2$ **51.** $(10.6 - 9.8)^2 \div 0.01 + 8.6$ **52.** $49.1 - (101.3 - 100.6)^2 \div 0.1$

53. $6 \div 0.1 + 8.9 \times 10 - 4.6$ **54.** $8 \div 10 + 7.6 \times 0.1 - (0.1)^2$

Review and Preview

Perform each operation. See Sections 2.4, 2.5, and 3.3.

55. $\dfrac{3}{4} \cdot \dfrac{5}{12}$

56. $\dfrac{56}{23} \cdot \dfrac{46}{8}$

57. $\dfrac{36}{56} \div \dfrac{30}{35}$

58. $\dfrac{5}{7} \div \dfrac{14}{10}$

59. $\dfrac{5}{12} - \dfrac{1}{3}$

60. $\dfrac{12}{15} + \dfrac{5}{21}$

Combining Concepts

Simplify each expression. Then estimate to see whether the proposed result is reasonable.

61. $1.96(7.852 - 3.147)^2$

62. $(6.02)^2 + (2.06)^2$

63. When simplifying $5 + 0.1(1.26 - 0.23)$, do you add, multiply, or subtract first? Explain your answer.

64. When simplifying $3(1.5)^2$, do you square first or multiply by 3 first? Explain your answer.

4.7 Fractions and Decimals

OBJECTIVES

Ⓐ Write fractions as decimals.

Ⓑ Compare fractions and decimals.

Ⓒ Solve area problems containing fractions and decimals.

SSM
TUTOR CENTER   SG   CD & VIDEO   MATH PRO   WEB

Ⓐ Writing Fractions as Decimals

To write a fraction as a decimal, we interpret the fraction bar to mean division and find the quotient.

Writing Fractions as Decimals

To write a fraction as a decimal, divide the numerator by the denominator.

EXAMPLE 1   Write $\frac{1}{4}$ as a decimal.

Solution: $\frac{1}{4} = 1 \div 4$

$$
\begin{array}{r}
0.25 \\
4\overline{)1.00} \\
-8 \\
\hline
20 \\
-20 \\
\hline
0
\end{array}
$$

Thus, $\frac{1}{4}$ written as a decimal is 0.25.

Practice Problem 1

a. Write $\frac{2}{5}$ as a decimal.

b. Write $\frac{9}{40}$ as a decimal.

EXAMPLE 2   Write $\frac{2}{3}$ as a decimal.

Solution:
$$
\begin{array}{r}
0.666\ldots \\
3\overline{)2.000} \\
-18 \\
\hline
20 \\
-18 \\
\hline
20 \\
-18 \\
\hline
2
\end{array}
$$
This pattern will continue because $\frac{2}{3} = 0.6666\ldots$.

We place a bar over the digit 6 to indicate that it repeats.

$$\frac{2}{3} = 0.666\ldots = 0.\overline{6}$$

We can also write a decimal approximation for $\frac{2}{3}$. For example, $\frac{2}{3}$ rounded to the nearest hundredth is 0.67. This can be written as $\frac{2}{3} \approx 0.67$.

Practice Problem 2

a. Write $\frac{5}{6}$ as a decimal.

b. Write $\frac{2}{9}$ as a decimal.

EXAMPLE 3

Write $\frac{22}{7}$ as a decimal. (The fraction $\frac{22}{7}$ is an approximation for π.) Round to the nearest hundredth.

Practice Problem 3

Write $\frac{1}{9}$ as a decimal. Round to the nearest thousandth.

Answers

1. a. 0.4, **b.** 0.225, **2. a.** $0.8\overline{3} \approx 0.83$,
b. $0.\overline{2} \approx 0.22$, **3.** $0.\overline{1} \approx 0.11$

Solution:

$$\begin{array}{r} 3.142 \approx 3.14 \\ 7\overline{)22.000} \\ -21 \\ \hline 1\,0 \\ -7 \\ \hline 30 \\ -28 \\ \hline 20 \\ -14 \\ \hline 6 \end{array}$$

Carry the division out to the thousandths place.

The fraction $\dfrac{22}{7}$ in decimal form is approximately 3.14.

B **Comparing Decimals and Fractions**

Now we can compare decimals and fractions by writing fractions as equivalent decimals.

Practice Problem 4

Insert $<$, $>$, or $=$ to form a true statement.

$\dfrac{1}{5}$ 0.25

EXAMPLE 4   Insert $<$, $>$, or $=$ to form a true statement.

$\dfrac{1}{8}$ 0.12

Solution: First we write $\dfrac{1}{8}$ as an equivalent decimal. Then we compare decimal places.

$$\begin{array}{r} 0.125 \\ 8\overline{)1.000} \\ -8 \\ \hline 20 \\ -16 \\ \hline 40 \\ -40 \\ \hline 0 \end{array}$$

Original numbers	$\dfrac{1}{8}$	0.12
Decimals	0.125	0.120
Compare	0.125 $>$ 0.12	
Thus,	$\dfrac{1}{8} > 0.12$	

Practice Problem 5

Insert $<$, $>$, or $=$ to form a true statement.

a. $\dfrac{1}{2}$ 0.54   b. $0.\overline{6}$ $\dfrac{2}{3}$

c. $\dfrac{5}{7}$ 0.72

EXAMPLE 5   Insert $<$, $>$, or $=$ to form a true statement.

$0.\overline{7}$ $\dfrac{7}{9}$

Solution: We write $\dfrac{7}{9}$ as a decimal and then compare.

$$\begin{array}{r} 0.77\ldots = 0.\overline{7} \\ 9\overline{)7.00} \\ -6\,3 \\ \hline 70 \\ -63 \\ \hline 7 \end{array}$$

Original numbers	$0.\overline{7}$	$\dfrac{7}{9}$
Decimals	$0.\overline{7}$	$0.\overline{7}$
Compare	$0.\overline{7} = 0.\overline{7}$	
Thus,	$0.\overline{7} = \dfrac{7}{9}$	

Answers

4. $<$, **5. a.** $<$, **b.** $=$, **c.** $<$

EXAMPLE 6   Write the numbers in order from smallest to largest.

$$\frac{9}{20}, \frac{4}{9}, 0.456$$

Solution:

Original numbers	$\frac{9}{20}$	$\frac{4}{9}$	0.456
Decimals	0.450	0.444…	0.456
Compare in order	2nd	1st	3rd

Written in order, we have

$$\frac{4}{9}, \frac{9}{20}, 0.456$$

●

C   **Solving Area Problems Containing Fractions and Decimals**

Sometimes real-life problems contain both fractions and decimals. In this section, we solve such problems concerning area.

△ **EXAMPLE 7**

The area of a triangle is Area $= \frac{1}{2} \cdot$ base $\cdot$ height. Find the area of the triangle shown.

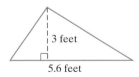

3 feet

5.6 feet

Solution:

$$\text{Area} = \frac{1}{2} \cdot \text{base} \cdot \text{height}$$

$$= \frac{1}{2} \cdot 5.6 \cdot 3$$

$$= 0.5 \cdot 5.6 \cdot 3 \qquad \text{Write } \frac{1}{2} \text{ as the decimal } 0.5.$$

$$= 8.4$$

The area of the triangle is 8.4 square feet.

●

Practice Problem 6

Write the numbers in order from smallest to largest.

a. $\frac{1}{3}, 0.302, \frac{3}{8}$ b. $1.26, 1\frac{1}{4}, 1\frac{2}{5}$

c. $0.4, 0.41, \frac{5}{7}$

Practice Problem 7   △

Find the area of the triangle.

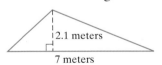

2.1 meters

7 meters

Answers

6. a. $0.302, \frac{1}{3}, \frac{3}{8}$,   **b.** $1\frac{1}{4}, 1.26, 1\frac{2}{5}$,   **c.** $0.4, 0.41, \frac{5}{7}$,

7. 7.35 sq m

STUDY SKILLS REMINDER

Are you prepared for a test on Chapter 4?

Below I have listed some *common trouble areas* for students in Chapter 4. After studying for your test—but before taking your test—read these.

- Don't forget the order of operations. To simplify $0.7 + 1.3(5 - 0.1)$, should you add, subtract, or multiply first? First, perform the subtraction within parentheses, then multiply, and finally add.

$$0.7 + 1.3(5 - 0.1) = 0.7 + 1.3(4.9) \quad \text{Subtract.}$$
$$= 0.7 + 6.37 \quad \text{Multiply.}$$
$$= 7.07 \quad \text{Add.}$$

- If you are having trouble with ordering or operations on decimals, don't forget that you can insert 0s after the last digit to the right of the decimal point as needed.

Addition	Addition with Zeros Inserted	Subtraction	Subtraction with Zeros Inserted
			$\overset{6\ \ 9\ 10}{\cancel{7.00}}$
8.1	8.100	7	7.00
0.6	0.600	$-\,0.28$	$-\,0.28$
$+\,23.003$	$+\,23.003$		6.72
	31.703		

Place in order from smallest to largest: $0.108, 0.18, 0.0092$

If we insert zeros, we have: $0.1080, 0.1800, 0.0092$

The decimals in order are: $0.0092, 0.1080, 0.1800$ or $0.0092, 0.108, 0.18$

EXERCISE SET 4.7

A *Write each fraction as a decimal. See Examples 1 through 3.*

1. $\dfrac{1}{5}$ **2.** $\dfrac{2}{5}$ **3.** $\dfrac{4}{8}$ **4.** $\dfrac{6}{8}$ **5.** $\dfrac{3}{4}$ **6.** $\dfrac{6}{5}$

7. $\dfrac{2}{25}$ **8.** $\dfrac{3}{25}$ **9.** $\dfrac{3}{8}$ **10.** $\dfrac{1}{4}$ **11.** $\dfrac{11}{12}$ **12.** $\dfrac{19}{25}$

13. $\dfrac{17}{40}$ **14.** $\dfrac{5}{12}$ **15.** $\dfrac{9}{20}$ **16.** $\dfrac{31}{40}$ **17.** $\dfrac{1}{3}$ **18.** $\dfrac{7}{9}$

19. $\dfrac{7}{16}$ **20.** $\dfrac{27}{16}$ **21.** $\dfrac{2}{9}$ **22.** $\dfrac{9}{11}$ **23.** $\dfrac{5}{3}$ **24.** $\dfrac{4}{11}$

Round each number as indicated.

25. Round your decimal answer to Exercise 17 to the nearest hundredth.

26. Round your decimal answer to Exercise 18 to the nearest hundredth.

27. Round your decimal answer to Exercise 19 to the nearest hundredth.

28. Round your decimal answer to Exercise 20 to the nearest hundredth.

29. Round your decimal answer to Exercise 21 to the nearest tenth.

30. Round your decimal answer to Exercise 22 to the nearest tenth.

31. Round your decimal answer to Exercise 23 to the nearest tenth.

32. Round your decimal answer to Exercise 24 to the nearest tenth.

Write each fraction as a decimal. If necessary, round to the nearest hundredth.

33. During the 2001 Boston Marathon, $\dfrac{17}{25}$ of the starting runners over the age of 70 finished the race. (*Source:* Boston Athletic Association)

34. About $\dfrac{21}{50}$ of all blood donors have type A blood. (*Source:* American Red Cross Biomedical Services)

35. On a typical day, $\frac{4}{10}$ of American adults eat out at a restaurant. (*Source*: National Restaurant Association)

36. By October 2000, $\frac{29}{46}$ of all individuals who had flown in space were citizens of the United States. Round to the nearest hundredth. (*Source*: Congressional Research Service)

37. Of the U.S. mountains that are over 14,000 feet in elevation, $\frac{56}{91}$ are located in Colorado. (*Source*: U.S. Geological Survey)

38. In 2000, about $\frac{701}{1048}$ Americans used the Internet in some form. (*Source*: UCLA Center for Communication Policy)

B *Insert* <, >, *or* = *to form a true statement. See Examples 4 and 5.*

39. 0.562　　0.569

40. 0.983　　0.988

41. 0.823　　0.813

42. 0.824　　0.821

43. 0.0923　　0.0932

44. 0.00536　　0.00563

45. $\frac{2}{3}$　$\frac{5}{6}$

46. $\frac{1}{9}$　$\frac{2}{17}$

47. $\frac{5}{9}$　$\frac{51}{91}$

48. $\frac{7}{12}$　$\frac{6}{11}$

49. $\frac{4}{7}$　0.14

50. $\frac{5}{9}$　0.557

 51. 1.38　$\frac{18}{13}$

52. 0.372　$\frac{22}{59}$

53. 7.123　$\frac{456}{64}$

54. 12.713　$\frac{89}{7}$

Write the numbers in order from smallest to largest. See Example 6.

55. 0.34, 0.35, 0.32

56. 0.47, 0.42, 0.40

57. 0.49, 0.491, 0.498

58. 0.72, 0.727, 0.728

59. $\frac{3}{4}$, 0.78, 0.73

60. $\frac{2}{5}$, 0.49, 0.42

61. $\frac{4}{7}$, 0.453, 0.412

62. $\frac{6}{9}$, 0.663, 0.668

63. 5.23, $\frac{42}{8}$, 5.34

64. 7.56, $\frac{67}{9}$, 7.562

65. $\frac{12}{5}$, 2.37, $\frac{17}{8}$

66. $\frac{29}{16}$, 1.75, $\frac{58}{32}$

C *Find the area of each triangle or rectangle. See Example 7.*

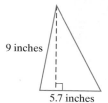 **67.**

9 inches

5.7 inches

△ **68.**

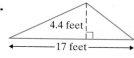

4.4 feet

17 feet

△ **69.**

3.6 centimeters

← 5.2 → centimeters

△ **70.**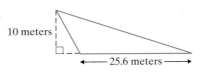

10 meters

← 25.6 meters →

△ **71.**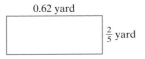

0.62 yard

$\frac{2}{5}$ yard

△ **72.**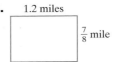

1.2 miles

$\frac{7}{8}$ mile

Review and Preview

Simplify. See Sections 1.9 and 3.5.

73. 2^3

74. 5^4

75. $6^2 \cdot 2$

76. $4 \cdot 3^4$

77. $\left(\frac{1}{3}\right)^4$

78. $\left(\frac{4}{5}\right)^3$

79. $\left(\frac{3}{5}\right)^2$

80. $\left(\frac{7}{2}\right)^2$

81. $\left(\frac{2}{5}\right)\left(\frac{5}{2}\right)^2$

82. $\left(\frac{2}{3}\right)^2\left(\frac{3}{2}\right)^3$

 Combining Concepts

In 2000, there were 10,716 commercial radio stations in the United States. The most popular formats are listed in the table along with their counts for 2000. Use the table to answer Exercises 83 through 86.

83. Write as a decimal the fraction of radio stations with a country music format. Round to the nearest thousandth.

84. Write as a decimal the fraction of radio stations with a talk format. Round to the nearest hundredth.

85. Round each number in the table to the nearest hundredth. Use the rounded numbers to estimate the total number of stations with the top five formats in 2000.

Top Commercial Radio Station Formats in 2000	
Format	**Number of Stations**
Country	2249
Adult contemporary	1557
News/talk/business/sports	1426
Oldies/classic hits	1135
Religion	1118

(*Source:* M Street Corporation)

86. Use your estimate from Exercise 85 to write as a decimal the fraction of radio stations accounted for by the top five formats. Round to the nearest thousandth.

87. Suppose that you are writing a paper on a word processor. You have been instructed to use $\frac{5}{8}$-inch margins. To change margins in the word processing software, you must enter margin widths in inches as a decimal. What number should you enter?

88. Suppose that you are using a word processor to format a table for a presentation. The first column of the table must be $3\frac{7}{16}$-inches wide. To adjust the width of the column in the word processing software, you must enter the column width in inches as a decimal. What number should you enter?

89. Describe how to determine the larger of two fractions.

281

This activity may be completed by working in groups or individually.

A checking account is a convenient way of handling money and paying bills. To open a checking account, the bank or savings and loan association requires a customer to make a deposit. Then the customer receives a checkbook that contains checks, deposit slips, and a register for recording checks written and deposits made. It is important to record all payments and deposits that affect the account. It is also important to keep the checkbook balance current by subtracting checks written and adding deposits made.

About once a month checking customers receive a statement from the bank listing all activity that the account has had in the last month. The statement lists a beginning balance, all checks and deposits, any service charges made against the account, and an ending balance. Because it may take several days for checks that a customer has written to clear the banking system, the check register may list checks that do not appear on the monthly bank statement. These checks are called **outstanding checks**. Deposits that are recorded in the check register but do not appear on the statement are called **deposits in transit**. Because of these differences, it is important to balance, or reconcile, the checkbook against the monthly statement. The steps for doing so are listed at the right.

BALANCING OR RECONCILING A CHECKBOOK

STEP 1. Place a check mark in the checkbook register next to each check and deposit listed on the monthly bank statement. Any entries in the register without a check mark are outstanding checks or deposits in transit.

STEP 2. Find the ending checkbook register balance and add to it any outstanding checks and any interest paid on the account.

STEP 3. From the total in Step 2, subtract any deposits in transit and any service charges.

STEP 4. Compare the amount found in Step 3 with the ending balance listed on the bank statement. If they are the same, the checkbook balances with the bank statement. Be sure to update the check register with service charges and interest.

STEP 5. If the checkbook does not balance, recheck the balancing process. Next, make sure that the running checkbook register balance was calculated correctly. Finally, compare the checkbook register with the statement to make sure that each check was recorded for the correct amount.

For the checkbook register and monthly bank statement given:

a. *update the checkbook register*   **b.** *list the outstanding checks and deposits in transit*
c. *balance the checkbook—be sure to update the register with any interest or service fees*

Checkbook Register						Balance
#	**Date**	**Description**	**Payment**	**✔**	**Deposit**	**425.86**
114	4/1	Market Basket	30.27			
115	4/3	May's Texaco	8.50			
	4/4	Cash at ATM	50.00			
116	4/6	UNO Bookstore	121.38			
	4/7	Deposit			100.00	
117	4/9	MasterCard	84.16			
118	4/10	Blockbuster	6.12			
119	4/12	Kroger	18.72			
120	4/14	Parking sticker	18.50			
	4/15	Direct deposit			294.36	
121	4/20	Rent	395.00			
122	4/25	Student fees	20.00			
	4/28	Deposit			75.00	

FIRST NATIONAL BANK
Monthly Statement 4/30

BEGINNING BALANCE:		425.86
Date	Number	Amount
CHECKS AND ATM		
WITHDRAWALS		
4/3	114	30.27
4/4	ATM	50.00
4/11	117	84.16
4/13	115	8.50
4/15	119	18.72
4/22	121	395.00
DEPOSITS		
4/7		100.00
4/15	Direct deposit	294.36
SERVICE CHARGES		
Low balance fee		7.50
INTEREST		
Credited 4/30 1.15		
ENDING BALANCE:		227.22

Chapter 4 VOCABULARY CHECK

Fill in each blank with one of the words listed below.

vertically   decimal   and   sum
denominator   numerator

1. Like fractional notation, _____ notation is used to denote a part of a whole.

2. To write fractions as decimals, divide the _____ by the _____ .

3. To add or subtract decimals, write the decimals so that the decimal points line up _____ .

4. When writing decimals in words, write "____" for the decimal point.

5. When multiplying decimals, the decimal point in the product is placed so that the number of decimal places in the product is equal to the ____ of the number of decimal places in the factors.

CHAPTER 4 Highlights

DEFINITIONS AND CONCEPTS	EXAMPLES
Section 4.1 Introduction to Decimals	

PLACE-VALUE CHART

hundreds	tens	4 ones	↑ decimal point	2 tenths	6 hundredths	5 thousandths	ten-thousandths	hundred-thousandths
100	10	1		$\frac{1}{10}$	$\frac{1}{100}$	$\frac{1}{1000}$	$\frac{1}{10,000}$	$\frac{1}{100,000}$

4.265 means

$$4 \cdot 1 + 2 \cdot \frac{1}{10} + 6 \cdot \frac{1}{100} + 5 \cdot \frac{1}{1000}$$

or

$$4 + \frac{2}{10} + \frac{6}{100} + \frac{5}{1000}$$

WRITING (OR READING) A DECIMAL IN WORDS

Step 1. Write the whole number part in words.

Step 2. Write "and" for the decimal point.

Step 3. Write the decimal part in words as though it were a whole number, followed by the place value of the last digit.

A decimal written in words can be written in standard form by reversing the above procedure.

Write 3.08 in words.
Three and eight hundredths

Write "four and twenty-one thousandths" in standard form.

4.021

DEFINITIONS AND CONCEPTS	EXAMPLES

Section 4.2   Order and Rounding

To compare decimals, compare digits in the same place from left to right. When two digits are not equal, the number with the larger digit is the larger decimal.

$3.0261 < 3.0186$ because

$$\underset{2}{\uparrow} \quad < \quad \underset{1}{\uparrow}$$

TO ROUND DECIMALS TO A PLACE VALUE TO THE RIGHT OF THE DECIMAL POINT

Step 1. Locate the digit to the right of the given place value.

Step 2. If this digit is 5 or greater, add 1 to the digit in the given place value and drop all digits to its right.

Step 3. If this digit is less than 5, drop all digits to the right of the given place value.

Round 86.1256 to the nearest hundredth.

$$86.12 \underset{\text{5 or greater}}{\overset{\text{hundredths place}}{\textcircled{5}}} 6$$

rounds to 86.13

Section 4.3   Adding and Subtracting Decimals

TO ADD OR SUBTRACT DECIMALS

Step 1. Write the decimals so that the decimal points line up vertically.

Step 2. Add or subtract as with whole numbers.

Step 3. Place the decimal point in the sum or difference so that it lines up vertically with the decimal points in the problem.

Add: 4.6 + 0.28

$$\begin{array}{r} 4.60 \\ + 0.28 \\ \hline 4.88 \end{array}$$

Subtract: 2.8 − 1.04

$$\begin{array}{r} 2.\overset{7\ 10}{8\cancel{0}} \\ - 1.04 \\ \hline 1.76 \end{array}$$

Section 4.4   Multiplying Decimals

TO MULTIPLY DECIMALS

Step 1. Multiply the decimals as though they are whole numbers.

Step 2. The decimal point in the product is placed so that the number of decimal places in the product is equal to the **sum** of the number of decimal places in the factors.

△ The circumference of a circle is the distance around the circle.

$C = \pi \cdot$ diameter or

$C = 2 \cdot \pi \cdot$ radius,

where $\pi \approx 3.14$ or $\dfrac{22}{7}$.

Multiply: 1.48 × 5.9

$$\begin{array}{r} 1.4\,8 \quad \leftarrow \text{2 decimal places} \\ \times \ 5.9 \quad \leftarrow \text{1 decimal place} \\ \hline 1\,3\,3\,2 \\ 7\,4\,0\,0 \\ \hline 8.7\,3\,2 \quad \leftarrow \text{3 decimal places} \end{array}$$

△ Find the exact circumference and an approximation by using 3.14 for π.

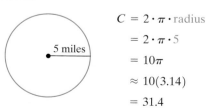

$C = 2 \cdot \pi \cdot$ radius

$\quad = 2 \cdot \pi \cdot 5$

$\quad = 10\pi$

$\quad \approx 10(3.14)$

$\quad = 31.4$

The circumference is exactly 10π miles and approximately 31.4 miles.

DEFINITIONS AND CONCEPTS	EXAMPLES

Section 4.5 Dividing Decimals

TO DIVIDE DECIMALS

Step 1. Move the decimal point in the divisor to the right until the divisor is a whole number.

Step 2. Move the decimal point in the dividend to the right the **same number of places** as the decimal point was moved in step 1.

Step 3. Divide. The decimal point in the quotient is directly over the moved decimal point in the dividend.

Divide: $1.118 \div 2.6$

$$
\begin{array}{r}
0.43 \\
2.6\overline{)1.118} \\
-1\ 04 \\
\hline
78 \\
-78 \\
\hline
0
\end{array}
$$

Section 4.6 Estimating and Order of Operations

ORDER OF OPERATIONS

1. Do all operations within grouping symbols such as parentheses or brackets.
2. Evaluate any expressions with exponents.
3. Multiply or divide in order from left to right.
4. Add or subtract in order from left to right.

Simplify.

$$
\begin{aligned}
1.9(12.8 - 4.1) &= 1.9(8.7) \quad \text{Subtract.} \\
&= 16.53 \quad \text{Multiply.}
\end{aligned}
$$

Section 4.7 Fractions and Decimals

To write fractions as decimals, divide the numerator by the denominator.

Write $\frac{3}{8}$ as a decimal.

$$
\begin{array}{r}
0.375 \\
8\overline{)3.000} \\
-2\ 4 \\
\hline
60 \\
-56 \\
\hline
40 \\
-40 \\
\hline
0
\end{array}
$$

FOCUS ON **History**

THE GOLDEN RECTANGLE IN ART

The golden rectangle is a rectangle whose length is approximately 1.6 times its width. The early Greeks thought that a rectangle with these dimensions was the most pleasing to the eye. Examples of the golden rectangle are found in many ancient, as well as modern, works of art. For example, the Parthenon in Athens, Greece, shows the golden rectangle in many aspects of its design. Modern-era artists, including Piet Mondrian (1872–1944) and Georges Seurat (1859–1891), also frequently used the proportions of a golden rectangle in their paintings.

To test whether a rectangle is a golden rectangle, divide the rectangle's length by its width. If the result is approximately 1.6, we can consider the rectangle to be a golden rectangle. For instance, consider Mondrian's *Composition with Gray and Light Brown*, which was painted on an 80.2 × 49.9 cm canvas.

Because $\frac{80.2}{49.9} \approx 1.6$, the dimensions of the canvas form a golden rectangle. In what other ways are golden rectangles connected with this painting?

Examples of golden rectangles can be found in the designs of many everyday objects. Visual artists, from architects to product and package designers, use the golden rectangle shape in such things as the face of a building, the floor of a room, the front of a food package, the front cover of a book, and even the shape of a credit card.

GROUP ACTIVITY

Find an example of a golden rectangle in a building or an everyday object. Use a ruler to measure its dimensions and verify that the length is approximately 1.6 times the width.

Chapter 4 Review

(4.1) *Determine the place value of the digit 4 in each decimal.*

1. 23.45

2. 0.000345

Write each decimal in words.

3. 23.45

4. 0.00345

5. 109.23

6. 200.000032

Write each decimal in standard form.

7. Two and fifteen hundredths

8. Five hundred three and one hundred two thousandths

9. Sixteen thousand twenty-five and fourteen ten-thousandths

Write the decimal as a fraction or a mixed number.

10. 0.16

11. 12.023

12. 1.0045

13. 0.00231

14. 25.25

Write each fraction as a decimal.

15. $\dfrac{9}{10}$

16. $\dfrac{25}{100}$

17. $\dfrac{45}{1000}$

18. $\dfrac{7}{100}$

(4.2) *Insert $<$, $>$, or $=$ to make a true statement.*

19. 0.49   0.43

20. 0.973   0.9730

21. 402.00032   402.000032

22. 0.230505   0.23505

Round each decimal to the given place value.

23. 0.623, nearest tenth

24. 0.9384, nearest hundredth

25. 42.895, nearest hundredth

26. 16.34925, nearest thousandth

Round each money amount to the nearest cent or dollar, as indicated.

27. $0.259, nearest cent

28. $12.461, nearest cent

29. $123.46, nearest dollar

30. $3,645.52, nearest dollar

31. Every day in America an average of 13,490.5 people get married. Round this number to the nearest hundred.

32. A certain kind of chocolate candy bar contains 10.75 teaspoons of sugar. Convert this number to a mixed number.

(4.3) *Add or subtract as indicated.*

33. $2.4 + 7.1$

34. $3.9 + 1.2$

35. $4.9 - 3.2$

36. $5.23 - 2.74$

37. $6.4 + 0.88$

38. $19.02 + 6.98$

39. $200.49 + 16.82 + 103.002$

40. $0.00236 + 100.45 + 48.29$

Subtract.

41. $892.1 - 432.4$

42. $100.342 - 0.064$

43. Subtract 34.98 from 100.

44. Subtract 10.02 from 200.

45. The price of oil was $17.02 per barrel on October 23. It was $17.46 on October 24. Find by how much the price of oil increased from the 23rd to the 24th.

△ **46.** Find the perimeter.

6.2 inches

Rectangle   4.9 inches

(4.4) *Multiply.*

47. 7.2 × 10

48. 9.345 × 1000

49. $\begin{array}{r} 34.02 \\ \times 2.3 \\ \hline \end{array}$

50. $\begin{array}{r} 839.02 \\ \times 87.3 \\ \hline \end{array}$

△ **51.** Find the exact circumference of the circle. Then use the approximation 3.14 for π and approximate the circumference.

7 meters

52. A kilometer is approximately 0.625 mile. It is 102 kilometers from Hays to Colby. Write 102 kilometers in miles to the nearest tenth of a mile.

(4.5) *Divide. Round the quotient to the nearest thousandth if necessary.*

53. 21 ÷ 0.3

54. 0.0063 ÷ 0.03

55. 0.005$\overline{)24.5}$

56. 2.3$\overline{)54.98}$

57. 0.34$\overline{)2.74}$

58. 20$\overline{)316.5}$

59. $\dfrac{2.67}{100}$

60. $\dfrac{93}{10}$

61. There are approximately 3.28 feet in 1 meter. Find how many meters are in 24 feet to the nearest tenth of a meter.

←——1 meter——→
←—— ≈3.28 feet ——→

62. George Strait pays $69.71 per month to pay back a loan of $3136.95. In how many months will the loan be paid off?

(4.6) *Perform each indicated operation. Then estimate to see whether each proposed result is reasonable.*

63. 2.4 + 6.7 + 9.1

64. 15.9 + 34.1

65. 340.03 − 240.98

66. 100 − 45.9

67. 6.02 × 5.91

68. 0.205 × 1.72

69. 62.13 ÷ 1.9

70. 601.92 ÷ 19.8

△ 71. Tomaso is going to fertilize his lawn, a rectangle that measures 77.3 feet by 115.9 feet. Approximate the area of the lawn by rounding each measurement to the nearest foot.

77.3 feet

115.9 feet

72. Estimate the cost of the items to see whether the groceries can be purchased with a $5 bill.

$1.89

BREAD

$1.07   3 cans for $0.99

Simplify each expression.

73. $7.6 \times 1.9 + 2.5$

74. $2.3^2 - 1.4$

75. $\dfrac{(3.2)^2}{100}$

76. $(2.6 + 1.4)(4.5 - 3.6)$

(4.7) *Write each fraction as a decimal. Round to the nearest thousandth if necessary.*

77. $\dfrac{4}{5}$

78. $\dfrac{12}{13}$

79. $\dfrac{3}{7}$

80. $\dfrac{13}{60}$

81. $\dfrac{9}{80}$

82. $\dfrac{8935}{175}$

Insert $<$, $>$, or $=$ to make a true statement.

83. $0.392 \quad 0.392$

84. $\dfrac{4}{7} \quad \dfrac{5}{8}$

85. $0.293 \quad \dfrac{5}{17}$

86. $\dfrac{6}{11} \quad 0.55$

Write the numbers in order from smallest to largest.

87. $0.837, 0.839, 0.832$

88. $\dfrac{3}{7}, 0.42, 0.43$

89. $\dfrac{18}{11}, 1.63, \dfrac{19}{12}$

90. $\dfrac{6}{7}, \dfrac{8}{9}, \dfrac{3}{4}$

Find each area.

△ 91.

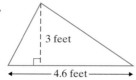

3 feet

4.6 feet

△ 92.

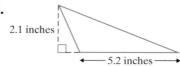

2.1 inches

5.2 inches

Chapter 4 Test

Write the decimal as indicated.

1. 45.092, in words

2. Three thousand and fifty-nine thousandths, in standard form

Round the decimal to the indicated place value.

3. 34.8923, nearest tenth

4. 0.8623, nearest thousandth

Insert <, >, or = to make a true statement.

5. 25.0909   25.9090

6. $\dfrac{4}{9}$ 0.445

Write the decimal as a fraction or a mixed number.

7. 0.345

8. 24.73

Write the fraction as a decimal. If necessary, round to the nearest thousandth.

9. $\dfrac{13}{26}$

10. $\dfrac{16}{17}$

Perform the indicated operations. Round the result to the nearest thousandth if necessary.

11. $2.893 + 4.2 + 10.49$

12. Subtract 8.6 from 20.

13. $\begin{array}{r} 10.2 \\ \times\ \ 4.3 \\ \hline \end{array}$

14. $0.23\overline{)12.88}$

15. $7\overline{)46.71}$

16. 126.9×100

17. $\dfrac{473}{10}$

18. $1.57 - (0.6)^2$

19. $\dfrac{0.23 + 1.63}{0.3}$

1. _____

2. _____

3. _____

4. _____

5. _____

6. _____

7. _____

8. _____

9. _____

10. _____

11. _____

12. _____

13. _____

14. _____

15. _____

16. _____

17. _____

18. _____

19. _____

Find the area

△ **20.**

1.1 miles
4.2 miles

△ **21.** Vivian Thomas is going to put insecticide on her lawn to control grubworms. The lawn is a rectangle that measures 123.8 feet by 80 feet. The amount of insecticide required is 0.02 ounces per square foot. Find how much insecticide Vivian needs to purchase.

△ **22.** Find the exact circumference of the circle. Then use the approximation 3.14 for π and approximate the circumference.

9 miles

23. A CD (compact disc) holds 700 megabytes of data. A DVD (digital versatile disc) holds 8740 megabytes of data. How many CDs does it take to hold as much data as a DVD? Round to the nearest whole CD.

24. Estimate the total distance from Bayette to Center City by rounding each distance to the nearest mile.

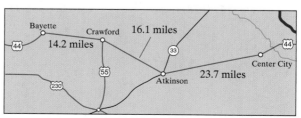

20. _____

21. _____

22. _____

23. _____

24. _____

Cumulative Review

1. Write 106,052,447 in words.

2. Alan Mayfield collects baseball cards. He has 109 cards for the New York Yankees, 96 for the Chicago White Sox, 79 for the Kansas City Royals, 42 for the Seattle Mariners, 67 for the Oakland Athletics, and 52 for the California Angels. How many cards does he have in total?

3. Subtract $900 - 174$. Then check by adding.

4. Round each number to the nearest hundred to find an estimated sum.
$$294$$
$$625$$
$$1071$$
$$+ \ \ 349$$

5. A single-density computer disk can hold 370 thousand bytes of information. How many total bytes can 42 such disks hold?

6. Divide: $6819 \div 17$

7. Simplify:
$$4^3 + [3^2 - (10 \div 2)] - 7 \cdot 3$$

8. Identify the numerator and the denominator: $\dfrac{3}{7}$

9. Write $\dfrac{6}{60}$ in simplest form.

10. Multiply: $\dfrac{3}{4} \cdot 20$

11. Divide: $\dfrac{7}{8} \div \dfrac{2}{9}$

12. Multiply: $1\dfrac{2}{3} \cdot 2\dfrac{1}{4}$

13. Divide: $\dfrac{3}{4} \div 5$

Subtract and simplify.

14. $\dfrac{8}{9} - \dfrac{1}{9}$

15. $\dfrac{7}{8} - \dfrac{5}{8}$

1. _____
2. _____
3. _____
4. _____
5. _____
6. _____
7. _____
8. _____
9. _____
10. _____
11. _____
12. _____
13. _____
14. _____
15. _____

16.

17.

18.

19.

20.

21.

22.

23.

24.

25.

16. Write an equivalent fraction with the indicated denominator: $\dfrac{3}{4} = \dfrac{}{20}$

17. Add: $\dfrac{2}{15} + \dfrac{3}{10}$

18. Sarah Grahamm purchases two packages of ground round. One package weighs $2\dfrac{3}{8}$ pounds and the other $1\dfrac{4}{5}$ pounds. What is the combined weight of the ground round?

Evaluate each expression.

19. $\left(\dfrac{1}{4}\right)^2$

20. $\left(\dfrac{1}{6}\right)^2 \cdot \left(\dfrac{3}{4}\right)^3$

21. Write 0.43 as a fraction.

22. Insert $<$, $>$, or $=$ to form a true statement.

0.378   0.368

23. Subtract: $35.218 - 23.65$

Multiply.

24. 23.702×100

25. $76,805 \times 0.01$

Ratio and Proportion

Having studied fractions in Chapters 2 and 3, we are ready to explore using fractions to express a relationship between two quantities. We often make comparisons between like quantities in real life. When we say there is a 2-to-1 ratio of men to women at a party, we mean that for every two men, there is one woman. When we find gas mileage as miles per gallon or compare two unlike quantities, we are finding a rate. We can apply both of these concepts when finding a proportion. A proportion compares two equal rates or ratios. As soon as we know how far a car can go on 1 gallon of gasoline (a rate), we can use a proportion to compute the number of gallons of gasoline needed for a 200-mile trip.

Lawn care is a more-than-$25-billion-per-year business. About one in five households in the United States hire professional services to care for and landscape their lawns. The sales of lawn care products alone account for roughly one-third of all money spent on gardening in this country. Why is there so much interest in maintaining lawns? Lawns have many benefits. They are aesthetically pleasing. They provide a cushioned play area for children. They can increase the selling price of a home. However, one of the most important benefits of lawns is their environmental impact. Turf grasses help clean the air with their oxygen production and also help control pollution. In Exercise 27 on page 306 and Exercise 27 on page 322, we will use rates and proportions to compute the effect lawns have on trapping dust and on oxygen production.

Name _____ Section _____ Date _____

Chapter 5 Pretest

1. _____

2. _____

3. _____

4. _____

5. _____

6. _____

7. _____

8. _____

9. _____

10. _____

11. _____

12. _____

13. _____

14. _____

15. _____

16. _____

17. _____

18. _____

19. _____

20. _____

Write each ratio as a fraction in simplest form.

1. 27 to 8   **2.** 4 to 37   **3.** $2\frac{1}{2}$ to 6

4. 8.4 to 6.3   **5.** 12 pints to 38 pints   **6.** 200 years to 550 years

Write each rate as a fraction in simplest form.

7. 140 students for 8 instructors   **8.** 21 cups of flour for 6 cakes

Write each phrase as a unit rate.

9. 945 kilometers in 12 hours   **10.** 780 cookies for 260 children

Compare the unit prices and decide which is the better buy.

11. Coffee:
8 ounces for $3.69
12 ounces for $5.25

12. Doughnuts:
12 doughnuts for $4.20
30 doughnuts for $10.80

13. Write the following sentence as a proportion: 5 incorrect answers is to 80 math problems as 15 incorrect answers is to 240 math problems.

Determine whether the proportions are true or false.

14. $\frac{8}{22} = \frac{24}{72}$ **15.** $\frac{4.8}{6.4} = \frac{3}{4}$

Find the unknown number n.

16. $\frac{n}{4} = \frac{28}{7}$ **17.** $\frac{3}{74} = \frac{n}{111}$ **18.** $\frac{9\frac{1}{2}}{n} = \frac{\frac{3}{7}}{6}$

Personal property taxes on a vehicle are figured at a rate of $3.05 tax for every $100 of assessed vehicle value.

19. If Tony pays $366 in personal property taxes for his car, find the assessed value of his car.

20. Find the personal property taxes on a truck with an assessed value of $15,500.

5.1 Ratios

A Writing Ratios as Fractions

A **ratio** is the quotient of two quantities. A ratio, in fact, is no different from a fraction, except that a ratio is sometimes written using notation other than fractional notation. For example, the ratio of 1 to 2 can be written as

$$1 \text{ to } 2 \quad \text{or} \quad \frac{1}{2} \quad \text{or} \quad 1:2$$

fractional notation   colon notation

These ratios are all read as, "the ratio of 1 to 2."

Try the Concept Check in the margin.

In this section, we write ratios using fractional notation.

> ### Writing a Ratio as a Fraction
>
> The order of the quantities is important when writing ratios. To write a ratio as a fraction, write the *first number* of the ratio as the *numerator* of the fraction and the *second number* as the *denominator*.

For example, the ratio of 6 to 11 is $\dfrac{6}{11}$, *not* $\dfrac{11}{6}$.

EXAMPLE 1   Write the ratio of 12 to 17 using fractional notation.

Solution: The ratio is $\dfrac{12}{17}$.

Helpful Hint

Don't forget that order is important when writing ratios. The ratio $\dfrac{17}{12}$ is *not* the same as the ratio $\dfrac{12}{17}$.

EXAMPLES   Write each ratio using fractional notation.

2. The ratio of 2.6 to 3.1 is $\dfrac{2.6}{3.1}$.

3. The ratio of $1\dfrac{1}{2}$ to $7\dfrac{3}{4}$ is $\dfrac{1\frac{1}{2}}{7\frac{3}{4}}$.

B Writing Ratios in Simplest Form

To simplify a ratio, we just write the fraction in simplest form. Common factors as well as common units can be divided out.

OBJECTIVES

A Write ratios as fractions.

B Write ratios in simplest form.

SSM   TUTOR CENTER   SG   CD & VIDEO   MATH PRO   WEB

Concept Check

How should each ratio be read aloud?

a. $\dfrac{8}{5}$ b. $\dfrac{5}{8}$

Practice Problem 1

Write the ratio of 20 to 23 using fractional notation.

Practice Problems 2–3

Write each ratio using fractional notation.

2. The ratio of 10.3 to 15.1

3. The ratio of $3\dfrac{1}{3}$ to $12\dfrac{1}{5}$

Answers

1. $\dfrac{20}{23}$, **2.** $\dfrac{10.3}{15.1}$, **3.** $\dfrac{3\frac{1}{3}}{12\frac{1}{5}}$

Concept Check: a. "eight to five",   b. "five to eight"

Practice Problem 4

Write the ratio of $8 to $6 as a fraction in simplest form.

Practice Problem 5

Write the ratio of 1.71 to 4.56 as a fraction in simplest form.

Practice Problem 6

Use the circle graph below to write the ratio of work miles to total miles as a fraction in simplest form.

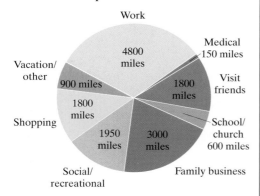

Work

4800 miles

Medical 150 miles

Vacation/ other

900 miles

1800 miles

Visit friends

1800 miles

Shopping

1950 miles

3000 miles

School/ church 600 miles

Social/ recreational

Family business

Total yearly mileage: 15,000

Sources: The American Automobile Manufacturers Association and The National Automobile Dealers Association.

Practice Problem 7 △

Given the triangle shown:

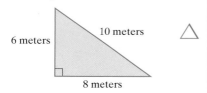

6 meters

10 meters

8 meters

a. Find the ratio of the length of the shortest side to the length of the longest side.
b. Find the ratio of the length of the longest side to the perimeter of the triangle.

Concept Check

Explain why the answer $\dfrac{7}{5}$ would be incorrect for part (a) of Example 7.

Answers

4. $\dfrac{4}{3}$, 5. $\dfrac{3}{8}$, 6. $\dfrac{8}{25}$, 7. a. $\dfrac{3}{5}$, b. $\dfrac{5}{12}$

Concept Check: $\dfrac{7}{5}$ would be the ratio of the rectangle's length to its width.

EXAMPLE 4 Write the ratio of $15 to $10 as a fraction in simplest form.

Solution:

$$\frac{\$15}{\$10} = \frac{15}{10} = \frac{3 \cdot \cancel{5}^{1}}{2 \cdot \cancel{5}_{1}} = \frac{3}{2}$$ ●

In the example above, although the fraction $\dfrac{3}{2}$ is equal to the mixed number $1\dfrac{1}{2}$, a ratio is a quotient of *two* quantities. For that reason, ratios are not written as mixed numbers.

If a ratio compares two decimal numbers, we write the ratio as a ratio of whole numbers.

EXAMPLE 5 Write the ratio of 2.6 to 3.1 as a fraction in simplest form.

Solution: The ratio is

$$\frac{2.6}{3.1}$$

Now let's clear the ratio of decimals.

$$\frac{2.6}{3.1} = \frac{2.6 \cdot 10}{3.1 \cdot 10} = \frac{26}{31} \quad \text{Simplest form}$$ ●

EXAMPLE 6 Writing a Ratio from a Circle Graph

The circle graph in the margin shows the part of a car's total mileage that falls into a particular category. Write the ratio of medical miles to total miles as a fraction in simplest form.

Solution:

$$\frac{\text{medical miles}}{\text{total miles}} = \frac{150 \text{ miles}}{15,000 \text{ miles}} = \frac{150}{15,000} = \frac{\cancel{150}^{1}}{\cancel{150}_{1} \cdot 100} = \frac{1}{100}$$ ●

EXAMPLE 7 Given the rectangle shown:

a. Find the ratio of its width to its length.
b. Find the ratio of its length to its perimeter.

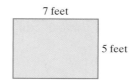

7 feet

5 feet

Solution:

a. The ratio of its width to its length is

$$\frac{\text{width}}{\text{length}} = \frac{5 \,\cancel{\text{feet}}}{7 \,\cancel{\text{feet}}} = \frac{5}{7}$$

b. Recall that the perimeter of the rectangle is the distance around the rectangle: $7 + 5 + 7 + 5 = 24$ feet. The ratio of its length to its perimeter is

$$\frac{\text{length}}{\text{perimeter}} = \frac{7 \,\cancel{\text{feet}}}{24 \,\cancel{\text{feet}}} = \frac{7}{24}$$ ●

Try the Concept Check in the margin.

Name _____ Section _____ Date _____

EXERCISE SET 5.1

A *Write each ratio using fractional notation. Do not simplify. See Examples 1 through 3.*

1. 11 to 14   **2.** 7 to 12   **3.** 23 to 10   **4.** 8 to 5   **5.** 151 to 201   **6.** 673 to 1000

7. 2.8 to 7.6   **8.** 3.9 to 4.2   **9.** 5 to $7\frac{1}{2}$ **10.** $5\frac{3}{4}$ to 3   **11.** $3\frac{3}{4}$ to $1\frac{2}{3}$ **12.** $2\frac{2}{5}$ to $6\frac{1}{2}$

B *Write each ratio as a ratio of whole numbers using fractional notation. Write the fraction in simplest form. See Examples 4 and 5.*

 13. 16 to 24   **14.** 25 to 150   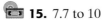 **15.** 7.7 to 10   **16.** 8.1 to 10

17. 4.63 to 8.21   **18.** 9.61 to 7.62   **19.** 9 inches to 12 inches   **20.** 14 centimeters to 20 centimeters

21. 10 hours to 24 hours   **22.** 18 quarts to 30 quarts   **23.** $32 to $100   **24.** $46 to $102

25. 24 days to 14 days   **26.** 80 miles to 120 miles   **27.** 32,000 bytes to 46,000 bytes   **28.** 600 copies to 150 copies

29. 8 inches to 20 inches   **30.** 9 yards to 2 yards

Find the ratio described in each problem. See Example 7.

△ **31.** Find the ratio of the length to the width of the swimming pool.

30 feet

18 feet

△ **32.** Find the ratio of the base to the height of the triangular mainsail.

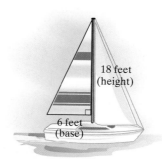

18 feet (height)

6 feet (base)

△ **33.** Find the ratio of the longest side to the perimeter of the right-triangular-shaped billboard.

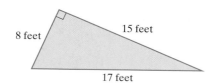

8 feet

15 feet

17 feet

△ **34.** Find the ratio of the width to the perimeter of the rectangular vegetable garden.

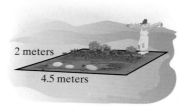

2 meters

4.5 meters

At the Honey Island College Glee Club meeting one night, there were 125 women and 100 men present.

35. Find the ratio of women to men.

36. Find the ratio of men to the total number of people present.

A poll at State University revealed that 4500 students out of 6000 students are single, and the rest are married.

37. Find the ratio of single students to married students.

38. Find the ratio of married students to the total student population.

Blood contains three types of cells: red blood cells, white blood cells, and platelets. For approximately every 600 red blood cells in healthy humans, there are 40 platelets and 1 white blood cell. (Source: American Red Cross Biomedical Services)

39. Write the ratio of red blood cells to platelet cells.

40. Write the ratio of white blood cells to red blood cells.

41. Of the U.S. mountains that are over 14,000 feet in elevation, 57 are located in Colorado and 19 are located in Alaska. Find the ratio of the number of mountains over 14,000 feet found in Alaska to the number of mountains over 14,000 feet found in Colorado. (*Source:* U.S. Geological Survey)

42. Citizens of the United States eat an average of 25 pints of ice cream per year. Residents of the New England states eat an average of 39 pints of ice cream per year. Find the ratio of the amount of ice cream eaten by New Englanders to the amount eaten by the average U.S. citizen. (*Source:* International Dairy Foods Association)

43. At the Summer Olympics in Sydney, Australia, U.S. athletes won a total of 40 gold medals, and Russian Federation athletes won a total of 32 gold medals. Find the ratio of gold medals won by the United States to the gold medals won by the Russian Federation. (*Source:* International Olympic Committee)

44. For the 2001 Boston Marathon, 8594 males and 4814 females finished the race. Find the ratio of female finishers to male finishers. (*Source:* Boston Athletic Association)

Review and Preview

Divide. See Section 4.5.

45. $9\overline{)20.7}$

46. $7\overline{)60.2}$

47. $3.7\overline{)0.555}$

48. $4.6\overline{)1.15}$

 Combining Concepts

49. As of 2001, Target stores operate in 46 states. Find the ratio of states without Target stores to states with Target stores. (*Source:* Target Corporation)

50. A total of 17 states have 200 or more public libraries. Find the ratio of states with 200 or more public libraries to states with fewer than 200 public libraries. (*Source:* U.S. Department of Education)

51. Write the ratio $2\frac{1}{2}$ to $5\frac{3}{4}$ as a fraction in simplest form.

52. A panty hose manufacturing machine will be repaired if the ratio of defective panty hose to good panty hose is at least 1 to 20. A quality control engineer found 10 defective panty hose in a batch of 200. Determine whether the machine should be repaired.

53. A grocer will refuse a shipment of tomatoes if the ratio of bruised tomatoes to the total batch is at least 1 to 10. A sample is found to contain 3 bruised tomatoes and 33 good tomatoes. Determine whether the shipment should be refused.

54. Is the ratio $\frac{5}{7}$ the same as the ratio $\frac{7}{5}$? Explain your answer.

55. In 2001, 17 states have mandatory helmet laws. (*Source:* Bicycle Helmet Safety Institute)
   a. Find the ratio of states with mandatory helmet laws to total U.S. states.

   b. Find the ratio of states with mandatory helmet laws to states without mandatory helmet laws.

   c. Are your ratios for parts **a** and **b** the same? Explain why or why not.

Internet Excursions

Go To: http://www.prenhall.com/martin-gay_basic What's Related

A scale model of an object is one in which there is a particular ratio between each measurement of the model and the corresponding measurement of the object being modeled. For instance, model trains are scale models of actual trains. Model train enthusiasts know that model trains are available in several different scales. By going to the World Wide Web site listed above, you will be directed to the Lionel Glossary in the Lionel model trains Web site, or a related site, where you can look up information to help you answer the questions below.

56. The two most popular model train scales are the HO scale and the N scale. What ratio of measurements is used in HO scale? What ratio of measurements is used in N scale? Describe what each of these ratios means and give an example.

57. Other model train scales include the O scale and the S scale. What ratio of measurements is used in O scale? What ratio of measurements is used in S scale? Describe what each of these ratios means and give an example.

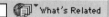

STUDY SKILLS REMINDER

How well do you know your textbook?

See if you can answer the questions below.

1. What does the 📷 icon mean?

2. What does the ✎ icon mean?

3. What does the △ icon mean?

4. Where can you find a review for each chapter? What answers to this review can be found in the back of your text?

5. Each chapter contains an overview of the chapter along with examples. What is this feature called?

6. Does this text contain any solutions to exercises? If so, where?

5.2 Rates

Ⓐ Writing Rates as Fractions

A special type of ratio is a rate. **Rates** are used to compare *different* kinds of quantities. For example, suppose that a recreational runner can run 3 miles in 33 minutes. If we write this rate as a fraction, we have

$$\frac{3 \text{ miles}}{33 \text{ minutes}} = \frac{1 \text{ mile}}{11 \text{ minutes}} \quad \text{In simplest form}$$

> **Helpful Hint**
>
> When comparing quantities with different units, write the units as part of the comparison. They do not divide out.
>
> **Same Units:** $\dfrac{3 \text{ \sout{inches}}}{12 \text{ \sout{inches}}} = \dfrac{1}{4}$
>
> **Different Units:** $\dfrac{2 \text{ miles}}{20 \text{ minutes}} = \dfrac{1 \text{ mile}}{10 \text{ minutes}}$ Units are still written.

EXAMPLE 1

Write the rate as a fraction in simplest form: 10 nails every 6 feet

Solution:

$$\frac{10 \text{ nails}}{6 \text{ feet}} = \frac{5 \text{ nails}}{3 \text{ feet}}$$

EXAMPLES   Write each rate as a fraction in simplest form.

2. $2160 for 12 weeks is $\dfrac{2160 \text{ dollars}}{12 \text{ weeks}} = \dfrac{180 \text{ dollars}}{1 \text{ week}}$

3. 360 miles on 16 gallons of gasoline is $\dfrac{360 \text{ miles}}{16 \text{ gallons}} = \dfrac{45 \text{ miles}}{2 \text{ gallons}}$

Try the Concept Check in the margin.

Ⓑ Finding Unit Rates

A **unit rate** is a rate with a denominator of 1. A familiar example of a unit rate is 55 mph, read as "55 **miles per hour**." This means 55 miles per 1 hour or

$$\frac{55 \text{ miles}}{1 \text{ hour}} \quad \text{Denominator of 1}$$

> **Writing a Rate as a Unit Rate**
>
> To write a rate as a unit rate, divide the numerator of the rate by the denominator.

Practice Problem 1

Write the rate as a fraction in simplest form: 12 commercials every 45 minutes

Practice Problems 2–3

Write each rate as a fraction in simplest form.

2. $1680 for 8 weeks

3. 236 miles on 12 gallons of gasoline

Concept Check

True or false? $\dfrac{16 \text{ gallons}}{4 \text{ gallons}}$ is a rate.

Explain.

Answers

1. $\dfrac{4 \text{ commercials}}{15 \text{ min}}$, **2.** $\dfrac{\$210}{1 \text{ wk}}$, **3.** $\dfrac{59 \text{ mi}}{3 \text{ gal}}$

Concept Check: false; a rate compares different kinds of quantities

Practice Problem 4

Write as a unit rate: 3600 feet every 12 seconds

EXAMPLE 4   Write as a unit rate: $27,000 every 6 months

Solution:

$$\frac{27,000 \text{ dollars}}{6 \text{ months}} \qquad 6\overline{)27,000} \; = 4,500$$

The unit rate is

$$\frac{4500 \text{ dollars}}{1 \text{ month}} \text{ or } 4500 \text{ dollars/month} \quad \text{Read as, "4500 dollars per month."} \quad \bullet$$

Practice Problem 5

Write as a unit rate: 52 bushels of fruit from 8 trees

EXAMPLE 5   Write as a unit rate: 318.5 miles every 13 gallons of gas

Solution:

$$\frac{318.5 \text{ miles}}{13 \text{ gallons}} \qquad 13\overline{)318.5} \; = 24.5$$

The unit rate is

$$\frac{24.5 \text{ miles}}{1 \text{ gallon}} \text{ or } 24.5 \text{ miles/gallon} \quad \text{Read as, "24.5 miles per gallon."} \quad \bullet$$

C Finding Unit Prices

Rates are used extensively in sports, business, medicine, and science. One of the most common uses of rates is in consumer economics. When a unit rate is "money per item," it is also called a **unit price**.

Practice Problem 6

An automobile rental agency charges $170 for 5 days for a certain model car. What is the unit price in dollars per day?

EXAMPLE 6   Finding Unit Price

A store charges $3.36 for a 16-ounce jar of picante sauce. What is the unit price in dollars per ounce?

Solution:

$$\frac{\text{unit}}{\text{price}} = \frac{\$3.36}{16 \text{ ounces}} = \frac{\$0.21}{1 \text{ ounce}} \text{ or } \$0.21 \text{ per ounce} \quad \bullet$$

Practice Problem 7

Approximate each unit price to decide which is the better buy for a bag of nacho chips: 11 ounces for $2.32 or 16 ounces for $3.59.

EXAMPLE 7   Finding the Best Buy

Approximate each unit price to decide which is the better buy: $0.99 for 4 bars of soap or $1.19 for 5 bars of soap.

Solution:

$$\frac{\text{unit}}{\text{price}} = \frac{\$0.99}{4 \text{ bars}} \approx \frac{\$0.25 \text{ per bar}}{\text{of soap}} \qquad 4\overline{)0.990} = 0.247 \approx 0.25 \quad \text{("is approximately")}$$

$$\frac{\text{unit}}{\text{price}} = \frac{\$1.19}{5 \text{ bars}} \approx \frac{\$0.24 \text{ per bar}}{\text{of soap}} \qquad 5\overline{)1.190} = 0.238 \approx 0.24$$

Thus, the 5-bar package is the better buy. $\quad \bullet$

Answers

4. $\dfrac{300 \text{ ft}}{1 \text{ sec}}$ or 300 ft/sec,

5. $\dfrac{6.5 \text{ bushels}}{1 \text{ tree}}$ or 6.5 bushels/tree,

6. $34 per day,   7. 11-oz bag

EXERCISE SET 5.2

A *Write each rate as a fraction in simplest form. See Examples 1 through 3.*

1. 5 shrubs every 15 feet

2. 14 lab tables for 28 students

3. 15 returns for 100 sales

4. 150 graduate students for 8 advisors

5. 8 phone lines for 36 employees

6. 6 laser printers for 28 computers

7. 18 gallons of pesticide for 4 acres of crops

8. 4 inches of rain in 18 hours

9. 6 flight attendants for 200 passengers

10. 240 pounds of grass seed for 9 lawns

11. 355 calories in a 10-fluid-ounce chocolate milkshake (*Source: Home and Garden Bulletin No. 72*, U.S. Department of Agriculture)

12. 160 calories in an 8-fluid-ounce serving of cream of tomato soup (*Source: Home and Garden Bulletin No. 72*, U.S. Department of Agriculture)

B *Write each rate as a unit rate. See Examples 4 and 5.*

13. 375 riders in 5 subway cars

14. 18 campaign yard signs in 6 blocks

15. 330 calories in a 3-ounce serving

16. 275 miles in 11 hours

17. 144 diapers for 24 babies

18. 420 feet in 3 seconds

19. $1,000,000 lottery paid over 20 years

20. 400,000 library books for 8000 students

21. 600 kilometers in 90 minutes

22. 5000 registered vehicles for 2000 campus parking spaces

23. The state of Arizona has approximately 114,000 square miles of land for 15 counties. (*Source:* U.S. Bureau of the Census)

24. The state of Louisiana has approximately 4,468,800 residents for 64 parishes. (*Note:* Louisiana is the only U.S. state with parishes instead of counties. *Source:* U.S. Bureau of the Census)

25. 12,000 good assembly line products to 40 defective products

26. 5,000,000 lottery tickets for 4 lottery winners

27. 12 million tons of dust and dirt are trapped by the 25 million acres of lawns in the United States each year. (*Source:* Professional Lawn Care Association of America)

28. Approximately 65,000,000,000 checks are written each year by a total of approximately 260,000,000 Americans. (*Source:* Board of Governors of the Federal Reserve System)

29. The National Zoo in Washington, D.C., has an annual budget of $29,000,000 for its 500 different species. (*Source:* Smithsonian Institution)

30. On average, it costs $36,120 for 5 hours of flight in a B747-100 aircraft. (*Source:* Air Transport Association of America)

31. On average, it costs $1,165,000 to build 25 Habitat for Humanity houses in the United States. (*Source:* Habitat for Humanity International)

32. The top-grossing concert tour in North America was the 1994 Rolling Stones tour, which grossed $121,200,000 for 60 shows. (*Source:* Pollstar)

33. Greer Krieger can assemble 250 computer boards in an 8-hour shift while Lamont Williams can assemble 400 computer boards in a 12-hour shift.
a. Find the unit rate of Greer.

b. Find the unit rate of Lamont.

c. Who can assemble computer boards faster, Greer or Lamont?

34. Jerry Stein laid 713 bricks in 46 minutes while his associate, Bobby Burns, laid 396 bricks in 30 minutes.
a. Find the unit rate of Jerry.

b. Find the unit rate of Bobby.

c. Who is the faster bricklayer?

C *Find each unit price. See Example 6.*

35. $57.50 for 5 compact discs

36. $0.87 for 3 apples

37. $1.19 for 7 bananas

38. $73.50 for 6 lawn chairs

Find each unit price and decide which is the better buy. Round to 3 decimal places. Assume that we are comparing different sizes of the same brand. See Examples 6 and 7.

39. Crackers:
$1.19 for 8 ounces
$1.59 for 12 ounces

40. Pickles:
$1.89 for 32 ounces
$0.89 for 18 ounces

41. Frozen orange juice:
$1.69 for 16 ounces

$0.69 per dozen

42. Eggs:
$0.69 for 6 ounces
$2.10 for a flat $\left(2\frac{1}{2} \text{ dozen} \right)$

43. Soy sauce:
$2.29 for 12 ounces
$1.49 for 8 ounces

44. Shampoo:
$1.89 for 20 ounces
$3.19 for 32 ounces

45. Napkins:
100 for $0.59
180 for $0.93

46. Crackers:
$2.39 for 20 ounces
$0.99 for 8 ounces

Multiply or divide as indicated. See Sections 4.4 and 4.5.

47. 1.7
       $\underline{\times\ \ \ 6}$

48. 2.3
       $\underline{\times\ \ \ 9}$

49. 3.7
       $\underline{\times\ 1.2}$

50. 6.6
       $\underline{\times 2.5}$

51. $2.3\overline{)4.37}$

52. $3.5\overline{)22.75}$

Combining Concepts

53. Fill in the table to calculate miles per gallon.

Beginning Odometer Reading	Ending Odometer Reading	Miles Driven	Gallons of Gas Used	Miles per Gallon (round to the nearest tenth)
79,286	79,543		13.4	
79,543	79,895		15.8	
79,895	80,242		16.1	

Find each unit rate.

54. Sammy Joe Wingfield from Arlington, Texas, is the fastest bricklayer on record. On May 20, 1994, he laid 1048 bricks in 60 minutes. Find his unit rate of bricks per minute rounded to the nearest tenth. (*Source: The Guinness Book of Records*, 1996)

55. The longest stairway is the service stairway for the Niesenbahn Cable railway near Spiez, Switzerland. It has 11,674 steps and rises to a height of 7759 feet. Find the unit rate of steps per foot rounded to the nearest tenth of a step. (*Source: The Guinness Book of Records*, 1996)

56. In the United States, 1,900,000 bank workers are employed by 9905 banking institutions. Write a unit rate in bank workers per bank. Round to the nearest whole. (*Source:* American Bankers Association)

57. In the United States, the total number of students enrolled in public schools is 45,000,000. There are 78,300 public schools. Write a unit rate in students per school. Round to the nearest whole. (*Source:* National Center for Education Statistics)

58. In your own words, define the term unit rate.

59. Should the rate $\dfrac{3\text{ lights}}{2\text{ feet}}$ be written as $\dfrac{3}{2}$? Explain why or why not?

Name _____ Section _____ Date _____

Integrated Review—Ratio and Rate

Write each ratio as a ratio of whole numbers using fractional notation. Write the fraction in simplest form.

1. 18 to 20   **2.** 36 to 100   **3.** 8.6 to 10   **4.** 1.6 to 4.6

5. 8.65 to 6.95   **6.** 7.2 to 8.4   **7.** $3\frac{1}{2}$ to 13   **8.** $1\frac{2}{3}$ to $2\frac{3}{4}$

9. 8 inches to 12 inches   **10.** 3 hours to 24 hours

Find the ratio described in each problem.

11. In 2000, the average weekly earnings for male university teachers was $1020. Average weekly earnings for female university teachers was $805. Find the ratio of the male university teachers' average earnings to the female university teachers' average earnings. (*Source:* Bureau of Labor Statistics)

12. At the end of 2000, Eastman Kodak had $14,212 million in assets and $4508 million in long-term debt. Find the ratio of assets to long-term debt. (*Source:* Eastman Kodak Company)

13. In 2000, 761 films were rated. Of these, 36 were rated G. Find the ratio of G-rated films to total films for that year.

2000 Films Rated

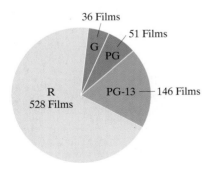

Source: Motion Picture Association of America

14. Find the ratio of the width to the length of the sign below.

1. _____

2. _____

3. _____

4. _____

5. _____

6. _____

7. _____

8. _____

9. _____

10. _____

11. _____

12. _____

13. _____

14. _____

Write each rate as a fraction in simplest form.

15. 5 offices for every 20 graduate assistants

16. 6 lights every 15 feet

17. 100 U.S. senators for 50 states

18. 5 teachers for every 140 students

19. 64 households with computers for every 100 households

20. 538 electoral votes for 50 states

Write each rate as a unit rate.

21. 165 miles in 3 hours

22. 560 feet in 4 seconds

23. 63 employees per 3 fax lines

24. 85 phone calls for 5 teenagers

25. 115 miles per 5 gallons

26. 112 teachers for 7 computers

27. 7524 books for 1254 college sudents

28. 2002 pounds for 13 adults

Write each unit price and decide which is the better buy.

29. Dog food:
8 pounds for $2.16
18 pounds for $4.99

30. Paper plates:
100 for $1.98
500 for $8.99

31. Microwave popcorn:
3 packs for $2.39
8 packs for $5.99

32. AA Batteries:
4 for $3.69
10 for $9.89

5.3 Proportions

Ⓐ Writing Proportions

A **proportion** is a statement that 2 ratios or rates are equal. For example,

$$\frac{5}{6} = \frac{10}{12}$$

is a proportion. We can read this as, "5 is to 6 as 10 is to 12."

EXAMPLE 1   Write each sentence as a proportion.

a. 12 diamonds is to 15 rubies as 4 diamonds is to 5 rubies.

b. 5 hits is to 9 at bats as 20 hits is to 36 at bats.

Solution:

a.   diamonds → $\dfrac{12}{15} = \dfrac{4}{5}$ ← diamonds
     rubies →   ← rubies

b.   hits → $\dfrac{5}{9} = \dfrac{20}{36}$ ← hits
     at bats →   ← at bats

> **Helpful Hint**
>
> Notice in the above examples of proportions that the numerators contain the same units and the denominators contain the same units. In this text, proportions will be written so that this is the case.

Ⓑ Determining Whether Proportions Are True

Like other mathematical statements, a proportion may be either true or false. A proportion is true if its ratios are equal. Since ratios are fractions, one way to determine whether a proportion is true is to write both fractions, in simplest form and compare them.

Another way is to compare cross products as we did in Section 2.3.

Determining Whether Proportions Are True or False

If cross products are *equal*, the proportion is *true*.
If cross products are *not equal*, the proportion is *false*.

EXAMPLE 2   Is $\dfrac{2}{3} = \dfrac{4}{6}$ a true proportion?

Solution:

$$\frac{2}{3} \diagdown\!\!\!\!\!\diagup \frac{4}{6} \qquad \left.\begin{array}{l} 3 \cdot 4 = 12 \\ 2 \cdot 6 = 12 \end{array}\right\}$$ Cross products are equal.

Since the cross products are equal, the proportion is true.

OBJECTIVES

Ⓐ Write sentences as proportions.

Ⓑ Determine whether proportions are true.

Ⓒ Find an unknown number in a proportion.

SSM   TUTOR CENTER   SG   CD & VIDEO   MATH PRO   WEB

Practice Problem 1

Write each sentence as a proportion.

a. 24 right is to 6 wrong as 4 right is to 1 wrong.

b. 32 Cubs fans is to 18 Mets fans as 16 Cubs fans is to 9 Mets fans.

Practice Problem 2

Is $\dfrac{3}{6} = \dfrac{4}{8}$ a true proportion?

Answers

1. a. $\dfrac{24}{6} = \dfrac{4}{1}$,   **b.** $\dfrac{32}{18} = \dfrac{16}{9}$,   **2.** yes

Practice Problem 3

Is $\dfrac{3.6}{6} = \dfrac{5.4}{8}$ a true proportion?

Practice Problem 4

Is $\dfrac{4\frac{1}{5}}{2\frac{1}{3}} = \dfrac{3\frac{3}{10}}{1\frac{5}{6}}$ a true proportion?

Concept Check

Write the proportion $\dfrac{5}{8} = \dfrac{10}{16}$ in two other ways so that it remains a true proportion.

EXAMPLE 3   Is $\dfrac{4.1}{7} = \dfrac{2.9}{5}$ a true proportion?

Solution:

$$\dfrac{4.1}{7} = \dfrac{2.9}{5} \qquad \left. \begin{array}{l} 7 \cdot 2.9 = 20.3 \\ 4.1 \cdot 5 = 20.5 \end{array} \right\} \quad \text{Cross products are not equal.}$$

Since cross products are not equal, $\dfrac{4.1}{7} \neq \dfrac{2.9}{5}$. The proportion is not a true proportion.   ●

EXAMPLE 4   Is $\dfrac{1\frac{1}{6}}{10\frac{1}{2}} = \dfrac{\frac{1}{2}}{4\frac{1}{2}}$ a true proportion?

Solution:

$$\dfrac{1\frac{1}{6}}{10\frac{1}{2}} = \dfrac{\frac{1}{2}}{4\frac{1}{2}} \qquad \left. \begin{array}{l} 10\frac{1}{2} \cdot \frac{1}{2} = \frac{21}{2} \cdot \frac{1}{2} = \frac{21}{4} \text{ or } 5\frac{1}{4} \\ 1\frac{1}{6} \cdot 4\frac{1}{2} = \frac{7}{6} \cdot \frac{9}{2} = \frac{63}{12} = \frac{21}{4} \text{ or } 5\frac{1}{4} \end{array} \right\} \quad \begin{array}{l}\text{Cross}\\\text{products}\\\text{are equal.}\end{array}$$

Since cross products are equal, the proportion is true.   ●

Try the Concept Check in the margin.

C   Finding Unknown Numbers in Proportions

When one number of a proportion is unknown, we can use cross products to find the unknown number. For example, to find the unknown number n in the proportion $\dfrac{2}{3} = \dfrac{n}{30}$, we cross multiply.

$$\dfrac{2}{3} = \dfrac{n}{30} \qquad \left. \begin{array}{l} 3 \cdot n \\ 2 \cdot 30 \end{array} \right\} \quad \text{Find the cross products.}$$

If the proportion is true, then cross products are equal.

$3 \cdot n = 2 \cdot 30$ Set the cross products equal to each other.
$3 \cdot n = 60$ Write $2 \cdot 30$ as 60.

To find the unknown number n, we ask ourselves, "3 times what number is 60?" The number is 20 and can be found by dividing 60 by 3.

$n = \dfrac{60}{3}$ Divide 60 by the number multiplied by n.

$n = 20$ Simplify.

Thus, the unknown number is 20.
To *check*, let's replace n with this value, 20, and verify that a true proportion results.

$\dfrac{2}{3} \overset{?}{=} \dfrac{20}{30}$ ← Let n be 20.

Answers

3. no,   **4.** yes

Concept Check: possible answers: $\dfrac{8}{5} = \dfrac{16}{10}$ and $\dfrac{5}{10} = \dfrac{8}{16}$

$$\frac{2}{3} \overset{?}{=} \frac{20}{30}$$

$$\left.\begin{array}{l} 3 \cdot 20 = 60 \\ 2 \cdot 30 = 60 \end{array}\right\}$$ Cross products are equal.

Finding an Unknown Value *n* in a Proportion

Step 1. Find cross products.
Step 2. Set the cross products equal to each other.
Step 3. Divide the number not multiplied by *n* by the number multiplied by *n*.

EXAMPLE 5   Find the value of the unknown number *n*.

$$\frac{34}{51} = \frac{n}{3}$$

Solution:

Step 1.

$$\frac{34}{51} = \frac{n}{3}$$

$$\left.\begin{array}{l} 51 \cdot n \\ 34 \cdot 3 = 102 \end{array}\right\}$$ Find the cross products.

Step 2.
$$51 \cdot n = 102$$ Set the cross products equal to each other.

Step 3.
$$n = \frac{102}{51}$$ Divide 102 by 51, the number multiplied by *n*.

$$n = 2$$ Simplify.

Check: $$\frac{34}{51} \overset{?}{=} \frac{2}{3}$$ Replace *n* with its value, 2.

$$\frac{34}{51} \overset{?}{=} \frac{2}{3}$$

$$\left.\begin{array}{l} 51 \cdot 2 = 102 \\ 34 \cdot 3 = 102 \end{array}\right\}$$ Cross products are equal, so the proportion is true.

EXAMPLE 6   Find the unknown number *n*.

$$\frac{6}{5} = \frac{7}{n}$$

Solution:

Step 1.

$$\frac{6}{5} = \frac{7}{n}$$

$$\left.\begin{array}{l} 5 \cdot 7 = 35 \\ 6 \cdot n \end{array}\right\}$$ Find the cross products.

Step 2.
$$6 \cdot n = 35$$ Set the cross products equal to each other.

Step 3.
$$n = \frac{35}{6}$$ Divide 35 by 6, the number multiplied by *n*.

$$n = 5\frac{5}{6}$$

Check to see that $5\frac{5}{6}$ is the unknown number.

Practice Problem 5

Find the value of the unknown number *n*.

$$\frac{2}{15} = \frac{n}{60}$$

Practice Problem 6

Find the unknown number *n*.

$$\frac{8}{n} = \frac{5}{9}$$

Answers

5. $n = 8$,   **6.** $n = 14\frac{2}{5}$

Practice Problem 7

Find the unknown number n.

$$\frac{n}{6} = \frac{0.7}{1.2}$$

Practice Problem 8

Find the unknown number n.

$$\frac{n}{4\frac{1}{3}} = \frac{4\frac{1}{2}}{1\frac{3}{4}}$$

EXAMPLE 7   Find the unknown number n.

$$\frac{n}{3} = \frac{0.8}{1.5}$$

Solution:

Step 1.

$$\frac{n}{3} = \frac{0.8}{1.5} \qquad \left. \begin{array}{l} 3 \cdot 0.8 = 2.4 \\ n \cdot 1.5 \end{array} \right\} \text{ Find the cross products.}$$

Step 2.

$n \cdot 1.5 = 2.4$ Set the cross products equal to each other.

Step 3.

$$n = \frac{2.4}{1.5} \qquad \text{Divide 2.4 by 1.5, the number multiplied by } n.$$

$$n = 1.6 \qquad \text{Simplify.}$$

Check to see that 1.6 is the unknown number.

EXAMPLE 8   Find the unknown number n.

$$\frac{1\frac{2}{3}}{3\frac{1}{4}} = \frac{n}{2\frac{3}{5}}$$

Solution:

Step 1.

$$\frac{1\frac{2}{3}}{3\frac{1}{4}} = \frac{n}{2\frac{3}{5}} \qquad \begin{array}{l} 3\frac{1}{4} \cdot n \\[6pt] 1\frac{2}{3} \cdot 2\frac{3}{5} = \frac{5}{3} \cdot \frac{13}{5} = \frac{\overset{1}{\cancel{5}} \cdot 13}{3 \cdot \underset{1}{\cancel{5}}} = \frac{13}{3} \end{array} \left. \right\} \begin{array}{l} \text{Find the} \\ \text{cross} \\ \text{products.} \end{array}$$

Step 2.

$$3\frac{1}{4} \cdot n = \frac{13}{3} \qquad \text{Set the cross products equal to each other.}$$

$$\frac{13}{4} \cdot n = \frac{13}{3} \qquad \text{Write } 3\frac{1}{4} \text{ as } \frac{13}{4}.$$

Step 3.

$$n = \frac{13}{3} \div \frac{13}{4} \qquad \text{Divide } \frac{13}{3} \text{ by } \frac{13}{4}, \text{ the number multiplied by } n.$$

$$n = \frac{13}{3} \cdot \frac{4}{13} = \frac{4}{3} \text{ or } 1\frac{1}{3} \qquad \text{Divide by multiplying by the reciprocal.}$$

Check to see that $1\frac{1}{3}$ is the unknown number.

Answers

7. $n = 3.5$,   **8.** $n = 11\frac{1}{7}$

Name _____ Section _____ Date _____

Mental Math

B *State whether each proportion is true or false.*

1. $\dfrac{2}{1} = \dfrac{6}{3}$ 2. $\dfrac{3}{1} = \dfrac{15}{5}$ 3. $\dfrac{1}{2} = \dfrac{3}{5}$ 4. $\dfrac{2}{11} = \dfrac{1}{5}$ 5. $\dfrac{2}{3} = \dfrac{4}{6}$ 6. $\dfrac{3}{4} = \dfrac{6}{8}$

EXERCISE SET 5.3

A *Write each sentence as a proportion. See Example 1.*

 1. 10 diamonds is to 6 opals as 5 diamonds is to 3 opals.

2. 8 books is to 6 courses as 4 books is to 3 courses.

3. 3 printers is to 12 computers as 1 printer is to 4 computers.

4. 4 hit songs is to 16 releases as 1 hit song is to 4 releases.

5. 6 eagles is to 58 sparrows as 3 eagles is to 29 sparrows.

6. 12 errors is to 8 pages as 1.5 errors is to 1 page.

7. $2\dfrac{1}{4}$ cups of flour is to 24 cookies as $6\dfrac{3}{4}$ cups of flour is to 72 cookies.

8. $1\dfrac{1}{2}$ cups milk is to 10 bagels as $\dfrac{3}{4}$ cup milk is to 5 bagels.

9. 22 vanilla wafers is to 1 cup of cookie crumbs as 55 vanilla wafers is to 2.5 cups of cookie crumbs. (*Source:* Based on data from *Family Circle* magazine)

10. 1 cup of instant rice is to 1.5 cups cooked rice as 1.5 cups of instant rice is to 2.25 cups of cooked rice. (*Source:* Based on data from *Family Circle* magazine)

B *Determine whether each proportion is a true proportion. See Examples 2 through 4.*

11. $\dfrac{15}{9} = \dfrac{5}{3}$ 12. $\dfrac{8}{6} = \dfrac{20}{15}$ 13. $\dfrac{8}{6} = \dfrac{9}{7}$ 14. $\dfrac{7}{12} = \dfrac{4}{7}$ 15. $\dfrac{9}{36} = \dfrac{2}{8}$ 16. $\dfrac{8}{24} = \dfrac{3}{9}$

17. $\dfrac{5}{8} = \dfrac{625}{1000}$ 18. $\dfrac{30}{50} = \dfrac{600}{1000}$ 19. $\dfrac{0.8}{0.3} = \dfrac{0.2}{0.6}$ 20. $\dfrac{0.7}{0.4} = \dfrac{0.3}{0.1}$ 21. $\dfrac{4.2}{8.4} = \dfrac{5}{10}$ 22. $\dfrac{8}{10} = \dfrac{5.6}{0.7}$

23. $\dfrac{\frac{3}{4}}{\frac{4}{3}} = \dfrac{\frac{1}{2}}{\frac{8}{9}}$ 24. $\dfrac{\frac{2}{5}}{\frac{2}{7}} = \dfrac{\frac{1}{10}}{\frac{1}{3}}$ 25. $\dfrac{2\frac{2}{5}}{\frac{2}{3}} = \dfrac{\frac{10}{9}}{\frac{1}{4}}$ 26. $\dfrac{5\frac{5}{8}}{\frac{5}{3}} = \dfrac{4\frac{1}{2}}{1\frac{1}{5}}$

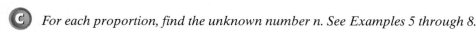

C *For each proportion, find the unknown number n. See Examples 5 through 8.*

27. $\dfrac{n}{5} = \dfrac{6}{10}$ **28.** $\dfrac{n}{3} = \dfrac{12}{9}$ **29.** $\dfrac{30}{10} = \dfrac{15}{n}$ **30.** $\dfrac{25}{100} = \dfrac{7}{n}$ **31.** $\dfrac{n}{8} = \dfrac{50}{100}$

32. $\dfrac{12}{18} = \dfrac{n}{21}$ **33.** $\dfrac{n}{6} = \dfrac{8}{15}$ **34.** $\dfrac{24}{n} = \dfrac{60}{96}$ **35.** $\dfrac{12}{10} = \dfrac{n}{16}$ **36.** $\dfrac{18}{54} = \dfrac{3}{n}$

37. $\dfrac{\frac{1}{3}}{\frac{3}{8}} = \dfrac{\frac{2}{5}}{n}$ **38.** $\dfrac{\frac{7}{9}}{\frac{8}{27}} = \dfrac{\frac{1}{4}}{n}$ **39.** $\dfrac{8}{\frac{1}{3}} = \dfrac{24}{n}$ **40.** $\dfrac{\frac{3}{4}}{12} = \dfrac{n}{48}$ **41.** $\dfrac{\frac{2}{3}}{\frac{6}{9}} = \dfrac{12}{n}$

42. $\dfrac{n}{24} = \dfrac{\frac{5}{8}}{3}$ **43.** $\dfrac{n}{\frac{6}{5}} = \dfrac{4\frac{1}{6}}{6\frac{2}{3}}$ **44.** $\dfrac{\frac{11}{4}}{\frac{25}{8}} = \dfrac{7\frac{3}{5}}{n}$ **45.** $\dfrac{n}{0.6} = \dfrac{0.05}{12}$ **46.** $\dfrac{0.2}{0.7} = \dfrac{8}{n}$

47. $\dfrac{3.5}{12.5} = \dfrac{7}{n}$ **48.** $\dfrac{7.8}{13} = \dfrac{n}{2.6}$

Review and Preview

Insert < or > to form a true statement. See Sections 2.1 and 4.2.

49. $8.01 \quad 8.1$ **50.** $7.26 \quad 7.026$ **51.** $2\frac{1}{2} \quad 2\frac{1}{3}$ **52.** $9\frac{1}{5} \quad 9\frac{1}{4}$ **53.** $5\frac{1}{3} \quad 6\frac{2}{3}$ **54.** $1\frac{1}{2} \quad 2\frac{1}{2}$

Combining Concepts

For each proportion, find the unknown number n.

55. $\dfrac{n}{7} = \dfrac{0}{8}$ **56.** $\dfrac{0}{2} = \dfrac{n}{3.5}$ **57.** $\dfrac{n}{1150} = \dfrac{588}{483}$

58. $\dfrac{585}{n} = \dfrac{117}{474}$ **59.** $\dfrac{222}{1515} = \dfrac{37}{n}$ **60.** $\dfrac{1425}{1062} = \dfrac{n}{177}$

61. Explain the difference between a ratio and a proportion.

62. Explain how to find the unknown number in a proportion such as $\dfrac{n}{18} = \dfrac{12}{8}$.

5.4 Proportions and Problem Solving

Ⓐ **Solving Problems by Writing Proportions**

Writing proportions is a powerful tool for solving problems in almost every field, including business, chemistry, biology, health sciences, and engineering, as well as in daily life. Given a specified ratio (or rate) of two quantities, a proportion can be used to determine an unknown quantity.

EXAMPLE 1   Determining Distances from a Map

On a chamber of commerce map of Abita Springs, 5 miles corresponds to 2 inches. How many miles correspond to 7 inches?

Solution:

1. UNDERSTAND. Read and reread the problem. You may want to draw a diagram.

$$\underbrace{\frac{5\ miles}{2\ inches}\ \frac{5\ miles}{2\ inches}\ \frac{5\ miles}{2\ inches}\ \frac{?}{1\ inch}}\ \begin{array}{l}= \text{a little over 15 miles}\\= 7\ inches\end{array}$$

From the diagram we can see that our solution should be a little over 15 miles.

2. TRANSLATE. We will let n represent our unknown number. Since 5 miles corresponds to 2 inches as n miles corresponds to 7 inches, we have the proportion

$$\begin{array}{ll}\text{miles} \rightarrow\\ \text{inches} \rightarrow\end{array} \frac{5}{2}=\frac{n}{7} \begin{array}{ll}\leftarrow \text{miles}\\ \leftarrow \text{inches}\end{array}$$

3. SOLVE:

$$\frac{5}{2}=\frac{n}{7} \qquad \left.\begin{array}{l}2\cdot n\\ 5\cdot 7=35\end{array}\right\} \text{ Find the cross products.}$$

$2\cdot n=35$ Set the cross products equal to each other.

$n=\dfrac{35}{2}$ Divide 35 by 2, the number multiplied by n.

$n=17.5$ Simplify.

4. INTERPRET. *Check* your work. This result is reasonable since it is a little over 15 miles. *State* your conclusion: 7 inches corresponds to 17.5 miles.

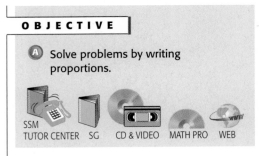

OBJECTIVE

Ⓐ Solve problems by writing proportions.

SSM TUTOR CENTER   SG   CD & VIDEO   MATH PRO   WEB

Practice Problem 1

On an architect's blueprint, 1 inch corresponds to 12 feet. How long is a wall represented by a $3\frac{1}{2}$-inch line on the blueprint?

Answer
1. 42 ft

Copyright 2003 Prentice-Hall, Inc.

Helpful Hint

We can also solve Example 1 by writing the proportion

$$\frac{2 \text{ inches}}{5 \text{ miles}} = \frac{7 \text{ inches}}{n \text{ miles}}$$

Although other proportions may be used to solve Example 1, we will solve by writing proportions so that the numerators have the same unit measures and the denominators have the same unit measures.

Practice Problem 2

An auto mechanic recommends that 3 ounces of isopropyl alcohol be mixed with a tankful of gas (14 gallons) to increase the octane of the gasoline for better engine performance. At this rate, how many gallons of gas can be treated with a 16-ounce bottle of alcohol?

EXAMPLE 2   Finding Medicine Dosage

The standard dose of an antibiotic is 4 cc (cubic centimeters) for every 25 pounds (lb) of body weight. At this rate, find the standard dose for a 140-lb woman.

Solution:

1. UNDERSTAND. Read and reread the problem. You may want to draw a diagram to estimate a reasonable solution.

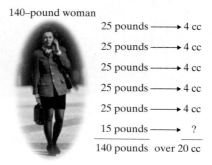

140–pound woman

25 pounds ⟶	4 cc
25 pounds ⟶	4 cc
25 pounds ⟶	4 cc
25 pounds ⟶	4 cc
25 pounds ⟶	4 cc
15 pounds ⟶	?
140 pounds	over 20 cc

From the diagram, we can see that a reasonable solution is a little over 20 cc.

2. TRANSLATE. We will let n represent the unknown number. From the problem, we know that 4 cc is to 25 pounds as n cc is to 140 pounds, or

$$\begin{array}{ll} \text{cubic centimeters} & \rightarrow \\ \text{pounds} & \rightarrow \end{array} \quad \frac{4}{25} = \frac{n}{140} \quad \begin{array}{ll} \leftarrow & \text{cubic centimeters} \\ \leftarrow & \text{pounds} \end{array}$$

3. SOLVE:

$$\frac{4}{25} \diagdown \frac{n}{140} \qquad \left. \begin{array}{l} 25 \cdot n \\ 4 \cdot 140 = 560 \end{array} \right\} \quad \text{Find the cross products.}$$

$25 \cdot n = 560$ Set the cross products equal to each other.

$n = \dfrac{560}{25}$ Divide 560 by 25, the number multiplied by n.

$n = 22.4$ Simplify.

4. INTERPRET. *Check* your work. This result is reasonable since it is a little over 20 cc. *State* your conclusion: The standard dose for a 140-lb woman is 22.4 cc.

Answer

2. $74\frac{2}{3}$ gal

△ **EXAMPLE 3**   **Calculating Supplies Needed to Fertilize a Lawn**

A 50-pound bag of fertilizer covers 2400 square feet of lawn. How many bags of fertilizer are needed to cover a town square containing 15,360 square feet of lawn? Round the answer up to the nearest whole bag.

Practice Problem 3   △

If a gallon of paint covers 400 square feet, how many gallons are needed to paint a retaining wall that is 260 feet long and 4 feet high? Round the answer up to the nearest whole gallon.

Solution:

1.   UNDERSTAND. Read and reread the problem. Draw a picture.

50 pounds   How many
cover   bags cover

2400 square feet	15,360 square feet

2.   TRANSLATE. We'll let n represent the unknown number. From the problem, we know that 1 bag is to 2400 square feet as n bags is to 15,360 square feet.

$$\begin{array}{ll} \text{bags} \quad \rightarrow \\ \text{square feet} \quad \rightarrow \end{array} \frac{1}{2400} = \frac{n}{15{,}360} \begin{array}{l} \leftarrow \quad \text{bags} \\ \leftarrow \quad \text{square feet} \end{array}$$

3.   SOLVE:

$$\frac{1}{2400} = \frac{n}{15{,}360} \qquad \left. \begin{array}{l} 2400 \cdot n \\ 1 \cdot 15{,}360 = 15{,}360 \end{array} \right\} \quad \begin{array}{l} \text{Find the cross} \\ \text{products.} \end{array}$$

$2400 \cdot n = 15{,}360$ Set the cross products equal to each other.

$n = \dfrac{15{,}360}{2400}$ Divide 15,360 by 2400, the number multiplied by n.

$n = 6.4$ Simplify.

4.   INTERPRET. *Check* that replacing n with 6.4 makes the proportion true. Is the answer reasonable?

Answer

3. 3 gal

1 bag covers:

15,360 square feet

6 bags

2400 square feet

+ 1 part of another bag / 7 bags

960 square feet

Yes. Since we must buy whole bags of fertilizer, 7 bags are needed. *State your conclusion:* To cover 15,360 square feet of lawn, 7 bags are needed.

Concept Check

You are told that 12 ounces of ground coffee will brew enough coffee to serve 20 people. How could you estimate how much ground coffee will be needed to serve 95 people?

Answer

Concept Check: Find how much will be needed for 100 people (20 × 5) by multiplying 12 ounces by 5, which is 60 ounces.

Try the Concept Check in the margin.

FOCUS ON **Real Life**

BODY DIMENSIONS

Have you ever noticed how large a baby's head looks compared to the rest of its body? We tend to notice this because we are used to seeing adults with very different "proportions" or dimensions. Try the following group activity to see whether there are any patterns in adult body dimensions from person to person.

GROUP ACTIVITY

Have another group member help you measure the height of your head from under your chin to the top of your skull. (Remember that the top of your head is round. You may find it helpful to hold a flat surface on top of your head to more accurately locate and measure to the "top.") Use your head measurement to make a measuring stick from a piece of cardboard. You will now use this measuring stick to measure other parts of your body in units equal to the height of your head. With the help of another group member as needed, measure the following distances with your "head" measuring stick:

- Your overall height
- Your arm length from shoulder to elbow
- Your arm length from elbow to longest fingertip
- Your leg length from hip to ankle
- Your leg length from knee to ankle
- The distance from your chin to waistline

(*Note:* Each group member must make his or her own measurements in relation to the measurement of his or her own head height.)

1. For each of the above measurements, find the ratio of the measurement to the height measurement of your head.
2. As a group, compile a table summarizing the ratios for each group member by category. (You might list the six measurement categories along the side of the table and group members' names along the top of the table.)
3. Analyze the table. Do you see any patterns? Explain.
4. About how many "heads tall" would you say a normal adult is?

EXERCISE SET 5.4

A Solve. See Examples 1 through 3.

The ratio of a quarterback's completed passes to attempted passes is 4 to 9.

1. If he attempted 27 passes, find how many passes he completed.

2. If he completed 20 passes, find how many passes he attempted.

It takes Sandra Hallahan 30 minutes to word process and spell check 4 pages.

 3. Find how long it takes her to word process and spell check 22 pages.

4. Find how many pages she can word process and spell check in 4.5 hours.

University Law School accepts 2 out of every 7 applicants.

5. If the school received 630 applications, find how many students were accepted.

6. If the school accepted 150 students, find how many applications were received.

On an architect's blueprint, 1 inch corresponds to 8 feet.

7. Find the length of a wall represented by a line $2\frac{7}{8}$ inches long on the blueprint.

8. If an exterior wall is 42 feet long, find how long the blueprint measurement should be.

A human-factors expert recommends that there be at least 9 square feet of floor space in a college classroom for every student in the class.

△ **9.** Find the minimum floor space that 30 students require.

△ **10.** Due to a space crunch, a university converts a 21-by-15-foot conference room into a classroom. Find the maximum number of students the room can accommodate.

A Honda Civic averages 450 miles on a 12-gallon tank of gas.

11. If Dave Smythe runs out of gas in a Honda Civic and AAA comes to his rescue with $1\frac{1}{2}$ gallons of gas, determine how far he can go. Round to the nearest mile.

12. Find how many gallons of gas Denise Wolcott can expect to burn on a 2000-mile vacation trip in a Honda Civic. Round to the nearest gallon.

The scale on an Italian map states that 1 centimeter corresponds to 30 kilometers.

13. Find how far apart Milan and Rome are if their corresponding points on the map are 15 centimeters apart.

14. On the map, a small Italian village is located 0.4 centimeter from the Mediterranean Sea. Find the actual distance.

A drink called Sea Breeze Punch is made by mixing 3 parts of grapefruit juice with 4 parts of cranberry juice.

15. Find how much grapefruit juice should be mixed with 32 ounces of cranberry juice.

16. For a party, 6 quarts of grapefruit juice have been purchased to make Sea Breeze Punch. Find how much cranberry juice should be purchased.

A bag of Scott fertilizer covers 3000 square feet of lawn.

△ **17.** Find how many bags of fertilizer should be purchased to cover a rectangular lawn 260 feet by 180 feet.

△ **18.** Find how many bags of fertilizer should be purchased to cover a square lawn measuring 160 feet on each side.

Yearly homeowner property taxes are figured at a rate of $1.45 tax for every $100 of house value.

19. If Janet Blossom pays $2349 in property taxes, find the value of her home.

20. Find the property taxes on a condominium valued at $72,000.

A Cubs baseball player makes 3 hits in every 8 times at bat.

21. If this Cubs player comes up to bat 40 times in the World Series, find how many hits he would be expected to make.

22. At this rate, if he made 12 hits, find how many times he batted.

A survey reveals that 2 out of 3 people prefer Coke to Pepsi.

23. In a room of 40 people, how many people are likely to prefer Coke? Round the answer to the nearest person.

24. In a college class of 36 students, find how many students are likely to prefer Pepsi.

An office uses 5 boxes of envelopes every 3 weeks.

25. Find how long a gross of envelope boxes is likely to last. (A gross of boxes is 144 boxes.) Round to the nearest week.

26. Find how many boxes should be purchased to last a month. Round to the nearest box.

 27. The daily supply of oxygen for one person is provided by 625 square feet of lawn. A total of 3750 square feet of lawn would provide the daily supply of oxygen for how many people? (*Source:* Professional Lawn Care Association of America)

28. In the United States, approximately 71 million of the 200 million cars and light trucks in service have driver-side air bags. In a parking lot containing 800 cars and light trucks, how many would be expected to have driver-side air bags? (*Source:* Insurance Institute for Highway Safety)

29. A student would like to estimate the height of the Statue of Liberty in New York City's harbor. The length of the Statue of Liberty's right arm is 42 feet. The student's right arm is 2 feet long and her height is $5\frac{1}{3}$ feet. Use this information to estimate the height of the Statue of Liberty. How close is your estimate to the statue's actual height of 111 feet, 1 inch from heel to top of head? (*Source:* National Park Service)

30. There are 76 milligrams of cholesterol in a 3-ounce serving of skinless chicken. How much cholesterol is in 8 ounces of chicken? (*Source:* USDA)

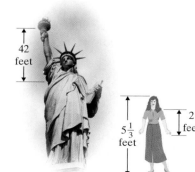

322

31. One pound of firmly-packed brown sugar yields $2\frac{1}{4}$ cups. How many pounds of brown sugar will be required in a recipe that calls for 6 cups of firmly packed brown sugar? (*Source:* Based on data from *Family Circle* magazine)

32. Eleven out of every 25 greeting cards sold in the United States are Hallmark brand cards. If a consumer purchased 75 greeting cards in the past year, how many do you expect would have been Hallmark brand cards? (*Source:* Hallmark Cards, Inc.)

33. One out of 3 American adults has worked in the restaurant industry at some point during his or her life. In an office of 84 workers, how many of these people would you expect to have worked in the restaurant industry at some point? (*Source:* National Restaurant Association)

34. According to the 2000 census, of people age 65 and older, there are 143 women for every 100 men. At a retirement community with 400 men over the age of 65, how many female residents over the age of 65 would be expected? (*Source:* U.S. Census Bureau)

Review and Preview

Find the prime factorization of each number. See Section 2.2.

35. 15   **36.** 21   **37.** 20   **38.** 24

39. 200   **40.** 300   **41.** 32   **42.** 81

 Combining Concepts

A board such as the one pictured below will balance if the following proportion is true:

$$\frac{\text{first weight}}{\text{second distance}} = \frac{\text{second weight}}{\text{first distance}}$$

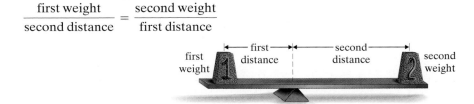

Use this proportion to solve Exercises 43 and 44.

43. Find the distance n that will allow the board to balance.

44. Find the length n needed to lift the weight below.

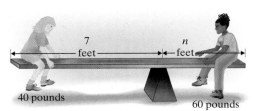

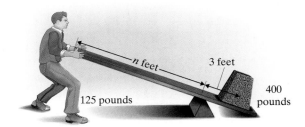

45. Describe a situation in which writing a proportion might solve a problem related to driving a car.

Investigating Scale Drawings

MATERIALS:

■ ruler

■ tape measure

■ grid paper (optional)

This activity may be completed by working in groups or individually.

Scale drawings are used by architects, engineers, interior designers, ship builders, and others. In a scale drawing, each unit measurement on the drawing represents a fixed length on the object being drawn. For instance, in an architect's scale drawing, 1 inch on the drawing may represent 10 feet on a building. The scale describes the relationship between the measurements. If the measurements have the same units, the scale can be expressed as a ratio. In this case, the ratio would be 1:120, representing 1 inch to 120 inches (or 10 feet).

Use a ruler and the scale drawing of an elementary school below to answer the following questions.

1. How wide are each of the front doors of the school?

2. How long is the front of the school?

3. How tall is the front of the school?

Now you will draw your own scale floor plan. First choose a room to draw—it can be your math classroom, your living room, your dormitory room, or any room that can be easily measured. Start by using a tape measure to measure the distances around the base of the walls in the room you are drawing.

4. Choose a scale for your floor plan.

5. Convert each measurement in the room you are drawing to the corresponding lengths needed for the scale drawing.

6. Complete your floor plan (you may find it helpful to use grid paper). Mark the locations of doors and windows on your floor plan. Be sure to indicate on the drawing the scale used in your floor plan.

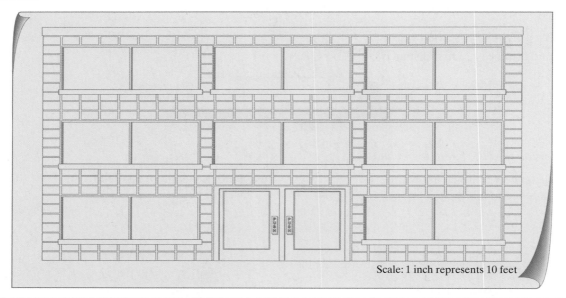

Scale: 1 inch represents 10 feet

Chapter 5   VOCABULARY CHECK

Fill in each blank with one of the words or phrases listed below.

rate   unit rate   ratio   unit price   proportion

1. A _____ is the quotient of two numbers. It can be written as a fraction, using a colon, or using the word *to*.

2. $\frac{x}{2} = \frac{7}{16}$ is an example of a _____ .

3. A _____ is a rate with a denominator of 1.

4. A _____ is a "money per item" unit rate.

5. A _____ is used to compare different kinds of quantities.

C H A P T E R

Highlights

DEFINITIONS AND CONCEPTS	EXAMPLES

Section 5.1   Ratios

A **ratio** is the quotient of two quantities.	The ratio of 3 to 4 can be written as $$\frac{3}{4} \quad \text{or} \quad 3:4$$ fraction notation colon notation

Section 5.2   Rates

Rates are used to compare different kinds of quantities.	Write the rate 12 spikes every 8 inches as a fraction in simplest form. $$\frac{12 \text{ spikes}}{8 \text{ inches}} = \frac{3 \text{ spikes}}{2 \text{ inches}}$$
A **unit rate** is a rate with a denominator of 1.	Write as a unit rate: 117 miles on 5 gallons of gas $$\frac{117 \text{ miles}}{5 \text{ gallons}} = \frac{23.4 \text{ miles}}{1 \text{ gallon}} \quad \text{or 23.4 miles per gallon} \\ \text{or 23.4 miles/gallon}$$
A **unit price** is a "money per item" unit rate.	Write as a unit price: $5.88 for 42 ounces of detergent $$\frac{\$5.88}{42 \text{ ounces}} = \frac{\$0.14}{1 \text{ ounce}} = \$0.14 \text{ per ounce}$$

DEFINITIONS AND CONCEPTS	**EXAMPLES**

Section 5.3   Proportions

A **proportion** is a statement that two ratios or rates are equal.
If cross products are equal, the proportion is true.
If cross products are not equal, the proportion is false.

$\dfrac{1}{2} = \dfrac{4}{8}$ is a proportion.

Is $\dfrac{6}{10} = \dfrac{9}{15}$ a true proportion?

$$\dfrac{6}{10} \bowtie \dfrac{9}{15} \qquad \left.\begin{array}{l} 10 \cdot 9 = 90 \\ 6 \cdot 15 = 90 \end{array}\right\} \begin{array}{l}\text{Cross products}\\\text{are equal.}\end{array}$$

Since cross products are equal, the proportion is a true proportion.

FINDING AN UNKNOWN VALUE *N* IN A PROPORTION

Find n: $\dfrac{n}{7} = \dfrac{5}{8}$

Step 1.   Find the cross products.

Step 1.

$$\dfrac{n}{7} \bowtie \dfrac{5}{8} \qquad \left.\begin{array}{l} 7 \cdot 5 = 35 \\ n \cdot 8 \end{array}\right\} \begin{array}{l}\text{Find the cross}\\\text{products.}\end{array}$$

Step 2.   Set the cross products equal to each other.

Step 2.

$n \cdot 8 = 35$ Set the cross products equal to each other.

Step 3.   Divide the number not multiplied by *n* by the number multiplied by *n*.

Step 3.

$n = \dfrac{35}{8}$ Divide 35 by 8, the number multiplied by *n*.

$n = 4\dfrac{3}{8}$

Section 5.4   Proportions and Problem Solving

Given a specified ratio (or rate) of two quantities, a proportion can be used to determine an unknown quantity.

On a map, 50 miles corresponds to 3 inches. How many miles correspond to 10 inches?

1. UNDERSTAND. Read and reread the problem.

2. TRANSLATE. We let *n* represent the unknown number. We are given that 50 miles is to 3 inches as *n* miles is to 10 inches.

$$\begin{array}{l}\text{miles} \rightarrow \\ \text{inches} \rightarrow\end{array} \quad \dfrac{50}{3} = \dfrac{n}{10} \quad \begin{array}{l}\leftarrow \text{miles} \\ \leftarrow \text{inches}\end{array}$$

3. SOLVE:

$$\dfrac{50}{3} \bowtie \dfrac{n}{10} \qquad \left.\begin{array}{l} 3 \cdot n \\ 50 \cdot 10 = 500 \end{array}\right\} \begin{array}{l}\text{Find the cross}\\\text{products.}\end{array}$$

$3 \cdot n = 500$ Set the cross products equal to each other.

$n = \dfrac{500}{3}$ Divide 500 by 3, the number multiplied by *n*.

$n = 166\dfrac{2}{3}$

4. INTERPRET. *Check* your work. *State* your conclusion:
On the map, $166\dfrac{2}{3}$ miles corresponds to 10 inches.

Chapter 5 Review

(5.1) *Write each ratio as a fraction in simplest form.*

1. 6000 people to 4800 people

2. 121 births to 143 births

3. $2\frac{1}{4}$ days to 10 days

4. 14 quarters to 5 quarters

5. 4 weeks to 15 weeks

6. 4 yards to 8 yards

7. $3\frac{1}{2}$ dollars to 7 dollars

8. 3.5 centimeters to 75 centimeters

(5.2) *Write each rate as a fraction in simplest form.*

9. 8 stillborn births to 1000 live births

10. 6 professors for 20 graduate research assistants

11. 15 word processing pages printed in 6 minutes

12. 8 computers assembled in 6 hours

Find each unit rate.

13. 468 miles in 9 hours

14. 180 feet in 12 seconds

15. $0.93 for 3 pears

16. $6.96 for 4 diskettes

17. 260 kilometers in 4 hours

18. 8 gallons of pesticide for 6 acres of crops

19. $184 for books for 5 college courses

20. 52 bushels of fruit from 4 trees

Find each unit price and decide which is the better buy. Assume that we are comparing different sizes of the same brand.

21. Taco sauce: 8 ounces for $0.99 or 12 ounces for $1.69

22. Peanut butter: 18 ounces for $1.49 or 28 ounces for $2.39

23. 2% milk: 16 ounces for $0.59, $\frac{1}{2}$ gallon for $1.69, or 1 gallon for $2.29 (1 gallon = 128 fluid ounces)

24. Coca-Cola: 12 ounces for $0.59, 16 ounces for $0.79, or 32 ounces for $1.19

(5.3) *Write each sentence as a proportion.*

25. 20 men is to 14 women as 10 men is to 7 women.

26. 50 tries is to 4 successes as 25 tries is to 2 successes.

27. 16 sandwiches is to 8 players as 2 sandwiches is to 1 player.

28. 12 tires is to 3 cars as 4 tires is to 1 car.

Determine whether each proportion is true.

29. $\dfrac{21}{8} = \dfrac{14}{6}$

30. $\dfrac{3}{5} = \dfrac{60}{100}$

31. $\dfrac{3.1}{6.2} = \dfrac{0.8}{0.16}$

32. $\dfrac{3.75}{3} = \dfrac{7.5}{6}$

Find the unknown number n in each proportion.

33. $\dfrac{n}{6} = \dfrac{15}{18}$

34. $\dfrac{n}{9} = \dfrac{5}{3}$

35. $\dfrac{4}{13} = \dfrac{10}{n}$

36. $\dfrac{8}{5} = \dfrac{9}{n}$

37. $\dfrac{16}{3} = \dfrac{n}{6}$

38. $\dfrac{n}{3} = \dfrac{9}{2}$

39. $\dfrac{n}{5} = \dfrac{27}{2\frac{1}{4}}$

40. $\dfrac{2\frac{1}{2}}{6} = \dfrac{3}{n}$

41. $\dfrac{n}{0.4} = \dfrac{4.7}{2}$

42. $\dfrac{6}{0.3} = \dfrac{7.2}{n}$

(5.4) *Solve.*

The ratio of a quarterback's completed passes to attempted passes is 3 to 7.

43. If he attempts 32 passes, find how many passes he completed. Round to the nearest whole pass.

44. If he completed 15 passes, find how many passes he attempted.

One bag of pesticide covers 4000 square feet of crops.

△ **45.** Find how many bags of pesticide should be purchased to cover a rectangular garden that is 180 feet by 175 feet.

△ **46.** Find how many bags of pesticide should be purchased to cover a square garden that is 250 feet on each side.

An owner of a Ford Escort can drive 420 miles on 11 gallons of gas.

47. If Tom Aloiso runs out of gas in an Escort and AAA comes to his rescue with $1\frac{1}{2}$ gallons of gas, determine whether Tom can then drive to a gas station 65 miles away.

48. Find how many gallons of gas Tom can expect to burn on a 3000-mile trip. Round to the nearest gallon.

Yearly homeowner property taxes are figured at a rate of $1.15 tax for every $100 of house value.

49. If a homeowner pays $627.90 in property taxes, find the value of his home.

50. Find the property taxes on a town house valued at $89,000.

On an architect's blueprint, 1 inch = 12 feet.

51. Find the length of a wall represented by a $3\frac{3}{8}$-inch line on the blueprint.

52. If an exterior wall is 99 feet long, find how long the blueprint measurement should be.

STUDY SKILLS REMINDER

Are you satisfied with your performance in this course thus far?

If not, ask yourself the following questions:

- Am I attending all class periods and arriving on time?
- Am I working and checking my homework assignments?
- Am I getting help when I need it?
- In addition to my instructor, am I using the supplements to this text that could help me? For example, the tutorial video lessons? Math-Pro, the tutorial software?
- Am I satisfied with my performance on quizzes and tests?

If you answered no to *any* of these questions, read or reread Section 1.1 for suggestions in these areas. Also, you may want to contact your instructor for additional feedback.

FOCUS ON **Business and Career**

CONSUMER PRICE INDEX

Do you remember when the regular price of a candy bar was 5¢, 10¢, or 25¢? It is certainly difficult to find a candy bar for that price these days. The reason is inflation: The tendency for the price of a given product to increase over time. Businesses and government agencies use the Consumer Price Index (CPI) to track inflation. The CPI measures the change in prices over time of basic consumer goods and services. It is calculated by the Bureau of Labor Statistics, part of the U.S. Department of Labor.

The CPI is very useful for comparing the prices of fixed items in various years. For instance, suppose an insurance company customer submits a claim for the theft of a fishing boat purchased in 1975. Because the customer's policy includes replacement cost coverage, the insurance company must calculate how much it would cost to replace the boat at today's prices. (Let's assume the theft took place in 1997.) The customer has a receipt for the boat showing that it cost $598 in 1975. The insurance company can use the following proportion to calculate the replacement cost:

$$\frac{\text{price in earlier year}}{\text{price in later year}} = \frac{\text{CPI value in earlier year}}{\text{CPI value in later year}}$$

Because the CPI value is 53.8 for 1975 and 160.5 for 1997, the insurance company would use the following proportion for this situation. (We will let n represent the unknown price in 1997).

$$\frac{\text{price in 1975}}{\text{price in 1997}} = \frac{\text{CPI value in 1975}}{\text{CPI value in 1997}}$$

$$\frac{598}{n} = \frac{53.8}{160.5}$$

$$53.8 \cdot n = 598(160.5)$$

$$53.8 \cdot n = 95,979$$

$$\frac{53.8 \cdot n}{53.8} = \frac{95,979}{53.8}$$

$$n \approx 1784$$

The replacement cost of the fishing boat at 1997 prices is $1784.

CRITICAL THINKING

1. What trends do you see in the CPI values in the table? Do you think these trends make sense? Explain.

2. A piece of jewelry cost $800 in 1975. What is its 2000 replacement value?

3. In 1995, the cost of a loaf of bread was about $1.89. What would an equivalent loaf of bread cost in 1950?

4. Suppose a couple purchased a house for $22,000 in 1920. At what price could they have expected to sell the house in 1990?

5. An original Ford Model T cost about $850 in 1915. What is the equivalent cost of a Model T in 2000 dollars?

Consumer Price Index	
Year	**CPI**
1915	10.1
1920	20.0
1925	17.5
1930	16.7
1935	13.7
1940	14.0
1945	18.0
1950	24.1
1955	26.8
1960	29.6
1965	31.5
1970	38.8
1975	53.8
1980	82.4
1985	107.6
1990	130.7
1995	152.4
1997	160.5
1998	163.0
1999	166.6
2000	172.2

(*Source:* Bureau of Labor Statistics, U.S. Department of Labor)

Chapter 5 Test

Write each ratio as a fraction in simplest form.

1. 4500 trees to 6500 trees

2. $75 to $10

3. 28 men to every 4 women

4. 9 inches of rain in 30 days

Find each unit rate.

5. 650 kilometers in 8 hours

6. 8 inches of rain in 12 hours

7. 140 students for 5 teachers

Find each unit price and decide which is the better buy.

8. Steak sauce:
8 ounces for $1.19
12 ounces for $1.89

9. Jelly:
16 ounces for $1.49
24 ounces for $2.39

Determine whether each proportion is true.

10. $\dfrac{28}{16} = \dfrac{14}{8}$

11. $\dfrac{3.6}{2.2} = \dfrac{1.9}{1.2}$

Answers

1. _____

2. _____

3. _____

4. _____

5. _____

6. _____

7. _____

8. _____

9. _____

10. _____

11. _____

12.

13.

14.

15.

16.

17.

18.

19.

20.

332

Find the unknown number n in each proportion.

12. $\dfrac{n}{3} = \dfrac{15}{9}$ **13.** $\dfrac{8}{n} = \dfrac{11}{6}$ **14.** $\dfrac{\frac{15}{12}}{\frac{3}{7}} = \dfrac{n}{\frac{4}{5}}$ **15.** $\dfrac{1.5}{5} = \dfrac{2.4}{n}$

Solve.

16. On an architect's drawing, 2 inches corresponds to 9 feet. Find the length of a home represented by a line that is 11 inches long.

17. If a car can be driven 80 miles in 3 hours, how long will it take to travel 100 miles?

18. The standard dose of medicine for a dog is 10 grams for every 15 pounds of body weight. What is the standard dose for a dog that weighs 80 pounds?

19. Jerome Grant worked 6 hours and packed 86 cartons of books. At this rate, how many cartons can he pack in 8 hours?

20. Currently 27 out of every 50 American adults drink coffee every day. In a town with a population of 7900 adults, how many of these adults would you expect to drink coffee every day? (*Source:* National Coffee Association)

Cumulative Review

1. Subtract. Check each answer by adding.
   a. $12 - 9$
   b. $11 - 6$
   c. $5 - 5$
   d. $7 - 0$

2. Round 248,982 to the nearest hundred.

3. Multiply: $\begin{array}{r} 25 \\ \times\ 8 \\ \hline \end{array}$

4. The director of a learning lab at a local community college is working on next year's budget. Thirty-three new video players are needed at a cost of $540 each. What is the total cost of these video players?

5. Find the prime factorization of 45.

6. Write $\dfrac{12}{20}$ in simplest form.

Multiply.

7. $\dfrac{3}{4} \cdot \dfrac{8}{5}$

8. $\dfrac{6}{13} \cdot \dfrac{26}{30}$

Add and simplify.

9. $\dfrac{2}{7} + \dfrac{3}{7}$

10. $\dfrac{7}{8} + \dfrac{6}{8} + \dfrac{3}{8}$

11. Find the LCM of 6 and 9.

12. Write an equivalent fraction with the indicated denominator. $\dfrac{1}{2} = \dfrac{}{14}$

1. a. _____

   b. _____

   c. _____

   d. _____

2. _____

3. _____

4. _____

5. _____

6. _____

7. _____

8. _____

9. _____

10. _____

11. _____

12. _____

13. _____

14. _____

15. _____

16. _____

17. _____

18. _____

19. _____

20. _____

21. _____

22. _____

23. _____

24. _____

25. _____

334

13. Subtract: $\dfrac{10}{11} - \dfrac{2}{3}$

14. A flight from Tucson to Phoenix, Arizona, requires $\dfrac{5}{12}$ of an hour. If the plane has been flying $\dfrac{1}{4}$ of an hour, find how much time remains before it will land.

15. Add: $2\dfrac{1}{3} + 5\dfrac{3}{8}$

16. Insert $<$ or $>$ to form a true statement.

$\dfrac{3}{10} \quad \dfrac{2}{7}$

17. Write the decimal 1.3 in words.

18. Round 736.2359 to the nearest tenth.

19. Add: $23.85 + 1.604$

20. Multiply: 0.283×0.3

21. Divide: $0.5 \div 4$

22. Simplify: $0.5(8.6 - 1.2)$

23. Write the following numbers in order from smallest to largest.

$\dfrac{9}{20}, \quad \dfrac{4}{9}, 0.456$

Write each ratio using fractional notation.

24. The ratio of 2.6 to 3.1

25. The ratio of $1\dfrac{1}{2}$ to $7\dfrac{3}{4}$

Percent

This chapter is devoted to percent, a concept used virtually every day in ordinary and business life. Understanding percent and using it efficiently depends on understanding ratios because a percent is a ratio whose denominator is 100. We present techniques to write percents as fractions and as decimals and then solve problems relating to sales tax, commission, discounts, interest, and other real-life situations that use percents.

The Model T, developed by Henry Ford in 1908, was the world's first mass-produced automobile. It sold for $850. In 1913, Ford Motor Company introduced interchangeable parts and the moving assembly line, and the automobile industry as we know it today was born. For almost half a century, American automobile manufacturers produced the majority of the world's motor vehicles. However, in the 1970s and 1980s, the sales of foreign auto imports in the United States began to rise. Imports reached a high in 1987, when 31.3% of all auto sales in the United States were imports. However, American auto producers started to make a comeback, reducing the sales of imported vehicles to a low of 14.9% of all such sales in 1996. Sales of imports have rebounded since then, reaching 28.2% of all auto sales in the United States in the first half of 2001. In Exercises 57, 58, 61 and 62 on page 351 we see some of the ways percents are used by the automobile manufacturing industry.

Name _____ Section _____ Date _____

Chapter 6 Pretest

1. _____

2. _____

3. _____

4. _____

5. _____

6. _____

7. _____

8. _____

9. _____

10. _____

11. _____

12. _____

13. _____

14. _____

15. _____

16. _____

17. _____

18. _____

19. _____

20. _____

1. In a classroom of 100 students, 48 are female. What percent of the students in the classroom are female?

Write each percent as a decimal.

2. 73%

3. 6.8%

Write each decimal as a percent.

4. 0.03

5. 2.1

Write each percent as a fraction in simplest form.

6. 22%

7. 1.5%

Write each fraction or mixed number as a percent.

8. $\dfrac{9}{10}$

9. $2\dfrac{1}{5}$

10. A salesman receives a commission of 3.5% of his total sales. Write 3.5% as a decimal.

11. Translate the following to a percent equation: 4 is what percent of 28?

12. Translate to a proportion: 18% of what number is 70?

Solve.

13. 6 is 15% of what number?

14. What number is 40% of 16?

15. A $230 CD player is on sale at 20% off. What is the discount and what is the sale price?

16. In hopes of increasing sales, a local fast-food chain decreased the price of its hamburger meal from $2.39 to $1.89. What is the percent decrease? Round to the nearest whole percent.

17. Find the sales tax and the total price on a purchase of a $42 dress in a city where the sales tax rate is 6%.

18. A house sold for $125,000, and the real estate agent earned a commission of $5000. Find the rate of commission.

19. Find the simple interest after 3 years on $600 at an interest rate of 8%.

20. Find the monthly payment on a $4000 loan for 4 years. The interest on the 4-year loan is $960.

6.1 Introduction to Percent

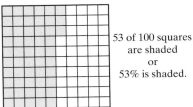
A Understanding Percent

The word **percent** comes from the Latin phrase *per centum*, which means "**per 100**." For example, 53% (percent) means 53 per 100. In the square below, 53 of the 100 squares are shaded. Thus, 53% of the figure is shaded.

53 of 100 squares
are shaded
or
53% is shaded.

Since 53% means 53 per 100, 53% is the ratio of 53 to 100, or $\frac{53}{100}$.

$$53\% = \frac{53}{100}$$

Also,

$$7\% = \frac{7}{100} \qquad \text{7 parts per 100 parts}$$

$$73\% = \frac{73}{100} \qquad \text{73 parts per 100 parts}$$

$$109\% = \frac{109}{100} \qquad \text{109 parts per 100 parts}$$

Percent

Percent means **per one hundred**. The "%" symbol is used to denote percent.

Percent is used in a variety of everyday situations. For example:

The interest rate is 5.7%.
45.4% of U.S. adults accessed the
    Internet in the last 30 days.
The store is having a 25%-off sale.
78% of us trust our local fire department.
Air traveler complaints increased 14%
    this year.
89% of California drivers use their
    seat belts.

EXAMPLE 1

In a survey of 100 people, 17 people drive blue cars. What percent of people drive blue cars?

Solution: Since 17 people out of 100 drive blue cars, the fraction is $\frac{17}{100}$.

Then

$$\frac{17}{100} = 17\%$$

Practice Problem 1

Of 100 students in a club, 23 are freshmen. What percent of the students are freshmen?

Answer
1. 23%

Practice Problem 2

29 out of 100 executives are in their forties. What percent of executives are in their forties?

Practice Problem 3

Write 89% as a decimal.

Practice Problems 4–7

Write each percent as a decimal.

4. 2.7%   5. 150%   6. 0.69%
7. 500%

Concept Check

Why is it incorrect to write the percent 0.033% as 3.3 in decimal form?

Answers

2. 29%,   **3.** 0.89,   **4.** 0.027,   **5.** 1.5,   **6.** 0.0069,
7. 5

Concept Check: To write a percent as a decimal, the decimal point should be moved two places to the left, not to the right. So the correct answer is 0.00033.

EXAMPLE 2

46 out of every 100 college students live at home. What percent of students live at home? (*Source:* Independent Insurance Agents of America)

Solution:

$$\frac{46}{100} = 46\%$$

B Writing Percents as Decimals

Since percent means "per hundred," we have that

$$1\% = \frac{1}{100} = 0.01$$

To write a percent as a decimal, we first write the percent as a fraction.

$$53\% = \frac{53}{100}$$

Now we write the fraction as a decimal as we did in Section 4.7.

$$\frac{53}{100} = 53(0.01) = 0.53 \qquad \text{(53-hundredths)}$$

Notice that the result is

$$53.\% = 53(0.01) = 0.53 \qquad \text{Replace the percent symbol with 0.01. Then multiply.}$$

Writing a Percent as a Decimal

Replace the percent symbol with its decimal equivalent, 0.01; then multiply.

$$43\% = 43(0.01) = 0.43$$

EXAMPLE 3   Write 23% as a decimal.

Solution:

$$23\% = 23(0.01) \qquad \text{Replace the percent symbol with 0.01.}$$
$$= 0.23 \qquad \text{Multiply.}$$

EXAMPLES   Write each percent as a decimal.

4. $4.6\% = 4.6(0.01) = 0.046$ Replace the percent symbol with 0.01. Then multiply.

5. $190\% = 190.(0.01) = 1.90$ or   1.9

6. $0.74\% = 0.74(0.01) = 0.0074$

7. $100\% = 100(0.01) = 1.00$ or   1

Try the Concept Check in the margin.

Ⓒ Writing Decimals as Percents

To write a decimal as a percent, we use the result of Example 7 above. In this example, we found that $1 = 100\%$.

$$0.38 = 0.38(1) = 0.38(100\%) = 38\%$$

Notice that the result is

$$0.38 = 0.38(100\%) = 38.\%$$ Multiply by 1 in the form of 100%.

Writing a Decimal as a Percent

Multiply by 1 in the form of 100%.

$$0.27 = 0.27(100\%) = 27.\%$$

EXAMPLE 8   Write 0.65 as a percent.

Solution:

$$0.65 = 0.65(100\%) = 65.\%$$ Multiply by 100%.
$$= 65\%$$

EXAMPLES   Write each decimal as a percent.

9. $1.25 = 1.25(100\%) = 125.\%$ or   125%

10. $0.012 = 0.012(100\%) = 001.2\%$ or   1.2%

11. $0.6 = 0.6(100\%) = 060.\%$ or   60%

Try the Concept Check in the margin.

Practice Problem 8

Write 0.19 as a percent.

●

Practice Problems 9–11

Write each decimal as a percent.
9. 1.75   10. 0.044   11. 0.7

●

Concept Check

Why is it incorrect to write the decimal 0.0345 as 34.5% in percent form?

Answers

8 19%,   **9.** 175%,   **10.** 4.4%,   **11.** 70%

Concept Check: To change a decimal to a percent, the decimal point should be moved *only* two places to the right. So the correct answer is 3.45%.

FOCUS ON **The Real World**

HOW MUCH CAN YOU AFFORD FOR A HOUSE?

When a home buyer takes out a mortgage to buy a house, the loan is generally repaid on a monthly basis with a monthly mortgage payment. (Some banks also offer biweekly payment programs.) An important consideration in choosing a house is the amount of the monthly payment. Usually, the amount that a home buyer can afford to make as a monthly payment will dictate the house purchase price that can be afforded.

The first step in deciding how much can be afforded for a house is finding out how much income the household has each month before taxes. The Mortgage Bankers Association of American (MBAA) suggests that the monthly mortgage payment be between 25% and 28% of the total monthly income. If other long-term debts exist (such as car or education loans and long-term credit card debt repayment), the MBAA further recommends that the total of housing costs and other monthly debt payments not exceed 36% of the total monthly income.

Once the size of the monthly payment that can be afforded has been found, a mortgage payment calculator can be used to work backward to estimate the mortgage amount that will give that desired monthly payment. For example, the Interest.com Web site includes a mortgage payment calculator at http://www.interest.com/calculators/monthly-payment.shtml. (Alternatively, visit www.interest.com and navigate to "Use our mortgage calculators." Look for the calculator to calculate the monthly payment for a particular mortgage loan.) With this mortgage payment calculator, the user can input the interest rate (as a percent), the term of the loan (in years), and total home loan amount (in dollars). This information is then used to calculate the associated monthly payment. To work backward with this mortgage payment calculator to find the total loan amount that can be afforded:

- Enter the interest rate that is likely for your loan and the term of the loan in which you are interested.
- Then make a guess (perhaps $100,000?) for the total home loan amount that can be afforded.
- Have the mortgage calculator calculate the monthly payment.
- If the monthly payment that is calculated is higher than the range that can be afforded, repeat the calculation using the same interest rate and loan term but a lower value for the total home loan amount.
- If the monthly payment that is calculated is lower than the range that can be afforded, repeat the calculation using the same interest rate and loan term but a higher value for the total home loan amount.
- Repeat these calculations methodically until a monthly payment is obtained that is in the range that can be afforded. The initial principal value that gave this monthly payment amount is an estimate of the mortgage amount that can be afforded to buy a home.

GROUP ACTIVITY

1. Research current interest rates on 30-year mortgages.
2. Use the method described above to find the size of mortgages that can be afforded by households with the following total monthly incomes before taxes. (Assume in each case that the household has no other debts.) Use a loan term of 30 years and a current interest rate on a 30-year mortgage.

a. $1500	**b.** $2000	**c.** $2500
d. $3000	**e.** $3500	**f.** $4000

3. Create a table of your results.

Copyright 2003 Prentice-Hall, Inc.

EXERCISE SET 6.1

A *Solve. See Examples 1 and 2.*

1. A basketball player makes 81 out of 100 attempted free throws. What percent of free throws was made?

2. In a survey of 100 people, 54 preferred chocolate syrup on their ice cream. What percent preferred chocolate syrup?

Adults were asked what type of cookie was their favorite. The circle graph below shows the results for every 100 people. Use this graph to answer Exercises 3 through 6. See Examples 1 and 2.

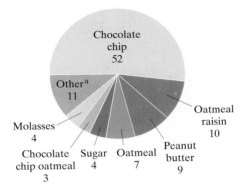

[a]all 1% or less
Source: *USA Today*

3. What percent preferred peanut butter cookies?

4. What percent preferred oatmeal raisin cookies?

5. What type of cookie was preferred by most adults? What percent preferred this type of cookie?

6. What two types of cookies were preferred by the same number of adults? What percent preferred each type?

7. Michigan leads the United States in tart cherry production, producing 75 out of every 100 tart cherries each year. What percent is this? (*Source:* Cherry Marketing Institute)

8. 51 out of 100 adults ages 30 to 49 say the best way to meet a potential date is through volunteer activities. What percent is this? (*Source:* NFO Research for Combe)

Write each percent as a decimal. See Examples 3 through 7.

9. 48%   **10.** 64%   **11.** 6%   **12.** 9%

13. 100%   **14.** 136%   **15.** 61.3%   **16.** 52.7%

17. 2.8%   **18.** 1.7%   **19.** 0.6%   **20.** 0.9%

21. 300%   **22.** 500%   **23.** 32.58%   **24.** 72.18%

Write each percent as a decimal. See Examples 3 through 7.

25. 67% of American men are happy with their current weight. (*Source:* Gallup for Wheat Foods Council)

26. About 95% of tableservice restaurants in the United States include appetizers on their menus. (*Source:* National Restaurant Association)

27. In June 2001, the unemployment rate was 4.5%. (*Source:* Bureau of Labor Statistics)

28. Approximately 14.7% of new compact or sports cars are silver, making silver the most popular new car color for that class. (*Source:* Ward's Communications)

29. Video games made up 21.2% of the total toy market in the United States in 2000. (*Source:* The NPD Group Worldwide)

30. In 2000, rock music accounted for 24.8% of recorded music sales in the United States. (*Source:* Recording Industry Association of America)

Write each decimal as a percent. See Examples 8 through 11.

31. 0.98   **32.** 0.75   **33.** 3.1   **34.** 4.8   **35.** 29.00

36. 56.00   **37.** 0.003   **38.** 0.006   **39.** 0.22   **40.** 0.45

41. 5.3   **42.** 1.6   **43.** 0.056   **44.** 0.027   **45.** 0.3328

46. 0.1115   **47.** 3.00   **48.** 5.00   **49.** 0.7   **50.** 0.8

Write each decimal as a percent. See Examples 8 through 11.

51. The Munoz family saves 0.10 of their take-home pay.

52. The cost of an item for sale is 0.7 of the sale price.

53. Nearly 0.093 of people in the United States are affected by pollen allergies. (*Source:* National Institute of Allergy and Infectious Diseases)

54. About 0.25 of the world's automobiles are produced in North America. (*Source:* Automotive Intelligence)

55. People take aspirin for a variety of reasons. The most common use of aspirin is to prevent heart disease, accounting for 0.38 of all aspirin use. (*Source:* Bayer Market Research)

56. According to the 2000 census, 0.509 of the American population is female. (*Source:* U.S. Census Bureau)

Review and Preview

Write each fraction as a decimal. See Section 4.7.

57. $\dfrac{1}{4}$ **58.** $\dfrac{3}{5}$ **59.** $\dfrac{13}{20}$ **60.** $\dfrac{11}{40}$ **61.** $\dfrac{9}{10}$ **62.** $\dfrac{7}{10}$

The bar graph shows the predicted fastest-growing occupations. Use this graph for Exercises 63 through 66.

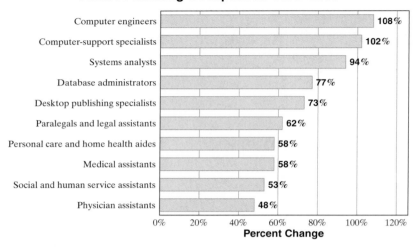

Fastest-Growing Occupations 1998–2008

Source: Bureau of Labor Statistics

63. What occupation is predicted to be the fastest growing?

64. What occupation is predicted to be the second fastest growing?

65. Write the percent change for database administrators as a decimal.

66. Write the percent change for medical assistants as a decimal.

 67. In your own words, explain how to write a percent as a decimal.

68. In your own words, explain how to write a decimal as a percent.

STUDY SKILLS REMINDER

Is your notebook still organized?

Is your notebook still organized? If it's not, it's not too late to start organizing it. Start writing your notes and completing your homework assignment in a notebook with pockets (spiral or ring binder). Take class notes in this notebook, and then follow the notes with your completed homework assignment. When you receive graded papers or handouts, place them in the notebook pocket so that you will not lose them.

Remember to mark (possibly with an explanation point) any note(s) that seem extra important to you. Also remember to mark (possibly with a question mark) any notes or homework that you are having trouble with. Don't forget to see your instructor or a math tutor to help you with the concepts or exercises that you are having trouble understanding.

Also—don't forget to write neatly and keep a positive attitude.

6.2 Percents and Fractions

A   Writing Percents as Fractions

When we write a percent as a fraction, we usually then write the fraction in simplest form. For example, recall from the previous section that

$$50\% = \frac{50}{100}$$

Then we write the fraction in simplest form:

$$\frac{50}{100} = \frac{\overset{1}{\cancel{50}}}{2 \cdot \underset{1}{\cancel{50}}} = \frac{1}{2}$$

OBJECTIVES

A   Write percents as fractions.
B   Write fractions as percents.
C   Convert percents, decimals, and fractions.

SSM   TUTOR CENTER   SG   CD & VIDEO   MATH PRO   WEB

Writing a Percent as a Fraction

Replace the percent symbol with its fraction equivalent, $\frac{1}{100}$; then multiply. Don't forget to simplify the fraction if possible.

$$7\% = 7 \cdot \frac{1}{100} = \frac{7}{100}$$

EXAMPLES

Write each percent as a fraction or mixed number in simplest form.

1. $40\% = 40 \cdot \frac{1}{100} = \frac{40}{100} = \frac{2 \cdot \overset{1}{\cancel{20}}}{5 \cdot \underset{1}{\cancel{20}}} = \frac{2}{5}$

2. $1.9\% = 1.9 \cdot \frac{1}{100} = \frac{1.9}{100}$. Next we multiply the numerator and denominator of 1.9 by 10.

$$= \frac{1.9 \cdot 10}{1 \cdot 10} \cdot \frac{1}{100} = \frac{19}{10} \cdot \frac{1}{100} = \frac{19}{1000}$$

3. $125\% = 125 \cdot \frac{1}{100} = \frac{125}{100} = \frac{5 \cdot \overset{1}{\cancel{25}}}{4 \cdot \underset{1}{\cancel{25}}} = \frac{5}{4} \text{ or } 1\frac{1}{4}$

4. $33\frac{1}{3}\% = \underbrace{33\frac{1}{3}}_{\substack{\rightarrow\text{Write as}}} \cdot \frac{1}{100} = \frac{100}{3} \cdot \frac{1}{100} = \frac{\overset{1}{\cancel{100}} \cdot 1}{3 \cdot \underset{1}{\cancel{100}}} = \frac{1}{3}$

  → Write as an improper fraction.

5. $100\% = 100 \cdot \frac{1}{100} = \frac{100}{100} = 1$

B   Writing Fractions as Percents

Recall that to write a percent as a fraction, we drop the percent symbol and divide by 100. We reverse these steps to write a fraction as a percent.

Practice Problems 1–5

Write each percent as a fraction in simplest form.

1.   25%
2.   2.3%
3.   150%
4. $66\frac{2}{3}\%$
5.   8%

Answers

1. $\frac{1}{4}$,   **2.** $\frac{23}{1000}$,   **3.** $\frac{3}{2}$,   **4.** $\frac{2}{3}$,   **5.** $\frac{2}{25}$

Writing a Fraction as a Percent

Multiply by 1 in the form of 100%.

$$\frac{1}{8} = \frac{1}{8} \cdot 100\% = \frac{1}{8} \cdot \frac{100}{1}\% = \frac{100}{8}\% = 12\frac{1}{2}\% \quad \text{or} \quad 12.5\%$$

Helpful Hint

From Example 5, we know that

$$100\% = 1$$

Recall that when we multiply a number by 1, we are not changing the value of that number. This means that when we multiply a number by 100%, we are not changing its value but rather writing the number as an equivalent percent.

Practice Problems 6–8

Write each fraction or mixed number as a percent.

6. $\frac{1}{2}$ 7. $\frac{7}{40}$ 8. $2\frac{1}{4}$

Concept Check

Which digit in the percent 76.4582% represents

a. A tenth percent?
b. A thousandth percent?
c. A hundredth percent?
d. A whole percent?

Practice Problem 9

Write $\frac{3}{17}$ as a percent. Round to the nearest hundredth percent.

EXAMPLES Write each fraction or mixed number as a percent.

6. $\frac{9}{20} = \frac{9}{20} \cdot 100\% = \frac{9}{20} \cdot \frac{100}{1}\% = \frac{900}{20}\% = 45\%$

7. $\frac{2}{3} = \frac{2}{3} \cdot 100\% = \frac{2}{3} \cdot \frac{100}{1}\% = \frac{200}{3}\% = 66\frac{2}{3}\%$

8. $1\frac{1}{2} = \frac{3}{2} \cdot 100\% = \frac{3}{2} \cdot \frac{100}{1}\% = \frac{300}{2}\% = 150\%$

Try the Concept Check in the margin.

EXAMPLE 9

Write $\frac{1}{12}$ as a percent. Round to the nearest hundredth percent.

Solution:

$$\frac{1}{12} = \frac{1}{12} \cdot 100\% = \frac{1}{12} \cdot \frac{100\%}{1} = \frac{100}{12}\% \approx 8.33\%$$

$$\begin{array}{r} 8.333 \approx 8.33 \\ 12\overline{)100.000} \\ -96 \\ \hline 4\,0 \\ -3\,6 \\ \hline 40 \\ -36 \\ \hline 40 \\ -36 \\ \hline 4 \end{array}$$

Thus, $\frac{1}{12}$ is approximately 8.33%.

Answers

6. 50%,   7. $17\frac{1}{2}\%$,   8. 225%,   9. 17.65%

Concept Check: a. 4,   b. 8,   c. 5,   d. 6

Ⓒ Converting Percents, Decimals, and Fractions

Let's summarize what we have learned so far about percents, decimals, and fractions:

Summary of Converting Percents, Decimals, and Fractions

- *To write a percent as a decimal,* replace the % symbol with its decimal equivalent, 0.01; then multiply.
- *To write a percent as a fraction,* replace the % symbol with its fraction/equivalent, $\frac{1}{100}$; then multiply.
- *To write a decimal or fraction as a percent,* multiply by 100%.

EXAMPLE 10

17.8% of automobile thefts in the continental United States occur in the Midwest. Write this percent as a decimal. (*Source:* The American Automobile Manufacturers Association)

Solution:

$$17.8\% = 17.8(0.01) = 0.178.$$

Thus, 17.8% written as a decimal is 0.178.

EXAMPLE 11

An advertisement for a stereo system reads "$\frac{1}{4}$ off." What percent off is this?

Solution: Write $\frac{1}{4}$ as a percent.

$$\frac{1}{4} = \frac{1}{4} \cdot 100\% = \frac{1}{4} \cdot \frac{100\%}{1} = \frac{100}{4}\% = 25\%$$

Thus, "$\frac{1}{4}$ off" is the same as "25% off."

It is helpful to know a few basic percent conversions. Appendix D contains a handy reference of percent, decimal, and fraction equivalencies.

Practice Problem 10

A family decides to spend no more than 25% of its monthly income on rent. Write 25% as a decimal.

Practice Problem 11

Provincetown's budget for waste disposal increased by $1\frac{1}{4}$ times over the budget from last year. What percent increase is this?

Answers

10. 0.25,   **11.** 125%

Tips for studying for an exam

To prepare for an exam, try the following study techniques.

- Start the study process days before your exam.

- Make sure that you are current and up to date on your assignments.

- If there is a topic that you are unsure of, use one of the many resources that are available to you. For example,

  See your instructor.

  Visit a learning resource center on campus where math tutors are available.

  Read the textbook material and examples on the topic.

  View a videotape on the topic.

- Reread your notes and carefully review the Chapter Highlights at the end of the chapter.

- Work the review exercises at the end of the chapter and check your answers. Make sure that you correct any missed exercises. If you have trouble on a topic, use a resource listed above.

- Find a quiet place to take the Chapter Test found at the end of the chapter. Do not use any resources when taking this sample test. This way you will have a clear indication of how prepared you are for your exam. Check your answers and make sure that you correct any missed exercises.

- Get lots of rest the night before the exam. It's hard to show how well you know the material if your brain is foggy from lack of sleep.

Good luck and keep a positive attitude.

Mental Math

Write each fraction as a percent.

1. $\dfrac{13}{100}$

2. $\dfrac{92}{100}$

3. $\dfrac{87}{100}$

4. $\dfrac{71}{100}$

5. $\dfrac{1}{100}$

6. $\dfrac{2}{100}$

EXERCISE SET 6.2

A *Write each percent as a fraction or mixed number in simplest form. See Examples 1 through 5.*

1. 12%

2. 24%

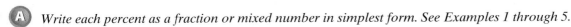 3. 4%

4. 2%

5. 4.5%

6. 7.5%

7. 175%

8. 250%

9. 73%

10. 86%

11. 12.5%

12. 62.5%

13. 6.25%

14. 37.5%

15. 8%

16. 16%

17. $10\dfrac{1}{3}\%$

18. $7\dfrac{3}{4}\%$

19. $22\dfrac{3}{8}\%$

20. $15\dfrac{5}{8}\%$

B *Write each fraction or mixed number as a percent. See Examples 6 through 8.*

21. $\dfrac{3}{4}$

22. $\dfrac{1}{2}$

23. $\dfrac{7}{10}$

24. $\dfrac{3}{10}$

25. $\dfrac{2}{5}$

26. $\dfrac{4}{5}$

27. $\dfrac{59}{100}$

28. $\dfrac{73}{100}$

29. $\dfrac{17}{50}$

30. $\dfrac{47}{50}$

31. $\dfrac{3}{8}$

32. $\dfrac{5}{8}$

33. $\dfrac{5}{16}$

34. $\dfrac{7}{16}$

35. $1\dfrac{3}{5}$

36. $1\dfrac{3}{4}$

37. $\dfrac{2}{3}$

38. $\dfrac{1}{3}$

39. $\dfrac{13}{20}$ **40.** $\dfrac{3}{20}$ **41.** $2\dfrac{1}{2}$ **42.** $2\dfrac{1}{5}$ **43.** $1\dfrac{9}{10}$ **44.** $2\dfrac{7}{10}$

Write each fraction as a percent. Round to the nearest hundredth percent. See Example 9.

45. $\dfrac{7}{11}$ **46.** $\dfrac{5}{12}$ **47.** $\dfrac{4}{15}$ **48.** $\dfrac{10}{11}$

49. $\dfrac{1}{7}$ **50.** $\dfrac{1}{9}$ **51.** $\dfrac{11}{12}$ **52.** $\dfrac{5}{6}$

C *Complete each table. See Examples 10 and 11.*

53.

Percent	Decimal	Fraction
35%		
		$\dfrac{1}{5}$
	0.5	
70%		
		$\dfrac{3}{8}$

54.

Percent	Decimal	Fraction
	0.525	
		$\dfrac{3}{4}$
$66\dfrac{2}{3}\%$		
		$\dfrac{5}{6}$
100%		

55.

Percent	Decimal	Fraction
40%		
	0.235	
		$\dfrac{4}{5}$
$33\dfrac{1}{3}\%$		
		$\dfrac{7}{8}$
7.5%		

56.

Percent	Decimal	Fraction
50%		
		$\dfrac{2}{5}$
	0.25	
12.5%		
		$\dfrac{5}{8}$
		$\dfrac{7}{50}$

Solve. See Examples 10 and 11.

57. Approximately 14.8% of new luxury cars are silver, making silver the most popular new vehicle color for that class. Write this percent as a fraction. (*Source:* Ward's Communications)

58. In 1950, the United States produced 75.7% of all motor vehicles made worldwide. Write this percent as a decimal. (*Source:* American Automobile Manufacturers Association)

59. In 2000, 40.2% of Americans' meat expenditures were on beef products. Write this percent as a decimal. (*Source:* National Cattlemen's Beef Association)

60. 52% of Americans say that their ideal family size is fewer than three children. Write this percent as a fraction. (*Source:* Gallup)

61. In the first half of 2001, $\frac{141}{500}$ of all new cars sold in the United States were imports. Write this fraction as a percent. (*Source:* Ward's AutoInfoBank)

62. In 1997, $\frac{41}{250}$ of all new cars sold in the United States were imports. Write this fraction as a percent. (*Source:* Ward's Communications)

63. The sales tax in Slidell, Louisiana, is 8.25%. Write this percent as a decimal.

64. A real estate agent receives a commission of 3% of the sale price of a house. Write this percent as a decimal.

65. In the 2000/2001 television season, one of the top-rated shows was *Who Wants to Be a Millionaire?*, which had an average audience share of $\frac{11}{50}$ of all those watching television during that time slot. Write this fraction as a percent. (*Source:* Nielsen Media Research)

66. The 2000 National Assessment of Educational Progress showed that $\frac{13}{50}$ of U.S. fourth-graders were proficient in math. Write this fraction as a percent. (*Source:* National Center for Education Statistics)

Review and Preview

Find the value of n. See Section 5.3.

67. $3 \cdot n = 45$

68. $7 \cdot n = 48$

69. $8 \cdot n = 80$

70. $2 \cdot n = 16$

71. $6 \cdot n = 72$

72. $5 \cdot n = 35$

Combining Concepts

Write each fraction as a decimal and then write each decimal as a percent. Round the decimal to three decimal places and the percent to the nearest tenth of a percent.

 73. $\dfrac{21}{79}$

74. $\dfrac{56}{102}$

75. $\dfrac{850}{736}$

76. $\dfrac{506}{248}$

Fill in the blanks.

77. A fraction written as a percent is greater than 100% when the numerator is _____ than the denominator.
(greater/less)

78. A decimal written as a percent is less than 100% when the decimal is _____ than 1.
(greater/less)

79. In your own words, explain how to write a percent as a fraction.

80. In your own words, explain how to write a fraction as a decimal.

Internet Excursions

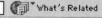

In basketball, one statistic that can be computed is the field goal percentage (FG%). This is computed by dividing total field goals made (FGM) by total field goals attempted (FGA).

This World Wide Web address will provide you with access to the official Web site of the Women's National Basketball Association (WNBA), or a related site. At this site, choose two WNBA players and locate the career stats of each.

81. Name the player you have chosen and choose a year in her career. What is her field goal percentage written as a decimal? Rewrite this statistic as a percent. Now use the data "field goals made" and "field goals attempted" to write this statistic as a fraction.

82. Repeat Exercise 81 for the second WNBA player you have chosen.

6.3 Solving Percent Problems Using Equations

Throughout this text, we have written mathematical statements such as $3 + 10 = 13$, or area = length·width. These statements are called "equations." An equation is simply a statement that contains an equal sign. To solve percent problems, we translate the problems into such mathematical statements, or equations.

OBJECTIVES

Ⓐ Write percent problems as equations.

Ⓑ Solve percent problems

SSM
TUTOR CENTER   SG   CD & VIDEO   MATH PRO   WEB

Ⓐ Writing Percent Problems as Equations

Recognizing key words in a percent problem is helpful in writing the problem as an equation. Three key words in the statement of a percent problem and their meanings are as follows:

of means **multiplication** (·)
is means **equals** (=)
what (or some equivalent) means **the unknown number**

In our examples, we let the letter n stand for the unknown number.

EXAMPLE 1   Translate to an equation.

　　　5 is what percent of 20?

Solution:　5 is what percent of 20?
　　　　　↓ ↓　　　↓　　　↓ ↓
　　　　　5 =　　　n　　　· 20

Practice Problem 1

Translate: 6 is what percent of 24?

Helpful Hint

Remember that an equation is simply a mathematical statement that contains an equal sign (=).

　　　$5 = n \cdot 20$
　　　　　↑
　　　equal sign

EXAMPLE 2   Translate to an equation.

　　　1.2 is 30% of what number?

Solution:　1.2 is 30% of what number?
　　　　　↓　↓　↓　↓　　　↓
　　　　　1.2 = 30% ·　　　n

Practice Problem 2

Translate: 1.8 is 20% of what number?

Practice Problem 3

Translate: What number is 40% of 3.6?

EXAMPLE 3   Translate to an equation.

　　　What number is 25% of 0.008?

Solution:　What number is 25% of 0.008?
　　　　　　↓　　　↓ ↓　↓　↓
　　　　　n　　　= 25% · 0.008

EXAMPLES   Translate each of the following to an equation:

4. 38% of 200 is what number?
　　　↓　　↓　↓ ↓　　　↓
　　　38% · 200 =　　　n

Practice Problems 4–6

Translate each to an equation.

4. 42% of 50 is what number?
5. 15% of what number is 9?
6. What percent of 150 is 90?

Answers

1. $6 = n \cdot 24$,　**2.** $1.8 = 20\% \cdot n$,
3. $n = 40\% \cdot 3.6$,　**4.** $42\% \cdot 50 = n$,
5. $15\% \cdot n = 9$,　**6.** $n \cdot 150 = 90$

5. 40% of what number is 80?

$$40\% \cdot n = 80$$

6. What percent of 85 is 34?

$$n \cdot 85 = 34$$

Concept Check

In the equation $2 \cdot n = 10$, what step should be taken to solve the equation?

Try the Concept Check in the margin.

B   **Solving Percent Problems**

You may have noticed by now that each percent problem has contained three numbers—in our examples, two are known and one is unknown. Each of these numbers is given a special name.

$$
\begin{array}{ccccc}
15\% & \text{of} & 60 & \text{is} & 9
\end{array}
$$

$$
\underset{\text{percent}}{15\%} \cdot \underset{\text{base}}{60} = \underset{\text{amount}}{9}
$$

We call this equation the **percent equation**.

> **Percent Equation**
>
> percent · base = amount

> **Helpful Hint**
>
> Notice that the percent equation given above is a true statement. To see this, simplify the left side as shown:
>
> $15\% \cdot 60 = 9$
> $0.15 \cdot 60 = 9$ Write 15% as 0.15.
> $9 = 9$ Multiply.
>
> The statement $9 = 9$ is true.

After a percent problem has been written as a percent equation, we can use the equation to find the unknown number. This is called **solving** the equation.

Solving Percent Equations for the Amount

EXAMPLE 7

Practice Problem 7

What number is 20% of 85?

What number is 35% of 40?

Solution:

$$
\begin{array}{ll}
n = 35\% \cdot 40 & \text{Translate to an equation.} \\
n = 0.35 \cdot 40 & \text{Write 35\% as 0.35.} \\
n = 14 & \text{Multiply:}
\end{array}
$$

$$
\begin{array}{r}
40 \\
0.35 \\
\hline
2\,00 \\
12\,00 \\
\hline
14.00
\end{array}
$$

Thus, 14 is 35% of 40.

Answers

7. 17

Concept Check: If $2 \cdot n = 10$, then $n = \dfrac{10}{2}$, or $n = 5$.

Helpful Hint

When solving a percent equation, write the percent as a decimal or fraction.

EXAMPLE 8

85% of 300 is what number?

Solution:

85%	$\cdot$	300	$=$	n	Translate to an equation.
0.85	$\cdot$	300	$=$	n	Write 85% as 0.85.
		255	$=$	n	Multiply: $0.85 \cdot 300 = 255$.

Thus, 85% of 300 is 255.

Solving Percent Equations for the Base

EXAMPLE 9

12% of what number is 0.6?

Solution:

12%	$\cdot$	n	$= 0.6$	Translate to an equation.
0.12	$\cdot$	n	$= 0.6$	Write 12% as 0.12.

Recall from Section 5.3 that if "0.12 times some number is 0.6," then the number is 0.6 divided by 0.12.

$$n = \frac{0.6}{0.12} \quad \text{Divide 0.6 by 0.12, the number multiplied by } n.$$

$$n = 5$$

$$\begin{array}{r} 5. \\ 0.12\overline{)0.60} \\ \underline{60} \\ 0 \end{array}$$

Thus, 12% of 5 is 0.6.

EXAMPLE 10

13 is $6\frac{1}{2}\%$ of what number?

Solution:

$$13 = 6\frac{1}{2}\% \cdot n \quad \text{Translate to an equation.}$$

$$13 = 0.065 \cdot n \quad 6\frac{1}{2}\% = 6.5\% = 0.065.$$

$$\frac{13}{0.065} = n \quad \begin{array}{l} \text{Divide 13 by 0.065, the} \\ \text{number multiplied by } n. \end{array}$$

$$200 = n$$

$$\begin{array}{r} 200. \\ 0.065\overline{)13.000} \\ \underline{130} \\ 0 \end{array}$$

Thus, 13 is $6\frac{1}{2}\%$ of 200.

Practice Problem 8

90% of 150 is what number?

Practice Problem 9

15% of what number is 1.2?

Practice Problem 10

27 is $4\frac{1}{2}\%$ of what number?

Answers

8. 135, **9.** 8, **10.** 600

Practice Problem 11

What percent of 80 is 8?

Solving Percent Equations for Percent

EXAMPLE 11

$$\underbrace{\text{What percent}}_{\downarrow} \quad \text{of} \quad \underset{\downarrow}{12} \quad \underset{\downarrow}{\text{is}} \quad \underset{\downarrow}{9}?$$

Solution:

$$n \quad \cdot \quad 12 \quad = \quad 9 \qquad \text{Translate to an equation.}$$

$$n \qquad\qquad = \frac{9}{12} \qquad \begin{array}{l}\text{Divide 9 by 12,}\\ \text{the number multiplied by } n.\end{array}$$

$$n \qquad\qquad = 0.75$$

Next, since we are looking for percent, we write 0.75 as a percent.

$$n = 75\%$$

So, 75% of 12 is 9.

> **Helpful Hint**
>
> If your unknown in the percent equation is the percent, don't forget to convert your answer to a percent.

Practice Problem 12

35 is what percent of 25?

EXAMPLE 12

$$\underset{\downarrow}{78} \quad \underset{\downarrow}{\text{is}} \quad \underbrace{\text{what percent}}_{\downarrow} \quad \text{of} \quad \underset{\downarrow}{65}?$$

Solution:

$$78 \quad = \quad n \quad \cdot \quad 65 \qquad \text{Translate to an equation.}$$

$$\frac{78}{65} \quad = \quad n \qquad\qquad \begin{array}{l}\text{Divide 78 by 65,}\\ \text{the number multiplied by } n.\end{array}$$

$$1.2 \quad = \quad n$$

$$120\% = \quad n \qquad\qquad \text{Write 1.2 as a percent.}$$

So, 78 is 120% of 65.

Answers

11. 10%, **12.** 140%

Name _____ Section _____ Date _____

Mental Math

Identify the percent, the base, and the amount in each equation. Recall that percent · base = amount.

1. $42\% \cdot 50 = 21$

2. $30\% \cdot 65 = 19.5$

3. $107.5 = 125\% \cdot 86$

4. $99 = 110\% \cdot 90$

EXERCISE SET 6.3

 **A** *Translate each to an equation. Do not solve. See Examples 1 through 6.*

1. 15% of 72 is what number?

2. What number is 25% of 55?

3. 30% of what number is 80?

4. 0.5 is 20% of what number?

5. What percent of 90 is 20?

6. 8 is 50% of what number?

7. 1.9 is 40% of what number?

8. 72% of 63 is what number?

9. What number is 9% of 43?

10. 4.5 is what percent of 45?

B *Solve. See Examples 7 and 8.*

11. 10% of 35 is what number?

12. 25% of 60 is what number?

13. What number is 14% of 52?

14. What number is 30% of 17?

Solve. See Examples 9 and 10.

15. 30 is 5% of what number?

16. 25 is 25% of what number?

17. 1.2 is 12% of what number?

18. 0.22 is 44% of what number?

Solve. See Examples 11 and 12.

19. 66 is what percent of 60?

20. 30 is what percent of 20?

21. 16 is what percent of 50?

22. 27 is what percent of 50?

Solve. See Examples 7 through 12.

23. 0.1 is 10% of what number?

24. 0.5 is 5% of what number?

25. 125% of 36 is what number?

26. 200% of 13.5 is what number?

27. 82.5 is $16\frac{1}{2}$% of what number?

28. 7.2 is $6\frac{1}{4}$% of what number?

29. 126 is what percent of 31.5?

30. 264 is what percent of 33?

31. What number is 42% of 60?

32. What number is 36% of 80?

33. What percent of 150 is 67.5?

34. What percent of 105 is 88.2?

35. 120% of what number is 42?

36. 160% of what number is 40?

Review and Preview

Find the value of n in each proportion. See Section 5.3.

37. $\dfrac{27}{n} = \dfrac{9}{10}$

38. $\dfrac{35}{n} = \dfrac{7}{5}$

39. $\dfrac{n}{5} = \dfrac{8}{11}$

40. $\dfrac{n}{3} = \dfrac{6}{13}$

Write each phrase as a proportion.

41. 17 is to 12 as n is to 20

42. 20 is to 25 as n is to 10

43. 8 is to 9 as 14 is to n

44. 5 is to 6 as 15 is to n

 Combining Concepts

Solve.

45. 1.5% of 45,775 is what number?

46. What percent of 75,528 is 27,945.36?

47. 22,113 is 180% of what number?

48. In your own words, explain how to solve a percent equation.

6.4 Solving Percent Problems Using Proportions

There is more than one method that can be used to solve percent problems. In the last section, we used the percent equation. In this section, we will use proportions.

A   Writing Percent Problems as Proportions

To understand the proportion method, recall that 70% means the ratio of 70 to 100, or $\frac{70}{100}$.

$$70\% = \frac{70}{100} = \frac{7}{10}$$

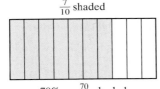

$\frac{7}{10}$ shaded

70% or $\frac{70}{100}$ shaded

Since the ratio $\frac{70}{100}$ is equal to the ratio $\frac{7}{10}$, we have the proportion

$$\frac{7}{10} = \frac{70}{100}.$$

We call this proportion the "percent proportion." In general, we can name the parts of this proportion as follows:

When we translate percent problems to proportions, the **percent**, p, can be identified by looking for the symbol % or the word *percent*. The **base**, b, usually follows the word *of*. The **amount**, a, is the part compared to the whole.

Percent Proportion

$$\frac{\text{amount}}{\text{base}} = \frac{\text{percent}}{100} \quad \leftarrow \text{always } 100$$

or

$$\text{amount} \rightarrow \frac{a}{b} = \frac{p}{100} \quad \leftarrow \text{percent}$$
$$\text{base} \rightarrow$$

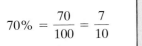 Helpful Hint

Part of Proportion	How It's Identified
Percent	% or percent
Base	Appears after *of*
Amount	Part compared to whole

EXAMPLE 1   Translate to a proportion.

12% of what number is 47?

Solution:

percent

base
It appears after the word *of*.

amount
It is the part compared to the whole.

$$\text{amount} \rightarrow \frac{47}{b} = \frac{12}{100} \quad \leftarrow \text{percent}$$
$$\text{base} \rightarrow$$

OBJECTIVES

A Write percent problems as proportions.

B Solve percent problems.

SSM   TUTOR CENTER   SG   CD & VIDEO   MATH PRO   WEB

Practice Problem 1

Translate to a proportion.
15% of what number is 55?

Answer

1. $\dfrac{15}{100} = \dfrac{55}{b}$

Practice Problem 2

Translate to a proportion.
35 is what percent of 70?

Practice Problem 3

Translate to a proportion.
What number is 25% of 68?

Practice Problem 4

Translate to a proportion.
520 is 65% of what number?

Practice Problem 5

Translate to a proportion.
65 is what percent of 50?

Practice Problem 6

Translate to a proportion.
36% of 80 is what number?

Concept Check

When solving a percent problem by using a proportion, describe how you can check the result.

Answers

2. $\dfrac{35}{70} = \dfrac{p}{100}$,   3. $\dfrac{a}{68} = \dfrac{25}{100}$,   4. $\dfrac{520}{b} = \dfrac{65}{100}$,

5. $\dfrac{65}{50} = \dfrac{p}{100}$,   6. $\dfrac{a}{80} = \dfrac{36}{100}$

Concept Check: by putting the result into the proportion and checking that the proportion is true

---

EXAMPLE 2   Translate to a proportion.

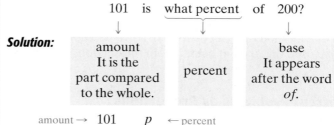

Solution:

$$\text{amount} \rightarrow \frac{101}{200} = \frac{p}{100} \leftarrow \text{percent}$$
$$\text{base} \rightarrow$$

EXAMPLE 3   Translate to a proportion.

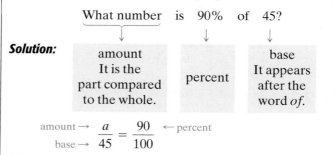

Solution:

$$\text{amount} \rightarrow \frac{a}{45} = \frac{90}{100} \leftarrow \text{percent}$$
$$\text{base} \rightarrow$$

EXAMPLE 4   Translate to a proportion.

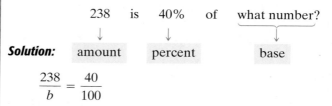

Solution:

$$\frac{238}{b} = \frac{40}{100}$$

EXAMPLE 5   Translate to a proportion.

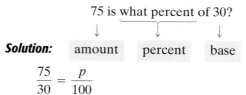

Solution:

$$\frac{75}{30} = \frac{p}{100}$$

EXAMPLE 6   Translate to a proportion.

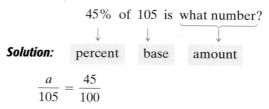

Solution:

$$\frac{a}{105} = \frac{45}{100}$$

Try the Concept Check in the margin.

B   Solving Percent Problems

The proportions that we have written in this section contain three values that can change: The percent number, the base, and the amount. If any two of these values are known, we can find the third (the unknown value). To do this, we write a percent proportion and find the unknown value as we did in Section 5.3.

EXAMPLE 7   Solving Percent Proportion for the Amount

What number is 30% of 9?

Solution: amount   percent   base

$$\frac{a}{9} = \frac{30}{100}$$

To solve, we set cross products equal to each other.

$$\frac{a}{9} = \frac{30}{100}$$ $9 \cdot 30 = 270$
$a \cdot 100$

$a \cdot 100 = 270$ Set cross products equal.

Recall from Section 5.3 that if "some number times 100 is 270," then the number is 270 divided by 100.

$$a = \frac{270}{100}$$ Divide 270 by 100, the number multiplied by a.

$a = 2.7$ Simplify.

Thus, 2.7 is 30% of 9.

Try the Concept Check in the margin.

Helpful Hint

The proportion in Example 7 contains the ratio $\frac{30}{100}$. A ratio in a proportion may be simplified before solving the proportion. The unknown number in both

$$\frac{a}{9} = \frac{30}{100} \quad \text{and} \quad \frac{a}{9} = \frac{3}{10} \quad \text{is 2.7}$$

EXAMPLE 8   Solving Percent Problems for the Base

150% of what number is 30?

Solution: percent   base   amount

$$\frac{30}{b} = \frac{150}{100}$$ Write the proportion.

$$\frac{30}{b} = \frac{3}{2}$$ Simplify $\frac{150}{100}$.

$30 \cdot 2 = b \cdot 3$ Set cross products equal.
$60 = b \cdot 3$ Write $30 \cdot 2$ as 60.

$$\frac{60}{3} = b$$ Divide 60 by 3, the number multiplied by b.

$20 = b$ Simplify.

Thus, 150% of 20 is 30.

Practice Problem 7

What number is 8% of 120?

Concept Check

In the statement "78 is what percent of 350?", which part of the percent proportion is unknown: a. the amount, b. the base, or c. the percent number?

Practice Problem 8

75% of what number is 60?

Answers
7. 9.6,   **8.** 80

Concept Check: c

Practice Problem 9

15 is 5% of what number?

EXAMPLE 9

$$\underset{\downarrow}{20.8} \quad is \quad \underset{\downarrow}{40\%} \quad of \quad \underset{\downarrow}{\underline{what\ number?}}$$

Solution: amount   percent   base

$$\frac{20.8}{b} = \frac{40}{100} \quad or \quad \frac{20.8}{b} = \frac{2}{5} \qquad \text{Write the proportion and simplify } \frac{40}{100}.$$

$$20.8 \cdot 5 = b \cdot 2 \qquad \text{Set cross products equal.}$$

$$104 = b \cdot 2 \qquad \text{Multiply.}$$

$$\frac{104}{2} = b \qquad \text{Divide 104 by 2, the number multiplied by } b.$$

$$52 = b \qquad \text{Simplify.}$$

So, 20.8 is 40% of 52.

Practice Problem 10

What percent of 40 is 5?

EXAMPLE 10   Solving Percent Problems for the Percent

$$\underset{\downarrow}{\underline{What\ percent}} \quad of \quad \underset{\downarrow}{50} \quad is \quad \underset{\downarrow}{8}?$$

Solution: percent   base   amount

$$\frac{8}{50} = \frac{p}{100} \quad or \quad \frac{4}{25} = \frac{p}{100} \qquad \text{Write the proportion and simplify } \frac{8}{50}.$$

$$4 \cdot 100 = 25 \cdot p \qquad \text{Set cross products equal.}$$

$$400 = 25 \cdot p \qquad \text{Multiply.}$$

$$\frac{400}{25} = p \qquad \text{Divide 400 by 25, the number multiplied by } p.$$

$$16 = p \qquad \text{Simplify.}$$

So, 16% of 50 is 8.

> **Helpful Hint**
>
> Recall from our percent proportion that this number already is a percent. Just keep the number as is and attach a % symbol.

Practice Problem 11

What percent of 160 is 336?

EXAMPLE 11

$$\underset{\downarrow}{504} \quad is \quad \underset{\downarrow}{\underline{what\ percent}} \quad of \quad \underset{\downarrow}{360}?$$

Solution: amount   percent   base

$$\frac{504}{360} = \frac{p}{100}$$

Let's choose not to simplify the ratio $\frac{504}{360}$.

$$504 \cdot 100 = 360 \cdot p \qquad \text{Set cross products equal.}$$

$$50,400 = 360 \cdot p \qquad \text{Multiply.}$$

$$\frac{50,400}{360} = p \qquad \text{Divide 50, 400 by 360, the number multiplied by } p.$$

$$140 = p \qquad \text{Simplify.}$$

Notice that by choosing not to simplify $\frac{504}{360}$, we had larger numbers in our equation. Either way, we find that 504 is 140% of 360.

Answers

9. 300,   **10.** 12.5%,   **11.** 210%

Mental Math

Identify the amount, the base, and the percent in each equation. Recall that $\dfrac{\text{amount}}{\text{base}} = \dfrac{\text{percent}}{100}$.

1. $\dfrac{12.6}{42} = \dfrac{30}{100}$

2. $\dfrac{201}{300} = \dfrac{67}{100}$

3. $\dfrac{20}{100} = \dfrac{102}{510}$

4. $\dfrac{40}{100} = \dfrac{248}{620}$

EXERCISE SET 6.4

Ⓐ *Translate each to a proportion. Do not solve. See Examples 1 through 6.*

1. 32% of 65 is what number?

2. What number is 5% of 125?

3. 40% of what number is 75?

4. 1.2 is 47% of what number?

5. What percent of 200 is 70?

6. 520 is 85% of what number?

7. 2.3 is 58% of what number?

8. 92% of 30 is what number?

9. What number is 19% of 130?

10. 8.2 is what percent of 82?

Ⓑ *Solve. See Example 7.*

11. 10% of 55 is what number?

12. 25% of 84 is what number?

13. What number is 18% of 105?

14. What number is 40% of 29?

Solve. See Examples 8 and 9.

15. 60 is 15% of what number?

16. 75 is 75% of what number?

17. 7.8 is 78% of what number?

18. 1.1 is 44% of what number?

Solve. See Examples 10 and 11.

19. 105 is what percent of 84?

20. 77 is what percent of 44?

21. 14 is what percent of 50?

22. 37 is what percent of 50?

Solve. See Examples 7 through 11.

23. 2.9 is 10% of what number?

24. 6.2 is 5% of what number?

25. 2.4% of 80 is what number?

26. 6.5% of 120 is what number?

27. 160 is 16% of what number?

28. 30 is 6% of what number?

29. 348.6 is what percent of 166?

30. 262.4 is what percent of 82?

31. What number is 89% of 62?

32. What number is 53% of 130?

33. What percent of 8 is 3.6?

34. What percent of 5 is 1.6?

35. 140% of what number is 119?

36. 170% of what number is 221?

Review and Preview

Add or subtract the fractions. See Sections 3.1, 3.3, and 3.4.

37. $\dfrac{11}{16} + \dfrac{3}{16}$

38. $\dfrac{5}{8} - \dfrac{7}{12}$

39. $3\dfrac{1}{2} - \dfrac{11}{30}$

40. $2\dfrac{2}{3} + 4\dfrac{1}{2}$

Add or subtract the decimals. See Section 4.3.

41.
$$\begin{array}{r} 0.41 \\ +0.29 \\ \hline \end{array}$$

42.
$$\begin{array}{r} 10.78 \\ 4.3 \\ +\ 0.21 \\ \hline \end{array}$$

43.
$$\begin{array}{r} 2.38 \\ -0.19 \\ \hline \end{array}$$

44.
$$\begin{array}{r} 16.37 \\ -\ 2.61 \\ \hline \end{array}$$

Combining Concepts

Solve. Round to the nearest tenth, if necessary.

45. What number is 22.3% of 53,862?

46. What percent of 110,736 is 88,542?

47. 8652 is 119% of what number?

48. In your own words, explain how to use a proportion to solve a percent problem.

Integrated Review–Percent and Percent Problems

Write each number as a percent.

1. 0.12   **2.** 0.68   **3.** $\frac{1}{4}$ **4.** $\frac{1}{2}$

5. 5.2   **6.** 7.8   **7.** $\frac{3}{50}$ **8.** $\frac{11}{25}$

9. $2\frac{1}{2}$ **10.** $3\frac{1}{4}$ **11.** 0.03   **12.** 0.05

Write each percent as a decimal.

13. 65%   **14.** 31%   **15.** 8%   **16.** 7%

17. 142%   **18.** 538%   **19.** 2.9%   **20.** 6.6%

Write each percent as a fraction or mixed number in simplest form.

21. 3%   **22.** 8%   **23.** 5.25%   **24.** 12.75%

Answers

1. _____
2. _____
3. _____
4. _____
5. _____
6. _____
7. _____
8. _____
9. _____
10. _____
11. _____
12. _____
13. _____
14. _____
15. _____
16. _____
17. _____
18. _____
19. _____
20. _____
21. _____
22. _____
23. _____
24. _____

25. _____

26. _____

27. _____

28. _____

29. _____

30. _____

31. _____

32. _____

33. _____

34. _____

35. _____

36. _____

37. _____

38. _____

39. _____

40. _____

25. 38%   **26.** 45%   **27.** $12\frac{1}{3}$%   **28.** $16\frac{2}{3}$%

Solve each percent problem.

29. 12% of 70 is what number?   **30.** 36 is 36% of what number?

31. 212.5 is 85% of what number?   **32.** 66 is what percent of 55?

33. 23.8 is what percent of 85?   **34.** 38% of 200 is what number?

35. What number is 25% of 44?   **36.** What percent of 99 is 128.7?

37. What percent of 250 is 215?   **38.** What number is 45% of 84?

39. 63 is 42% of what number?   **40.** 58.9 is 95% of what number?

6.5 Applications of Percent

A Solving Applications Involving Percent

Percent is used in a variety of everyday situations. The next examples show just a few ways that percent occurs in real-life settings. (Each of these examples shows two ways of solving these problems. If you studied Section 6.3 only, see *Method 1*. If you studied Section 6.4 only, see *Method 2*.)

OBJECTIVES

A Solve applications involving percent.
B Find percent increase and percent decrease.

SSM   TUTOR CENTER   SG   CD & VIDEO   MATH PRO   WEB

EXAMPLE 1   Finding Totals Using Percents

Mr. Buccaran, the principal at Slidell High School, counted 31 freshmen absent during a particular day. If this is 4% of the total number of freshmen, how many freshmen are there at Slidell High School?

Solution: *Method 1.* First we state the problem in words; then we translate.

In words: 31 is 4% of what number?

Translate: $31 = 4\% \cdot n$

Next, we solve for *n*.

$31 = 0.04 \cdot n$ Write 4% as a decimal.

$\dfrac{31}{0.04} = n$ Divide 31 by 0.04, the number multiplied by *n*.

$775 = n$ Simplify.

There are 775 freshmen at Slidell High School.

Method 2. First we state the problem in words; then we translate.

In words: 31 is 4% of what number?

            amount   percent   base

Translate: $\underset{\text{base} \,\rightarrow}{\overset{\text{amount} \,\rightarrow}{\dfrac{31}{b}}} = \dfrac{4}{100} \leftarrow \text{percent}$

Next, we solve for *b*.

$31 \cdot 100 = b \cdot 4$ Set cross products equal.

$3100 = b \cdot 4$ Multiply.

$\dfrac{3100}{4} = b$ Divide 3100 by 4, the number multiplied by *b*.

$775 = b$ Simplify.

There are 775 freshmen at Slidell High School.

Practice Problem 1

The freshmen class of 775 students is 31% of all students at Euclid University. How many students go to Euclid University?

Answer

1. 2500

Practice Problem 2

The nutrition label below is from a can of cashews. Find what percent of total calories are from fat. Round to the nearest tenth of a percent.

Nutrition Facts

Serving Size $\frac{1}{4}$ cup (33g)
Servings Per Container About 9

Amount Per Serving

Calories 190 Calories from Fat 130

	% Daily Value
Total Fat 16g	**24%**
Saturated Fat 3g	**16%**
Cholesterol 0mg	**0%**
Sodium 135mg	**6%**
Total Carbohydrate 9g	**3%**
Dietary Fiber 1g	**5%**
Sugars 2g	
Protein 5g	

Vitamin A 0% • Vitamin C 0%
Calcium 0% • Iron 8%

EXAMPLE 2 Finding Percents

Standardized nutrition labels like the one shown have been on foods since 1994. It is recommended that no more than 30% of your calorie intake be from fat. Find what percent of the total calories shown are fat.

Nutrition Facts

Serving Size 1 pouch (20g)
Servings Per Container 6

Amount Per Serving

Calories	80
Calories from fat	10

	% Daily Value*
Total Fat 1g	**2%**
Sodium 45mg	**2%**
Total Carbohydrate 17g	**6%**
Sugars 9g	
Protein 0g	

Vitamin C	25%

Not a significant source of saturated fat, cholesterol, dietary fiber, vitamin A, calcium and iron.

*Percent Daily Values are based on a 2,000 calorie diet.

Fruit snacks nutrition label

Solution: *Method 1.*

In words: 10 is what percent of 80?

Translate: $10 = n \cdot 80$

Next, we solve for n.

$$\frac{10}{80} = n \qquad \text{Divide 10 by 80, the number multiplied by } n.$$

$$0.125 = n \qquad \text{Simplify.}$$

$$12.5\% = n \qquad \text{Write 0.125 as a percent.}$$

12.5% of this food's total calories are from fat.

Method 2.

In words: 10 is what percent of 80?

amount   percent   base

Translate: amount → $\dfrac{10}{80} = \dfrac{p}{100}$ ← percent
base →

Next, we solve for p.

$$10 \cdot 100 = 80 \cdot p \qquad \text{Set cross products equal.}$$

$$1000 = 80 \cdot p \qquad \text{Multiply.}$$

$$\frac{1000}{80} = p \qquad \text{Divide 1000 by 80, the number multiplied by } p.$$

$$12.5 = p \qquad \text{Simplify.}$$

12.5% of this food's total calories are from fat.

Answer

2. 68.4%

Ⓑ Finding Percent Increase and Percent Decrease

We often use percents to show how much an amount has increased or decreased.

EXAMPLE 3   Finding an Increase

The state of Nevada had the largest percent increase in population, about 66%, from the 1990 census to the 2000 census. In 1990, the population of Nevada was about 1202 thousand. Find the population of Nevada in 2000. (*Source:* U.S. Census Bureau)

Nevada

Arizona

Solution: *Method 1.* First we find the increase in population.

In words: What number is 66% of 1202?

Translate: n $= 66\%$ · 1202

Next, we solve for n.

$n = 0.66 \cdot 1202$ Write 66% as a decimal.

$n = 793.32$ Multiply.

The increase in population is 793.32 thousand. This means that the population of Nevada in 2000 was

1202 thousand + 793.32 thousand = 1995.32 thousand

Method 2. First we find the increase in population.

In words: What number is 66% of 1202?

$$\underbrace{\text{amount}} \quad \underbrace{\text{percent}} \quad \underbrace{\text{base}}$$

Translate: $\begin{array}{c}\text{amount} \rightarrow \\ \text{base} \rightarrow\end{array} \dfrac{a}{1202} = \dfrac{66}{100} \; \leftarrow \text{percent}$

Next, we solve for a.

$a \cdot 100 = 1202 \cdot 66$ Set cross products equal.

$a \cdot 100 = 79{,}332$ Multiply.

$a = \dfrac{79{,}332}{100}$ Divide 79,332 by 100, the number multiplied by a.

$a = 793.32$ Simplify.

The increase in population is 793.32 thousand. This means that the population of Nevada in 2000 was

1202 thousand + 793 thousand = 1995.32 thousand   ●

Suppose that the population of a town is 10,000 people and then it increases by 2000 people. The **percent of increase** is

$$\begin{array}{c}\text{amount of increase} \rightarrow \\ \text{original amount} \rightarrow\end{array} \dfrac{2000}{10{,}000} = 0.2 = 20\%$$

Practice Problem 3

The state of Arizona had the second-largest percent increase in population, 40%, from the 1990 census to the 2000 census. In 1990, the population of Arizona was about 3665 thousand. Find the population of Arizona in 2000. (*Source:* U.S. Census Bureau)

Answer

3. 5131 thousand

In general, we have the following.

Percent of Increase

$$\text{percent of increase} = \frac{\text{amount of increase}}{\text{original amount}}$$

Then write the quotient as a percent.

Practice Problem 4

Saturday's attendance at the play *Peter Pan* increased to 333 people over Friday's attendance of 285 people. What was the percent increase in attendance? Round to the nearest tenth of a percent.

EXAMPLE 4 Finding Percent Increase

The number of applications for a mathematics scholarship at Yale increased from 34 to 45 in one year. What is the percent increase? Round to the nearest whole percent.

Solution: First we find the amount of increase by subtracting the original number of applicants from the new number of applicants.

$$\text{amount of increase} = 45 - 34 = 11$$

The amount of increase is 11 applicants. To find the percent of increase,

$$\text{percent of increase} = \frac{\text{amount of increase}}{\text{original amount}} = \frac{11}{34} \approx 0.32 = 32\%$$

Helpful Hint

Make sure that this number is the original number and not the new number.

The number of applications increased by about 32%.

Try the Concept Check in the margin.

Concept Check

A student is calculating the percent increase in enrollment from 180 students one year to 200 students the next year. Explain what is wrong with the following calculations:

$$\begin{aligned} \text{Amount of increase} &= 200 - 180 = 20 \\ \text{Percent of increase} &= \frac{20}{200} = 0.1 = 10\% \end{aligned}$$

Suppose that your income was $300 a week and then it decreased by $30. The percent decrease is

$$\begin{aligned} \text{amount of decrease} \rightarrow \\ \text{original amount} \rightarrow \end{aligned} \quad \frac{\$30}{\$300} = 0.1 = 10\%$$

Percent of Decrease

$$\text{percent of decrease} = \frac{\text{amount of decrease}}{\text{original amount}}$$

Then write the quotient as a percent.

Practice Problem 5

A town with a population of 20,145 decreased to 18,430 over a 10-year period. What was the percent decrease? Round to the nearest tenth of a percent.

EXAMPLE 5 Finding Percent Decrease

In response to a decrease in sales, a company with 1500 employees reduces the number of employees to 1230. What is the percent decrease?

Solution: First we find the amount of decrease by subtracting 1230 from 1500.

$$\text{amount of decrease} = 1500 - 1230 = 270$$

The amount of decrease is 270. To find the percent of decrease,

$$\text{percent of decrease} = \frac{\text{amount of decrease}}{\text{original amount}} = \frac{270}{1500} = 0.18 = 18\%$$

The number of employees decreased by 18%.

Answers

4. 16.8%, **5.** 8.5%

Concept Check: To find the percent of increase, you have to divide the amount of increase by the original amount $\left(\dfrac{20}{180}\right)$.

EXERCISE SET 6.5

A *Solve. See Example 1.*

1. An inspector found 24 defective bolts during an inspection. If this is 1.5% of the total number of bolts inspected, how many bolts were inspected?

2. A day care worker found 28 children absent one day during an epidemic of chicken pox. If this was 35% of the total number of children attending the day care center, how many children attend this day care center?

3. An owner of a repair service company estimates that for every 40 hours a repairperson is on the job, he can bill for only 75% of the hours. The remaining hours, the repairperson is idle or driving to or from a job. Determine the number of hours per 40-hour week the owner can bill for a repairperson.

4. The Hodder family paid 20% of the purchase price of a $75,000 home as a down payment. Determine the amount of the down payment.

5. Vera Faciane earns $2000 per month and budgets $300 per month for food. What percent of her monthly income is spent on food?

6. Last year, Mai Toberlan bought a share of stock for $83. She was paid a dividend of $4.15. Determine what percent of the stock price is the dividend.

7. A manufacturer of electronic components expects 1.04% of its products to be defective. Determine the number of defective components expected in a batch of 28,350 components. Round to the nearest whole component.

8. 18% of Frank's wages are withheld for income tax. Find the amount withheld from Frank's wages of $3680 per month.

For each food described, find what percent of total calories is from fat. If necessary, round to the nearest tenth of a percent. See Example 2.

9.

Nutrition Facts

Serving Size 18 crackers (29g)
Servings Per Container About 9

Amount Per Serving

Calories 120 Calories from Fat 35

	% Daily Value*
Total Fat 4g	**6%**
Saturated Fat 0.5g	**3%**
Polyunsaturated Fat 0g	
Monounsaturated Fat 1.5g	
Cholesterol 0mg	**0%**
Sodium 220mg	**9%**
Total Carbohydrate 21g	**7%**
Dietary Fiber 2g	**7%**
Sugars 3g	
Protein 2g	

Vitamin A 0% • Vitamin C 0%
Calcium 2% • Iron 4%
Phosphorus 10%

Snack Crackers

10.

Nutrition Facts

Serving Size 28 crackers (31g)
Servings Per Container About 6

Amount Per Serving

Calories 130 Calories from Fat 35

	% Daily Value*
Total Fat 4g	**6%**
Saturated Fat 2g	**10%**
Polyunsaturated Fat 1g	
Monounsaturated Fat 1g	
Cholesterol 0mg	**0%**
Sodium 470mg	**20%**
Total Carbohydrate 23g	**8%**
Dietary Fiber 1g	**4%**
Sugars 4g	
Protein 2g	

Vitamin A 0% • Vitamin C 0%
Calcium 0% • Iron 2%

B *Solve. Round money amounts to the nearest cent and all other amounts to the nearest tenth. See Example 3.*

11. Ace Furniture Company currently produces 6200 chairs per month. If production increases 8%, find the increase and the new number of chairs produced each month.

12. The enrollment at a local college increased 5% over last year's enrollment of 7640. Find the increase in enrollment and the current enrollment.

13. By carefully planning their meals, a family was able to decrease their weekly grocery bill by 20%. Their weekly grocery bill used to be $170. What is their new weekly grocery bill?

14. The profit of Ramone Company last year was $175,000. This year's profit decreased by 11%. Find this year's profit.

15. A car manufacturer announced that next year the price of a certain model of car would increase 4.5%. This year the price is $19,286. Find the increase in price and the new price.

16. A union contract calls for a 6.5% salary increase for all employees. Determine the increase and the new salary that a worker currently making $28,500 under this contract can expect.

17. From 1998 to 2008, the number of people employed as physician assistants in the United States is expected to increase 48%. The number of people employed as physician assistants in 1998 was 66,000. Find the predicted number of physician assistants in 2008. (*Source:* Bureau of Labor Statistics)

18. The state of North Dakota had the smallest percent increase in population, 0.5%, from the 1990 census to the 2000 census. In 1990, the population of North Dakota was 638,800. What was the population of North Dakota in 2000? (*Source:* U.S. Census Bureau)

Find the amount of increase and the percent increase. See Example 4.

	Original Amount	New Amount	Amount of Increase	Percent Increase
19.	40	50	_____	_____
20.	10	15	_____	_____
21.	85	187	_____	_____
22.	78	351	_____	_____

Find the amount of decrease and the percent decrease. See Example 5.

	Original Amount	New Amount	Amount of Decrease	Percent Decrease
23.	8	6	_____	_____
24.	25	20	_____	_____
25.	160	40	_____	_____
26.	200	162	_____	_____

Solve. Round percents to the nearest tenth, if necessary. See Examples 3 through 5.

27. There are 150 calories in a cup of whole milk and only 84 in a cup of skim milk. In switching to skim milk, find the percent decrease in number of calories per cup.

28. In reaction to a slow economy, the number of employees at a soup company decreased from 530 to 477. What was the percent decrease in employees?

29. By changing his driving routines, Alan Miller increased his car's rate of miles per gallon from 19.5 to 23.7. Find the percent increase.

30. John Smith decided to decrease the number of calories in his diet from 3250 to 2100. Find the percent decrease.

 31. The price of a loaf of bread decreased from $1.59 to $1.39. Find the percent decrease.

32. Before taking a typing course, Geoffry Landers could type 32 words per minute. By the end of the course, he was able to type 76 words per minute. Find the percent increase.

33. In 1940, the average size of a U.S. farm was 174 acres. By 2000, the average size of a U.S. farm had increased to 434 acres. What was the percent increase? (*Source:* National Agricultural Statistics Service)

34. In 1995, 272.6 million recorded music cassettes were shipped to retailers in the United States. By 2000, this number had decreased to 76.0 million cassettes. What was the percent decrease? (*Source:* Recording Industry Association of America)

35. In 1994, approximately 16,000,000 Americans subscribed to cellular phone service. By 2000, this number had increased to about 110,000,000 American subscribers. What was the percent increase? (*Source:* Network World, Inc.)

36. In 1970, there were 1754 deaths from boating accidents in the United States. By 2000, the number of deaths from boating accidents had decreased to 698. What was the percent decrease? (*Source:* U.S. Coast Guard)

Review and Preview

Perform each indicated operation. See Sections 4.3 and 4.4.

37.   0.12
      $\times$ 38

38.   42
       $\times$ 0.7

39. 9.20 + 1.98

40. 46 + 7.89

41. 78 − 19.46

42. 64.80 − 10.72

Combining Concepts

Solve. Round percents to the nearest tenth.

43. The population of Tokyo is expected to increase from 26,518 thousand in 1994 to 28,700 thousand in 2015. Find the percent increase. (*Source:* United Nations, Department for Economic and Social Information and Policy Analysis)

Japan

Tokyo

44. In 1998, approximately 299,000 computer engineers were employed in the United States. By 2008, this number is expected to increase to 622,000 computer engineers. What is the percent increase? (*Source:* Bureau of Labor Statistics)

45. In your own words, explain what is wrong with the following statement. "Last year we had 80 students attend. This year we have a 50% increase or a total of 160 students attend."

MORTGAGES

Buying a house may be one of the most expensive purchases we make. The amount borrowed from a lending institution for real estate is called a **mortgage**. The lending institutions that normally make mortgage loans include banks, savings and loan associations, credit unions, and mortgage companies.

There are basically three items that define a mortgage loan: the principal, the loan term, and the interest rate. The principal is the dollar amount being borrowed or financed by the home buyers. The loan term is the length of the loan, or how long it will take to pay off the loan. The interest rate, normally expressed as a percent, governs how much must be paid for the privilege of borrowing the money.

Mortgages come in all shapes and sizes. Loan terms can range anywhere from 10 years to 15, 20, 25, 30, or even 40 years. The interest rates on shorter loans are generally lower than the interest rates on longer loans. For instance, the interest rate on a 15-year loan might be 7.25% while the interest rate on a 30-year loan is 7.5%. Interest rates also tend to be lower on loans with a smaller principal as compared with a large principal. For example, many banks offer a jumbo mortgage loan that applies only to principals over a certain limit, generally over $227,150. The interest rates on jumbo mortgages are higher (often by about 0.25%) than on other mortgage programs.

Most lending institutions require a home buyer to make a **down payment** in cash on a home. The size of the down payment usually depends on the buyer's circumstances and the specific mortgage program chosen, but down payments generally range from 3% to 20% of the home's value. A typical down payment on a house is 10% of the purchase price. After a down payment has been chosen, the mortgage amount can be calculated by subtracting the amount of the down payment from the purchase price:

mortgage = purchase price − down payment

Besides the down payment, there are a number of initial costs related to the mortgage that must be paid. These costs are called **closing costs**. Two expensive items on the list of closing costs are the **loan origination fee** and **loan discount points**. Both of these items are generally given in "points," where each point is equal to 1% of the mortgage amount. For example, "3 points" means 3% of the mortgage amount. The loan origination fee is the fee charged by the lender to cover the costs of preparing all the loan documents. Loan discount points are prepaid interest paid at closing. Home buyers generally can choose whether or not they will pay loan discount points. Doing so lowers the interest rate on the mortgage.

loan origination fee = mortgage · points

loan discount points = mortgage · points

CRITICAL THINKING

Suppose you are considering buying a house with a purchase price of $140,000.

1. Find the amount of the down payment if you plan to make a 10% down payment.
2. Find the mortgage amount.
3. Calculate the loan origination fee of 1 point.
4. To get a lower interest rate on your loan, suppose you choose to pay 2.5 loan discount points at closing. How much will you be paying in loan discount points?

6.6 Percent and Problem Solving: Sales Tax, Commission, and Discount

OBJECTIVES

Ⓐ Calculate sales tax and total price.
Ⓑ Calculate commissions.
Ⓒ Calculate discount and sale price.

SSM
TUTOR CENTER   SG   CD & VIDEO   MATH PRO   WEB

Ⓐ Calculating Sales Tax and Total Price

Percents are frequently used in the retail trade. For example, most states charge a tax on certain items when purchased. This tax is called a **sales tax**, and retail stores collect it for the state. Sales tax is almost always stated as a percent of the purchase price.

A 6% sales tax rate on a purchase of a $10 item gives a sales tax of

$$\text{sales tax} = 6\% \text{ of } \$10 = 0.06 \cdot \$10.00 = \$0.60$$

The total price to the customer would be

$$\underbrace{\text{purchase price}}_{} \text{ plus } \underbrace{\text{sales tax}}_{}$$

$$\$10.00 \quad + \quad \$0.60 = \$10.60$$

This example suggests the following equations:

> **Sales Tax and Total Price**
>
> $$\text{sales tax} = \text{tax rate} \cdot \text{purchase price}$$
> $$\text{total price} = \text{purchase price} + \text{sales tax}$$

In this section we round dollar amounts to the nearest cent.

EXAMPLE 1   Finding Sales Tax and Purchase Price

Find the sales tax and the total price on the purchase of an $85.50 trench coat in a city where the sales tax rate is 7.5%.

SALE
$85.50
+ 7.5% tax

Solution: The purchase price is $85.50 and the tax rate is 7.5%.

$$\underbrace{\text{sales tax}} = \underbrace{\text{tax rate}} \cdot \underbrace{\text{purchase price}}$$

$$\text{sales tax} = 7.5\% \quad \cdot \quad \$85.50$$

$$= 0.075 \quad \cdot \quad \$85.5 \qquad \text{Write 7.5\% as a decimal.}$$

$$\approx \$6.41 \qquad \text{Rounded to the nearest cent.}$$

Thus,

$$\underbrace{\text{total price}} = \underbrace{\text{purchase price}} + \underbrace{\text{sales tax}}$$

$$\text{total price} = \$85.50 + \$6.41$$

$$= \$91.91$$

The sales tax on $85.50 is $6.41, and the total price is $91.91.

Practice Problem 1

If the sales tax rate is 6%, what is the sales tax and the total amount due on a $29.90 Goodgrip tire?

Answer

1. tax: $1.79; total: $31.69

Concept Check

The purchase price of a textbook is $50 and sales tax is 10%. If you are told by the cashier that the total price is $75, how can you tell that a mistake has been made?

Practice Problem 2

The sales tax on a $13,500 automobile is $1080.00. Find the sales tax rate.

Try the Concept Check in the margin.

EXAMPLE 2   Finding a Sales Tax Rate

The sales tax on a $300 printer is $22.50. Find the sales tax rate.

SALE
$300
+$22.50 sales tax

Solution: Let r be the unknown sales tax rate. Then

sales tax = tax rate · purchase price

$$\$22.50 = r \cdot \$300$$

$$\frac{22.50}{300} = r \qquad \text{Divide 22.50 by 300, the number multiplied by } r.$$

$$0.075 = r \qquad \text{Simplify.}$$

$$7.5\% = r \qquad \text{Write 0.075 as a percent.}$$

The sales tax rate is 7.5%.

B   Calculating Commissions

A **wage** is payment for performing work. Hourly wage, commissions, and salary are some of the ways wages can be paid. Many people who work in sales are paid a commission. An employee who is paid a **commission** is paid a percent of his or her total sales.

Commission

$$\text{commission} = \text{commission rate} \cdot \text{sales}$$

EXAMPLE 3   Finding a Commission Rate

Sherry Souter, a real estate broker for Wealth Investments, sold a house for $114,000 last week. If her commission is 1.5% of the selling price of the home, find the amount of her commission.

Solution:

commission = commission rate · sales

$$\text{commission} = 1.5\% \cdot \$114,000$$

$$= 0.015 \cdot \$114,000 \qquad \text{Write 1.5\% as 0.015.}$$

$$= \$1710 \qquad \text{Multiply.}$$

Practice Problem 3

Mr. Olsen is a sales representative for Miko Copiers. Last month he sold $37,632 worth of copy equipment and supplies. What is his commission for the month if he is paid a commission of 6.6% of his total sales for the month?

Answers

2. 8%, **3.** $2483.71

Concept Check: Since $10\% = \frac{1}{10}$, the sales tax is $\frac{\$50}{10} = \$5$. The total price should have been $55.

Her commission on the house is $1710.

EXAMPLE 4   Finding a Commission Rate

A salesperson earned $1560 for selling $13,000 worth of television and stereo systems. Find the commission rate.

Solution: Let r stand for the unknown commission rate. Then

$$\text{commission} = \text{commission rate} \cdot \text{sales}$$

$$\$1560 = r \cdot \$13,000$$

$$\frac{1560}{13,000} = r \qquad \text{Divide 1560 by 13,000, the number multiplied by } r.$$

$$0.12 = r \qquad \text{Simplify.}$$

$$12\% = r \qquad \text{Write 0.12 as a percent.}$$

The commission rate is 12%.

(C) Calculating Discount and Sale Price

Suppose that an item that normally sells for $40 is on sale for 25% off. This means that the **original price** of $40 is reduced, or **discounted**, by 25% of $40, or $10. The **discount rate** is 25%, the **amount of discount** is $10, and the **sale price** is $40 − $10, or $30.

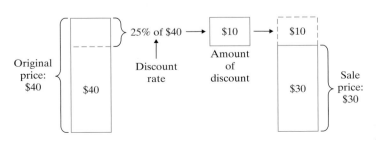

Original price: $40 — $40 — Discount rate — 25% of $40 → $10 — Amount of discount → $10 — $30 — Sale price: $30

Practice Problem 4

A salesperson earns $1290 for selling $8600 worth of appliances. Find the commission rate.

Answer

4. 15%

To calculate discounts and sale prices, we can use the following equations:

Discount and Sale Price

amount of discount = discount rate · original price

sale price = original price − amount of discount

Practice Problem 5

A Panasonic TV is advertised on sale for 15% off the regular price of $700. Find the discount and the sale price.

EXAMPLE 5 Finding a Discount and a Sale Price

A speaker that normally sells for $65 is on sale for 25% off. What is the discount, and what is the sale price?

Solution: First we find the discount.

amount of discount = discount rate · original price

amount of discount = 25% · $65

= 0.25 · $65   Write 25% as 0.25.

= $16.25   Multiply.

The discount is $16.25. Next, find the sale price.

sale price = original price − discount

sale price = $65 − $16.25

= $48.75   Subtract.

The sale price is $48.75.

Answer

5. $105; $595

Name _____ Section _____ Date _____

EXERCISE SET 6.6

A *Solve. See Examples 1 and 2.*

1. What is the sales tax on a suit priced at $150 if the sales tax rate is 5%?

2. If the sales tax rate is 6%, find the sales tax on a microwave oven priced at $188.

3. The purchase price of a camcorder is $799. What is the total price if the sales tax rate is 7.5%?

4. A stereo system has a purchase price of $426. What is the total price if the sales tax rate is 8%?

5. A chair and ottoman have a purchase price of $600. If the sales tax on this purchase is $54, find the sales tax rate.

6. The sales tax on the purchase of a $2500 computer is $162.50. Find the sales tax rate.

7. A table saw sells for $120. With a sales tax rate of 8.5%, find the total price.

$120
+8.5% tax

8. A one-half carat diamond ring is priced at $800. The sales tax rate is 9.5%. Find the total price.

$800
+9.5% tax

379

9. A gold and diamond bracelet sells for $1800. Find the total price if the sales tax rate is 6.5%.

10. The purchase price of a personal computer is $1890. If the sales tax rate is 8%, what is the total price?

11. The sales tax on the purchase of a truck is $920. If the tax rate is 8%, find the purchase price of the truck.

12. The sales tax on the purchase of a desk is $27.50. If the tax rate is 5%, find the purchase price of the desk.

13. A cordless phone costs $90 and a battery recharger costs $15. What is the total price for purchasing these items if the sales tax rate is 7%?

14. Ms. Warner bought a blouse for $35, a skirt for $55, and a blazer for $95. Find the total price she paid, given a sales tax rate of 6.5%.

15. The sales tax is $98.70 on a stereo sound system purchase of $1645. Find the sales tax rate.

16. The sales tax is $103.50 on a necklace purchase of $1150. Find the sales tax rate.

Solve. See Examples 3 and 4.

17. Jane Moreschi, a sales representative for a large furniture warehouse, is paid a commission rate of 4%. Find her commission if she sold $1,236,856 worth of furniture last month.

18. Rosie Davis-Smith is a beauty consultant for a home cosmetic business. She is paid a commission rate of 4.8%. Find her commission if she sold $1638 in cosmetics last month.

19. A salesperson earned a commission of $1380.40 for selling $9860 worth of paper products. Find the commission rate.

20. A salesperson earned a commission of $3575 for selling $32,500 worth of books to various bookstores. Find the commission rate.

21. How much commission will Jack Pruet make on the sale of a $125,900 house if he receives 1.5% of the selling price?

22. Frankie Lopez sold $9638 of jewelry this week. Find her commission for the week if she receives a commission rate of 5.6%.

23. A house sold for $85,500, and the real estate agent earned a commission of $2565. Find the rate of commission.

24. A salesperson earned $1750 for selling $25,000 worth of fertilizer. Find the commission rate.

Find the amount of discount and the sale price. See Example 5.

	Original Price	Discount Rate	Amount of Discount	Sale Price
25.	$68	10%	_____	_____
26.	$47	20%	_____	_____
27.	$96.50	50%	_____	_____
28.	$110.60	40%	_____	_____
29.	$215	35%	_____	_____
30.	$370	25%	_____	_____
31.	$21,700	15%	_____	_____
32.	$17,800	12%	_____	_____

 33. A $300 fax machine is on sale for 15% off. Find the discount and the sale price.

34. A $2000 designer dress is on sale for 30% off. Find the discount and the sale price.

Review and Preview

Multiply. See Sections 4.4 and 4.7.

35. $2000 \cdot 0.3 \cdot 2$

36. $500 \cdot 0.08 \cdot 3$

37. $400 \cdot 0.03 \cdot 11$

38. $1000 \cdot 0.05 \cdot 5$

39. $600 \cdot 0.04 \cdot \dfrac{2}{3}$

40. $6000 \cdot 0.06 \cdot \dfrac{3}{4}$

Combining Concepts

 41. A diamond necklace sells for $24,966. If the tax rate is 7.5%, find the total price.

42. A house recently sold for $562,560. The commission rate on the sale is 5.5%. If the real estate agent is to receive 60% of the commission, find the amount received by the agent.

43. Which is better, a 30% discount followed by an additional 25% off or a 20% discount followed by an additional 40% off? To see, suppose an item costs $100 and calculate each discounted price. Explain your answer.

6.7 Percent and Problem Solving: Interest

O B J E C T I V E S

Ⓐ Calculate simple interest.
Ⓑ Calculate compound interest.
Ⓒ Calculate monthly payments.

SSM
TUTOR CENTER   SG   CD & VIDEO   MATH PRO   WEB

Ⓐ Calculating Simple Interest

Interest is money charged for using other people's money. When you borrow money, you pay interest. When you loan or invest money, you earn interest. The money borrowed, loaned, or invested is called the **principal amount**, or simply **principal**. Interest is normally stated in terms of a percent of the principal for a given period of time. The **interest rate** is the percent used in computing the interest. Unless stated otherwise, *the rate is understood to be per year*. When the interest is computed on the original principal, it is called **simple interest**. Simple interest is calculated using the following equation:

> **Simple Interest**
>
> simple interest = principal · rate · time
>
> where the rate is understood to be per year and time is in years.

EXAMPLE 1 Finding Simple Interest

Find the simple interest after 2 years on $500 at an interest rate of 12%.

Solution: In this example, the principal is $500, the rate is 12%, and the time is 2 years.

$$\text{simple interest} = \text{principal} \cdot \text{rate} \cdot \text{time}$$

$$
\begin{aligned}
\text{simple interest} &= \$500 \cdot 12\% \cdot 2 \\
&= \$500 \cdot 0.12 \cdot 2 \quad \text{Write 12\% as 0.12.} \\
&= \$120 \quad\quad\quad\quad\quad\ \text{Multiply.}
\end{aligned}
$$

The simple interest is $120.

●

If time is not given in years, we need to convert the given time to years.

EXAMPLE 2 Finding Simple Interest

Ivan Borski borrowed $2400 at 10% simple interest for 8 months to buy a used Chevy S-10. Find the simple interest he paid.

Solution: Since there are 12 months in a year, we first find what part of a year 8 months is.

$$8 \text{ months} = \frac{8}{12} \text{ year} = \frac{2}{3} \text{ year}$$

Now we find the simple interest.

$$\text{simple interest} = \text{principal} \cdot \text{rate} \cdot \text{time}$$

$$
\begin{aligned}
\text{simple interest} &= \$2400 \cdot 10\% \cdot \frac{2}{3} \\
&= \$2400 \cdot 0.10 \cdot \frac{2}{3} \\
&= \$160
\end{aligned}
$$

The interest on Ivan's loan is $160.

Practice Problem 1

Find the simple interest after 3 years on $750 at an interest rate of 8%.

Practice Problem 2

Juanita Lopez borrowed $800 for 9 months at a simple interest rate of 20%. How much interest did she pay?

Answers

1. $180,   **2.** $120

●

When money is borrowed, the borrower pays the original amount borrowed, or the principal, as well as the interest. When money is invested, the investor receives the original amount invested, or the principal, as well as the interest. In either case, the **total amount** is the sum of the principal and the interest.

> **Finding the Total Amount of a Loan or Investment**
>
> total amount (paid or received) = principal + interest

Practice Problem 3

If $500 is borrowed at a simple interest rate of 12% for 6 months, find the total amount paid.

EXAMPLE 3   Finding the Total Amount of an Investment

An accountant invested $2000 at a simple interest rate of 10% for 2 years. What total amount of money will she have from her investment in 2 years?

Solution: First we find her interest.

$$
\begin{aligned}
\text{simple interest} &= \text{principal} \cdot \text{rate} \cdot \text{time} \\
\text{simple interest} &= \$2000 \cdot 10\% \cdot 2 \\
&= \$2000 \cdot 0.10 \cdot 2 \\
&= \$400
\end{aligned}
$$

The interest is $400.

Next, we add the interest to the principal.

$$
\begin{aligned}
\text{total amount} &= \text{principal} + \text{interest} \\
\text{total amount} &= \$2000 + \$400 \\
&= \$2400
\end{aligned}
$$

After 2 years, she will have a total amount of $2400.

Try the Concept Check in the margin.

Concept Check

Which investment would earn more interest: An amount of money invested at 8% interest for 2 years, or the same amount of money invested at 8% for 3 years? Explain.

B   Calculating Compound Interest

Recall that simple interest depends on the original principal only. Another type of interest is compound interest. **Compound interest** is computed on not only the principal, but also on the interest already earned in previous compounding periods. Compound interest is used more often than simple interest.

Let's see how compound interest differs from simple interest. Suppose that $2000 is invested at 7% interest **compounded annually** for 3 years. This means that interest is added to the principal at the end of each year and that next year's interest is computed on this new amount. In this section, we round dollar amounts to the nearest cent.

Answers

3. $530

Concept Check: 8% for 3 years. Since the interest rate is the same, the longer you keep the money invested, the more interest you earn.

	Amount at Beginning of Year	Principal	·	Rate	·	Time	= Interest	Amount at End of Year
1st year	$2000	$2000	·	0.07	·	1	= $140	$2000 + 140 = $2140
2nd year	$2140	$2140	·	0.07	·	1	= $149.80	$2140 + 149.80 = $2289.80
3rd year	$2289.80	$2289.80	·	0.07	·	1	= $160.29	$2289.80 + 160.29 = $2450.09

The compound interest earned can be found by

total amount	−	original principal	=	compound interest
↓		↓		↓
$2450.09	−	$2000	=	$450.09

The simple interest earned would have been

principal	·	rate	·	time	=	interest
↓		↓		↓		↓
$2000	·	0.07	·	3	=	$420

Since compound interest earns "interest on interest," compound interest earns more than simple interest.

Computing compound interest using the method above can be tedious. We can use a **compound interest table** to compute interest more quickly. The compound interest table in this textbook is found in Appendix F. This table gives the total compound interest and principal paid on $1 for given rates and numbers of years. Then we can use the following equation to find the total amount of interest and principal:

Finding Total Amounts with Compound Interest

$$\text{total amount} = \text{original principal} \cdot \begin{array}{c} \text{compound interest factor} \\ \text{(from table)} \end{array}$$

EXAMPLE 4   Finding Total Amount Received on an Investment

$4000 is invested at 8% compounded semiannually for 10 years. Find the total amount at the end of 10 years.

Solution: Look in Appendix F. The compound interest factor for 10 years at 8% in the Compounded Semiannually section is 2.19112.

total amount	=	original principal	·	compound interest factor
↓		↓		↓
total amount	=	$4000	·	2.19112
	=	$8764.48		

Therefore, the total amount at the end of 10 years is $8764.48. ●

EXAMPLE 5   Finding Compound Interest Earned

In Example 4 we found that the total amount for $4000 invested at 8% compounded semiannually for 10 years is $8764.48. Find the compound interest earned.

Solution:

interest earned	=	total amount	−	original principal
↓		↓		↓
interest earned	=	$8764.48	−	$4000
	=	$4764.48		

The compound interest earned is $4764.48.

Practice Problem 4

$5500 is invested at 7% compounded daily for 5 years. Find the total amount at the end of 5 years.

Practice Problem 5

If the total amount is $9933.14 when $5500 is invested, find the compound interest earned.

Answers
● | **4.** $7804.61,   **5.** $4433.14

(c) Calculating a Monthly Payment

We conclude this section with a method to find the monthly payment on a loan.

> **Finding the Monthly Payment of a Loan**
>
> $$\text{monthly payment} = \frac{\text{principal} + \text{interest}}{\text{total number of payments}}$$

EXAMPLE 6   Finding a Monthly Payment

Find the monthly payment on a $2000 loan for 2 years. The interest on the 2-year loan is $435.88.

Solution: First we determine the total number of monthly payments. The loan is for 2 years. Since there are 12 months per year, the number of payments is $2 \cdot 12$, or 24. Now we calculate the monthly payment.

$$\text{monthly payment} = \frac{\text{principal} + \text{interest}}{\text{total number of payments}}$$

$$\text{monthly payment} = \frac{\$2000 + \$435.88}{24}$$

$$\approx \$101.50$$

The monthly payment is about $101.50.

Practice Problem 6

Find the monthly payment on a $3000 3-year loan if the interest on the loan is $1123.58.

Answer

6. $114.54

CALCULATOR EXPLORATIONS

Compound Interest Factor

A compound interest factor may be found by using your calculator and evaluating the formula

$$\textbf{compound interest factor} = \left(1 + \frac{r}{n}\right)^{nt}$$

where r is the interest rate, t is the time in years, and n is the number of times compounded per year. For example, we stated earlier that the compound interest factor for 10 years at 8% compounded semiannually is 2.19112. Let's find this factor by evaluating the compound interest factor formula when $r = 8\%$ or 0.08, $t = 10$, and $n = 2$ (compounded semiannually means 2 times per year). Thus,

$$\text{compound interest factor} = \left(1 + \frac{0.08}{2}\right)^{2 \cdot 10}$$

$$\text{or} \quad \left(1 + \frac{0.08}{2}\right)^{20}$$

To evaluate, press the keys

The display will read $\boxed{2.1911231}$. Rounded to 5 decimal places, this is 2.19112.

Find the compound interest factors. Use the table in Appendix F to check your answers.

1. 5 years, 9%, compounded quarterly
2. 15 years, 14%, compounded daily
3. 20 years, 11%, compounded annually
4. 1 year, 7%, compounded semiannually
5. Find the total amount after 4 years when $500 is invested at 6% compounded quarterly.
6. Find the total amount for 19 years when $2500 is invested at 5% compounded daily.

EXERCISE SET 6.7

(A) *Find the simple interest. See Examples 1 and 2.*

	Principal	Rate	Time
1.	$200	8%	2 years
3.	$160	11.5%	4 years
5.	$5000	10%	$1\frac{1}{2}$ years
7.	$375	18%	6 months
9.	$2500	16%	21 months

	Principal	Rate	Time
2.	$800	9%	3 years
4.	$950	12.5%	5 years
6.	$1500	14%	$2\frac{1}{4}$ years
8.	$1000	10%	18 months
10.	$775	15%	8 months

Solve. See Examples 1 through 3.

11. A company borrows $62,500 for 2 years at a simple interest of 12.5% to buy an airplane. Find the total amount paid on the loan.

12. $65,000 is borrowed to buy a house. If the simple interest rate on the 30-year loan is 10.25%, find the total amount paid on the loan.

 13. A money market fund advertises a simple interest rate of 9%. Find the total amount received on an investment of $5000 for 15 months.

14. The Real Service Company takes out a 270-day (9-month) short-term, simple interest loan of $4500 to finance the purchase of some new equipment. If the interest rate is 14%, find the total amount that the company pays back.

15. Marsha Waide borrows $8500 and agrees to pay it back in 4 years. If the simple interest rate is 12%, find the total amount she pays back.

16. Ms. Lapchinski gives her 18-year-old daughter a graduation gift of $2000. If this money is invested at 8% simple interest for 5 years, find the total amount.

(B) *Find the total amount in each compound interest account. See Example 4.*

17. $6150 is compounded semiannually at a rate of 14% for 15 years.

18. $2060 is compounded annually at a rate of 15% for 10 years.

19. $1560 is compounded daily at a rate of 8% for 5 years.

20. $1450 is compounded quarterly at a rate of 10% for 15 years.

21. $10,000 is compounded semiannually at a rate of 9% for 20 years.

22. $3500 is compounded daily at a rate of 8% for 10 years.

Find the total amount of compound interest earned. See Example 5.

23. $2675 is compounded annually at a rate of 9% for 1 year.

24. $6375 is compounded semiannually at a rate of 10% for 1 year.

25. $2000 is compounded annually at a rate of 8% for 5 years.

26. $2000 is compounded semiannually at a rate of 8% for 5 years.

27. $2000 is compounded quarterly at a rate of 8% for 5 years.

28. $2000 is compounded daily at a rate of 8% for 5 years.

C *Solve. See Example 6.*

29. A college student borrows $1500 for 6 months to pay for a semester of school. If the interest is $61.88, find the monthly payment.

30. Jim Tillman borrows $1800 for 9 months. If the interest is $148.90, find his monthly payment.

 31. $20,000 is borrowed for 4 years. If the interest on the loan is $10,588.70, find the monthly payment.

32. $105,000 is borrowed for 15 years. If the interest on the loan is $181,125, find the monthly payment.

Review and Preview

Find the perimeter of each figure. See Section 1.3.

△ **33.**

Rectangle 6 yards
10 yards

△ **34.**

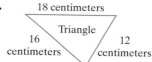

18 centimeters
Triangle
16 centimeters
12 centimeters

△ **35.**

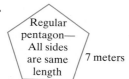

Regular pentagon— All sides are same length
7 meters

△ **36.**

Square 21 miles

Combining Concepts

37. Explain how to look up the compound interest factor in the compound interest table.

38. Explain how to find the amount of interest in a compounded account.

39. Compare the following accounts: Account 1: $1000 is invested for 10 years at a simple interest rate of 6%. Account 2: $1000 is compounded semiannually at a rate of 6% for 10 years. Discuss how the interest is computed for each account. Determine which account earns more interest. Why?

MATERIALS:

■ calculator

This activity may be completed by working in groups or individually.

Suppose you have just moved to a new city and must open new bank accounts with $5000. You would like to open a checking account, a savings account, and a certificate of deposit (CD). After doing some research, you find the following five options for CDs at five local banks:

Bank	Minimum Deposit	Interest Rate	Number of Compoundings per Year
A	$1000	6.50%	26
B	$1000	6.55%	365
C	$1000	6.60%	4
D	$1000	6.65%	2
E	$1000	6.70%	1

1. Without making any calculations, which CD investment option do you think would make the best choice? Explain your reasoning.

2. For each CD option, find the values of a CD opened with the minimum deposit after an investment of 2, 5, and 10 years (use the compound interest table on the next page). Create a table of your results.

3. Based on your table from Question 2, which CD option would be the best choice? Explain your reasoning. Does this surprise you? What other factors should you consider when choosing a savings tool such as a CD?

4. You may have noticed that it can sometimes be difficult to compare options when different interest rates and numbers of compoundings are involved. When describing investment options, banks often use *annual percentage yield* (APY) instead to help eliminate confusion. The APY of an investment represents the percent increase from the original principal to the ending value of the investment after 1 year. Find the APY of each CD option. Do you think that the APY makes it easier to compare the options? Explain.

CHAPTER 6 ACTIVITY # Compound Interest Table

Compounded Annually

Years	6.50%	6.55%	6.60%	6.65%	6.70%	6.75%	6.80%	6.85%	6.90%	6.95%	7.00%
2	1.13423	1.13529	1.13636	1.13742	1.13849	1.13956	1.14062	1.14169	1.14276	1.14383	1.14490
5	1.37009	1.37331	1.37653	1.37976	1.38300	1.38624	1.38949	1.39275	1.39601	1.39928	1.40255
10	1.87714	1.88597	1.89484	1.90374	1.91269	1.92167	1.93069	1.93975	1.94884	1.95798	1.96715

Compounded Semiannually

Years	6.50%	6.55%	6.60%	6.65%	6.70%	6.75%	6.80%	6.85%	6.90%	6.95%	7.00%
2	1.13648	1.13758	1.13868	1.13978	1.14089	1.14199	1.14309	1.14420	1.14531	1.14641	1.14752
5	1.37689	1.38023	1.38358	1.38693	1.39029	1.39365	1.39703	1.40041	1.40380	1.40720	1.41060
10	1.89584	1.90504	1.91428	1.92357	1.93290	1.94227	1.95169	1.96115	1.97065	1.98020	1.98979

Compounded Quarterly

Years	6.50%	6.55%	6.60%	6.65%	6.70%	6.75%	6.80%	6.85%	6.90%	6.95%	7.00%
2	1.13764	1.13876	1.13988	1.14100	1.14212	1.14325	1.14437	1.14550	1.14663	1.14775	1.14888
5	1.38042	1.38382	1.38723	1.39064	1.39407	1.39750	1.40094	1.40439	1.40784	1.41131	1.41478
10	1.90556	1.91496	1.92440	1.93389	1.94342	1.95300	1.96263	1.97230	1.98202	1.99179	2.00160

Compounded Monthly

Years	6.50%	6.55%	6.60%	6.65%	6.70%	6.75%	6.80%	6.85%	6.90%	6.95%	7.00%
2	1.13843	1.13956	1.14070	1.14183	1.14297	1.14410	1.14524	1.14638	1.14752	1.14866	1.14981
5	1.38282	1.38626	1.38971	1.39317	1.39664	1.40011	1.40360	1.40709	1.41060	1.41411	1.41763
10	1.91218	1.92172	1.93130	1.94092	1.95060	1.96032	1.97009	1.97991	1.98978	1.99970	2.00966

Compounded Biweekly

Years	6.50%	6.55%	6.60%	6.65%	6.70%	6.75%	6.80%	6.85%	6.90%	6.95%	7.00%
2	1.13864	1.13978	1.14092	1.14206	1.14320	1.14434	1.14548	1.14662	1.14777	1.14891	1.15006
5	1.38347	1.38692	1.39039	1.39386	1.39734	1.40083	1.40432	1.40783	1.41134	1.41487	1.41840
10	1.91399	1.92356	1.93317	1.94284	1.95255	1.96232	1.97213	1.98199	1.99189	2.00185	2.01186

Compounded Daily

Years	6.50%	6.55%	6.60%	6.65%	6.70%	6.75%	6.80%	6.85%	6.90%	6.95%	7.00%
2	1.13882	1.13995	1.14109	1.14224	1.14338	1.14452	1.14567	1.14681	1.14796	1.14911	1.15026
5	1.38399	1.38745	1.39093	1.39441	1.39790	1.40140	1.40490	1.40842	1.41194	1.41548	1.41902
10	1.91543	1.92503	1.93468	1.94437	1.95412	1.96391	1.97375	1.98364	1.99359	2.00358	2.01362

Chapter 6   VOCABULARY CHECK

Fill in each blank with one of the words or phrases listed below.

percent	of	amount	100%	compound interest
base	is	0.01	$\frac{1}{100}$	

1. In a mathematical statement, _____ usually means "multiplication."
2. In a mathematical statement, _____ means "equals."
3. _____ means "per hundred."
4. _____ is computed not only on the principal, but also on interest already earned in previous compounding periods.
5. In the percent proportion $\dfrac{\overline{}}{\underline{}} = \dfrac{\text{percent}}{100}$
6. To write a decimal or fraction as a percent, multiply by _____.
7. The decimal equivalent of the % symbol is _____.
8. The fraction equivalent of the % symbol is _____.

C H A P T E R | # Highlights

DEFINITIONS AND CONCEPTS	**EXAMPLES**
Section 6.1 Introduction to Percent	

Percent means "per hundred." The % symbol denotes percent.	$51\% = \dfrac{51}{100}$ 51 per 100 $7\% = \dfrac{7}{100}$ 7 per 100
To write a percent as a decimal, replace the % symbol with its decimal equivalent, 0.01, and multiply.	$32\% = 32(0.01) = 0.32$
To write a decimal as a percent, multiply by 100%.	$0.08 = 0.08(100\%) = 08\% = 8\%$

Section 6.2 Percents and Fractions	
To write a percent as a fraction, replace the % symbol with its fraction equivalent, $\dfrac{1}{100}$, and multiply.	$25\% = \dfrac{25}{100} = \dfrac{\overset{1}{\cancel{25}}}{4 \cdot \cancel{25}} = \dfrac{1}{4}$
To write a fraction as a percent, multiply by 100%.	$\dfrac{1}{6} = \dfrac{1}{6} \cdot 100\% = \dfrac{1}{6} \cdot \dfrac{100}{1}\% = \dfrac{100}{6}\% = 16\dfrac{2}{3}\%$

Section 6.3 Solving Percent Problems Using Equations	

Three key words in the statement of a percent problem are
 	
  of, which means **multiplication** (·)
  is, which means **equals** (=)
  what (or some equivalent word or phrase), which stands
    for **the unknown number**
	Solve:
	
	6 is 12% of what number?
	↓ ↓ ↓ ↓ ↓
	$6 = 12\% \cdot n$
	$6 = 0.12 \cdot n$ Write 12% as a decimal.
	$\dfrac{6}{0.12} = n$ Divide 6 by 0.12, the number multiplied by n.
	$50 = n$
	
	Thus, 6 is 12% of 50.

DEFINITIONS AND CONCEPTS	EXAMPLES

Section 6.4   Solving Percent Problems Using Proportions

PERCENT PROPORTION

$$\frac{\text{amount}}{\text{base}} = \frac{\text{percent}}{100} \quad \leftarrow \text{always } 100$$

or

$$\text{amount} \rightarrow \frac{a}{b} = \frac{p}{100} \quad \leftarrow \text{percent}$$
$$\text{base} \rightarrow$$

Solve:

$$20.4 \text{ is what percent of } 85?$$
$$\downarrow \qquad\qquad \downarrow \qquad\quad \downarrow$$
$$\boxed{\text{amount}} \qquad \boxed{\text{percent}} \quad \boxed{\text{base}}$$

$$\text{amount} \rightarrow \frac{20.4}{85} = \frac{p}{100} \quad \leftarrow \text{percent}$$
$$\text{base} \rightarrow$$

$$20.4 \cdot 100 = 85 \cdot p \qquad \text{Set cross products equal.}$$
$$2040 = 85 \cdot p \qquad \text{Multiply.}$$
$$\frac{2040}{85} = p \qquad \text{Divide 2040 by 85, the number multiplied by } p.$$
$$24 = p \qquad \text{Simplify.}$$

Thus, 20.4 is 24% of 85.

Section 6.5   Applications of Percent

PERCENT OF INCREASE

$$\text{percent of increase} = \frac{\text{amount of increase}}{\text{original amount}}$$

PERCENT OF DECREASE

$$\text{percent of decrease} = \frac{\text{amount of decrease}}{\text{original amount}}$$

A town with a population of 16,480 decreased to 13,870 over a 12-year period. Find the percent decrease. Round to the nearest whole percent.

$$\text{amount of decrease} = 16{,}480 - 13{,}870$$
$$= 2610$$
$$\text{percent of decrease} = \frac{\text{amount of decrease}}{\text{original amount}}$$
$$= \frac{2610}{16{,}480} \approx 0.16$$
$$= 16\%$$

The town's population decreased by 16%.

Section 6.6   Percent and Problem Solving: Sales Tax, Commission, and Discount

SALES TAX

$$\text{sales tax} = \text{sales tax rate} \cdot \text{purchase price}$$
$$\text{total price} = \text{purchase price} + \text{sales tax}$$

Find the sales tax and the total price of a purchase of $42 if the sales tax rate is 9%.

$$\boxed{\text{sales tax}} \quad = \quad \boxed{\text{sales tax rate}} \quad \cdot \quad \boxed{\text{purchase price}}$$
$$\downarrow \qquad\qquad\qquad \downarrow \qquad\qquad\qquad \downarrow$$
$$\text{sales tax} \quad = \qquad\quad 9\% \qquad\quad \cdot \qquad\quad \$42$$
$$= \quad 0.09 \cdot \$42$$
$$= \quad \$3.78$$

The total price is

$$\boxed{\text{total price}} \quad = \quad \boxed{\text{purchase price}} \quad + \quad \boxed{\text{sales tax}}$$
$$\downarrow \qquad\qquad\qquad \downarrow \qquad\qquad\qquad \downarrow$$
$$\text{total price} \quad = \qquad \$42 \qquad\quad + \qquad \$3.78$$
$$= \quad \$45.78$$

DEFINITIONS AND CONCEPTS	**EXAMPLES**

Section 6.6   Percent and Problem Solving: Sales Tax, Commission, and Discount *(continued)*

COMMISSION

commission = commission rate · sales

A salesperson earns a commission of 3%. Find the commission from sales of $12,500 worth of appliances.

commission	=	commission rate	·	sales
↓		↓		↓
commission	=	3%	·	$12,500
	=	0.03 · 12,500		
	=	$375		

DISCOUNT AND SALE PRICE

amount of discount = discount rate · original price

sale price = original price − amount of discount

A suit is priced at $320 and is on sale today for 25% off. What is the sale price?

amount of discount	=	discount rate	·	original price
↓		↓		↓
amount of discount	=	25%	·	$320
	=	0.25 · 320		
	=	$80		

sale price	=	original price	−	amount of discount
↓		↓		↓
sale price	=	$320	−	$80
	=	$240		

The sale price is $240.

Section 6.7   Percent and Problem Solving: Interest

SIMPLE INTEREST

interest = principal · rate · time

where the rate is understood to be per year.

Find the simple interest after 3 years on $800 at an interest rate of 5%.

interest	=	principal	·	rate	·	time
↓		↓		↓		↓
interest	=	$800	·	5%	·	3
	=	$800 · 0.05 · 3				Write 5% as 0.05.
	=	$120				Multiply.

The interest is $120.

Compound interest is computed not only on the principal, but also on interest already earned in previous compounding periods. (See Appendix F.)

total amount = original principal · compound interest factor

$800 is invested at 5% compounded quarterly for 10 years. Find the total amount at the end of 10 years.

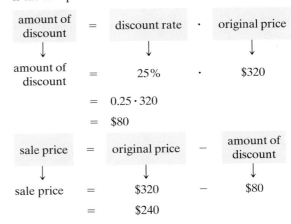

total amount	=	original principal	·	compound interest factor
↓		↓		↗
total amount	=	$800 · 1.64362		
	≈	$1314.90		

Are you prepared for a test on Chapter 6?

Below I have listed some *common trouble areas* for students in Chapter 6. After studying for your test—but before taking your test—read these.

- Can you convert from percents to fractions or decimals and from fractions or decimals to percents?

    Percent to decimal: $7.5\% = 7.5(0.01) = 0.075$

    Percent to fraction: $11\% = 11 \cdot \dfrac{1}{100} = \dfrac{11}{100}$

    Decimal to percent: $0.36 = 0.36(100\%) = 36\%$

    Fraction to percent: $\dfrac{6}{7} = \dfrac{6}{7} \cdot 100\% = \dfrac{6}{7} \cdot \dfrac{100}{1}\% = \dfrac{600}{7}\%$
    $$= 85\dfrac{5}{7}\%$$

- Do you remember how to find percent increase or percent decrease? The number of CDs increased from 40 to 48. Find the percent increase.

    $$\dfrac{\text{percent}}{\text{increase}} = \dfrac{\text{increase}}{\text{original number}} = \dfrac{8}{40} = 0.20 = 20\%$$

Chapter 6 Review

(6.1) *Solve.*

1. In a survey of 100 adults, 37 preferred pepperoni on their pizzas. What percent preferred pepperoni?

2. A basketball player made 77 out 100 attempted free throws. What percent of free throws was made?

Write each percent as a decimal.

3. 83%

4. 75%

5. 73.5%

6. 1.5%

7. 125%

8. 145%

9. 0.5%

10. 0.7%

11. 200%

12. 400%

13. 26.25%

14. 85.34%

Write each decimal as a percent.

15. 2.6

16. 0.055

17. 0.35

18. 1.02

19. 0.725

20. 0.25

21. 0.076

22. 0.085

23. 0.75

24. 0.65

25. 4.00

26. 9.00

(6.2) *Write each percent as a fraction or mixed number in simplest form.*

27. 1%

28. 10%

29. 25%

30. 8.5%

31. 10.2%

32. $16\frac{2}{3}$%

33. $33\frac{1}{3}$%

34. 110%

Write each fraction or mixed number as a percent.

35. $\frac{1}{5}$

36. $\frac{7}{10}$

37. $\frac{5}{6}$

38. $\frac{5}{8}$

39. $1\frac{2}{3}$

40. $1\frac{1}{4}$

41. $\frac{3}{5}$

42. $\frac{1}{16}$

(6.3) *Translate each to an equation and solve.*

43. 1250 is 1.25% of what number?

44. What number is $33\frac{1}{3}$% of 24,000?

45. 124.2 is what percent of 540?

46. 22.9 is 20% of what number?

47. What number is 40% of 7500?

48. 693 is what percent of 462?

(6.4) *Translate each to a proportion and solve.*

49. 104.5 is 25% of what number?

50. 16.5 is 5.5% of what number?

51. What number is 36% of 180?

52. 63 is what percent of 35?

53. 93.5 is what percent of 85?

54. What number is 33% of 500?

(6.5) *Solve.*

55. In a survey of 2000 people, it was found that 1320 have a microwave oven. Find the percent of people who own microwaves.

56. Of the 12,360 freshmen entering County College, 2000 are enrolled in basic college mathematics. Find the percent of entering freshmen who are enrolled in basic college mathematics. Round to the nearest whole percent.

57. The current charge for dumping waste in a local landfill is $16 per cubic foot. To cover new environmental costs, the charge will increase to $33 per cubic foot. Find the percent increase.

58. The number of violent crimes in a city decreased from 675 to 534. Find the percent decrease. Round to the nearest tenth.

59. This year the fund drive for a charity collected $215,000. Next year, a 4% decrease is expected. Find how much is expected to be collected in next year's drive.

60. A local union negotiated a new contract that increases the hourly pay 15% over last year's pay. The old hourly rate was $11.50. Find the new hourly rate rounded to the nearest cent.

61. If the sales tax rate is 5.5%, what is the total amount charged for a $250 coat?

62. Find the sales tax paid on a $25.50 purchase if the sales tax rate is 4.5%.

63. Russ James is a sales representative for a chemical company and is paid a commission rate of 5% on all sales. Find his commission if he sold $100,000 worth of chemicals last month.

64. Carol Sell is a sales clerk in a clothing store. She receives a commission of 7.5% on all sales. Find her commission for the week if her sales for the week were $4005. Round to the nearest cent.

65. A $3000 mink coat is on sale for 30% off. Find the discount and the sale price.

66. A $90 calculator is on sale for 10% off. Find the discount and the sale price.

(6.7) *Solve.*

67. Find the simple interest due on $4000 loaned for 3 months at 12% interest.

68. Find the total amount due on an 8-month loan of $1200 at a simple interest rate of 15%.

69. Find the total amount in an account if $5500 is compounded annually at 12% for 15 years.

70. Find the total amount in an account if $6000 is compounded semiannually at 11% for 10 years.

71. Find the compound interest earned if $100 is compounded quarterly at 12% for 5 years.

72. Find the compound interest earned if $1000 is compounded quarterly at 18% for 20 years.

Chapter 6 Test

Write each percent as a decimal.

1. 85%   **2.** 500%   **3.** 0.6%

Write each decimal as a percent.

4. 0.056   **5.** 6.1   **6.** 0.35

Write each percent as a fraction or mixed number in simplest form.

7. 120%   **8.** 38.5%   **9.** 0.2%

Write each fraction or mixed number as a percent.

10. $\frac{11}{20}$ **11.** $\frac{3}{8}$ **12.** $1\frac{3}{4}$

Solve.

13. What number is 42% of 80?   **14.** 0.6% of what number is 7.5?

15. 567 is what percent of 756?

Answers

1. _____

2. _____

3. _____

4. _____

5. _____

6. _____

7. _____

8. _____

9. _____

10. _____

11. _____

12. _____

13. _____

14. _____

15. _____

Solve. Round all dollar amounts to the nearest cent.

16. An alloy is 12% copper. How much copper is contained in 320 pounds of this alloy?

17. A farmer in Nebraska estimates that 20% of his potential crop, or $11,350, has been lost to a hard freeze. Find the total value of his potential crop.

18. If the local sales tax rate is 1.25%, find the total amount charged for a stereo system priced at $354.

19. A town's population increased from 25,200 to 26,460. Find the percent increase.

20. A $120 framed picture is on sale for 15% off. Find the discount and the sale price.

21. Randy Nguyen is paid a commission rate of 4% on all sales. Find Randy's commission if his sales were $9875.

22. A sales tax of $1.53 is added to an item's price of $152.99. Find the sales tax rate. Round to the nearest whole percent.

23. Find the simple interest earned on $2000 saved for $3\frac{1}{2}$ years at an interest rate of 9.25%.

24. $1365 is compounded annually at 8%. Find the total amount in the account after 5 years.

25. A couple borrowed $400 from a bank at 13.5% for 6 months for car repairs. Find the total amount due the bank at the end of the 6-month period.

Cumulative Review

1. How many cases can be filled with 9900 cans of jalapeños if each case holds 48 cans? How many cans will be left over? Will there be enough cases to fill an order for 200 cases?

2. Write each fraction as a mixed number or a whole number.

   a. $\dfrac{30}{7}$

   b. $\dfrac{16}{15}$

   c. $\dfrac{84}{6}$

3. Use a factor tree to find the prime factorization of 24.

4. Write $\dfrac{10}{27}$ in simplest form.

5. Multiply and simplify: $\dfrac{23}{32} \cdot \dfrac{4}{7}$

6. Find the reciprocal of $\dfrac{11}{8}$.

△ **7.** Find the perimeter of the rectangle.

$\frac{2}{15}$ inch

$\frac{4}{15}$ inch

8. Find the LCM of 12 and 20.

9. Add: $\dfrac{2}{5} + \dfrac{4}{15}$

10. Subtract: $7\dfrac{3}{14} - 3\dfrac{6}{7}$

Perform each indicated operation.

11. $\dfrac{1}{2} \div \dfrac{8}{7}$

12. $\dfrac{2}{9} \cdot \dfrac{3}{11}$

1. _____

2. a. _____

  b. _____

  c. _____

3. _____

4. _____

5. _____

6. _____

7. _____

8. _____

9. _____

10. _____

11. _____

12. _____

13. _____

14. _____

15. _____

16. _____

17. _____

18. _____

19. _____

20. _____

21. _____

22. _____

23. _____

24. _____

25. _____

Write each fraction as a decimal.

13. $\dfrac{8}{10}$

14. $\dfrac{87}{10}$

15. The price of a gallon of gasoline in Aimsville is currently $1.0279. Round this to the nearest cent.

16. Add: $763.7651 + 22.001 + 43.89$

17. Multiply: $23.6 \cdot 0.78$

Divide.

18. $\dfrac{786}{10,000}$

19. $\dfrac{0.12}{10}$

20. Simplify: $(1.3)^2$

21. Write $\dfrac{1}{4}$ as a decimal.

22. Write the following rate as a fraction in simplest form:

10 nails every 6 feet

23. Is $\dfrac{4.1}{7} = \dfrac{2.9}{5}$ a true proportion?

24. On a chamber of commerce map of Abita Springs, 5 miles corresponds to 2 inches. How many miles correspond to 7 inches?

25. Translate to an equation: What number is 25% of 0.008?

Measurement

The use of measurements is common in everyday life. A sales representative records the number of miles she has driven when she submits her travel expense report. A respiratory therapist measures the volume of air exhaled by a patient. A measurement is necessary in each case.

The Statue of Liberty, standing on Liberty Island in New York City's harbor, is one of the largest statues in the world. Originally named the "Statue of Liberty Enlightening the World," this symbol of freedom was a gift from the people of France. U.S. citizens raised money for the pedestal on which the statue stands and for the installation of the statue. After the pedestal had been prepared in 1885, the statue arrived dismantled in 214 packing cases of iron framework and copper sheeting and was assembled using rivets. Over a year later it was dedicated by President Grover Cleveland. The statue alone weighs 225 tons and is over 150 feet tall. For its 100th anniversary in 1986, the Statue of Liberty received extensive renovations to its framework and torch. In Exercises 39 and 40 on page 414, we will explore some of the measurements of the Statue of Liberty.

Name _____ Section _____ Date _____

Chapter 7 Pretest

Convert.

1. 6 ft to inches

2. 98 in. = ____ ft ____ in.

3. 2.8 mi to feet

4. 9.8 m to centimeters

5. 7 lb to ounces

6. 29 dg to grams

7. 0.6 kg to grams

8. 18 qt to gallons

9. 6280 L to kiloliters

10. 45°C to Fahrenheit

11. 77°F to Celsius

12. 15 BTU to foot-pounds

Perform the indicated operation.

13. 9 ft 10 in. + 3 ft 7 in.

14. 6 yd 1 ft − 1 yd 2 ft

15. 5.4 mm × 9

16. 5 tons 1300 lb ÷ 4

17. 20.6 g + 391 mg

18. 9 qt 1 pt × 7

19. 3 c 2 fl oz + 5 c 6 fl oz

20. Jay Dolesky paid $13 to fill his car with 40 liters of gasoline. Find the price per liter of gasoline to the nearest tenth of a cent.

Answers column:

1. _____
2. _____
3. _____
4. _____
5. _____
6. _____
7. _____
8. _____
9. _____
10. _____
11. _____
12. _____
13. _____
14. _____
15. _____
16. _____
17. _____
18. _____
19. _____
20. _____

7.1 Length: U.S. and Metric Systems of Measurement

OBJECTIVES

Ⓐ Define U.S. units of length and convert from one unit to another.

Ⓑ Use mixed units of length.

Ⓒ Perform arithmetic operations on U.S. units of length.

Ⓓ Define metric units of length and convert from one unit to another.

Ⓔ Perform arithmetic operations on metric units of length.

SSM   TUTOR CENTER   SG   CD & VIDEO   MATH PRO   WEB

Ⓐ Defining and Converting U.S. System Units of Length

In the United States, two systems of measurement are commonly used. They are the **United States (U.S.), or English, measurement system** and the **metric system**. The U.S. measurement system is familiar to most Americans. Units such as feet, miles, ounces, and gallons are used. However, the metric system is also commonly used in fields such as medicine, sports, international marketing, and certain physical sciences. We are accustomed to buying 2-liter bottles of soft drinks, watching televised coverage of the 100-meter dash at the Olympic Games, or taking a 200-milligram dose of pain reliever.

The U.S. system of measurement uses the **inch**, **foot**, **yard**, and **mile** to measure **length**. The following is a summary of equivalencies between units of length:

U.S. Units of Length	Unit Fractions
12 inches (in.) = 1 foot (ft)	$\dfrac{12 \text{ in.}}{1 \text{ ft}} = \dfrac{1 \text{ ft}}{12 \text{ in.}} = 1$
3 feet = 1 yard (yd)	$\dfrac{3 \text{ ft}}{1 \text{ yd}} = \dfrac{1 \text{ yd}}{3 \text{ ft}} = 1$
5280 feet = 1 mile (mi)	$\dfrac{5280 \text{ ft}}{1 \text{ mi}} = \dfrac{1 \text{ mi}}{5280 \text{ ft}} = 1$

To convert from one unit of length to another, **unit fractions** may be used. A unit fraction is a fraction that equals 1. For example, since 12 in. = 1 ft, we have the unit fractions

$$\frac{12 \text{ in.}}{1 \text{ ft}} = \frac{1 \text{ ft}}{12 \text{ in.}} = 1$$

For example, to convert 48 inches to feet, we *multiply by a unit fraction* that relates feet to inches. The unit fraction should be written so that *the units we are converting to*, feet, *are in the numerator and the original units*, inches, *are in the denominator*. We do this so that like units will divide out, as shown next:

$$48 \text{ in.} = \frac{48 \text{ in.}}{1} \cdot \overbrace{\frac{1 \text{ ft}}{12 \text{ in.}}}^{\text{Unit fraction}} \quad \leftarrow \text{ Units to convert to}$$
$$\qquad\qquad\qquad\qquad \leftarrow \text{ Original units}$$

$$= \frac{48 \cdot 1 \text{ ft}}{1 \cdot 12}$$

$$= \frac{48 \text{ ft}}{12}$$

$$= 4 \text{ ft}$$

Therefore, 48 inches equals 4 feet, as seen in the diagram:

12 in.	12 in.	12 in.	12 in.
1 ft	1 ft	1 ft	1 ft

48 in. = 4 ft

Practice Problem 1

Convert 5 feet to inches.

Practice Problem 2

Convert 7 yards to feet.

EXAMPLE 1 Convert 8 feet to inches.

Solution: We multiply 8 feet by a unit to convert fraction that compares 12 inches to 1 foot. The unit fraction should be $\dfrac{\text{units to convert to}}{\text{original units}}$ or $\dfrac{12 \text{ inches}}{1 \text{ foot}}$.

$$8 \text{ ft} = \frac{8 \, \cancel{\text{ft}}}{1} \cdot \frac{\overbrace{12 \text{ in.}}^{\text{Unit fraction}}}{1 \, \cancel{\text{ft}}}$$

$$= 8 \cdot 12 \text{ in.}$$

$$= 96 \text{ in.} \qquad \text{Multiply.}$$

Thus, 8 ft = 96 in., as shown in the diagram:

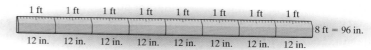

EXAMPLE 2 Convert 7 feet to yards.

Solution: We multiply by a unit fraction that compares 1 yard to 3 feet.

$$7 \text{ ft} = \frac{7 \, \cancel{\text{ft}}}{1} \cdot \frac{1 \text{ yd}}{3 \, \cancel{\text{ft}}} \quad \begin{array}{l} \leftarrow \text{ Units to convert to} \\ \leftarrow \text{ Original units} \end{array}$$

$$= \frac{7 \text{ yd}}{3}$$

$$= 2\frac{1}{3} \text{ yd} \qquad \text{Divide.}$$

Thus, $7 \text{ ft} = 2\frac{1}{3}$ yd.

B Using Mixed U.S. System Units of Length

Sometimes it is more meaningful to express a measurement of length with mixed units such as 1 ft and 5 in. We usually condense this and write 1 ft 5 in.

In Example 2, we found that 7 feet was the same as $2\frac{1}{3}$ yards. The measurement can also be written as a mixture of yards and feet. That is,

$$7 \text{ ft} = \underline{\quad} \text{ yd} \underline{\quad} \text{ft}$$

Because 3 ft = 1 yd, we divide 3 into 7 to see how many whole yards are in 7 feet. The quotient is the number of yards, and the remainder is the number of feet.

$$\begin{array}{r} 2 \text{ yd } 1 \text{ ft} \\ 3\overline{)7} \\ \underline{-6} \\ 1 \end{array}$$

Thus, 7 ft = 2 yd 1 ft, as seen in the diagram:

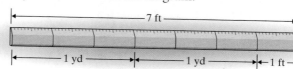

EXAMPLE 3   Convert: 134 in. = _____ ft _____ in.

Solution: Because 12 in. = 1 ft, we divide 12 into 134. The quotient is the number of feet. The remainder is the number of inches. To see why we divide 12 into 134, notice that

$$134 \text{ in.} = \frac{134 \text{ in.}}{1} \cdot \frac{1 \text{ ft}}{12 \text{ in.}} = \frac{134}{12} \text{ ft}$$

$$\begin{array}{r} 11 \text{ ft } 2 \text{ in.} \\ 12\overline{)134} \\ -12 \\ \hline 14 \\ -12 \\ \hline 2 \end{array}$$

Thus, 134 in. = 11 ft 2 in.

EXAMPLE 4   Convert 3 feet 7 inches to inches.

Solution: First, we convert 3 feet to inches. Then we add 7 inches.

$$3 \text{ ft} = \frac{3 \text{ ft}}{1} \cdot \frac{12 \text{ in.}}{1 \text{ ft}} = 36 \text{ in.}$$

Then

$$3 \text{ ft } 7 \text{ in.} = 36 \text{ in.} + 7 \text{ in.} = 43 \text{ in.}$$

C   Performing Operations on U.S. System Units of Length

Finding sums or differences of measurements often involves converting units, as shown in the next example. Just remember that, as usual, only like units can be added or subtracted.

EXAMPLE 5   Add 3 ft 2 in. and 5 ft 11 in.

Solution: To add, we line up the similar units.

$$\begin{array}{r} 3 \text{ ft } \ 2 \text{ in.} \\ + 5 \text{ ft } 11 \text{ in.} \\ \hline 8 \text{ ft } 13 \text{ in.} \end{array}$$

Since 13 inches is the same as 1 ft 1 in., we have

$$8 \text{ ft } 13 \text{ in.} = 8 \text{ ft} + 1 \text{ ft } 1 \text{ in.}$$
$$= 9 \text{ ft } 1 \text{ in.}$$

Try the Concept Check in the margin.

EXAMPLE 6   Multiply 8 ft 9 in. by 3.

Solution: By the distributive property, we multiply 8 ft by 3 and 9 in. by 3.

$$\begin{array}{r} 8 \text{ ft } \ 9 \text{ in.} \\ \times \qquad\quad 3 \\ \hline 24 \text{ ft } 27 \text{ in.} \end{array}$$

Practice Problem 3

Convert: 68 in. = _____ ft _____ in.

Practice Problem 4

Convert 5 yards 2 feet to feet.

Practice Problem 5

Add 4 ft 8 in. to 8 ft 11 in.

Concept Check

How could you estimate the following sum?

$$\begin{array}{r} 7 \text{ yd } \ 4 \text{ in.} \\ + 3 \text{ yd } 27 \text{ in.} \end{array}$$

Practice Problem 6

Multiply 4 ft 7 in. by 4.

Answers

3. 5 ft 8 in.,   **4.** 17 ft,   **5.** 13 ft 7 in.,
6. 18 ft 4 in.

Concept Check: round each to the nearest yard:
7 yd + 4 yd = 11 yd

Since 27 in. is the same as 2 ft 3 in., we simplify the product as

$$24 \text{ ft } 27 \text{ in.} = 24 \text{ ft } + 2 \text{ ft } 3 \text{ in.}$$
$$= 26 \text{ ft } 3 \text{ in.}$$

Practice Problem 7

Divide 18 ft 6 in. by 2.

EXAMPLE 7 Divide 24 yd 6 in. by 3.

Solution: We divide each of the units by 3.

$$
\begin{array}{r}
8 \text{ yd } 2 \text{ in.} \\
3\overline{)24 \text{ yd } 6 \text{ in.}} \\
-24 \text{ yd} \\
\hline
6 \text{ in.} \\
-6 \text{ in.} \\
\hline
0
\end{array}
$$

The quotient is 8 yd 2 in.

To check, see that 8 yd 2 in. multiplied by 3 is 24 yd 6 in.

Practice Problem 8

A carpenter cuts 1 ft 9 in. from a board of length 5 ft 8 in. Find the remaining length of the board.

EXAMPLE 8 Finding the Length of a Piece of Rope

A rope of length 6 yd 1 ft has 2 yd 2 ft cut from one end. Find the length of the remaining rope.

Solution: Subtract 2 yd 2 ft from 6 yd 1 ft.

$$
\begin{array}{lcr}
\text{beginning length} & \rightarrow & 6 \text{ yd } 1 \text{ ft} \\
- \quad \text{amount cut} & \rightarrow & -2 \text{ yd } 2 \text{ ft} \\
\hline
\text{remaining length} & &
\end{array}
$$

We cannot subtract 2 ft from 1 ft, so we borrow 1 yd from the 6 yd. One yard is converted to 3 ft and combined with the 1 ft already there.

Borrow 1 yd = 3 ft   The problem now reads:

$$5 \text{ yd } + \boxed{1 \text{ yd}} \quad \boxed{3 \text{ ft}}$$

$$
\begin{array}{r}
\cancel{6 \text{ yd}} 1 \text{ ft} \\
-2 \text{ yd } 2 \text{ ft} \\
\end{array}
\qquad
\begin{array}{r}
5 \text{ yd } 4 \text{ ft} \\
-2 \text{ yd } 2 \text{ ft} \\
\hline
3 \text{ yd } 2 \text{ ft}
\end{array}
$$

The remaining rope is 3 yd 2 ft long.

D Defining and Converting Metric System Units of Length

The basic unit of length in the metric system is the **meter**. A meter is slightly longer than a yard. It is approximately 39.37 inches long. Recall that a yard is 36 inches long.

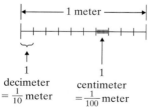

$$
\begin{array}{c}
1 \text{ decimeter} = \tfrac{1}{10} \text{ meter} \\
1 \text{ centimeter} = \tfrac{1}{100} \text{ meter}
\end{array}
$$

All units of length in the metric system are based on the meter. The following is a summary of the prefixes used in the metric system. Also shown are equivalencies between units of length. Like the decimal system, the metric system uses powers of 10 to define units.

Answers

7. 9 ft 3 in.,   **8.** 3 ft 11 in.

Prefix	Meaning	Metric Unit of Length
kilo	1000	1 **kilo**meter (km) = 1000 meters (m)
hecto	100	1 **hecto**meter (hm) = 100 m
deka	10	1 **deka**meter (dam) = 10 m
		1 meter (m) = 1 m
deci	1/10	1 **deci**meter (dm) = 1/10 m or 0.1 m
centi	1/100	1 **centi**meter (cm) = 1/100 m or 0.01 m
milli	1/1000	1 **milli**meter (mm) = 1/1000 m or 0.001 m

These same prefixes are used in the metric system for mass and capacity. The most commonly used measurements of length in the metric system are the **meter**, **millimeter**, **centimeter**, and **kilometer**.

Being comfortable with the metric units of length means gaining a "feeling" for metric lengths, just as you have a "feeling" for the length of an inch, a foot, and a mile. To help you accomplish this, study the following examples:

A millimeter is about the thickness of a large paper clip.

A centimeter is about the width of a large paper clip.

1 mm

1 cm

A meter is slightly longer than a yard.

A kilometer is about two-thirds of a mile.

The length of this book is approximately 27.5 centimeters.

The width of this book is approximately 21.5 centimeters.

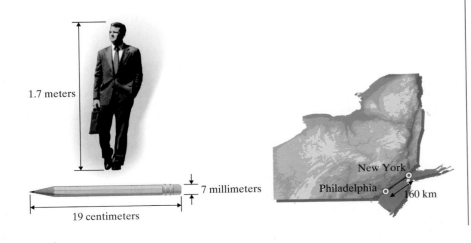

1.7 meters

7 millimeters

19 centimeters

New York

Philadelphia

160 km

As with the U.S. system of measurement, unit fractions may be used to convert from one unit of length to another. The metric system does, however, have a distinct advantage over the U.S. system of measurement: The ease of converting from one unit of length to another. Since all units of length are powers of 10 of the meter, converting from one unit of length to another is as simple as moving the decimal point. Listing units of length in order from largest to smallest helps to keep track of how many places to move the decimal point when converting.

For example, let's convert 1200 meters to kilometers. To convert from meters to kilometers, we move along the chart shown 3 units to the left, from meters to kilometers. This means that we move the decimal point 3 places to the left.

km	hm	dam	**m**	dm	cm	mm

3 units to the left

1200 m $= 1.200$ km

	1000 m	200 m

3 places to the left | ←——— 1 km ———→ | ← 0.2 km → |

The same conversion can be made using unit fractions.

$$1200 \text{ m} = \frac{1200 \text{ m}}{1} \cdot \overbrace{\frac{1 \text{ km}}{1000 \text{ m}}}^{\text{Unit fraction}} = \frac{1200 \text{ km}}{1000} = 1.2 \text{ km}$$

EXAMPLE 9   Convert 2.3 m to centimeters.

Solution: First we will convert by using a unit fraction.

$$2.3 \text{ m} = \frac{2.3 \text{ m}}{1} \cdot \overbrace{\frac{100 \text{ cm}}{1 \text{ m}}}^{\text{Unit fraction}} = 230 \text{ cm}$$

Now we will convert by listing the units of length in a chart and moving from meters to centimeters.

km	hm	dam	m	dm	cm	mm

2 units to the right

2.30 m $= 230.$ cm

2 places to the right

With either method, we get 230 cm.   ●

EXAMPLE 10   Convert 450,000 mm to meters.

Solution: We list the units of length in a chart and move from millimeters to meters.

km	hm	dam	m	dm	cm	mm

3 units to the left

$450,000$ mm $= 450.000$ m   or 450 m   ●

Try the Concept Check in the margin.

Performing Operations on Metric System Units of Length

To add, subtract, multiply, or divide with metric measurements of length, we write all numbers using the same unit of length and then add, subtract, multiply, or divide as with decimals.

EXAMPLE 11   Subtract 430 m from 1.3 km.

Solution: First we convert both measurements to kilometers or both to meters.

$$430 \text{ m} = 0.43 \text{ km} \quad \text{or} \quad 1.3 \text{ km} = 1300 \text{ m}$$

$$
\begin{array}{r}
1.30 \text{ km} \\
- 0.43 \text{ km} \\
\hline
0.87 \text{ km}
\end{array}
\qquad
\begin{array}{r}
1300 \text{ m} \\
- 430 \text{ m} \\
\hline
870 \text{ m}
\end{array}
$$

The difference is 0.87 km or 870 m.   ●

EXAMPLE 12   Multiply 5.7 mm by 4.

Solution: Here we simply multiply the two numbers. Note that the unit of measurement remains the same.

$$
\begin{array}{r}
5.7 \text{ mm} \\
\times \quad 4 \\
\hline
22.8 \text{ mm}
\end{array}
$$

EXAMPLE 13   Finding a Person's Height

Fritz Martinson was 1.2 meters tall on his last birthday. Since then, he has grown 14 centimeters. Find his current height in meters.

Solution:

$$
\begin{array}{lll}
\text{original height} & \rightarrow & 1.20 \text{ m} \\
\underline{+ \text{ height grown}} & \rightarrow & \underline{+ 0.14 \text{ m}} \quad \text{(Since 14 cm} = 0.14 \text{ m)} \\
\text{current height} & & 1.34 \text{ m}
\end{array}
$$

Fritz is now 1.34 meters tall.   ●

Practice Problem 11

Subtract 640 m from 2.1 km.

Practice Problem 12

Multiply 18.3 hm by 5.

Practice Problem 13

Doris Blackwell is knitting a scarf that is currently 0.8 meter long. If she knits an additional 45 centimeters, how long will the scarf be?

Answers

11. 1.46 km or 1460 m,   **12.** 91.5 hm,
13. 125 cm or 1.25 m

CALCULATOR EXPLORATIONS

Metric to U.S. System Conversions in Length

To convert between the two systems of measurement in length, the following *approximations* may be used:

meters $\times$ 3.28 $\approx$ feet	feet $\times$ 0.305 $\approx$ meters
meters $\times$ 1.09 $\approx$ yards	yards $\times$ 0.914 $\approx$ meters
centimeters $\times$ 0.39 $\approx$ inches	inches $\times$ 2.54 = centimeters
kilometers $\times$ 0.62 $\approx$ miles	miles $\times$ 1.609 $\approx$ kilometers

EXAMPLE

The distance from New Orleans, Louisiana, to Pensacola, Florida, is about 400 miles. How many kilometers is this?

Solution: From the above approximations,

$$\text{miles} \times 1.609 \approx \text{kilometers}$$
$$\downarrow$$
$$400 \times 1.609 \approx \text{kilometers}$$

To multiply on your calculator, press the keys

The display will read 643.6 . Thus, 400 miles $\approx$ 643.6 kilometers.
("is approximately")

Convert as indicated.

1. 7 meters to feet.
2. 11.5 yards to meters.
3. 8.5 inches to centimeters.
4. 15 kilometers to miles.
5. A 5-kilometer race is being held today. How many miles is this?
6. A 100-meter dash is being held today. How many yards is this?

Name _____ Section _____ Date _____

Mental Math

Convert as indicated.

1. 12 inches to feet

2. 6 feet to yards

3. 24 inches to feet

4. 36 inches to feet

5. 36 inches to yards

6. 2 yards to inches

Determine whether the measurement in each statement is reasonable.

7. The screen of a home television set has a 30-meter diagonal.

8. A window measures 1 meter by 0.5 meter.

9. A drinking glass is made of glass 2 millimeters thick.

10. A paper clip is 4 kilometers long.

11. The distance across the Colorado River is 50 kilometers.

12. A model's hair is 30 centimeters long.

EXERCISE SET 7.1

Ⓐ *Convert each measurement as indicated. See Examples 1 and 2.*

1. 60 in. to feet

2. 84 in. to feet

3. 12 yd to feet

4. 18 yd to feet

5. 42,240 ft to miles

6. 36,960 ft to miles

7. 102 in. to feet

8. 150 in. to feet

9. 10 ft to yards

10. 25 ft to yards

11. 6.4 mi to feet

12. 3.8 mi to feet

Ⓑ *Convert each measurement as indicated. See Examples 3 and 4.*

13. 40 ft = ____ yd ____ ft

14. 100 ft = ____ yd ____ ft

15. 41 in. = ____ ft ____ in.

16. 75 in. = ____ ft ____ in.

17. 10,000 ft = ____ mi ____ ft

18. 25,000 ft = ____ mi ____ ft

19. 5 ft 2 in. = ____ in.

20. 4 ft 11 in. = ____ in.

21. 5 yd 2 ft = ____ ft

22. 7 yd 1 ft = ____ ft

23. 2 yd 1 ft = ____ in.

24. 1 yd 2 ft = ____ in.

Ⓒ *Perform each indicated operation. Simplify the result if possible. See Examples 5 through 7.*

25. 5 ft 8 in. + 6 ft 7 in.

26. 9 ft 10 in. + 8 ft 4 in.

27. 12 yd 2 ft + 9 yd 2 ft

28. 16 yd 2 ft + 8 yd 1 ft

29. 24 ft 8 in. − 16 ft 3 in.

30. 15 ft 5 in. − 8 ft 2 in.

31. 16 ft 3 in. − 10 ft 9 in.

32. 14 ft 8 in. − 3 ft 11 in.

33. 6 ft 8 in. ÷ 2

34. 26 ft 10 in. ÷ 2

35. 12 yd 2 ft × 4

36. 15 yd 1 ft × 8

Solve. Remember to insert units when writing your answers. See Example 8.

37. The National Zoo maintains a small patch of bamboo, which it grows as a food supply for its pandas. Two weeks ago, the bamboo was 6 ft 10 in. tall. Since then, the bamboo has grown 3 ft 8 in. taller. How tall is the bamboo now?

38. While exploring in the Marianas Trench, a submarine probe was lowered to a point 1 mile 1400 feet below the ocean's surface. Later it was lowered an additional 1 mile 4000 feet below this point. How far is the probe below the surface of the Pacific?

39. The length of one of the Statue of Liberty's hands is 16 ft 5 in. One of the Statue's eyes is 2 ft 6 in. across. How much longer is a hand than the width of an eye? (*Source:* National Park Service)

40. The width of the Statue of Liberty's head from ear to ear is 10 ft. The height of the Statue's head from chin to cranium is 17 ft 3 in. How much taller is the Statue's head than its width? (*Source:* National Park Service)

41. The Amana Corporation stacks up its microwave ovens in a distribution warehouse. Each stack is 1 ft 9 in. wide. How far from the wall would 9 of these stacks extend?

1ft 9 in.

?

42. The highway commission is installing concrete sound barriers along a highway. Each barrier is 1 yd 2 ft long. How far will 25 barriers in a row reach?

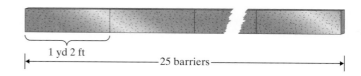

1 yd 2 ft

25 barriers

43. A carpenter needs to cut a board into thirds. If the board is 9 ft 3 in. long originally, how long will each cut piece be?

9 feet 3 inches

44. A wall is erected exactly halfway between two buildings that are 192 ft 8 in. apart. If the wall is 8 in. wide, how far is it from the wall to either of the buildings?

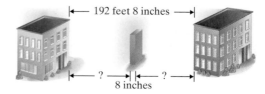

192 feet 8 inches

? 8 inches ?

△**45.** Evelyn Pittman plans to erect a fence around her garden to keep the rabbits out. If the garden is a rectangle 24 ft 9 in. long by 18 ft 6 in. wide, what is the length of the fencing material she must purchase?

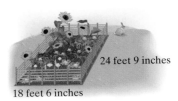

24 feet 9 inches

18 feet 6 inches

△**46.** Ronnie Hall needs new gutters for the front and *both sides* of his home. The front of the house is 50 ft. 8 in., and each side is 22 ft 9 in. wide. What length of gutter must he buy?

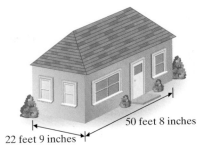

50 feet 8 inches

22 feet 9 inches

47. The world's longest Coca-Cola truck is in Sweden and is 79 feet long. How many *yards* long are 4 of these trucks? (*Source: Coca-Cola Today*)

△**48.** The world's largest Coca-Cola sign is in Arica, Chile. It is in the shape of a rectangle whose length is 400 feet and whose width is 131 feet. Find the area of the sign. (*Source: Coca-Cola Today*) (*Hint*: Recall that area of a rectangle is the product: length times width.)

D *Convert as indicated. See Examples 9 and 10.*

49. 40 m to centimeters

50. 18 m to centimeters

51. 40 mm to centimeters

52. 18 mm to centimeters

53. 300 m to kilometers

54. 400 m to kilometers

55. 1400 mm to meters

56. 6400 mm to meters

57. 1500 cm to meters

58. 6400 cm to meters

59. 8.3 cm to millimeters

60. 4.6 cm to millimeters

61. 20.1 mm to decimeters

62. 140.2 mm to decimeters

63. 0.04 m to millimeters

64. 0.2 m to millimeters

E *Perform each indicated operation. See Examples 11 and 12.*

65. $8.6 \, m + 0.34 \, m$

66. $14.1 \, cm + 3.96 \, cm$

67. $2.9 \, m + 40 \, mm$

68. $30 \, cm + 8.9 \, m$

69. $24.8 \, mm - 1.19 \, cm$

70. $45.3 \, m - 2.16 \, dam$

71. $15 \, km - 2360 \, m$

72. $14 \, cm - 15 \, mm$

73. $18.3 \, m \times 3$

74. $14.1 \, m \times 4$

75. $6.2 \, km \div 4$

76. $9.6 \, m \div 5$

Solve. Remember to insert units when writing your answers. See Example 13.

77. A 3.4-m rope is attached to a 5.8-m rope. However, when the ropes are tied, 8 cm of length is lost to form the knot. What is the length of the tied ropes?

78. A 2.15-m-long sash cord has become frayed at both ends so that 1 cm is trimmed from each end. How long is the remaining cord?

79. The ice on Doc Miller's pond is 5.33 cm thick. For safe skating, Doc insists that it must be 80 mm thick. How much thicker must the ice be before Doc goes skating?

80. The sediment on the bottom of the Towamencin Creek is normally 14 cm thick, but the recent flood washed away 22 mm of sediment. How thick is it now?

81. An art class is learning how to make kites. The two sticks used for each kite have lengths of 1 m and 65 cm. What total length of wood must be ordered for the sticks if 25 kites are to be built?

82. The total pages of a hardbound economics text are 3.1 cm thick. The front and back covers are each 2 mm thick. How high would a stack of 10 of these texts be?

83. A logging firm needs to cut a 67-m-long redwood log into 20 equal pieces before loading it onto a truck for shipment. How long will each piece be?

84. An 18.3-m-tall flagpole is mounted on a 65-cm-high pedestal. How far is the top of the flagpole from the ground?

85. At one time it was believed that the fort of Basasi, on the Indian-Tibetan border, was the highest located structure at an elevation of 5.988 km above sea level. However, a settlement has been located that is 21 m higher than the fort. What is the elevation of this settlement?

86. The average American male at age 35 is 1.75 m tall. The average 65-year-old male is 48 mm shorter. How tall is the average 65-year-old male?

87. A floor tile is 22.86 cm wide. How many tiles in a row are needed to cross a room 3.429 m wide?

△ **88.** A standard postcard is 1.6 times longer than it is wide. If it is 9.9 cm wide, what is its length?

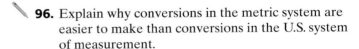

Jim Doe
99 Red Rd
Austin, TX 23942

9.9 centimeters

Review and Preview

Write each decimal or fraction as a percent. See Sections 6.1 and 6.2.

89. 0.21

90. 0.86

91. $\dfrac{13}{100}$

92. $\dfrac{47}{100}$

93. $\dfrac{1}{4}$

94. $\dfrac{3}{20}$

Combining Concepts

95. To convert from meters to centimeters, the decimal point is moved two places to the right. Explain how this relates to the fact that the prefix *centi* means $\dfrac{1}{100}$.

96. Explain why conversions in the metric system are easier to make than conversions in the U.S. system of measurement.

△ **97.** Anoa Longway plans to use 26.3 meters of leftover fencing material to enclose a square garden plot for her daughter. How long will each side of the garden be?

98. A marathon is a running race over a distance of 26 mi 385 yd. If a runner runs five marathons in a year, what is the total distance he or she has run? (*Source: Microsoft Encarta 97 Encyclopedia*)

FOCUS ON **the Real World**

COMPUTER STORAGE CAPACITY

Have you ever heard someone describe a computer as having a 4 gig hard drive and wondered what exactly that means? In this chapter, we focused on measuring basic physical characteristics such as length, weight/mass, and capacity. However, another type of measurement that we frequently encounter in the real world deals with computer storage capacity; that is, how much information can be stored on a computer device such as a floppy diskette, data cartridge, CD-ROM, hard drive, or computer memory (such as RAM).

To understand computer storage capacities, we must first understand the building blocks of information in the world of computers. The smallest piece of information handled by a computer is called a **bit** (abbreviated "b"). Bit is short for "binary digit" and consists of either a 0 or a 1. A collection of 8 bits is called a **byte** (abbreviated "B"). A single byte represents a standard character such as a letter of the alphabet, a number, or a punctuation mark. The following table shows the relationships among various terms used to describe computer storage capacities:

$$1 \text{ nibble } = 4 \text{ bits}$$

$$1 \text{ byte (B) } = 8 \text{ bits}$$

$$1 \text{ kilobyte (KB) } = 1024 \text{ bytes } = 2^{10} \text{ bytes}$$

$$1 \text{ megabyte (MB) } = 1024 \text{ kilobytes } = 1024^2 \text{ bytes } = 2^{20} \text{ bytes}$$

$$1 \text{ gigabyte (GB) } = 1024 \text{ megabytes } = 1024^3 \text{ bytes } = 2^{30} \text{ bytes}$$

$$1 \text{ terabyte (TB) } = 1024 \text{ gigabytes } = 1024^4 \text{ bytes } = 2^{40} \text{ bytes}$$

Note: Sometimes a megabyte is referred to as a "meg" and a gigabyte as a "gig."

GROUP ACTIVITY

Using advertisements for computer systems, find five different examples of uses of any of the computer storage capacity terms defined above. Then convert each example to both bytes and bits. Make a table to organize your results. Be sure to include an explanation of the original use of each example.

7.2 Weight and Mass: U.S. and Metric Systems of Measurement

A Defining and Converting U.S. System Units of Weight

Whenever we talk about how heavy an object is, we are concerned with the object's **weight**. We discuss weight when we refer to a 12-ounce box of Rice Krispies, an overweight 19-pound tabby cat, or a barge hauling 24 tons of garbage.

The most common units of weight in the U.S. measurement system are the **ounce**, the **pound**, and the **ton**. The following is a summary of equivalencies between units of weight:

U.S. Units of Weight	Unit Fractions
16 ounces (oz) = 1 pound (lb)	$\dfrac{16\text{ oz}}{1\text{ lb}} = \dfrac{1\text{ lb}}{16\text{ oz}} = 1$
2000 pounds = 1 ton	$\dfrac{2000\text{ lb}}{1\text{ ton}} = \dfrac{1\text{ ton}}{2000\text{ lb}} = 1$

Try the Concept Check in the margin.

Unit fractions that equal 1 are used to convert between units of weight in the U.S. system. When converting using unit fractions, recall that the numerator of a unit fraction should contain the units we are converting to and the denominator should contain the original units.

To convert 40 ounces to pounds, multiply by $\dfrac{1\text{ lb}}{16\text{ oz}}$ ← Units to convert to
                                                    ← Original units

$$40 \text{ oz} = \frac{40 \cancel{\text{ oz}}}{1} \cdot \overbrace{\frac{1 \text{ lb}}{16 \cancel{\text{ oz}}}}^{\text{Unit fraction}}$$

$$= \frac{40 \text{ lb}}{16} \qquad \text{Multiply.}$$

$$= \frac{5}{2} \text{ lb} \quad \text{or} \quad 2\frac{1}{2} \text{ lb, as a mixed number}$$

EXAMPLE 1   Convert 9000 pounds to tons.

Solution: We multiply 9000 lb by the unit fraction

$$\frac{1 \text{ ton}}{2000 \text{ lb}} \qquad \begin{array}{l}\leftarrow \text{Units to convert to} \\ \leftarrow \text{Original units}\end{array}$$

Concept Check

If you were describing the weight of a semitrailer, which type of unit would you use: ounce, pound, or ton? Why?

Practice Problem 1

Convert 4500 pounds to tons.

Answers

1. $2\frac{1}{4}$ tons

Concept Check: ton

$$9000 \text{ lb} = \frac{9000 \text{ lb}}{1} \cdot \frac{1 \text{ ton}}{2000 \text{ lb}} = \frac{9000 \text{ tons}}{2000} = \frac{9}{2} \text{ tons or} 4\frac{1}{2} \text{ tons}$$

2000 lb — 1 ton   2000 lb — 1 ton   2000 lb — 1 ton   2000 lb — 1 ton   1000 lb — $\frac{1}{2}$ ton   9000 lb = $4\frac{1}{2}$ tons

Practice Problem 2

Convert 56 ounces to pounds.

EXAMPLE 2   Convert 3 pounds to ounces.

Solution: We multiply by the unit fraction $\frac{16 \text{ oz}}{1 \text{ lb}}$ to convert from pounds to ounces.

$$3 \text{ lb} = \frac{3 \text{ lb}}{1} \cdot \frac{16 \text{ oz}}{1 \text{ lb}} = 3 \cdot 16 \text{ oz} = 48 \text{ oz}$$

1 lb — 16 oz   1 lb — 16 oz   1 lb — 16 oz   3 lb = 48 oz

As with length, it is sometimes useful to simplify a measurement of weight by writing it in terms of mixed units.

$$33 \text{ ounces} = \underline{\hspace{1cm}} \text{ lb} \underline{\hspace{1cm}} \text{ oz}$$

Because 16 oz = 1 lb, divide 16 into 33 to see how many pounds are in 33 ounces. The quotient is the number of pounds, and the remainder is the number of ounces. To see why we divide 16 into 33, notice that

$$33 \text{ oz} = 33 \text{ oz} \cdot \frac{1 \text{ lb}}{16 \text{ oz}} = \frac{33}{16} \text{ lb}$$

$$\begin{array}{r} 2 \text{ lb } 1 \text{ oz} \\ 16 \overline{)33} \\ -32 \\ \hline 1 \end{array}$$

Thus, 33 ounces is the same as 2 lb 1 oz.

16 oz — 1 lb   16 oz — 1 lb   1 oz — 1 oz   33 oz = 2 lb 1 oz

B   Performing Operations on U.S. System Units of Weight

Performing arithmetic operations on units of weight works the same way as performing arithmetic operations on units of length.

Practice Problem 3

Subtract 5 tons 1200 lb from 8 tons 100 lb.

EXAMPLE 3   Subtract 3 tons 1350 lb from 8 tons 1000 lb.

Solution: To subtract, we line up similar units.

$$\begin{array}{r} 8 \text{ tons } 1000 \text{ lb} \\ - 3 \text{ tons } 1350 \text{ lb} \end{array}$$

Since we cannot subtract 1350 lb from 1000 lb, we borrow 1 ton from the 8 tons. To do so, we write 1 ton as 2000 lb and combine it with the 1000 lb.

$$7 \text{ tons} + \overset{\curvearrowright}{\textcircled{1 \text{ ton}}} 2000 \text{ lb}$$

$$\begin{array}{r} \cancel{8} \text{ tons } 1000 \text{ lb} \\ - 3 \text{ tons } 1350 \text{ lb} \end{array} \qquad \text{becomes} \qquad \begin{array}{r} 7 \text{ tons } 3000 \text{ lb} \\ - 3 \text{ tons } 1350 \text{ lb} \\ \hline 4 \text{ tons } 1650 \text{ lb} \end{array}$$

To check, see that the sum of 4 tons 1650 lb and 3 tons 1350 lb is 8 tons 1000 lb. ●

EXAMPLE 4   Multiply 5 lb 9 oz by 6.

Solution: We multiply 5 lb by 6 and 9 oz by 6.

$$\begin{array}{r} 5 \text{ lb } 9 \text{ oz} \\ \times \qquad 6 \\ \hline 30 \text{ lb } 54 \text{ oz} \end{array}$$

To write 54 oz as mixed units, we divide by 16 (1 lb = 16 oz).

$$\begin{array}{r} 3 \text{ lb } 6 \text{ oz} \\ 16 \overline{)54} \\ - 48 \\ \hline 6 \end{array}$$

Thus,

$$30 \text{ lb } 54 \text{ oz} = 30 \text{ lb} + 3 \text{ lb } 6 \text{ oz} = 33 \text{ lb } 6 \text{ oz}$$ ●

EXAMPLE 5   Divide 9 lb 6 oz by 2.

Solution: We divide each of the units by 2.

$$\begin{array}{r} 4 \text{ lb} \qquad 11 \text{ oz} \\ 2 \overline{)9 \text{ lb}} \qquad 6 \text{ oz} \\ - 8 \\ \hline 1 \text{ lb} = 16 \text{ oz} \\ \hline 22 \text{ oz} \end{array} \qquad \text{Divide 2 into 22 oz to get 11 oz.}$$

To check, multiply 4 pounds 11 ounces by 2. The result will be 9 pounds 6 ounces. ●

EXAMPLE 6   Finding the Weight of a Child

Bryan weighed 8 lb 8 oz at birth. By the time he was 1 year old, he had gained 11 lb 14 oz. Find his weight at age 1 year.

Solution:

$$\begin{array}{r} \text{birth weight} \rightarrow \qquad 8 \text{ lb} \quad 8 \text{ oz} \\ + \text{ weight gained} \rightarrow + 11 \text{ lb } 14 \text{ oz} \\ \hline \text{total weight} \rightarrow \qquad 19 \text{ lb } 22 \text{ oz} \end{array}$$

Since 22 oz equals 1 lb 6 oz.,

$$19 \text{ lb } 22 \text{ oz} = 19 \text{ lb} + 1 \text{ lb } 6 \text{ oz}$$
$$= 20 \text{ lb } 6 \text{ oz}$$

Bryan weighed 20 lb 6 oz on his first birthday. ●

Practice Problem 4

Multiply 4 lb 11 oz by 8.

Practice Problem 5

Divide 5 lb 8 oz by 4.

Practice Problem 6

A 5-lb 14-oz batch of cookies is packed into a 6-oz container before it is mailed. Find the total weight.

Answers

4. 37 lb 8 oz,   **5.** 1 lb 6 oz,   **6.** 6 lb 4 oz

Ⓒ Defining and Converting Metric System Units of Mass

In scientific and technical areas, a careful distinction is made between **weight** and **mass**. **Weight** is really a measure of the pull of gravity. The farther from Earth an object gets, the less it weighs. However, **mass** is a measure of the amount of substance in the object and does not change. Astronauts orbiting Earth weigh much less than they weigh on Earth, but they have the same mass in orbit as they do on Earth. Here on Earth weight and mass are the same, so either term may be used.

The basic unit of mass in the metric system is the **gram**. It is defined as the mass of water contained in a cube 1 centimeter (cm) on each side.

1 cm

1 cm

1 cm

The following examples may help you get a feeling for metric masses:

A tablet contains 200 milligrams of ibuprofen.

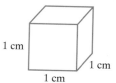

A large paper clip weighs approximately 1 gram.

A box of crackers weighs 453 grams.

A kilogram is slightly over 2 pounds. An adult woman may weigh 60 kilograms.

The prefixes for units of mass in the metric system are the same as for units of length, as shown in the following table:

Prefix	Meaning	Metric Unit of Mass
kilo	1000	1 kilogram (kg) = 1000 grams (g)
hecto	100	1 hectogram (hg) = 100 g
deka	10	1 dekagram (dag) = 10 g
		1 gram (g) = 1 g
deci	1/10	1 decigram (dg) = 1/10 g or 0.1 g
centi	1/100	1 centigram (cg) = 1/100 g or 0.01 g
milli	1/1000	1 milligram (mg) = 1/1000 g or 0.001 g

Try the Concept Check in the margin.

The **milligram**, the **gram**, and the **kilogram** are the three most commonly used units of mass in the metric system.

As with lengths, all units of mass are powers of 10 of the gram, so converting from one unit of mass to another involves moving only the decimal point. To convert from one unit of mass to another in the metric system, list the units of mass in order from largest to smallest.

Let's convert 4300 milligrams to grams. To convert from milligrams to grams, we move along the table 3 units to the left.

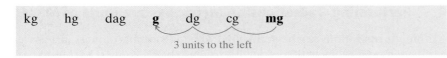

kg   hg   dag   **g**   dg   cg   **mg**

3 units to the left

This means that we move the decimal point 3 places to the left to convert from milligrams to grams.

4300 mg = 4.3 g

The same conversion can be done with unit fractions.

$$4300 \text{ mg} = \frac{4300 \text{ mg}}{1} \cdot \frac{0.001 \text{ g}}{1 \text{ mg}}$$

$$= 4300 \cdot 0.001 \text{ g}$$

$$= 4.3 \text{ g} \quad \text{To multiply by 0.001, move the decimal point 3 places to the left.}$$

To see that this is reasonable, study the diagram:

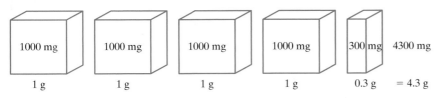

| 1000 mg | 1000 mg | 1000 mg | 1000 mg | 300 mg | 4300 mg |
| 1 g | 1 g | 1 g | 1 g | 0.3 g | = 4.3 g |

Thus, 4300 mg = 4.3 g

EXAMPLE 7   Convert 3.2 kg to grams.

Solution: First we convert by using a unit fraction.

Unit fraction

$$3.2 \text{ kg} = 3.2 \text{ kg} \cdot \frac{1000 \text{ g}}{1 \text{ kg}} = 3200 \text{ g}$$

Now let's list the units of mass in a chart and move from kilograms to grams.

kg   hg   dag   g   dg   cg   mg

3 units to the right

3.200 kg = 3200. g

3 places to the right

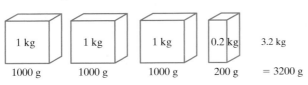

| 1 kg | 1 kg | 1 kg | 0.2 kg | 3.2 kg |
| 1000 g | 1000 g | 1000 g | 200 g | = 3200 g |

Concept Check

True or false? A decigram is larger than a dekagram. Explain.

Practice Problem 7

Convert 3.41 g to milligrams.

Answers

7. 3410 mg

Concept Check: false

Practice Problem 8

Convert 56.2 cg to grams.

EXAMPLE 8 Convert 2.35 cg to grams.

Solution: We list the units of mass in a chart and move from centigrams to grams.

kg   hg   dag   g   dg   cg   mg

2 units to the left

02.35 cg = 0.0235 g

2 places to the left

D Performing Operations on Metric System Units of Mass

Arithmetic operations can be performed with metric units of mass just as we performed operations with metric units of length. We convert each number to the same unit of mass and add, subtract, multiply, or divide as with decimals.

Practice Problem 9

Subtract 3.1 dg from 2.5 g.

EXAMPLE 9 Subtract 5.4 dg from 1.6 g.

Solution: We convert both numbers to decigrams or to grams before subtracting.

$$5.4 \text{ dg} = 0.54 \text{ g}$$
$$1.60 \text{ g}$$
$$- 0.54 \text{ g}$$
$$\overline{1.06 \text{ g}}$$

or

$$1.6 \text{ g} = 16 \text{ dg}$$
$$16.0 \text{ dg}$$
$$- 5.4 \text{ dg}$$
$$\overline{10.6 \text{ dg}}$$

The difference is 1.06 g or 10.6 dg.

Practice Problem 10

Multiply 12.6 kg by 4.

EXAMPLE 10 Multiply 15.4 kg by 5.

Solution: We multiply the two numbers together.

$$15.4 \text{ kg}$$
$$\times \quad 5$$
$$\overline{77.0 \text{ kg}}$$

The result is 77.0 kg.

Practice Problem 11

Twenty-four bags of cement weigh a total of 550 kg. Find the average weight of 1 bag, rounded to the nearest kilogram.

EXAMPLE 11 Calculating Allowable Weight in an Elevator

An elevator has a weight limit of 1400 kg. A sign posted in the elevator indicates that the maximum capacity of the elevator is 17 persons. What is the average allowable weight for each passenger, rounded to the nearest kilogram?

Solution: To solve, notice that the total weight of 1400 kilograms ÷ 17 = average weight

$$
\begin{array}{r}
82.3 \text{ kg} \approx 82 \text{ kg} \\
17\overline{)1400.0 \text{ kg}} \\
- 136 \\
\hline
40 \\
- 34 \\
\hline
60 \\
- 51 \\
\hline
9
\end{array}
$$

Answers

8. 0.562 g,   **9.** 2.19 g or 21.9 dg,   **10.** 50.4 kg,
11. 23 kg

Each passenger can weigh an average of 82 kg. (Recall that a kilogram is slightly over 2 pounds, so 82 kilograms is over 164 pounds. For a better approximation, see the Calculator Explorations box.) ●

CALCULATOR EXPLORATIONS

Metric to U.S. System Conversions in Weight

To convert between the two systems of measurement in weight, the following *approximations* can be used:

grams × 0.035 ≈ ounces   ounces × 28.35 ≈ grams

kilograms × 2.20 ≈ pounds   pounds × 0.454 ≈ kilograms

grams × 0.0022 ≈ pounds   pounds × 454 ≈ grams

EXAMPLE

A bulldog weighs 35 pounds. How many kilograms is this?

Solution: From the above approximations,

pounds × 0.454 ≈ kilograms
↓
35 × 0.454 ≈ kilograms

To multiply on your calculator, press the keys

[35] [×] [0.454] [=] or [ENTER].

The display will read [15.89]. Thus, 35 pounds ≈ 15.89 kilograms. ●

Convert as indicated.

1. 15 ounces to grams
2. 11.2 grams to ounces
3. 7 kilograms to pounds
4. 23 pounds to kilograms
5. A piece of candy weighs 5 grams. How many ounces is this?
6. If a person weighs 82 kilograms, how many pounds is this?

THE DEVELOPMENT OF UNITS OF MEASURE

The earliest units of measure were based on the human body. The ancient Egyptians, Babylonians, Hebrews, and Mesopotamians used a unit of length called the **cubit**, which represents the distance between a human elbow and fingertips. For instance, in the book of Genesis in the Bible, Noah's ark is described as having a length of "three hundred cubits, its width fifty cubits, and its height thirty cubits." Other commonly used measures found in documents of these ancient cultures include the digit, hand, span, and foot. These measures are shown at the right.

Several thousand years later, the English system of measurement also consisted of a mixed bag of body- and nature-related units of measure. A rod was the combined length of the left feet of 16 men. An inch was the distance spanned by three grains of barley. A foot was the length of the foot of the king currently in power. Around A.D. 1100 King Henry I of England decreed that a yard was the distance between the king's nose and the thumb of his outstretched arm.

Although the English system became somewhat standardized by the 13th century, the problem with it and the ancient systems of measurement was that the relationship between the various units was not necessarily easy to remember. This made it difficult to convert from one type of unit to another.

The metric system grew out of the French Revolution during the 1790s. As a reaction against the lack of consistency and utility of existing systems of measurement, the French Academy of Sciences set out to develop an easy-to-use, internationally standardized system of measurements. Originally the basic unit of the metric system, the meter, was to be one ten-millionth of the distance between the North Pole and the Equator on a line through Paris. However, it was soon discovered that this distance was nearly impossible to measure accurately. The meter has been redefined several times since the end of the French Revolution, most recently in 1983 as the distance that light travels in a vacuum during $\frac{1}{299,792,458}$ of a second.

Although the definition of the meter has evolved over time, the original idea of developing an easy-to-use system of measurements has endured. The metric system uses a standard set of prefixes for all basic unit types (length, mass, capacity, etc.) and all conversions within a unit type are based on powers of 10.

CRITICAL THINKING

Develop your own unit of measure for (a) length and (b) area. Explain how you chose your units and for what types of relative sizes of measurements they would be most useful. Discuss the advantages and disadvantages of your units of measurement. Then use your units to measure the width of your classroom desk and the area of the front cover of this textbook.

digit

hand

span

foot

cubit

Name _____ Section _____ Date _____

Mental Math

Convert.

1. 16 ounces to pounds   **2.** 32 ounces to pounds   **3.** 1 ton to pounds   **4.** 3 tons to pounds

5. 1 pound to ounces   **6.** 3 pounds to ounces   **7.** 2000 pounds to tons   **8.** 4000 pounds to tons

Determine whether the measurement in each statement is reasonable.

9. The doctor prescribed a pill containing 2 kg of medication.

10. A full-grown cat weighs approximately 15 g.

11. A bag of flour weighs 4.5 kg.

12. A staple weighs 15 mg.

13. A professor weighs less than 150 g.

14. A car weighs 2000 mg.

EXERCISE SET 7.2

A *Convert as indicated. See Examples 1 and 2.*

1. 2 pounds to ounces   **2.** 3 pounds to ounces   **3.** 5 tons to pounds   **4.** 3 tons to pounds

5. 12,000 pounds to tons   **6.** 32,000 pounds to tons   **7.** 60 ounces to pounds   **8.** 90 ounces to pounds

9. 3500 pounds to tons   **10.** 9000 pounds to tons   **11.** 16.25 pounds to ounces   **12.** 14.5 pounds to ounces

 13. 4.9 tons to pounds   **14.** 8.3 tons to pounds   **15.** $4\frac{3}{4}$ pounds to ounces   **16.** $9\frac{1}{8}$ pounds to ounces

17. 2950 pounds to the nearest tenth of a ton   **18.** 51 ounces to the nearest tenth of a pound

427

19. 34 lb 12 oz + 18 lb 14 oz

20. 6 lb 10 oz + 10 lb 8 oz

21. 6 tons 1540 lb + 2 tons 850 lb

22. 2 tons 1575 lb + 1 ton 480 lb

23. 5 tons 1050 lb − 2 tons 875 lb

24. 4 tons 850 lb − 1 ton 260 lb

25. 12 lb 4 oz − 3 lb 9 oz

26. 45 lb 6 oz − 26 lb 10 oz

27. 5 lb 3 oz × 6

28. 2 lb 5 oz × 5

29. 6 tons 1500 lb ÷ 5

30. 5 tons 400 lb ÷ 4

Solve. Remember to insert units when writing your answers. See Example 6.

31. Doris Johnson has two open containers of Uncle Ben's rice. If she combines 1 lb 10 oz from one container with 3 lb 14 oz from the other container, how much total rice does she have?

32. Dru Mizel maintains the records of the amount of coal delivered to his department in the steel mill. In January, 3 tons 1500 lb were delivered. In February, 2 tons 1200 lb were delivered. Find the total amount delivered in these two months.

33. Carla Hamtini was amazed when she grew a 28 lb 10 oz zucchini in her garden, but later she learned that the heaviest zucchini ever grown weighed 64 lb 8 oz in Llanharry, Wales, by B. Lavery in 1990. How far below the record weight was Carla's zucchini? (*Source: The Guinness Book of Records*, 1996)

34. The heaviest baby born in good health weighed an incredible 22 lb 8 oz. He was born in Italy in September, 1955. How much heavier is this than a 7 lb 12 oz baby? (*Source: The Guinness Book of Records*, 1996)

35. The Shop 'n Bag supermarket chain ships hamburger meat by placing 10 packages of hamburger in a box, with each package weighing 3 lb 4 oz. How much will 4 boxes of hamburger weigh?

36. The Quaker Oats Company ships its 1-lb 2-oz boxes of oatmeal in cartons containing 12 boxes of oatmeal. How much will 3 such cartons weigh?

37. A carton of Del Monte Pineapple weighs 55 lb 4 oz, but 2 lb 8 oz of this weight is due to packaging. Subtract the weight of the packaging to find the actual weight of the pineapple in 4 cartons.

38. The Hormel Corporation ships cartons of canned ham weighing 43 lb 2 oz each. Of this weight, 3 lb 4 oz is due to packaging. Find the actual weight of the ham found in 3 cartons.

39. One bag of Pepperidge Farm Bordeaux cookies weighs $6\frac{3}{4}$ ounces. How many pounds will a dozen bags weigh?

40. One can of Payless Red Beets weighs $8\frac{1}{2}$ ounces. How much will eight cans weigh?

C *Convert as indicated. See Examples 7 and 8.*

41. 500 g to kilograms **42.** 650 g to kilograms **43.** 4 g to milligrams **44.** 9 g to milligrams

45. 25 kg to grams **46.** 18 kg to grams **47.** 48 mg to grams **48.** 112 mg to grams

49. 6.3 g to kilograms **50.** 4.9 g to kilograms **51.** 15.14 g to milligrams **52.** 16.23 g to milligrams

53. 4.01 kg to grams **54.** 3.16 kg to grams

D *Perform each indicated operation. See Examples 9 and 10.*

55. $3.8\,\text{mg} + 9.7\,\text{mg}$ **56.** $41.6\,\text{g} + 9.8\,\text{g}$ **57.** $205\,\text{mg} + 5.61\,\text{g}$ **58.** $2.1\,\text{g} + 153\,\text{mg}$

59. $9\,\text{g} - 7150\,\text{mg}$ **60.** $4\,\text{kg} - 2410\,\text{g}$ **61.** $1.61\,\text{kg} - 250\,\text{g}$ **62.** $6.13\,\text{g} - 418\,\text{mg}$

63. $5.2\,\text{kg} \times 2.6$ **64.** $4.8\,\text{kg} \times 9.3$ **65.** $17\,\text{kg} \div 8$ **66.** $8.25\,\text{g} \div 6$

Solve. Remember to insert units when writing your answers. See Example 11.

67. A can of 7-Up weighs 336 grams. Find the weight in kilograms of 24 cans.

68. Guy Green normally weighs 73 kg, but he lost 2800 grams after being sick with the flu. Find Guy's new weight.

69. Sudafed is a decongestant that comes in two strengths. Regular strength contains 60 mg of medication. Extra strength contains 0.09 g of medication. How much extra medication is in the extra-strength tablet?

70. A small can of Planters sunflower seeds weighs 177 g. If each can contains 6 servings, find the weight of one serving.

71. Tim Caucutt's doctor recommends that Tim limit his daily intake of sodium to 0.6 gram. A one-ounce serving of Cheerios with $\frac{1}{2}$ cup of fortified skim milk contains 350 mg of sodium. How much more sodium can Tim have after he eats a bowl of Cheerios for breakfast, assuming he intends to follow the doctor's orders?

72. A large bottle of Hire's Root Beer weighs 1900 grams. If a carton contains 6 large bottles of root beer, find the weight in kilograms of 5 cartons.

73. Three milligrams of preservatives are added to a 0.5-kg box of dried fruit. How many milligrams of preservatives are in 3 cartons of dried fruit if each carton contains 16 boxes?

74. One box of Swiss Miss Cocoa Mix weighs 0.385 kg, but 39 grams of this weight is the packaging. Find the actual weight of the cocoa in 8 boxes.

75. A carton of 12 boxes of Quaker Oats Oatmeal weighs 6.432 kg. Each box includes 26 grams of packaging material. What is the actual weight of the oatmeal in the carton?

76. The supermarket prepares hamburger in 85-gram market packages. When Leo Gonzalas gets home, he divides the package in half before refrigerating the meat. How much will each package weigh?

77. A package of Trailway's Gorp, a high-energy hiking trail mix, contains 0.3 kg of nuts, 0.15 kg of chocolate bits, and 400 grams of raisins. Find the total weight of the package.

78. The manufacturer of Anacin wants to reduce the caffeine content of its aspirin by $\frac{1}{4}$. Currently, each regular tablet contains 32 mg of caffeine. How much caffeine should be removed from each tablet?

79. A regular-size bag of Lay's potato chips weighs 198 grams. Find the weight of a dozen bags, rounded to the nearest hundredth of a kilogram.

80. Clarence Patterson's cat weighs a hefty 9 kg. The vet has recommended that the cat lose 1500 grams. How much should the cat weigh?

Review and Preview

Write each fraction as a decimal. See Section 4.7.

81. $\frac{1}{4}$

82. $\frac{1}{20}$

83. $\frac{4}{25}$

84. $\frac{3}{5}$

85. $\frac{7}{8}$

86. $\frac{3}{16}$

Combining Concepts

 87. Why is the decimal point moved to the right when grams are converted to milligrams?

88. To change 8 pounds to ounces, multiply by 16. Why is this the correct procedure?

7.3 Capacity: U.S. and Metric Systems of Measurement

OBJECTIVES

Ⓐ Define U.S. units of capacity and convert from one unit to another.

Ⓑ Perform arithmetic operations on U.S. units of capacity.

Ⓒ Define metric units of capacity and convert from one unit to another.

Ⓓ Perform arithmetic operations on metric units of capacity.

SSM   TUTOR CENTER   SG   CD & VIDEO   MATH PRO   WEB

Ⓐ Defining and Converting U.S. System Units of Capacity

Units of **capacity** are generally used to measure liquids. The number of gallons of gasoline needed to fill a gas tank in a car, the number of cups of water needed in a bread recipe, and the number of quarts of milk sold each day at a supermarket are all examples of using units of capacity. The following summary shows equivalencies between units of capacity:

U.S. Units of Capacity

$$8 \text{ fluid ounces (fl oz)} = 1 \text{ cup (c)}$$
$$2 \text{ cups} = 1 \text{ pint (pt)}$$
$$2 \text{ pints} = 1 \text{ quart (qt)}$$
$$4 \text{ quarts} = 1 \text{ gallon (gal)}$$

Just as with units of length and weight, we can form unit fractions to convert between different units of capacity. For instance,

$$\frac{2 \text{ c}}{1 \text{ pt}} = \frac{1 \text{ pt}}{2 \text{ c}} = 1 \quad \text{and} \quad \frac{2 \text{ pt}}{1 \text{ qt}} = \frac{1 \text{ qt}}{2 \text{ pt}} = 1$$

EXAMPLE 1   Convert 9 quarts to gallons.

Solution: We multiply by the unit fraction $\frac{1 \text{ gal}}{4 \text{ qt}}$.

$$9 \text{ qt} = \frac{9 \, \cancel{\text{qt}}}{1} \cdot \frac{1 \text{ gal}}{4 \, \cancel{\text{qt}}}$$

$$= \frac{9 \text{ gal}}{4}$$

$$= 2\frac{1}{4} \text{ gal}$$

Thus, 9 quarts is the same as $2\frac{1}{4}$ gallons, as shown in the diagram:

9 quarts

Practice Problem 1

Convert 43 pints to quarts.

EXAMPLE 2   Convert 14 cups to quarts.

Solution: Our equivalency table contains no direct conversion from cups to quarts. However, from this table we know that

$$1 \text{ qt} = 2 \text{ pt} = 4 \text{ c}$$

so 1 qt = 4c. Now we have the unit fraction $\frac{1 \text{ qt}}{4 \text{ c}}$. Thus,

$$14 \text{ c} = \frac{14 \, \cancel{\text{c}}}{1} \cdot \frac{1 \text{ qt}}{4 \, \cancel{\text{c}}} = \frac{7}{2} \text{ qt} \quad \text{or} \quad 3\frac{1}{2} \text{ qt}$$

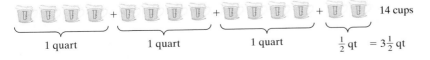

14 cups

1 quart   1 quart   1 quart $\frac{1}{2}$ qt $= 3\frac{1}{2}$ qt

Practice Problem 2

Convert 26 quarts to cups.

Answers

1. $21\frac{1}{2}$ qt,   2. 104 c

Concept Check

If 50 cups are converted to quarts, will the equivalent number of quarts be less than or greater than 50? Explain.

Practice Problem 3

Subtract 2 qt from 1 gal 1 qt.

Practice Problem 4

Multiply 2 gal 3 qt by 2.

Practice Problem 5

Divide 6 gal 1 qt by 2.

Try the Concept Check in the margin.

B Performing Operations on U.S. System Units of Capacity

As is true of units of length and weight, units of capacity can be added, subtracted, multiplied, and divided.

EXAMPLE 3 Subtract 3 qt from 4 gal 2 qt.

Solution: To subtract, we line up similar units.

$$\begin{array}{r} 4 \text{ gal } 2 \text{ qt} \\ - \quad\quad 3 \text{ qt} \\ \hline \end{array}$$

We cannot subtract 3 qt from 2 qt. We need to borrow 1 gallon from the 4 gallons, convert it to 4 quarts, and then combine it with the 2 quarts.

Borrow 1 gal = 4 qt

3 gal + (1 gal) 4 qt

$$\begin{array}{r} 4 \text{ gal } 2 \text{ qt} \\ - \quad\quad 3 \text{ qt} \\ \hline \end{array} \quad \text{or} \quad \begin{array}{r} 3 \text{ gal } 6 \text{ qt} \\ - \quad\quad 3 \text{ qt} \\ \hline 3 \text{ gal } 3 \text{ qt} \end{array}$$

To check, see that the sum of 3 gal 3 qt and 3 qt is 4 gal 2 qt.

EXAMPLE 4 Multiply 3 qt 1 pt by 3.

Solution: We multiply each of the units of capacity by 3.

$$\begin{array}{r} 3 \text{ qt } 1 \text{ pt} \\ \times \quad\quad 3 \\ \hline 9 \text{ qt } 3 \text{ pt} \end{array}$$

Since 3 pints is the same as 1 quart and 1 pint, we have

$$9 \text{ qt } 3 \text{ pt} = 9 \text{ qt} + 1 \text{ qt } 1 \text{ pt} = 10 \text{ qt } 1 \text{ pt}$$

The 10 quarts can be changed to gallons by dividing by 4, since there are 4 quarts in a gallon. To see why we divide, notice that

$$10 \text{ qt} = \frac{10 \text{ qt}}{1} \cdot \frac{1 \text{ gal}}{4 \text{ qt}} = \frac{10}{4} \text{ gal}$$

$$\begin{array}{r} 2 \text{ gal } 2 \text{ qt} \\ 4)\overline{10 \text{ qt}} \\ -8 \\ \hline 2 \end{array}$$

Thus, the product is 10 qt 1 pt or 2 gal 2 qt 1 pt.

EXAMPLE 5 Divide 3 gal 2 qt by 2.

Solution: We divide each unit of capacity by 2.

$$\begin{array}{r} 1 \text{ gal} \quad 3 \text{ qt} \\ 2)\overline{3 \text{ gal}} \quad 2 \text{ qt} \\ -2 \\ \hline 1 \text{ gal} = 4 \text{ qt} \\ 6 \text{ qt} \quad\quad 6 \text{ qt} \div 2 = 3 \text{ qt} \end{array}$$

EXAMPLE 6   Finding the Amount of Water in an Aquarium

An aquarium contains 6 gal 3 qt of water. If 2 gal 2 qt of water is added, what is the total amount of water in the aquarium?

Solution:

beginning water	→	6 gal 3 qt
+ water added	→	+ 2 gal 2 qt
total water	→	8 gal 5 qt

Since 5 qt = 1 gal 1 qt, we have

$$= \overbrace{8 \text{ gal}}^{8 \text{ gal}} + \overbrace{1 \text{ gal } 1 \text{ qt}}^{5 \text{ qt}}$$
$$= 9 \text{ gal } 1 \text{ qt}$$

The total amount of water is 9 gal 1 qt. ●

C   Defining and Converting Metric System Units of Capacity

Thus far, we know that the basic unit of length in the metric system is the meter and that the basic unit of mass in the metric system is the gram. What is the basic unit of capacity? The **liter**. By definition, a **liter** is the capacity or volume of a cube measuring 10 centimeters on each side.

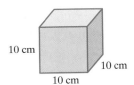

10 cm 10 cm 10 cm

The following examples may help you get a feeling for metric capacities:

One liter of liquid is slightly more than one quart.

1 quart   1 liter

Many soft drinks are packaged in 2-liter bottles.

The metric system was designed to be a consistent system. Once again, the prefixes for metric units of capacity are the same as for metric units of length and mass, as summarized in the following table:

Practice Problem 6

A large oil drum contains 15 gal 3 qt of oil. How much will be in the drum if an additional 4 gal 3 qt of oil is poured into it?

Answer
6. 20 gal 2 qt

Prefix	Meaning	Metric Unit of Capacity	
kilo	1000	1 kiloliter	(kl) = 1000 liters (L)
hecto	100	1 hectoliter	(hl) = 100 L
deka	10	1 dekaliter	(dal) = 10 L
		1 liter (L) = 1 L	
deci	1/10	1 deciliter	(dl) = 1/10 L or 0.1 L
centi	1/100	1 centiliter	(cl) = 1/100 L or 0.01 L
milli	1/1000	1 milliliter	(ml) = 1/1000 L or 0.001 L

The **milliliter** and the **liter** are the two most commonly used metric units of capacity.

Converting from one unit of capacity to another involves multiplying by powers of 10 or moving the decimal point to the left or to the right. Listing units of capacity in order from largest to smallest helps to keep track of how many places to move the decimal point when converting.

Let's convert 2.6 liters to milliliters. To convert from liters to milliliters, we move along the chart 3 units to the right.

kl   hl   dal   **L**   dl   cl   ml

3 units to the right

This means that we move the decimal point 3 places to the right to convert from liters to milliliters.

$$2.600 \text{ L} = 2600. \text{ ml}$$

This same conversion can be done with unit fractions.

$$2.6 \text{ L} = \frac{2.6 \, \cancel{L}}{1} \cdot \frac{1000 \text{ ml}}{1 \, \cancel{L}}$$
$$= 2.6 \cdot 1000 \text{ ml}$$
$$= 2600 \text{ ml}$$

To multiply by 1000, move the decimal point 3 places to the right.

To visualize the result, study the diagram below:

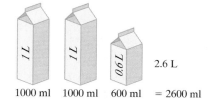

1 L   1 L   600   2.6 L

1000 ml   1000 ml   600 ml   = 2600 ml

Thus, 2.6 L = 2600 ml.

Practice Problem 7

Convert 2100 ml to liters.

Answer

7. 2.1 L

EXAMPLE 7   Convert 3210 ml to liters.

Solution: Let's use the unit fraction method first.

Unit fraction

$$3210 \text{ ml} = 3210 \, \cancel{ml} \cdot \frac{1 \text{ L}}{1000 \, \cancel{ml}} = 3.21 \text{ L}$$

Now let's list the unit measures in a chart and move from milliliters to liters.

kl   hl   dal   L   dl   cl   ml

3 units to the left

$$3210 \text{ ml} = 3.210 \text{ L}$$

3 places to the left

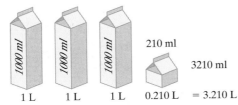

210 ml

3210 ml

1000 ml   1000 ml   1000 ml

1 L   1 L   1 L   0.210 L   = 3.210 L

EXAMPLE 8   Convert 0.185 dl to milliliters.

Solution: We list the unit measures in a chart and move from deciliters to milliliters.

kl   hl   dal   L   dl   cl   ml

2 units to the right

$$0.185 \text{ dl} = 18.5 \text{ ml}$$

2 places to the right

(D) Performing Operations on Metric System Units of Capacity

As was true for length and weight, arithmetic operations involving metric units of capacity can also be performed. Make sure that the metric units of capacity are the same before adding, subtracting, multiplying, or dividing.

EXAMPLE 9   Add 2400 ml to 8.9 L.

Solution: We must convert both to liters or both to milliliters before adding the capacities together.

2400 ml = 2.4 L   or   8.9 L = 8900 ml

$$\begin{array}{r} 2.4 \text{ L} \\ + 8.9 \text{ L} \\ \hline 11.3 \text{ L} \end{array} \qquad \begin{array}{r} 2400 \text{ ml} \\ + 8900 \text{ ml} \\ \hline 11{,}300 \text{ ml} \end{array}$$

The total is 11.3 L or 11,300 ml. They both represent the same capacity.

Try the Concept Check in the margin.

EXAMPLE 10   Divide 18.08 ml by 16.

Solution:

$$\begin{array}{r} 1.13 \text{ ml} \\ 16{\overline{)18.08 \text{ ml}}} \\ -16\phantom{.08 \text{ ml}} \\ \hline 2\,0\phantom{8 \text{ ml}} \\ -1\,6\phantom{8 \text{ ml}} \\ \hline 48\phantom{ \text{ ml}} \\ -48\phantom{ \text{ ml}} \\ \hline 0\phantom{ \text{ ml}} \end{array}$$

The solution is 1.13 ml.

Practice Problem 8

Convert 2.13 dal to liters.

Practice Problem 9

Add 1250 ml to 2.9 L.

Concept Check

How could you estimate the following operation? Subtract 950 ml from 7.5 L.

Practice Problem 10

Divide 146.9 L by 13.

Answers

8. 21.3 L,   **9.** 4150 ml or 4.15 L,   **10.** 11.3 L

Concept Check: 950 ml = 0.95 L; round 0.95 to 1;
7.5 − 1 = 6.5 L

Practice Problem 11

If 28.6 L of water can be pumped every minute, how much water can be pumped in 85 minutes?

Answer

11. 2431 L

EXAMPLE 11 Finding the Amount of Medication a Person Has Received

A patient hooked up to an IV unit in the hospital is to receive 12.5 ml of medication every hour. How much medication does the patient receive in 3.5 hours?

Solution: We multiply 12.5 ml by 3.5.

$$
\begin{array}{r}
\text{medication per hour} \quad \rightarrow \quad 12.5 \text{ ml} \\
\times \qquad\qquad \text{hours} \quad \rightarrow \quad \times \; 3.5 \\
\hline
\text{total medication} \qquad\qquad 625 \\
375 \\
\hline
43.75 \text{ ml}
\end{array}
$$

The patient receives 43.75 ml of medication.

CALCULATOR EXPLORATIONS

Metric To U.S. System Conversions in Capacity

To convert between the two systems of measurement in capacity, the following *approximations* can be used:

$$\text{liters} \times 1.06 \approx \text{quarts} \qquad \text{quarts} \times 0.946 \approx \text{liters}$$
$$\text{liters} \times 0.264 \approx \text{gallons} \qquad \text{gallons} \times 3.785 \approx \text{liters}$$

EXAMPLE How many quarts are there in a 2-liter bottle of cola?

Solution: From the above approximations,

$$
\begin{array}{c}
\text{liters} \times 1.06 \approx \text{quarts} \\
\downarrow \\
2 \times 1.06 \approx \text{quarts}
\end{array}
$$

To multiply on your calculator, press the keys

$\boxed{2}\;\boxed{\times}\;\boxed{1.06}\;\boxed{=}$ or $\boxed{\text{ENTER}}$.

The display should read $\boxed{2.12}$.

$$2 \text{ liters} \approx 2.12 \text{ quarts}$$

Convert as indicated.

1. 5 quarts to liters
2. 26 gallons to liters
3. 17.5 liters to gallons
4. 7.8 liters to quarts
5. A 1-gallon container holds how many liters?
6. How many quarts are contained in a 3-liter bottle of cola?

Mental Math

Convert as indicated.

1. 2 c to pints

2. 4 c to pints

3. 4 qt to gallons

4. 8 qt to gallons

5. 2 pt to quarts

6. 6 pt to quarts

7. 8 fl oz to cups

8. 24 fl oz to cups

9. 1 pt to cups

10. 3 pt to cups

11. 1 gal to quarts

12. 2 gal to quarts

Determine whether the measurement in each statement is reasonable.

13. Clair took a dose of 2 L of cough medicine to cure her cough.

14. John drank 250 ml of milk for lunch.

15. Jeannie likes to relax in a tub filled with 3000 ml of hot water.

16. Sarah pumped 20 L of gasoline into her car yesterday.

EXERCISE SET 7.3

 Convert each measurement as indicated. See Examples 1 and 2.

1. 32 fluid ounces to cups

2. 16 quarts to gallons

 3. 8 quarts to pints

4. 9 pints to quarts

5. 10 quarts to gallons

6. 15 cups to pints

7. 80 fluid ounces to pints

8. 18 pints to gallons

9. 2 quarts to cups

10. 3 pints to fluid ounces

11. 120 fluid ounces to quarts

12. 20 cups to gallons

13. 6 gallons to fluid ounces

14. 5 quarts to cups

15. $4\frac{1}{2}$ pints to cups

16. $6\frac{1}{2}$ gallons to quarts

17. $2\frac{3}{4}$ gallons to pints   **18.** $3\frac{1}{4}$ quarts to cups

B *Perform each indicated operation. See Examples 3 through 5.*

19. 4 gal 3 qt + 5 gal 2 qt   **20.** 2 gal 3 qt + 8 gal 3 qt   **21.** 1 c 5 fl oz + 2 c 7 fl oz   **22.** 2 c 3 fl oz + 2 c 6 fl oz

23. 3 gal − 1 gal 3 qt   **24.** 2 pt − 1 pt 1 c   **25.** 3 gal 1 qt − 1 qt 1 pt   **26.** 3 qt 1 c − 1 c 4 fl oz

27. 1 pt 1 c × 3   **28.** 1 qt 1 pt × 2   **29.** 8 gal 2 qt × 2   **30.** 6 gal 1 pt × 2

31. 9 gal 2 qt ÷ 2   **32.** 5 gal 6 fl oz ÷ 2

Solve. Remember to insert units when writing your answers. See Example 6.

33. A can of Hawaiian Punch holds $1\frac{1}{2}$ quarts of liquid. How many fluid ounces is this?

34. Weight Watchers Double Fudge bars contain 21 fluid ounces of ice cream. How many cups of ice-cream is this?

35. Many diet experts advise individuals to drink 64 ounces of water each day. How many quarts of water is this?

36. A recipe for walnut fudge cake calls for $1\frac{1}{4}$ cups of water. How many fluid ounces is this?

37. Can 5 pt 1 c of fruit punch and 2 pt 1 c of ginger ale be poured into a 1-gal container without it overflowing?

38. Three cups of prepared Jell-O are poured into 6 dessert dishes. How many fluid ounces of Jell-O are in each dish?

39. How much punch has been prepared if 1 qt 1 pt of Ocean Spray Cranapple drink is mixed with 1 pt 1 c of ginger ale?

40. Henning's Supermarket sells homemade soup in 1-qt-1-pt containers. How much soup is contained in three such containers?

438

41. A case of Pepsi Cola holds 24 cans, each of which contains 12 ounces of Pepsi Cola. How many *quarts* are there in a case of Pepsi?

42. Manuela's Service Station has a drum that holds 40 gallons of oil. If 6 gallons and 3 quarts have been used, how much oil remains?

C *Convert as indicated. See Examples 7 and 8.*

43. 5 L to milliliters

44. 8 L to milliliters

45. 4500 ml to liters

46. 3100 ml to liters

47. 410 L to kiloliters

48. 250 L to kiloliters

49. 64 ml to liters

50. 39 ml to liters

51. 0.16 kl to liters

52. 0.48 kl to liters

53. 3.6 L to milliliters

54. 1.9 L to milliliters

55. 0.16 L to kiloliters

56. 0.127 L to kiloliters

D *Perform each indicated operation. See Examples 9 and 10.*

57. 2.9 L + 19.6 L

58. 18.5 L + 4.6 L

59. 2700 ml + 1.8 L

60. 4.6 L + 1600 ml

61. 8.6 L − 190 ml

62. 4.8 L − 283 ml

63. 11,400 ml − 0.8 L

64. 6850 ml − 0.3 L

65. 480 ml × 8

66. 290 ml × 6

67. 81.2 L ÷ 0.5

68. 5.4 L ÷ 3.6

Solve. Remember to insert units when writing your answers. See Example 11.

69. Mike Schaferkotter drank 410 ml of Mountain Dew from a 2-liter bottle. How much Mountain Dew remains in the bottle?

70. The Werners' Volvo has a 54.5-L gas tank. Only 3.8 liters of gasoline still remain in the tank. How much is needed to fill it?

71. Margie Phitts added 354 ml of Prestone dry gas to the 18.6 L of gasoline in her car's tank. Find the total amount of gasoline in the tank.

72. Chris Peckaitis wishes to share a 2-L bottle of Coca Cola equally with 7 of his friends. How much will each person get?

73. Stanley Fisher paid $14 to fill his car with 44.3 liters of gasoline. Find the price per liter of gasoline to the nearest tenth of a cent.

74. A student carelessly misread the scale on a cylinder in the chemistry lab and added 40 cl of water to a mixture instead of 40 ml. Find the excess amount of water.

75. A large bottle of Ocean Spray Cranicot drink contains 1.42 L of beverage. The smaller bottle contains only 946 ml. How much more is in the larger bottle?

76. In a lab experiment, Melissa Martin added 400 ml of salt water to 1.65 L of water. Later 320 ml of the solution was drained off. How much of the solution still remained?

Review and Preview

Write each decimal as a fraction. See Section 4.7.

77. 0.7   **78.** 0.9   **79.** 0.03   **80.** 0.007   **81.** 0.006   **82.** 0.08

 Combining Concepts

83. Explain how to borrow in order to subtract 1 gal 2 qt from 3 gal 1 qt.

Internet Excursions

  Go To: http://www.prenhall.com/martin-gay_basic 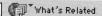 What's Related

By going to the World Wide Web address listed above, you will be directed to a site called A Dictionary of Units, or a related site, that will help you answer the questions below.

84. There are more units of capacity in use than the ones mentioned in this section. For instance, there are special measures for the capacity of dry items such as grains or fruits. Visit this Web site and locate the "U.S. System of Measurements" area. List the measure equivalencies for dry capacity. Then convert 38 pecks of apples to bushels.

85. There are also special apothecaries' measures for capacity of liquid drugs and medicines. Although metric measures are commonly used in the pharmacy industry today, awareness of these measures is still useful. Within the "U.S. System of Measurements" area of this Web site, find and list the apothecaries' measures equivalencies. Then convert 2 pints to fl drams.

Integrated Review—Length, Weight, and Capacity

Convert each measurement as indicated.

Length

1. 36 in. = _____ ft

2. 10,560 ft = _____ mi

3. 20 ft = _____ yd

4. 6 yd = _____ ft

5. 2.1 mi = _____ ft

6. 3.2 ft = _____ in.

7. 30 m = _____ cm

8. 24 mm = _____ cm

9. 2000 mm = _____ m

10. 1800 cm = _____ m

11. 7.2 cm = _____ mm

12. 600 m = _____ km

Weight or Mass

13. 5 tons = _____ lb

14. 11,000 lb = _____ tons

15. 8.5 lb = _____ oz

16. 40 oz = _____ lb

Answers

1. _____

2. _____

3. _____

4. _____

5. _____

6. _____

7. _____

8. _____

9. _____

10. _____

11. _____

12. _____

13. _____

14. _____

15. _____

16. _____

17. 56 oz = _____ lb

18. 5 lb = _____ oz

19. 28 kg = _____ g

20. 1400 mg = _____ g

21. 5.6 g = _____ kg

22. 6 kg = _____ g

23. 670 mg = _____ g

24. 3.6 g = _____ kg

Capacity

25. 6 qt = _____ pt

26. 5 pt = _____ qt

27. 14 qt = _____ gal

28. 17 c = _____ pt

29. $3\frac{1}{2}$ pt = _____ c

30. 26 qt = _____ gal

31. 7 L = _____ ml

32. 350 L = _____ kl

33. 47 ml = _____ L

34. 0.97 kl = _____ L

35. 0.126 kl = _____ L

36. 75 ml = _____ L

7.4 Temperature: U.S. and Metric Systems of Measurement

When Gabriel Fahrenheit and Anders Celsius independently established units for temperature scales, each based his unit on the heat of water the moment it boils compared to the moment it freezes. One degree Celsius is 1/100 of the difference in heat. One degree Fahrenheit is 1/180 of the difference in heat. Celsius arbitrarily labeled the temperature at the freezing point at 0°C, making the boiling point 100°C; Fahrenheit labeled the freezing point 32°F, making the boiling point 212°F. Water boils at 212°F and 100°C.

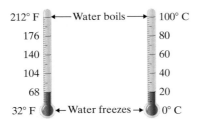

By comparing the two scales in the figure, we see that a 20°C day is as warm as a 68°F day. Similarly, a sweltering 104°F day in the Mojave Desert corresponds to a 40°C day.

Try the Concept Check in the margin.

A Converting Degrees Celsius to Degrees Fahrenheit

To convert from Celsius temperatures to Fahrenheit temperatures, see the box below. In this box, we use the symbol F to represent degrees Fahrenheit and the symbol C to represent degrees Celsius.

Converting Celsius to Fahrenheit

$$F = \frac{9}{5} \cdot C + 32 \qquad \text{or} \qquad F = 1.8 \cdot C + 32$$

(To convert to Fahrenheit temperature, multiply the Celsius temperature by $\frac{9}{5}$ or 1.8, and then add 32.)

Concept Check

Which of the following statements is correct? Explain.

a. 6°C is below the freezing point of water.

b. 6°F is below the freezing point of water.

Answers

Concept Check: a. false,   b. true

Practice Problem 1

Convert 50°C to degrees Fahrenheit.

EXAMPLE 1 Convert 15°C to degrees Fahrenheit.

Solution:

$$F = \frac{9}{5} \cdot C + 32$$

$$= \frac{9}{5} \cdot 15 + 32 \qquad \text{Replace C with 15.}$$

$$= 27 + 32 \qquad \text{Simplify.}$$

$$= 59 \qquad \text{Add.}$$

Thus, 15°C is equivalent to 59°F. ●

Practice Problem 2

Convert 18°C to degrees Fahrenheit.

EXAMPLE 2 Convert 29°C to degrees Fahrenheit.

Solution:

$$F = 1.8 \cdot C + 32$$

$$= 1.8 \cdot 29 + 32 \qquad \text{Replace C with 29.}$$

$$= 52.2 + 32 \qquad \text{Multiply 1.8 by 29.}$$

$$= 84.2 \qquad \text{Add.}$$

Therefore, 29°C is the same as 84.2°F. ●

B **Converting Degrees Fahrenheit to Degrees Celsius**

To convert from Fahrenheit temperatures to Celsius temperatures, see the box below. The symbol C represents degrees Celsius, and the symbol F represents degrees Fahrenheit.

Converting Fahrenheit to Celsius

$$C = \frac{5}{9} \cdot (F - 32)$$

(To convert to Celsius temperature, subtract 32 from the Fahrenheit temperature, and then multiply by $\frac{5}{9}$.)

Practice Problem 3

Convert 86°F to degrees Celsius.

EXAMPLE 3 Convert 59°F to degrees Celsius.

Solution: We evaluate the formula $C = \frac{5}{9} \cdot (F - 32)$ when F is 59.

$$C = \frac{5}{9} \cdot (F - 32)$$

$$= \frac{5}{9} \cdot (59 - 32) \qquad \text{Replace F with 59.}$$

$$= \frac{5}{9} \cdot (27) \qquad \text{Subtract inside parentheses.}$$

$$= 15 \qquad \text{Multiply.}$$

Answers

1. 122°F, **2.** 64.4°F, **3.** 30°C

Therefore, 59°F is the same temperature as 15°C. ●

EXAMPLE 4

Convert 114°F to degrees Celsius. If necessary, round to the nearest tenth of a degree.

Solution:

$$C = \frac{5}{9} \cdot (F - 32)$$

$$= \frac{5}{9} \cdot (114 - 32) \qquad \text{Replace F with 114.}$$

$$= \frac{5}{9} \cdot (82) \qquad \text{Subtract inside parentheses.}$$

$$\approx 45.6 \qquad \text{Multiply.}$$

Therefore, 114°F is approximately 45.6°C.   ●

EXAMPLE 5

Normal body temperature is 98.6°F. What is this temperature in degrees Celsius?

Solution: We evaluate the formula $C = \frac{5}{9} \cdot (F - 32)$ when F is 98.6.

$$C = \frac{5}{9} \cdot (F - 32)$$

$$= \frac{5}{9} \cdot (98.6 - 32) \qquad \text{Replace F with 98.6.}$$

$$= \frac{5}{9} \cdot (66.6) \qquad \text{Subtract inside parentheses.}$$

$$= 37 \qquad \text{Multiply.}$$

Therefore, normal body temperature is 37°C.   ●

Try the Concept Check in the margin.

Practice Problem 4

Convert 113°F to degrees Celsius. If necessary, round to the nearest tenth of a degree.

Practice Problem 5

During a bout with the flu, Albert's temperature reaches 102.8°F. What is his temperature measured in degrees Celsius? Round to the nearest tenth of a degree.

Concept Check

Clarissa must convert 40°F to degrees Celsius. What is wrong with her work shown below?

$$F = 1.8 \cdot C + 32$$
$$F = 1.8 \cdot 40 + 32$$
$$F = 72 + 32$$
$$F = 104$$

Answers

4. 45°C,   **5.** 39.3°C

Concept Check: She used the conversion for Celsius to Fahrenheit instead of Fahrenheit to Celsius.

Do you remember what to do the day of an exam?

On the day of an exam, don't forget to try the following:

- Allow yourself plenty of time to arrive.
- Read the directions on the test carefully.
- Read each problem carefully as you take your test. Make sure that you answer the question asked.
- Watch your time and pace yourself so that you may attempt each problem on your test.
- If you have time, check your work and answers.
- Do not turn your test in early. If you have extra time, spend it double-checking your work.

Good luck!

Mental Math

Determine whether the measurement in each statement is reasonable.

1. A 72°F room feels comfortable.

2. Water heated to 110°F will boil.

3. Josiah has a fever if a thermometer shows his temperature to be 40°F.

4. An air temperature of 20°F on a Vermont ski slope can be expected in the winter.

5. When the temperature is 30°C outside, an overcoat is needed.

6. An air-conditioned room at 60°C feels quite chilly.

7. Barbara has a fever when a thermometer records her temperature at 40°C.

8. Water cooled to 32°C will freeze.

EXERCISE SET 7.4

Ⓐ Ⓑ *Convert as indicated. When necessary, round to the nearest tenth of a degree. See Examples 1 through 5.*

1. 41°F to degrees Celsius

2. 68°F to degrees Celsius

3. 104°F to degrees Celsius

4. 86°F to degrees Celsius

5. 60°C to degrees Fahrenheit

6. 80°C to degrees Fahrenheit

7. 115°C to degrees Fahrenheit

8. 35°C to degrees Fahrenheit

9. 62°F to degrees Celsius

10. 182°F to degrees Celsius

11. 142.1°F to degrees Celsius

12. 43.4°F to degrees Celsius

13. 92°C to degrees Fahrenheit

14. 75°C to degrees Fahrenheit

15. 16.3°C to degrees Fahrenheit

16. 48.6°C to degrees Fahrenheit

17. The hottest temperature ever recorded in New Mexico was 122°F. Convert this temperature to degrees Celsius. (*Source*: National Climatic Data Center)

18. The hottest temperature ever recorded in Rhode Island was 104°F. Convert this temperature to degrees Celsius. (*Source*: National Climatic Data Center)

19. A weather forecaster in Caracas predicts a high temperature of 27°C. Find this measurement in degrees Fahrenheit.

20. While driving to work, Alan Olda notices a temperature of 18°C flash on the local bank's temperature display. Find the corresponding temperature in degrees Fahrenheit.

21. At Mack Trucks' headquarters, the room temperature is to be set at 70°F, but the thermostat is calibrated in degrees Celsius. Find the temperature to be set.

22. The computer room at Merck, Sharp, and Dohm is normally cooled to 66°F. Find the corresponding temperature in degrees Celsius.

23. Najib Tan is running a fever of 100.2°F. Find his temperature as it would be shown on a Celsius thermometer.

24. William Saylor generally has a subnormal temperature of 98.2°F. Find what this temperature would be on a Celsius thermometer.

25. In a European cookbook, a recipe requires the ingredients for caramels to be heated to 118°C, but the cook has access only to a Fahrenheit thermometer. Find the temperature in degrees Fahrenheit that should be used to make the caramels.

26. The ingredients for divinity should be heated to 127°C, but the candy thermometer that Myung Kim has is calibrated to degrees Fahrenheit. Find how hot he should heat the ingredients.

27. Mark Tabbey's recipe for Yorkshire pudding calls for a 500°F oven. Find the temperature setting he should use with an oven having Celsius controls.

28. The temperature of Earth's core is estimated to be 4000°C. Find the corresponding temperature in degrees Fahrenheit.

29. The surface temperature of Venus can reach 864°F. Find this temperature in degrees Celsius.

Review and Preview

Find the perimeter of each figure. See Section 1.3.

△ **30.**
3 in.

3 in. Square

△ **31.**
25 m

6 m Rectangle

△ **32.**
4 cm   3 cm
Triangle
5 cm

△ **33.**
3 ft   3 ft
Pentagon
3 ft   3 ft
3 ft

△ **34.**
2 ft 8 in.

1 ft 6 in. Rectangle

△ **35.**
2.6 m

2.6 m Square

◆ Combining Concepts

36. On February 17, 1995, in the Tokamak Fusion Test Reactor at Princeton University, the highest temperature produced in a laboratory was achieved. This temperature was 918,000,000°F. Convert this temperature to degrees Celsius. (*Source: Guinness Book of Records*)

37. The hottest-burning substance known is carbon subnitride. Its flame at one atmospheric pressure reaches 9010°F. Convert this temperature to degrees Celsius. (*Source: Guinness Book of Records*)

38. In your own words, describe how to convert from degrees Celsius to degrees Fahrenheit.

7.5 Energy: U.S. and Metric Systems of Measurement

Many people think of energy as a concept that involves movement or activity. However, **energy** is defined as "the capacity to do work." Often energy is stored, awaiting use at some later point in time.

OBJECTIVES

Ⓐ Define and use U.S. units of energy and convert from one unit to another.

Ⓑ Define and use metric units of energy.

SSM   SG   CD & VIDEO   MATH PRO   WEB
TUTOR CENTER

Ⓐ Defining and Using the U.S. System Units of Energy

In the U.S. system of measurement, energy is commonly measured in foot-pounds. One **foot-pound (ft-lb)** is the amount of energy needed to lift a 1-pound object a distance of 1 foot. To determine the amount of energy necessary to move a 50-pound weight a distance of 100-feet, we simply multiply these numbers. That is,

50 pounds · 100 feet = 5000 ft-lb of energy

EXAMPLE 1   Finding the Amount of Energy Needed to Move a Carton

An employee for the Jif Peanut Butter company must lift a carton of peanut butter jars 16 feet to the top of the warehouse. In the carton are 24 jars, each of which weighs 1.125 pounds. How much energy is required to lift the carton?

Solution: First we determine the weight of the carton.

$$\begin{aligned} \text{weight of carton} &= \text{weight of a jar} \cdot \text{number of jars} \\ &= 1.125 \text{ pounds} \cdot 24 \\ &= 27 \text{ pounds} \end{aligned}$$

Thus, the carton weighs 27 pounds.
To find the energy needed to lift the 27-pound carton, we multiply the weight times the distance.

$$\text{energy} = 27 \text{ pounds} \cdot 16 \text{ feet} = 432 \text{ ft-lb}$$

Thus, 432 ft-lb of energy are required to lift the carton.

Try the Concept Check in the margin.

Another form of energy is heat. In the U.S. system of measurement, heat is measured in **British Thermal Units (BTU)**. A BTU is the amount of heat required to raise the temperature of 1 pound of water 1 degree Fahrenheit. To relate British Thermal Units to foot-pounds, we need to know that

1 BTU = 778 ft-lb

EXAMPLE 2   Converting BTU to Foot-Pounds

The Raywall Company produces several different furnace models. Their FC-4 model requires 13,652 BTU every hour to operate. Convert the required energy to foot-pounds.

Solution:

To convert BTU to foot-pounds, we multiply by the unit fraction $\dfrac{778 \text{ ft-lb}}{1 \text{ BTU}}$.

$$13{,}652 \text{ BTU} = 13{,}652 \ \cancel{\text{BTU}} \cdot \overset{\text{Unit fraction}}{\dfrac{778 \text{ ft-lb}}{1 \ \cancel{\text{BTU}}}}$$

$$= 10{,}621{,}256 \text{ ft-lb}$$

Thus, 13,652 BTU is equivalent to 10,621,256 ft-lb.

Practice Problem 1

Three bales of cardboard must be lifted 340 feet. If each bale weighs 63 pounds, find the amount of work required to lift the cardboard.

Concept Check

Suppose you would like to find how many foot-pounds of energy are needed to lift an object weighing 12 ounces a total of 14 yards. What adjustments should you make before computing the answer?

Practice Problem 2

The FC-5 model furnace produced by Raywall uses 17,065 BTU every hour. Convert this energy requirement to foot-pounds.

Answers

1. 64,260 ft-lb,   **2.** 13,276,570 ft-lb

Concept Check: convert 12 ounces to .75 pound and 14 yards to 42 feet; 31.5 ft-lb

70 calories

B Defining and Using the Metric System Units of Energy

In the metric system, heat is measured in calories. A **calorie (cal)** is the amount of heat required to raise the temperature of 1 kilogram of water 1 degree Celsius.

The fact that an apple contains 70 calories means that 70 calories of heat energy are stored in our bodies whenever we eat an apple. This energy is stored in fat tissue and is burned (or "oxidized") by our bodies when we require energy to do work. We need 20 calories each hour just to stand still. This means that 20 calories of heat energy will be burned by our bodies each hour that we spend standing.

Practice Problem 3

It takes 30 calories each hour for Alan to fly a kite. How many calories will he use if he flies his kite for 2 hours?

EXAMPLE 3   Finding the Number of Calories Needed

It takes 20 calories for Jim to stand for 1 hour. How many calories does he use when standing for 3 hours at a crowded party?

Solution: We multiply the number of calories used in 1 hour by the number of hours spent standing.

$$\text{total calories} = 20 \cdot 3 = 60 \text{ calories}$$

Therefore, Jim uses 60 calories to stand for 3 hours at the party.   ●

Practice Problem 4

It takes 200 calories for Melanie to play Frisbee for an hour. How many calories will she use playing Frisbee for an hour each day for 5 days?

EXAMPLE 4   Finding the Number of Calories Needed

It takes 115 calories for Kathy to walk slowly for 1 hour. How many calories does she use walking slowly for 1 hour a day for 6 days?

Solution: We multiply the total number of calories used in 1 hour each day by the number of days.

$$\text{total calories} = 115 \cdot 6 = 690 \text{ calories}$$

Therefore, Kathy uses 690 calories walking slowly for 1 hour for 6 days. ●

Practice Problem 5

To play volleyball for an hour requires 300 calories. If Martha plays volleyball 1.25 hours each day for 4 days, how many calories does Martha use?

EXAMPLE 5   Finding the Number of Calories Needed

It requires 100 calories to play a game of cards for an hour. If Jason plays poker for 1.5 hours each day for 5 days, how many calories are required?

Solution: We first determine the number of calories Jason uses each day to play poker.

$$\text{calories used each day} = 100(1.5) = 150 \text{ calories}$$

Then we multiply the number of calories used each day by the number of days.

$$\text{calories used for 5 days} = 150 \cdot 5 = 750 \text{ calories}$$

Thus, Jason uses 750 calories to play poker for 1.5 hours each day for 5 days.
●

Answers

3. 60 cal, **4.** 1000 cal, **5.** 1500 cal

Mental Math

Solve.

1. How many foot-pounds of energy are needed to lift a 6-pound object 5 feet?

2. How many foot-pounds of energy are needed to lift a 10-pound object 4 feet?

3. How many foot-pounds of energy are needed to lift a 3-pound object 20 feet?

4. How many foot-pounds of energy are needed to lift a 5-pound object 9 feet?

5. If 30 calories are burned by the body in 1 hour, how many calories are burned in 3 hours?

6. If 15 calories are burned by the body in 1 hour, how many calories are burned in 2 hours?

7. If 20 calories are burned by the body in 1 hour, how many calories are burned in $\frac{1}{4}$ of an hour?

8. If 50 calories are burned by the body in 1 hour, how many calories are burned in $\frac{1}{2}$ of an hour?

EXERCISE SET 7.5

Ⓐ *Solve. See Examples 1 and 2.*

 1. How much energy is required to lift a 3-pound math textbook 380 feet up a hill?

2. How much energy is required to lift a 20-pound sack of potatoes 55 feet?

3. How much energy is required to lift a 168-pound person 22 feet?

4. How much energy is needed to lift a 2250-pound car a distance of 45 feet?

5. How many foot-pounds of energy are needed to take 2.5 tons of topsoil 85 feet from the pile delivered by the nursery to the garden?

6. How many foot-pounds of energy are needed to lift 4.25 tons of coal 16 feet into a new coal bin?

7. Convert 30 BTU to foot-pounds.

8. Convert 50 BTU to foot-pounds.

9. Convert 1000 BTU to foot-pounds.

10. Convert 10,000 BTU to foot-pounds.

11. A 20,000 BTU air conditioner requires how many foot-pounds of energy to operate?

12. A 24,000 BTU air conditioner requires how many foot-pounds of energy to operate?

13. The Raywall model FC-10 heater uses 34,130 BTU each hour to operate. How many foot-pounds of energy does it use each hour?

14. The Raywall model FC-12 heater uses 40,956 BTU each hour to operate. How many foot-pounds of energy does it use each hour?

15. 8,000,000 ft-lb is equivalent to how many BTU, rounded to the nearest whole number?

16. 450,000 ft-lb is equivalent to how many BTU, rounded to the nearest whole number?

Solve. See Examples 3 through 5.

17. While walking slowly, Janie Gaines burns 115 calories each hour. How many calories does she burn if she walks slowly for an hour every day of the week?

18. Dancing burns 270 calories per hour. How many calories are needed to go dancing an hour a night for 3 nights?

 19. Approximately 300 calories are burned each hour skipping rope. How many calories are required to skip rope $\frac{1}{2}$ of an hour each day for 5 days?

20. Ebony Jordan burns 360 calories per hour while riding her stationary bike. How many calories does she burn when she rides her bicycle $\frac{1}{4}$ of an hour each day for 6 days?

21. Julius Davenport goes through a rigorous exercise routine each day. He burns calories at a rate of 720 calories per hour. How many calories does he need to exercise 20 minutes per day, 6 days a week?

22. A roller skater can easily use 325 calories per hour while skating. How many calories are needed to roller skate 75 minutes per day for 3 days?

23. A casual stroll burns 165 calories per hour. How long will it take to stroll off the 425 calories contained in a hamburger, to the nearest tenth of an hour?

24. Even when asleep, the body burns 15 calories per hour. How long must a person sleep to burn off the calories in an 80-calorie orange, to the nearest tenth of an hour?

25. One pound of body weight is lost whenever 3500 calories are burned off. If walking briskly burns 200 calories for each mile walked, how far must Sheila Osby walk to lose 1 pound?

26. Bicycling can burn as much as 500 calories per hour. How long must a person ride a bicycle at this rate to use up the 3500 calories needed to lose 1 pound?

Review and Preview

Write each fraction in simplest form. See Section 2.3.

27. $\frac{20}{25}$ **28.** $\frac{75}{100}$ **29.** $\frac{27}{45}$ **30.** $\frac{56}{60}$ **31.** $\frac{72}{80}$ **32.** $\frac{18}{20}$

Combining Concepts

33. A 123.9-pound pile of prepacked canned goods must be lifted 9 inches to permit a door to close. How much energy is needed to do the job?

34. A 14.3-pound framed picture must be lifted 6 feet 3 inches. How much energy is required to move the picture?

35. 6400 ft-lb of energy were needed to lift an anvil 25 feet. Find the weight of the anvil.

36. 825 ft-lb of energy were needed to lift 40 pounds of apples to a new container. How far were the apples moved?

Map Reading

MATERIALS:

- ruler
- string
- calculator

This activity may be completed by working in groups or individually.

Investigate the route you would take from Santa Rosa, New Mexico, to San Antonio, New Mexico. Use the map in the figure to answer the following questions. You may find that using string to match the roads on the map is useful when measuring distances.

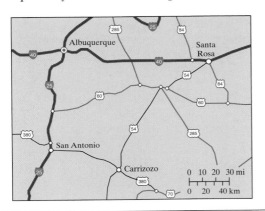

1. How many miles is it from Santa Rosa to San Antonio via Interstate 40 and Interstate 25? Convert this distance to kilometers.

2. How many miles is it from Santa Rosa to San Antonio via U.S. 54 and U.S. 380? Convert this distance to kilometers.

3. Assume that the speed limit on Interstates 40 and 25 is 65 miles per hour. How long would the trip take if you took this route and traveled 65 miles per hour the entire trip?

4. At what average speed would you have to travel on the U.S. routes to make the trip from Santa Rosa to San Antonio in the same amount of time that it would take on the interstate routes? Do you think this speed is reasonable on this route? Explain your reasoning.

5. Discuss in general the factors that might affect your decision among the different routes.

6. Explain which route you would choose in this case and why.

How are your homework assignments going?

By now, you should have good homework habits. If not, it's never too late to begin. Why is it so important in mathematics to keep up with homework? You probably now know the answer to that question. You have probably realized by now that many concepts in mathematics build on each other. Your understanding of one chapter in mathematics usually depends on your understanding of the previous chapter's material.

Don't forget that completing your homework assignment involves a lot more than attempting a few of the problems assigned.

To complete a homework assignment, remember these four things:

1. Attempt all of it.

2. Check it.

3. Correct it.

4. If needed, ask questions about it.

Chapter 7   VOCABULARY CHECK

Fill in each blank with one of the words or phrases listed below.

mass	unit fractions	gram	energy	weight
meter	liter	calorie	British Thermal Unit	

1. _____ is a measure of the pull of gravity.
2. _____ is a measure of the amount of substance in an object. This measure does not change.
3. The basic unit of length in the metric system is the _____.
4. To convert from one unit of length to another, _____ may be used.
5. A _____ is the basic unit of mass in the metric system.
6. _____ is the capacity to do work.
7. In the U.S. system of measurement, a _____ is the amount of heat required to raise the temperature of 1 pound of water 1 degree Fahrenheit.
8. The _____ is the basic unit of capacity in the metric system.
9. In the metric system, a _____ is the amount of heat required to raise the temperature of 1 kilogram of water 1 degree Celsius.

C H A P T E R

Highlights

DEFINITIONS AND CONCEPTS	**EXAMPLES**

Section 7.1   Length: U.S. and Metric Systems of Measurement

To convert from one unit of length to another, **unit fractions** may be used.
The unit fraction should be in the form

$$\frac{\text{units to convert to}}{\text{original units}}$$

$$\frac{12 \text{ inches}}{1 \text{ foot}}, \frac{1 \text{ foot}}{12 \text{ inches}}, \frac{3 \text{ feet}}{1 \text{ yard}}$$

Convert 6 feet to inches.

$$6 \text{ ft} = \frac{6 \text{ ft}}{1} \cdot \frac{12 \text{ in.}}{1 \text{ ft}} \quad \leftarrow \text{units to convert to}$$
$$\leftarrow \text{original units}$$
$$= 6 \cdot 12 \text{ in.}$$
$$= 72 \text{ in.}$$

Convert 3650 centimeters to meters.

$$3650 \text{ cm} = \frac{3650 \text{ cm}}{1} \cdot \frac{0.01 \text{ m}}{1 \text{ cm}} = 36.5 \text{ m}$$

or

km   hm   dam   m   dm   cm   mm

2 units to the left

LENGTH: U.S. SYSTEM OF MEASUREMENT

$$12 \text{ inches (in.)} = 1 \text{ foot (ft)}$$
$$3 \text{ feet} = 1 \text{ yard (yd)}$$
$$5280 \text{ feet} = 1 \text{ mile (mi)}$$

The basic unit of length in the metric system is the **meter**. A meter is slightly longer than a yard.

$$3650 \text{ cm} = 36.5 \text{ m}$$

2 places to the left

LENGTH: METRIC SYSTEM OF MEASUREMENT

Prefix	Meaning	Metric Unit of Length
kilo	1000	1 kilometer (km) = 1000 meters (m)
hecto	100	1 hectometer (hm) = 100 m
deka	10	1 dekameter (dam) = 10 m
		1 meter (m) = 1 m
deci	1/10	1 decimeter (dm) = 1/10 m or 0.1 m
centi	1/100	1 centimeter (cm) = 1/100 m or 0.01 m
milli	1/1000	1 millimeter (mm) = 1/1000 m or 0.001 m

DEFINITIONS AND CONCEPTS	EXAMPLES

Section 7.2 Weight and Mass: U.S. and Metric Systems of Measurement

Weight is really a measure of the pull of gravity. **Mass** is a measure of the amount of substance in an object and does not change.

Convert 5 pounds to ounces.

$$5\,\text{lb} = \frac{5\,\text{lb}}{1} \cdot \frac{16\,\text{oz}}{1\,\text{lb}} = 80\,\text{oz}$$

WEIGHT: U.S. SYSTEM OF MEASUREMENT

16 ounces (oz) = 1 pound (lb)

2000 pounds = 1 ton

A **gram** is the basic unit of mass in the metric system. It is the mass of water contained in a cube 1 centimeter on each side. A paper clip weighs about 1 gram.

Convert 260 grams to kilograms.

$$260\,\text{g} = \frac{260\,\text{g}}{1} \cdot \frac{1\,\text{kg}}{1000\,\text{g}} = 0.26\,\text{kg}$$

or

kg hg dag g dg cg mg

3 units to the left

260 g = 0.260 kg

3 places to the left

MASS: METRIC SYSTEM OF MEASUREMENT

Prefix	Meaning	Metric Unit of Mass
kilo	1000	1 kilogram (kg) = 1000 grams (g)
hecto	100	1 hectogram (hg) = 100 g
deka	10	1 dekagram (dag) = 10 g
		1 gram (g) = 1 g
deci	1/10	1 decigram (dg) = 1/10 g or 0.1 g
centi	1/100	1 centigram (cg) = 1/100 g or 0.01 g
milli	1/1000	1 milligram (mg) = 1/1000 g or 0.001 g

Section 7.3 Capacity: U.S. and Metric Systems of Measurement

CAPACITY: U.S. SYSTEM OF MEASUREMENT

8 fluid ounces (fl oz) = 1 cup (c)

2 cups = 1 pint (pt)

2 pints = 1 quart (qt)

4 quarts = 1 gallon (gal)

The **liter** is the basic unit of capacity in the metric system. It is the capacity or volume of a cube measuring 10 centimeters on each side. A liter of liquid is slightly more than 1 quart.

Convert 5 pints to gallons.

1 gal = 4 qt = 8 pt

$$5\,\text{pt} = \frac{5\,\text{pt}}{1} \cdot \frac{1\,\text{gal}}{8\,\text{pt}} = \frac{5}{8}\,\text{gal}$$

Convert 1.5 liters to milliliters.

$$1.5\,\text{L} = \frac{1.5\,\text{L}}{1} \cdot \frac{1000\,\text{ml}}{1\,\text{L}} = 1500\,\text{ml}$$

or

kl hl dal L dl cl ml

3 units to the right

1.500 L = 1500 ml

3 places to the right

Section 7.3   Capacity: U.S. and Metric Systems of Measurement *(continued)*

CAPACITY: METRIC SYSTEM OF MEASUREMENT

Prefix	Meaning	Metric Unit of Capacity
kilo	1000	1 kiloliter (kl) $=$ 1000 liters (L)
hecto	100	1 hectoliter (hl) $=$ 100 L
deka	10	1 dekaliter (dal) $=$ 10 L
		1 liter (L) $=$ 1 L
deci	1/10	1 deciliter (dl) $=$ 1/10 L or 0.1 L
centi	1/100	1 centiliter (cl) $=$ 1/100 L or 0.01 L
milli	1/1000	1 milliliter (ml) $=$ 1/1000 L or 0.001 L

Section 7.4   Temperature: U.S. and Metric Systems of Measurement

CELSIUS TO FAHRENHEIT

$$F = \frac{9}{5} \cdot C + 32 \text{ or } F = 1.8 \cdot C + 32$$

FAHRENHEIT TO CELSIUS

$$C = \frac{5}{9} \cdot (F - 32)$$

Convert 35°C to degrees Fahrenheit.

$$F = \frac{9}{5} \cdot 35 + 32 = 63 + 32 = 95$$

$$35°C = 95°F$$

Convert 50°F to degrees Celsius.

$$C = \frac{5}{9} \cdot (50 - 32) = \frac{5}{9} \cdot 18 = 10$$

$$50°F = 10°C$$

Section 7.5   Energy: U.S. and Metric Systems of Measurement

Energy is the capacity to do work.
In the U.S. system of measurement, a **foot-pound** (ft-lb) is the amount of energy needed to lift a 1-pound object a distance of 1 foot.

In the U.S. system of measurement, a **British Thermal Unit** (BTU) is the amount of heat required to raise the temperature of 1 pound of water 1 degree Fahrenheit. (1 BTU = 778 ft-lb)

In the metric system, a **calorie** is the amount of heat required to raise the temperature of 1 kilogram of water 1 degree Celsius.

How much energy is needed to lift a 20-pound object 35 feet?

$$20 \text{ pounds} \cdot 35 \text{ feet} = 700 \text{ foot-pounds of energy}$$

Convert 50 BTU to foot-pounds.

$$50 \text{ BTU} = 50 \text{ } \cancel{BTU} \cdot \frac{778 \text{ ft-lb}}{1 \text{ } \cancel{BTU}} = 38,900 \text{ ft-lb}$$

A stationary bicyclist uses 350 calories in 1 hour.
How many calories will the bicyclist use in $\frac{1}{2}$ of an hour?

$$\text{calories used} = 350 \cdot \frac{1}{2} = 175 \text{ calories}$$

SPEEDS

A speed measures how far something travels in a given unit of time. You already learned in Section 5.2 that the speed 55 miles per hour is a unit rate that can be written as $\dfrac{55 \text{ miles}}{1 \text{ hour}}$. Just as there are different units of measurement for length or distance, there are different units of measurement for speed as well. It is also possible to perform unit conversions on speeds. Before we learn about converting speeds, we will review units of time. The following is a summary of equivalencies between various units of time.

Units of Time	Unit Fractions
60 seconds (s) = 1 minute (min)	$\dfrac{60 \text{ s}}{1 \text{ min}} = \dfrac{1 \text{ min}}{60 \text{ s}} = 1$
60 minutes = 1 hour (h)	$\dfrac{60 \text{ min}}{1 \text{ h}} = \dfrac{1 \text{ h}}{60 \text{ min}} = 1$
3600 seconds = 1 hour	$\dfrac{3600 \text{ s}}{1 \text{ h}} = \dfrac{1 \text{ h}}{3600 \text{ s}}$

Here are some common speeds.

Speeds

Miles per hour (mph)
Miles per minute (mi/min)
Miles per second (mi/s)
Feet per second (ft/s)
Feet per minute (ft/min)
Kilometers per hour (kmph or km/h)
Kilometers per second (kmps or km/s)
Meters per second (m/s)
Knots

Helpful Hint

A **knot** is 1 nautical mile per hour and is a measure of speed used for ships.
1 nautical mile (nmi) ≈ 1.15 miles (mi)
1 nautical mile (nmi) ≈ 6076.12 feet (ft)

To convert from one speed to another, unit fractions may be used. To convert from mph to ft/s first write the original speed as a unit rate. Then multiply by a unit fraction that relates miles to feet and by a unit fraction that relates hours to seconds. The unit fractions should be written so that like units will divide out. For example, to convert 55 mph to ft/s:

$$55 \text{ mph} = \frac{55 \text{ miles}}{1 \text{ hour}} = \frac{55 \text{ miles}}{1 \text{ hour}} \cdot \frac{5280 \text{ ft}}{1 \text{ mile}} \cdot \frac{1 \text{ hour}}{3600 \text{ s}}$$

$$= \frac{55 \cdot 5280 \text{ ft}}{3600 \text{ s}}$$

$$= \frac{290{,}400 \text{ ft}}{3600 \text{ s}}$$

$$= 80\frac{2}{3} \text{ ft/s}$$

GROUP ACTIVITY

1. Research the current world land speed record. Convert the speed from mph to feet per second.
2. Research the current world water speed record. Convert from mph to knots.
3. Research and then describe the Beaufort Wind Scale, its origins, and how it is used. Give the scale keyed to both miles per hour and knots. Why would both measures be useful?

Chapter 7 Review

(7.1) *Convert.*

1. 108 in. to feet

2. 72 ft to yards

3. 2.5 mi to feet

4. 6.25 ft to inches

5. 52 ft = _____ yd _____ ft

6. 46 in. = _____ ft _____ in.

7. 42 m to centimeters

8. 82 cm to millimeters

9. 12.18 mm to meters

10. 2.31 m to kilometers

Perform each indicated operation.

11. 4 yd 2 ft + 16 yd 2 ft

12. 12 ft 1 in. − 4 ft 8 in.

13. 8 ft 3 in. × 5

14. 7 ft 4 in. ÷ 2

15. 8 cm + 15 mm

16. 4 m + 126 cm

17. 9.3 km − 183 m

18. 4100 mm − 3 m

Solve.

19. A bolt of cloth contains 333 yd 1 ft of cotton ticking. Find the amount of material that remains after 163 yd 2 ft is removed from the bolt.

20. The local ambulance corps plans to award 20 framed certificates of valor to some of its outstanding members. If each frame requires 6 ft 4 in. of framing material, how much material is needed for all the frames?

459

21. The trip from Philadelphia to Washington, D.C., is 217 km. Four friends agree to share the driving equally. How far must each drive on this round-trip vacation?

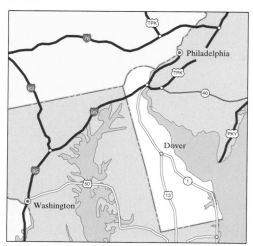

△ **22.** The college has ordered that NO SMOKING signs be placed above the doorway of each classroom. Each sign is 0.8 m long and 30 cm wide. Find the area of each sign. (*Hint*: Recall that the area of a rectangle = width · length.)

30 centimeters

0.8 meter

(7.2) *Convert.*

23. 66 oz to pounds

24. 2.3 tons to pounds

25. 52 oz = _____ lb _____ oz.

26. 8200 lb = _____ tons _____ lb

27. 1400 mg to grams

28. 40 kg to grams

29. 2.1 hg to dekagrams

30. 0.03 mg to decigrams

Perform each indicated operation.

31. 6 lb 5 oz − 2 lb 12 oz

32. 5 tons 1600 lb + 4 tons 1200 lb

33. 6 tons 2250 lb ÷ 3

34. 8 lb 6 oz × 4

35. 1300 mg + 3.6 g

36. 4.8 kg + 4200 g

37. 9.3 g − 1200 mg

38. 6.3 kg × 8

Solve

39. Donshay Berry ordered 1 lb 12 oz of soft-center candies and 2 lb 8 oz of chewy-center candies for his party. Find the total weight of the candy ordered.

40. Four local townships jointly purchase 38 tons 300 lb of cinders to spread on their roads during an ice storm. Determine the weight of the cinders each township receives if they share the purchase equally.

41. Linda Holden ordered 8.3 kg of whole wheat flour from the health store, but she received 450 g less. How much flour did she actually receive?

42. Eight friends spent a weekend in the Poconos tapping maple trees and preparing 9.3 kg of maple syrup. Find the weight each friend receives if they share the syrup equally.

(7.3) *Convert.*

43. 16 pints to quarts

44. 40 fluid ounces to cups

45. 6.75 gallons to quarts

46. 8.5 pints to cups

47. 9 pt = _____ qt _____ pt

48. 15 qt = _____ gal _____ qt

49. 3.8 L to milliliters

50. 4.2 ml to deciliters

51. 14 hl to kiloliters

52. 30.6 L to centiliters

Perform each indicated operation.

53. 1 qt 1 pt + 3 qt 1 pt

54. 3 gal 2 qt 1 pt $\times$ 2

55. 0.946 L − 210 ml

56. 6.1 L + 9400 ml

Solve.

57. Carlos Perez prepares 4 gal 2 qt of iced tea for a block party. During the first 30 minutes of the party, 1 gal 3 qt of the tea is consumed. How much iced tea remains?

58. A recipe for soup stock calls for 1 c 4 fl oz of beef broth. How much should be used if the recipe is cut in half?

59. Each bottle of Kiwi liquid shoe polish holds 85 ml of the polish. Find the number of liters of shoe polish contained in 8 boxes if each box contains 16 bottles.

60. Ivan Miller wants to pour three separate containers of saline solution into a single vat with a capacity of 10 liters. Will 6 liters of solution in the first container combined with 1300 milliliters in the second container and 2.6 liters in the third container fit into the larger vat?

(7.4) *Convert. Round to the nearest tenth of a degree, if necessary.*

61. 245°C to degrees Fahrenheit

62. 160°C to degrees Fahrenheit

63. 42°C to degrees Fahrenheit

64. 86°C to degrees Fahrenheit

65. 93.2°F to degrees Celsius

66. 51.8°F to degrees Celsius

67. 41.3°F to degrees Celsius

68. 80°F to degrees Celsius

Solve. Round to the nearest tenth of a degree, if necessary.

69. A sharp dip in the jet stream caused the temperature in New Orleans to drop to 35°F. Find the corresponding temperature in degrees Celsius.

70. The recipe for meat loaf calls for a 165°C oven. Find the setting used if the oven has a Fahrenheit thermometer.

(7.5) *Solve.*

71. How many foot-pounds of energy are needed to lift a 5.6-pound radio a distance of 12 feet?

72. How much energy is required to lift a 21-pound carton of Rice-A-Roni a distance of 6.5 feet?

73. How much energy is used when a 1.2-ton pile of sand is lifted 15 yards?

74. The energy required to operate a 12,000-BTU air conditioner is equivalent to how many foot-pounds?

75. Convert 2,000,000 foot-pounds to BTU, rounded to the nearest hundred.

76. Kip Yates burns off 450 calories each hour he plays handball. How many calories does he use to play handball for $2\frac{1}{2}$ hours?

77. Qwanetta Sesson uses 210 calories each hour she spends mowing the grass. Find the number of calories needed to mow the grass if she spends 3 hours mowing each week for 24 weeks.

78. Four ounces of sirloin steak contain 420 calories. If Edith Lutrell uses 180 calories each hour she walks, how long must she walk to burn off the calories from this steak?

Chapter 7 Test

Convert.

1. 280 in. to feet and inches

2. $2\frac{1}{2}$ gal to quarts

3. 30 oz to pounds

4. 2.8 tons to pounds

5. 38 pt to gallons

6. 40 mg to grams

7. 2.4 kg to grams

8. 3.6 cm to millimeters

9. 4.3 dg to grams

10. 0.83 L to milliliters

Perform each indicated operation.

11. 3 qt 1 pt + 2 qt 1 pt

12. 8 lb 6 oz − 4 lb 9 oz

13. 2 ft 9 in. × 3

14. 5 gal 2 qt ÷ 2

15. 8 cm − 14 mm

16. 1.8 km + 456 m

Convert. Round to the nearest tenth of a degree, if necessary.

17. 84°F to degrees Celsius

18. 12.6°C to degrees Fahrenheit

19. The sugar maples in front of Bette MacMillan's house are 8.4 meters tall. Because they interfere with the phone lines, the telephone company plans to remove the top third of the trees. How tall will the maples be after they are shortened?

20. A total of 15 gal 1 qt of oil has been removed from a 20-gallon drum. How much oil still remains in the container?

21. The doctors are quite concerned about Lucia Gillespie, who is running a 41°C fever. Find Lucia's temperature in degrees Fahrenheit.

22. Gordan Cooper, the engineer in charge of bridge construction, said that the span of a certain bridge would be 88 m. But the actual construction required it to be 340 cm longer. Find the span of the bridge.

Answers

1. _____
2. _____
3. _____
4. _____
5. _____
6. _____
7. _____
8. _____
9. _____
10. _____
11. _____
12. _____
13. _____
14. _____
15. _____
16. _____
17. _____
18. _____
19. _____
20. _____
21. _____
22. _____

23. _____

24. _____

25. _____

26. _____

27. _____

28. _____

29. _____

23. If 2 ft 9 in. of material is used to manufacture one scarf, how much material is needed for 6 scarves?

24. Phillipe Jordaine must lift a 48.5-pound carton of canned tomatoes a distance of 14 feet. Find the energy required to move the carton.

25. Energy used by a 26,000-BTU heater is equivalent to how many foot-pounds?

26. Robin Nestle burns 180 calories each hour when she swims. How many calories does she use if she swims 1 hour per day for 5 days?

27. The Vietnam Veterans Memorial, inscribed with the names of 58,226 deceased and missing U.S. soldiers from the Vietnam War, is located on the National Mall in Washington, D.C. This memorial is formed from two straight sections of wall that meet at an angle at the center of the monument. Each wall is 246 ft 9 in. long. What is the total length of the Vietnam Veterans Memorial's wall? (*Source*: National Park Service)

28. Each panel making up the wall of the Vietnam Veterans Memorial is 101.6 cm wide. There are a total of 148 panels making up the wall. What is the total length of the wall in meters? (*Source*: National Park Service)

29. During the 2001 Harrah's 500 auto race at Michigan International Speedway, the maximum racetrack temperature during the race was 129°F. Find this temperature in degrees Celsius. Round to the nearest tenth. (*Source*: Championship Auto Racing Teams)

Cumulative Review

1. Find the sum: $1647 + 246 + 32 + 85$

2. Find the prime factorization of 80.

3. Find the LCM of 11 and 33.

4. Add: $3\frac{4}{5} + 1\frac{4}{15}$

Write each decimal as a fraction or mixed number in simplest form.

5. 0.125

6. 105.083

7. Insert $<, >$, or $=$ to form a true statement.

0.052   0.236

8. Subtract $85 - 17.31$. Check your answer.

Multiply.

9. 42.1×0.1

10. 9.2×0.001

11. Divide $60.24 \div 8$. Check your answer.

12. Estimate the distance in miles between Garden City, Kansas, and Wichita, Kansas, by rounding each given distance to the nearest ten.

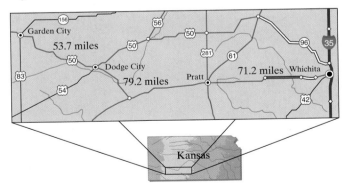

13. Write $\frac{2}{3}$ as a decimal.

14. Write the ratio of 12 to 17 using fractional notation.

Answers

1. _____

2. _____

3. _____

4. _____

5. _____

6. _____

7. _____

8. _____

9. _____

10. _____

11. _____

12. _____

13. _____

14. _____

15. _____

16. _____

17. _____

18. _____

19. _____

20. _____

21. _____

22. _____

23. _____

24. _____

25. _____

15. Write "318.5 miles every 13 gallons of gas" as a unit rate.

16. Find the unknown number n.

$$\frac{6}{5} = \frac{7}{n}$$

△ **17.** A 50-pound bag of fertilizer covers 2400 square feet of lawn. How many bags of fertilizer are needed to cover a town square containing 15,360 square feet of lawn? Round the answer up to the nearest whole bag.

18. Write 23% as a decimal.

19. Write $\frac{1}{12}$ as a percent. Round to the nearest hundredth percent.

20. What number is 35% of 40?

21. Translate to a proportion. 75 is what percent of 30?

22. In response to a decrease in sales, a company with 1500 employees reduces the number of employees to 1230. What is the percent decrease?

23. A speaker that normally sells for $65 is on sale at 25% off. What is the discount and what is the sale price?

24. Find the simple interest after 2 years on $500 at an interest rate of 12%.

25. Convert 9000 pounds to tons.

Geometry

The word *geometry* is formed from the Greek words *geo*, meaning Earth, and *metron*, meaning measure. Geometry literally means to measure the Earth. In this chapter we learn about various geometric figures and their properties such as perimeter, area, and volume. Knowledge of geometry can help us solve practical problems in real-life situations. For instance, knowing certain measures of a circular swimming pool allows us to calculate how much water it can hold.

Just outside of Cairo, Egypt, is a famous plateau called Giza. This is the home of the Great Pyramids, the only surviving entry on the list of the Seven Wonders of the Ancient World. There are three pyramids at Giza. The largest, and oldest, was built as the tomb of the Pharaoh Khufu around 2550 B.C. This pyramid is made from over 2,300,000 blocks of stone weighing a total of 6.5 million tons. It took about 30 years to build this monument. Prior to the 20th century, Khufu's pyramid was the tallest building in the world. Khufu's son Khafre is responsible for the second-largest pyramid at Giza during his rule as pharaoh between 2520 to 2494 B.C. The smallest of the pyramids at Giza is credited to Menkaure, believed to be the son of Khafre and grandson of Khufu. In Exercises 35–38 on page 514, we use some of the measurements of the Great Pyramids to calculate their volumes.

Name _____ Section _____ Date _____

Chapter 8 Pretest

1. _____

2. _____

3. _____

4. _____

5. _____

6. _____

7. _____

8. _____

9. _____

10. _____

11. _____

12. _____

13. _____

14. _____

15. _____

Classify each angle as acute, right, obtuse, or straight.

1.

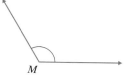

2.

3. Find the complement of a 54° angle.

4. Find the measures of angles x, y, and z.

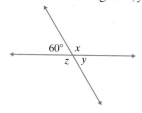

5. Find the measure of $\angle x$.

6. The radius of a sphere is 12.2 inches. Find its diameter.

7. Find the perimeter of the given rectangle.

8. Find the circumference of the given circle. Use $\pi \approx 3.14$.

9. Find the area of the given triangle.

10. Find the volume of a rectangular box 10 inches by 17 inches by 3 inches.

11. Find the exact volume of a sphere with a radius of 8 centimeters.

Simplify.

12. $\sqrt{36}$

13. $\sqrt{\dfrac{81}{49}}$

14. Find the length of the hypotenuse of a right triangle with leg lengths 8 yards and 3 yards. Approximate the length to the nearest thousandth.

15. Given that the triangles are similar, find the length of the side labeled n.

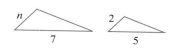

△ 8.1 Lines and Angles

Ⓐ Identify Lines, Line Segments, Rays, and Angles

Let's begin with a review of two important concepts—plane and space.

A **plane** is a flat surface that extends indefinitely. Surfaces like a plane are a classroom floor or a blackboard or whiteboard.

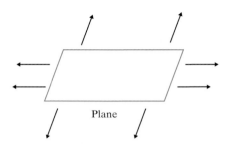

Plane

OBJECTIVES

Ⓐ Identify lines, line segments, rays, and angles.

Ⓑ Classify angles as acute, right, obtuse, or straight.

Ⓒ Identify complementary and supplementary angles.

Ⓓ Find measures of angles.

SSM TUTOR CENTER   SG   CD & VIDEO   MATH PRO   WEB

Space extends in all directions indefinitely. Examples of objects in space are houses, grains of salt, bushes, your *Basic College Mathematics* text-book, and you.

The most basic concept of geometry is the idea of a point in space. A **point** has no length, no width, and no height, but it does have location. We represent a point by a dot, and we label points with letters.

$$P$$
•
Point P

A **line** is a set of points extending indefinitely in two directions. A line has no width or height, but it does have length. We name a line by any two of its points. A **line segment** is a piece of a line with two end points.

Line AB or $\overleftrightarrow{AB}$ Line segment AB or $\overline{AB}$

A **ray** is a part of a line with one end point. A ray extends indefinitely in one direction. An **angle** is made up of two rays that share the same end point. The common end point is called the **vertex**.

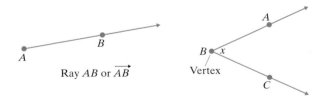

Ray AB or $\overrightarrow{AB}$ Vertex

The angle in the figure above can be named

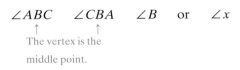

$$\angle ABC \qquad \angle CBA \qquad \angle B \qquad \text{or} \qquad \angle x$$

The vertex is the middle point.

Helpful Hint

Use the vertex alone to name an angle only when there is no confusion as to what angle is being named.

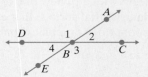

Name of ∠B is all right.
There is no confusion. ∠B means ∠1.

Name of ∠B is *not* all right.
There is confusion. Does ∠B mean ∠1, ∠2, ∠3, or ∠4?

Rays *BA* and *BC* are **sides** of the angle.

Practice Problem 1

Identify each figure as a line, a ray, a line segment, or an angle. Then name the figure using the given points.

a. **b.**

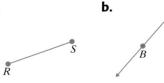

c. **d.**

Practice Problem 2

Use the figure in Example 2 to list other ways to name ∠z.

EXAMPLE 1

Identify each figure as a line, a ray, a line segment, or an angle. Then name the figure using the given points.

a. **b.**

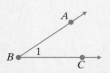

c. **d.**

Solution: Figure (a) extends indefinitely in two directions. It is line *CD* or $\overleftrightarrow{CD}$.
Figure (b) has two end points. It is line segment *EF* or $\overline{EF}$.
Figure (c) has two rays with a common end point. It is ∠*MNO*, ∠*ONM*, or ∠*N*.
Figure (d) is part of a line with one end point. It is ray *PT* or $\overrightarrow{PT}$. ●

EXAMPLE 2 List other ways to name ∠y.

Solution: Two other ways to name ∠y are ∠*QTR* and ∠*RTQ*. We may *not* use the vertex alone to name this angle because three different angles have *T* as their vertex.

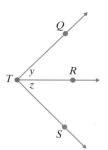

Answers

1. a. line segment; line segment *RS* or $\overline{RS}$,
b. ray; ray *AB* or $\overrightarrow{AB}$, **c.** line; line *EF* or $\overleftrightarrow{EF}$,
d. angle; ∠*TVH*, or ∠*HVT* or ∠*V*,
2. ∠*RTS*, ∠*STR*

B Classifying Angles as Acute, Right, Obtuse, or Straight

An angle can be measured in **degrees**. The symbol for degrees is a small, raised circle, °. There are 360° in a full revolution, or a full circle.

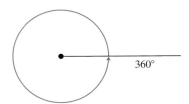

$\frac{1}{2}$ of a revolution measures $\frac{1}{2}(360°) = 180°$. An angle that measures 180° is called a **straight angle**.

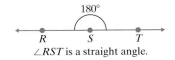

∠RST is a straight angle.

$\frac{1}{4}$ of a revolution measures $\frac{1}{4}(360°) = 90°$. An angle that measures 90° is called a **right angle**. The symbol ∟ is used to denote a right angle.

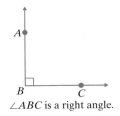

∠ABC is a right angle.

An angle whose measure is between 0° and 90° is called an **acute angle**.

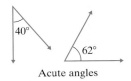

Acute angles

An angle whose measure is between 90° and 180° is called an **obtuse angle**.

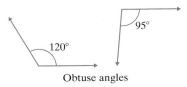

Obtuse angles

EXAMPLE 3   Classify each angle as acute, right, obtuse, or straight.

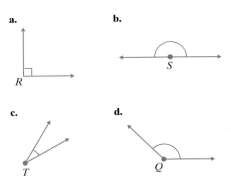

Practice Problem 3

Classify each angle as acute, right, obtuse, or straight.

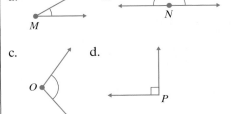

Answers

3. a. acute,   **b.** straight,   **c.** obtuse,   **d.** right

Solution:

a. ∠R is a right angle, denoted by ∟ .

b. ∠S is a straight angle.

c. ∠T is an acute angle. It measures between 0° and 90°.

d. ∠Q is an obtuse angle. It measures between 90° and 180°. ●

C **Identifying Complementary and Supplementary Angles**

Two angles that have a sum of 90° are called **complementary angles**. We say that each angle is the **complement** of the other.

∠R and ∠S are complementary angles because

60° + 30° = 90°

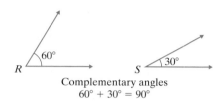

Complementary angles
60° + 30° = 90°

Two angles that have a sum of 180° are called **supplementary angles**. We say that each angle is the **supplement** of the other.

∠M and ∠N are supplementary angles because

125° + 55° = 180°

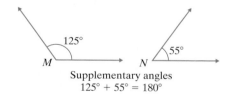

Supplementary angles
125° + 55° = 180°

Practice Problem 4

Find the complement of a 36° angle.

EXAMPLE 4   Find the complement of a 48° angle.

Solution: The complement of an angle that measures 48° is an angle that measures 90° − 48° = 42°. ●

Practice Problem 5

Find the supplement of an 88° angle.

Concept Check

True or false? The supplement of a 48° angle is 42°. Explain.

EXAMPLE 5   Find the supplement of a 107° angle.

Solution: The supplement of an angle that measures 107° is an angle that measures 180° − 107° = 73°. ●

Try the Concept Check in the margin.

Answers

4. 54°, **5.** 92°

Concept Check: false; the complement of a 48° angle is 42°; the supplement of a 48° angle is 132°

D **Finding Measures of Angles**

Measures of angles can be added or subtracted to find measures of related angles.

EXAMPLE 6   Find the measure of ∠x.

Solution: ∠x = 87° − 52° = 35°

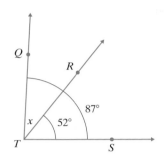

Practice Problem 6

Find the measure of ∠y.

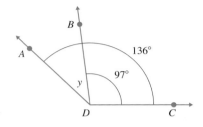

●

Two lines in a plane can be either parallel or intersecting. **Parallel lines** never meet. **Intersecting lines** meet at a point. The symbol ‖ is used to indicate "is parallel to." For example, in the figure $p\|q$.

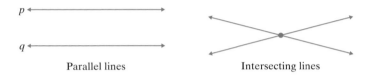

Parallel lines   Intersecting lines

Some intersecting lines are perpendicular. Two lines are **perpendicular** if they form right angles when they intersect. The symbol ⊥ is used to denote "is perpendicular to." For example, in the figure below, $n \perp m$.

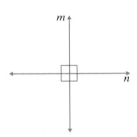

Perpendicular lines

When two lines intersect, four angles are formed. Two of these angles that are opposite each other are called **vertical angles**. Vertical angles have the same measure. Two angles that share a common side are called **adjacent angles**. Adjacent angles formed by intersecting lines are supplementary. That is, they have a sum of 180°.

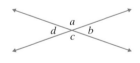

Vertical angles:
∠a and ∠c
∠d and ∠b

Adjacent angles:
∠a and ∠b
∠b and ∠c
∠c and ∠d
∠d and ∠a

Answer

6. 39°

Practice Problem 7

Find the measure of ∠a, ∠b, and ∠c.

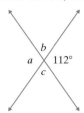

EXAMPLE 7

Find the measure of ∠x, ∠y, and ∠z if the measure of ∠t is 42°.

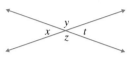

Solution: Since ∠t and ∠x are vertical angles, they have the same measure, so ∠x measures 42°.

Since ∠t and ∠y are adjacent angles, their measures have a sum of 180°. So ∠y measures 180° − 42° = 138°.

Since ∠y and ∠z are vertical angles, they have the same measure. So ∠z measures 138°. ●

A line that intersects two or more lines at different points is called a **transversal**. Line *l* is a transversal that intersects lines *m* and *n*. The eight angles formed have special names. Some of these names are:

Corresponding angles: ∠a and ∠e,
   ∠c and ∠g, ∠b and ∠f, ∠d and ∠h
Alternate interior angles: ∠c and ∠f,
   ∠d and ∠e

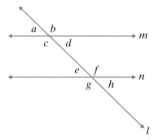

When two lines cut by a transversal are *parallel*, the following are true:

Parallel Lines Cut by a Transversal

If two parallel lines are cut by a transversal, then the measures of **corresponding angles are equal** and **alternate interior angles are equal**.

Practice Problem 8

Given that *m*‖*n* and that the measure of ∠w = 40°, find the measures of ∠x, ∠y, and ∠z.

EXAMPLE 8

Given that *m*‖*n* and that the measure of ∠w is 100°, find the measures of ∠x, ∠y, and ∠z.

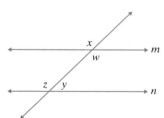

Solution:

The measure of ∠x = 100°. ∠x and ∠w are vertical angles.

The measure of ∠z = 100°. ∠x and ∠z are corresponding angles.

The measure of ∠y = 180° − 100° = 80°. ∠z and ∠y are supplementary angles. ●

Answers

7. ∠a = 112°; ∠b = 68°; ∠c = 68°,
8. ∠x = 40°; ∠y = 40°; ∠z = 140°

EXERCISE SET 8.1

 A *Identify each figure as a line, a ray, a line segment, or an angle. Then name the figure using the given points. See Example 1.*

1.

2.

3.

4.

5.

6.

7.

8.

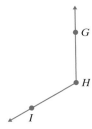

B *Find the measure of each angle in the following figure:*

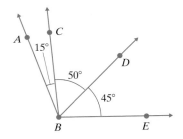

 9. ∠ABC

10. ∠EBD

 11. ∠CBD

12. ∠CBA

13. ∠DBA

14. ∠EBC

15. ∠CBE

16. ∠ABE

Fill in each blank. See Example 3.

17. A right angle has a measure of _____ .

18. A straight angle has a measure of _____ .

19. An acute angle measures between _____ and _____ .

20. An obtuse angle measures between _____ and _____ .

Classify each angle as acute, right, obtuse, or straight. See Example 3.

21.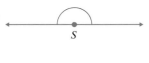

22.

23.

24.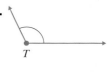

25.

26.

27.

28.

 C *Find each complementary or supplementary angle as indicated. See Examples 4 and 5.*

29. Find the complement of a 17° angle.

30. Find the complement of an 87° angle.

31. Find the supplement of a 17° angle.

32. Find the supplement of an 87° angle.

33. Find the complement of a 48° angle.

34. Find the complement of a 22° angle.

35. Find the supplement of a 125° angle.

36. Find the supplement of a 155° angle.

37. Identify the pairs of complementary angles.

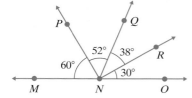

38. Identify the pairs of complementary angles.

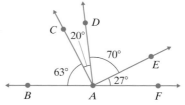

39. Identify the pairs of supplementary angles.

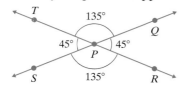

40. Identify the pairs of supplementary angles.

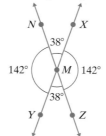

D *Find the measure of ∠x in each figure. See Example 6.*

41.

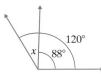

120°

x 88°

42.

35° 82°
x

43.

x
15°

44.

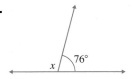

x 76°

Find the measures of angles x, y, and z in each figure. See Examples 7 and 8.

45.

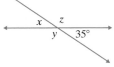

x z
y 35°

46.

x 75°
y z

47.

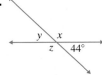

x 103°
y z

48.

y x
z 44°

49. *m∥n*

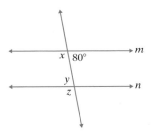

x 80°
m
y n
z

50. *m∥n*

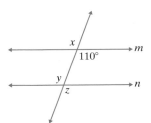

x 110°
m
y n
z

51. *m∥n*

x 46°
m
z y n

52. *m∥n*

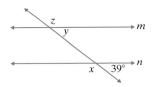

z m
y
x 39° n

Review and Preview

Perform each indicated operation. See Sections 2.4, 2.5, 3.3, and 3.4.

53. $\dfrac{7}{8} + \dfrac{1}{4}$

54. $\dfrac{7}{8} - \dfrac{1}{4}$

55. $\dfrac{7}{8} \cdot \dfrac{1}{4}$

56. $\dfrac{7}{8} \div \dfrac{1}{4}$

57. $3\dfrac{1}{3} - 2\dfrac{1}{2}$

58. $3\dfrac{1}{3} + 2\dfrac{1}{2}$

59. $3\dfrac{1}{3} \div 2\dfrac{1}{2}$

60. $3\dfrac{1}{3} \cdot 2\dfrac{1}{2}$

Combining Concepts

61. The angle between the two walls of the Vietnam Veterans Memorial in Washington, D.C., is 125.2°. Find the supplement of this angle. (*Source*: National Park Service)

62. The faces of Khafre's Pyramid at Giza, Egypt, are inclined at an angle of 53.13°. Find the complement of this angle. (*Source*: PBS *NOVA* Online)

63. If lines *m* and *n* are parallel, find the measures of angles *a* through *e*.

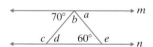

64. In your own words, describe how to find the complement and the supplement of a given angle.

△ 8.2 Plane Figures and Solids

In order to prepare for the sections ahead in this chapter, we first review plane figures and solids.

A Identifying Plane Figures

Recall from Section 8.1 that a **plane** is a flat surface that extends indefinitely.

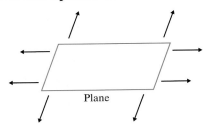

Plane

A **plane figure** is a figure that lies on a plane. Plane figures, like planes, have length and width but no thickness or depth.

A **polygon** is a closed plane figure that basically consists of three or more line segments that meet at their end points.

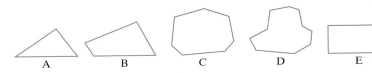

A **regular polygon** is one whose sides are all the same length and whose angles are the same measure.

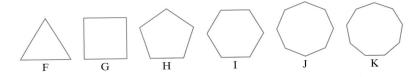

A polygon is named according to the number of its sides.

Some triangles and quadrilaterals are given special names, so let's study these special polygons further. We begin with triangles. The sum of the measures of the angles of a triangle is 180°.

EXAMPLE 1   Find the measure of ∠a.

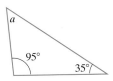

Solution: Since the sum of the measures of the three angles is 180°, we have

measure of ∠a = 180° − 95° − 35° = 50°

To check, see that 95° + 35° + 50° = 180°.

●

We can classify triangles according to the lengths of their sides. (We will use tick marks to denote the sides and angles of a figure that are equal.)

OBJECTIVES

A Identify plane figures.
B Identify solid figures.

SSM TUTOR CENTER   SG   CD & VIDEO   MATH PRO   WEB

Number of Sides	Name	Figure Examples
3	Triangle	A, F
4	Quadrilateral	B, E, G
5	Pentagon	H
6	Hexagon	I
7	Heptagon	C
8	Octagon	J
9	Nonagon	K
10	Decagon	D

Practice Problem 1

Find the measure of ∠x.

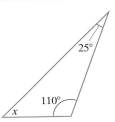

Answer

1. 45°

Equilateral triangle	Isoceles triangle	Scalene triangle
All three sides are the same length. Also, all three angles have the same measure.	Two sides are the same length. Also, the angles opposite the equal sides have equal measures.	No sides are the same length. No angles have the same measure.

One other important type of triangle is a right triangle. A **right triangle** is a triangle with a right angle. The side opposite the right angle is called the **hypotenuse**, and the other two sides are called **legs**.

Practice Problem 2

Find the measure of ∠ y.

EXAMPLE 2   Find the measure of ∠ b.

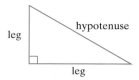

Solution: We know that the measure of the right angle, ⌐, is 90°. Since the sum of the measures of the angles is 180°, we have

measure of ∠b = 180° − 90° − 30° = 60°   ●

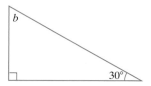

Helpful Hint

From the previous example, can you see that in a right triangle, the sum of the other two acute angles is 90°? This is because

$$90° \quad + \quad 90° \quad = \quad 180°$$

| right angle's measure | sum of other two angles' measures | sum of angles' measures |

Now we review some special quadrilaterals. A **parallelogram** is a special quadrilateral with opposite sides parallel and equal in length.

A **rectangle** is a special **parallelogram** that has four right angles.

A **square** is a special **rectangle** that has all four sides equal in length.

A **rhombus** is a special **parallelogram** that has all four sides equal in length.

A **trapezoid** is a quadrilateral with exactly one pair of opposite sides parallel.

Try the Concept Check in the margin.

In addition to triangles and quadrilaterals, circles are common plane figures. A **circle** is a plane figure that consists of all points that are the same fixed distance from a point c. The point c is called the **center** of the circle. The **radius** of a circle is the distance from the center of the circle to any point on the circle. The **diameter** of a circle is the distance across the circle passing through the center. The diameter is twice the radius, and the radius is half the diameter.

$$\text{diameter} = 2 \cdot \text{radius} \qquad \text{radius} = \frac{\text{diameter}}{2}$$

$$d = 2 \cdot r \qquad r = \frac{d}{2}$$

EXAMPLE 3   Find the diameter of the circle.

Solution: The diameter is twice the radius.

$$d = 2 \cdot r$$
$$d = 2 \cdot 5 \text{ cm} = 10 \text{ cm}$$

The diameter is 10 centimeters.

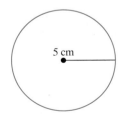

5 cm

B   Identifying Solid Figures

Recall from Section 8.1 that space extends in all directions indefinitely.
   A **solid** is a figure that lies in space. Solids have length, width, and height or depth.

Concept Check

True or false? All quadrilaterals are parallelograms. Explain.

Practice Problem 3

Find the radius of the circle.

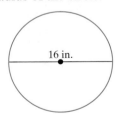

16 in.

Answers

8. 8 in.

Concept Check: false

A **rectangular solid** is a solid that consists of six sides, or faces, all of which are rectangles.

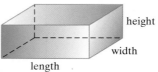

A **cube** is a rectangular solid whose six sides are squares.

A **pyramid** is shown below.

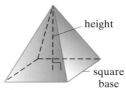

A **sphere** consists of all points in space that are the same distance from a point *c*. The point *c* is called the **center** of the sphere. The **radius** of a sphere is the distance from the center to any point on the sphere. The **diameter** of a sphere is the distance across the sphere passing through the center.

The radius and diameter of a sphere are related in the same way as the radius and diameter of a circle.

$$d = 2 \cdot r \quad \text{or} \quad r = \frac{d}{2}$$

Practice Problem 4

Find the diameter of the sphere.

7 mi

EXAMPLE 4   Find the radius of the sphere.

Solution: The radius is half the diameter.

$$r = \frac{d}{2}$$

$$r = \frac{36 \text{ feet}}{2} = 18 \text{ feet}$$

The radius is 18 feet.

The **cylinders** we will study have bases that are in the shape of circles and are perpendicular to their height.

The **cones** we will study have bases that are circles and are perpendicular to their height.

Answer

4. 14 mi

EXERCISE SET 8.2

A *Classify each triangle as equilateral, isosceles, or scalene.*

1.

2.

3.

4.

5.

6.

Find the measure of ∠x in each figure. See Examples 1 and 2.

 7.

8.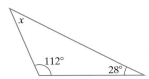

9.

10.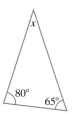

 11.

12.

Fill in each blank.

13. Twice the radius of a circle is its
_____ .

14. A rectangle with all four sides equal is
a _____ .

15. A parallelogram with four right angles is a _____.

16. Half the diameter of a circle is its _____.

17. A quadrilateral with opposite sides parallel is a _____.

18. A quadrilateral with exactly one pair of opposite sides parallel is a _____.

19. The side opposite the right angle of a right triangle is called the _____.

20. A triangle with no equal sides is a _____.

Determine whether each statement is true or false.

21. A square is also a rhombus.

22. A square is also a rectangle.

 23. A rectangle is also a parallelogram.

24. A trapezoid is also a parallelogram.

25. A pentagon is also a quadrilateral.

26. A rhombus is also a parallelogram.

Find the unknown diameter or radius in each figure. See Example 3.

27.
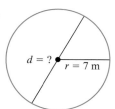

28.
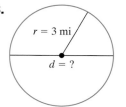

29.
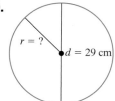

30.
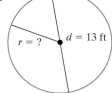

Identify each regular polygon.

31.

32.

33.

34.

B *Identify each solid.*

 35.

36.

 37.

38.

39.

40.

Find each unknown radius or diameter. See Example 4.

41. The radius of a sphere is 7.4 inches. Find its diameter.

42. The radius of a sphere is 5.8 meters. Find its diameter.

43. The diameter of a sphere is 26 miles. Find its radius.

44. The diameter of a sphere is 78 centimeters. Find its radius.

Identify the shape of each item.

45.

46.

47.

48.

49. **50.**

51.

52.

Review and Preview

Perform each indicated operation. See Sections 4.3 and 4.4.

53. 4(28.6) **54.** 2(7.8) + 2(9.6) **55.** 2(18) + 2(36) **56.** 4(87)

 Combining Concepts

57. Saturn has a radius of approximately 36,184 miles. What is its diameter?

58. Is an isosceles right triangle possible? If so, draw one.

59. The following demonstration is credited to the mathematician Pascal, who is said to have developed it as a young boy.

   Cut a triangle from a piece of paper. The length of the sides and the size of the angles is unimportant. Tear the points off the triangle.

Place the points of the triangle together. Notice that a straight line is formed. What was Pascal trying to show?

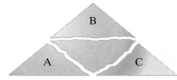

60. In your own words, explain whether a square is also a rhombus.

△ **8.3 Perimeter**

Ⓐ **Using Formulas to Find Perimeters**

Recall from Section 1.3 that the perimeter of a polygon is the distance around the polygon. This means that the perimeter of a polygon is the sum of the lengths of its sides.

EXAMPLE 1   Find the perimeter of the rectangle below.

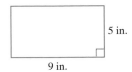

5 in.

9 in.

Solution:

$$\text{perimeter} = 9 \text{ inches} + 9 \text{ inches} + 5 \text{ inches} + 5 \text{ inches}$$
$$= 28 \text{ inches}$$

Notice that the perimeter of the rectangle in Example 1 can be written as $2 \cdot (9 \text{ inches}) + 2 \cdot (5 \text{ inches})$.

↑ length   ↑ width

In general, we can say that the perimeter of a rectangle is always

$$2 \cdot \text{length} + 2 \cdot \text{width}$$

As we have just seen, the perimeter of some special figures such as rectangles form patterns. These patterns are given as **formulas**. The formula for the perimeter of a rectangle is shown next:

Perimeter of a Rectangle

$$\text{perimeter} = 2 \cdot \text{length} + 2 \cdot \text{width}$$

In symbols, this can be written as

$$P = 2 \cdot l + 2 \cdot w$$

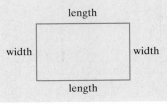

length

width   width

length

EXAMPLE 2

Find the perimeter of a rectangle with a length of 11 inches and a width of 3 inches.

11 in.

3 in.

Solution: We use the formula for perimeter and replace the letters by their known lengths.

$$P = 2 \cdot l + 2 \cdot w$$
$$= 2 \cdot 11 \text{ in.} + 2 \cdot 3 \text{ in.} \quad \text{Replace } l \text{ with 11 in. and } w \text{ with 3 in.}$$
$$= 22 \text{ in.} + 6 \text{ in.}$$
$$= 28 \text{ in.}$$

The perimeter is 28 inches.

O B J E C T I V E S

Ⓐ Use formulas to find perimeter.

Ⓑ Use formulas to find circumferences

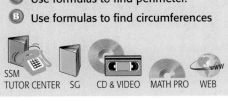

SSM   TUTOR CENTER   SG   CD & VIDEO   MATH PRO   WEB

Practice Problem 1

Find the perimeter of the rectangular lot shown below:

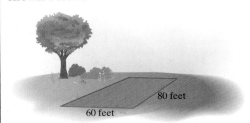

80 feet

60 feet

Practice Problem 2

Find the perimeter of a rectangle with a length of 22 centimeters and a width of 10 centimeters.

Answers

1. 280 ft,   **2.** 64 cm

Recall that a square is a special rectangle with all four sides the same length. The formula for the perimeter of a square is shown next:

Perimeter of a Square

$$\text{Perimeter} = \text{side} + \text{side} + \text{side} + \text{side}$$
$$= 4 \cdot \text{side}$$

In symbols,

$$P = 4 \cdot s$$

side

side   side

side

Practice Problem 3

How much fencing is needed to enclose a square field 50 yards on a side?

50 yards

EXAMPLE 3 Finding the Perimeter of a Tabletop

Find the perimeter of a square tabletop if each side is 5 feet long.

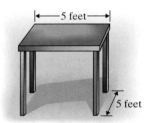

5 feet

5 feet

Solution: The formula for the perimeter of a square is $P = 4 \cdot s$. We use this formula and replace s by 5 feet.

$$P = 4 \cdot s$$
$$= 4 \cdot 5 \text{ feet}$$
$$= 20 \text{ feet}$$

The perimeter of the square tabletop is 20 feet.   ●

The formula for the perimeter of a triangle with sides of lengths $a, b,$ and c is given next:

Perimeter of a Triangle

$$\text{Perimeter} = \text{side } a + \text{side } b + \text{side } c$$

In symbols,

$$P = a + b + c$$

side a side b

side c

Answer

3. 200 yd

EXAMPLE 4

Find the perimeter of a triangle when the sides are 3 inches, 7 inches, and 6 inches.

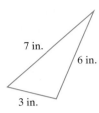

Solution: The formula is $P = a + b + c$, where $a, b,$ and c are the lengths of the sides. Thus,

$$P = a + b + c$$
$$= 3 \text{ in.} + 7 \text{ in.} + 6 \text{ in.}$$
$$= 16 \text{ in.}$$

The perimeter of the triangle is 16 inches.

Recall that to find the perimeter of other polygons, we find the sum of the lengths of their sides.

EXAMPLE 5   Find the perimeter of the trapezoid shown below:

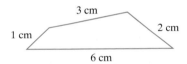

Solution: To find the perimeter, we find the sum of the lengths of its sides.

perimeter $= 3 \text{ cm} + 2 \text{ cm} + 6 \text{ cm} + 1 \text{ cm} = 12 \text{ cm}$

The perimeter is 12 centimeters.

EXAMPLE 6   Finding the Perimeter of a Room

Find the perimeter of the room shown below:

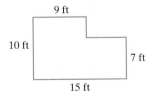

Solution: To find the perimeter of the room, we first need to find the lengths of all sides of the room.

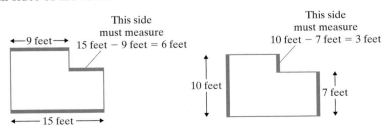

Practice Problem 4

Find the perimeter of a triangle when the sides are 5 centimeters, 9 centimeters, and 7 centimeters in length.

Practice Problem 5

Find the perimeter of the trapezoid shown.

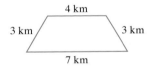

Practice Problem 6

Find the perimeter of the room shown.

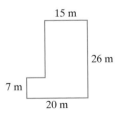

Answers

4. 21 cm,   **5.** 17 km,   **6.** 92 m

Now that we know the measures of all sides of the room, we can add the measures to find the perimeter.

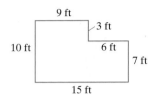

$$\text{perimeter} = 10\,\text{ft} + 9\,\text{ft} + 3\,\text{ft} + 6\,\text{ft} + 7\,\text{ft} + 15\,\text{ft}$$
$$= 50\,\text{ft}$$

The perimeter of the room is 50 feet.   ●

Practice Problem 7

A rectangular lot measures 60 feet by 120 feet. Find the cost to install fencing around the lot if the cost of fencing is $1.90 per foot.

EXAMPLE 7   Calculating the Cost of Baseboard

A rectangular room measures 10 feet by 12 feet. Find the cost to install new baseboard around the room if the cost of the baseboard is $0.66 per foot.

Solution: First we find the perimeter of the room.

$$P = 2 \cdot l + 2 \cdot w$$
$$= 2 \cdot 12 \text{ feet} + 2 \cdot 10 \text{ feet} \qquad \text{Replace } l \text{ with 12 feet and } w \text{ with 10 feet.}$$
$$= 24 \text{ feet} + 20 \text{ feet}$$
$$= 44 \text{ feet}$$

The cost of the baseboard is

$$\text{cost} = 0.66 \cdot 44 = 29.04$$

The cost of the baseboard is $29.04.   ●

B   Using Formulas to Find Circumferences

Recall from Section 4.4 that the distance around a circle is called the **circumference**. This distance depends on the radius or the diameter of the circle.
    The formulas for circumference are shown next:

Circumference of a Circle

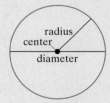

Circumference $= 2 \cdot \pi \cdot$ radius or Circumference $= \pi \cdot$ diameter
In symbols,

$$C = 2 \cdot \pi \cdot r \quad \text{or} \quad C = \pi \cdot d,$$

where $\pi \approx 3.14$ or $\pi \approx \dfrac{22}{7}$.

Answer

7. $684

To better understand circumference and π (pi), try the following experiment. Take any can and measure its circumference and its diameter.

The can in the figure above has a circumference of 23.5 centimeters and a diameter of 7.5 centimeters. Now divide the circumference by the diameter.

$$\frac{\text{circumference}}{\text{diameter}} = \frac{23.5 \text{ cm}}{7.5 \text{ cm}} \approx 3.1$$

Try this with other sizes of cylinders and circles—you should always get a number close to 3.1. The exact ratio of circumference to diameter is π. (Recall that $\pi \approx 3.14$ or $\approx \frac{22}{7}$).

EXAMPLE 8

Mary Catherine Dooley plans to install a border of new tiling around the circumference of her circular spa. If her spa has a diameter of 14 feet, find its circumference.

Solution: Because we are given the diameter, we use the formula $C = \pi \cdot d$.

$$C = \pi \cdot d$$
$$= \pi \cdot 14 \text{ ft} \qquad \text{Replace } d \text{ with 14 feet.}$$
$$= 14\pi \text{ ft}$$

The circumference of the spa is *exactly* 14π feet. By replacing π with the *approximation* 3.14, we find that the circumference is *approximately* 14 feet $\cdot$ 3.14 = 43.96 feet.

Try the Concept Check in the margin.

Practice Problem 8

An irrigation device waters a circular region with a diameter of 20 yards. What is the circumference of the watered region?

Concept Check

The distance around which figure is greater: a square with side length 5 inches or a circle with radius 3 inches?

Answers

8. 62.8 yd

Concept Check: a square with length 5 in.

STUDY SKILLS REMINDER

How are you doing?

If you haven't done so yet, take a few moments and think about how you are doing in this course. Are you working toward your goal of successfully completing this course? Is your performance on homework, quizzes, and tests satisfactory? If not, you might want to see your instructor to see if he/she has any suggestions on how you can improve your performance. Let me once again remind you that, in addition to your instructor, there are many places to get help with your mathematics course. A few suggestions are below.

- This text has an accompanying video lesson for every text section.
- The back of this book contains answers to odd-numbered exercises and selected solutions.
- MathPro is available with this text. It is a tutorial software program with lessons corresponding to each text section.
- There is a student solutions manual available that contains worked-out solutions to odd-numbered exercises as well as solutions to every exercise in the Chapter Pretests, Integrated Reviews, Chapter Reviews, Chapter Tests, and Cumulative Reviews.
- Don't forget to check with your instructor for other local resources available to you, such as a tutor center.

Name _____ Section _____ Date _____

EXERCISE SET 8.3

A *Find the perimeter of each figure. See Examples 1 through 6.*

 1.
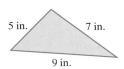
15 ft   Rectangle
17 ft

2.
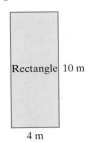
Rectangle   10 m
4 m

3.

Square
9 cm

4.

Square
46 mi

 5.
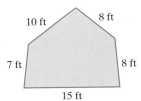
5 in.   7 in.
9 in.

6.
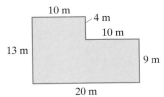
4 units   10 units
9 units

7.
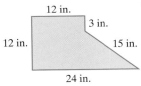
Parallelogram   25 cm
35 cm

8.
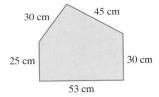
Parallelogram
3 yd
2 yd

9.
10 ft   8 ft
7 ft   8 ft
15 ft

10.
10 m   4 m
10 m
13 m   9 m
20 m

11.
12 in.   3 in.
12 in.   15 in.
24 in.

12.
30 cm   45 cm
25 cm   30 cm
53 cm

Solve. See Example 7.

13. A polygon has sides of length 5 feet, 3 feet, 2 feet, 7 feet, and 4 feet. Find its perimeter.

14. A triangle has sides of length 8 inches, 12 inches, and 10 inches. Find its perimeter.

15. Baseboard is to be installed in a square room that measures 15 feet on one side. Find how much baseboard is needed.

16. Find how much fencing is needed to enclose a rectangular rose garden 85 feet by 15 feet.

17. If a football field is 53 yards wide and 120 yards long, what is the perimeter?

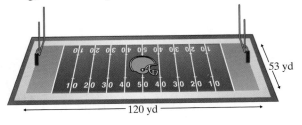
53 yd
120 yd

18. A stop sign has eight equal sides of length 12 inches. Find its perimeter.

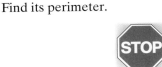

19. A metal strip is being installed around a workbench that is 8 feet long and 3 feet wide. Find how much stripping is needed.

20. Find how much fencing is needed to enclose a rectangular garden 70 feet by 21 feet.

21. If the stripping in Exercise 19 costs $3 per foot, find the total cost of the stripping.

22. If the fencing in Exercise 20 costs $2 per foot, find the total cost of the fencing.

23. A regular hexagon has a side length of 6 inches. Find its perimeter.

24. A regular pentagon has a side length of 14 meters. Find its perimeter.

25. Find the perimeter of the top of a square compact disc case if the length of one side is 7 inches.

26. Find the perimeter of a square ceramic tile with a side of length 5 inches.

27. A rectangular room measures 6 feet by 8 feet. Find the cost of installing a strip of wallpaper around the room if the wallpaper costs $0.86 per foot.

28. A rectangular house measures 75 feet by 60 feet. Find the cost of installing gutters around the house if the cost is $2.36 per foot.

Find the perimeter of each figure. See Example 6.

29.

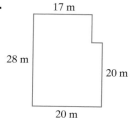

30.

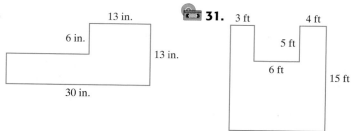

31.

32.

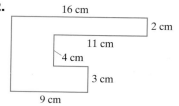

33.

34.

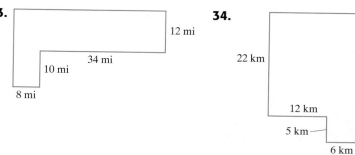

494

B *Find the circumference of each circle. Give the exact circumference and then an approximation. Use π ≈ 3.14. See Example 8.*

35.

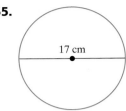

17 cm

36.

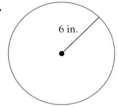

6 in.

37.

8 mi

38.

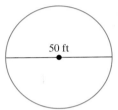

50 ft

39.

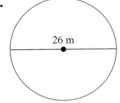

26 m

40.

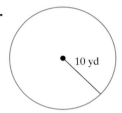

10 yd

41. A circular fountain has a radius of 5 feet. Approximate the distance around the fountain. Use $\frac{22}{7}$ for π.

42. A circular walkway has a radius of 40 meters. Approximate the distance around the walkway. Use 3.14 for π.

43. Meteor Crater, near Winslow, Arizona, is 4000 feet in diameter. Approximate the distance around the crater. Use 3.14 for π. (*Source: The Handy Science Answer Book*)

44. The largest pearl, the *Pearl of Lao-tze*, has a diameter of $5\frac{1}{2}$ inches. Approximate the distance around the pearl. Use $\frac{22}{7}$ for π. (*Source: The Guinness Book of Records*)

Review and Preview

Simplify. See Section 1.9.

45. $5 + 6 \cdot 3$

46. $25 - 3 \cdot 7$

47. $(20 - 16) \div 4$

48. $6 \cdot (8 + 2)$

49. $(18 + 8) - (12 + 4)$

50. $72 \div (2 \cdot 6)$

51. $(72 \div 2) \cdot 6$

52. $4^1 \cdot (2^3 - 8)$

Recall from Section 1.6 that area measures the amount of surface of a region. Given the following situations, tell whether you are more likely to be concerned with area or perimeter.

53. ordering fencing to fence a yard

54. ordering grass seed to plant in a yard

55. buying carpet to install in a room

56. buying gutters to install on a house

57. ordering paint to paint a wall

58. ordering baseboards to install in a room

59. buying a wallpaper border to go on the walls around a room

60. buying fertilizer for your yard

61. a. Find the circumference of each circle. Approximate the circumference by using 3.14 for π.

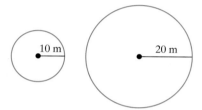

b. If the radius of a circle is doubled, is its corresponding circumference doubled?

62. a. Find the circumference of each circle. Approximate the circumference by using 3.14 for π.

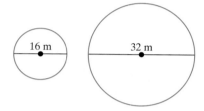

b. If the diameter of a circle is doubled, is its corresponding circumference doubled?

63. Find the perimeter of the skating rink.

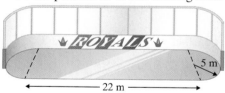

64. In your own words, explain how to find the perimeter of any polygon.

8.4 Area

 Finding Area of Geometric Figures

Recall that area measures the amount of surface of a region. Thus far, we know how to find the area of a rectangle and a square. These formulas, as well as formulas for finding the areas of other common geometic figures, are given next:

Area Formulas of Common Geometric Figures

Geometric Figure	Area Formula
RECTANGLE 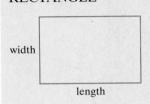	Area of a rectangle: **A**rea = **l**ength · **w**idth $A = lw$
SQUARE 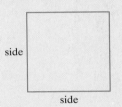	Area of a square: **A**rea = **s**ide · **s**ide $A = s \cdot s = s^2$
TRIANGLE 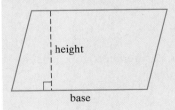	Area of a triangle: **A**rea = $\dfrac{1}{2}$ · **b**ase · **h**eight $A = \dfrac{1}{2} \cdot b \cdot h$
PARALLELOGRAM 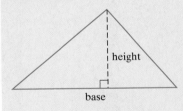	Area of a parallelogram: **A**rea = **b**ase · **h**eight $A = b \cdot h$
TRAPEZOID 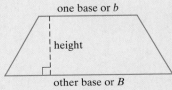	Area of a trapezoid: **area** = $\dfrac{1}{2}$ · (one **b**ase + other **B**ase) · **height** $A = \dfrac{1}{2} \cdot (b + B) \cdot h$

Use these formulas for the following examples.

> **Helpful Hint**
> Area is always measured in square units.

Practice Problem 1

Find the area of the triangle.

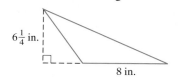

EXAMPLE 1 Find the area of the triangle.

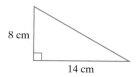

Solution:

$$A = \frac{1}{2} \cdot b \cdot h$$

$$= \frac{1}{2} \cdot 14\,cm \cdot 8\,cm$$

$$= \frac{\overset{1}{2} \cdot 7 \cdot 8}{\underset{1}{2}}\ \text{square centimeters}$$

$$= 56 \text{ square centimeters}$$

The area is 56 square centimeters.

Practice Problem 2

Find the area of the square.

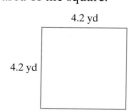

EXAMPLE 2 Find the area of the parallelogram.

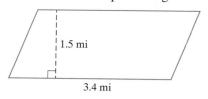

Solution:

$$A = b \cdot h$$
$$= 3.4 \text{ miles} \cdot 1.5 \text{ miles}$$
$$= 5.1 \text{ square miles}$$

The area is 5.1 square miles.

> **Helpful Hint**
> When finding the area of figures, be sure all measurements are changed to the same unit before calculations are made.

Practice Problem 3

Find the area of the figure.

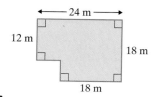

EXAMPLE 3 Find the area of the figure.

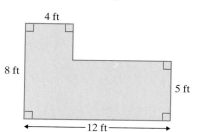

Answers

1. 25 sq in., **2.** 17.64 sq yd, **3.** 396 sq m

Solution: Split the figure into two rectangles. To find the area of the figure, we find the sum of the areas of the two rectangles.

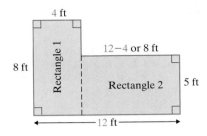

Area of Rectangle 1 $= l \cdot w$

$\qquad = 8 \, \text{feet} \cdot 4 \, \text{feet}$

$\qquad = 32 \, \text{square feet}$

Notice that the length of Rectangle 2 is 12 feet $-$ 4 feet, or 8 feet.

Area of Rectangle 2 $= l \cdot w$

$\qquad = 8 \, \text{feet} \cdot 5 \, \text{feet}$

$\qquad = 40 \, \text{square feet}$

Area of the Figure $=$ Area of Rectangle 1 $+$ Area of Rectangle 2

$\qquad = 32 \, \text{square feet} + 40 \, \text{square feet}$

$\qquad = 72 \, \text{square feet}$

⬤

Helpful Hint

The figure in Example 3 can also be split into two rectangles as shown:

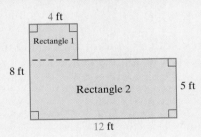

To better understand the formula for area of a circle, try the following. Cut a circle into many pieces as shown:

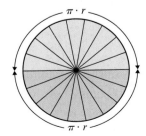

The circumference of a circle is $2 \cdot \pi \cdot r$. This means that the circumference of half a circle is half of $2 \cdot \pi \cdot r$, or $\pi \cdot r$.

Then unfold the two halves of the circle and place them together as shown:

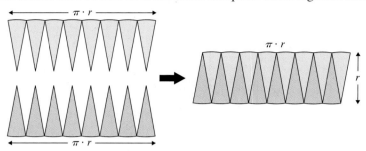

The figure on the right is almost a parallelogram with a base of $\pi \cdot r$ and a height of r. The area is

$$A = \boxed{\text{base}} \cdot \boxed{\text{height}}$$
$$ = (\pi \cdot r) \cdot \quad r$$
$$ = \pi \cdot r^2$$

This is the formula for area of a circle.

Area Formula of a Circle

CIRCLE

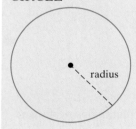

Area of a circle:

Area $= \pi \cdot (\text{radius})^2$

$A = \pi \cdot r^2$

(A fraction approximation for π is $\dfrac{22}{7}$.)

(A decimal approximation for π is 3.14.)

Practice Problem 4

Find the area of the given circle. Find the exact area and an approximation. Use 3.14 as an approximation for π.

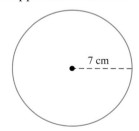

7 cm

Concept Check

Use estimation to decide which figure would have a larger area: a circle of diameter 10 inches or a square 10 inches long on each side.

Answers

4. 49π sq cm ≈ 153.86 sq cm

Concept Check: a square 10 in. long on each side

EXAMPLE 4

Find the area of a circle with a radius of 3 feet. Find the exact area and an approximation. Use 3.14 as an approximation for π.

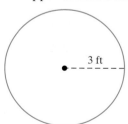

3 ft

Solution: We let $r = 3$ feet and use the formula.

$$A = \pi \cdot r^2$$
$$ = \pi \cdot (3 \text{ feet})^2$$
$$ = 9 \cdot \pi \text{ square feet}$$

To approximate this area, we substitute 3.14 for π.

$$9 \cdot \pi \text{ square feet} \approx 9 \cdot 3.14 \text{ square feet}$$
$$= 28.26 \text{ square feet}$$

The *exact* area of the circle is 9π square feet, which is *approximately* 28.26 square feet. ●

Try the Concept Check in the margin.

Name _____ Section _____ Date _____

EXERCISE SET 8.4

 A *Find the area of the geometric figure. If the figure is a circle, give an exact area and then use the given* ***approximation*** *for* π *to approximate the area. See Examples 1 through 4.*

 1.

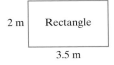

2 m | Rectangle
3.5 m

2.

2.75 ft | Rectangle
7 ft

 **3.**

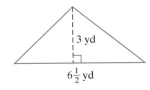

3 yd

$6\frac{1}{2}$ yd

4.

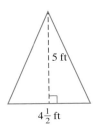

5 ft

$4\frac{1}{2}$ ft

5.

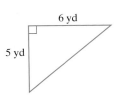

6 yd

5 yd

6.

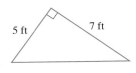

5 ft   7 ft

7. Use 3.14 for π.

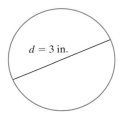

d = 3 in.

8. Use $\frac{22}{7}$ for π.

r = 2 cm

9.

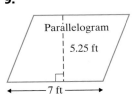

Parallelogram
5.25 ft
◄――― 7 ft ―――►

10.

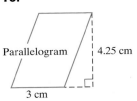

Parallelogram   4.25 cm

3 cm

11.

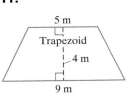

5 m
Trapezoid
4 m
9 m

12.

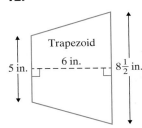

Trapezoid
5 in.   6 in. $8\frac{1}{2}$ in.

13.

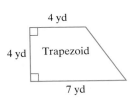

4 yd
4 yd | Trapezoid
7 yd

14.

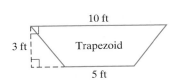

10 ft
3 ft | Trapezoid
5 ft

15.

10 ft

7 ft
Parallelogram
$5\frac{1}{4}$ ft

16.

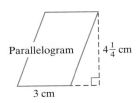

Parallelogram $4\frac{1}{4}$ cm

3 cm

501

17.

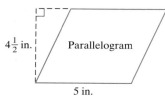

4½ in. Parallelogram

5 in.

18.

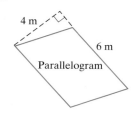

4 m

6 m

Parallelogram

19.

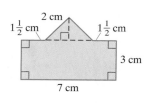

2 cm

1½ cm   1½ cm

3 cm

7 cm

20.

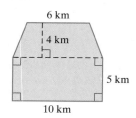

6 km

4 km

5 km

10 km

21.

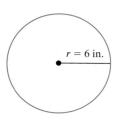

5 mi

10 mi

3 mi

17 mi

22.

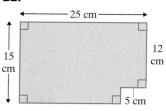

25 cm

15 cm

12 cm

5 cm

23.

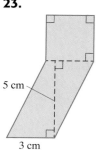

5 cm

3 cm

24.

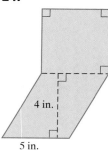

4 in.

5 in.

25. Use $\dfrac{22}{7}$ for π.

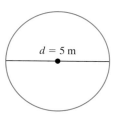

$r = 6$ in.

26. Use 3.14 for π.

$d = 5$ m

Solve. See Examples 1 through 4.

27. A $10\dfrac{1}{2}$-foot by 16-foot concrete wall is to be built using concrete blocks. Find the area of the wall.

28. The floor of Terry's attic is 24 feet by 35 feet. Find how many square feet of insulation are needed to cover the attic floor.

29. The world's largest flag measures 505 feet by 225 feet. It's the U.S. "Super flag" owned by "Ski" Demski of Long Beach, California. Find its area. (*Source: Guinness World Records*, 2001)

505 feet

225 feet

30. The longest illuminated sign is in Ramat Gan, Israel, and measures 197 feet by 66 feet. Find its area. (*Source: The Guinness Book of Records*)

66 ft

197 ft

502

31. One side of a concrete block measures 8 inches by 16 inches. Find the area of the side in square inches. Find the area in square feet (144 sq in. = 1 sq ft).

32. A standard *double* roll of wallpaper is $6\frac{5}{6}$ feet wide and 33 feet long. Find the area of the *double* roll.

33. A picture frame measures 20 inches by $25\frac{1}{2}$ inches. Find how many square inches of glass the frame requires.

34. A mat to go under a tablecloth is made to fit a round dining table with a 4-foot diameter. Approximate how many square feet of mat there are. Use 3.14 as an approximation for π.

35. A drapery panel measures 6 feet by 7 feet. Find how many square feet of material are needed for *four* panels.

36. A page in this book measures 27.5 centimeters by 20.5 centimeters. Find its area.

37. Find how many square feet of land are in the following plot:

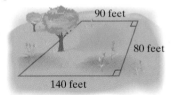

38. For Gerald Gomez to determine how much grass seed he needs to buy, he must know the size of his yard. Use the drawing to determine how many square feet are in his yard.

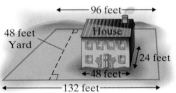

39. The shaded part of the roof shown is in the shape of a trapezoid and needs to be shingled. The number of packages of shingles to buy depends on the area. Use the dimensions given to find the area of the shaded part of the roof to the nearest whole square foot.

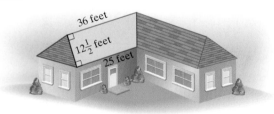

40. The end of the building shaded in the drawing is to be bricked. The number of bricks to buy depends on the area.
a. Find the area.

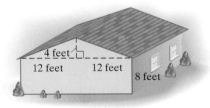

b. If the side area of each brick (including mortar room) is $\frac{1}{6}$ square feet, find the number of bricks needed to buy.

Review and Preview

Find the perimeter or circumference of each geometric figure. See Section 8.3.

41. Use 3.14 for π.

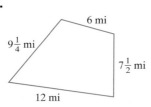

14 in.

42.

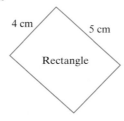

4 cm   5 cm

Rectangle

43.

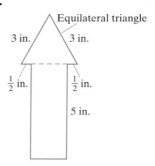

3 ft $3\frac{1}{2}$ ft

4 ft

6 ft

$8\frac{1}{2}$ ft

44.

6 mi

$9\frac{1}{4}$ mi

$7\frac{1}{2}$ mi

12 mi

45.

$2\frac{1}{8}$ ft

Regular hexagon

46.

Equilateral triangle

3 in.   3 in.

$\frac{1}{2}$ in. $\frac{1}{2}$ in.

5 in.

◆ Combining Concepts

47. A pizza restaurant recently advertised two specials. The first special was a 12-inch pizza for $10. The second special was two 8-inch pizzas for $9. Determine the better buy. (*Hint:* First compare the areas of the two specials and then find a price per square inch for both specials.)

48. Find the approximate area of the state of Utah.

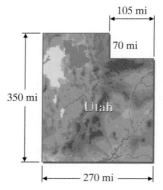

105 mi

70 mi

350 mi

Utah

270 mi

49. Find the area of a rectangle that measures 2 *feet* by 8 *inches*. Give the area in square feet and in square inches.

50. In your own words, explain why perimeter is measured in units and area is measured in square units. (*Hint:* See Section 1.6 for an introduction on the meaning of area.)

Internet Excursions

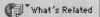
The World Wide Web address listed above will direct you to the official Web site of the U.S. Department of Defense and the Pentagon in Washington, D.C., or a related site. You will find information that helps you complete the questions below.

51. Visit this Web site and locate information on the length of each outer wall of the Pentagon. Then calculate the outer perimeter of the Pentagon.

52. Notice that each of the five outer walls of the Pentagon is in the shape of a rectangle. Visit this Web site and locate information on the height of the building. Then use this height information along with the length information found in Exercise 51 to calculate the area of each outer wall of the Pentagon.

THE COST OF ROAD SIGNS

There are nearly 4 million miles of streets and roads in the United States. With streets, roads, and highways comes the need for traffic control, guidance, warning, and regulation. Road signs perform many of these tasks. Just in our routine travels, we see a wide variety of road signs every day. Think how many road signs must exist on the 4 million miles of roads in the United States. Have you ever wondered how much signs like these cost?

The cost of a road sign generally depends on the type of sign. Costs for several types of signs and signposts are listed in the table. Examples of various types of signs are shown below.

Road Sign Costs	
Type of Sign	**Cost**
Regulatory, warning, marker	$15–$18 per square foot
Large guide	$20–$25 per square foot
Type of Post	**Cost**
U-channel	$125–$200 each
Square tube	$10–$15 per foot
Steel breakaway posts	$15–$25 per foot

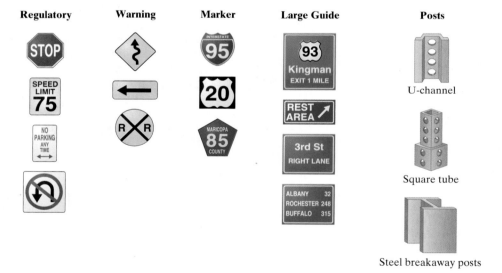

The cost of a sign is based on its area. For diamond, square, or rectangular signs, the area is found by multiplying the length (in feet) times the width (in feet). Then the area is multiplied by the cost per square foot. For signs with irregular shapes, costs are generally figured *as if* the sign were a rectangle, multiplying the height and width at the tallest and widest parts of the sign.

GROUP ACTIVITY

Locate four different kinds of road signs on or near your campus. Measure the dimensions of each sign, including the height of the post on which it is mounted. Using the cost data given in the table, find the minimum and maximum costs of each sign, including its post. Summarize your results in a table, and include a sketch of each sign.

8.5 Volume

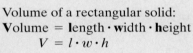 **Finding Volume of Solids**

Volume is a measure of the space of a region. The volume of a box or can, for example, is the amount of space inside. Volume can be used to describe the amount of juice in a pitcher or the amount of concrete needed to pour a foundation for a house.

The volume of a solid is the number of **cubic units** in the solid. A cubic centimeter and a cubic inch are illustrated.

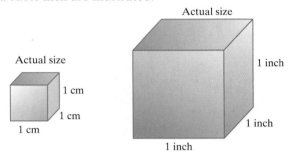

Actual size

1 cm
1 cm
1 cm

Actual size

1 inch
1 inch
1 inch

1 cubic centimeter   **1 cubic inch**

Formulas for finding the volumes of some common solids are given next:

Volume Formulas of Common Solids

Solid	Volume Formulas
RECTANGULAR SOLID 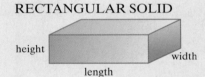 height, width, length	Volume of a rectangular solid: **Volume** = **length · width · height** $V = l \cdot w \cdot h$
CUBE 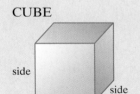 side, side, side	Volume of a cube: **Volume** = **side · side · side** $V = s^3$
SPHERE 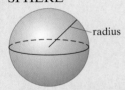 radius	Volume of a sphere: **Volume** = $\dfrac{4}{3} \cdot \pi \cdot (\textbf{radius})^3$ $V = \dfrac{4}{3} \cdot \pi \cdot r^3$
CIRCULAR CYLINDER 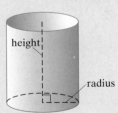 height, radius	Volume of a circular cylinder: **Volume** = $\pi \cdot (\textbf{radius})^2 \cdot \textbf{height}$ $V = \pi \cdot r^2 \cdot h$

Volume Formulas of Common Solids

Solid	**Volume Formulas**

CONE

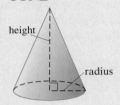

Volume of a cone:

$$\mathbf{V}\text{olume} = \frac{1}{3} \cdot \pi \cdot (\mathbf{radius})^2 \cdot \mathbf{height}$$

$$V = \frac{1}{3} \cdot \pi \cdot r^2 \cdot h$$

SQUARE-BASED PYRAMID

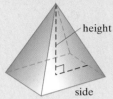

Volume of a square-based pyramid:

$$\mathbf{V}\text{olume} = \frac{1}{3} \cdot (\mathbf{side})^2 \cdot \mathbf{height}$$

$$V = \frac{1}{3} \cdot s^2 \cdot h$$

Practice Problem 1

Draw a diagram and find the volume of a rectangular box that is 5 feet long, 2 feet wide, and 4 feet deep.

Concept Check

Juan is calculating the volume of the following rectangular solid. Find the error in his calculation.

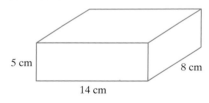

Volume = $l + w + h$
= 14 + 8 + 5
= 27 cu cm

Practice Problem 2

Draw a diagram and approximate the volume of a ball of radius $\frac{1}{2}$ centimeters. Use $\frac{22}{7}$ for π.

Answers

1. 40 cu ft,   **2.** $\frac{11}{21}$ cu cm

Concept Check: Volume = $l \cdot w \cdot h$
= 14 · 8 · 5
= 560 cu cm

Helpful Hint

Volume is always measured in cubic units.

EXAMPLE 1

Find the volume of a rectangular box that is 12 inches long, 6 inches wide, and 3 inches high.

Solution:

$$V = l \cdot w \cdot h$$
$$V = 12 \text{ inches} \cdot 6 \text{ inches} \cdot 3 \text{ inches} = 216 \text{ cubic inches}$$

The volume of the rectangular box is 216 cubic inches.

Try the Concept Check in the margin.

EXAMPLE 2   Approximate the volume of a ball of radius 3 inches.

Use the approximation $\frac{22}{7}$ for π.

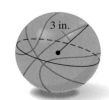

3 in.

Solution:

$$V = \frac{4}{3} \cdot \pi \cdot r^3$$

$$\approx \frac{4}{3} \cdot \frac{22}{7} \, (3 \text{ inches})^3$$

$$= \frac{4}{3} \cdot \frac{22}{7} \cdot 27 \text{ cubic inches}$$

$$= \frac{4 \cdot 22 \cdot \overset{1}{\cancel{3}} \cdot 9}{\underset{1}{\cancel{3}} \cdot 7} \text{ cubic inches}$$

$$= \frac{792}{7} \quad \text{or} \quad 113\frac{1}{7} \text{ cubic inches}$$

The volume is *approximately* $113\frac{1}{7}$ cubic inches.

EXAMPLE 3

Approximate the volume of a can that has a $3\frac{1}{2}$-inch radius and a height of 6 inches. Use $\frac{22}{7}$ for π.

$3\frac{1}{2}$ in.

6 in.

Solution: Using the formula for a circular cylinder, we have

$$V = \pi \cdot r^2 \cdot h$$

$$3\frac{1}{2} = \frac{7}{2}$$

$$= \pi \cdot \left(\frac{7}{2} \text{ inches}\right)^2 \cdot 6 \text{ inches}$$

or approximately

$$\approx \frac{22}{7} \cdot \frac{49}{4} \cdot 6 \text{ cubic inches}$$

$$= 231 \text{ cubic inches}$$

The volume is approximately 231 cubic inches.

Practice Problem 3

Approximate the volume of a cylinder of radius 5 inches and height 7 inches. Use 3.14 for π.

Answer

3. 549.5 cu in.

Practice Problem 4

Find the volume of a square-based pyramid that has a 3-meter side and a height of 5.1 meters.

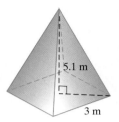

EXAMPLE 4

Approximate the volume of a cone that has a height of 14 centimeters and a radius of 3 centimeters. Use $\frac{22}{7}$ for π.

Solution: Using the formula for volume of a cone, we have

$$V = \frac{1}{3} \cdot \pi \cdot r^2 \cdot h$$

$$= \frac{1}{3} \cdot \pi \cdot (3 \text{ centimeters})^2 \cdot 14 \text{ centimeters} \quad \text{Replace } r, \text{ with 3 cm and } h \text{ with 14 cm.}$$

$$= 42\pi$$

or approximately

$$\approx 42 \cdot \frac{22}{7} \text{ cubic centimeters}$$

$$= 132 \text{ cubic centimeters}$$

The volume is approximately 132 cubic centimeters. ●

Answer

4. 15.3 cu m

Name _____ Section _____ Date _____

EXERCISE SET 8.5

A *Find the volume of each solid. See Examples 1 through 4. Use $\frac{22}{7}$ for π.*

1.
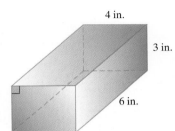
4 in.
3 in.
6 in.

2.
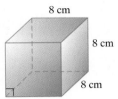
3 mi

3.
8 cm
8 cm
8 cm

4.
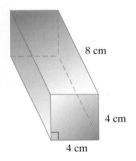
8 cm
4 cm
4 cm

5.
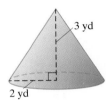
3 yd
2 yd

6.

10 ft
6 ft

7.

10 in.

8.

$1\frac{3}{4}$ in.
9 in.

9.
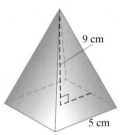
9 cm
5 cm

10.

1 ft

511

Solve.

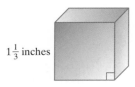 **11.** Find the volume of a cube with edges of $1\frac{1}{3}$ inches.

$1\frac{1}{3}$ inches

12. A water storage tank is in the shape of a cone with the pointed end down. If the radius is 14 feet and the depth of the tank is 15 feet, approximate the volume of the tank in cubic feet. Use $\frac{22}{7}$ for π.

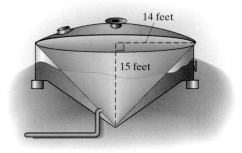

14 feet

15 feet

13. Find the volume of a rectangular box 2 feet by 1.4 feet by 3 feet.

14. Find the volume of a box in the shape of a cube that is 5 feet on each side.

15. Find the volume of a pyramid with a square base 5 inches on a side and a height of 1.3 inches.

16. Approximate to the nearest hundredth the volume of a sphere with a radius of 2 centimeters. Use 3.14 for π.

17. A paperweight is in the shape of a square-based pyramid 20 centimeters tall. If an edge of the base is 12 centimeters, find the volume of the paperweight.

18. A birdbath is made in the shape of a hemisphere (half-sphere). If its radius is 10 inches, approximate the volume. Use $\frac{22}{7}$ for π.

10 inches

19. Find the exact volume of a sphere with a radius of 7 inches.

20. A tank is in the shape of a cylinder 8 feet tall and 3 feet in radius. Find the exact volume of the tank.

21. Find the volume of a rectangular block of ice 2 feet by $2\frac{1}{2}$ feet by $1\frac{1}{2}$ feet.

22. Find the capacity (volume in cubic feet) of a rectangular ice chest with inside measurements of 3 feet by $1\frac{1}{2}$ feet by $1\frac{3}{4}$ feet.

23. An ice cream cone with a 4-centimeter diameter and 3-centimeter depth is filled exactly level with the top of the cone. Approximate how much ice cream (in cubic centimeters) is in the cone. Use $\frac{22}{7}$ for π.

24. A child's toy is in the shape of a square-based pyramid 10 inches tall. If an edge of the base is 7 inches, find the volume of the toy.

25. Ball lightning is a rare form of lightning in which a moving white or colored luminous sphere is seen. It can last from a few seconds to a few minutes and travels at about walking pace. An average sphere size is 6 inches in diameter. Find the exact volume of a sphere with this diameter and then approximate the volume using 3.14 for π.

26. A monkey ball tree produces large green fruit in the shape of spheres. These fruits are approximately 4 inches (or 10 centimeters) in diameter and have a coarse surface. Find the exact volume of a sphere with diameter 4 inches and then approximate the volume using 3.14 for π. (Round to the nearest tenth.)

Review and Preview

Evaluate. See Section 1.9.

27. 5^2

28. 7^2

29. 3^2

30. 20^2

31. $1^2 + 2^2$

32. $5^2 + 3^2$

33. $4^2 + 2^2$

34. $1^2 + 6^2$

35. The Great Pyramid of Khufu at Giza is the largest of the ancient Egyptian pyramids. Its original height was 146.5 meters. The length of each side of its square base was originally 230 meters. Find the volume of the Great Pyramid of Khufu as it was originally built. Round to the nearest whole cubic meter. (*Source*: PBS *NOVA* Online)

36. The second-largest pyramid at Giza is Khafre's Pyramid. Its original height was 471 feet. The length of each side of its square base was originally 704 feet. Find the volume of Khafre's Pyramid as it was originally built. (*Source*: PBS *NOVA* Online)

37. Menkaure's Pyramid, the smallest of the three Great Pyramids at Giza, was originally 65.5 meters tall. Each of the sides of its square base was originally 344 meters long. What was the volume of Menkaure's Pyramid as it was originally built? Round to the nearest whole cubic meter. (*Source*: PBS *NOVA* Online)

38. Due to factors such as weathering and loss of outer stones, the Great Pyramid of Khufu now stands only 137 meters tall. Its square base is now only 227 meters on a side. Find the current volume of the Great Pyramid of Khufu to the nearest whole cubic meter. How much has its volume decreased since it was built? See Exercise 35 for comparison. (*Source*: PBS *NOVA* Online)

39. The centerpiece of the New England Aquarium in Boston is its Giant Ocean Tank. This exhibit is a four-story cylindrical saltwater tank containing sharks, sea turtles, stingrays, and tropical fish. The radius of the tank is 16.3 feet and its height is 32 feet (assuming that a story is 8 feet). What is the volume of the Giant Ocean Tank? Use $\pi \approx 3.14$ and round to the nearest tenth of a cubic foot. (*Source*: New England Aquarium)

40. Except for service dogs for guests with disabilities, Walt Disney World does not allow pets in its parks or hotels. However, the resort does make pet-boarding services available to guests. The pet-care kennels at Walt Disney World offer three different sizes of indoor kennels. Of these, the smaller two kennels measure
a. $2'1'' \times 1'8'' \times 1'7''$ and
b. $1'1'' \times 2' \times 8''$
What is the volume of each kennel rounded to the nearest cubic foot? Which is larger? (*Source*: Walt Disney World Resort)

41. Can you compute the volume of a rectangle? Why or why not?

Integrated Review—Geometry Concepts

△ **1.** Find the supplement and the complement of a 27° angle.

Find the measures of angles x, y, and z in each figure.

2.

3. *m∥n*

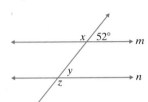

4. Find the measure of ∠x.

Find the perimeter (or circumference) and area of each figure. For the circle give an exact circumference and area. Then use π ≈ 3.14 to approximate each. Don't forget to attach correct units.

5.

Square | 5 m

6.

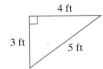

4 ft

3 ft   5 ft

7.

3 cm

8.

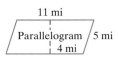

11 mi

Parallelogram | 5 mi
4 mi

9. The smallest cathedral is in High-landville, Missouri. The floor of the cathedral measures 14 feet by 17 feet. Find its perimeter and its area. (*Source: The Guinness Book of Records*)

Find the volume of each solid. Don't forget to attach correct units.

10. A cube with edges of 4 inches each.

11. A rectangular box 2 feet by 3 feet by 5.1 feet.

12. A pyramid with a square base 10 centimeters on a side and a height of 12 centimeters.

13. A sphere with a diameter of 16 miles. Give the exact volume and then use $\pi \approx \frac{22}{7}$ to approximate.

7. _____

8. _____

9. _____

10. _____

11. _____

12. _____

13. _____

△ **8.6** Square Roots and the Pythagorean Theorem

O B J E C T I V E S

A Find the square root of a number.

B Approximate square roots.

C Use the Pythagorean theorem.

SSM
TUTOR CENTER   SG   CD & VIDEO   MATH PRO   WEB

A Finding Square Roots

The square of a number is the number times itself. For example:

The square of 5 is 25 because 5^2 or $5 \cdot 5 = 25$.
The square of 3 is 9 because 3^2 or $3 \cdot 3 = 9$.
The square of 10 is 100 because 10^2 or $10 \cdot 10 = 100$.

The reverse process of squaring is finding a **square root**. For example:

A square root of 9 is 3 because $3^2 = 9$.
A square root of 25 is 5 because $5^2 = 25$.
A square root of 100 is 10 because $10^2 = 100$.

We use the symbol $\sqrt{}$, called a **radical sign**, to name square roots. For example:

$$\sqrt{9} = 3 \quad \text{because } 3^2 = 9$$
$$\sqrt{25} = 5 \quad \text{because } 5^2 = 25$$

Square Root of a Number

A square root of a number a is a number b whose square is a. We use the radical sign $\sqrt{}$ to name square roots.

EXAMPLE 1 Find each square root.

a. $\sqrt{49}$ **b.** $\sqrt{36}$ **c.** $\sqrt{1}$ **d.** $\sqrt{81}$

Solution:

a. $\sqrt{49} = 7$ because $7^2 = 49$
b. $\sqrt{36} = 6$ because $6^2 = 36$
c. $\sqrt{1} = 1$ because $1^2 = 1$
d. $\sqrt{81} = 9$ because $9^2 = 81$

EXAMPLE 2 Find: $\sqrt{\dfrac{1}{36}}$

Solution: $\sqrt{\dfrac{1}{36}} = \dfrac{1}{6}$ because $\dfrac{1}{6} \cdot \dfrac{1}{6} = \dfrac{1}{36}$

EXAMPLE 3 Find: $\sqrt{\dfrac{4}{25}}$

Solution: $\sqrt{\dfrac{4}{25}} = \dfrac{2}{5}$ because $\dfrac{2}{5} \cdot \dfrac{2}{5} = \dfrac{4}{25}$

B Approximating Square Roots

Thus far, we have found square roots of perfect squares. Numbers like $\dfrac{1}{4}, 36, \dfrac{4}{25}$, and 1 are called **perfect squares** because their square root is a whole number or a fraction. A square root such as $\sqrt{5}$ cannot be written as a whole number or a fraction since 5 is not a perfect square.

Practice Problem 1

Find each square root.

a. $\sqrt{100}$
b. $\sqrt{64}$
c. $\sqrt{121}$
d. $\sqrt{0}$

Practice Problem 2

Find: $\sqrt{\dfrac{1}{4}}$

Practice Problem 3

Find: $\sqrt{\dfrac{9}{16}}$

Answers

1. a. 10, **b.** 8, **c.** 11, **d.** 0, **2.** $\dfrac{1}{2}$, **3.** $\dfrac{3}{4}$

Although $\sqrt{5}$ cannot be written as a whole number or a fraction, it can be approximated by estimating, by using a table (as in Appendix E), or by using a calculator.

Practice Problem 4

Use Appendix E or a calculator to approximate the square root of 11 to the nearest thousandth.

EXAMPLE 4

Use Appendix E or a calculator to approximate the square root of 43 to the nearest thousandth.

Solution: $\sqrt{43} \approx 6.557$ ●

$\sqrt{43}$ is *approximately* 6.557. This means that if we multiply 6.557 by 6.557, the product is *close* to 43.

$$6.557 \times 6.557 = 42.994249$$

Practice Problem 5

Approximate $\sqrt{29}$ to the nearest thousandth.

EXAMPLE 5   Approximate $\sqrt{32}$ to the nearest thousandth.

Solution: $\sqrt{32} \approx 5.657$ ●

C Using the Pythagorean Theorem

One important application of square roots has to do with right triangles. Recall that a **right triangle** is a triangle in which one of the angles is a right angle, or measures 90°. The **hypotenuse** of a right triangle is the side opposite the right angle. The **legs** of a right triangle are the other two sides. These are shown in the following figure. The right angle in the triangle is indicated by the small square drawn in that angle.

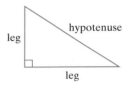

The following theorem is true for all right triangles:

Pythagorean Theorem

In any **right triangle**,

$$(\text{leg})^2 + (\text{other leg})^2 = (\text{hypotenuse})^2$$

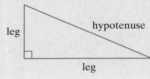

Using the Pythagorean theorem, we can use one of the following formulas to find an unknown length of a right triangle:

Finding an Unknown Length of a Right Triangle

$$\text{hypotenuse} = \sqrt{(\text{leg})^2 + (\text{other leg})^2}$$

or

$$\text{leg} = \sqrt{(\text{hypotenuse})^2 - (\text{other leg})^2}$$

Answers

4. 3.317, **5.** 5.385

EXAMPLE 6

Find the length of the hypotenuse of the given right triangle.

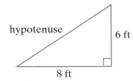

Solution: Since we are finding the hypotenuse, we use the formula

$$\text{hypotenuse} = \sqrt{(\text{leg})^2 + (\text{other leg})^2}$$

Putting the known values into the formula, we have

$$\text{hypotenuse} = \sqrt{(6)^2 + (8)^2} \quad \text{The legs are 6 feet and 8 feet.}$$
$$= \sqrt{36 + 64}$$
$$= \sqrt{100}$$
$$= 10$$

The hypotenuse is 10 feet long.

Practice Problem 6

Find the length of the hypotenuse of the given right triangle.

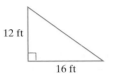

EXAMPLE 7

Approximate the length of the hypotenuse of the given right triangle. Round the length to the nearest whole unit.

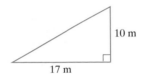

Solution:

$$\text{hypotenuse} = \sqrt{(\text{leg})^2 + (\text{other leg})^2}$$
$$= \sqrt{(17)^2 + (10)^2} \quad \text{The legs are 10 meters and 17 meters.}$$
$$= \sqrt{289 + 100}$$
$$= \sqrt{389}$$
$$\approx 20 \quad \text{From Appendix E or a calculator}$$

The hypotenuse is exactly $\sqrt{389}$ meters, which is approximately 20 meters.

Practice Problem 7

Approximate the length of the hypotenuse of the given right triangle. Round to the nearest whole unit.

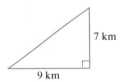

EXAMPLE 8

Find the length of the leg in the given right triangle. Give the exact length and a two-decimal-place approximation.

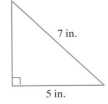

Solution: Notice that the hypotenuse measures 7 inches and that the length of one leg measures 5 inches. Since we are looking for the length of the other leg, we use the formula

$$\text{leg} = \sqrt{(\text{hypotenuse})^2 - (\text{other leg})^2}$$

Practice Problem 8

Find the length of the leg in the given right triangle. Give the exact length and a two-decimal-place approximation.

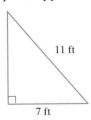

Answers

6. 20 ft,   **7.** 11 km,   **8.** $\sqrt{72}$ ft ≈ 8.49 ft

Putting the known values into the formula, we have

$$\text{leg} = \sqrt{(7)^2 - (5)^2} \qquad \text{The hypotenuse is 7 inches, and the other leg is 5 inches.}$$
$$= \sqrt{49 - 25}$$
$$= \sqrt{24}$$
$$\approx 4.90 \qquad \text{From Appendix E or a calculator}$$

The length of the leg is exactly $\sqrt{24}$ inches, which is approximately 4.90 inches.

Try the Concept Check in the margin.

EXAMPLE 9   Finding the Diagonal Length of a City Block

A standard city block is a square that measures 300 feet on a side. Find the length of the diagonal of a city block rounded to the nearest whole foot.

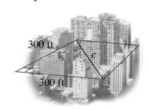

300 ft

300 ft

c

Solution: The diagonal is the hypotenuse of a right triangle, so we use the formula

$$\text{hypotenuse} = \sqrt{(\text{leg})^2 + (\text{other leg})^2}$$

Putting the known values into the formula we have

$$\text{hypotenuse} = \sqrt{(300)^2 + (300)^2} \qquad \text{The legs are both 300 feet.}$$
$$= \sqrt{90,000 + 90,000}$$
$$= \sqrt{180,000}$$
$$\approx 424 \qquad \text{From Appendix E or a calculator}$$

The length of the diagonal is approximately 424 feet.

Concept Check

The following lists are the lengths of the sides of two triangles. Which set forms a right triangle?

a. 8, 15, 17
b. 24, 30, 40

Practice Problem 9

A football field is a rectangle measuring 100 yards by 53 yards. Draw a diagram and find the length of the diagonal of a football field to the nearest yard.

Answers

9. 113 yd

Concept Check: a. yes,   b. no

 CALCULATOR EXPLORATIONS

Finding Square Roots

To simplify or approximate square roots using a calculator, locate the key marked $\boxed{\sqrt{\ }}$.

To simplify $\sqrt{64}$, for example, press the keys

$$\boxed{64} \; \boxed{\sqrt{\ }} \quad \text{or} \quad \boxed{\sqrt{\ }} \; \boxed{64}$$

The display should read $\boxed{\qquad 8}$. Then

$$\sqrt{64} = 8$$

To *approximate* $\sqrt{10}$, press the keys

$$\boxed{10} \; \boxed{\sqrt{\ }} \quad \text{or} \quad \boxed{\sqrt{\ }} \; \boxed{10}$$

The display should read $\boxed{3.16227766}$. This is an *approximation* for $\sqrt{10}$. A three-decimal-place approximation is

$$\sqrt{10} \approx 3.162$$

Is this answer reasonable? Since 10 is between perfect squares 9 and 16, $\sqrt{10}$ is between $\sqrt{9} = 3$ and $\sqrt{16} = 4$. Our answer is reasonable since 3.162 is between 3 and 4.

Simplify.

1. $\sqrt{1024}$ **2.** $\sqrt{676}$

Approximate each square root. Round each answer to the nearest thousandth.

3. $\sqrt{15}$ **4.** $\sqrt{19}$ **5.** $\sqrt{97}$ **6.** $\sqrt{56}$

EXERCISE SET 8.6

A *A Find each square root. See Examples 1 through 3.*

 1. $\sqrt{4}$

 2. $\sqrt{9}$

3. $\sqrt{625}$

4. $\sqrt{16}$

 5. $\sqrt{\dfrac{1}{81}}$

6. $\sqrt{\dfrac{1}{64}}$

7. $\sqrt{\dfrac{144}{64}}$

8. $\sqrt{\dfrac{36}{81}}$

9. $\sqrt{256}$

10. $\sqrt{144}$

11. $\sqrt{\dfrac{9}{4}}$

12. $\sqrt{\dfrac{121}{169}}$

B *Use Appendix E or a calculator to approximate each square root. Round the square root to the nearest thousandth. See Examples 4 and 5.*

13. $\sqrt{3}$

14. $\sqrt{5}$

 15. $\sqrt{15}$

16. $\sqrt{17}$

17. $\sqrt{14}$

18. $\sqrt{18}$

19. $\sqrt{47}$

20. $\sqrt{85}$

21. $\sqrt{8}$

22. $\sqrt{10}$

23. $\sqrt{26}$

24. $\sqrt{35}$

25. $\sqrt{71}$

26. $\sqrt{62}$

27. $\sqrt{7}$

28. $\sqrt{2}$

C *Find the unknown length in each right triangle. If necessary, approximate the length to the nearest thousandth. See Examples 6 through 8.*

29.

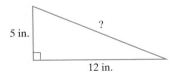

30.

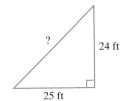

31.

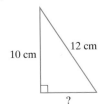

32.

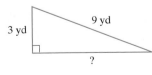

Sketch each right triangle and find the length of the side not given. If necessary, approximate the length to the nearest thousandth. See Examples 6 through 8.

33. leg = 3, leg = 4

34. leg = 9, leg = 12

35. leg = 6, hypotenuse = 10

36. leg = 48, hypotenuse = 53

37. leg = 10, leg = 14

38. leg = 32, leg = 19

39. leg = 2, leg = 16

40. leg = 27, leg = 36

41. leg = 5, hypotenuse = 13

42. leg = 45, hypotenuse = 117

43. leg = 35, leg = 28

44. leg = 30, leg = 15

45. leg = 30, leg = 30

46. leg = 110, leg = 132

47. hypotenuse = 2, leg = 1

48. hypotenuse = 7, leg = 6

Solve. See Example 9

49. A standard city block is a square with each side measuring 100 yards. Find the length of the diagonal of a city block to the nearest hundredth yard.

100 yd
100 yd
?

50. A section of land is a square with each side measuring 1 mile. Find the length of the diagonal of the section of land to the nearest thousandth mile.

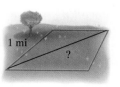

1 mi
?

51. Find the height of the tree. Round the height to one decimal place.

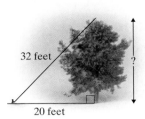

32 feet

?

20 feet

52. Find the height of the antenna. Round the height to one decimal place.

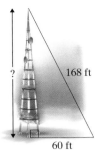

?

168 ft

60 ft

53. A football field is a rectangle that is 300 feet long by 160 feet wide. Find, to the nearest foot, the length of a straight-line run that started at one corner and went diagonally to end at the opposite corner.

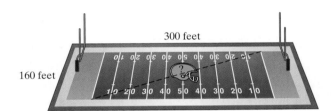

300 feet

160 feet

54. A baseball diamond is in the shape of a square and has sides of length 90 feet. Find the distance across the diamond from third base to first base, to the nearest tenth of a foot.

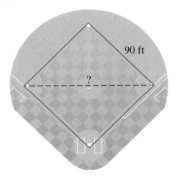

90 ft

?

Review and Preview

Find the value of n in each proportion. See Section 5.3.

55. $\dfrac{n}{6} = \dfrac{2}{3}$

56. $\dfrac{8}{n} = \dfrac{4}{8}$

57. $\dfrac{9}{11} = \dfrac{n}{55}$

58. $\dfrac{5}{6} = \dfrac{35}{n}$

59. $\dfrac{3}{n} = \dfrac{7}{14}$

60. $\dfrac{n}{9} = \dfrac{4}{6}$

Combining Concepts

Determine what two whole numbers each square root is between without using a calculator or table. Then use a calculator or table to check.

61. $\sqrt{38}$

62. $\sqrt{27}$

63. $\sqrt{101}$

64. $\sqrt{85}$

65. Without using a calculator, explain how you know that $\sqrt{105}$ is *not* approximately 9.875.

Are you preparing for a test on Chapter 8?

Below I have listed some common trouble areas for students in Chapter 8. After studying for your test—but before taking your test—read these.

- Don't forget the difference between complementary and supplementary angles.

    Complementary angles have a sum of 90°.

    Supplementary angles have a sum of 180°.

    The complement of a 15° angle measures 90° − 15° = 75°.

    The supplement of a 15° angle measures 180° − 15° = 165°.

- Remember:

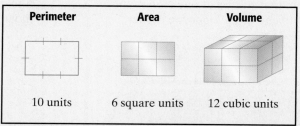

Perimeter	Area	Volume
10 units	6 square units	12 cubic units

- Remember that, for example, that since

$$18 \text{ is between } \quad 16 \quad \text{and} \quad 25,$$
$$\sqrt{18} \text{ is between } \quad \sqrt{16} \quad \text{and} \quad \sqrt{25} \text{ or}$$
$$\downarrow \qquad\qquad \downarrow$$
$$\sqrt{18} \text{ is between } \quad 4 \quad \text{and} \quad 5.$$

Remember: This is simply a checklist of common trouble areas. For a review of Chapter 8, see the Highlights and Chapter Review at the end of this chapter.

8.7 Congruent and Similar Triangles

OBJECTIVES

Ⓐ Decide whether two triangles are congruent.

Ⓑ Find the ratio of corresponding sides in similar triangles.

Ⓒ Find unknown lengths of sides in similar triangles.

SSM   TUTOR CENTER   SG   CD & VIDEO   MATH PRO   WEB

Ⓐ Deciding Whether Two Triangles Are Congruent

Two triangles are **congruent** when they have the same shape and the same size. In congruent triangles, the measures of corresponding angles are equal and the lengths of corresponding sides are equal. The following triangles are congruent:

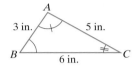
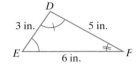

Since these triangles are congruent, the measures of corresponding angles are equal.

Angles with equal measure: ∠A and ∠D, ∠B and ∠E, ∠C and ∠F

Also, the lengths of corresponding sides are equal.

Equal corresponding sides: $\overline{AB}$ and $\overline{DE}$, $\overline{BC}$ and $\overline{EF}$, $\overline{CA}$ and $\overline{FD}$

Any one of the following may be used to determine whether two triangles are congruent:

Congruent Triangles

Angle-Side-Angle (ASA)

If the measures of two angles of a triangle equal the measures of two angles of another triangle, and the lengths of the sides between each pair of angles are equal, the triangles are congruent.

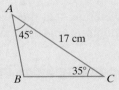
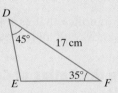

For example, these two triangles are congruent by Angle-Side-Angle.

Side-Side-Side (SSS)

If the lengths of the three sides of a triangle equal the lengths of the corresponding sides of another triangle, the triangles are congruent.

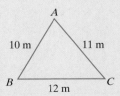
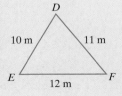

For example, these two triangles are congruent by Side-Side-Side.

Congruent Triangles, continued

Side-Angle-Side (SAS)

If the lengths of two sides of a triangle equal the lengths of corresponding sides of another triangle, and the measures of the angles between each pair of sides are equal, the triangles are congruent.

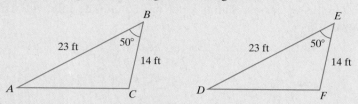

For example, these two triangles are congruent by Side-Angle-Side.

Practice Problem 1

Determine whether triangle *MNO* is congruent to triangle *RQS*.

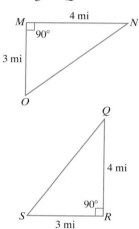

EXAMPLE 1

Determine whether triangle *ABC* is congruent to triangle *DEF*.

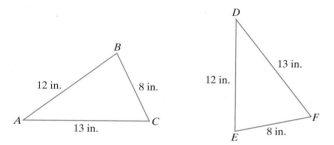

Solution: Since the lengths of all three sides of triangle *ABC* equal the lengths of all three sides of triangle *DEF*, the triangles are congruent. ●

In Example 1, notice that as soon as we know that the two triangles are congruent, we know that all three corresponding angles are congruent.

B **Finding the Ratios of Corresponding Sides in Similar Triangles**

Two triangles are **similar** when they have the same shape but not necessarily the same size. In similar triangles, the measures of corresponding angles are equal and corresponding sides are in proportion. The following triangles are similar:

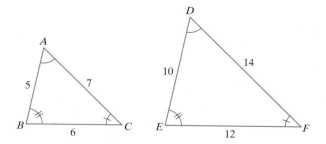

Since these triangles are similar, the measures of corresponding angles are equal.

Answer

1. congruent

Angles with equal measure: $\angle A$ and $\angle D$, $\angle B$ and $\angle E$, $\angle C$ and $\angle F$
Also, the lengths of corresponding sides are in proportion.

Sides in proportion: $\dfrac{AB}{DE} = \dfrac{5}{10} = \dfrac{1}{2}, \dfrac{BC}{EF} = \dfrac{6}{12} = \dfrac{1}{2}, \dfrac{CA}{FD} = \dfrac{7}{14} = \dfrac{1}{2}$

The ratio of corresponding sides is $\dfrac{1}{2}$.

EXAMPLE 2

Find the ratio of corresponding sides for the similar triangles ABC and DEF.

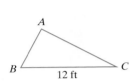 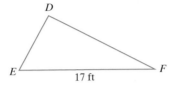

Solution: We are given the lengths of two corresponding sides. Their ratio is

$$\frac{12 \ \text{feet}}{17 \ \text{feet}} = \frac{12}{17}$$

Finding Unknown Lengths of Sides in Similar Triangles

Because the ratios of lengths of corresponding sides are equal, we can use proportions to find unknown lengths in similar triangles.

EXAMPLE 3

Given that the triangles are similar, find the missing length n.

Solution: Since the triangles are similar, corresponding sides are in proportion. Thus, the ratio of 2 to 3 is the same as the ratio of 10 to n, or

$$\frac{2}{3} = \frac{10}{n}$$

To find the unknown length n, we set cross products equal.

$$\frac{2}{3} = \frac{10}{n}$$

$2 \cdot n = 30$ Set cross products equal.

$n = \dfrac{30}{2}$

$n = 15$

The missing length is 15 units.

Practice Problem 2

Find the ratio of corresponding sides for the similar triangles QRS and XYZ.

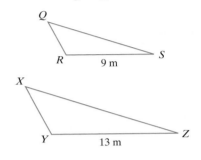

Practice Problem 3

Given that the triangles are similar, find the missing length n.

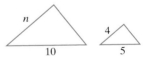

Answers

2. $\dfrac{9}{13}$,   **3.** $n = 8$

Concept Check

The following two triangles are similar. Which vertices of the first triangle appear to correspond to which vertices of the second triangle?

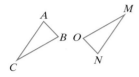

Practice Problem 4

Tammy Shultz, a firefighter, needs to estimate the height of a burning building. She estimates the length of her shadow to be 8 feet long and the length of the building's shadow to be 60 feet long. Find the approximate height of the building if she is 5 feet tall.

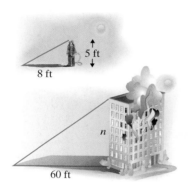

Try the Concept Check in the margin.

Many applications involve a diagram containing similar triangles. Surveyors, astronomers, and many other professionals continually use ratios of similar triangles in their work.

EXAMPLE 4   Finding the Height of a Tree

Mel Rose is a 6-foot-tall park ranger who needs to know the height of a particular tree. He measures the shadow of the tree to be 69 feet long when his own shadow is 9 feet long. Find the height of the tree.

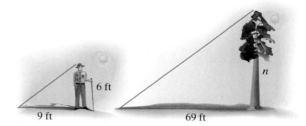

Solution:

1. UNDERSTAND. Read and reread the problem. Notice that the triangle formed by the Sun's rays, Mel, and his shadow is similar to the triangle formed by the Sun's rays, the tree, and its shadow.

2. TRANSLATE. Write a proportion from the similar triangles formed.

    $$\begin{array}{ll} \text{Mel's height} \rightarrow \\ \text{height of tree} \rightarrow \end{array} \dfrac{6}{n} = \dfrac{9}{69} \begin{array}{ll} \leftarrow \text{length of Mel's shadow} \\ \leftarrow \text{length of tree's shadow} \end{array}$$

    $$\text{or } \dfrac{6}{n} = \dfrac{3}{23} \text{ (ratio in lowest terms)}$$

3. SOLVE for n:

    $$\dfrac{6}{n} = \dfrac{3}{23}$$

    $$6 \cdot 23 = n \cdot 3 \qquad \text{Set cross products equal.}$$
    $$138 = n \cdot 3$$
    $$\dfrac{138}{3} = n$$
    $$46 = n$$

4. INTERPRET. *Check* to see that replacing n with 46 in the proportion makes the proportion true. *State* your conclusion: The height of the tree is 46 feet.   ●

Answers

4. approximately 37.5 ft

Concept Check: *A* corresponds to *O*;
B corresponds to *N*; *C* corresponds to *M*

EXERCISE SET 8.7

Ⓐ *Determine whether each pair of triangles is congruent. See Example 1.*

1.

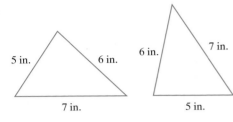

5 in. 6 in. 7 in. 6 in. 7 in. 5 in.

2.

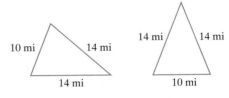

10 mi 14 mi 14 mi 14 mi 14 mi 10 mi

3.

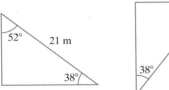

52° 21 m 38° 52° 21 m 38°

4.

2 yd 45° 55° 55° 45° 2 yd

Ⓑ *Find each ratio of the corresponding sides of the given similar triangles. See Example 2.*

5.

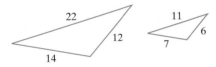

22 12 14 11 6 7

6.

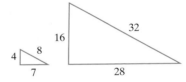

16 32 4 8 7 28

7.

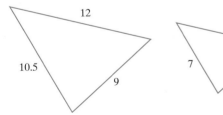

12 10.5 9 8 7 6

8.

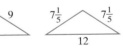

9 9 15 $7\frac{1}{5}$ $7\frac{1}{5}$ 12

Ⓒ *Given that the pairs of triangles are similar, find the length of the side labeled n. See Example 3.*

9.

3 6 n 9

10.

5 3 2 n 60 24

11.

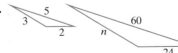

12 18 4 n

12.
4
7
n
14

13.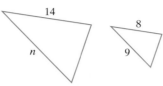
n   12
3.75   9

14.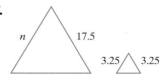
9   15
n   22.5

15.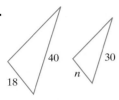
40
30
18
n

16.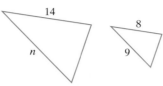
14
8
n
9

17.
n   17.5
3.25   3.25

18.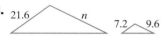
21.6   n
7.2   9.6

19.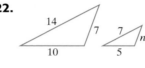
$8\frac{1}{2}$
$2\frac{1}{8}$
n
n   2

20.
9
n
6
9

21.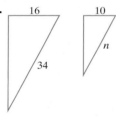
16   10
n
34

22.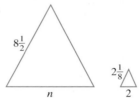
14
7   7   n
10   5

Solve. See Example 4.

23. Given the following diagram, approximate the height of the First National Center in Oklahoma City, OK. (*Source: The World Almanac*, 2001)

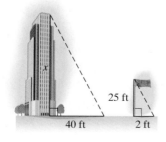

25 ft
40 ft   2 ft

24. The tallest tree standing today is a redwood located near Ukiah, California. Given the following diagram, approximate its height. (*Source: Guinness World Record*, 2001)

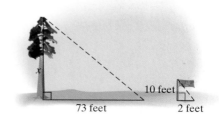

10 feet
73 feet   2 feet

25. Samantha Black, a 5-foot-tall park ranger, needs to know the height of a tree. She notices that when the shadow of the tree is 48 feet long, her shadow is 4 feet long. Find the height of the tree.

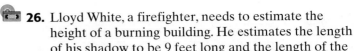

26. Lloyd White, a firefighter, needs to estimate the height of a burning building. He estimates the length of his shadow to be 9 feet long and the length of the building's shadow to be 75 feet long. Find the approximate height of the building if he is 6 feet tall.

27. If a 30-foot tree casts an 18-foot shadow, find the length of the shadow cast by a 24-foot tree.

28. If a 24-foot flagpole casts a 32-foot shadow, find the length of the shadow cast by a 44-foot antenna. Round to the nearest tenth.

29. The print on a particular page measures 7 inches by 9 inches. A printing shop is to copy the page and reduce the print so that its length is 5 inches. What will its width be? Will the print now fit on a 3-/by-5-inch index card?

30. Ben and Joyce Lander draw a triangular deck on their house plans. Joyce measures the deck drawing on the plans to be 3 inches by $4\frac{1}{2}$ inches. If the scale on the drawing is $\frac{1}{4}$ in. = 1 foot, find the dimensions of the deck they want built.

Review and Preview

Find the average of each list of numbers. See Section 1.7.

31. 14, 17, 21, 18

32. 87, 84, 93

33. 76, 79, 88

34. 7, 8, 4, 6, 3, 8

 Combining Concepts

Given that the pairs of triangles are similar, find the length of the side labeled n. Round your results to 1 decimal place.

35.

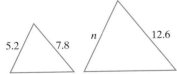

36.

37. In your own words, describe any differences in similar triangles and congruent triangles.

MATERIALS:

■ Cardboard cut into 1-inch squares

This activity may be completed by working in groups or individually. You will explore the perimeters, areas, and diagonal lengths of patterns formed by flooring tiles and sidewalk tiles. Complete each of the following tables. You will analyze the data in your tables and identify the numerical patterns that are present. Note: A diagonal is a line that extends from the upper-left corner of the figure to the lower-right corner.

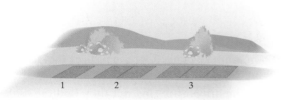

Figure 1 Sidewalk Tile Patterns

Sidewalk Pattern Number	1	2	3	4	5	6
Perimeter						
Area						
Length of diagonal						

1. In the table, record the perimeter, area, and length of diagonal for each of the three sidewalk tile patterns shown in Figure 1. (*Hint*: For the length of the diagonal, use the Pythagorean theorem.)

2. Form the appropriate sidewalk tile patterns for 4, 5, and 6 sidewalk tiles with cardboard squares. Record the perimeters, areas, and diagonal lengths in the table.

3. Study the results for perimeter, area, and length of diagonal. What patterns do you notice? Describe each pattern.

4. Use the patterns you observed in Question 3 to predict the perimeter, area, and diagonal length for a sidewalk made of 7 tiles. Use the cardboard tiles to form this pattern and verify your predictions.

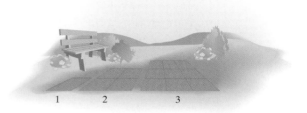

Figure 2 Floor Tile Patterns

Floor Pattern Number	1	2	3	4	5	6
Perimeter						
Area						
Length of diagonal						

5. In the table, record the perimeter, area, and length of diagonal for each of the three floor tile patterns shown in Figure 2.

6. Form the appropriate floor tile patterns with cardboard squares for 4, 5, and 6. Record the perimeters, areas, and diagonal lengths in the table.

7. Study the results for perimeter, area, and length of diagonal. What patterns do you notice? Describe each pattern.

8. Use the patterns you observed in Question 7 to predict the perimeter, area, and diagonal length for the floor tile pattern corresponding to 7 floor tiles. Use the cardboard tiles to form this pattern and verify your predictions.

Chapter 8 VOCABULARY CHECK

Fill in each blank with one of the words or phrases listed below.

transversal	line segment	obtuse	straight	adjacent
right	volume	area	legs	acute
right triangle	perimeter	hypotenuse	vertical	supplementary
similar	congruent	square root	ray	angle
		line	complementary	vertex

1. A _____ is a triangle with a right angle. The side opposite the right angle is called the _____, and the other two sides are called _____ .
2. A _____ is a piece of a line with two end points.
3. Two angles that have a sum of 90° are called _____ angles.
4. A _____ is a set of points extending indefinitely in two directions.
5. The _____ of a polygon is the distance around the polygon.
6. An _____ is made up of two rays that share the same end point. The common end point is called the _____ .
7. _____ triangles have the same shape and the same size.
8. _____ measures the amount of surface of a region.
9. A _____ is a part of a line with one end point. A ray extends indefinitely in one direction.
10. A _____ of a number a is a number b whose square is a.
11. A line that intersects two or more lines at different points is called a _____ .
12. An angle that measures 180° is called a _____ angle.
13. The measure of the space of a solid is called its _____ .
14. When two lines intersect, four angles are formed. Two of these angles that are opposite each other are called _____ angles.
15. Two of these angles that share a common side are called _____ angles.
16. An angle whose measure is between 90° and 180° is called an _____ angle.
17. An angle that measures 90° is called a _____ angle.
18. An angle whose measure is between 0° and 90° is called an _____ angle.
19. Two angles that have a sum of 180° are called _____ angles.
20. _____ triangles have exactly the same shape but not necessarily the same size.

C H A P T E R | # Highlights

DEFINITIONS AND CONCEPTS	**EXAMPLES**
Section 8.1 Lines and Angles	

A **line** is a set of points extending indefinitely in two directions. A line has no width or height, but it does have length. We name a line by any two of its points. A **line segment** is a piece of a line with two end points.	Line AB or $\overleftrightarrow{AB}$ Line segment AB or $\overline{AB}$
A **ray** is a part of a line with one end point. A ray extends indefinitely in one direction.	Ray AB or $\overrightarrow{AB}$
An **angle** is made up of two rays that share the same end point. The common end point is called the **vertex**.
Vertex |

An angle that measures 180° is called a **straight angle**.

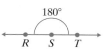

∠*RST* is a straight angle.

An angle that measures 90° is called a **right angle**. The symbol ∟ is used to denote a right angle.

∠*ABC* is a right angle.

An angle whose measure is between 0° and 90° is called an **acute angle**.

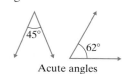

Acute angles

An angle whose measure is between 90° and 180° is called an **obtuse angle**.

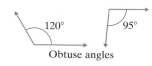

Obtuse angles

Two angles that have a sum of 90° are called **complementary angles**. We say that each angle is the **complement** of the other.

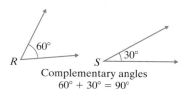

Complementary angles
60° + 30° = 90°

Two angles that have a sum of 180° are called **supplementary angles**. We say that each angle is the **supplement** of the other.

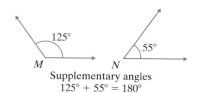

Supplementary angles
125° + 55° = 180°

When two lines intersect, four angles are formed. Two of these angles that are opposite each other are called **vertical angles**. Vertical angles have the same measure.

Two of these angles that share a common side are called **adjacent angles**. Adjacent angles formed in intersecting lines are supplementary.

A line that intersects two or more lines at different points is called a **transversal**. Line *l* is a transversal that intersects lines *m* and *n*. The eight angles formed have special names. Some of these names are:

Corresponding angles: ∠*a* and ∠*e*, ∠*c* and ∠*g*, ∠*b* and ∠*f*, ∠*d* and ∠*h*

Alternate interior angles: ∠*c* and ∠*f*, ∠*d* and ∠*e*

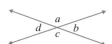

Vertical angles:
∠*a* and ∠*c*
∠*d* and ∠*b*
Adjacent angles:
∠*a* and ∠*b*
∠*b* and ∠*c*
∠*c* and ∠*d*
∠*d* and ∠*a*

PARALLEL LINES CUT BY A TRANSVERSAL

If two parallel lines are cut by a transversal, then the measures of **corresponding angles are equal** and the measures of **alternate interior angles are equal**.

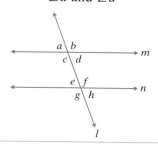

DEFINITIONS AND CONCEPTS

EXAMPLES

Section 8.2 Plane Figures and Solids

The **sum of the measures** of the angles of a triangle is 180°.

Find the measure of $\angle x$.

The measure of $\angle x = 180° - 85° - 45° = 50°$

A **right triangle** is a triangle with a right angle. The side opposite the right angle is called the **hypotenuse**, and the other two sides are called **legs**.

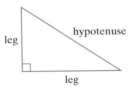

For a circle or a sphere:

$$\text{diameter} = 2 \cdot \text{radius}$$
$$d = 2 \cdot r$$

$$\text{radius} = \frac{\text{diameter}}{2}$$

$$r = \frac{d}{2}$$

Find the diameter of the circle.

$$d = 2 \cdot r$$
$$= 2 \cdot 6 \text{ feet} = 12 \text{ feet}$$

Section 8.3 Perimeter

PERIMETER FORMULAS

Rectangle:
$$P = 2 \cdot l + 2 \cdot w$$
Square:
$$P = 4 \cdot s$$
Triangle:
$$P = a + b + c$$
Circumference of a Circle:
$$C = 2 \cdot \pi \cdot r \quad \text{or} \quad C = \pi \cdot d,$$
where $\pi \approx 3.14 \quad$ or $\quad \pi \approx \frac{22}{7}$

Find the perimeter of the rectangle.

$$P = 2 \cdot l + 2 \cdot w$$
$$= 2 \cdot 28 \text{ m} + 2 \cdot 15 \text{ m}$$
$$= 56 \text{ m} + 30 \text{ m}$$
$$= 86 \text{ m}$$

The perimeter is 86 meters.

Section 8.4 Area

AREA FORMULAS

Rectangle:
$$A = l \cdot w$$
Square:
$$A = s^2$$
Triangle:
$$A = \frac{1}{2} \cdot b \cdot h$$
Parallelogram:
$$A = b \cdot h$$
Trapezoid:
$$A = \frac{1}{2} \cdot (b + B) \cdot h$$
Circle:
$$A = \pi \cdot r^2$$

Find the area of the square.

$$A = s^2$$
$$= (8 \text{ cm})^2$$
$$= 64 \text{ square centimeters}$$

The area of the square is 64 square centimeters.

Section 8.5   Volume

VOLUME FORMULAS

Rectangular Solid:

$$V = l \cdot w \cdot h$$

Cube:

$$V = s^3$$

Sphere:

$$V = \frac{4}{3} \cdot \pi \cdot r^3$$

Right Circular Cylinder:

$$V = \pi \cdot r^2 \cdot h$$

Cone:

$$V = \frac{1}{3} \cdot \pi \cdot r^2 \cdot h$$

Square-Based Pyramid:

$$V = \frac{1}{3} \cdot s^2 \cdot h$$

Find the volume of the sphere. Use $\frac{22}{7}$ for π.

$$V = \frac{4}{3} \cdot \pi \cdot r^3$$

$$\approx \frac{4}{3} \cdot \frac{22}{7} \cdot (4 \text{ inches})^3$$

$$= \frac{4 \cdot 22 \cdot 64}{3 \cdot 7} \text{ cubic inches}$$

$$= \frac{5632}{21} \quad \text{or} \quad 268\frac{4}{21} \text{ cubic inches}$$

Section 8.6   Square Roots and the Pythagorean Theorem

SQUARE ROOT OF A NUMBER

A **square root** of a number a is a number b whose square is a. We use the radical sign $\sqrt{}$ to name square roots.

PYTHAGOREAN THEOREM

$$(\text{leg})^2 + (\text{other leg})^2 = (\text{hypotenuse})^2$$

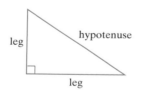

TO FIND AN UNKNOWN LENGTH OF A RIGHT TRIANGLE

$$\text{hypotenuse} = \sqrt{(\text{leg})^2 + (\text{other leg})^2}$$
$$\text{leg} = \sqrt{(\text{hypotenuse})^2 - (\text{other leg})^2}$$

$$\sqrt{9} = 3, \sqrt{100} = 10, \sqrt{1} = 1$$

Find the hypotenuse of the given triangle.

3 in.   hypotenuse   8 in.

$$\text{hypotenuse} = \sqrt{(\text{leg})^2 + (\text{other leg})^2}$$
$$= \sqrt{(3)^2 + (8)^2} \quad \text{The legs are 3 and 8 inches.}$$
$$= \sqrt{9 + 64}$$
$$= \sqrt{73} \text{ inches}$$
$$\approx 8.5 \text{ inches}$$

Section 8.7   Congruent and Similar Triangles

Congruent triangles have the same shape and the same size. Corresponding angles are equal, and corresponding sides are equal.

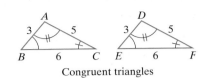

Congruent triangles

Similar triangles have exactly the same shape but not necessarily the same size. Corresponding angles are equal, and the ratios of the lengths of corresponding sides are equal.

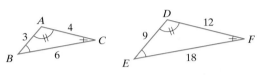

$$\frac{AB}{DE} = \frac{3}{9} = \frac{1}{3}, \frac{BC}{EF} = \frac{6}{18} = \frac{1}{3},$$

$$\frac{CA}{FD} = \frac{4}{12} = \frac{1}{3}$$

FOCUS ON **History**

THE PYTHAGOREAN THEOREM

Pythagoras was a Greek teacher who lived from about 580 B.C. to 501 B.C. He founded his school at Croton in southern Italy. The school was a tightly-knit community of eager young students, with the motto "All is number." Pythagoras himself taught by lecturing on the subjects of arithmetic, music, geometry, and astronomy—all based on mathematics. In fact, Pythagoras and his students held numbers to be sacred and tried to find mathematical order in all parts of life and nature. Pythagoras' followers, known as Pythagoreans, were sworn not to reveal any of the Master's teachings to outsiders. It is for this reason that very little of his life or teachings are known today—Pythagoreans were forbidden from recording any of the Master's teachings in written form, and Pythagoras left none of his own writings.

Traditionally, the so-called Pythagorean theorem is attributed to Pythagoras himself. However, some scholars question whether Pythagoras ever gave any rigorous proof of the theorem at all. There is ample evidence that several ancient cultures had knowledge of this important theorem and used its results in practical matters well before the time of Pythagoras.

■ A Babylonian clay tablet (#322 in the Plimpton collection at Columbia University) dating from 1900 B.C. gives numerical evidence that the Babylonians were well aware of what we today call the Pythagorean theorem.

■ The ancient Chinese knew of a very simple geometric proof of the Pythagorean theorem, which can be seen in *Arithmetic Classics of the Gnomon and the Circular Paths of Heaven*, dating from about 600 B.C.

■ The ancient Egyptians used the Pythagorean theorem to form right angles when they needed to measure a plot of land. They put equally-spaced knots in pieces of rope so that by stretching the rope out on the ground they could form a triangle with a right angle. They used equally-spaced knots to ensure that the lengths of the measuring ropes were in the ratio $3:4:5$.

CRITICAL THINKING

In the Egyptian use of the Pythagorean theorem, if three pieces of rope were used with one piece having a total of 10 knots equally spaced and another piece having a total of 13 knots with the same equal spacing, how many knots with the same equal spacing would a third piece of rope need so that the three pieces would form a right triangle with the first two pieces as the legs of the triangle? Draw a figure and explain your answer. (Assume that each piece of rope was knotted at both ends.)

538

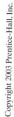

Chapter 8 Review

(8.1) *Classify each angle as acute, right, obtuse, or straight.*

1.

2.

3.

4.

5. Find the complement of a 25° angle.

6. Find the supplement of a 105° angle.

7. Find the supplement of a 72° angle.

8. Find the complement of a 1° angle.

Find the measure of x in each figure.

9.

10.

11.

12.

13. Identify the pairs of supplementary angles.

14. Identify the pairs of complementary angles.

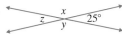

Find the measures of angles x, y, and z in each figure.

15.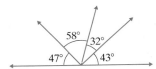

16.

17.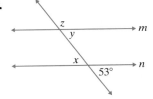

18.

19.

32°   45°   x

20.

62°   x   58°

21.

x   30°

22.
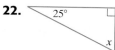
25°   x

Find the unknown diameter or radius as indicated.

23.

d = ?   r = 2.1 m

24.

d = 14 ft   r = ?

25.

d = 19 m   r = ?

26.
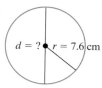
d = ?   r = 7.6 cm

Identify each solid.

27.

28.

29.

30.

Find the unknown radius or diameter as indicated.

31. The radius of a sphere is 9 inches. Find its diameter.

32. The diameter of a sphere is 4.7 meters. Find its radius.

Identify each regular polygon.

33.
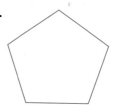

34.

(8.3) *Find the perimeter of each figure.*

35.
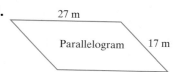
27 m   Parallelogram   17 m

36.
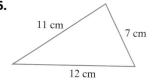
11 cm   7 cm   12 cm

37.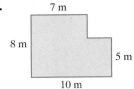
7 m
8 m
5 m
10 m

38.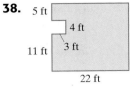
5 ft
4 ft
11 ft   3 ft
22 ft

Solve.

39. Find the perimeter of a rectangular sign that measures 6 feet by 10 feet.

40. Find the perimeter of a town square that measures 110 feet on a side.

Find the circumference of each circle. Use $\pi \approx 3.14$.

41.
1.7 in.

42.
5 yd

(8.4) *Find the area of each figure. For the circles, find the exact area and then use $\pi \approx 3.14$ to approximate the area.*

43.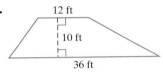
12 ft
10 ft
36 ft

44.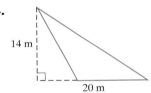
14 m
20 m

45.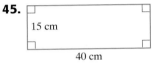
15 cm
40 cm

46.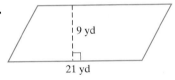
9 yd
21 yd

47.
7 ft

48.
2 in.

49.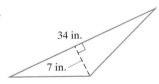
34 in.
7 in.

50.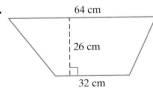
64 cm
26 cm
32 cm

51.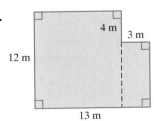
4 m
3 m
12 m
13 m

52. The amount of sealer necessary to seal a driveway depends on the area. Find the area of a rectangular driveway 36 feet by 12 feet.

53. Find how much carpet is necessary to cover the floor of the room shown.

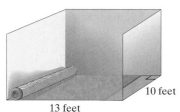

10 feet

13 feet

(8.5) *Find the volume of each solid. For Exercises 56 and 57, use* $\pi \approx \dfrac{22}{7}$.

54.

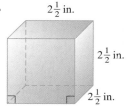

$2\frac{1}{2}$ in.

$2\frac{1}{2}$ in.

$2\frac{1}{2}$ in.

55.

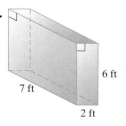

6 ft

7 ft

2 ft

56.

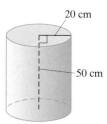

20 cm

50 cm

57.

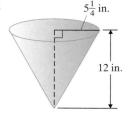

$5\frac{1}{4}$ in.

12 in.

58. Find the volume of a pyramid with a square base 2 feet on a side and a height of 2 feet.

59. Approximate the volume of a tin can 8 inches high and 3.5 inches in radius. Use 3.14 for π.

60. A chest has 3 drawers. If each drawer has inside measurements of $2\frac{1}{2}$ feet by $1\frac{1}{2}$ feet by $\frac{2}{3}$ foot, find the total volume of the 3 drawers.

61. A cylindrical canister for a shop vacuum is 2 feet tall and 1 foot in *diameter*. Find its exact volume.

62. Find the volume of air in a rectangular room 15 feet by 12 feet with a 7-foot ceiling.

63. A mover has two boxes left for packing. Both are cubical, one 3 feet on a side and the other 1.2 feet on a side. Find their combined volume.

(8.6) *Simplify.*

64. $\sqrt{64}$

65. $\sqrt{144}$

66. $\sqrt{36}$

67. $\sqrt{1}$

68. $\sqrt{\dfrac{4}{25}}$

69. $\sqrt{\dfrac{1}{100}}$

Find the unknown length of each given right triangle. If necessary, round to the nearest tenth.

70. leg = 12, leg = 5

71. leg = 20, leg = 21

72. leg = 9, hypotenuse = 14

73. leg = 124, hypotenuse = 155

74. leg = 66, leg = 56

75. Find the length to the nearest hundredth of the diagonal of a square that has a side of length 20 centimeters.

76. Find the height of the building rounded to the nearest tenth.

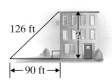

(8.7) *Given that the pairs of triangles are similar, find the unknown length n.*

77.

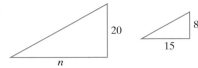

78.

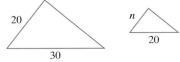

79.

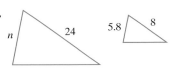

80.

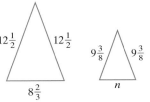

Solve.

81. A housepainter needs to estimate the height of a condominium. He estimates the length of his shadow to be 7 feet long and the length of the building's shadow to be 42 feet long. Find the approximate height of the building if the housepainter is $5\frac{1}{2}$ feet tall.

82. Santa's elves are making a triangular sail for a toy sailboat. The toy sail is to be the same shape as a real sailboat's sail. Use the following diagram to find the unknown lengths *x* and *y*.

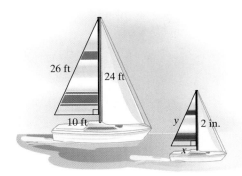

PRODUCT PACKAGING

Suppose you have just developed a new product that you would like to market. You will need to think about who would like to buy it, where and how it should be sold, for how much it will sell, how to package it, and other pressing concerns. Although all of these items are important to think through, many package designers believe that the packaging in which a product is sold is at least as important as the product itself.

Product packaging contains the product, keeping it from leaking out, keeping it fresh if perishable, providing protective cushioning against breakage, and keeping all the pieces together as a bundle. Product packaging also provides a way to give information about the product: What it is, how to use it, for whom it is designed, how it is beneficial or advantageous, whom to contact if more information is needed, what other products are necessary to use with the product, etc. Product packaging must be pleasing and eye-catching to the product's audience.

It must be capable of selling its contents without further assistance.

CRITICAL THINKING

1. How can a knowledge of geometry be helpful in the packaging design process?

2. Design two different packages for the same product that have roughly the same volume. Does one package "look" larger than the other? How could this be useful to a package designer?

Name _____ Section _____ Date _____

Chapter 8 Test

1. Find the complement of a 78° angle.   **2.** Find the supplement of a 124° angle.

3. Find the measure of ∠x.

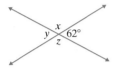

Find the measure of x, y, and z in each figure.

4.

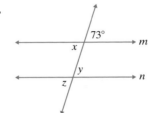

5.

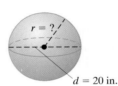

Find the unknown diameter or radius as indicated.

6.

7.

8. Find the measure of ∠x.

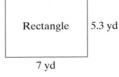

Find the perimeter (or circumference) and area of each figure. For the circle, give the exact value and then use π ≈ 3.14 for an approximation.

9.

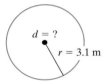

10.

Rectangle   5.3 yd

7 yd

11.

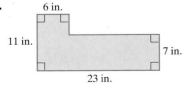

6 in.

11 in.

7 in.

23 in.

12. _____

13. _____

14. _____

15. _____

16. _____

17. _____

18. _____

19. _____

20. _____

21. _____

22. _____

23. _____

Find the volume of each solid. For the cylinder, use $\pi \approx \dfrac{22}{7}$.

12.

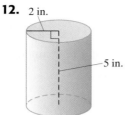

2 in.

5 in.

13.

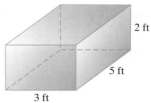

2 ft

5 ft

3 ft

Find each square root and simplify. Round the square root to the nearest thousandth if necessary.

14. $\sqrt{49}$

15. $\sqrt{157}$

16. $\sqrt{\dfrac{64}{100}}$

Solve.

17. Find the perimeter of a square photo with a side length of 4 inches.

4 in.

18. How much soil is needed to fill a rectangular hole 3 feet by 3 feet by 2 feet?

19. Find how much baseboard is needed to go around a rectangular room that measures 18 feet by 13 feet.

20. Approximate to the nearest hundredth of a centimeter the length of the missing side of a right triangle with legs of 4 centimeters each.

21. Vivian Thomas is going to put insecticide on her lawn to control grubworms. The lawn is a rectangle measuring 123.8 feet by 80 feet. The amount of insecticide required is 0.02 ounces per square foot. Find how much insecticide Vivian needs to purchase.

22. Given that the following triangles are similar, find the missing length n.

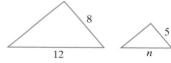

8

12

5

n

23. Tamara Watford, a surveyor, needs to estimate the height of a tower. She estimates the length of her shadow to be 4 feet long and the length of the tower's shadow to be 48 feet long. Find the height of the approximate tower if she is $5\dfrac{3}{4}$ feet tall.

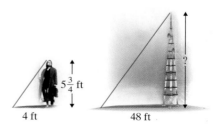

$5\dfrac{3}{4}$ ft

4 ft

48 ft

546

Cumulative Review

1. Write the decimal 19.5023 in words.

2. Round 736.2359 to the nearest hundredth.

3. Add: 45 + 2.06

Multiply.

4. 7.68 × 10

5. 76.3 × 1000

6. Divide: 270.2 ÷ 7

7. Simplify: $\dfrac{0.7 + 1.84}{0.4}$

8. Insert < , > , or = to form a true statement. $\dfrac{1}{8}$ 0.12

9. Write the ratio of 2.6 to 3.1 as a fraction in simplest form.

10. Is $\dfrac{2}{3} = \dfrac{4}{6}$ a true proportion?

11. In a survey of 100 people, 17 people drive blue cars. What percent drive blue cars?

Write each percent as a fraction in simplest form.

12. 1.9%

13. 125%

14. 85% of 300 is what number?

15. 20.8 is 40% of what?

1. _____

2. _____

3. _____

4. _____

5. _____

6. _____

7. _____

8. _____

9. _____

10. _____

11. _____

12. _____

13. _____

14. _____

15. _____

16. Mr. Buccaran, the principal at Slidell High School, counted 31 freshmen absent during a particular day. If this is 4% of the total number of freshmen, how many freshmen are there at Slidell High School?

17. Sherry Souter, a real estate broker for Wealth Investments, sold a house for $114,000 last week. If her commission is 1.5% of the selling price of the home, find the amount of her commission.

18. Convert 8 feet to inches.

19. Convert 3.2 kilograms to grams.

20. Subtract 3 quarts from 4 gallons 2 quarts.

21. Convert 29°C to degrees Fahrenheit.

22. Find the measure of ∠a.

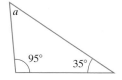

23. Find the perimeter of the rectangle below:

24. Find $\sqrt{\dfrac{4}{25}}$.

25. Find the ratio of corresponding sides for triangles *ABC* and *DEF*.

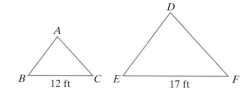

16. _____

17. _____

18. _____

19. _____

20. _____

21. _____

22. _____

23. _____

24. _____

25. _____

Statistics and Probability

We often need to make decisions based on known statistics or the probability of an event occurring. For example, we decide whether or not to bring an umbrella to work based on the probability of rain. We choose an investment based on its mean, or average, return. We can predict which football team will win based on the trend in its previous wins and losses. This chapter reviews presenting data in a usable form on a graph and the basic ideas of statistics and probability.

A tornado is a violent, whirling column of air that is often spawned by the unstable weather conditions that occur during thunderstorms. Although tornadoes are capable of sustaining wind speeds of 250 to more than 300 mph, most tornadoes have wind speeds under 110 mph. The average forward speed of a tornado is 30 mph, but some tornadoes have been known to travel over land at speeds up to 70 mph. The path of a tornado can extend anywhere from a few feet to 100 miles long. Each year in the United States, an average of 800 tornadoes occur, causing an average of 80 deaths. The deadliest tornado in the United, States was the Tri-State Tornado Outbreak on March 18, 1925, which killed 689 people and injured over 2000 more in Missouri, Illinois, and Indiana. In Exercises 17–22 on page 558 and the Chapter Highlights on page 594, we will see how graphs can be used to summarize data about tornadoes.

Name _____ Section _____ Date _____

Chapter 9 Pretest

1. _____

2. _____

3. _____

4.   see graph

5. _____

6. _____

7. _____

8. _____

9.   see graph

10. _____

11. _____

12. _____

13. _____

14. _____

550

The line graph below shows the number of burglaries in a town during the months of March through September.

Mar Apr May June July Aug Sept

1. During which month, between March and September, did the fewest number of burglaries occur?

2. During which month, between March and September, were there 400 burglaries?

3. How many burglaries were there in September?

4. The following table shows a breakdown of an average day for Dawn Miller.

Attending college classes	4 hours
Studying	3 hours
Working	5 hours
Sleeping	8 hours
Driving	1 hour
Other	3 hours

Draw a circle graph showing this data.

Below is a list of scores from the final exam given in Mrs. Maxwell's basic college mathematics class. Use this list to complete the table below:

| 76 | 71 | 94 | 73 | 81 | 78 | 96 | 65 |
| 95 | 80 | 90 | 86 | 98 | 88 | 62 | 91 |

	Class Intervals (Scores)	Tally	Class Frequency (Number of Exams)
5.	60–69		
6.	70–79		
7.	80–89		
8.	90–99		

9. Use the table from Exercises 5 through 8 to draw a histogram.

10. Find the grade point average if the following grades were earned in one semester.

Grade	Credit Hours
B	4
B	3
A	3
C	5
D	2

11. Find the median of the following list of numbers: 28, 36, 81, 64, 73, 31, 25, 92, 74

A single die is tossed. Find the probability that the die is each of the following:

12. a 4   **13.** a number greater than 3   **14.** a 3 or a 5

9.1 Reading Pictographs, Bar Graphs, and Line Graphs

Often data is presented visually in a graph. In this section, we practice reading several kinds of graphs including pictographs, bar graphs, and line graphs.

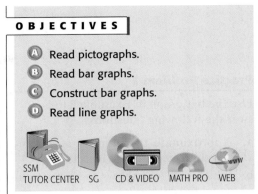

OBJECTIVES

Ⓐ Read pictographs.
Ⓑ Read bar graphs.
Ⓒ Construct bar graphs.
Ⓓ Read line graphs.

SSM   SG   CD & VIDEO   MATH PRO   WEB
TUTOR CENTER

Ⓐ Reading Pictographs

A **pictograph** such as the one below is a graph in which pictures or symbols are used. This type of graph contains a key that explains the meaning of the symbol used. An advantage of using a pictograph to display information is that comparisons can easily be made. A disadvantage of using a pictograph is that it is often hard to tell what fractional part of a symbol is shown. For example, in the pictograph below, Germany shows a part of a symbol, but it's hard to read with any accuracy what fractional part of a symbol is shown.

EXAMPLE 1

The following pictograph shows the approximate amount of nuclear energy generated by selected countries in the year 2000. Use this pictograph to answer the questions.

Nuclear Energy Generated by Selected Countries (2000)

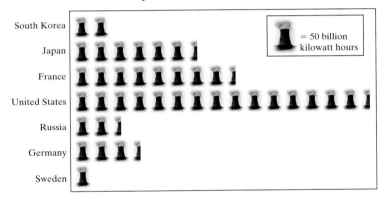

Source: Energy Information Administration

a. Approximate the amount of nuclear energy that is generated in South Korea.
b. Approximate how much more nuclear energy is generated in France than in South Korea.

Solution:

a. South Korea corresponds to 2 symbols, and each symbol represents 50 billion kilowatt hours of energy. This means that South Korea generates approximately $2 \cdot (50 \text{ billion})$ or 100 billion kilowatt hours of energy.
b. France shows $6\frac{1}{2}$ more symbols than South Korea. This means that France generates $6\frac{1}{2} \cdot (50 \text{ billion})$ or 325 billion more kilowatt hours of nuclear energy than South Korea.   ●

Ⓑ Reading Bar Graphs

Another way to present data by a graph is with a **bar graph**. Bar graphs can appear with vertical bars or horizontal bars. Although we have studied bar graphs in previous sections, we now practice reading the height of the bars contained in a bar graph. An advantage to using bar graphs is that a scale is usually included for greater accuracy. Care must be taken when reading bar

Practice Problem 1

Use the pictograph shown in Example 1 to answer the following questions:

a. Approximate the amount of nuclear energy that is generated in Sweden.
b. Approximate the total nuclear energy generated in Sweden and Russia.

Answers

1. a. 50 billion kilowatt hours,   **b.** 175 billion kilowatt hours

graphs, as well as other types of graphs—they may be misleading, as shown later in this section.

EXAMPLE 2

The following bar graph shows the number of endangered species in 2001. Use this graph to answer the questions.

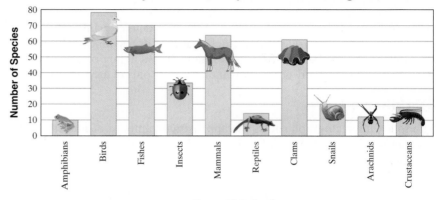

How Many U.S. Animal Species Are Endangered?

Source: U.S. Fish and Wildlife Service

a. Approximate the number of endangered species that are reptiles.

b. Which category has the most endangered species?

Solution:

a. To approximate the number of endangered species that are reptiles, we go to the top of the bar that represents reptiles. From the top of this bar, we move horizontally to the left until the scale is reached. We read the height of the bar on the scale as approximately 15. There are approximately 15 reptile species that are endangered, as shown in the next figure.

b. The most endangered species is represented by the longest bar. The longest bar corresponds to birds.

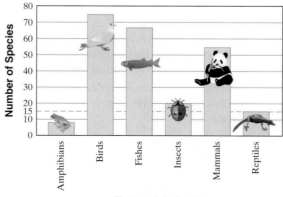

How Many Species Are Endangered?

Source: U.S. Fish and Wildlife Service

Practice Problem 2

Use the bar graph in Example 2 to answer the following questions:

a. Approximate the number of endangered species that are insects.

b. Which category shows the fewest endangered species?

Answers

2. a. 33, **b.** amphibians

As mentioned previously, graphs can be misleading. Both graphs below show the same information, but with different scales. Special care should be taken when forming conclusions from the appearance of a graph.

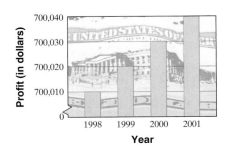

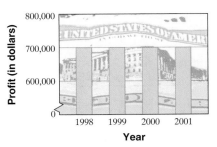

Are profits shown in the graphs above greatly increasing, or are they remaining about the same?

C Constructing Bar Graphs

Next, we practice constructing a bar graph.

EXAMPLE 3

Draw a vertical bar graph using the information in the table about the caffeine content of selected foods.

Average Caffeine Content of Selected Foods			
Food	**Milligrams**	**Food**	**Milligrams**
Brewed coffee (percolator, 8 ounces)	124	Instant coffee (8 ounces)	104
Brewed decaffeinated coffee (8 ounces)	3	Brewed tea (U.S. brands, 8 ounces)	64
Coca-Cola Classic (8 ounces)	31	Mr. Pibb (8 ounces)	27
Dark chocolate (semi sweet, $1\frac{1}{2}$ ounces)	30	Milk chocolate (8 ounces)	9

(*Sources:* International Food Information Council and the Coca-Cola Company)

Solution: We draw and label a vertical line and a horizontal line as shown on the next page in the bar graph on the left. We place the different food categories along the horizontal line. Along the vertical line, we place a scale. The scale shown next starts at 0 and then shows multiples of 20 placed at equally distant intervals. It may also be helpful to draw horizontal lines along the scale markings to help draw the vertical bars at the correct height. The finished bar graph is shown on the next page on the right.

Practice Problem 3

Draw a vertical bar graph using the information in the table about electoral votes for selected states.

Electoral Votes for President by Selected States	
State	**Electoral Votes**
Texas	34
California	55
Florida	27
Louisiana	9
Illinois	21
Massachusetts	12

(*Source:* National Archives and Records Administration)

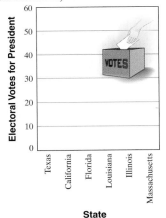

Answer

3.

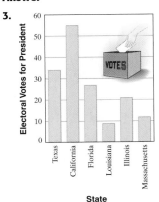

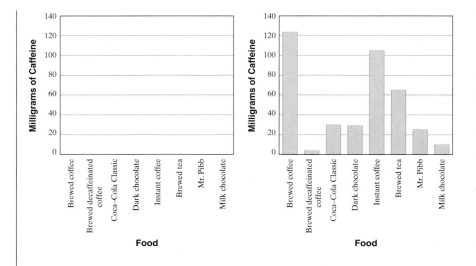

D Reading Line Graphs

Another common way to display information with a graph is by using a **line graph**. An advantage of a line graph is that it can be used to visualize relationships between two quantities. A line graph can also be very useful in showing a change over time.

Practice Problem 4

Use the temperature graph in Example 4 to answer the following questions:

a. During what month is the average daily temperature the lowest?

b. During what month is the average daily temperature 25°F?

c. During what months is the average daily temperature greater than 70°F?

EXAMPLE 4

The following line graph shows the average daily temperature for each month for Omaha, Nebraska. Use this graph to answer the questions.

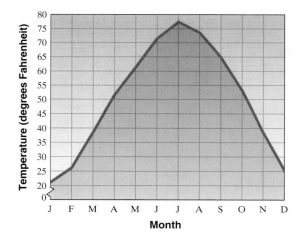

Source: National Climatic Data Center

a. During what month is the average daily temperature the highest?

b. During what month is the average daily temperature 65°F?

c. During what months is the average daily temperature less than 30°F?

Solution:

a. The month with the highest temperature corresponds to the highest point. We follow the highest point downward to the horizontal month scale and see that this point corresponds to July.

b. We find the 65°F mark on the vertical scale and move to the right until a darkened point on the graph is reached. From that point, we move downward to the Month scale and read the corresponding month. During the month of September, the average daily temperature was 65°F.

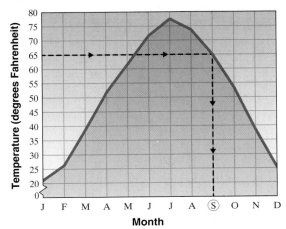

Source: National Climatic Data Center

c. To see what months the temperature is less than 30°F, we find what months correspond to darkened points that fall below the 30°F mark on the vertical scale. These months are January, February, and December.

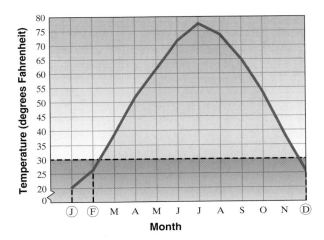

Source: National Climatic Data Center

FOCUS ON **Business and Career**

SURVEYS

How often have you read an article in a newspaper or a magazine that included results from a survey or a poll? Surveys have become very popular ways for businesses and organizations to get feedback on a variety of topics. A political organization may hire a polling group to gauge the public's response to a political candidate. A food company may send out surveys to customers to gather market research on a new food product. A health club may ask its patrons to fill out a brief comment card to report their views on the services offered at the club. Surveys are useful for collecting information needed to make a decision: Should we do a media blitz to increase public awareness of our candidate? Do we need to change the recipe for our new food product to make it more appealing to a wider audience? Should we add a new service to attract new customers and retain our current clientele?

After data have been collected from a survey, they must be summarized to be useful in decision making. Survey results can be summarized in any of the types of graphs presented in this chapter. Survey results can also be summarized by reporting average responses or with a combination of basic statistics and graphs.

GROUP ACTIVITY

1. Conduct a survey of 30 students in one of your classes. Ask each student to report his or her age.
2. Find the difference between the ages of the youngest and oldest survey respondents (this difference between the largest and smallest value is called the "range"). Divide the range into five or six equal age categories. Tally the number of your respondents that fall into each category. Make a histogram of your results. What does this graph tell you about the ages of your survey respondents?
3. Find the average age of your survey respondents.
4. Find the median age of your survey respondents.
5. Find the mode of your survey respondents.
6. Compare the mean, median, and mode of your age data. Are these measures similar? Which is the largest? Which is the smallest? If there is a noticeable difference between any of these measures, can you explain why?

EXERCISE SET 9.1

A *The following pictograph shows the annual automobile production by one plant for the years 1995–2001. Use this graph to answer Exercises 1 through 8. See Example 1.*

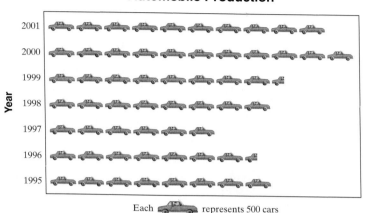

Automobile Production

Each represents 500 cars

1. In what year was the greatest number of cars manufactured?

2. In what year was the least number of cars manufactured?

3. Approximate the number of cars manufactured in the year 1998.

4. Approximate the number of cars manufactured in the year 1999.

5. In what year(s) did the production of cars decrease from the previous year?

6. In what year(s) did the production of cars increase from the previous year?

7. In what year(s) were 4000 cars manufactured?

8. In what year(s) were 5500 cars manufactured?

The following pictograph shows the average number of ounces of chicken consumed per person per week in the United States. Use this graph to answer Exercises 9 through 16. See Example 1.

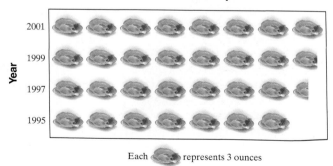

Chicken Consumption

Each represents 3 ounces

Source: National Agricultural Statistics Service

9. Approximate the number of ounces of chicken consumed per week in 1997.

10. Approximate the number of ounces of chicken consumed per week in 2001.

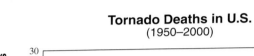

11. In what year(s) was the number of ounces of chicken consumed per week greater than 21 ounces?

12. In what year(s) was the number of ounces of chicken consumed per week 21 ounces or less?

13. What was the increase in average chicken consumption from 1995 to 2001?

14. What was the increase in average chicken consumption from 1997 to 2001?

15. Describe a trend in eating habits shown by this graph.

16. In 2001, did you eat less than, greater than, or about the same as the U.S. average number of ounces consumed per week?

B *The following bar graph shows the average number of people killed by tornadoes during the months of the year. Use this graph to answer Exercises 17 through 22. See Example 2.*

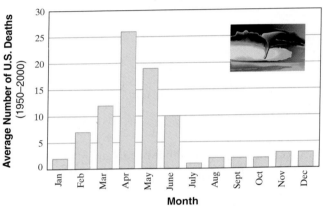

Tornado Deaths in U.S.
(1950–2000)

Source: Storm Prediction Center

17. In which month(s) did the most tornado-related deaths occur?

18. In which month(s) did the fewest tornado-related deaths occur?

19. Approximate the number of tornado-related deaths that occurred in May.

20. Approximate the number of tornado-related deaths that occurred in April.

21. In which month(s) did over 5 deaths occur?

22. In which month(s) did over 15 deaths occur?

The following horizontal bar graph shows the 2001 population of the world's largest agglomerations (cities plus their suburbs). Use this graph to answer Exercises 23 through 28. See Example 2.

World's Largest Agglomerations

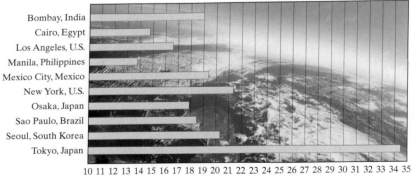

Source: Thomas Brinkhoff: *The Principal Agglomerations of the World*,
http://www.citypopulation.de, 13.05.2001

23. Name the city with the largest population and estimate its population.

24. Name the city whose population is between 16 and 17 million and estimate its population.

25. Name the city in the United States with the largest population and estimate its population.

26. Name the city with the smallest population and estimate its population.

27. How much larger is Tokyo than Sao Paulo?

28. How much larger is Bombay than Manila?

C *Use the information given to draw a vertical bar graph. Clearly label the bars. See Example 3.*

29.

Fiber Content of Selected Foods	
Food	**Grams of Total Fiber**
Kidney beans $\left(\frac{1}{2}c\right)$	4.5
Oatmeal, cooked $\left(\frac{3}{4}c\right)$	3.0
Peanut butter, chunky (2 tbsp)	1.5
Popcorn (1 c)	1.0
Potato, baked with skin (1 med)	4.0
Whole wheat bread (1 slice)	2.5

(*Sources:* American Dietetic Association and National Center for Nutrition and Dietetics)

Fiber Content of Selected Foods

Grams of Total Fiber

2
1
0

Food

30.

U.S. Restaurant Industry Annual Food and Beverage Sales	
Year	**Sales in Billions of Dollars**
1970	43
1980	120
1990	239
2001	399

(*Source:* National Restaurant Association)

U.S. Restaurant Industry Annual Food and Beverage Sales

Sales (in billions of dollars)

200
100

Year

31.

Best-selling Albums of All Time (U.S. Sales)	
Album	**Multiplatinum Level**
Pink Floyd: *The Wall* (1979)	23
Michael Jackson: *Thriller* (1982)	26
AC/DC: *Back in Black* (1980)	19
Billy Joel: *Greatest Hits Volumes I & II* (1985)	21
Eagles: *Their Greatest Hits* (1976)	27
Led Zeppelin: *Led Zeppelin IV* (1971)	22

(*Source:* Recording Industry Association of America)

Best-selling Albums of All Time (U.S. sales)

Multiplatinum Level

20
16

32.

Fuel Economy of the Top-Selling Vehicles in the United States for 2000	
Vehicle (sales rank)	Highway Fuel Economy* (in miles per gallon)
Ford F-Series (1)	21
Chevrolet Silverado (2)	23
Ford Explorer (3)	23
Toyota Camry (4)	32
Honda Accord (5)	31
Ford Taurus (6)	28

*Maximum fuel economy available among all model trims
(*Sources:* Edmunds.com, Inc. and U.S. Environmental Protection Agency)

Fuel Economy of the Top-Selling Vehicles in the United States for 2000

Highway Fuel Economy (in miles per gallon)

22
20

Vehicle

Ⓓ *The following line graph shows the average number of goals scored per Major League Soccer game during the years shown. Use this graph to answer Exercises 33 through 38. See Example 4.*

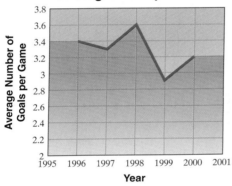

Major League Soccer Average Goals per Game

Source: Major League Soccer

33. Approximate the average number of goals per game in 1996.

34. Approximate the average number of goals per game in 2000.

35. During what year shown was the average number of goals per game the highest?

36. During what year shown was the average number of goals per game the lowest?

37. Between 1996 and 1997, did the average number of goals per game increase or decrease?

38. Between 1999 and 2000, did the average number of goals per game increase or decrease?

Review and Preview

Find each percent. See Sections 6.3 and 6.4.

39. 30% of 12

40. 45% of 120

41. 10% of 62

42. 95% of 50

Write each fraction as a percent. See Section 6.2.

43. $\dfrac{1}{4}$

44. $\dfrac{2}{5}$

45. $\dfrac{17}{50}$

46. $\dfrac{9}{10}$

Combining Concepts

The following double-line graph shows temperature highs and lows for a week. Use this graph to answer Exercises 47 through 52.

47. What was the high temperature reading on Thursday?

48. What was the low temperature reading on Thursday?

49. What day was the temperature the lowest? What was this low temperature?

50. What day of the week was the temperature the highest? What was this high temperature?

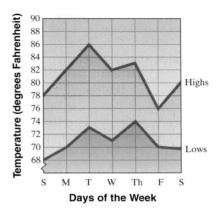

51. On what day of the week was the difference between the high temperature and the low temperature the greatest? What was this difference in temperature?

52. On what day of the week was the difference between the high temperature and the low temperature the least? What was this difference in temperature?

Internet Excursions

Go To: http://www.prenhall.com/martin-gay_basic 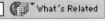 What's Related

The Bureau of Labor Statistics, within the U.S. Department of Labor, is the principal fact-finding agency for the Federal Government in the broad field of labor economics and statistics. The World Wide Web address listed here will provide you with access to the "U.S. Economy at a Glance" Web site of the Bureau of Labor Statistics, or a related site. You will find links to graphs of various data series.

53. Visit this Web site and view the graph of "Unemployment Rate." What type of graph is this? Use the graph to estimate when the highest unemployment rate occurred during the period of time covered by the graph, and estimate that unemployment rate.

54. Visit this Web site and view the graph of "Average Hourly Earnings." What type of graph is this? Describe any trends that you see in the graph.

FOCUS ON **the Real World**

MISLEADING GRAPHS

Graphs are very common in magazines and in newspapers such as *USA Today*. Graphs can be a convenient way to get an idea across because, as the old saying goes, "A picture is worth a thousand words." However, some graphs can be deceptive, which may or may not be intentional. It is important to know some of the ways that graphs can be misleading.

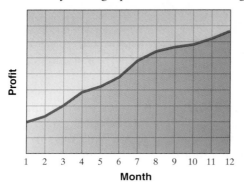

Beware of graphs like the one above. Notice that the graph shows a company's profit for various months. It appears that profit is growing quite rapidly. However, this impressive picture tells us little without knowing what units of profit are being graphed. Does the graph show profit in dollars or millions of dollars? An unethical company with profit increases of only a few pennies could use a graph like this one to make the profit increase seem much more substantial than it really is. A truthful graph describes the size of the units used along the vertical axis.

Another type of graph to watch for is one that misrepresents relationships. This can occur in both bar graphs and circle graphs. For example, the bar

graph below shows the number of men and women employees in the accounting and shipping departments of a certain company. In the accounting department, the bar representing the number of women is twice as tall as the bar representing the number of men. However, the number of women (13) is not twice the number of men (10). This set of bars misrepresents the relationship between the number of men and women. Do you see how the relationship between the number of men and women in the shipping department is distorted by the heights of the bars used? A truthful graph will use bar heights or circle sectors that are proportional in size to the numbers they represent.

We have already seen that the impression a graph can give also depends on its vertical scale. Here is another example: The two graphs below represent exactly the same data. The only difference between the two graphs is the vertical scale—one shows enrollments from 246 to 260 students, and the other shows enrollments between 0 and 300 students. If you were trying to convince readers that algebra enrollment at UPH had changed drastically over the period 1996–2000, which graph would you use? Which graph do you think gives the more honest representation?

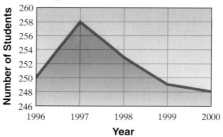

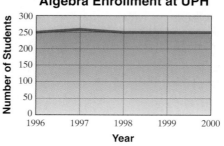

9.2 **Reading Circle Graphs**

A **Reading Circle Graphs**

In Section 6.1, the following circle graph was shown. This particular graph shows the favorite cookie for every 100 people.

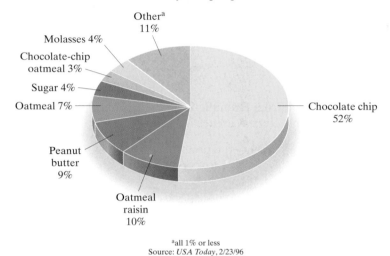

Other[a]
11%

Molasses 4%

Chocolate-chip oatmeal 3%

Sugar 4%

Oatmeal 7%

Chocolate chip
52%

Peanut butter
9%

Oatmeal raisin
10%

[a]all 1% or less
Source: *USA Today*, 2/23/96

Each sector of the graph (shaped like a piece of pie) shows a category and the relative size of the category. In other words, the most popular cookie is the chocolate chip cookie, and it is represented by the largest sector.

EXAMPLE 1

Find the ratio of people preferring chocolate chip cookies to total people. Write the ratio as a fraction in simplest form.

Solution: The ratio is

$$\frac{52 \text{ people preferring chocolate chip}}{100 \text{ people}} = \frac{52}{100} = \frac{13}{25}$$

●

A circle graph is often used to show percents in different categories, with the whole circle representing 100%. For example, in 2000 the population of the United States was about 281,400,000. The following circle graph shows the percent of Americans with various numbers of working computers at home. Notice that the percents in each category sum to 100%.

Number of Working Computers at Home

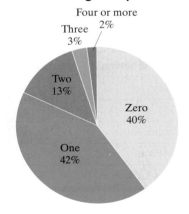

Four or more
2%

Three
3%

Two
13%

Zero
40%

One
42%

Source: UCLA Center for Communication Policy

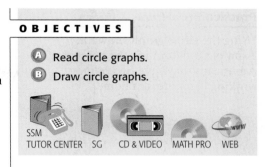

OBJECTIVES

A Read circle graphs.

B Draw circle graphs.

SSM
TUTOR CENTER   SG   CD & VIDEO   MATH PRO   WEB

Practice Problem 1

Find the ratio of people preferring oatmeal raisin cookies to total people. Write the ratio as a fraction in simplest form.

Answer

1. $\frac{1}{10}$

Practice Problem 2

Using the circle graph shown in Example 2, determine the percent of Americans that have two or more working computers at home.

Practice Problem 3

Using the circle graph from Example 2, find the number of Americans that have four or more working computers at home.

Concept Check

Can the following data be represented by a circle graph? Why or why not?

Responses to the question, "In which activities are you involved?"	
Intramural sports	60%
On-campus job	42%
Fraternity/sorority	27%
Academic clubs	21%
Music programs	14%

Practice Problem 4

Use the data shown to draw a circle graph.

Freshmen	30%
Sophomores	27%
Juniors	25%
Seniors	18%

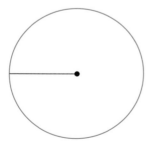

Answers

2. 18%,   **3.** 5,628,000 Americans,   **4.** see page 565

Concept Check: no; the percents add up to more than 100%

EXAMPLE 2

Using the circle graph shown, determine the percent of Americans that have one or more working computers at home.

Solution: To find this percent, we add the percents corresponding to one, two, three, and four or more working computers at home. The percent of Americans that have one or more working computers at home is

$$42\% + 13\% + 3\% + 2\% = 60\%$$ ●

EXAMPLE 3

Using the circle graph from Example 2, find the number of Americans that have no working computers at home.

Solution: We use the percent equation.

amount	=	percent	·	base

$$\text{amount} = 0.40 \cdot 281{,}400{,}000$$
$$= 0.40(281{,}400{,}000) = 112{,}560{,}000$$

Thus, 112,560,000 Americans have no working computer at home.   ●

Try the Concept Check in the margin.

Ⓑ Drawing Circle Graphs

To draw a circle graph, we use the fact that a whole circle contains 360° (degrees).

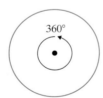

EXAMPLE 4

The following table shows the percent of U.S. armed forces personnel that are in each branch of service. (*Source:* U.S. Department of Defense)

Branch of Service	Percent
Army	33%
Navy	27%
Marine Corps	12%
Air Force	25%
Coast Guard	3%

Draw a circle graph showing this data.

Solution: First we find the number of degrees in each sector representing each branch of service. Remember that the whole circle contains 360°. (We will round degrees to the nearest whole.)

Sector	Degrees in Each Sector
Army	33% × 360° = 118.8° ≈ 119°
Navy	27% × 360° = 97.2° ≈ 97°
Marine Corps	12% × 360° = 43.2° ≈ 43°
Air Force	25% × 360° = 90° = 90°
Coast Guard	3% × 360° = 10.8° ≈ 11°

Next we draw a circle and mark its center. Then we draw a line from the center of the circle to the circle itself.

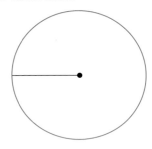

To construct the sectors, we will use a **protractor**. We place the hole in the protractor over the center of the circle. Then we adjust the protractor so that 0° on the protractor is aligned with the line that we drew.

It makes no difference which sector we draw first. To construct the "Army" sector, we find 119° on the protractor and mark our circle. Then we remove the protractor and use this mark to draw a second line from the center to the circle itself.

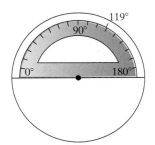
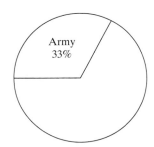

To construct the "Navy" sector, we follow the same procedure as above, except that we line up 0° with the second line we drew and mark the protractor at 97°.

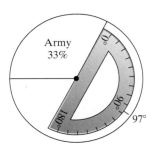
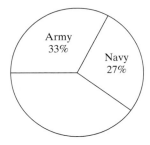

Answer

4.

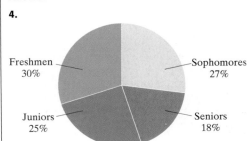

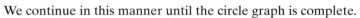

We continue in this manner until the circle graph is complete.

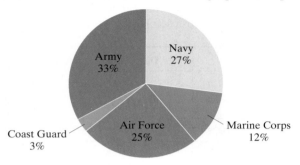

Concept Check

True or false? The larger a sector in a circle graph, the larger the percent of the total it represents. Explain your answer.

Try the Concept Check in the margin.

Answer

Concept Check: true

EXERCISE SET 9.2

A *The following circle graph is a result of surveying 700 college students. They were asked where they live while attending college. Use this graph to answer Exercises 1 through 6. Write all ratios as fractions in simplest form. See Example 1.*

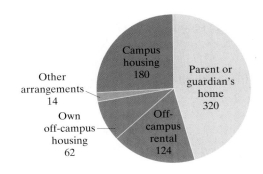

1. Where do most of these college students live?

2. Besides the category "Other Arrangements," where do least of these college students live?

3. Find the ratio of students living in campus housing to total students.

4. Find the ratio of students living in off-campus rentals to total students.

5. Find the ratio of students living in campus housing to students living at home.

6. Find the ratio of students living in off-campus rentals to students living at home.

The following circle graph shows the relative sizes of the continents of Earth. Use this graph for Exercises 7 through 14. See Examples 2 and 3.

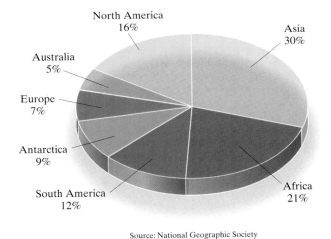

Source: National Geographic Society

7. Which continent is the largest?

8. Which continent is the smallest?

9. What percent of the land on Earth is accounted for by Asia and Europe together?

10. What percent of the land on Earth is accounted for by North and South America?

The total amount of land on Earth is approximately 57,000,000 square miles. Use the graph to find the area of the continents given in Exercises 11 through 14.

11. Asia   **12.** South America   **13.** Australia   **14.** Europe

The following circle graph shows the percent of the types of books available at Midway Memorial Library. Use this graph for Exercises 15 through 24. See Examples 2 and 3.

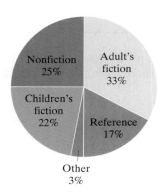

 15. What percent of books are classified as some type of fiction?

16. What percent of books are nonfiction or reference?

17. What is the second-largest category of books?

18. What is the third-largest category of books?

If this library has 125,600 books, find how many books are in each category given in Exercises 19 through 24.

 19. Nonfiction

20. Reference

21. Children's fiction

22. Adult's fiction

23. Reference or other

24. Nonfiction or other

B *Draw a circle graph to represent the information given in each table. See Example 4.*

25.

First Half of 2001 Light Vehicle Sales by Vehicle Origin	
Country of Origin	**Percent**
United States	65%
Asia	30%
Europe	5%

(*Source:* Ward's AutoInfoBank)

26.

Size of Kellogg's Business Segments after Acquiring Keebler	
Business Segment	**Percent of Annual Sales**
U.S. cereal	27%
U.S. convenience foods	43%
International	30%

(*Source:* Kellogg Company's 2000 Annual Report)

Review and Preview

Write the prime factorization of each number. See Section 2.2.

27. 20   **28.** 25   **29.** 40   **30.** 16   **31.** 85   **32.** 105

Combining Concepts

The following circle graph shows the relative sizes of the great oceans. These oceans together make up 264,489,800 square kilometers of the Earth's surface. Find the square kilometers for each ocean.

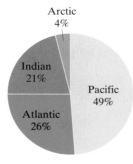

Arctic
4%

Indian
21%

Pacific
49%

Atlantic
26%

Source: *Philip's World Atlas*

33. Before actually calculating, determine which ocean is the largest. How can you answer this question by looking at the circle graph?

34. Before calculating, determine which ocean is the smallest. How can you answer this question by looking at the circle graph?

35. Pacific Ocean   **36.** Atlantic Ocean   **37.** Indian Ocean   **38.** Arctic Ocean

The following circle graph summarizes the results of a survey of 2711 Internet users who make purchases online. Use this graph for Exercises 39 through 41. Round to the nearest whole.

Online Spending per Month

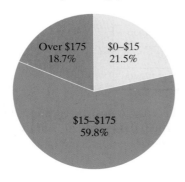

Over $175
18.7%

$0–$15
21.5%

$15–$175
59.8%

Source: UCLA Center for Communication Policy

39. How many of the survey respondents said that they spend $0–$15 online each month?

40. How many of the survey respondents said that they spend $15–$175 online each month?

41. How many of the survey respondents said that they spend at least $15 or over $175 online each month?

STEM-AND-LEAF DISPLAYS

Stem-and-leaf displays are another way to organize data. After data are logically organized, it can be much easier to draw conclusions from them.

Suppose we have collected the following set of data. It could represent the test scores for an algebra class or the pulse rates of a group of small children.

90	73	93	99	79	95	69	78	93	80
89	85	97	78	75	79	72	76	97	88
83	98	72	94	92	79	70	98	85	99

In a stem-and-leaf display, the last digit of each number forms the *leaf*, and the remaining digits to the left form the *stem*. For the first number in the list, 90, 9 is the stem and 0 is the leaf. To make the stem-and-leaf display, we write all of the stems in numerical order in a column. Then we write each leaf on the horizontal line next to its stem, aligning leaves in vertical columns. In this case, because the data range from 69 to 99, we use the stems 6, 7, 8, and 9. Each line of the display represents an interval of data; for instance, the line corresponding to the stem **7** represents all data that fall in the interval **70** to **79**, inclusive. After the data have been divided into stems and leaves on the display as shown in the table on the left, we simply rearrange the leaves on each line to appear in numerical order, as shown in the table on the right.

Stem	Leaf		Stem	Leaf
6	9		6	9
7	39885926290	⟶	7	02235688999
8	095835		8	035589
9	039537784289		9	023345778899

Now that the data have been organized into a stem-and-leaf display, it is easy to answer questions about the data such as: What are the least and greatest values in the set of data? Which data interval contains the most items from the data set? How many values fall between 74 and 84? Which data value occurs most frequently in the data set? What patterns or trends do you see in the data?

CRITICAL THINKING

Make a stem-and-leaf display of the weekend emergency room admission data shown on the right. Then answer the following questions:

1. What is the difference between the least number and the greatest number of weekend ER admissions?
2. How many weekends had between 125 and 165 ER admissions?
3. What number of weekend ER admissions occurred most frequently?
4. Which interval contains the most weekend ER admissions?

Number of Emergency Room Admissions on Weekends				
198	168	117	185	159
160	177	169	112	175
170	188	137	117	145
198	169	154	163	192
167	179	155	133	121
162	188	124	145	146
128	181	198	149	140
122	162	161	180	177

9.3   Reading Histograms

Ⓐ   Reading Histograms

Suppose that the test scores of 36 students are summarized in the table below:

Student Scores	Frequency (number of students)
40–49	1
50–59	3
60–69	2
70–79	10
80–89	12
90–99	8

The results in the table can be displayed in a histogram. A **histogram** is a special bar graph. The width of each bar represents a range of numbers called a **class interval**. The height of each bar corresponds to how many times a number in the class interval occurred and is called the **class frequency**.

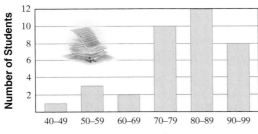

Number of Students / Student Test Scores

EXAMPLE 1

Use the preceding histogram to determine how many students scored 50–59 on the test.

Solution: We find the bar representing 50–59. The height of this bar is 3, which means 3 students scored 50–59 on the test.   ●

EXAMPLE 2

Use the preceding histogram to determine how many students scored 80 or above on the test.

Solution: We see that two different bars fit this description. There are 12 students who scored 80–89 and 8 students who scored 90–99. The sum of these two categories is 12 + 8 or 20 students. Thus, 20 students scored 80 or above on the test.   ●

Ⓑ   Constructing Histograms

The daily high temperatures for 1 month in New Orleans, Louisiana, are recorded in the following list:

85°	90°	95°	89°	88°	94°
87°	90°	95°	92°	95°	94°
82°	92°	96°	91°	94°	92°
89°	89°	90°	93°	95°	91°
88°	90°	88°	86°	93°	89°

Practice Problem 1

Use the histogram for Examples 1 and 2 to determine how many students scored 70–79 on the test.

Practice Problem 2

Use the histogram for Examples 1 and 2 to determine how many students scored less than 60 on the test.

Answers

1. 10, **2.** 4

Practice Problem 3

Complete the frequency distribution table for the data below. Each number represents a credit card owner's unpaid balance each month.

0	53	89	125
265	161	37	76
62	201	136	42

Class Intervals (Credit Card Balances)	Tally	Class Frequency (Number of Months)
$0–$49	___	___
$50–$99	___	___
$100–$149	___	___
$150–$199	___	___
$200–$249	___	___
$250–$299	___	___

Practice Problem 4

Construct a histogram from the frequency distribution table above.

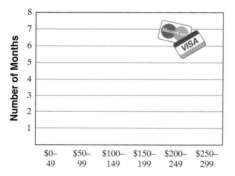

Concept Check

Which of the following sets of data is better suited to representation by a histogram? Explain.

Grade on Final	# of Students	Section Number	Avg. Grade on Final
51–60	12	150	78
61–70	18	151	83
71–80	29	152	87
81–90	23	153	73
91–100	25		

Answers

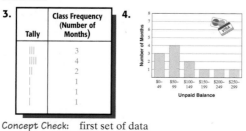

3.

Tally	Class Frequency (Number of Months)
				3
					4
			2
		1
		1
		1

4.

Concept Check: first set of data

The data in this list have not been organized and can be hard to interpret. One way to organize the data is to place it in a **frequency distribution table**. We will do this in Example 3.

EXAMPLE 3

Complete the frequency distribution table for the preceding temperature data.

Solution: Go through the data and place a tally mark next to the class interval (in the second table column). Then count the tally marks and write each total in the third table column.

Class Intervals (Temperatures)	Tally	Class Frequency (Number of Days)
82°–84°	|	1
85°–87°	|||	3
88°–90°	‖‖ ‖‖ |	11
91°–93°	‖‖ ||	7
94°–96°	‖‖ |||	8

EXAMPLE 4

Construct a histogram from the frequency distribution table in Example 3.

Solution:

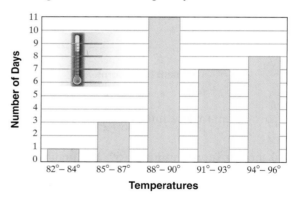

Try the Concept Check in the margin.

EXERCISE SET 9.3

A *The following histogram shows the number of miles that each adult, from a survey of 100 adults, drives per week. Use this histogram to answer Exercises 1 through 10. See Examples 1 and 2.*

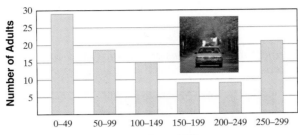

1. How many adults drive 100–149 miles per week?

2. How many adults drive 200–249 miles per week?

3. How many adults drive fewer than 150 miles per week?

4. How many adults drive 200 miles or more per week?

5. How many adults drive 100–199 miles per week?

6. How many adults drive 0–149 miles per week?

7. How many more adults drive 250–299 miles per week than 200–249 miles per week?

8. How many more adults drive 0–49 miles per week than 50–99 miles per week?

9. What is the ratio of adults who drive 150–199 miles per week to the total number of adults surveyed?

10. What is the ratio of adults who drive 50–99 miles per week to the total number of adults surveyed?

The following histogram shows the projected ages of householders for the year 2005. Use this histogram to answer Exercises 11 through 18. See Examples 1 and 2.

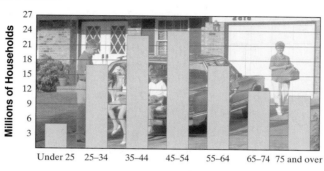

Source: U.S. Bureau of the Census, *Current Population Reports*

11. The most householders will be in what age range?

12. The least householders will be in what age range?

13. How many householders will be 55–64 years old?

14. How many householders will be 35–44 years old?

15. How many householders will be 44 years old or younger?

16. How many householders will be 55 years old or older?

17. Which bar represents the household you expect to be in, in the year 2005?

18. How many more householders will be 45–54 years old than 55–64 years old?

B *The following list shows the golf scores for an amateur golfer. Use this list to complete the frequency distribution table below. See Example 3.*

78	84	91	93	97
97	95	85	95	96
101	89	92	89	100

	Class Intervals (Scores)	Tally	Class Frequency (Number of Games)
19.	70–79	_____	_____
20.	80–89	_____	_____
21.	90–99	_____	_____
22.	100–109	_____	_____

Twenty-five people in a survey were asked to give their current checking account balances. Use the balances shown in the following list to complete the frequency distribution table below. See Example 3.

$53	$105	$162	$443	$109
$468	$47	$259	$316	$228
$207	$357	$15	$301	$75
$86	$77	$512	$219	$100
$192	$288	$352	$166	$292

	Class Intervals (Account Balances)	Tally	Class Frequency (Number of People)
23.	$0–$99	_____	_____
24.	$100–$199	_____	_____
25.	$200–$299	_____	_____
26.	$300–$399	_____	_____
27.	$400–$499	_____	_____
28.	$500–$599	_____	_____

29. Use the table from Exercises 19 through 22 to construct a histogram. See Example 4.

30. Use the table from Exercises 23 through 28 to construct a histogram. See Example 4.

Number of Games

2
1

Golf Scores

Number of People

2
1

Account Balances

Review and Preview

Find the average of each list of numbers. See Section 1.7. (Recall that the average of a list of numbers is the sum of the numbers divided by the number of numbers.)

31. 86, 94   **32.** 75, 87   **33.** 12, 28, 20   **34.** 19, 10, 22   **35.** 30, 22, 23, 33   **36.** 39, 25, 31, 37

Combining Concepts

The following graph is called a "double bar graph." Study this graph and use it to answer Exercises 37 through 39.

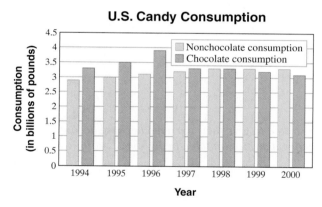

U.S. Candy Consumption

Sources: National Confectioners Association and Chocolate Manufacturers Association

37. In which year was the difference between consumption of chocolate candy and consumption of nonchocolate candy the greatest?

38. In which year was the difference between consumption of chocolate candy and consumption of nonchocolate candy the least?

39. In which year did consumption of chocolate candy surpass the consumption of nonchocolate candy for the first time?

40. How might a graph of this type be helpful?

41. Use the data and answers to Exercises 39 through 41 of Section 9.2 to construct a histogram.

Name _____ Section _____ Date _____

Integrated Review–Reading Graphs

The following pictograph shows the average number of pounds of beef and veal consumed per person per year in the United States. Use this graph to answer Exercises 1 through 4.

Beef and Veal Consumption

2000
1995
1990
1985
1980

Each 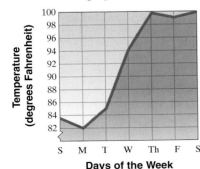 represents 10 pounds

Source: U.S. Department of Agriculture

1. Approximate the number of pounds of beef and veal consumed per person in 1995.

2. Approximate the number of pounds of beef and veal consumed per person in 1980.

3. In what year(s) was the number of pounds consumed the greatest?

4. In what year(s) was the number of pounds consumed the least?

The following bar graph shows the highest U.S. dams. Use this graph to answer Exercises 5 through 8.

Highest U.S. Dams

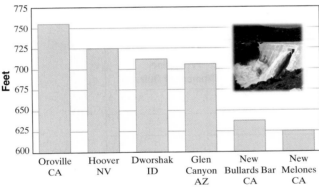

Feet

775
750
725
700
675
650
625
600

Oroville CA · Hoover NV · Dworshak ID · Glen Canyon AZ · New Bullards Bar CA · New Melones CA

Source: Committee on Register of Dams

5. Name the U.S. dam with the greatest height and estimate its height.

6. Name the U.S. dam whose height is between 625 and 650 feet and estimate its height.

7. Estimate how much higher the Hoover Dam is than the Glen Canyon Dam.

8. How many U.S. dams have heights over 700 feet?

The following line graph shows the daily high temperatures for 1 week in Annapolis, Maryland. Use this graph to answer Exercises 9 through 12.

Temperature (degrees Fahrenheit)

100
98
96
94
92
90
88
86
84
82

S M T W Th F S

Days of the Week

9. Name the day(s) of the week with the highest temperature and give that high temperature.

10. Name the day(s) of the week with the lowest temperature and give that low temperature.

11. On what days of the week was the temperature less than 90° Fahrenheit?

12. On what days of the week was the temperature greater than 90° Fahrenheit?

13. _____

14. _____

15. _____

16. _____

17. see table _____

18. see table _____

19. see table _____

20. see table _____

21. see table _____

22. see graph _____

The following circle graph shows the type of beverage milk consumed in the United States. Use this graph for Exercises 13 through 16. If a store in Kerrville, Texas, sells 200 quart containers of milk per week, estimate how many quart containers are sold in each category below.

Types of Beverage Milk Consumed

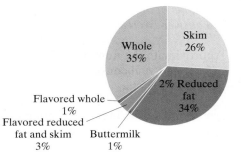

Whole 35%
Skim 26%
2% Reduced fat 34%
Flavored whole 1%
Flavored reduced fat and skim 3%
Buttermilk 1%

Source: U.S. Department of Agriculture

13. Whole milk

14. Skim milk

15. Buttermilk

16. Flavored reduced fat and skim milk

The following list shows weekly quiz scores for a student in basic college mathematics. Use this list to complete the frequency distribution table.

50	80	71	83	86
67	89	93	88	97
	53	90		
75	80	78	93	99

	Class Intervals (Scores)	Tally	Class Frequency (Number of Quizzes)
17.	50–59	_____	_____
18.	60–69	_____	_____
19.	70–79	_____	_____
20.	80–89	_____	_____
21.	90–99	_____	_____

22. Use the table from Exercises 17 through 21 to construct a histogram.

Number of Quizzes

2
1

Quiz Scores

9.4 Mean, Median, and Mode

OBJECTIVES

Ⓐ Find the mean of a list of numbers.

Ⓑ Find the median of a list of numbers.

Ⓒ Find the mode of a list of numbers.

SSM
TUTOR CENTER   SG   CD & VIDEO   MATH PRO   WEB

Ⓐ Finding the Mean

Sometimes we want to summarize data by displaying them in a graph, but sometimes it is also desirable to be able to describe a set of data, or a set of numbers, by a single "middle" number. Three such **measures of central tendency** are the **mean**, the **median**, and the **mode**.

The most common measure of central tendency is the mean (sometimes called the "arithmetic mean" or the "average"). Recall that we first introduced finding the average of a list of numbers in Section 1.7.

> The **mean (average)** of a set of number items is the sum of the items divided by the number of items.

EXAMPLE 1

Seven students in a psychology class conducted an experiment on mazes. Each student was given a pencil and asked to successfully complete the same maze. The timed results are below:

Student	Ann	Thanh	Carlos	Jesse	Melinda	Ramzi	Dayni
Time (seconds)	13.2	11.8	10.7	16.2	15.9	13.8	18.5

a. Who completed the maze in the shortest time? Who completed the maze in the longest time?

b. Find the mean time.

c. How many students took longer than the mean time? How many students took shorter than the mean time?

Solution:

a. Carlos completed the maze in 10.7 seconds, the shortest time. Dayni completed the maze in 18.5 seconds, the longest time.

b. To find the mean (or average), we find the sum of the number items and divide by 7, the number of items.

$$\text{mean} = \frac{13.2 + 11.8 + 10.7 + 16.2 + 15.9 + 13.8 + 18.5}{7}$$

$$= \frac{100.1}{7} = 14.3$$

c. Three students, Jesse, Melinda, and Dayni, had times longer than the mean time. Four students, Ann, Thanh, Carlos, and Ramzi, had times shorter than the mean time.   ●

Try the Concept Check in the margin.

Often in college, the calculation of a **grade point average** (GPA) is a **weighted mean** and is calculated as shown in Example 2.

Practice Problem 1

Find the mean of the following test scores: 77, 85, 86, 91, and 88.

Concept Check

Estimate the mean of the following set of data:

5, 10, 10, 10, 10, 15

Answers

1. 85.4

Concept Check: 10

Practice Problem 2

Find the grade point average if the following grades were earned in one semester.

Grade	Credit Hours
A	2
C	4
B	5
D	2
A	2

EXAMPLE 2

The following grades were earned by a student during one semester. Find the student's grade point average.

Course	Grade	Credit Hours
College mathematics	A	3
Biology	B	3
English	A	3
PE	C	1
Social studies	D	2

Solution: To calculate the grade point average, we need to know the point values for the different possible grades. The point values of grades commonly used in colleges and universities are given below:

A: 4, B: 3, C: 2, D: 1, F: 0

Now, to find the grade point average, we multiply the number of credit hours for each course by the point value of each grade. The grade point average is the sum of these products divided by the sum of the credit hours.

Course	Grade	Point Value of Grade	Credit Hours	Point Value · Credit Hours
College mathematics	A	4	3	12
Biology	B	3	3	9
English	A	4	3	12
PE	C	2	1	2
Social studies	D	1	2	2
			Totals:	12	37

$$\text{grade point average} = \frac{37}{12} \approx 3.08 \text{ rounded to two decimal places}$$

The student earned a grade point average of 3.08.

B Finding the Median

You may have noticed that a very low number or a very high number can affect the mean of a list of numbers. Because of this, you may sometimes want to use another measure of central tendency. A second measure of central tendency is called the **median**. The median of a list of numbers is not affected by a low or high number in the list.

The **median** of an ordered set of numbers is the middle number. If the number of items is even, the median is the mean of the two middle numbers.

Practice Problem 3

Find the median of the list of numbers: 7, 9, 13, 23, 24, 35, 38, 41, 43

Answers

2. 2.73, **3.** 24

EXAMPLE 3 Find the median of the following list of numbers:

25, 54, 56, 57, 60, 71, 98

Solution: Because this list is in numerical order, the median is the middle number, 57.

EXAMPLE 4

Find the median of the following list of scores: 67, 91, 75, 86, 55, 91

Solution: First we list the scores in numerical order and then find the middle number.

55, 67, 75, 86, 91, 91

Since there is an even number of scores, there are two middle numbers. The median is the mean of the two middle numbers.

$$\text{median} = \frac{75 + 86}{2} = 80.5$$

The median is 80.5.

C Finding the Mode

The last common measure of central tendency is called the **mode**.

The **mode** of a set of numbers is the number that occurs most often. (It is possible for a set of numbers to have more than one mode or to have no mode.)

EXAMPLE 5   Find the mode of the list of numbers:

11, 14, 14, 16, 31, 56, 65, 77, 77, 78, 79

Solution: There are two numbers that occur the most often. They are 14 and 77. This list of numbers has two modes, 14 and 77.

EXAMPLE 6

Find the median and the mode of the following set of numbers. These numbers were high temperatures for 14 consecutive days in a city in Montana.

76, 80, 85, 86, 89, 87, 82, 77, 76, 79, 82, 89, 89, 92

Solution: First we write the numbers in numerical order.

76, 76, 77, 79, 80, 82, 82, 85, 86, 87, 89, 89, 89, 92

Since there is an even number of items, the median is the mean of the two middle numbers.

$$\text{median} = \frac{82 + 85}{2} = 83.5$$

The mode is 89, since 89 occurs most often.

Try the Concept Check in the margin.

> **Helpful Hint**
>
> Don't forget that it is possible for a list of numbers to have no mode. For example, the list
>
> 2, 4, 5, 6, 8, 9
>
> has no mode. There is no number or numbers that occur more often than the others.

Find the median of the list of scores:

43, 89, 78, 65, 95, 95, 88, 71

Find the mode of the list of numbers:

9, 10, 10, 13, 15, 15, 15, 17, 18, 18, 20

Find the median and the mode of the list of numbers:

26, 31, 15, 15, 26, 30, 16, 18, 15, 35

Concept Check

True or false? Every set of numbers *must* have a mean, median, and mode. Explain your answer.

Answers

4. 83,   **5.** 15,   **6.** median: 22; mode: 15
Concept Check: false; a set of numbers may have no mode

STUDY SKILLS REMINDER

Tips for studying for an exam

To prepare for an exam, try the following study techniques.

- Start the study process days before your exam.
- Make sure that you are current and up-to-date on your assignments.
- If there is a topic that you are unsure of, use one of the many resources that are available to you. For example,

    See your instructor.

    Visit a learning resource center on campus where math tutors are available.

    Read the textbook material and examples on the topic.

    View a videotape on the topic.

- Reread your notes and carefully review the Chapter Highlights at the end of the chapter.
- Work the review exercises at the end of the chapter and check your answers. Make sure that you correct any missed exercises. If you have trouble on a topic, use a resource listed above.
- Find a quiet place to take the Chapter Test found at the end of the chapter. Do not use any resources when taking this sample test. This way you will have a clear indication of how prepared you are for your exam. Check your answers and make sure that you correct any missed exercises.
- Get lots of rest the night before the exam. It's hard to show how well you know the material if your brain is foggy from lack of sleep.

Good luck and keep a positive attitude.

Name _____ Section _____ Date _____

Mental Math

State the mean for each list of numbers.

1. 3, 5   **2.** 10, 20   **3.** 1, 3, 5   **4.** 7, 7, 7

EXERCISE SET 9.4

    For each set of numbers, find the mean, the median, and the mode. If necessary, round the mean to one decimal place. See Examples 1 and 3 through 6.

1. 21, 28, 16, 42, 38   **2.** 42, 35, 36, 40, 50   **3.** 7.6, 8.2, 8.2, 9.6, 5.7, 9.1   **4.** 4.9, 7.1, 6.8, 6.8, 5.3, 4.9

5. 0.2, 0.3, 0.5, 0.6, 0.6, 0.9, 0.2, 0.7, 1.1   **6.** 0.6, 0.6, 0.8, 0.4, 0.5, 0.3, 0.7, 0.8, 0.1

7. 231, 543, 601, 293, 588, 109, 334, 268   **8.** 451, 356, 478, 776, 892, 500, 467, 780

The eight tallest buildings in the world are listed in the following table. Use this table to answer Exercises 9 through 12. If necessary, round results to one decimal place. See Examples 1 and 3 through 6.

9. Find the mean height of the five tallest buildings.

10. Find the median height of the five tallest buildings.

11. Find the median height of the eight tallest buildings.

12. Find the mean height of the eight tallest buildings.

Building	Height (in feet)
Petronas Tower 1, Kuala Lumpur	1483
Petronas Tower 2, Kuala Lumpur	1483
Sears Tower, Chicago	1450
Jin Mao Building, Shanghai	1381
Citic Plaza, Guangzhou	1283
Shun Hing Square, Shenzhen	1260
Empire State Building, New York	1250
Central Plaza, Hong Kong	1227

(*Source:* Council on Tall Buildings and Urban Habitat)

13. Given the building heights, explain how you know, without calculating, that the answer to Exercise 10 is more than the answer to Exercise 11.

14. Given the building heights, explain how you know, without calculating, that the answer to Exercise 12 is less than the answer to Exercise 9.

For Exercises 15 through 18, the grades are given for a student for a particular semester. Find the grade point average. If necessary, round the grade point average to the nearest hundredth. See Example 2.

 15.

Grade	Credit Hours
B	3
C	3
A	4
C	4

16.

Grade	Credit Hours
D	1
F	1
C	4
B	5

17.

Grade	Credit Hours
A	3
A	3
B	4
B	1
B	2

18.

Grade	Credit Hours
B	2
B	2
A	3
C	3
B	3

During an experiment, the following times (in seconds) were recorded:

7.8, 6.9, 7.5, 4.7, 6.9, 7.0.

19. Find the mean. Round to the nearest tenth.

20. Find the median.

21. Find the mode.

In a mathematics class, the following test scores were recorded for a student: 86, 95, 91, 74, 77, 85.

22. Find the mean. Round to the nearest hundredth.

23. Find the median.

24. Find the mode.

The following pulse rates were recorded for a group of 15 students: 78, 80, 66, 68, 71, 64, 82, 71, 70, 65, 70, 75, 77, 86, 72.

25. Find the mean.

26. Find the median.

27. Find the mode.

28. How many rates were higher than the mean?

29. How many rates were lower than the mean?

Review and Preview

Write each fraction in simplest form. See Section 2.3.

30. $\dfrac{12}{20}$

31. $\dfrac{6}{18}$

32. $\dfrac{4}{36}$

33. $\dfrac{18}{30}$

34. $\dfrac{35}{100}$

35. $\dfrac{55}{75}$

Combining Concepts

Find the missing numbers in each set of numbers.

36. 16, 18, __, __, __. The mode is 21. The median is 20.

37. __, __, __, 40, __. The mode is 35. The median is 37. The mean is 38.

38. Write a list of numbers for which you feel the median would be a better measure of central tendency than the mean.

584

9.5   Counting and Introduction to Probability

OBJECTIVES

Ⓐ Use a tree diagram to count outcomes.

Ⓑ Find the probability of an event.

SSM   TUTOR CENTER   SG   CD & VIDEO   MATH PRO   WEB

Ⓐ Using a Tree Diagram

In our daily conversations, we often talk about the likelihood or the probability of a given result occurring. For example:

The *chance* of thundershowers is 70 percent.

What are the *odds* that the Saints will go to the Super Bowl?

What is the *probability* that you will finish cleaning your room today?

Each of these chance happenings—thundershowers, the Saints playing in the Super Bowl, and cleaning your room today—is called an **experiment**. The possible results of an experiment are called **outcomes**. For example, flipping a coin is an experiment, and the possible outcomes are heads (H) or tails (T).

One way to picture the outcomes of an experiment is to draw a tree diagram. Each outcome is shown on a separate branch. For example, the outcomes of flipping a coin are

Heads   Tails

EXAMPLE 1

Draw a tree diagram for tossing a coin twice. Then use the diagram to find the number of possible outcomes.

Solution: There are 4 possible outcomes when tossing a coin twice.

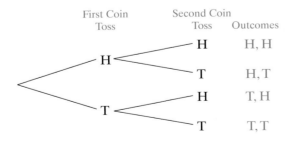

First Coin Toss	Second Coin Toss	Outcomes
H	H	H, H
	T	H, T
T	H	T, H
	T	T, T

Practice Problem 1

Draw a tree diagram for tossing a coin three times. Then use the diagram to find the number of possible outcomes.

Answer

1.

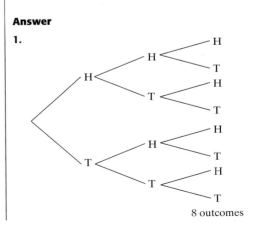

8 outcomes

Practice Problem 2

Draw a tree diagram for an experiment consisting of tossing a coin and then rolling a die. Then use the diagram to find the number of possible outcomes.

EXAMPLE 2

Draw a tree diagram for an experiment consisting of rolling a die and then tossing a coin. Then use the diagram to find the number of possible outcomes.

Die

Solution: Recall that a die has six sides and that each side represents a number, 1 through 6.

Roll a Die	Toss a coin	Outcomes
1	H	1, H
	T	1, T
2	H	2, H
	T	2, T
3	H	3, H
	T	3, T
4	H	4, H
	T	4, T
5	H	5, H
	T	5, T
6	H	6, H
	T	6, T

There are 12 possible outcomes for rolling a die and then tossing a coin. ●

Any number of outcomes considered together are called an **event**. For example, when tossing a coin twice, H, H is an event. The event is tossing heads first and tossing heads second. Another event would be tossing tails first and then heads (T, H), and so on.

Answer

2.

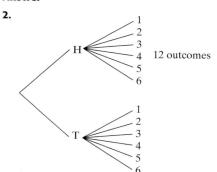

12 outcomes

B   **Finding the Probability of an Event**

As we mentioned earlier, the probability of an event is a measure of the chance or likelihood of it occurring. For example, if a coin is tossed, what is the probability that heads occurs? Since one of two equally likely possible outcomes is heads, the probability is $\frac{1}{2}$.

The Probability of an Event

$$\text{probability of an event} = \frac{\text{number of ways that the event can occur}}{\text{number of possible outcomes}}$$

Helpful Hint

Note from the definition of probability that the probability of an event is always between 0 and 1, inclusive (i.e., including 0 and 1). A probability of 0 means that an event won't occur, and a probability of 1 means that an event is certain to occur.

EXAMPLE 3

If a coin is tossed twice, find the probability of tossing heads and then heads (H, H).

Solution: 1 way the event can occur

H, T,   H, H,   T, H,   T, T

4 possible outcomes

$$\text{probability} = \frac{1}{4} \quad \begin{array}{l}\text{Number of ways the event can occur}\\ \text{Number of possible outcomes}\end{array}$$

The probability of tossing heads and then heads is $\frac{1}{4}$.

EXAMPLE 4

If a die is rolled one time, find the probability of rolling a 3 or a 4.

Solution: Recall that there are 6 possible outcomes when rolling a die.

2 ways that the event can occur

possible outcomes: 1,   2,   3,   4,   5,   6

6 possible outcomes

$$\text{probability of a 3 or a 4} = \frac{2}{6} \quad \begin{array}{l}\text{Number of ways the event can occur}\\ \text{Number of possible outcomes}\end{array}$$

$$= \frac{1}{3} \quad \text{Simplest form}$$

Try The Concept Check in the margin.

Practice Problem 3

If a coin is tossed three times, find the probability of tossing heads, then tails, then tails (H, T, T).

Practice Problem 4

If a die is rolled one time, find the probability of rolling a 1 or a 2.

Concept Check

Suppose you have calculated a probability of $\frac{11}{9}$. How do you know that you have made an error in your calculation?

Answers

3. $\frac{1}{8}$,   **4.** $\frac{1}{3}$

Concept Check: The number of ways an event can occur can't be larger than the number of possible outcomes.

Practice Problem 5

Use the diagram from Example 5 and find the probability of choosing a blue marble from the box.

Answer

5. $\frac{1}{2}$

EXAMPLE 5

Find the probability of choosing a red marble from a box containing 1 red, 1 yellow, and 2 blue marbles.

Solution:

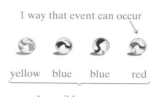

1 way that event can occur

yellow   blue   blue   red

4 possible outcomes

probability $= \dfrac{1}{4}$

STUDY SKILLS REMINDER

Are you satisfied with your performance on a particular quiz or exam?

If not, don't forget to analyze your quiz or exam and look for common errors.
Were most of your errors a result of

- *Carelessness*? If your errors were careless, did you turn in your work before the allotted time expired? If so, resolve next time to use the entire time allotted. Any extra time can be spent checking your work.

- *Running out of time*? If so, make a point to better manage your time on your next exam. A few suggestions are to work any questions that you are unsure of last and to check your work after all questions have been answered.

- *Not understanding a concept*? If so, review that concept and correct your work. Remember next time to make sure that all concepts on a quiz or exam are understood before the exam.

- *Test conditions*? When studying for your test, are you placing yourself in conditions similar to test conditions? In other words, once you feel that you know the material, use a few sheets of blank paper and take a sample test. (A sample test can be one provided by your instructor or you may use the Chapter Test found at the end of each chapter, for example.)

Name _____ Section _____ Date _____

Mental Math

If a coin is tossed once, find the probability of each event.

1. The coin lands heads up.

2. The coin lands tails up.

If the spinner shown is spun once, find the probability of each event.

3. The spinner stops on red.

4. The spinner stops on blue.

EXERCISE SET 9.5

Ⓐ *Draw a tree diagram for each experiment. Then use the diagram to find the number of possible outcomes. See Examples 1 and 2.*

1. Choosing a vowel (a, e, i, o, u) and then a number (1, 2, or 3)

2. Choosing a number (1 or 2) and then a vowel (a, e, i, o, u)

Spinner A

Spinner B

3. Spinning Spinner A once

4. Spinning Spinner B once

5. Spinning Spinner B twice

6. Spinning Spinner A twice

7. Spinning Spinner A and then Spinner B

8. Spinning Spinner B and then Spinner A

9. Tossing a coin and then spinning Spinner B

10. Tossing a coin and then spinning Spinner A

B *If a single die is tossed once, find the probability of each event. See Examples 3 through 5.*

11. A 5

12. A 7

13. A 1 or a 4

14. A 2 or a 3

15. An even number

16. An odd number

Suppose the spinner shown is spun once. Find the probability of each event. See Examples 3 through 5.

17. The result of the spin is 2.

18. The result of the spin is 3.

19. The result of the spin is an odd number.

20. The result of the spin is an even number.

If a single choice is made from the bag of marbles shown, find the probability of each event. See Examples 3 through 5.

21. A red marble is chosen.

22. A blue marble is chosen.

23. A yellow marble is chosen.

24. A green marble is chosen.

A new drug is being tested that is supposed to lower blood pressure. This drug was given to 200 people and the results are below.

Lower Blood Pressure	Higher Blood Pressure	Blood Pressure Not Changed
152	38	10

25. If a person is testing this drug, what is the probability that their blood pressure will be higher?

26. If a person is testing this drug, what is the probability that their blood pressure will be lower?

27. If a person is testing this drug, what is the probability that their blood pressure will not change?

28. What is the sum of the answers to exercises 25, 26, and 27? In your own words, explain why.

Review and Preview

Perform each indicated operation. See Sections 2.4, 2.5, and 3.3.

29. $\dfrac{1}{2} + \dfrac{1}{3}$ **30.** $\dfrac{7}{10} - \dfrac{2}{5}$ **31.** $\dfrac{1}{2} \cdot \dfrac{1}{3}$ **32.** $\dfrac{7}{10} \div \dfrac{2}{5}$ **33.** $5 \div \dfrac{3}{4}$ **34.** $\dfrac{3}{5} \cdot 10$

Combining Concepts

Recall that a deck of cards contains 52 cards. These cards consist of four suits (hearts, spades, clubs, and diamonds) of each of the following: 2, 3, 4, 5, 6, 7, 8, 9, 10, jack, queen, king, and ace. If a card is chosen from a deck of cards, find the probability of each event.

35. The king of hearts

36. The 10 of spades

37. A king

38. A 10

39. A heart

40. A club

Two dice are tossed. Find the probability of each sum of the dice. (Hint: Draw a tree diagram of the possibilities of two tosses of a die, and then find the sum of the numbers on each branch.)

41. A sum of 4   **42.** A sum of 11   **43.** A sum of 13   **44.** A sum of 2

45. In your own words, explain why the probability of an event cannot be greater than 1.

46. In your own words, explain when the probability of an event is 0.

CHAPTER 9 ACTIVITY **Investigating Probability**

MATERIALS:

■ paper or foam cup
■ 30 thumbtacks

This activity may be completed by working in groups or individually. Recall that probability is a measure of how likely it is that an event will occur. We can report probabilities as fractions or percents because any fraction can be written as a percent. In this activity you will investigate one way that probabilities can be estimated.

Number of Tacks	Number of Tacks Landing Point Up	Fraction of Point-Up Tacks	Percent of Point-Up Tacks
30			
60			
90			
120			
150			

1. Place the thumbtacks in the cup. Shake the cup and toss out the thumbtacks onto a flat surface. Count the number of tacks that land point up, and record this number in the table. Complete the experiment for 30, 60, 90, 120, and 150 tacks. (*Hint:* For 60 thumbtacks, count the number of tacks landing point up in two tosses of the 30 thumbtacks, etc.)

2. For each row of the table, find the fraction of tacks that landed point up. Then express each fraction as a percent. Add these values to the table in the columns labeled "Fraction of Point-Up Tacks" and "Percent of Point-Up Tacks."

3. Each of the percents you computed in Question 2 is an *estimate* of the probability that a single thumbtack will land point up when tossed. When you estimate a probability experimentally, the larger the number of trials used (i.e., the number of tacks tossed), the better the estimate of the actual probability. What do you suppose the value of the actual probability is? Explain your reasoning.

4. Combine your results for all 450 of your tack tosses with the other students' results. Of this total number of tacks, compute the percent of tacks that landed point up. This is your best estimate of the probability that a tack will land point up when tossed.

5. If you tossed 200 thumbtacks, what percent would you expect to land point up? How many tacks would you expect to land point up? Use the percent (probability) you computed in Question 4 to make this calculation. What if you tossed 300 thumbtacks?

Chapter 9   VOCABULARY CHECK

Fill in each blank with one of the words or phrases listed below.

outcomes	bar	experiment	mean	tree diagram
pictograph	line	circle	median	probability
histogram		class frequency	mode	class interval

1. A _____ graph presents data using vertical or horizontal bars.

2. The _____ of a set of number items is

$$\frac{\text{sum of items}}{\text{number of items}}.$$

3. The possible results of an experiment are the _____.

4. A _____ is a graph in which pictures or symbols are used to visually present data.

5. The _____ of a set of numbers is the number that occurs most often.

6. A _____ graph displays information with a line that connects data points.

7. The _____ of an ordered set of numbers is the middle number.

8. A _____ is one way to picture and count outcomes.

9. An _____ is an activity being considered, such as tossing a coin or rolling a die.

10. In a _____ graph, each section (shaped like a piece of pie) shows a category and the relative size of the category.

11. The _____ of an event is

$$\frac{\text{number of ways that the event can occur}}{\text{number of possible outcomes}}.$$

12. A _____ is a special bar graph in which the width of each bar represents a _____ and the height of each bar represents the _____.

CHAPTER 9

Highlights

DEFINITIONS AND CONCEPTS	EXAMPLES

Section 9.1   Reading Pictographs, Bar Graphs, and Line Graphs

A **pictograph** is a graph in which pictures or symbols are used to visually present data.

A **line graph** displays information with a line that connects data points.

A **bar graph** presents data using vertical or horizontal bars.

The bar graph on the right shows the number of acres of wheat harvested in 1996 for leading states.

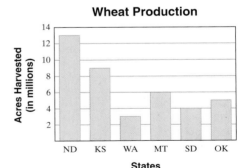

Source: U.S. Department of Agriculture

1. Approximately how many acres of wheat were harvested in Kansas?

   9,000,000 acres

2. About how many more acres of wheat were harvested in North Dakota than South Dakota?

   $$\begin{array}{r} 13 \text{ million} \\ - 4 \text{ million} \\ \hline 9 \text{ million} \end{array} \quad \text{or } 9,000,000 \text{ acres}$$

DEFINITIONS AND CONCEPTS	EXAMPLES

Section 9.2   Reading Circle Graphs

In a **circle graph**, each section (shaped like a piece of pie) shows a category and the relative size of the category.

The circle graph on the right classifies tornadoes by wind speed.

Tornado Wind Speeds

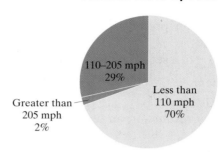

110–205 mph
29%

Greater than
205 mph
2%

Less than
110 mph
70%

Source: National Oceanic and Atmospheric Administration

1. What percent of tornadoes have wind speeds of 110 mph or greater?
   29% + 2% = 31%

2. If there were 1235 tornadoes in the United States in 1995, how many of these might we expect to have had wind speeds less than 110 mph? Find 70% of 1235.

   $70\%(1235) = 0.70(1235) = 864.5 \approx 865$

   Around 865 tornadoes would be expected to have had wind speeds of less than 110 mph.

Section 9.3   Reading Histograms

A **histogram** is a special bar graph in which the width of each bar represents a **class interval** and the height of each bar represents the **class frequency**. The histogram on the right shows student quiz scores.

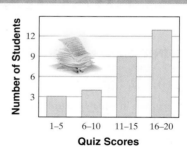

Number of Students

12

9

6

3

1–5   6–10   11–15   16–20

Quiz Scores

1. How many students received a score of 6–10?
   4 students

2. How many students received a score of 11–20?
   9 + 13 = 22 students

Section 9.4   Mean, Median, and Mode

The **mean** (or **average**) of a set of number items is

$$\text{mean} = \frac{\text{sum of items}}{\text{number of items}}$$

The **median** of an ordered set of numbers is the middle number. If the number of items is even, the median is the mean of the two middle numbers.

The **mode** of a set of numbers is the number that occurs most often. (A set of numbers may have no mode or more than one mode.)

Find the mean, median, and mode of the following set of numbers: 33, 35, 35, 43, 68, 68

$$\text{mean} = \frac{33 + 35 + 35 + 43 + 68 + 68}{6} = 47$$

The median is the mean of the two middle numbers:

$$\text{median} = \frac{35 + 43}{2} = 39$$

There are two modes because there are two numbers that occur twice:

35 and 68

DEFINITIONS AND CONCEPTS	EXAMPLES

Section 9.5   Counting and Introduction to Probability

An **experiment** is an activity being considered, such as tossing a coin or rolling a die. The possible results of an experiment are the **outcomes**. A **tree diagram** is one way to picture and count outcomes.

Draw a tree diagram for tossing a coin and then choosing a number from 1 to 4.

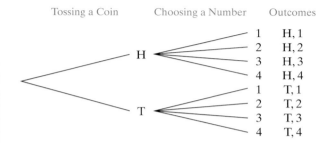

Any number of outcomes considered together is called an **event**. The **probability** of an event is a measure of the chance or likelihood of it occurring.

$$\text{probability of an event} = \frac{\text{number of ways that the event can occur}}{\text{number of possible outcomes}}$$

Find the probability of tossing a coin twice and tails occurring each time.

1 way the event can occur

HH,   HT,   TH,   TT

4 possible outcomes

$$\text{probability} = \frac{1}{4}$$

Are you prepared for a test on Chapter 9?

Below I have listed some *common trouble areas* for students in chapter 9. After studying for your test—but before taking your test—read these.

- Do you remember that a set of numbers can have no mode, 1 mode, or even more than 1 mode?

$2, 5, 8, 9$	no mode
$2, 2, 8, 9$	mode: 2
$2, 2, 3, 3, 5, 7, 7$	mode: 2, 3, 7

- Do you remember how to find the median of an even-numbered set of numbers?

  $$2, \overbrace{5, 8}, 9 \qquad \frac{5 + 8}{2} = 6.5$$

  The median is the average of the two "middle" numbers.

- Don't forget that the probability of an event is always between 0 and 1 inclusive (including 0 and 1).

- What is the probability of an event that won't occur? 0

  What is the probability of an event that is certain to occur? 1

Chapter 9 Review

(9.1) *The following pictograph shows the number of new homes constructed, by state. Use this graph to answer Exercises 1 through 6.*

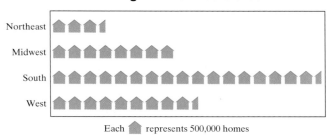

**2000 Housing Starts by
Region of United States**

Each 🏠 represents 500,000 homes

Source: U.S. Census Bureau

1. How many housing starts were there in the Midwest in 2000?

2. How many housing starts were there in the Northeast in 2000?

3. Which region had the most housing starts?

4. Which region had the fewest housing starts?

5. Which region(s) had 4,000,000 or more housing starts?

6. Which region(s) had fewer than 4,000,000 housing starts?

The following bar graph shows the percent of persons age 25 or over who completed four or more years of college. Use this graph to answer Exercises 7 through 10.

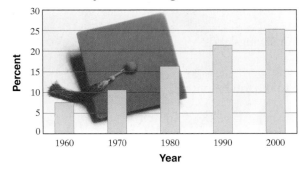

**Four or More Years of College
by Persons Age 25 or Over**

Source: U.S. Census Bureau

7. Approximate the percent of persons who completed four or more years of college in 1960.

8. What year shown had the greatest percent of persons completing four or more years of college?

9. What years shown had 15% or more of persons completing four or more years of college?

10. Describe any patterns you notice in this graph.

The following line graph shows the average price of a 30-second television advertisement during the Super Bowl for the years shown. Use this graph to answer Exercises 11 through 15.

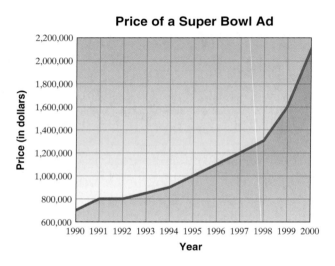

Price of a Super Bowl Ad

Sources: Nielsen Media Research and *Advertising Age* research

11. Approximate the price of a Super Bowl ad in 2000.

12. Approximate the price of a Super Bowl ad in 1997.

13. Between which two years did the price of a Super Bowl ad *not* increase?

14. Between which two years did the price of a Super Bowl ad increase the most?

15. During which years was the price of a Super Bowl ad *less than* $1,000,000?

(9.2) *The following circle graph shows a family's $4000 monthly budget. Use this graph to answer Exercises 16 through 22. Write all ratios as fractions in simplest form.*

16. What is the largest budget item?

17. What is the smallest budget item?

18. How much money is budgeted for the mortgage payment and utilities?

19. How much money is budgeted for savings and contributions?

20. Find the ratio of the house note to the total monthly budget.

21. Find the ratio of food to the total monthly budget.

22. Find the ratio of car expenses to food.

598

The following circle graph shows the percent of states with various rural interstate highway speed limits in 2000. Use this graph to determine the number of states with each speed limit in Exercises 23 through 26.

Percent of States with Rural Interstate Highway Speed Limit

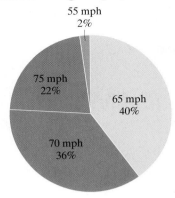

55 mph
2%

75 mph
22%

65 mph
40%

70 mph
36%

Source: Insurance Institute for Highway Safety

23. How many states have a rural interstate highway speed limit of 65 mph?

24. How many states have a rural interstate highway speed limit of 75 mph?

25. How many states have a rural interstate highway speed limit of 55 mph?

26. How many states have a rural interstate highway speed limit of 70 mph or 75 mph?

(9.3) *The following histogram shows the hours worked per week by the employees of Southern Star Furniture. Use this histogram to answer Exercises 27 through 30.*

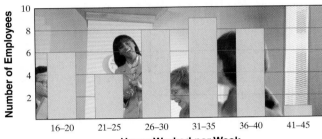

Number of Employees

10
8
6
4
2

16–20 21–25 26–30 31–35 36–40 41–45
Hours Worked per Week

27. How many employees work 21–25 hours per week?

28. How many employees work 41–45 hours per week?

29. How many employees work 36 hours or more per week?

30. How many employees work 30 hours or less per week?

Following is a list of monthly record high temperatures for New Orleans, Louisiana. Use this list to complete the frequency distribution table below.

83   96   101   92
85   100   92   102
89   101   87   84

	Class Intervals (Temperatures)	Tally	Class Frequency (Number of Months)
31.	80°–89°	_____	_____
32.	90°–99°	_____	_____
33.	100°–109°	_____	_____

34. Use the table from Exercises 31, 32, and 33 to draw a histogram.

Record Highs

2
1

Temperatures

(9.4) *Find the mean, median, and any mode(s) for each list of numbers.*

35. 13, 23, 33, 14, 6

36. 45, 21, 60, 86, 64

37. $14,000, $20,000, $12,000, $20,000, $36,000, $45,000

38. 560, 620, 123, 400, 410, 300, 400, 780, 430, 450

For Exercises 39 and 40, the grades are given for a student for a particular semester. Find each grade point average. If necessary, round the grade point average to the nearest hundredth.

39.

Grade	Credit Hours
A	3
A	3
C	2
B	3
C	1

40.

Grade	Credit Hours
B	3
B	4
C	2
D	2
B	3

(9.5) *Draw a tree diagram for each experiment. Then use the diagram to determine the number of outcomes.*

Spinner 1

Spinner 2

600

41. Tossing a coin and then spinning Spinner 1

42. Spinning Spinner 2 and then tossing a coin

43. Spinning Spinner 1 twice

44. Spinning Spinner 2 twice

45. Spinning Spinner 1 and then Spinner 2

Find the probability of each event.

Die

46. Rolling a 4 on a die

47. Rolling a 3 on a die

48. Spinning a 4 on Spinner 1

49. Spinning a 3 on Spinner 1

50. Spinning either a 1, 3, or 5 on Spinner 1

51. Spinning either a 2 or a 4 on Spinner 1

Name _____ Section _____ Date _____

Chapter 9 Test

The following pictograph shows the money collected each week from a wrapping paper fund-raiser. Use this graph to answer Exercises 1 through 3.

Weekly Wrapping Paper Sales

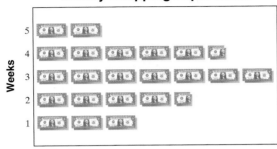

Each 🖸 represents $50

1. How much money was collected during the second week?

2. During which week was the most money collected? How much money was collected during that week?

3. What was the total money collected for the fundraiser?

The following bar graph shows the best-selling vehicles in the United States for 2000 and the number of each model that was sold. Use this graph to answer Exercises 4 through 6.

Best-Selling Vehicles in 2000

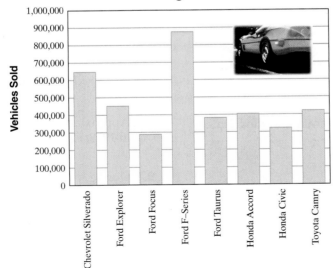

Source: Edmunds.com, Inc.

4. What was the best-selling vehicle for 2000?

5. Which of the models shown sold the least in 2000?

6. For which models were fewer than 400,000 vehicles sold during 2000?

7. Use the information in the following table to draw a bar graph. Clearly label each bar with the title of the album.

Best-Selling Albums of the 1990s (U.S. Sales)	
Album	**Multi-platinum Level***
Shania Twain: *Come on Over* (1997)	18
Whitney Houston & Various Artists: *The Bodyguard* (1992)	17
Alanis Morissette: *Jagged Little Pill* (1995)	16
Garth Brooks: *No Fences* (1990)	16
Hootie & the Blowfish: *Cracked Rear Window* (1995)	16
Backstreet Boys: *Backstreet Boys* (1997)	14
Santana: *Supernatural* (1999)	14

*As of July 2001
Source: Recording Industry Association of America

Best-selling Albums of the 1990s
(U.S. sales)

Multiplatinum Level

15
14

Album

The following line graph shows the annual inflation rate in the United States for the years 1990–2000. Use this graph to answer Exercises 8 through 10.

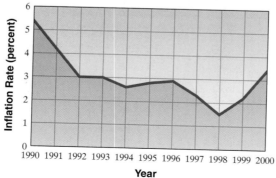

U.S. Annual Inflation Rate

Source: Bureau of Labor Statistics

8. Approximate the annual inflation rate in 1998.

9. During which of the years shown was the inflation rate greater than 3%?

10. During which sets of years was the inflation rate increasing?

6. _____

8. _____

9. _____

10. _____

The result of a survey of 200 people is shown in the following circle graph. Each person was asked to tell his or her favorite type of music. Use this graph to answer Exercises 11 and 12.

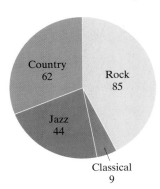

11. Find the ratio of those who prefer rock music to the total number surveyed.

12. Find the ratio of those who prefer country music to those who prefer jazz.

The following circle graph shows the U.S. labor force employment by industry for 2000. There were approximately 132,000,000 people employed by these industries in the United States in 2000. Use the graph to find how many people were employed by the industries given in Exercises 13 and 14.

U.S. Labor Force Employment by Industry

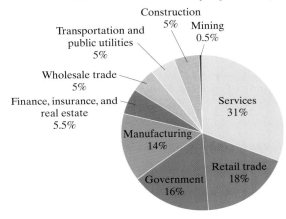

Source: Bureau of Labor Statistics

13. Services

14. Government

A professor measures the heights of the students in her class. The results are shown in the following histogram. Use this histogram to answer Exercises 15 and 16.

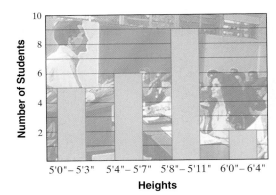

15. How many students are 5'8"–5'11" tall?

16. How many students are 5'7" or shorter?

11. _____

12. _____

13. _____

14. _____

15. _____

16. _____

17. see table

18. see graph

19.

20.

21.

22.

23.

24.

25.

17. The history test scores of 25 students are shown below. Use these scores to complete the frequency distribution table.

70	86	81	65	92
43	72	85	69	97
82	51	75	50	68
88	83	85	77	99
77	63	59	84	90

Class Intervals (Scores)	Tally	Class Frequency (Number of Students)
40–49	_____	_____
50–59	_____	_____
60–69	_____	_____
70–79	_____	_____
80–89	_____	_____
90–99	_____	_____

18. Use the results of Exercise 17 to draw a histogram.

Find the mean, median, and mode of each list of numbers.

19. 26, 32, 42, 43, 49

20. 8, 10, 16, 16, 14, 12, 12, 13

Find the grade point average. If necessary, round to the nearest hundredth.

21.

Grade	Credit Hours
A	3
B	3
C	3
B	4
A	1

22. Draw a tree diagram for the experiment of spinning the spinner twice.

23. Draw a tree diagram for the experiment of tossing a coin twice.

Suppose that the numbers 1 to 10 are each written on a scrap of paper and placed in a bag. You then select one number from the bag.

24. What is the probability of choosing a 6 from the bag?

25. What is the probability of choosing a 3 or a 4 from the bag?

Cumulative Review

1. Simplify: $(8 - 6)^2 + 2^3 \cdot 3$

2. Write $\dfrac{30}{108}$ in simplest form.

3. Add: $1\dfrac{4}{5} + 4 + 2\dfrac{1}{2}$

△ **4.** The formula for finding the area of a triangle is Area $= \dfrac{1}{2} \cdot$ base $\cdot$ height. Find the area of the triangle shown.

3 feet

5.6 feet

5. Write the ratio of $10 to $15 as a fraction in simplest form.

Write each rate as a fraction in simplest form.

6. $2160 for 12 weeks

7. 360 miles on 16 gallons of gasoline

8. Is $\dfrac{1\frac{1}{6}}{10\frac{1}{2}} = \dfrac{\frac{1}{2}}{4\frac{1}{2}}$ a true proportion?

9. The standard dose of an antibiotic is 4 cc (cubic centimeters) for every 25 pounds (lb) of body weight. At this rate, find the standard dose for a 140-lb woman.

Write each percent as a decimal.

10. 4.6%

11. 190%

12. _____	
13. _____	
14. _____	
15. _____	
16. _____	
17. _____	
18. _____	
19. _____	
20. _____	
21. _____	
22. _____	
23. _____	
24. _____	
25. _____	

Write each percent as a fraction in simplest form.

12. 40%

13. $33\frac{1}{3}\%$

14. Translate the following to an equation: Five is what percent of 20?

15. Find the sales tax and the total price on the purchase of an $85.50 trench coat in a city where the sales tax rate is 7.5%.

16. An accountant invested $2000 at a simple interest rate of 10% for 2 years. What total amount of money will she have from her investment in 2 years?

17. Convert 7 feet to yards.

18. Divide 9 lb 6 oz by 2.

19. Convert 2.35 cg to grams.

20. Convert 3210 ml to liters.

21. Convert 15°C to degrees Fahrenheit.

22. Find the complement of a 48° angle.

23. Find $\sqrt{\dfrac{1}{36}}$.

24. Find the mode of the following list of numbers:
11, 14, 14, 16, 31, 56, 65, 77, 77, 78, 79

25. If a coin is tossed twice, find the probability of tossing heads and then heads.

Signed Numbers

Thus far, we have studied whole numbers, fractions, and decimals. However, these numbers are not sufficient for representing many situations in real life. For example, to express 5 degrees below zero or $100 in debt, numbers less than zero are needed. This chapter is devoted to signed numbers, which include numbers less than zero, and to operations on these numbers.

G olf continues to increase in popularity. This game has its roots in Scotland in the 14th and 15th centuries. It became so popular that in 1457, the Scottish Parliament had to outlaw playing golf because it was keeping people from practicing their archery, which was extremely important from a military standpoint. The game of golf eventually spread across the European continent and then to North America and around the world. A variety of golf governing bodies and golfers associations now exist. One of these is the Ladies Professional Golf Association (LPGA). The LPGA Tour was founded in 1950, offering 14 events and a total of $50,000 in prize money in its first year. In 1996, LPGA rookie Karrie Webb became the first rookie in all of the golf world to earn $1 million on a single season of play. By 2000, the LPGA Tour offered 42 events and a total of more than $38 million in prize money. In Exercise 58 on page 623 and Exercises 65 and 66 on page 641, we will see how negative numbers are used in scoring in the game of golf.

Name _____ Section _____ Date _____

1. see number line

2. _____

3. _____

4. _____

5. _____

6. _____

7. _____

8. _____

9. _____

10. _____

11. _____

12. _____

13. _____

14. _____

15. _____

16. _____

17. _____

18. _____

19. _____

20. _____

Chapter 10 Pretest

1. Graph each signed number in the following list on the same number line:

$$-4, 1, 2\frac{1}{2}, -1\frac{1}{4}$$

Insert < or > between each pair of numbers to make a true statement.

2. -15 -17

3. $\frac{3}{5}$ $-\frac{1}{5}$

Find the absolute value.

4. $|-22|$

5. $|9.8|$

6. Find the opposite of -12.

Perform indicated operations.

7. $-31 + 50$ **8.** $-8 - 16$ **9.** $14 - 29$ **10.** $4 - 9 + 10$

11. $7 - (-41) + (-9)$ **12.** $-6(-9)$ **13.** $-3(-4)(10)$

14. $(-3)^4$ **15.** $\dfrac{-92}{-2}$ **16.** $\dfrac{1.44}{-1.2}$

17. $5 + 8(-3)$ **18.** $(-2 \div 2) - 6 \cdot 3 - 9$

19. Juanita has $275 in her checking account. She writes a check for $102, makes a deposit of $29, and then writes another check for $210. Find the amount left in her account. (Write the amount as an integer.)

20. A card player had a score of -24 for each of 3 games. Find his total score.

10.1 Signed Numbers

Ⓐ Representing Real-Life Situations

Thus far in this text, all numbers have been 0 or greater than 0. Numbers greater than 0 are called **positive numbers**. However, sometimes situations exist that cannot be represented by a number greater than 0. For example,

5 degrees below 0
20 feet below sea level

0° ↓ 5 degrees below 0

Sea level
20 feet below sea level

To represent these situations, we need numbers less than 0.

Extending the number line to the left of 0 allows us to picture **negative numbers**, numbers that are less than 0.

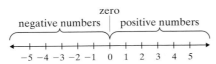

zero
negative numbers | positive numbers
−5 −4 −3 −2 −1 0 1 2 3 4 5

When a single + sign or no sign is in front of a number, the number is a positive number. When a single − sign is in front of a number, the number is a negative number. Together, we call positive numbers, negative numbers, and zero the **signed numbers**.

−5 indicates "negative five."

5 and +5 both indicate "positive five."

The number 0 is neither positive nor negative.

Some signed numbers are integers. The **integers** consist of the positive numbers, zero, and the negative numbers labeled on the number line above. The integers are

$$\ldots, -3, -2, -1, 0, 1, 2, 3, \ldots$$

Now we have numbers to represent the situations previously mentioned.

5 degrees below 0   −5°

20 feet below sea level   −20 feet

EXAMPLE 1   Representing Depth with a Signed Number

Jack Mayfield, a miner for the Molly Kathleen Gold Mine, is presently 150 feet below the surface of the Earth. Represent this position using a signed number.

Solution: If 0 represents the surface of the Earth, then 150 feet below the surface can be represented by −150.

Practice Problem 1

a. A deep-sea diver is 800 feet below the surface of the ocean. Represent this position using a signed number.

b. A company reports a $2 million loss for the year. Represent this amount using a signed number.

Answers

1. a. −800, **b.** −2 million

Practice Problem 2

Graph the signed numbers -5, 3, -3, $-1\frac{3}{4}$, and -4.5 on a number line.

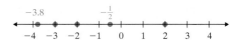

B Graphing Signed Numbers

EXAMPLE 2

Graph the signed numbers -3, 2 -2, $-\frac{1}{2}$, and -3.8 on a number line.

Solution:

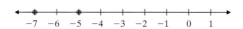

C Comparing Signed Numbers

For any two numbers graphed on a number line, the number to the **right** is the **greater number**, and the number to the **left** is the **smaller number**. Recall that the symbol $>$ means "is greater than" and the symbol $<$ means "is less than."

To illustrate, both -5 and -7 are graphed on the number line shown:

The graph of -7 is **to the left of** -5, so -7 **is less than** -5. We can write this as

$$-7 < -5$$

We can also write

$$-5 > -7$$

since -5 is **to the right** of -7, so -5 **is greater than** -7.

Concept Check

Is there a smallest negative number? Is there a largest positive number? Explain.

Try the Concept Check in the margin.

Practice Problems 3–9

Insert $<$ or $>$ between each pair of numbers to make a true statement.

3. 8 -8
4. -11 0
5. -15 -14
6. 3 -4.6
7. 0 -2
8. -7.9 7.9
9. $-\frac{3}{8}$ $-1\frac{1}{7}$

EXAMPLES

Insert $<$ or $>$ between each pair of numbers to make a true statement.

3. -7 7 -7 is to the left of 7, so $-7 < 7$.
4. 0 -4 0 is to the right of -4, so $0 > -4$.
5. -9 -11 -9 is to the right of -11, so $-9 > -11$.
6. -2.9 -1 -2.9 is to the left of -1, so $-2.9 < -1$.
7. -6 0 -6 is to the left of 0, so $-6 < 0$.
8. 8.6 -8.6 8.6 is to the right of -8.6, so $8.6 > -8.6$.
9. $-\frac{1}{4}$ $-2\frac{1}{2}$ $-\frac{1}{4}$ is to the right of $-2\frac{1}{2}$, so $-1\frac{1}{4} > -2\frac{1}{2}$.

Answers

2.

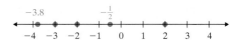

3. $>$, **4.** $<$, **5.** $<$, **6.** $>$, **7.** $>$,
8. $<$, **9.** $>$

Concept Check: no

> **Helpful Hint**
>
> If you think of $<$ and $>$ as arrowheads, notice that in a true statement the arrow always points to the smaller number.
>
> $$5 > -4 \qquad\qquad -3 < -1$$
>
> smaller   smaller
> number   number

D Finding the Absolute Value of a Number

The **absolute value** of a number is the number's distance from 0 on the number line. The symbol for absolute value is $|\ |$. For example, $|3|$ is read as "the absolute value of 3."

$|3| = 3$ because 3 is 3 units from 0.

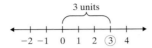

$|-3| = 3$ because -3 is 3 units from 0.

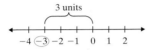

EXAMPLES   Find each absolute value.

10. $|-2| = 2$ because -2 is 2 units from 0.

11. $|5| = 5$ because 5 is 5 units from 0.

12. $|0| = 0$ because 0 is 0 units from 0.

13. $\left|-\dfrac{3}{4}\right| = \dfrac{3}{4}$ because $-\dfrac{3}{4}$ is $\dfrac{3}{4}$ unit from 0.

14. $|1.2| = 1.2$ because 1.2 is 1.2 units from 0.

●

> **Helpful Hint**
>
> Since the absolute value of a number is that number's *distance* from 0, the absolute value of a number is always 0 or positive. It is never negative.
>
> $|0| = 0$ $|-6| = 6$
> $\quad\;\uparrow$ $\quad\;\;\uparrow$
> $\;$ zero   a positive number

E Finding the Opposite of a Number

Two numbers that are the same distance from 0 on the number line but are on opposite sides of 0 are called **opposites**.

4 and -4 are opposites.

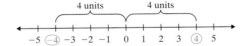

When two numbers are opposites, we say that each is the opposite of the other. Thus, **4 is the opposite of -4** and **-4 is the opposite of 4**.

The phrase "the opposite of" is written in symbols as "$-$". For example,

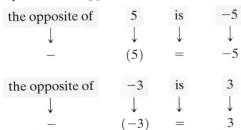

the opposite of   5   is -5
$\quad\downarrow$ $\downarrow$ $\downarrow$ $\downarrow$
$\quad-$ (5) $=$ -5

the opposite of -3 is   3
$\quad\downarrow$ $\downarrow$ $\downarrow$ $\downarrow$
$\quad-$ (-3) $=$ 3

Notice we just stated that

$-(-3) = 3$

Practice Problems 10–14

Find each absolute value.

10. $|6|$

11. $|-4|$

12. $|0|$

13. $\left|\dfrac{7}{8}\right|$

14. $|-3.4|$

Answers

10. 6,   **11.** 4,   **12.** 0,   **13.** $\dfrac{7}{8}$,   **14.** 3.4

In general, we have the following:

Opposites

If a is a number, then $-(-a) = a$.

Notice that because "the opposite of" is written as "$-$", to find the opposite of a number we place a "$-$" sign in front of the number.

Practice Problems 15–19

Find the opposite of each number.

15. -7 16. 0

17. $\dfrac{11}{15}$ 18. -9.6

19. 4

EXAMPLES   Find the opposite of each number.

15. 12   The opposite of 12 is -12.

16. 1.7   The opposite of 1.7 is -1.7.

17. -3 The opposite of -3 is $-(-3)$ or 3.

18. 0   The opposite of 0 is -0 or 0.

> **Helpful Hint**
> Remember that 0 is neither positive nor negative.

19. $-\dfrac{10}{13}$ The opposite of $-\dfrac{10}{13}$ is $-\left(-\dfrac{10}{13}\right)$ or $\dfrac{10}{13}$.

Practice Problems 20–22

Simplify.

20. $-(-11)$

21. $-|7|$

22. $-|-2|$

EXAMPLES   Simplify.

20. $-(-4) = 4$ The opposite of negative 4 is 4.

21. $-|6| = -6$ The opposite of the absolute value of 6 is the opposite of 6, which is -6.

22. $-|-5| = -5$ The opposite of the absolute value of -5 is the opposite of 5, which is -5.

Answers

15. 7, **16.** 0, **17.** $-\dfrac{11}{15}$, **18.** 9.6, **19.** -4,

20. 11, **21.** -7, **22.** -2

Name _____ Section _____ Date _____

A *Represent each quantity by a signed number. See Example 1.*

1. A worker in a silver mine in Nevada works 1445 feet underground.

2. A scuba diver is swimming 35 feet below the surface of the water in the Gulf of Mexico.

3. The peak of Mt. Rainier in Washington State is 14,410 feet above sea level. (*Source:* U.S. Geological Survey)

4. The average depth of the Atlantic Ocean, without its adjacent seas, is 12,877 feet below the surface of the ocean. (*Source:* Naval Meteorology and Oceanography Command)

5. The Virginia Cavaliers football team lost 15 yards on a play.

6. The record low temperature in Alaska is 80 degrees Fahrenheit below zero. (*Source:* National Climatic Data Center)

 7. The Dow Jones stock market average fell 317 points in one day.

8. The highest elevation in the United States is Mt. McKinley in Alaska at an elevation of 20,320 feet above sea level. (*Source:* U.S. Geological Survey)

9. Gateway, Inc., manufactures personal computers. In the second quarter of fiscal year 2001, Gateway posted a net loss of $20,786 thousand. (*Source:* Gateway, Inc.)

10. During the first quarter of fiscal year 2001, Apple Computer, Inc., reported a net loss of $195 million. (*Source:* Apple Computer, Inc.)

11. The temperature on one January day in Chicago was −10 degrees Celsius. Tell whether this temperature is cooler or warmer than −5 degrees Celsius.

12. Two divers are exploring the bottom of a trench in the Pacific Ocean. Joe is at 135 feet below the surface of the ocean, and Sara is at 157 feet below the surface. Determine who is deeper in the water.

13. In 2000, the number of music cassette singles shipped to retailers reflected a 91% loss from the previous year. Write a signed number to represent the percent loss in cassette singles shipped. (*Source:* Recording Industry Association of America)

14. In 2000, the number of music CD singles shipped to retailers reflected a 38.8% loss from the previous year. Write a signed number to represent the percent loss in CD singles shipped. (*Source:* Recording Industry Association of America)

B *Graph the signed numbers in each list on a number line. See Example 2.*

 15. $-3, 0, 4, -1\frac{1}{2}$

16. $-4, 0, 2, -3\frac{1}{4}$

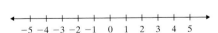

17. $5, -2, -4.7$

18. $3, -1, -4, -2.1$

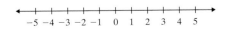

C *Insert < or > between each pair of numbers to make a true statement. See Examples 3 through 9.*

19. 5   7   **20.** 16   10   **21.** 4   0   **22.** 8   0

23. −5   −7

24. −12   −10

25. 0   −3

26. 0   −7

27. −26   26

28. 13   −13

29. −4.6   −2.7

30. 0 $-\dfrac{7}{8}$

31. $-1\dfrac{3}{4}$ 0

32. −8.4   −1.6

33. $\dfrac{1}{4}$ $-\dfrac{8}{11}$

34. −0.2   6

D *Find each absolute value. See Examples 10 through 14.*

35. $|5|$

36. $|7|$

37. $|-8|$

38. $|-19|$

39. $|0|$

40. $|100|$

41. $|-5|$

42. $|-10|$

43. $|-8.1|$

44. $\left|-\dfrac{1}{2}\right|$

45. $\left|\dfrac{9}{10}\right|$

46. $|-31.6|$

47. $\left|-\dfrac{3}{8}\right|$

48. $\left|\dfrac{20}{23}\right|$

49. $|7.6|$

50. $|-0.6|$

E *Find the opposite of each number. See Examples 15 through 19.*

51. 5

52. 8

53. −4

54. −6

55. 23.6

56. 123.9

57. $-\dfrac{9}{16}$

58. $-\dfrac{4}{9}$

59. −0.7

60. −4.4

61. $\dfrac{17}{18}$

62. $\dfrac{2}{3}$

Simplify. See Examples 20 through 22.

63. $|-7|$

64. $|-11|$

65. $-|20|$

66. $-|43|$

67. $-|-3|$

68. $-|-18|$

69. $-(-8)$

70. $-(-7)$

71. $|-14|$

72. $-(-14)$

73. $-(-29)$

74. $-|-29|$

Review and Preview

Add. See Section 1.3.

75. $0 + 13$

76. $9 + 0$

77. $15 + 20$

78. $20 + 15$

79. $47 + 236 + 77$

80. $362 + 37 + 90$

Combining Concepts

For Exercises 81 through 83, determine whether each statement is true or false.

81. A positive number is always greater than a negative number.

82. The absolute value of a number is *always* a positive number.

83. Zero is always less than a positive number.

84. Explain how to determine which of two signed numbers is larger.

85. Write in your own words how to find the absolute value of a signed number.

616

10.2 Adding Signed Numbers

Ⓐ Adding Signed Numbers

Adding signed numbers can be visualized by using a number line. A positive number can be represented on the number line by an arrow of appropriate length pointing to the right, and a negative number by an arrow of appropriate length pointing to the left.

Both arrows represent 2 or +2.

They both point to the right, and they are both 2 units long.

Both arrows represent −3.

They both point to the left, and they are both 3 units long.

To add signed numbers such as $5 + (-2)$ on a number line, we start at 0 on the number line and draw an arrow representing 5. From the tip of this arrow, we draw another arrow representing −2. The tip of the second arrow ends at their sum, 3.

$$5 + (-2) = 3$$

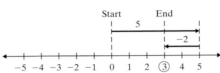

To add $-1 + (-4)$ on the number line, we start at 0 and draw an arrow representing −1. From the tip of this arrow, we draw another arrow representing −4. The tip of the second arrow ends at their sum, −5.

$$-1 + (-4) = -5$$

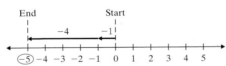

EXAMPLE 1   Add using a number line: $-3 + (-4)$

Solution:

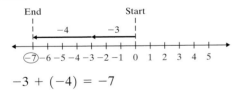

$$-3 + (-4) = -7$$

EXAMPLE 2   Add using a number line: $-7 + 3$

Solution:

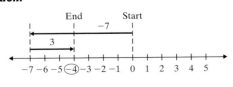

$$-7 + 3 = -4$$

Using a number line each time we add two numbers can be time consuming. Instead, we can notice patterns in the previous examples and write rules for adding signed numbers.

Practice Problem 1

Add using a number line: $5 + (-1)$

Practice Problem 2

Add using a number line: $-6 + (-2)$

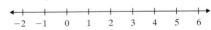

Answers

1.

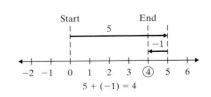

$$5 + (-1) = 4$$

2.

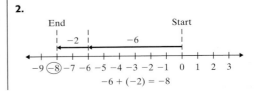

$$-6 + (-2) = -8$$

Rules for adding signed numbers depend on whether we are adding numbers with the same sign or different signs. When adding two numbers with the same sign, notice that the sign of the sum is the same as the sign of the addends.

Adding Two Numbers with the Same Sign

Step 1. Add their absolute values.

Step 2. Use their common sign as the sign of the sum.

Practice Problem 3

Add: $(-3) + (-9)$

EXAMPLE 3   Add: $-2 + (-21)$

Solution:

Step 1. $|-2| = 2, |-21| = 21$, and $2 + 21 = 23$.

Step 2. Their common sign is negative, so the sum is negative:

$$-2 + (-21) = -23$$

Practice Problems 4–7

Add.

4. $-12 + (-3)$

5. $9 + 5$

6. $-\dfrac{3}{7} + \left(-\dfrac{2}{7}\right)$

7. $-8.3 + (-5.7)$

EXAMPLES   Add.

4. $-5 + (-1) = -6$

5. $2 + 6 = 8$

6. $-\dfrac{3}{8} + \left(-\dfrac{1}{8}\right) = -\dfrac{\overset{1}{\cancel{4}}}{\underset{2}{\cancel{8}}} = -\dfrac{1}{2}$

7. $-1.2 + (-7.1) = -8.3$

The rule for adding two numbers with different signs follows:

Adding Two Numbers with Different Signs

Step 1. Find the larger absolute value minus the smaller absolute value.

Step 2. Use the sign of the number with the larger absolute value as the sign of the sum.

Practice Problem 8

Add: $-3 + 9$

EXAMPLE 8   Add: $-2 + 5$

Solution:

Step 1. $|-2| = 2, |5| = 5$, and $5 - 2 = 3$.

Step 2. 5 has the larger absolute value and its sign is an understood $+$:

$$-2 + 5 = +3 \text{ or } 3$$

Practice Problem 9

Add: $2 + (-8)$

Practice Problems 10–13

Add.

10. $-46 + 20$ 11. $8.6 + (-6.2)$

12. $-\dfrac{3}{4} + \dfrac{1}{8}$ 13. $-2 + 0$

EXAMPLE 9   Add: $3 + (-7)$

Solution:

Step 1. $|3| = 3, |-7| = 7$, and $7 - 3 = 4$.

Step 2. -7 has the larger absolute value and its sign is $-$:

$$3 + (-7) = -4$$

EXAMPLES   Add.

10. $-18 + 10 = -8$

11. $12.9 + (-8.6) = 4.3$

12. $-\dfrac{1}{2} + \dfrac{1}{6} = -\dfrac{3}{6} + \dfrac{1}{6} = -\dfrac{2}{6} = -\dfrac{1}{3}$

13. $0 + (-5) = -5$ The sum of 0 and any number is the number.

Answers

3. -12,   4. -15,   5. 14,   6. $-\dfrac{5}{7}$,   7. -14,

8. 6,   9. -6,   10. -26,   11. 2.4,   12. $-\dfrac{5}{8}$,

13. -2

Try the Concept Check in the margin.

When we add three or more numbers, we follow the order of operations and add from left to right. That is, we start at the left and add the first two numbers. Then we add their sum to the next number. We continue this process until the addition is completed.

EXAMPLE 14   Add: $(-3) + 4 + (-11)$

Solution:

$$(-3) + 4 + (-11) = 1 + (-11)$$
$$= -10$$

EXAMPLE 15   Add: $1 + (-10) + (-8) + 9$

Solution:

$$1 + (-10) + (-8) + 9 = -9 + (-8) + 9$$
$$= -17 + 9$$
$$= -8$$

Try the Concept Check in the margin.

B   **Solving Problems by Adding Signed Numbers**

Next, we practice solving problems that require adding signed numbers.

EXAMPLE 16   **Calculating Temperature**

On January 6 the temperature in Caribou, Maine, at 8 A.M. was $-12°$ Fahrenheit. By 9 A.M., the temperature had risen by 4 degrees, and by 10 A.M. it had risen 6 degrees from the 9 A.M. temperature. What was the temperature at 10 A.M.?

Solution:

1. UNDERSTAND. Read and reread the problem.
2. TRANSLATE.

In words:	temperature at 10 A.M.	=	8 A.M. temperature	+	rise of 4°	+	rise of 6°
	↓		↓		↓		↓
Translate:	temperature at 10 A.M.	=	-12	+	$(+4)$	+	$(+6)$

3. SOLVE:

$$\text{temperature at 10 A.M.} = -12 + (+4) + (+6)$$
$$= -8 + (+6)$$
$$= -2$$

4. INTERPRET. Check and state your conclusion: The temperature was $-2°F$ at 10 A.M.

Concept Check

What is wrong with the following calculation?

$$5 + (-13) = 8$$

Practice Problem 14

Add: $8 + (-3) + (-13)$

Practice Problem 15

Add: $5 + (-3) + 12 + (-14)$

Concept Check

Estimate the sum.

$$-8.19 + (-1.87) + 5.22$$

Practice Problem 16

If the temperature was $-8°$ Fahrenheit at 6 A.M., and it rose 4 degrees by 7 A.M., and then rose another 7 degrees in the hour from 7 A.M. to 8 A.M., what was the temperature at 8 A.M.?

Answers

14. -8,   **15.** 0,   **16.** 3°F

Concept Check: $5 + (-13) = -8$

Concept Check: $-8 + (-2) + 5 = -5$

CALCULATOR EXPLORATIONS

Entering Negative Numbers

To enter a negative number on a calculator, find the key marked $\boxed{+/-}$. (Some calculators have a key marked $\boxed{\text{CHS}}$, and some calculators have a special key $\boxed{(-)}$ for entering a negative sign.) To enter the number -2, for example, press the keys $\boxed{2}\ \boxed{+/-}$. The display will read $\boxed{\qquad -2}$.

To find $-32 + (-131)$, press the keys

$$\boxed{32}\ \boxed{+/-}\ \boxed{+}\ \boxed{131}\ \boxed{+/-}\ \boxed{=} \text{ or } \boxed{\text{ENTER}}.$$

The display will read $\boxed{\qquad -163}$. Thus, $-32 + (-131) = -163$.

Use a calculator to perform each indicated operation.

1. $-256 + 97$
2. $811 + (-1058)$
3. $6(15) + (-46)$
4. $-129 + 10(48)$
5. $-108.65 + (-786.205)$
6. $-196.662 + (-129.856)$

Add. See Examples 14 and 15.

 49. −4 + 2 + (−5)

50. −1 + 5 + (−8)

51. −5.2 + (−7.7) + (−11.7)

52. −10.3 + (−3.2) + (−2.7)

53. 12 + (−4) + (−4) + 12

54. 18 + (−9) + 5 + (−2)

55. (−10) + 14 + 25 + (−16)

56. 34 + (−12) + (−11) + 213

B *Solve. See Example 16.*

57. The temperature at 4 P.M. on February 2 was −10° Celsius. By 11 P.M. the temperature had risen 12 degrees. Find the temperature at 11 P.M.

58. Scores in golf can be positive or negative integers. For example, a score of 3 *over* par can be represented by +3, and a score of 5 *under* par can be represented by −5. If Fred Couples had scores of 3 over par, 6 under par, and 7 under par for three games of golf, what was his total score?

 59. Suppose a deep-sea diver dives from the surface to 165 feet below the surface. He then dives down 16 more feet. Use positive and negative numbers to represent this situation. Then find the diver's present depth.

60. Suppose a diver dives from the surface to 248 meters below the surface and then swims up 6 meters, down 17 meters, down another 24 meters, and then up 23 meters. Use positive and negative numbers to represent this situation. Then find the diver's depth after these movements.

In some card games, it is possible to have positive and negative scores. The table shows the scores for two teams playing a series of four card games. Use this table to answer Exercises 61 and 62.

	Game 1	Game 2	Game 3	Game 4
Team 1	−2	−13	20	2
Team 2	5	11	−7	−3

61. Find each team's total score after four games. If the winner is the team with the greater score, find the winning team.

62. Find each team's total score after three games. If the winner is the team with the greater score, which team was winning after three games?

63. The all-time record low temperature for Illinois is −36°F, which was recorded on January 5, 1999. South Carolina's all-time record low temperature is 17°F higher than Illinois' record low. What is South Carolina's record low temperature? (*Source:* National Climatic Data Center)

64. The all-time record low temperature for Texas is −23°F, which occurred on February 8, 1933. In Hawaii, the lowest temperature ever recorded is 35°F more than Texas' all-time low temperature. What is the all-time record low temperature for Hawaii? (*Source:* National Climatic Data Center)

65. The deepest spot in the Pacific Ocean is the Mariana Trench, which has an elevation of 10,924 meters below sea level. The bottom of the Pacific's Aleutian Trench has an elevation 3245 meters higher than that of the Mariana Trench. Use a negative number to represent the depth of the Aleutian Trench. (*Source:* Defense Mapping Agency)

66. The deepest spot in the Atlantic Ocean is the Puerto Rico Trench, which has an elevation of 8605 meters below sea level. The bottom of the Atlantic's Cayman Trench has an elevation 1070 meters above the level of the Puerto Rico Trench. Use a negative number to represent the depth of the Cayman Trench. (*Source:* Defense Mapping Agency)

Review and Preview

Subtract. See Section 1.4.

67. 44 − 0   **68.** 91 − 0   **69.** 52 − 52   **70.** 103 − 103   **71.** 87 − 59   **72.** 32 − 18

Combining Concepts

For Exercises 73 through 76, determine whether each statement is true or false.

73. The sum of two negative numbers is always a negative number.

74. The sum of two positive numbers is always a positive number.

75. The sum of a positive number and a negative number is always a negative number.

76. The sum of zero and a negative number is always a negative number.

 77. In your own words, explain how to add two negative numbers.

78. In your own words, explain how to add a positive number and a negative number.

10.3 Subtracting Signed Numbers

In Section 10.1 we discussed the opposite of a number.

The opposite of 3 is −3.
The opposite of −6.7 is 6.7.

Now let's find the sum of these opposites.

$$3 + (-3) = 0 \quad \text{and} \quad -6.7 + 6.7 = 0$$

From this, we see that the sum of a number and its opposite is 0. In this section, we use opposites to subtract signed numbers.

Ⓐ Subtracting Signed Numbers

To subtract signed numbers, we write the subtraction problem as an addition problem. To see how we do this, study the examples below:

$$10 - 4 = 6$$

$$10 + (-4) = 6$$

Since both expressions simplify to 6, this means that

$$10 - 4 = 10 + (-4) = 6$$

Also,

$$3 - 2 = 3 + (-2) = 1$$

$$15 - 1 = 15 + (-1) = 14$$

Thus, to subtract two numbers, we add the first number to the opposite (also called the **additive inverse**) of the second number.

EXAMPLES   Subtract.

subtraction	=	first number	+	opposite of the second number		
1. 8 − 5	=	8	+	(−5)	=	3
2. −4 − 10	=	−4	+	(−10)	=	−14
3. 6 − (−5)	=	6	+	5	=	11
4. −11 − (−7)	=	−11	+	7	=	−4

OBJECTIVES

Ⓐ Subtract signed numbers.

Ⓑ Add and subtract signed numbers.

Ⓒ Solve problems by subtracting signed numbers.

SSM
TUTOR CENTER   SG   CD & VIDEO   MATH PRO   WEB

Subtracting Two Numbers
If a and b are numbers, then
$a - b = a + (-b)$.

Practice Problems 1–4

Subtract.

1. 12 − 7   2. −6 − 4
3. 11 − (−14)   4. −9 − (−1)

Answers

1. 5,   **2.** −10,   **3.** 25,   **4.** −8

Practice Problems 5–9

Subtract.

5. $5 - 9$ 6. $-12 - 4$

7. $-2 - (-7)$ 8. $-10.5 - 14.3$

9. $\dfrac{5}{13} - \dfrac{12}{13}$

EXAMPLES Subtract.

5. $-10 - 5 = -10 + (-5) = -15$

6. $8 - 15 = 8 + (-15) = -7$

7. $-4 - (-5) = -4 + 5 = 1$

8. $-1.7 - 6.2 = -1.7 + (-6.2) = -7.9$

9. $-\dfrac{10}{11} - \left(-\dfrac{3}{11}\right) = -\dfrac{10}{11} + \dfrac{3}{11} = -\dfrac{7}{11}$

Try the Concept Check in the margin.

Concept Check

What is wrong with the following calculation? $-7 - (-3) = -10$

Practice Problem 10

Subtract 5 from -10.

EXAMPLE 10 Subtract 7 from -3.

Solution: To subtract 7 *from* -3, we find

$$-3 - 7 = -3 + (-7) = -10$$

B Adding and Subtracting Signed Numbers

If a problem involves adding or subtracting more than two signed numbers, we rewrite differences as sums and add from left to right.

Practice Problem 11

Simplify: $-4 - 3 - 7 - (-5)$

EXAMPLE 11 Simplify: $7 - 8 - (-5) - 1$

Solution:

$$
\begin{aligned}
7 - 8 - (-5) - 1 &= 7 + (-8) + 5 + (-1) \\
&= -1 + 5 + (-1) \\
&= 4 + (-1) \\
&= 3
\end{aligned}
$$

Practice Problem 12

Simplify: $3 + (-5) - 6 - (-4)$

EXAMPLE 12 Simplify: $7 + (-12) - 3 - (-8)$

Solution:

$$
\begin{aligned}
7 + (-12) - 3 - (-8) &= 7 + (-12) + (-3) + 8 \\
&= -5 + (-3) + 8 \\
&= -8 + 8 \\
&= 0
\end{aligned}
$$

Answers

5. -4, 6. -16, 7. 5, 8. -24.8, 9. $-\dfrac{7}{13}$

10. -15, 11. -9, 12. -4

Concept Check: $-7 - (-3) = -4$

C Solving Problems by Subtracting Signed Numbers

Solving problems often requires subtraction of signed numbers.

EXAMPLE 13   Finding a Change in Elevation

The highest point in the United States is the top of Mount McKinley, at a height of 20,320 feet above sea level. The lowest point is Death Valley, California, which is 282 feet below sea level. How much higher is Mount McKinley than Death Valley? (*Source:* U.S. Geological Survey)

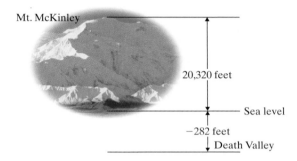

Practice Problem 13

The highest point in Asia is the top of Mount Everest, at a height of 29,028 feet above sea level. The lowest point is the Dead Sea, which is 1312 feet below sea level. How much higher is Mount Everest than the Dead Sea? (*Source:* National Geographic Society)

Solution:

1. UNDERSTAND. Read and reread the problem. To find "how much higher," we subtract. Don't forget that if Death Valley is 282 feet *below* sea level, we represent its height by -282. Draw a diagram to help visualize the problem.

2. TRANSLATE.

In words:	how much higher is Mt. McKinley	=	height of Mt. McKinley	minus	height of Death Valley
	↓	↓	↓	↓	↓
Translate:	How much higher is Mt. McKinley	=	20,320	−	(−282)

3. SOLVE:

$$20,320 - (-282) = 20,320 + 282 = 20,602$$

4. INTERPRET. Check and state your conclusion: Mount McKinley is 20,602 feet higher than Death Valley.

Answer

13. 30,340 ft

FOCUS ON **Mathematical Connections**

MODELING ADDING AND SUBTRACTING INTEGERS

Just as we used physical models to represent fractions and operations on whole numbers, we can use a physical model to help us add and subtract integers. In this model, we use objects called counters to represent numbers. Black counters ● represent positive numbers, and red counters ● represent negative numbers. The key to this model is remembering that taking a ● and ● together creates a neutral or zero pair. Once a neutral pair has been formed, it can be removed from or added to the model without changing the overall value.

Adding Integers

$5 + (-8)$

Begin with 5 black counters and add to that 8 red counters.

Next, form and remove neutral pairs.

   Now there are only 3 red counters left, so $5 + (-8) = -3$

$-4 + (-3)$

Begin with 4 red counters and add to that 3 red counters.

● ● ● ●
● ● ●

Because this group does not contain a mixture of red and black counters, there is no need to form neutral pairs. Simply find the total number of red counters and remember that red counters represent negative numbers. So,

$$-4 + (-3) = -7$$

Subtracting Integers

$6 - (-2)$

Begin with 6 black counters.

● ● ● ● ● ●

Subtracting a -2 indicates taking away 2 red counters. However, there are no red counters in the model at this point. Add enough neutral pairs to the model to obtain 2 red counters.

● ● ● ● ● ●   ● ●   ● ●

Now take away 2 red counters.

● ● ● ● ● ● ● ○ ● ○ ●

Because there are 8 black counters remaining, $6 - (-2) = 8$.

$-3 - 9$

Begin with 3 red counters.

● ● ●

Subtracting 9 indicates taking away 9 black counters, However, there are no black counters in the model at this point. Add enough neutral pairs to the model to obtain 9 black counters.

● ● ●   ● ● ●   ● ● ●   ● ● ●
● ● ●   ● ● ●   ● ● ●   ● ● ●
● ● ●   ● ● ●   ● ● ●   ● ● ●

Now take away 9 black counters.

Because there are 12 red counters remaining, $-3 - 9 = -12$.

CRITICAL THINKING

Use the counter model to perform each indicated operation.

1. $3 + 7$ **2.** $(-5) + 6$ **3.** $(-8) + (-4)$ **4.** $9 + (-11)$ **5.** $(-2) + 2$

6. $8 - 4$ **7.** $(-3) - (-1)$ **8.** $2 - (-5)$ **9.** $(-6) - 4$ **10.** $5 - 7$

Name _____ Section _____ Date _____

Mental Math

Subtract.

1. $5 - 5$ **2.** $7 - 7$ **3.** $6.2 - 6.2$ **4.** $1.9 - 1.9$

EXERCISE SET 10.3

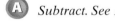 **A** *Subtract. See Examples 1 through 9.*

1. $-5 - (-5)$ **2.** $-6 - (-6)$ **3.** $8 - 3$ **4.** $5 - 2$ **5.** $3 - 8$

6. $2 - 5$ **7.** $7 - (-7)$ **8.** $12 - (-12)$ **9.** $-5 - (-8)$ **10.** $-25 - (-25)$

11. $-14 - 4$ **12.** $-2 - 42$ **13.** $2 - 16$ **14.** $8 - 9$ **15.** $2.2 - 5.5$

16. $1.7 - 6.3$ **17.** $3.62 - (-0.4)$ **18.** $8.44 - (-0.2)$ **19.** $-\dfrac{3}{10} - \left(-\dfrac{7}{10}\right)$ **20.** $-\dfrac{2}{11} - \left(-\dfrac{5}{11}\right)$

21. $\dfrac{2}{5} - \dfrac{7}{10}$ **22.** $\dfrac{3}{16} - \dfrac{7}{8}$ **23.** $\dfrac{1}{2} - \left(-\dfrac{1}{3}\right)$ **24.** $\dfrac{1}{7} - \left(-\dfrac{1}{4}\right)$

Solve. See Example 10.

25. Subtract 18 from -20.   **26.** Subtract 10 from -22.   **27.** Find the difference of -20 and -3.

28. Find the difference of -8 and -13.   **29.** Subtract -11 from 2.   **30.** Subtract -50 from -50.

B *Simplify: See Examples 11 and 12.*

31. $7 - 3 - 2$ **32.** $8 - 4 - 1$ **33.** $12 - 5 - 7$ **34.** $30 - 7 - 12$

35. $-5 - 8 - (-12)$ **36.** $-10 - 6 - (-9)$ **37.** $-10 + (-5) - 12$ **38.** $-15 + (-8) - 4$

39. $12 - (-34) + (-6)$ **40.** $23 - (-17) + (-9)$

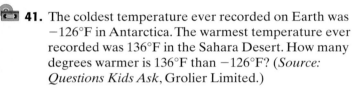

Solve. See Example 13.

41. The coldest temperature ever recorded on Earth was $-126°F$ in Antarctica. The warmest temperature ever recorded was $136°F$ in the Sahara Desert. How many degrees warmer is $136°F$ than $-126°F$? (*Source: Questions Kids Ask*, Grolier Limited.)

42. The coldest temperature ever recorded in South America was $-27°F$ in Sarmiento, Argentina. The warmest temperature ever recorded in South America was $120°F$ in Rivadavia, Argentina. How many degrees warmer is $120°F$ than $-27°F$? (*Source*: National Climatic Data Center)

43. Aaron Aiken has $125 in his checking account. He writes a check for $117, makes a deposit of $45, and then writes another check for $69. Find the amount left in his account. (Write the amount as an integer.)

44. In canasta, it is possible to have a negative score. If Juan Santanilla's score is 15, what is his new score if he loses 20 points?

45. The deepest point in the Atlantic Ocean lies in the Puerto Rico Trench at a depth of 28,374 feet below sea level. The Atlantic's Romanche Trench reaches a depth of 24,455 feet below sea level. What is the difference in elevation between these two points? (*Source*: Naval Meteorology and Oceanography Command)

46. The temperature on a February morning is $-6°$ Celsius at 6 A.M. If the temperature drops 3 degrees by 7 A.M., rises 4 degrees between 7 A.M. and 8 A.M., and then drops 7 degrees between 8 A.M. and 9 A.M., find the temperature at 9 A.M.

The bar graph shows heights of selected lakes. For Exercises 47 through 50, find the difference in elevation for the lakes listed. (Source: U.S. Geological Survey)

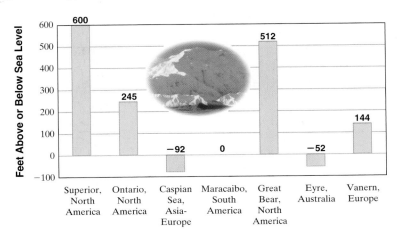

47. Lake Superior and Lake Eyre

48. Great Bear Lake and the Caspian Sea

49. Lake Maracaibo and Lake Vanern

50. Lake Eyre and the Caspian Sea

51. The average temperature on the surface of Mercury is 166.85°C. The average temperature on the surface of Neptune is −215.15°C. How many degrees warmer is the surface of Mercury than the surface of Neptune? (*Source*: National Space Science Data Center)

52. The average temperature on the surface of Earth is 14.85°C. The average temperature on the surface of Mars is −63.15°C. How many degrees warmer is the surface of Earth than the surface of Mars? (*Source*: National Space Science Data Center)

53. The average temperature on the surface of Saturn is −176°C. The average temperature on the surface of Uranus is −215°C. Which planet has the warmer average temperature? How much warmer is it? (*Source*: National Space Science Data Center)

54. The average temperature on the surface of Pluto is −223.15°C. The average temperature on the surface of Jupiter is −144°C. What is the difference between the average temperature on Jupiter and the average temperature on Pluto? (*Source*: National Space Science Data Center)

Review and Preview

Multiply. See Section 1.6.

55. $8 \cdot 0$

56. $0 \cdot 8$

57. $1 \cdot 8$

58. $8 \cdot 1$

59. 23
$\times 46$

60. 51
$\times 89$

Combining Concepts

Simplify. (Hint: Find the absolute values first.)

61. $|-3| - |-7|$

62. $|-12| - |-5|$

63. $|-6| - |6|$

64. $|-9| - |9|$

65. $|-17| - |-29|$

66. $|-23| - |-42|$

For Exercises 67 and 68, determine whether each statement is true or false.

67. $|-8 - 3| = 8 - 3$

68. $|-2 - (-6)| = |-2| - |-6|$

 69. In your own words, explain how to subtract one signed number from another.

Internet Excursions

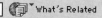
Go To: http://www.prenhall.com/martin-gay_basic   What's Related

The World Wide Web address listed here will direct you to the Web site of the CNN Financial Network, or a related site. You will be able to collect data on prices of stocks traded on the New York Stock Exchange and answer the questions below.

70. Pick stocks for any four companies. You can use the Web site to look up the stock market ticker symbol for the companies. Then get a stock quote for each company. Complete the table below. For each company in the table, show that the sum of the previous closing price and the change in price give the last price.

71. For the same four companies you used in Exercise 70, show how to use the last price and the change in price to calculate the previous closing price.

Date of stock quotes: _____   Time of stock quotes: _____

Company Name	Ticker Symbol	Previous Close	Change	Last

Name _____

Chapter 10 Integrated Review—Signed Numbers

Represent each quantity by a signed number.

1. The peak of Mount Everest in Asia is 29,028 feet above sea level. (*Source*: U.S. Geological Survey)

2. The Mariana Trench in the Pacific Ocean is 35,840 feet below sea level. (*Source: The World Almanac*, 1998)

3. The Peru-Chile Trench, off the west coast of South America in the Pacific Ocean, is 26,322 feet below sea level. (*Source*: Naval Meteorology and Oceanography Command)

4. The peak of K2 in Asia is 28,250 feet above sea level. (*Source*: National Geographic Society)

5. Graph the signed numbers on the given number line. $-4, 0, -1.5, 3\frac{1}{4}$

Insert < or > between each pair of numbers to make a true statement.

6. 0 -3

7. -15 -5

8. $-1\frac{1}{2}$ $1\frac{1}{4}$

9. -2 -7.6

10. 7   0

11. -4 -40

12. 3.9 -3.9

13. $-\frac{1}{2}$ $-\frac{1}{3}$

Simplify.

14. $|-1|$

15. $|1|$

16. $|0|$

17. $-|-4|$

18. $|-8.6|$

19. $|100.3|$

20. $-\left(-\frac{3}{4}\right)$

21. $-\left|-\frac{3}{4}\right|$

Answers

1. _____
2. _____
3. _____
4. _____
5. see number line
6. _____
7. _____
8. _____
9. _____
10. _____
11. _____
12. _____
13. _____
14. _____
15. _____
16. _____
17. _____
18. _____
19. _____
20. _____
21. _____

22. _____

23. _____

24. _____

25. _____

26. _____

27. _____

28. _____

29. _____

30. _____

31. _____

32. _____

33. _____

34. _____

35. _____

36. _____

37. _____

38. _____

39. _____

40. _____

41. _____

42. _____

43. _____

Find the opposite of each number.

22. 6   **23.** −3   **24.** 89.1   **25.** $-\dfrac{2}{9}$

Add or subtract as indicated.

26. −7 + 12   **27.** −9.2 + (−11.6)   **28.** $\dfrac{5}{9} + \left(-\dfrac{1}{3}\right)$

29. 1 − 3   **30.** $\dfrac{3}{8} - \left(-\dfrac{3}{8}\right)$ **31.** −2.6 − 1.4

32. −7 + (−2.6)   **33.** −14 + 8   **34.** −8 − (−20)

35. −18 − (−102)   **36.** 8.65 − 12.09

37. $-\dfrac{4}{5} - \dfrac{3}{10}$ **38.** −8 + (−6) + 20

39. −11 − 7 − (−19)   **40.** −4 + (−8) − 16 − (−9)

41. Subtract 14 from 26.   **42.** Subtract −8 from −12.

43. The coldest temperature ever recorded in Europe was −67°F in Ust'Shchugor, Russia. The warmest temperature ever recorded in Europe was 122°F in Seville, Spain. How many degrees warmer is 122°F than −67°F? (*Source:* National Climatic Data Center)

10.4 Multiplying and Dividing Signed Numbers

Multiplying and dividing signed numbers is similar to multiplying and dividing whole numbers. One difference is that we need to determine whether the result is a positive number or a negative number.

OBJECTIVES

A Multiply signed numbers.

B Divide signed numbers.

C Solve problems by multiplying or dividing signed numbers.

SSM   TUTOR CENTER   SG   CD & VIDEO   MATH PRO   WEB

A Multiplying Signed Numbers

Consider the following pattern of products:

First factor decreases by 1 each time.

$3 \cdot 2 = 6$

$2 \cdot 2 = 4$ Product decreases by 2 each time.

$1 \cdot 2 = 2$

$0 \cdot 2 = 0$

This pattern can be continued, as follows:

$-1 \cdot 2 = -2$

$-2 \cdot 2 = -4$

$-3 \cdot 2 = -6$

This suggests that the product of a negative number and a positive number is a negative number.

What is the sign of the product of two negative numbers? To find out, we form another pattern of products. Again, we decrease the first factor by 1 each time, but this time the second factor is negative.

$2 \cdot (-3) = -6$

$1 \cdot (-3) = -3$ Product increases by 3 each time.

$0 \cdot (-3) = 0$

This pattern continues as:

$-1 \cdot (-3) = 3$

$-2 \cdot (-3) = 6$

$-3 \cdot (-3) = 9$

This suggests that the product of two negative numbers is a positive number. Thus, we can determine the sign of a product when we know the signs of the factors.

Multiplying Signed Numbers

The product of two numbers having the same sign is a positive number. The product of two numbers having different signs is a negative number.

EXAMPLES   Multiply.

1. $-7 \cdot 3 = -21$

2. $-2(-5) = 10$

3. $0 \cdot (-4) = 0$

4. $\left(-\dfrac{1}{2}\right)\left(-\dfrac{2}{3}\right) = \dfrac{1 \cdot \overset{1}{\cancel{2}}}{\underset{1}{\cancel{2}} \cdot 3} = \dfrac{1}{3}$

5. $-3(1.2) = -3.6$

Practice Problems 1–5

Multiply.

1. $-2 \cdot 6$ 2. $-4(-3)$

3. $0 \cdot (-10)$ 4. $\left(\dfrac{3}{7}\right)\left(-\dfrac{1}{3}\right)$

5. $-4(-3.2)$

Answers

1. -12, **2.** 12, **3.** 0, **4.** $-\dfrac{1}{7}$, **5.** 12.8

Concept Check

Estimate the product

$$(3.92) \cdot \left(-9\frac{1}{8}\right)$$

Practice Problem 6

Multiply.

a. $(-2)(-1)(-2)(-1)$
b. $7(0)(-4)$

Concept Check

What is the sign of the product of five negative numbers? Explain.

Practice Problem 7

Evaluate: $(-3)^4$

Practice Problems 8–11

Divide.

8. $\dfrac{28}{-7}$ 9. $-18 \div (-2)$

9. $\dfrac{-4.6}{0.2}$ 11. $-\dfrac{3}{5} \div \dfrac{2}{7}$

Answers

6. a. 4, **b.** 0, **7.** 81, **8.** −4, **9.** 9,

10. −23, **11.** $-\dfrac{21}{10}$

Concept Check: $4 \cdot (-9) = -36$

Concept Check: negative

Try the Concept Check in the margin.

To find the product of more than two numbers, we multiply from left to right.

EXAMPLE 6 Multiply.

a. $7(-6)(-2)$ b. $(-2)(-3)(-4)$

Solution:

a. $7(-6)(-2) = -42(-2) = 84$

b. $(-2)(-3)(-4) = 6(-4) = -24$

Try the Concept Check in the margin.

Recall from our study of exponents that $2^3 = 2 \cdot 2 \cdot 2 = 8$. We can now work with bases that are negative numbers. For example,

$$(-2)^3 = (-2)(-2)(-2) = -8$$

EXAMPLE 7 Evaluate: $(-5)^2$

Solution: Remember that $(-5)^2$ means 2 factors of -5.

$$(-5)^2 = (-5)(-5) = 25$$

B **Dividing Signed Numbers**

Division of signed numbers is related to multiplication of signed numbers. The sign rules for division can be discovered by writing a related multiplication problem. For example,

$$\frac{6}{2} = 3 \qquad \text{because } 3 \cdot 2 = 6$$

$$\frac{-6}{2} = -3 \qquad \text{because } -3 \cdot 2 = -6$$

$$\frac{6}{-2} = -3 \qquad \text{because } -3 \cdot (-2) = 6$$

$$\frac{-6}{-2} = 3 \qquad \text{because } 3 \cdot (-2) = -6$$

Dividing Signed Numbers

The quotient of two numbers having the same sign is a positive number. The quotient of two numbers having different signs is a negative number.

EXAMPLES Divide.

8. $\dfrac{-12}{6} = -2$

9. $-20 \div (-4) = 5$

10. $\dfrac{1.2}{-0.6} = -2$

11. $-\dfrac{7}{9} \div \dfrac{2}{5} = -\dfrac{7}{9} \cdot \dfrac{5}{2} = -\dfrac{7 \cdot 5}{9 \cdot 2} = -\dfrac{35}{18}$

Try the Concept Check in the margin.

EXAMPLES Divide, if possible.

12. $\dfrac{0}{-5} = 0$ because $0 \cdot -5 = 0$

13. $\dfrac{-7}{0}$ is undefined because there is no number that gives a product of -7 when multiplied by 0 ●

(C) Solving Problems by Multiplying and Dividing Signed Numbers

Many real-life problems involve multiplication and division of signed numbers.

EXAMPLE 14 Calculating Total Golf Score

A professional golfer finished seven strokes under par (-7) for each of three days of a tournament. What was his total score for the tournament?

Solution:

1. UNDERSTAND. Read and reread the problem. Although the key word is "total," since this is repeated addition of the same number we multiply.

2. TRANSLATE.

In words: golfer's total score $=$ number of days $\cdot$ score each day

$\downarrow \qquad \downarrow \qquad \downarrow \qquad \downarrow \qquad \downarrow$

Translate: golfer's total $=$ 3 $\cdot$ (-7)

3. SOLVE: $3 \cdot (-7) = -21$

4. INTERPRET. Check and state your conclusion: The golfer's total score is -21, or 21 strokes under par. ●

Concept Check

Find the error in the following computation:

$$\frac{1}{2} \div (-3) = \frac{1}{2} \cdot \frac{1}{3} = \frac{1}{6}$$

Practice Problems 12–13

Divide, if possible.

12. $\dfrac{-1}{0}$

13. $\dfrac{0}{-2}$

Practice Problems 14

A card player had a score of -12 for each of four games. Find her total score.

Answers

12. undefined, **13.** 0, **14.** -48

Concept Check: $\dfrac{1}{2} \div (-3) = \dfrac{1}{2} \cdot \left(-\dfrac{1}{3}\right) = -\dfrac{1}{6}$

FOCUS ON History

MAGIC SQUARES

A magic square is a set of numbers arranged in a square table so that the sum of the numbers in each column, row, and diagonal is the same. For instance, in the magic square below, the sum of each column, row, and diagonal is 15. Notice that no number is used more than once in the magic square.

2	9	4
7	5	3
6	1	8

The properties of magic squares have been known for a very long time and once were thought to be good luck charms. The ancient Egyptians and Greeks understood their patterns. A magic square even made it into a famous work of art. The engraving titled *Melencolia I*, created by German artist Albrecht Dürer in 1514, features the following four-by-four magic square on the building behind the central figure.

16	3	2	13
5	10	11	8
9	6	7	12
4	15	14	1

CRITICAL THINKING

1. Verify that what is shown in the Dürer engraving is, in fact, a magic square. What is the common sum of the columns, rows, and diagonals?

2. Negative numbers can also be used in magic squares. Complete the following magic square:

		−2
	−1	
0		−4

3. Use the numbers −16, −12, −8, −4, 0, 4, 8, 12, and 16 to form a magic square:

Name _____ Section _____ Date _____

EXERCISE SET 10.4

 A *Multiply. See Examples 1 through 5.*

 1. $-2(-3)$ **2.** $5(-3)$ **3.** $-4(9)$ **4.** $-7(-2)$ **5.** $(2.6)(-1.2)$

6. $-0.3(5.6)$ **7.** $0(-14)$ **8.** $-6(0)$ **9.** $-\frac{3}{5}\left(-\frac{2}{7}\right)$ **10.** $-\frac{3}{4}\left(\frac{9}{10}\right)$

Multiply. See Example 6.

11. $6(-4)(2)$ **12.** $-2(3)(-7)$ **13.** $-1(-2)(-4)$ **14.** $8(-3)(3)$

15. $-4(4)(-5)$ **16.** $-2(-5)(-4)$ **17.** $10(-5)(0)$ **18.** $2(-1)(3)(-2)$

Evaluate. See Example 7.

19. $(-2)^2$ **20.** $(-2)^4$ **21.** $(-3)^3$ **22.** $(-1)^4$

23. $(-5)^2$ **24.** $(-4)^3$ **25.** $(-5)^3$ **26.** $(-3)^2$

Multiply. See Examples 1 through 7.

27. $-12(0)$ **28.** $0(-100)$ **29.** $\frac{3}{4}\left(-\frac{7}{8}\right)$ **30.** $-\frac{3}{11}\cdot\frac{1}{2}$

31. $-1 \cdot (-1) \cdot (-1) \cdot (-1)$

32. $-1 \cdot (-1) \cdot (-1)$

33. $-1(2)(7)(-3.1)$

34. $-2(3)(5)(-6.2)$

35. $(-2)^3$

36. $(-3)^5$

B *Divide. See Examples 8 through 13.*

37. $-24 \div 6$

38. $90 \div (-9)$

39. $\dfrac{-30}{6}$

40. $\dfrac{56}{-8}$

41. $\dfrac{-88}{-11}$

42. $\dfrac{-32}{4}$

43. $\dfrac{0}{14}$

44. $\dfrac{-13}{0}$

45. $\dfrac{39}{-3}$

46. $\dfrac{-24}{-12}$

47. $\dfrac{7.8}{-0.3}$

48. $\dfrac{1.21}{-1.1}$

49. $-\dfrac{7}{12} \div \left(-\dfrac{1}{6}\right)$

50. $-\dfrac{3}{8} \div \left(-\dfrac{2}{7}\right)$

51. $\dfrac{100}{-20}$

52. $\dfrac{45}{-9}$

53. $240 \div (-40)$

54. $480 \div (-8)$

55. $\dfrac{-12}{-4}$

56. $\dfrac{-36}{-3}$

57. $\dfrac{-120}{0.4}$

58. $\dfrac{-200}{2.5}$

59. $-\dfrac{8}{15} \div \dfrac{2}{3}$

60. $-\dfrac{1}{6} \div \dfrac{7}{18}$

61. A football team lost four yards on each of three consecutive plays. Represent the total loss as a product of signed numbers and find the total loss.

62. Joe Norstrom lost $400 on each of seven consecutive days in the stock market. Represent his total loss as a product of signed numbers and find his total loss.

63. A deep-sea diver must move up or down in the water in short steps in order to keep from getting a physical condition called the "bends." Suppose a diver moves down from the surface in five steps of 20 feet each. Represent his total movement as a product of signed numbers and find the product.

64. A weather forecaster predicts that the temperature will drop five degrees each hour for the next six hours. Represent this drop as a product of signed numbers and find the total drop in temperature.

65. During the 2001 U.S. Women's Open golf tournament, the winner, Karrie Webb, had scores of 0, −5, −1, and −1 in four rounds of golf. Find her average score per round. (*Source:* Ladies Professional Golf Association)

66. During the 2001 LPGA Jamie Farr Kroger Classic golf tournament, the winner, Se Ri Pak, had scores of −1, −9, −2, and −3 in four rounds of golf. Find her average score per round. (*Source:* Ladies Professional Golf Association)

67. During the first quarter of 2000, JCPenney posted a net income of −$118 million. If this continued, what would JCPenney's net income have been after four quarters? (*Source:* J. C. Penney Company, Inc.)

68. During the first quarter of 2001, Apple Computer posted a net income of −$195 million. If this continued, what would Apple's net income have been after four quarters? (*Source:* Apple Computer, Inc.)

69. In 1979, there were 35 California Condors in the entire world. By 1987, there were only 27 California Condors remaining. Thanks to conservation efforts, in 2001 there were 184 California Condors. (*Source:* California Department of Fish and Game)
 a. Find the change in the number of California Condors from 1979 to 1987.
 b. Find the average change per year in the California Condor population over the period in part a.
 c. Find the change in the number of California Condors from 1987 to 2001.
 d. Find the average change per year in the California Condor population over the period in part c. Round to the nearest whole.

70. In 1995, a total of 272.6 million music cassettes were shipped to retailers in the United States. In 2000, this number had dropped to 76.0 music cassettes. (*Source:* Recording Industry Association of America)
 a. Find the change in the number of music cassettes shipped to retailers from 1995 to 2000.
 b. Find the average change per year in the number of music cassettes shipped to retailers over this period.

Perform each indicated operation. See Section 1.9.

71. $(3 \cdot 5)^2$

72. $(12 - 3)^2(18 - 10)$

73. $90 + 12^2 - 5^3$

74. $3 \cdot (7 - 4) + 2 \cdot 5^2$

75. $12 \div 4 - 2 + 7$

76. $12 \div (4 - 2) + 7$

 Combining Concepts

In Exercises 77 through 79, determine whether each statement is true or false.

77. The product of two negative numbers is always a negative number.

78. The product of a positive number and a negative number is always a negative number.

79. The quotient of two negative numbers is always a positive number.

80. In 1999 there were 2321 commercial country music radio stations in the United States. By 2000, that number had declined to 2249. (*Source:* M Street Corporation)
   a. Find the change in the number of country music radio stations from 1999 to 2000.
   b. If this change continues, what will be the total change in the number of country music stations after another four years?
   c. Based on your answer to part b, how many country music radio stations will there be in 2004?

81. In 1999, a total of 29,380 Land Rovers were sold in the United States. By 2000, the number of Land Rovers sold in the United States had decreased to 27,148. (*Source:* Land Rover North America, Inc.)
   a. Find the change in the number of Land Rovers sold from 1999 to 2000.
   b. If this change continues, what will be the total change in the number of Land Rovers sold after another six years?
   c. Based on your answer to part b, how many Land Rovers will be sold in 2006?

82. In your own words, explain how to multiply two signed numbers.

83. In your own words, explain how to divide two signed numbers.

10.5 Order of Operations

A Simplifying Expressions

We first discussed the order of operations in Chapter 1. In this section, you are given an opportunity to practice using the order of operations when expressions contain signed numbers. The rules for the order of operations from Section 1.9 are repeated here.

If there are no other grouping symbols such as fraction bars or absolute value bars, perform operations in the following order:

Order of Operations

1. Perform all operations within parentheses or brackets.
2. Evaluate any expressions with exponents and find any square roots.
3. Multiply or divide in order from left to right.
4. Add or subtract in order from left to right.

EXAMPLES   Find the value of each expression.

1. $(-3)^2 = (-3)(-3) = 9$

2. $-3^2 = -(3)(3) = -9$

3. $\left(-\dfrac{1}{2}\right)^3 = \left(-\dfrac{1}{2}\right)\left(-\dfrac{1}{2}\right)\left(-\dfrac{1}{2}\right) = \dfrac{1}{4}\left(-\dfrac{1}{2}\right) = -\dfrac{1}{8}$

Practice Problems 1–3

Find the value of each expression.

1. $(-2)^4$
2. -2^4
3. $\left(-\dfrac{1}{3}\right)^3$

Helpful Hint

When simplifying expressions with exponents, notice that parentheses make an important difference.
$(-3)^2$ and -3^2 do not mean the same thing.
$(-3)^2$ means $(-3)(-3) = 9$.
-3^2 means the opposite of $3 \cdot 3$, or -9.
Only with parentheses is the -3 squared.

EXAMPLE 4   Simplify: $\dfrac{-6(2)}{-3}$

Solution: First we multiply -6 and 2. Then we divide.

$$\dfrac{-6(2)}{-3} = \dfrac{-12}{-3}$$
$$= 4$$

Practice Problem 4

Simplify: $\dfrac{25}{5(-1)}$

EXAMPLE 5   Simplify: $\dfrac{12 - 16}{-1 + 3}$

Solution: We simplify above and below the fraction bar separately. Then we divide.

$$\dfrac{12 - 16}{-1 + 3} = \dfrac{-4}{2}$$
$$= -2$$

Practice Problem 5

Simplify: $\dfrac{-18 + 6}{-3 - 1}$

Answers

1. 16,   **2.** -16,   **3.** $-\dfrac{1}{27}$,   **4.** -5,   **5.** 3

Practice Problem 6

Simplify: $20 + 50 + (-4)^3$

Practice Problem 7

Simplify: $-2^3 + (-4)^2 + 1^5$

Practice Problem 8

Simplify: $2(2 - 8) + (-12) - 3$

Practice Problem 9

Simplify: $(-5) \cdot |-4| + (-3) + 2^3$

Practice Problem 10

Simplify: $4(-6) \div [3(5 - 7)^2]$

Concept Check

True or false? Explain your answer. The result of

$-4 \cdot (3 - 7) - 8 \cdot (9 - 6)$

is positive because there are four negative signs.

Practice Problem 11

Simplify: $\dfrac{5}{8} \div \left(\dfrac{1}{5} + \dfrac{3}{4}\right)$

Answers

6. 6, **7.** 9, **8.** −27, **9.** −15, **10.** −2,
11. $\dfrac{25}{38}$

Concept Check: false;
$-4 \cdot (3 - 7) - 8 \cdot (9 - 6) = -8$

EXAMPLE 6 Simplify: $60 + 30 + (-2)^3$

Solution:

$$60 + 30 + (-2)^3 = 60 + 30 + (-8) \qquad \text{Write } (-2)^3 \text{ as } -8.$$
$$= 90 + (-8) \qquad \text{Add from left to right.}$$
$$= 82$$

EXAMPLE 7 Simplify: $-4^2 + (-3)^2 - 1^3$

Solution:

$$-4^2 + (-3)^2 - 1^3 = -16 + 9 - 1 \qquad \text{Simplify expressions with exponents.}$$
$$= -7 - 1 \qquad \text{Add or subtract from left to right.}$$
$$= -8$$

EXAMPLE 8 Simplify: $3(4 - 7) + (-2) - 5$

Solution:

$$3(4 - 7) + (-2) - 5 = 3(-3) + (-2) - 5 \qquad \text{Simplify inside parentheses.}$$
$$= -9 + (-2) - 5 \qquad \text{Multiply.}$$
$$= -11 - 5 \qquad \text{Add or subtract from left to right.}$$
$$= -16$$

EXAMPLE 9 Simplify: $(-3) \cdot |-5| - (-2) + 4^2$

Solution:

$$(-3) \cdot |-5| - (-2) + 4^2 = (-3) \cdot 5 - (-2) + 4^2 \qquad \text{Write } |-5| \text{ as 5.}$$
$$= (-3) \cdot 5 - (-2) + 16 \qquad \text{Write } 4^2 \text{ as 16.}$$
$$= -15 - (-2) + 16 \qquad \text{Multiply.}$$
$$= -13 + 16 \qquad \text{Add or subtract from left to right.}$$
$$= 3$$

EXAMPLE 10 Simplify: $-2[-3 + 2(-1 + 6)] - 5$

Solution: Here we begin with the innermost set of parentheses.

$$-2[-3 + 2(-1 + 6)] - 5 = -2[-3 + 2(5)] - 5 \qquad \text{Write } -1 + 6 \text{ as 5.}$$
$$= -2[-3 + 10] - 5 \qquad \text{Multiply.}$$
$$= -2(7) - 5 \qquad \text{Add.}$$
$$= -14 - 5 \qquad \text{Multiply.}$$
$$= -19 \qquad \text{Subtract.}$$

Try the Concept Check in the margin.

EXAMPLE 11 Simplify: $\left(\dfrac{1}{6} - \dfrac{5}{6}\right) \cdot \dfrac{3}{7}$

Solution:

$$\left(\frac{1}{6} - \frac{5}{6}\right) \cdot \frac{3}{7} = -\frac{2}{3} \cdot \frac{3}{7} \qquad \frac{1}{6} - \frac{5}{6} = -\frac{4}{6} = -\frac{2}{3}$$

$$= -\frac{2 \cdot \overset{1}{\cancel{3}}}{\underset{1}{\cancel{3}} \cdot 7} \qquad \text{Multiply.}$$

$$= -\frac{2}{7} \qquad \text{Simplify.}$$

CALCULATOR EXPLORATIONS

Simplifying an Expression Containing a Fraction Bar

Even though most calculators follow the order of operations, parentheses must sometimes be inserted. For example, to simplify $\dfrac{-8 + 6}{-2}$ on a calculator, enter parentheses around the expression above the fraction bar so that it is simplified separately.

To simplify $\dfrac{-8 + 6}{-2}$, press the keys

or ENTER .

The display will read [1].

Thus, $\dfrac{-8 + 6}{-2} = 1$.

Use a calculator to simplify.

1. $\dfrac{-12 - 36}{-10}$

2. $\dfrac{475}{-0.2 + (-1.7)}$

3. $\dfrac{-316 + (-458)}{28 + (-25)}$

4. $\dfrac{-234 + 86}{-18 + 16}$

NET INCOME AND NET LOSS

For most businesses, a financial goal is to "make money." But what does this mean from a mathematical point of view? To find out, we must first discuss some common business terms.

- **Revenue** is the amount of money a business takes in. A company's annual revenue is the amount of money it collects during its fiscal, or business, year. For most companies, the largest source of revenue is from the sales of their products or services. For instance, a grocery store's annual revenue is the amount of money it collects during the year from selling groceries to customers. Large companies may also have revenues from interest or rentals.

- **Expenses** are the costs of doing business. For instance, a large part of a grocery store's expenses includes the cost of the food items it buys from wholesalers to resell to customers. Other expenses include salaries, mortgage payments, equipment, taxes, advertising, and so on.

- **Net income/loss** is the difference between a company's annual revenues and expenses. If the company's revenues are larger than its expenses, the difference is a positive number, and the company posts a net income for the year. Posting a net income can be interpreted as "making money." If the company's revenues are smaller than its expenses, the difference is a negative number, and the company posts a net loss for the year. Posting a net loss can be interpreted as "losing money."

Net income can also be thought of as profit. A negative profit is a net loss.

GROUP ACTIVITY

Search for corporate annual reports or articles in financial newspapers and magazines that report a company's net income or net loss. Describe what the income or loss means for the company. What are some of the factors that contributed to the net income or loss?

EXERCISE SET 10.5

 Simplify. See Examples 1 through 11.

1. $-1(-2) + 1$

2. $3 + (-8) \div 2$

3. $3 - 6 + 2$

4. $5 - 9 + 2$

5. $9 - 12 - 4$

6. $10 - 23 - 12$

7. $4 + 3(-6)$

8. $8 + 4(-3)$

9. $\dfrac{4}{9}\left(\dfrac{2}{10} - \dfrac{7}{10}\right)$

10. $\dfrac{2}{5}\left(\dfrac{3}{8} - \dfrac{4}{8}\right)$

11. $(-10) + 4 \div 2$

12. $(-12) + 6 \div 3$

13. $25 \div (-5) + 12$

14. $28 \div (-7) + 10$

15. $\dfrac{16 - 13}{-3}$

16. $\dfrac{20 - 15}{-1}$

17. $\dfrac{24}{10 + (-4)}$

18. $\dfrac{88}{-8 - 3}$

19. $5(-3) - (-12)$

20. $7(-4) - (-6)$

21. $(-19) - 12(3)$

22. $(-24) - 14(2)$

23. $8 + 4^2$

24. $12 + 3^3$

25. $[8 + (-4)]^2$

26. $[9 + (-2)]^3$

27. $3^3 - 12$

28. $5^2 - 100$

29. $(3 - 12) \div 3$

30. $(12 - 19) \div 7$

31. $5 + 2^3 - 4^2$

32. $12 + 5^2 - 2^4$

33. $(5 - 9)^2 \div (4 - 2)^2$

34. $(2 - 7)^2 \div (4 - 3)^4$

35. $|8 - 24| \cdot (-2) \div (-2)$

36. $|3 - 15| \cdot (-4) \div (-16)$

37. $(-12 - 20) \div 16 - 25$

38. $(-20 - 5) \div 5 - 15$

39. $5(5 - 2) + (-5)^2 - 6$

40. $3 \cdot (8 - 3) + (-4) - 10$

41. $(0.2 - 0.7)(0.6 - 1.9)$

42. $(0.4 - 1.2)(0.8 - 1.7)$

43. $2 - 7 \cdot 6 - 19$

44. $4 - 12 \cdot 8 - 17$

45. $(-36 \div 6) - (4 \div 4)$

46. $(-4 \div 4) - (8 \div 8)$

47. $\left(\dfrac{1}{2}\right)^2 - \left(\dfrac{1}{3}\right)^2$

48. $\left(\dfrac{1}{4}\right)^2 - \left(\dfrac{1}{2}\right)^2$

49. $(-5)^2 - 6^2$

50. $(-4)^4 - (5)^4$

51. $(10 - 4^2)^2$

52. $(11 - 3^2)^3$

53. $2(8 - 10)^2 - 5(1 - 6)^2$

54. $-3(4 - 8)^2 + 5(14 - 16)^3$

55. $3(-10) \div [5(-3) - 7(-2)]$

56. $12 - [7 - (3 - 6)] + (2 - 3)^3$

57. $\dfrac{(-7)(-3) - (4)(3)}{3[7 \div (3 - 10)]}$

58. $\dfrac{10(-1) - (-2)(-3)}{2[-8 \div (-2 - 2)]}$

59. $(0.2)^2 - (1.5)^2$

60. $(1.3)^2 - (2.2)^2$

Review and Preview

Perform each indicated operation. See Sections 1.3, 1.4, 1.6, and 1.7.

61. $45 \cdot 90$

62. $90 \div 45$

63. $90 - 45$

64. $45 + 90$

Find the perimeter of each figure. See Section 1.3.

△ **65.** Square

8 in.

△ **66.** Parallelogram

5 cm

3 cm

△ **67.** Rectangle

6 ft

9 ft

△ **68.** Triangle

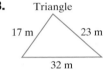

17 m   23 m

32 m

 Combining Concepts

Insert parentheses where needed so that each expression evaluates to the given number.

69. $2 \cdot 7 - 5 \cdot 3$; evaluates to 12

70. $7 \cdot 3 - 4 \cdot 2$; evaluates to 34

71. $-6 \cdot 10 - 4$; evaluates to -36

72. $2 \cdot 8 \div 4 - 20$; evaluates to -36

Evaluate.

73. $(-12)^4$

74. $(-17)^6$

75. Are parentheses necessary in the expression $3 + (4 \cdot 5)$? Explain your answer.

76. Are parentheses necessary in the expression $(3 + 4) \cdot 5$? Explain your answer.

Investigating Positive and Negative Numbers

MATERIALS:

- colored thumbtacks
- coin
- cardboard
- six-sided die
- tape

Work with a partner or a small group to try the following activity. The object is to have the largest absolute value at the end of the game.

1. Attach this page to a piece of cardboard with tape. Each person should choose a different colored thumbtack as his or her playing piece. Insert each thumbtack at the starting place 0 on the number line.

2. Each player takes a turn as follows: Roll the die and flip the coin. "Heads" on the coin makes the number that lands faceup on the die positive. "Tails" on the coin makes the number that lands faceup on the die negative. Record your number along with its sign (positive or negative) in the table below. Move your thumbtack on the number line according to your number.

3. Continue taking turns, until each person has taken five turns. Verify your final position on the number line by finding the total of the integers in the table. Do your total and final position agree?

4. The winner is the player having the final position with the largest absolute value. Find the absolute value of your final position. Create a table listing the absolute values of each person's final position. Who won? How could you tell who won just by looking at the number line?

5. Many board games include instructions for moving playing pieces forward or backward. Make a list of games that include such instructions. Then explain how these instructions for moving forward or backward are related to positive and negative numbers.

	Positive or Negative Number
Turn 1	
Turn 2	
Turn 3	
Turn 4	
Turn 5	
Total	

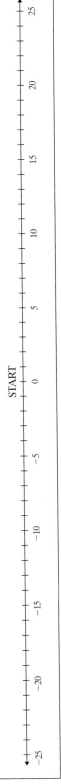

649

Are you prepared for a test on Chapter 10?

Below I have listed some *common trouble areas* for students in Chapter 10. After studying for your test—but before taking your test—read these.

■ Don't forget the difference between $-(-5)$ and $-|-5|$.

$-(-5) = 5$ The opposite of -5 is 5.

$-|-5| = -5$ The opposite of the absolute value of -5 is the opposite of 5, which is -5.

■ Remember how to simplify $(-7)^2$ and -7^2.

$$(-7)^2 = (-7)(-7) = 49$$

$$-7^2 = -(7)(7) = -49$$

■ Don't forget order of operations.

$$1 + 3(4 - 6) = 1 + 3(-2) \quad \text{Simplify inside parentheses.}$$
$$= 1 + (-6) \quad \text{Multiply.}$$
$$= -5 \quad \text{Add.}$$

Remember: This is simply a checklist of common trouble spots. For a review of Chapter 10, see the Highlights and Chapter Review at the end of this Chapter.

Chapter 10   VOCABULARY CHECK

Fill in each blank with one of the words or phrases listed below.

signed   opposites   absolute value   integers

1. Two numbers that are the same distance from 0 on the number line but are on opposite sides of 0 are called

_____.

2. Together, positive numbers, negative numbers, and 0 are called _____ numbers.

3. The _____ of a number is that number's distance from 0 on the number line.

4. The _____ are $\ldots, -3, -2, -1, 0, 1, 2, 3, \ldots$.

CHAPTER 10   Highlights

DEFINITIONS AND CONCEPTS	EXAMPLES

Section 10.1   Signed Numbers

Together, positive numbers, negative numbers, and 0 are called **signed numbers**.

$-432, -10, 0, 15$

The **integers** are $\ldots, -3, -2, -1, 0, 1, 2, 3, \ldots$.

The **absolute value** of a number is that number's distance from 0 on the number line. The symbol for absolute value is $|\ \ |$.

$|-2| = 2$

$|2| = 2$

Two numbers that are the same distance from 0 on the number line but are on opposite sides of 0 are called **opposites**.

5 and -5 are opposites.

If a is a number, then $-(-a) = a$.

$-(-11) = 11 \qquad -|-3| = -3$

Section 10.2   Adding Signed Numbers

ADDING TWO NUMBERS WITH THE SAME SIGN

Step 1. Add their absolute values.

Step 2. Use their common sign as the sign of the sum.

Add:

$$-3 + (-2) = -5$$
$$-7 + (-15) = -22$$
$$-1.2 + (-5.7) = -6.9$$

ADDING TWO NUMBERS WITH DIFFERENT SIGNS

Step 1. Find the larger absolute value minus the smaller absolute value.

Step 2. Use the sign of the number with the larger absolute value as the sign of the sum.

$$-6 + 4 = -2$$
$$17 + (-12) = 5$$
$$-\frac{4}{11} + \frac{1}{11} = -\frac{3}{11}$$
$$-32 + (-2) + 14 = -34 + 14$$
$$= -20$$

DEFINITIONS AND CONCEPTS	EXAMPLES

Section 10.3   Subtracting Signed Numbers

SUBTRACTING TWO NUMBERS

If a and b are numbers, then $a - b = a + (-b)$.

Subtract:

$$-35 - 4 = -35 + (-4) = -39$$
$$3 - 8 = 3 + (-8) = -5$$
$$-7.8 - (-10.2) = -7.8 + 10.2 = 2.4$$
$$7 - 20 - 18 - (-3) = 7 + (-20) + (-18) + 3$$
$$= -13 + (-18) + 3$$
$$= -31 + 3$$
$$= -28$$

Section 10.4   Multiplying and Dividing Signed Numbers

MULTIPLYING SIGNED NUMBERS

The product of two numbers having the same sign is a positive number.
The product of two numbers having different signs is a negative number.

Multiply:

$$(-7)(-6) = 42$$
$$9(-4) = -36$$
$$-3(0.7) = -2.1$$

Evaluate:

$$(-3)^2 = (-3)(-3) = 9$$
$$\left(-\frac{2}{3}\right)^2 = \left(-\frac{2}{3}\right)\left(-\frac{2}{3}\right) = \frac{4}{9}$$

DIVIDING SIGNED NUMBERS

The quotient of two numbers having the same sign is a positive number.
The quotient of two numbers having different signs is a negative number.

Divide:

$$-100 \div (-10) = 10$$
$$\frac{14}{-2} = -7, \frac{-3.6}{-0.3} = 12, \frac{0}{-3} = 0, \frac{22}{0} \text{ is undefined.}$$

Section 10.5   Order of Operations

ORDER OF OPERATIONS

1. Perform all operations within parentheses or brackets.
2. Evaluate any expressions with exponents.
3. Multiply or divide in order from left to right.
4. Add or subtract in order from left to right.

Simplify:

$$3 + 2 \cdot (-5) = 3 + (-10)$$
$$= -7$$
$$\frac{-2(5 - 7)}{-7 + |-3|} = \frac{-2(-2)}{-7 + 3}$$
$$= \frac{4}{-4}$$
$$= -1$$

Chapter 10 Review

(10.1) *Represent each quantity by a signed number.*

1. A gold miner is working 1435 feet down in a mine.

2. A mountain peak is 7562 meters above sea level.

Graph each number on a number line.

3. $-2, 4, -3.5, 0$

4. $-7, -1.6, 3, -4\dfrac{1}{3}$

Insert < or > between each pair of numbers to make a true statement.

5. $-18 \quad -20$

6. $-5 \quad 5$

7. $-12.3 \quad -19.8$

Find each absolute value.

8. $|-12|$

9. $|0|$

10. $\left|-\dfrac{7}{8}\right|$

Find the opposite of each number.

11. -12

12. $\dfrac{1}{2}$

Simplify.

13. $-(-7)$

14. $-|-7|$

Determine whether each statement is true or false.

15. A negative number is always less than a positive number.

16. The absolute value of a number is always 0 or a positive number.

(10.2) *Add.*

17. $5 + (-3)$

18. $18 + (-4)$

19. $-12 + 16$

20. $-23 + 40$

21. $-8 + (-15)$

22. $-5 + (-17)$

23. $-2.4 + 0.3$

24. $-8.9 + 1.9$

25. $\frac{2}{3} + \left(-\frac{2}{5}\right)$ **26.** $-\frac{8}{9} + \frac{1}{3}$ **27.** $-43 + (-108)$ **28.** $-100 + (-506)$

29. The temperature at 5 A.M. on a day in January was $-15°$ Celsius. By 6 A.M. the temperature had fallen 5 degrees. Use a signed number to represent the temperature at 6 A.M.

30. A diver starts out at 127 feet below the surface and then swims downward another 23 feet. Use a signed number to represent the diver's current depth.

31. During the 2001 PGA Masters Tournament, the winner, Tiger Woods, had scores of -2, -6, -4, and -4 over four rounds of golf. What was his total score for the tournament? (*Source*: Professional Golfer's Association)

32. During the 2001 British Open golf tournament, the winner, David Duval, had a score of -10. The second-place finisher, Niclas Fasth, had a score that was 3 points more than the winning score. What was Niclas Fasth's score in the British Open? (*Source*: Professional Golfer's Association)

(10.3) *Subtract.*

33. $12 - 4$ **34.** $-12 - 4$ **35.** $\frac{2}{5} - \frac{7}{10}$ **36.** $-8 - 19$

37. $7 - (-13)$ **38.** $-6 - (-14)$ **39.** $16 - 16$ **40.** $-16 - 16$

41. $-12 - (-12)$ **42.** $-0.5 - (-1.2)$ **43.** $-(-5) - 12 - (-3)$ **44.** $\frac{3}{7} - \frac{5}{7} - \left(-\frac{1}{14}\right)$

45. Josh Weidner has $142 in his checking account. He writes a check for $125, makes a deposit for $43, and then writes another check for $85. Represent the balance in his account by a signed number.

46. If the elevation of Lake Superior is 600 feet above sea level and the elevation of the Caspian Sea is 92 feet below sea level, find their difference in elevation. Represent the difference by a signed number.

Determine whether each statement is true or false.

47. $|-5| - |-6| = 5 - 6$

48. $|-5 - (-6)| = 5 + 6$

(10.4) *Multiply or divide.*

49. $(-3) \cdot (-7)$ **50.** $(-6) \cdot 3$ **51.** $-\frac{2}{3} \cdot \frac{7}{8}$ **52.** $(-0.5) \cdot (-12)$

53. $-2 \div 0$

54. $\dfrac{-38}{-1}$

55. $45 \div (-9)$

56. A football team lost five yards on each of two consecutive plays. Represent the total loss by a product of signed numbers and find the product.

57. A racehorse bettor lost $50 on each of four consecutive races. Represent the total loss by a product of signed numbers and find the product.

(10.5) *Simplify.*

58. $5 - 8 + 3$

59. $-3 + 12 + (-7) - 10$

60. $-10 + 3 \cdot (-2)$

61. $5 - 10 \cdot (-3)$

62. $16 \cdot (-2) + 4$

63. $3 \cdot (-12) - 8$

64. $5.2 + 6 \div (-0.3)$

65. $-6 + (-10) \div (-2)$

66. $16 + (-3) \cdot 12 \div 4$

67. $\dfrac{3}{4} \cdot \left(-\dfrac{1}{9}\right) - \left(\dfrac{3}{4} - \dfrac{5}{12}\right)$

68. $4^3 - (8 - 3)^2$

69. $\left(-\dfrac{1}{3}\right)^2 - \dfrac{8}{3}$

70. $-(-4) \cdot |-3| - 5$

71. $|5 - 1|^2 \cdot (-5)$

72. $\dfrac{(-4)(-3) - (-2)(-1)}{-10 + 5}$

73. $\dfrac{4(12 - 18)}{-10 \div (-2 - 3)}$

Chapter 10 Test

Simplify each expression.

1. $-5 + 8$

2. $18 - 24$

3. $0.5 \cdot (-20)$

4. $(-16) \div (-4)$

5. $-\dfrac{3}{11} + \left(-\dfrac{5}{22}\right)$

6. $-7 - (-19)$

7. $(-5) \cdot (-13)$

8. $\dfrac{-2.5}{-0.5}$

9. $|-25| + (-13)$

10. $14 - |-20|$

11. $|5| \cdot |-10|$

12. $\dfrac{|-10|}{-|-5|}$

13. $(-8) + 9 \div (-3)$

14. $-7 + (-32) - 12 + 5$

15. $(-5)^3 - 24 \div (-3)$

16. $(5 - 9)^2 \cdot (8 - 2)^3$

Answers

1. _____

2. _____

3. _____

4. _____

5. _____

6. _____

7. _____

8. _____

9. _____

10. _____

11. _____

12. _____

13. _____

14. _____

15. _____

16. _____

17. _____

18. _____

19. _____

20. _____

21. _____

22. _____

23. _____

24. _____

25. _____

17. $\left(\dfrac{5}{9} - \dfrac{7}{9}\right)^2 + \left(-\dfrac{2}{9}\right)$

18. $3 - (8 - 2)^3$

19. $-6 + (-15) \div (-3)$

20. $\dfrac{4}{2} - \dfrac{8^2}{16}$

21. $\dfrac{-3(-2) + 12}{-1(-4 - 5)}$

22. $\dfrac{|25 - 30|^2}{2(-6) + 7}$

23. At an elevation of 6684 feet above sea level, Mt. Mitchell in North Carolina is the highest U.S. mountain east of the Mississippi River. The South Sandwich Trench in the Atlantic Ocean reaches a depth of 27,651 feet below sea level. Find the difference in elevation between these two points. (*Sources*: U.S. Geological Survey and Naval Meteorology and Oceanography Command)

24. A mountain climber is at an elevation of 14,893 feet and moves down the mountain a distance of 147 feet. Represent his final elevation as a sum of signed numbers and find the sum.

25. Jane Hathaway has $237 in her checking account. She writes a check for $157, writes another check for $77, and deposits $38. Represent the balance in her account by a signed number.

Cumulative Review

1. Multiply: 631×125 **2.** Divide: $\dfrac{2}{5} \div \dfrac{1}{2}$ **3.** Add: $\dfrac{2}{3} + \dfrac{1}{7}$

Write each decimal in standard form.

4. Forty-eight and twenty-six-hundredths

5. Six and ninety-five-thousandths

6. Subtract: $3.5 - 0.068$

7. Simplify: $5.68 + (0.9)^2 \div 100$

8. Write "$27,000 every 6 months" as a unit rate.

9. Find the value of the unknown number n: $\dfrac{34}{51} = \dfrac{n}{3}$

10. 46 out of every 100 college students live at home. What percent of students live at home? (*Source*: Independent Insurance Agents of America)

Write each fraction or mixed number as a percent.

11. $\dfrac{9}{20}$

12. $1\dfrac{1}{2}$

13. 13 is $6\dfrac{1}{2}\%$ of what number?

14. Translate to a proportion. 101 is what percent of 200?

15. Ivan Borski borrowed $2400 at 10% simple interest for 8 months to buy a used Chevy S-10. Find the simple interest he paid.

Answers

1. _____

2. _____

3. _____

4. _____

5. _____

6. _____

7. _____

8. _____

9. _____

10. _____

11. _____

12. _____

13. _____

14. _____

15. _____

16. _____

17. _____

18. _____

19. _____

20. _____

21. _____

22. _____

23. _____

24. _____

25. _____

16. Add 3 ft 2 in. and 5 ft 11 in.

17. The normal body temperature is 98.6°F. What is this temperature in degrees Celsius?

△ **18.** Find the supplement of a 107° angle.

△ **19.** Find the measure of ∠b.

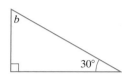

△ **20.** Find the area of the parallelogram:

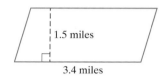

1.5 miles

3.4 miles

△ **21.** Approximate the volume of a ball of radius 3 inches. Use the approximation $\frac{22}{7}$ for π.

3 inches

22. Find the median of the list of numbers: 25, 54, 56, 57, 60, 71, 98

23. If a die is rolled, find the probability of rolling a 3 or a 4.

24. Add: $-2 + (-21)$

25. Simplify: $60 + 30 + (-2)^3$

Introduction to Algebra

In this chapter we make the transition from arithmetic to algebra. In algebra, letters are used to stand for unknown quantities. Using variables is a very powerful tool for solving problems that cannot be solved with arithmetic alone. This chapter introduces variables, algebraic expressions, and solving variable equations.

Accounting is a system of recording and reporting financial information about a business. Keeping track of numerical data can be traced to prehistoric times when primitive records were kept by scratching marks into branches, bones, or cave walls. The earliest known financial accounting documents are ancient Babylonian payroll records dating from about 3500 B.C. It wasn't until around A.D. 1494 that the first known text on accounting practices was published in Italy. As business has grown and changed and become more complicated, so has financial accounting. However, the basic relationships remain the same and can be described by very simple equations. In Exercises 49 and 50 on page 678, we will see one way that an equation is used in accounting.

Name _____ Section _____ Date _____

Chapter 11 Pretest

1. Evaluate: $3x - 2y$ when $x = -1$ and $y = 3$

Simplify by combining like terms.

2. $x - 11x$

3. $5x + 4 + 2x - 3y + 7 - y + 9$

4. Multiply: $-2(6m - 3)$

Decide whether the given number is a solution to the given equation.

5. Is -4 a solution to $x + 11 = 15$?

6. Is 7 a solution to $6 - n = -1$?

Solve.

7. $8 = 6 + y$

8. $4 - 9 = x + 3$

9. $-4z = 32$

10. $\frac{2}{5}a = 14$

11. $-12x = -4 + 28$

12. $-39 - 3 = -7x + x$

13. $9b - 12 = 6$

14. $4n + 2 = 10n - 16$

15. $3(y + 2) = 9$

16. $20 + 2(w - 7) = 3w + 4$

Translate each sentence into an equation. Use x to represent "a number."

17. The product of 3 and 12 is 36.

18. Twice the sum of 4 and 6 yields 20.

19. The quotient of 10 and three times a number equals 40.

Solve.

20. A number less 7 is 8. Find the number.

11.1 Introduction to Variables

 Evaluating Algebraic Expressions

OBJECTIVES

Ⓐ Evaluate algebraic expressions for given replacement values for the variables.

Ⓑ Use properties of numbers to combine like terms.

Ⓒ Use properties of numbers to multiply expressions.

SSM
TUTOR CENTER   SG   CD & VIDEO   MATH PRO   WEB

Perhaps the most important quality of mathematics is that it is a science of pattern. Communicating about patterns is made possible by using a letter to represent all the numbers fitting a pattern. We call such a letter a **variable**. For example, in Section 1.3 we presented the addition property of 0, which states that the sum of 0 and any whole number is that number. We might write

$$0 + 1 = 1$$
$$0 + 2 = 2$$
$$0 + 3 = 3$$
$$0 + 4 = 4$$
$$0 + 5 = 5$$
$$0 + 6 = 6$$
$$\vdots$$

continuing indefinitely. This is a pattern, and all whole numbers fit the pattern. We can communicate this pattern for all whole numbers by letting a letter, such as a, represent all whole numbers. We can then write

$$0 + a = a$$

Helpful Hint

Variables have been used in previous chapters, although we have not called them that. For example, in the ratio and proportion chapter, we wrote equations such as

$$\frac{4}{n} = \frac{6}{12}$$

Here, the letter n is a variable.

Learning to use variable notation is a primary goal of learning **algebra**. We now take some important beginning steps in learning to use variable notation.

A combination of operations on letters (variables) and numbers is called an **algebraic expression** or simply an **expression**.

$$3 + x \qquad 5 \cdot y \qquad 2 \cdot z - 1 + x$$

If two variables or a number and a variable are next to each other, with no operation sign between them, the indicated operation is multiplication. For example,

$$2x \qquad \text{means} \qquad 2 \cdot x$$

and

$$xy \text{ or } x(y) \qquad \text{means} \qquad x \cdot y$$

Also, the meaning of an exponent remains the same when the base is a variable. For example,

$$\underbrace{x^2 = x \cdot x}_{\text{2 factors of } x} \qquad \text{and} \qquad \underbrace{y^5 = y \cdot y \cdot y \cdot y \cdot y}_{\text{5 factors of } y}$$

Algebraic expressions have different values depending on the replacement values for x. Replacing a variable in an expression by a number and then finding the value of the expression is called **evaluating the expression** for the variable. When finding the value of an expression, remember to follow the order of operations.

Practice Problem 1

Evaluate: $5x - y$ when $x = 2$ and $y = 3$

EXAMPLE 1   Evaluate: $2x + y$ when $x = 8$ and $y = 7$

Solution: Replace x with 8 and y with 7 in $2x + y$.

$$2x + y = 2 \cdot 8 + 7 \qquad \text{Replace } x \text{ with 8 and } y \text{ with 7.}$$
$$= 16 + 7 \qquad \text{Multiply first because of the order of operations.}$$
$$= 23 \qquad \text{Add.}$$

Practice Problem 2

Evaluate: $x - y$ when $x = -12$ and $y = -2$

EXAMPLE 2   Evaluate: $x - y$ when $x = 14$ and $y = -3$

Solution:

$$x - y = 14 - (-3) \qquad \text{Replace } x \text{ with 14 and } y \text{ with } -3.$$
$$= 14 + (3) \qquad \text{To subtract, add the opposite of } -3.$$
$$= 17 \qquad \text{Add.}$$

Practice Problem 3

Evaluate: $\dfrac{5r - 2s}{3q}$ when $r = 3, s = 3,$ and $q = 1$

EXAMPLE 3   Evaluate: $\dfrac{3m - 2n}{2q}$ when $m = 8, n = 4,$ and $q = 1$

Solution:

$$\frac{3m - 2n}{2q} = \frac{3 \cdot 8 - 2 \cdot 4}{2 \cdot 1} \qquad \text{Replace } m \text{ with 8, } n \text{ with 4, and } q \text{ with 1.}$$
$$= \frac{24 - 8}{2} \qquad \text{Multiply.}$$
$$= \frac{16}{2} \qquad \text{Subtract in the numerator.}$$
$$= 8 \qquad \text{Divide.}$$

Practice Problem 4

Evaluate: $b^2 - (3a + c)$ when $a = 2, b = -3,$ and $c = -1$

EXAMPLE 4   Evaluate: $x^3 - (a - b)$ when $x = 2, a = -3,$ and $b = 5$

Solution:

$$x^3 - (a - b) = 2^3 - (-3 - 5) \qquad \text{Replace } x \text{ with 2, } a \text{ with } -3, \text{ and } b \text{ with 5.}$$
$$= 2^3 - (-8) \qquad \text{Simplify inside the parentheses.}$$
$$= 8 - (-8) \qquad \text{Evaluate the exponential expression.}$$
$$= 8 + 8$$
$$= 16 \qquad \text{Add.}$$

Practice Problem 5

The formula for finding the perimeter of a rectangle is $P = 2l + 2w$. Find the perimeter of a rectangular garden that is 25 meters wide and 40 meters long.

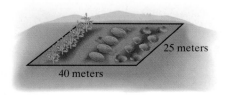

25 meters

40 meters

EXAMPLE 5   Finding the Area of a Rectangle

The formula for finding the area of a rectangle is $A = lw$, where l is the length of the rectangle and w is the width. Find the area of a rectangular floor that is 45 feet long and 30 feet wide.

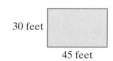

30 feet

45 feet

Answers

1. 7,   **2.** -10,   **3.** 3,   **4.** 4,   **5.** 130 m

Solution:

$$A = \quad l \quad \cdot \quad w$$

$$= (45\,\text{feet}) \cdot (30\,\text{feet}) \qquad \text{Replace } l \text{ with 45 feet and } w \text{ with 30 feet.}$$

$$= 1350\,\text{square feet}$$

The area of the floor is 1350 square feet. ●

B Combining Like Terms

The addends of an algebraic expression are called the **terms** of the expression.

$$\underbrace{x + 3}_{\text{2 terms}}$$

$$\underbrace{3y^2 + (-6y) + 4}_{\text{3 terms}}$$

A term that is only a number has a special name. It is called a **constant term**, or simply a **constant**. A term that contains a variable is called a **variable term**.

$$
\begin{array}{ccccc}
x & + & 3 & \qquad 3y^2 + (-6y) & + & 4 \\
\uparrow & & \uparrow & \uparrow \qquad\qquad \uparrow & & \uparrow \\
\text{variable} & & \text{constant} & \text{variable} & & \text{constant} \\
\text{term} & & \text{term} & \text{terms} & & \text{term}
\end{array}
$$

The number factor of a variable term is called the **numerical coefficient**. A numerical coefficient of 1 is usually not written.

$$
\begin{array}{cccc}
5x & x \text{ or } 1x & 3y^2 & -6y \\
\uparrow & \uparrow & \uparrow & \uparrow
\end{array}
$$

| Numerical coefficient is 5. | Understood numerical coefficient is 1. | Numerical coefficient is 3. | Numerical coefficient is -6. |

Terms that are exactly the same, except that they may have different numerical coefficients, are called **like terms**.

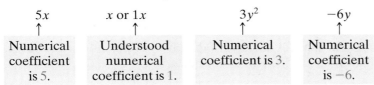

Like Terms	Unlike Terms
$3x, \dfrac{1}{2}x$	$5x, x^2$
$-6y, 2y, y$	$7x, 7y$

A sum or difference of like terms can be simplified using the **distributive property**. Recall from Chapter 1 that the distributive property says that multiplication distributes over addition (and subtraction). Using variables, we can write the distributive property as follows:

Distributive Property

If a, b, and c are numbers, then

$$ac + bc = (a + b)c$$

Also,

$$ac - bc = (a - b)c$$

The distributive property guarantees that, no matter what number x is, $7x + 5x$ (for example) has the same value as $(7 + 5)x$, or $12x$. We then have that

$$7x + 5x = (7 + 5)x = 12x$$

This is an example of **combining like terms**. An algebraic expression is **simplified** when all like terms have been combined.

EXAMPLE 6 Simplify each expression by combining like terms.

a. $3x + 2x$ **b.** $y - 7y$

Solution: We add or subtract like terms.

a. $3x + 2x = (3 + 2)x$
$$= 5x$$

$\overset{\text{Understood 1}}{\downarrow}$

b. $y - 7y = 1y - 7y$
$$= (1 - 7)y$$
$$= -6y$$ ●

The commutative and associative properties of addition and multiplication can also help us simplify expressions. We presented these properties in Sections 1.3 and 1.6 and state them again using variables.

> **Properties of Addition and Multiplication**
>
> If a, b, and c are numbers, then
>
> $a + b = b + a$ Commutative property of addition
> $a \cdot b = b \cdot a$ Commutative property of multiplication
>
> That is, the **order** of adding or multiplying two numbers can be changed without changing their sum or product.
>
> $(a + b) + c = a + (b + c)$ Associative property of addition
> $(a \cdot b) \cdot c = a \cdot (b \cdot c)$ Associative property of multiplication
>
> That is, the **grouping** of numbers in addition or multiplication can be changed without changing their sum or product.

EXAMPLE 7 Simplify: $2y - 6 + 4y + 8$

Solution: We begin by writing subtraction as the opposite of addition.

$$2y - 6 + 4y + 8 = 2y + (-6) + 4y + 8$$
$$= 2y + 4y + (-6) + 8 \quad \text{Apply the commutative property of addition.}$$
$$= (2 + 4)y + (-6) + 8 \quad \text{Apply the distributive property.}$$
$$= 6y + 2 \quad \text{Simplify.} \quad ●$$

EXAMPLES Simplify each expression by combining like terms.

8. $6x + 2x - 5 = 8x - 5$

9. $4x + 2 - 5x + 3 = 4x - 5x + 2 + 3$
$$= -1x + 5 \quad \text{or} \quad -x + 5$$

10. $1.2y + 10 - 5.7y - 9 = 1.2y - 5.7y + 10 - 9$
$$= -4.5y + 1$$

11. $2x - 5 + 3y + 4x - 10y + 11 = 6x - 7y + 6$ ●

Practice Problem 6

Simplify each expression by combining like terms.

a. $8m - 11m$
b. $5a + a$

Practice Problem 7

Simplify: $8m + 5 + m - 4$

Practice Problems 8–11

Simplify each expression by combining like terms.

8. $7y + 11y - 8$
9. $2y - 6 + y + 7$
10. $3.7x + 5 - 4.2x + 15$
11. $-9y + 2 - 4y - 8x + 12 - x$

Answers

6. a. $-3m$, **b.** $6a$, **7.** $9m + 1$, **8.** $18y - 8$,
9. $3y + 1$, **10.** $-0.5x + 20$,
11. $-13y - 9x + 14$

Copyright 2003 Prentice-Hall, Inc.

C Multiplying Expressions

We can also use properties of numbers to multiply expressions such as $3(2x)$. By the associative property of multiplication, we can write the product $3(2x)$ as $(3 \cdot 2)x$, which simplifies to $6x$.

EXAMPLES   Multiply.

12. $5(3y) = (5 \cdot 3)y$ Apply the associative property of multiplication.

$\qquad = 15y$ Multiply.

13. $-2(4x) = (-2 \cdot 4)x$ Apply the associative property of multiplication.

$\qquad = -8x$ Multiply.

Practice Problems 12–13

Multiply.

12. $7(8a)$

13. $-5(9x)$

We can use the distributive property to combine like terms, which we have done, and also to multiply expressions such as $2(3 + x)$. By the distributive property, we have that

$$2(3 + x) = 2 \cdot 3 + 2 \cdot x \qquad \text{Apply the distributive property.}$$

$$= 6 + 2x \qquad \text{Multiply.}$$

EXAMPLE 14   Use the distributive property to multiply: $6(x + 4)$

Solution: By the distributive property,

$$6(x + 4) = 6 \cdot x + 6 \cdot 4 \qquad \text{Apply the distributive property.}$$

$$= 6x + 24 \qquad \text{Multiply.}$$

Try the Concept Check in the margin.

Practice Problem 14

Use the distributive property to multiply: $7(y + 2)$

Concept Check

What's wrong with the following?

$8(a - b) = -8ab$

EXAMPLE 15   Multiply: $-3(5a + 2)$

Solution: By the distributive property,

$$-3(5a + 2) = -3(5a) + (-3)(2) \qquad \text{Apply the distributive property.}$$

$$= (-3 \cdot 5)a + (-6) \qquad \text{Multiply.}$$

$$= -15a - 6 \qquad \text{Multiply.}$$

Practice Problem 15

Multiply: $4(7a - 5)$

To simplify expressions containing parentheses, we first use the distributive property and multiply.

EXAMPLE 16   Simplify: $2(3 + x) - 15$

Solution: First we use the distributive property to remove parentheses.

$$2(3 + x) - 15 = 2(3) + 2(x) - 15 \qquad \text{Apply the distributive property.}$$

$$= 6 + 2x - 15 \qquad \text{Multiply.}$$

$$= 2x + (-9) \quad \text{or} \quad 2x - 9 \qquad \text{Combine like terms.}$$

Helpful Hint

2 is *not* distributed to the -15 since it is not within the parentheses.

Practice Problem 16

Simplify: $5(y - 3) - 8 + y$

Answers

12. $56a$,   **13.** $-45x$,   **14.** $7y + 14$,
15. $28a - 20$,   **16.** $6y - 23$

Concept Check: did not distribute the 8

Practice Problem 17

Simplify: $5(2x - 3) + 7(x - 1)$

EXAMPLE 17 Simplify: $-2(x - 5) + 4(2x + 2)$

Solution: First we use the distributive property to remove parentheses.

$$-2(x - 5) + 4(2x + 2) = -2(x) + (-2)(-5)$$
$$+ 4(2x) + 4(2) \quad \text{Apply the distributive property.}$$
$$= -2x + 10 + 8x + 8 \quad \text{Multiply.}$$
$$= 6x + 18 \quad \text{Combine like terms.}$$

Practice Problem 18

Find the area of the rectangular garden.

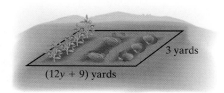

3 yards

$(12y + 9)$ yards

EXAMPLE 18 Finding the Area of a Deck

Find the area of the rectangular deck.

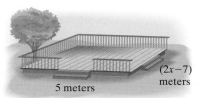

$(2x - 7)$ meters

5 meters

Solution: Recall how to find the area of a rectangle.

$$A = l \cdot w$$
$$= 5(2x - 7) \quad \text{Let length = 5 and width = } (2x - 7).$$
$$= 10x - 35 \quad \text{Multiply.}$$

The area is $(10x - 35)$ *square* meters.

Answers

17. $17x - 22$, **18.** $(36y + 27)$ sq yd

Name _____ Section _____ Date _____

Mental Math

Solve each equation.

1. $x - 2 = 0$

2. $x - 5 = 0$

3. $x + 1 = 0$

4. $x + 6 = 0$

EXERCISE SET 11.2

Ⓐ *Decide whether the given number is a solution of the given equation. See Examples 1 and 2.*

1. Is 10 a solution of $x - 8 = 2$?

2. Is 9 a solution of $y - 2 = 7$?

3. Is -5 a solution of $x + 12 = 17$?

4. Is -7 a solution of $a + 23 = -16$?

5. Is 8 a solution of $7f = 64 - f$?

6. Is 3 a solution of $12 - k = 9$?

7. Is 3 a solution of
$4c + 2 - 3c = -1 + 6$?

8. Is 1 a solution of $2(b - 3) = 10$?

Ⓑ *Solve. Check each solution. See Examples 3 through 7.*

9. $a + 5 = 23$

10. $s - 7 = 15$

11. $d - 9 = -17$

12. $f + 4 = -6$

13. $7 = y - 2$

14. $-10 = z - 15$

15. $-12 = x + 4$

16. $1 = y + 7$

17. $x + \dfrac{1}{2} = \dfrac{7}{2}$

18. $x + \dfrac{1}{3} = \dfrac{4}{3}$

19. $y - \dfrac{3}{4} = -\dfrac{5}{8}$

20. $y - \dfrac{5}{6} = -\dfrac{11}{12}$

21. $x - 3 = -1 + 4$

22. $y - 8 = -5 - 1$

23. $-7 + 10 = m - 5$

24. $1 - 8 = n + 2$

25. $x - 0.6 = 4.7$

26. $y - 1.2 = 7.5$

27. $-2 - 3 = -4 + x$

28. $7 - (-10) = x - 5$

29. $y + 2.3 = -9.2 - 8.6$

30. $x + 4.7 = -7.5 + 3.4$

31. $-8x + 4 + 9x = -1 + 7$

32. $3x - 2x + 5 = 5$

33. $2 - 2 = 5x - 4x$

34. $11 + (-15) = 6x - 4 - 5x$

35. $7x + 14 - 6x = -4 + (-10)$

36. $-10x + 11x + 5 = 9 + (-5)$

37. In your own words, explain what is meant by the phrase "a number is a solution of an equation."

38. In your own words, explain how to check a possible solution of an equation.

Review and Preview

Perform each indicated operation. See Section 10.4.

39. $\dfrac{-7}{-7}$

40. $\dfrac{4.2}{4.2}$

41. $\dfrac{1}{3} \cdot 3$

42. $\dfrac{1}{5} \cdot 5$

43. $-\dfrac{2}{3} \cdot -\dfrac{3}{2}$

44. $-\dfrac{7}{2} \cdot -\dfrac{2}{7}$

 Combining Concepts

Solve.

45. $x - 76{,}862 = 86{,}102$

46. $-968 + 432 = 86y - 508 - 85y$

A football team's total offense T is found by adding the total passing yardage P to the total rushing yardage R: T = P + R.

47. During the 2000 football season, the Baltimore Ravens' total offense was 5301 yards. The Ravens' passing yardage for the season was 3102 yards. How many yards did the Ravens gain by rushing during the season? (*Source:* National Football League)

48. During the 2000 football season, the San Francisco 49ers' total offense was 6201 yards. The 49ers' rushing yardage for the season was 1801 yards. How many yards did the 49ers gain by passing during the season? (*Source:* National Football League)

In accounting, a company's annual net income I can be computed using the relation I = R − E, where R is the company's total revenues for the year and E is the company's total expenses for the year.

49. At the end of fiscal year 2001, Best Buy had a net income of $604,308,000. During the year, Best Buy had total revenues of $15,326,552,000. What was Best Buy's total expenses for the year? (*Source:* Best Buy Co., Inc.)

50. At the end of fiscal year 2000, Kodak had a net income of $1,407,000,000. During the year, Kodak had total expenses of $12,683,000,000. What was Kodak's total revenues for the year? (*Source:* Eastman Kodak Company)

Internet Excursions

 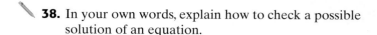

The World Wide Web address listed here will provide you with access to the Web site of the National Football League, or a related site. Team and individual player statistics are available that will help you complete the questions below:

51. Choose any NFL team. On its page of statistics, look for the listing of Season Stats. Look for statistics on total net yards (this is the team's total offense yardage), net yards rushing, and net yards passing. Use these statistics to write a problem similar to those in Exercises 47 and 48 about the team's total offense.

52. Choose another NFL team and write a similar problem about the team's total offense. Then trade your problems with another student in your class to solve.

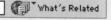

11.3 Solving Equations: The Multiplication Property

A Using the Multiplication Property to Solve Equations

Although the addition property of equality is a powerful tool for helping us solve equations, it cannot help us solve all types of equations. For example, it cannot help us solve an equation such as $2x = 6$. To solve this equation, we use a second property of equality called the **multiplication property of equality**.

Multiplication Property of Equality

Let a, b, and c represent numbers and let $c \neq 0$.

If $a = b$, then

$$a \cdot c = b \cdot c \qquad \text{and} \qquad \frac{a}{c} = \frac{b}{c}$$

In other words, both sides of an equation may be multiplied or divided by the same nonzero number without changing the solution of the equation.

Picturing again our balanced scale, if we multiply or divide the weight on each side by the same nonzero number, the scale (or equation) remains balanced.

To solve $2x = 6$ for x, we use the multiplication property of equality to divide both sides of the equation by 2, and simplify as follows:

$$2x = 6$$

$$\frac{\overset{1}{\cancel{2}} \cdot x}{\underset{1}{\cancel{2}}} = \frac{6}{2} \qquad \text{Divide both sides by 2.}$$

$$1 \cdot x = 3 \quad \text{or} \quad x = 3$$

EXAMPLE 1   Solve: $-5x = 15$

Solution: To get x by itself, we divide both sides by -5.

$$-5x = 15 \qquad \text{Original equation}$$

$$\frac{\overset{1}{\cancel{-5}}x}{\underset{1}{\cancel{-5}}} = \frac{15}{-5} \qquad \text{Divide both sides by } -5$$

$$1x = -3 \quad \text{or} \quad x = -3 \qquad \text{Simplify.}$$

Check: To check, we replace x with -3 in the original equation.

$$-5x = 15 \qquad \text{Original equation}$$

$$-5(-3) \overset{?}{=} 15 \qquad \text{Let } x = -3.$$

$$15 \overset{?}{=} 15 \qquad \text{True}$$

The solution is -3.

Practice Problem 1

Solve: $3y = -18$

Answer
1. -6

Practice Problem 2

Solve: $-16 = 8x$

Practice Problem 3

Solve: $-3y = -27$

Practice Problem 4

Solve: $\dfrac{5}{7}b = 25$

Copyright 2003 Prentice-Hall, Inc.

Answers

2. -2, **3.** 9, **4.** 35

EXAMPLE 2 Solve: $-8 = 2y$

Solution: To get y alone, we divide both sides of the equation by 2.

$$-8 = 2y$$

$$\frac{-8}{2} = \frac{\overset{1}{\cancel{2}}y}{\underset{1}{\cancel{2}}}\qquad\text{Divide both sides by 2.}$$

$$-4 = 1y\quad\text{or}\quad y = -4$$

Check to see that -4 is the solution. ●

EXAMPLE 3 Solve: $-1.2x = -36$

Solution: We divide both sides of the equation by the coefficient of x, which is -1.2.

$$-1.2x = -36$$

$$\frac{\overset{1}{\cancel{-1.2}}x}{\underset{1}{\cancel{-1.2}}} = \frac{-36}{-1.2}$$

$$x = 30$$

Check to see that 30 is the solution. ●

EXAMPLE 4 Solve: $\dfrac{3}{5}a = 9$

Solution: Recall that the product of a number and its reciprocal is 1. To get a alone then, we multiply both sides by $\dfrac{5}{3}$, the reciprocal of $\dfrac{3}{5}$.

$$\frac{3}{5}a = 9$$

$$\frac{\overset{1}{\cancel{5}}}{\underset{1}{\cancel{3}}}\cdot\frac{\overset{1}{\cancel{3}}}{\underset{1}{\cancel{5}}}a = \frac{5}{3}\cdot 9\qquad\text{Multiply both sides by }\frac{5}{3}.$$

$$1a = \frac{5\cdot\overset{3}{\cancel{9}}}{\underset{1}{\cancel{3}}\cdot 1}\qquad\text{Multiply.}$$

$$a = 15\qquad\text{Simplify.}$$

Check: To check, we replace a with 15 in the original equation.

$$\frac{3}{5}a = 9\qquad\text{Original equation}$$

$$\frac{3}{5}\cdot 15 \overset{?}{=} 9\qquad\text{Replace }a\text{ with 15.}$$

$$\frac{3}{\cancel{5}}\cdot\frac{\overset{3}{\cancel{15}}}{1} \overset{?}{=} 9\qquad\text{Multiply.}$$

$$9 \overset{?}{=} 9\qquad\text{True}$$

Since $9 = 9$ is true, 15 is the solution of $\dfrac{3}{5}a = 9$. ●

EXAMPLE 5   Solve: $\frac{1}{4}x = -\frac{1}{8}$

Solution: We multiply both sides of the equation by $\frac{4}{1}$, the reciprocal of $\frac{1}{4}$.

$$\frac{1}{4}x = -\frac{1}{8}$$

$$\frac{\overset{1}{\cancel{4}}}{1} \cdot \frac{1}{\underset{1}{\cancel{4}}}x = \frac{4}{1} \cdot -\frac{1}{8} \qquad \text{Multiply both sides by } \frac{4}{1}.$$

$$1x = -\frac{\overset{1}{\cancel{4}} \cdot 1}{1 \cdot \underset{2}{\cancel{8}}} \qquad \text{Multiply.}$$

$$x = -\frac{1}{2} \qquad \text{Simplify.}$$

Check to see that $-\frac{1}{2}$ is the solution.

Try the Concept Check in the margin.

   We often need to simplify one or both sides of an equation before applying the properties of equality to get the variable alone.

EXAMPLE 6   Solve: $3y - 7y = 12$

Solution: First we combine like terms.

$$3y - 7y = 12$$

$$-4y = 12 \qquad \text{Combine like terms.}$$

$$\frac{\overset{1}{\cancel{-4}}y}{\underset{1}{\cancel{-4}}} = \frac{12}{-4} \qquad \text{Divide both sides by } -4.$$

$$y = -3 \qquad \text{Simplify.}$$

Check: We replace y with -3.

$$3y - 7y = 12$$

$$3(-3) - 7(-3) \overset{?}{=} 12$$

$$-9 + 21 \overset{?}{=} 12$$

$$12 \overset{?}{=} 12 \qquad \text{True}$$

The solution is -3.

EXAMPLE 7   Solve: $-z - z = 11 - 5$

Solution: We simplify both sides of the equation first.

$$-z - z = 11 - 5$$

$$-2z = 6 \qquad \text{Combine like terms.}$$

$$\frac{\overset{1}{\cancel{-2}}z}{\underset{1}{\cancel{-2}}} = \frac{6}{-2} \qquad \text{Divide both sides by } -2.$$

$$z = -3 \qquad \text{Simplify.}$$

Check to see that -3 is the solution.

Practice Problem 5

Solve: $-\frac{7}{10}x = \frac{2}{5}$

Concept Check

Which operation is appropriate for solving each of the following equations, addition or multiplication?

a. $6 = -4x$
b. $6 = x - 4$

Practice Problem 6

Solve: $10 = 2m - 4m$

Practice Problem 7

Solve: $-8 + 6 = -3a + 2a$

Answers

5. $-\frac{4}{7}$,   **6.** -5,   **7.** 2

Concept Check: **a.** multiplication,   **b.** addition

STUDY SKILLS REMINDER

Are you prepared for a test on Chapter 11?

Below I have listed some *common trouble areas* for students in Chapter 11. After studying for your test, but before taking your test, read these.

- Be careful when evaluating expressions. For example, evaluate $3x - y$ when $x = -2$ and $y = -3$.

$$3x - y = 3(-2) - (-3) \qquad \text{Let } x = -2 \text{ and } y = -3.$$
$$= -6 - (-3) \qquad \text{Multiply.}$$
$$= -6 + 3$$
$$= -3 \qquad \text{Add.}$$

- Remember the distributive property.

$$5(4x - 3) + 2 = 5 \cdot 4x - 5 \cdot 3 + 2 \qquad \text{Use the distributive property.}$$
$$= 20x - 15 + 2 \qquad \text{Simplify.}$$
$$= 20x - 13 \qquad \text{Combine like terms.}$$

- Don't forget the steps for solving a linear equation.

$$2(3x - 2) + 16 = 6$$
$$6x - 4 + 16 = 6 \qquad \text{Apply the distributive property.}$$
$$6x + 12 = 6 \qquad \text{Combine like terms.}$$
$$6x + 12 - 12 = 6 - 12 \qquad \text{Subtract 12 from both sides.}$$
$$6x = -6 \qquad \text{Simplify.}$$
$$\frac{\overset{1}{\cancel{6}}x}{\underset{1}{\cancel{6}}} = \frac{-6}{6} \qquad \text{Divide both sides by 6.}$$
$$x = -1 \qquad \text{Simplify.}$$

Remember: This is simply a checklist of common trouble areas. For a review of Chapter 11 see the Highlights and Chapter Review at the end of this chapter.

EXERCISE SET 11.3

A *Solve. See Examples 1 through 5.*

1. $5x = 20$ **2.** $6y = 48$ **3.** $-3z = 12$ **4.** $-2x = 26$ **5.** $0.4y = 0$ **6.** $0.8x = -8$

7. $2z = -34$ **8.** $7y = -21$ **9.** $-0.3x = -15$ **10.** $-0.4z = -12$ **11.** $\frac{2}{5}x = 10$ **12.** $\frac{3}{7}x = 27$

13. $\frac{1}{6}y = -5$ **14.** $\frac{1}{8}y = -3$ **15.** $\frac{5}{6}x = \frac{5}{18}$ **16.** $\frac{4}{7}y = \frac{8}{21}$ **17.** $-\frac{2}{9}z = \frac{4}{27}$ **18.** $-\frac{3}{4}v = \frac{9}{14}$

 19. $\frac{8}{5}t = -\frac{3}{8}$ **20.** $\frac{4}{7}r = -\frac{7}{2}$ **21.** $-\frac{3}{5}x = -\frac{6}{15}$ **22.** $-\frac{6}{7}y = -\frac{1}{14}$

Solve. First combine any like terms on each side of the equation. See Examples 6 and 7.

23. $2w - 12w = 40$ **24.** $8y + y = 45$ **25.** $16 = 10t - 8t$ **26.** $100 = 15y + 5y$

27. $2z = 1.2 - 1.4$ **28.** $-3x = 1.1 - 0.2$ **29.** $4 - 10 = -3z$ **30.** $20 - 12 = -4x$

31. $-3x - 3x = 50 - 2$ **32.** $5y - 9y = -14 + (-14)$ **33.** $-36 = 9u + 3u$ **34.** $-50 = 4y - 14y$

35. $23x - 25x = 7 - 9$ **36.** $8x - 6x = 12 - 22$ **37.** $5 - 5 = 2x + 7x$

38. $7x + 8x = 12 + (-12)$ **39.** $-42 + 20 = -2x + 13x$ **40.** $4y - 9y = -20 + 15$

Review and Preview

Evaluate each expression when $x = 5$. See Section 11.1.

41. $3x + 10$ **42.** $40x$ **43.** $\dfrac{x - 3}{2}$ **44.** $7x - 20$ **45.** $\dfrac{3x + 5}{x - 7}$ **46.** $\dfrac{2x - 1}{x - 8}$

Combining Concepts

47. Why does the multiplication property of equality not allow us to divide both sides of an equation by zero?

48. Is the equation $-x = 6$ solved for the variable? Explain why or why not.

49. Solve: $-0.025x = 91.2$

50. Solve: $3.6y = -1.259 - 3.277$

The equation $d = r \cdot t$ describes the relationship between distance d in miles, rate r in miles per hour, and time t in hours. If necessary, round answers to the nearest tenth.

51. The distance between New Orleans, Louisiana, and Memphis, Tennessee by road is 390 miles. How long will it take to drive from New Orleans to Memphis if the driver maintains a speed of 60 miles per hour? (*Source: 2001 World Almanac*)

52. The distance between Boston, Massachusetts, and Milwaukee, Wisconsin, by road is 1050 miles. How long will it take to drive from Boston to Milwaukee if the driver maintains a speed of 55 miles per hour? (*Source: 2001 World Almanac*)

53. The distance between Cleveland, Ohio, and Indianapolis, Indiana, by road is 294 miles. At what speed should a driver drive if he or she would like to make the trip in 5 hours? (*Source: 2001 World Almanac*)

54. The distance between St. Louis, Missouri, and Minneapolis, Minnesota, by road is 552 miles. If it took 9 hours to drive from St. Louis to Minneapolis, what was the driver's average speed? (*Source: 2001 World Almanac*)

Integrated Review—Expressions and Equations

Evaluate each expression when $x = -1$ and $y = 3$.

1. $y - x$ **2.** $\dfrac{y}{x}$ **3.** $5x + 2y$ **4.** $\dfrac{y^2 + x}{2x}$

Simplify each expression by combining like terms.

5. $7x + x$ **6.** $6y - 10y$ **7.** $2a + 5a - 9a - 2$

8. $3x - y + 4 - 5x + 4y - 11$

Multiply. Simplify if possible.

9. $-2(4x)$ **10.** $5(y + 2)$

11. $3(x + 5) - 3$ **12.** $-4(x - 1) + 3(5x + 4)$

1. _____

2. _____

3. _____

4. _____

5. _____

6. _____

7. _____

8. _____

9. _____

10. _____

11. _____

12. _____

13. _____	
14. _____	
15. _____	
16. _____	
17. _____	
18. _____	
19. _____	
20. _____	
21. _____	
22. _____	
23. _____	
24. _____	
25. _____	
26. _____	

Find the area.

△ **13.**
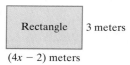
3 meters

(4x − 2) meters

△ **14.**

5y inches

Solve and check.

15. $x + 7 = 20$

16. $-11 = x - 2$

17. $11x = 55$

18. $-7y = 0$

19. $12 = 11x - 14x$

20. $\dfrac{3}{5}x = 15$

21. $x - 1.2 = -4.5 + 2.3$

22. $8y + 7y = -45$

23. $6 - (-5) = x + 5$

24. $-0.2m = -1.6$

25. $-\dfrac{2}{3}n = \dfrac{6}{11}$

26. $n - \dfrac{2}{5} = \dfrac{3}{10}$

11.4 Solving Equations Using Addition and Multiplication Properties

OBJECTIVES

Ⓐ Solve equations using addition and multiplication properties.

Ⓑ Solve equations containing parentheses.

Ⓒ Write sentences as equations.

SSM   TUTOR CENTER   SG   CD & VIDEO   MATH PRO   WEB

Ⓐ Solving Equations Using Addition and Multiplication Properties

We will now solve equations using more than one property of equality. To solve an equation such as $2x - 6 = 18$, we must first get the variable term $2x$ alone on one side of the equation.

EXAMPLE 1   Solve: $2x - 6 = 18$

Solution: We start by adding 6 to both sides to get the variable term $2x$ alone.

$$2x - 6 = 18$$
$$2x - 6 + 6 = 18 + 6 \qquad \text{Add 6 to both sides.}$$
$$2x = 24 \qquad \text{Simplify.}$$

To finish solving, we divide both sides by 2.

$$\frac{\overset{1}{\cancel{2}}x}{\underset{1}{\cancel{2}}} = \frac{24}{2} \qquad \text{Divide both sides by 2.}$$

$$x = 12 \qquad \text{Simplify.}$$

Check:
$$2x - 6 = 18$$
$$2(12) - 6 \overset{?}{=} 18 \qquad \text{Replace } x \text{ with 12 and simplify.}$$
$$24 - 6 \overset{?}{=} 18$$
$$18 \overset{?}{=} 18 \qquad \text{True}$$

The solution is 12. ●

Practice Problem 1

Solve: $5y + 2 = 17$

EXAMPLE 2   Solve: $20 - x = 21$

Solution: First we get the variable term alone on one side of the equation.

$$20 - x = 21$$
$$20 - x - 20 = 21 - 20 \qquad \text{Subtract 20 from both sides.}$$
$$-1x = 1 \qquad \text{Simplify. Recall that } -x \text{ means } -1x.$$

$$\frac{\overset{1}{\cancel{-1}}x}{\underset{1}{\cancel{-1}}} = \frac{1}{-1} \qquad \text{Divide both sides by } -1.$$

$$x = -1 \qquad \text{Simplify.}$$

Check:
$$20 - x = 21$$
$$20 - (-1) \overset{?}{=} 21$$
$$21 \overset{?}{=} 21 \qquad \text{True}$$

The solution is -1. ●

Practice Problem 2

Solve: $45 = -10 - y$

Answers

1. 3, **2.** -55

Practice Problem 3

Solve: $\dfrac{3}{4}y + 19 = 11$

Helpful Hint

Don't forget that we can get the variable alone on either side of the equation.

Practice Problem 4

Solve: $7x + 12 = 3x - 4$

Practice Problem 5

Solve: $8x + 4.2 = 10x + 11.6$

EXAMPLE 3   Solve: $1 = \dfrac{2}{3}x + 7$

Solution: Subtract 7 from both sides to get the variable term alone.

$$1 - 7 = \dfrac{2}{3}x + 7 - 7 \qquad \text{Subtract 7 from both sides.}$$

$$-6 = \dfrac{2}{3}x \qquad \text{Simplify.}$$

$$\dfrac{3}{2} \cdot -6 = \dfrac{\cancel{3}}{\cancel{2}} \cdot \dfrac{\cancel{2}}{\cancel{3}}x \qquad \text{Multiply both sides by } \dfrac{3}{2}.$$

$$\dfrac{3}{\cancel{2}} \cdot \dfrac{\cancel{-6}^{-3}}{1} = 1x \qquad \text{Simplify.}$$

$$-9 = x \qquad \text{Simplify.}$$

Check to see that the solution is -9.   ●

If an equation contains variable terms on both sides, we use the addition property of equality to get all the variable terms on one side and all the constants, or numbers, on the other side.

EXAMPLE 4   Solve: $3a - 6 = a + 4$

Solution:

$$3a - 6 = a + 4$$
$$3a - 6 + 6 = a + 4 + 6 \qquad \text{Add 6 to both sides.}$$
$$3a = a + 10 \qquad \text{Simplify.}$$
$$3a - a = a + 10 - a \qquad \text{Subtract } a \text{ from both sides.}$$
$$2a = 10 \qquad \text{Simplify.}$$
$$\dfrac{\cancel{2}a}{\cancel{2}} = \dfrac{10}{2} \qquad \text{Divide both sides by 2.}$$
$$a = 5 \qquad \text{Simplify.}$$

Check to see that the solution is 5.   ●

EXAMPLE 5   Solve: $7x + 3.2 = 4x - 1.6$

Solution:

$$7x + 3.2 = 4x - 1.6$$
$$7x + 3.2 - 3.2 = 4x - 1.6 - 3.2 \qquad \text{Subtract 3.2 from both sides.}$$
$$7x = 4x - 4.8 \qquad \text{Simplify.}$$
$$7x - 4x = 4x - 4.8 - 4x \qquad \text{Subtract } 4x \text{ from both sides.}$$
$$3x = -4.8 \qquad \text{Simplify.}$$
$$\dfrac{\cancel{3}x}{\cancel{3}} = -\dfrac{4.8}{3} \qquad \text{Divide both sides by 3.}$$
$$x = -1.6 \qquad \text{Simplify.}$$

Check to see that -1.6 is the solution.   ●

B **Solving Equations Containing Parentheses**

If an equation contains parentheses, we must first use the distributive property to remove them.

EXAMPLE 6   Solve: $7(x - 2) = 9x - 6$

Solution: First we apply the distributive property.

$$7(x - 2) = 9x - 6$$
$$7x - 14 = 9x - 6 \quad \text{Apply the distributive property.}$$

Next, we move variable terms to one side of the equation and constants to the other side.

$$7x - 14 - 9x = 9x - 6 - 9x \quad \text{Subtract } 9x \text{ from both sides.}$$
$$-2x - 14 = -6 \quad \text{Simplify.}$$
$$-2x - 14 + 14 = -6 + 14 \quad \text{Add 14 to both sides.}$$
$$-2x = 8 \quad \text{Simplify.}$$
$$\frac{-2x}{-2} = \frac{8}{-2} \quad \text{Divide both sides by } -2.$$
$$x = -4 \quad \text{Simplify.}$$

Check to see that -4 is the solution.   ●

You may want to use the steps in the margin to solve equations.

EXAMPLE 7   Solve: $3(2x - 6) + 6 = 0$

Solution:

$$3(2x - 6) + 6 = 0$$

Step 1. $6x - 18 + 6 = 0$ Apply the distributive property.

Step 2. $6x - 12 = 0$ Combine like terms on the left side of the equation.

Step 3. $6x - 12 + 12 = 0 + 12$ Add 12 to both sides.

$$6x = 12 \quad \text{Simplify.}$$

Step 4. $$\frac{6x}{6} = \frac{12}{6} \quad \text{Divide both sides by 6.}$$
$$x = 2 \quad \text{Simplify.}$$

Check:

Step 5. $3(2x - 6) + 6 = 0$
$3(2 \cdot 2 - 6) + 6 \stackrel{?}{=} 0$
$3(4 - 6) + 6 \stackrel{?}{=} 0$
$3(-2) + 6 \stackrel{?}{=} 0$
$-6 + 6 \stackrel{?}{=} 0$
$0 \stackrel{?}{=} 0 \quad \text{True}$

The solution is 2.   ●

(C)   Writing Sentences as Equations

Next, we practice translating sentences into equations. Below are key words and phrases that translate to an equal sign:

Practice Problem 6

Solve: $6(a - 5) = 4(a + 1)$

Steps for Solving an Equation

Step 1.   If parentheses are present, use the distributive property.

Step 2.   Combine any like terms on each side of the equation.

Step 3.   Use the addition property of equality to rewrite the equation so that variable terms are on one side of the equation and constant terms are on the other side.

Step 4.   Use the multiplication property of equality to divide both sides by the numerical coefficient of the variable to solve for.

Step 5.   Check the solution in the *original equation*.

Practice Problem 7

Solve: $4(x + 3) = 12$

Answers
6. 17,   **7.** 0

Key Words or Phrases	Examples	Symbols
equals	3 equals 2 plus 1	$3 = 2 + 1$
gives	the quotient of 10 and -5 gives -2	$\dfrac{10}{-5} = -2$
is/was	x is 5	$x = 5$
yields	y plus 2 yields 13	$y + 2 = 13$
amounts to	twice x amounts to -30	$2x = -30$
is equal to	-24 is equal to 2 times -12	$-24 = 2(-12)$

Practice Problem 8

Translate each sentence into an equation.

a. The difference of 110 and 80 is 30.

b. The product of 3 and the sum of -9 and 11 amounts to 6.

c. The quotient of twice 12 and -6 yields -4.

EXAMPLE 8 Translate each sentence into an equation.

a. The product of 7 and 6 is 42.

b. Twice the sum of 3 and 5 is equal to 16.

c. The quotient of -45 and 5 yields -9.

Solution:

a. In words: | the product of 7 and 6 | is | 42 |

Translate: $\quad 7 \cdot 6 \quad = \quad 42$

b. In words: | twice | the sum of 3 and 5 | is equal to | 16 |

Translate: $\quad 2 \quad (3 + 5) \quad = \quad 16$

c. In words: | the quotient of -45 and 5 | yields | -9 |

Translate: $\quad \dfrac{-45}{5} \quad = \quad -9$

Answers

8. a. $110 - 80 = 30$, **b.** $3(-9 + 11) = 6$,

c. $\dfrac{2(12)}{-6} = -4$

CALCULATOR EXPLORATIONS

Checking Equations

A calculator can be used to check possible solutions of equations. To do this, replace the variable by the possible solution and evaluate each side of the equation separately. For example, to see whether 7 is a solution of the equation $52x = 15x + 259$, replace x with 7 and use your calculator to evaluate each side separately.

Equation: $52x = 15x + 259$

$$52 \cdot 7 \overset{?}{=} 15 \cdot 7 + 259 \qquad \text{Replace } x \text{ with 7.}$$

Evaluate left side: $\boxed{52}\ \boxed{\times}\ \boxed{7}\ \boxed{=}$ or $\boxed{\text{ENTER}}$.   Display: $\boxed{364}$.

Evaluate right side: $\boxed{15}\ \boxed{\times}\ \boxed{7}\ \boxed{+}\ \boxed{259}\ \boxed{=}$ or $\boxed{\text{ENTER}}$.   Display: $\boxed{364}$.

Since the left side equals the right side, 7 is a solution of the equation $52x = 15x + 259$.

Use a calculator to determine whether the numbers given are solutions of each equation.

1. $76(x - 25) = -988;\quad 12$

2. $-47x + 862 = -783;\quad 35$

3. $x + 562 = 3x + 900;\quad -170$

4. $55(x + 10) = 75x + 910;\quad -18$

5. $29x - 1034 = 61x - 362;\quad -21$

6. $-38x + 205 = 25x + 120;\quad 25$

Name _____ Section _____ Date _____

EXERCISE SET 11.4

 A *Solve each equation. See Examples 1 through 5.*

1. $2x - 6 = 0$

2. $3y - 12 = 0$

3. $5n + 10 = 30$

4. $4z + 8 = 40$

5. $6 - n = 10$

6. $7 - y = 9$

7. $10x + 15 = 6x + 3$

8. $5x - 3 = 2x - 18$

9. $3x - 7 = 4x + 5$

10. $3x + 1 = 8x - 4$

11. $-\dfrac{2}{5}x + 19 = -21$

12. $-\dfrac{3}{7}y - 14 = 7$

13. $1.7 = 2y + 9.5$

14. $-5.1 = 3x + 2.4$

15. $9a + 29 = -7$

16. $10 + 4v = -6$

17. $8 - t = 3$

18. $6 - x = 4$

19. $0 = 4x - 4$

20. $0 = 5y + 5$

21. $2n + 8 = 0$

22. $8w - 40 = 0$

23. $7 = 4c - 1$

24. $9 = 2b - 5$

25. $3r + 4 = 19$ **26.** $5m + 1 = 46$ **27.** $2x - 1 = -7$ **28.** $3t - 2 = -11$

29. $2 = 3z - 4$ **30.** $4 = 4p - 12$ **31.** $5x - 2 = -12$ **32.** $7y - 3 = -24$

33. $-7c + 1 = -20$ **34.** $-2b + 5 = -7$ **35.** $-5 = -13 - 8k$ **36.** $-7 = -17 - 10d$

37. $4x + 3 = 2x + 11$ **38.** $6y - 8 = 3y + 7$ **39.** $-2y - 10 = 5y + 18$ **40.** $7n + 5 = 12n - 10$

41. $-8n + 1 = -6n - 5$ **42.** $10w + 8 = w - 10$ **43.** $9 - 3x = 14 + 2x$ **44.** $4 - 7m = -3m + 4$

45. $\frac{3}{8}x + 14 = \frac{5}{8}x - 2$ **46.** $\frac{2}{7}x - 9 = \frac{5}{7}x - 15$ **47.** $-1.4x - 2 = -1.2x + 7$ **48.** $5.7y + 14 = 5.4y - 10$

B *Solve each equation. See Examples 6 and 7.*

49. $3(x - 1) = 12$ **50.** $2(x + 5) = -8$ **51.** $-2(y + 4) = 2$ **52.** $-1(y + 3) = 10$

53. $35 = 17 + 3(x - 2)$ **54.** $22 - 42 = 4(x - 1)$ **55.** $2(y - 3) = y - 6$ **56.** $3(z + 2) = 5z + 6$

57. $2t - 1 = 3(t + 7)$ **58.** $4 + 3c = 2(c + 2)$ **59.** $3(5c - 1) - 2 = 13c + 3$

60. $4(3t + 4) - 20 = 3 + 5t$ **61.** $10 + 5(z - 2) = 4z + 1$ **62.** $14 + 4(w - 5) = 6 - 2w$

63. $7(6 + w) = 6(2 + w)$ **64.** $6(5 + c) = 5(c - 4)$

c *Write each sentence as an equation. See Example 8.*

65. The sum of -42 and 16 is -26.

66. The difference of -30 and 10 equals -40.

67. The product of -5 and -29 gives 145.

68. The quotient of -16 and 2 yields -8.

69. Three times the difference of -14 and 2 amounts to -48.

70. The product of -2 and the sum of 3 and 12 is -30.

71. The quotient of 100 and twice 50 is equal to 1.

72. Seventeen subtracted from −12 equals −29.

Review and Preview

The following bar graph shows the estimated number of U.S. federal individual income tax returns that will be filed electronically during the years shown. Electronically filed returns include Telefile and online returns. Use this graph to answer Exercises 73 through 76. See Section 9.1.

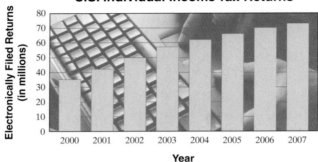

Total Electronically Filed U.S. Individual Income Tax Returns

Source: IRS Compliance Research Division

73. Approximate the number of electronically filed returns estimated for 2002.

74. Approximate the number of electronically filed returns estimated for 2005.

75. By how much is the number of electronically filed returns expected to increase from 2000 to 2007?

76. Describe any trends shown in this graph.

◆ Combining Concepts

The equation $C = \dfrac{5}{9}(F - 32)$ gives the relationship between Celsius temperatures C and Fahrenheit temperatures F.

77. The highest recorded temperature in Australia occurred in January 1960 at Oodnadatta, South Australia. The temperature reached 50.7°C. Use the given equation to convert this temperature to degrees Fahrenheit. (*Source:* World Weather Centre at Perth)

78. The highest recorded temperature in Africa occurred in September 1922 at Al'Aziziyah, Libya. The temperature reached 57.7°C. Use the given equation to convert this temperature to degrees Fahrenheit. (*Source:* World Weather Centre at Perth)

79. The lowest recorded temperature in Australia occurred in June 1994 at Charlotte Pass, New South Wales. The temperature plummeted to −23.0°C. Use the given equation to convert this temperature to degrees Fahrenheit. (*Source:* World Weather Centre at Perth)

80. The lowest recorded temperature in North America occurred in February 1947 at Snag, Canada. The temperature plummeted to −63.0°C. Use the given equation to convert this temperature to degrees Fahrenheit. (*Source:* World Weather Centre at Perth)

11.5   Equations and Problem Solving

(A)   Writing Phrases as Algebraic Expressions

Now that we have practiced solving equations for a variable, we can extend considerably our problem-solving skills. We begin by writing phrases as algebraic expressions using the following key words and phrases as a guide:

Addition	Subtraction	Multiplication	Division	Equal Sign
sum	difference	product	quotient	equals
plus	minus	times	divided by	gives
added to	subtracted from	multiply	into	is/was
more than	less than	twice	per	yields
increased by	decreased by	of		amounts to
total	less	double		is equal to

OBJECTIVES

(A)   Write phrases as algebraic expressions.

(B)   Write sentences as equations.

(C)   Use problem-solving steps to solve problems.

SSM
TUTOR CENTER   SG   CD & VIDEO   MATH PRO   WEB

EXAMPLE 1

Write each phrase as an algebraic expression. Use x to represent "a number."

a. 7 increased by a number
b. 15 decreased by a number
c. the product of 2 and a number
d. the quotient of a number and 5
e. 2 subtracted from a number

Solution:

a. In words:

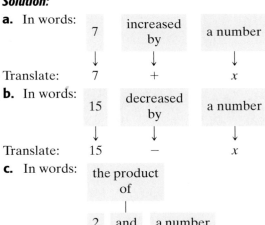

b. In words:

c. In words:

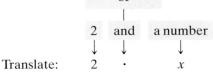

d. In words:

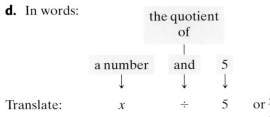

e. In words:

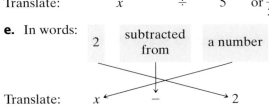

Practice Problem 1

Write each phrase as an algebraic expression. Use x to represent "a number."

a. twice a number
b. 8 increased by a number
c. 10 minus a number
d. 10 subtracted from a number
e. the quotient of 6 and a number

Answers

1. a. $2x$, **b.** $8 + x$, **c.** $10 - x$, **d.** $x - 10$,
e. $\dfrac{6}{x}$

B Writing Sentences as Equations

Now that we have practiced writing phrases as algebraic expressions, let's write sentences as equations. You may want to first study the key words and phrases chart to review some key words and phrases that translate to an equal sign.

Practice Problem 2

Write each sentence as an equation. Use x to represent "a number."

a. Five times a number is 20.

b. The sum of a number and -5 yields 14.

c. Ten subtracted from a number amounts to -23.

d. Five times a number added to 7 is equal to -8.

e. The quotient of 6 and the sum of a number and 4 gives 1.

EXAMPLE 2

Write each sentence as an equation. Use x to represent "a number."

a. Nine increased by a number is 5.

b. Twice a number equals -10.

c. A number minus 6 amounts to 168.

d. Three times the sum of a number and 5 is -30.

e. The quotient of 8 and twice a number is equal to 2.

Solution:

a. In words:

nine	increased by	a number	is	5
↓	↓	↓	↓	↓

Translate: $9 \quad + \quad x \quad = \quad 5$

b. In words:

twice a number	equals	-10
↓	↓	↓

Translate: $2x \quad = \quad -10$

c. In words:

a number	minus	6	amounts to	168
↓	↓	↓	↓	↓

Translate: $x \quad - \quad 6 \quad = \quad 168$

d. In words:

three times	the sum of a number and 5	is	-30
↓	↓	↓	↓

Translate: $3 \quad (x + 5) \quad = \quad -30$

e. In words:

the quotient of

8	and	twice a number	is equal to	2
↓	↓	↓	↓	↓

Translate: $8 \quad \div \quad 2x \quad = \quad 2$

or $\dfrac{8}{2x} = 2$

C Using Problem-Solving Steps to Solve Problems

Our main purpose for studying arithmetic and algebra is to solve problems. The same problem-solving steps that have been used throughout this text are used in this section also. Those steps are repeated on the next page.

The first problem that we solve consists of finding an unknown number.

EXAMPLE 3   Finding an Unknown Number

Twice a number added to 3 is the same as the number minus 6. Find the number.

Solution:

1. UNDERSTAND the problem. To do so, we read and reread the problem. Then we assign a variable to the unknown. We let x = the unknown number.

2. TRANSLATE the problem into an equation.

In words:

twice a number	added to 3	is the same as	the number minus 6
↓	↓	↓	↓
$2x$	$+ 3$	$=$	$x - 6$

Translate:

3. SOLVE the equation. To solve the equation, we first subtract x from both sides.

$$2x + 3 = x - 6$$
$$2x + 3 - x = x - 6 - x$$
$$x + 3 = -6 \qquad \text{Simplify.}$$
$$x + 3 - 3 = -6 - 3 \qquad \text{Subtract 3 from both sides.}$$
$$x = -9 \qquad \text{Simplify.}$$

4. INTERPRET the results. First, *Check* the proposed solution in the stated problem. Twice "-9" is -18 and $-18 + 3$ is -15. This is equal to the number minus 6, or "-9" -6, or -15. Then *state* your conclusion: The unknown number is -9.   ●

Try the Concept Check in the margin.

EXAMPLE 4   Determining Voter Counts

In the 2000 Senate election in Wyoming, incumbent Craig Thomas received 110,280 *more* votes than his challenger. If a total of 204,358 votes were cast, find how many votes Craig Thomas received. (*Source: 2001 World Almanac*)

Solution:

1. UNDERSTAND the problem. We read and reread the problem. Then we assign a variable to an unknown. We use this variable to represent any other unknown quantities. We let

    x = the number of challenger votes

    Then

    $x + 110{,}280$ = the number of incumbent votes
    since he received 110,280 more votes

2. TRANSLATE the problem into an equation.

In words:

challenger votes	+	incumbent votes	=	total votes
↓		↓		↓
x	$+$	$x + 110{,}280$	$=$	$204{,}358$

Translate:

Practice Problem 3

Translate "The sum of a number and 2 equals 6 added to three times the number" into an equation and solve.

Problem-Solving Steps

1. UNDERSTAND the problem. During this step, become comfortable with the problem. Some ways of doing this are:

    Read and reread the problem. Choose a variable to represent the unknown.

    Construct a drawing.

    Propose a solution and check. Pay careful attention to how you check your proposed solution. This will help when writing an equation to model the problem.

2. TRANSLATE the problem into an equation.

3. SOLVE the equation.

4. INTERPRET the results: *Check* the proposed solution in the stated problem and *state* your conclusion.

Concept Check

Suppose you have solved an equation involving perimeter to find the length of a rectangular table. Explain why you would want to recheck your math if you obtain the result of -5.

Practice Problem 4

At a recent U.S./Japan summit meeting, 121 delegates attended. If the United States sent 19 more delegates than Japan, find how many the United States sent.

Answers

3. -2,   **4.** 70 delegates

Concept Check: length cannot be negative

3. SOLVE the equation:

$$x + x + 110{,}280 = 204{,}358$$
$$2x + 110{,}280 = 204{,}358 \quad \text{Combine like terms.}$$
$$2x + 110{,}280 - 110{,}280 = 204{,}358 - 110{,}280 \quad \text{Subtract 110,280 from both sides.}$$
$$2x = 94{,}078 \quad \text{Simplify.}$$
$$\frac{2x}{2} = \frac{94{,}078}{2} \quad \text{Divide both sides by 2.}$$
$$x = 47{,}039 \quad \text{Simplify.}$$

4. INTERPRET the results. First *Check* the proposed solution in the stated problem. Since x represents the number of votes the challenger received, the challenger received 47,039 votes. The incumbent received $x + 110{,}280 = 47{,}039 + 110{,}280 = 157{,}319$ votes. To check, notice that the total number of challenger votes and incumbent votes is $47{,}039 + 157{,}319 = 204{,}358$ votes, the given total of votes cast. Also, 157,319 is 110,280 more votes than 47,039, so the solution checks. Then, *state* your conclusion: The incumbent, Craig Thomas, received 157,319 votes. ●

Practice Problem 5

A woman's $21,000 estate is to be divided so that her husband receives twice as much as her son. How much will each receive?

EXAMPLE 5   Calculating Separate Costs

Leo Leal sold a used computer system and software for $2100, receiving four times as much money for the computer system as for the software. Find the price of each.

Solution:

1. UNDERSTAND the problem. We read and reread the problem. Then we assign a variable to an unknown. We use this variable to represent any other unknown quantities. We let

$$x = \text{the software price}$$
$$4x = \text{the computer system price}$$

2. TRANSLATE the problem into an equation.

In words: software price and computer price is 2100

Translate: $x + 4x = 2100$

3. SOLVE the equation:

$$x + 4x = 2100$$
$$5x = 2100 \quad \text{Combine like terms.}$$
$$\frac{5x}{5} = \frac{2100}{5} \quad \text{Divide both sides by 5.}$$
$$x = 420 \quad \text{Simplify.}$$

4. INTERPRET the results. *Check* the proposed solution in the stated problem. The software sold for $420. The computer system sold for $4x = 4(\$420) = \$1680$. Since $\$420 + \$1680 = \$2100$, the total price, and $1680 is four times $420, the solution checks. *State* your conclusion: The software sold for $420, and the computer system sold for $1680. ●

Answer

5. husband: $14,000; son: $7,000

EXERCISE SET 11.5

A *Write each phrase as a variable expression. Use x to represent "a number." See Example 1.*

1. The sum of a number and five

2. Ten plus a number

3. The total of a number and eight

4. The difference of a number and five hundred

5. Twenty decreased by a number

6. A number less thirty

7. The product of 512 and a number

8. A number times twenty

9. A number divided by 2

10. The quotient of six and a number

11. The sum of seventeen and a number added to the product of five and the number

12. The difference of twice a number, and four

13. The product of five and a number

14. The quotient of twenty and a number, decreased by three

15. A number subtracted from 11

16. Twelve subtracted from a number

17. Fifty decreased by eight times a number

18. Twenty decreased by twice a number

B *Write each sentence as an equation. Use x to represent "a number." See Example 2.*

19. A number added to -5 is -7.

20. Five subtracted from a number equals 10.

21. Three times a number yields 27.

22. The quotient of 8 and a number is -2.

23. A number subtracted from -20 amounts to 104.

24. Two added to twice a number gives -14.

C *Solve. See Example 3.*

25. Three times a number added to 9 is 33. Find the number.

26. Twice a number subtracted from 60 is 20. Find the number.

27. The sum of 3, 4, and a number amounts to 16. Find the number.

28. A number less 5 is 11. Find the number.

29. Eight decreased by some number equals the quotient of 15 and 5. Find the number.

30. The product of some number plus 2 and 5 is 11 less than the number times 8. Find the number.

31. Five times a number less 40 is 8 more than the number.

32. The product of 4 and a number is the same as 30 less twice that same number.

33. Three times the difference of some number and 5 amounts to the quotient of 108 and 12. Find the number.

34. Thirty less a number is equal to the product of 3 and the sum of the number and 6.

35. The difference of a number and 3 is equal to the quotient of 10 and 5.

36. The product of a number and 3 is twice the sum of that number and 5.

Solve. See Examples 4 and 5.

37. Hearst Magazines, a leading publisher of monthly magazines worldwide, publishes titles such as *Popular Mechanics* and *Cosmopolitan*. Hearst publishes 86 more international editions than the number of U.S. titles it publishes. The total number of editions, U.S. and international combined, that Hearst publishes is 118. How many U.S. titles does Hearst Magazines publish? (*Source:* The Hearst Corporation)

38. Disneyland®, in Anaheim, California, has 12 more rides than its neighboring sister park, Disney's California Adventure, which opened in 2001. Together, the two parks have 36 rides. How many rides does each park have? (*Source:* The Walt Disney Company)

39. Mark and Stuart Martin collect comic books. Mark has twice the number of books Stuart has. Together they have 120 comic books. Find how many books Mark has.

40. Heather and Mary Gamber collect baseball cards. Heather's collection is three times as large as Mary's. Together they have 400 baseball cards. Find the number of baseball cards in each collection.

41. A Toyota Camry is traveling twice as fast as a Dodge truck. If their combined speed is 105 miles per hour, find the speed of the car and the speed of the truck.

42. A crow will eat five more ounces of food a day than a finch. If together they eat 13 ounces of food, find how many ounces of food the crow consumes and how many ounces of food the finch consumes.

43. Anthony Tedesco sold his used mountain bike and accessories for $270. If he received five times as much money for the bike as he did for the accessories, find how much money he received for the bike.

44. A tractor and a plow attachment are worth $1200. The tractor is worth seven times as much money as the plow. Find the value of the tractor and the value of the plow.

45. During the 2001 Women's NCAA Division I basketball championship game, the Notre Dame Fighting Irish scored 2 more points than the Purdue Boilermakers. Together, both teams scored a total of 134 points. How many points did the 2001 Champion Notre Dame Fighting Irish score during this game? (*Source:* National Collegiate Athletic Association)

46. During the 2001 Men's NCAA Division I basketball championship game, the Arizona Wildcats scored 10 fewer points than the Duke Blue Devils. Together, both teams scored a total of 154 points. How many points did the 2001 Champion Duke Blue Devils score during this game? (*Source:* National Collegiate Athletic Association)

Review and Preview

Round each number to the given place value. See Section 1.5.

47. 586 to the nearest ten

48. 82 to the nearest ten

49. 1026 to the nearest hundred

50. 52,333 to the nearest thousand

51. 2986 to the nearest thousand

52. 101,552 to the nearest hundred

 Combining Concepts

53. Solve Example 4 again, but this time let x be the number of incumbent votes. Did you get the same results? Explain why or why not.

This activity may be completed by working in groups or individually.

A couple is getting married, and they have decided to hold their wedding reception at a local hotel. The hotel charges a flat fee of $1200 for renting their large reception hall. This cost does not include food, wedding cake, or drinks. The costs of the food items will depend on the number of guests. The hotel offers the following buffet and cake packages:

Option	Description	Cost of Buffet per Person	Cost of Cake and Beverages per Person
A	Cold cuts, white cake, punch	$16	$2
B	Hot buffet, white cake, punch	$22	$2
C	Hot buffet, salad bar, white cake, mints, nuts, punch, coffee	$28	$3
D	Hot buffet, salad bar, white cake, mints, nuts, punch, coffee, open bar	$28	$10

1. The couple decides that they can afford to spend a total of $7000 on the reception. Translate each option into an algebraic equation that describes the total cost of the package in terms of the flat fee and the costs that depend on the number of guests. Let x represent the number of guests and assume that the couple will spend the entire $7000. For instance, the cost of Option A can be represented by the equation

    $7000 = 1200 + 16x + 2x$ or

    $7000 = 1200 + 18x.$

2. If the couple can spend only a total of $7000 on the reception, how many guests can the couple invite under each option? (Be sure to round your answers so that the couple would not spend over $7000.)

3. The couple prefers Option C, but they want to invite a total of 240 people. How much additional money will they need to cover the cost of the reception?

4. (Optional) Suppose as a member of a student organization, you have been asked to organize an awards banquet. Investigate the cost of a catered meal with three different caterers in your area. If the organization has a fixed budget of $2000 for the banquet, how many people could be invited in each scenario? Be sure to take any hall rentals or other fixed fees into consideration.

Chapter 11 VOCABULARY CHECK

Fill in each blank with one of the words or phrases listed below.

variable   simplified   numerical coefficient

terms   combined   algebraic expression

like   evaluating the expression

1. An algebraic expression is _____ when all like terms have been _____ .

2. Terms that are exactly the same, except that they may have different numerical coefficients, are called _____ terms.

3. A letter used to represent a number is called a _____ .

4. A combination of operations on variables and numbers is called an _____ .

5. The addends of an algebraic expression are called the _____ of the expression.

6. The number factor of a variable term is called the _____ .

7. Replacing a variable in an expression by a number and then finding the value of the expression is called _____ for the variable.

CHAPTER 11 Highlights

DEFINITIONS AND CONCEPTS	EXAMPLES

Section 11.1   Introduction to Variables

A letter used to represent a number is called a **variable**.

A combination of operations on variables and numbers is called an **algebraic expression**.

Replacing a variable in an expression by a number and then finding the value of the expression is called **evaluating the expression** for the variable.

x, y, z, a, b

$3 + x, 7y, x^3 + y - 10$

Evaluate: $2x + y$ when $x = 22$ and $y = 4$

$$2x + y = 2 \cdot 22 + 4 \quad \text{Replace } x \text{ with 22 and } y \text{ with 4.}$$
$$= 44 + 4 \quad \text{Multiply.}$$
$$= 48 \quad \text{Add.}$$

The addends of an algebraic expression are called the **terms** of the expression.

The number factor of a variable term is called the **numerical coefficient**.

Terms that are exactly the same, except that they may have different numerical coefficients, are called **like terms**.

$$5x^2 + (-4x) + (-2) \qquad \text{3 terms}$$

Term	Numerical Coefficient
$7x$	7
$-6y$	-6
x or $1x$	1

$$5x + 11x = (5 + 11)x = 16x$$
like terms
$$y - 6y = (1 - 6)y = -5y$$

An algebraic expression is **simplified** when all like terms have been **combined**.

Use the distributive property to multiply an algebraic expression by a term.

Simplify:

$$-4(x + 2) + 3(5x - 7)$$
$$= -4(x) + (-4)(2) + 3(5x) + 3(-7)$$
$$= -4x + (-8) + 15x + (-21)$$
$$= 11x + (-29) \quad \text{or} \quad 11x - 29$$

DEFINITIONS AND CONCEPTS	EXAMPLES

Section 11.2   Solving Equations: The Addition Property

ADDITION PROPERTY OF EQUALITY

Let a, b, and c represent numbers.

If $a = b$, then

$$a + c = b + c \quad \text{and} \quad a - c = b - c$$

In other words, the same number may be added to or subtracted from both sides of an equation without changing the solution of the equation.

Solve for x:

$$x + 8 = 2 + (-1)$$
$$x + 8 = 1$$

$x + 8 - 8 = 1 - 8$ Subtract 8 from both sides.

$x = -7$ Simplify.

The solution is -7.

Section 11.3   Solving Equations: The Multiplication Property

MULTIPLICATION PROPERTY OF EQUALITY

Let a, b, and c represent numbers and let $c \neq 0$.

If $a = b$, then

$$a \cdot c = b \cdot c \quad \text{and} \quad \frac{a}{c} = \frac{b}{c}$$

In other words, both sides of an equation may be multiplied or divided by the same nonzero number without changing the solution of the equation.

Solve: $-7x = 42$

$$\frac{-7x}{-7} = \frac{42}{-7} \quad \text{Divide both sides by } -7.$$

$x = -6$ Simplify.

Solve: $\frac{2}{3}x = -10$

$$\frac{3}{2} \cdot \frac{2}{3}x = \frac{3}{2} \cdot -10 \quad \text{Multiply both sides by } \frac{3}{2}.$$

$x = -15$ Simplify.

Section 11.4   Solving Equations Using Addition and Multiplication Properties

STEPS FOR SOLVING AN EQUATION

Step 1. If parentheses are present, use the distributive property.

Step 2. Combine any like terms on each side of the equation.

Step 3. Use the addition property of equality to rewrite the equation so that variable terms are on one side of the equation and constant terms are on the other side.

Step 4. Use the multiplication property of equality to divide both sides by the numerical coefficient of the variable to solve.

Step 5. Check the solution in the *original equation*.

Solve for x: $5(3x - 1) + 15 = -5$

Step 1. $15x - 5 + 15 = -5$ Apply the distributive property.

Step 2. $15x + 10 = -5$ Combine like terms.

Step 3. $15x + 10 - 10 = -5 - 10$ Subtract 10 from both sides.

$15x = -15$

Step 4. $\dfrac{15x}{15} = \dfrac{-15}{15}$ Divide both sides by 15.

$x = -1$

Step 5. Check to see that -1 is the solution.

Section 11.5   Equations and Problem Solving

PROBLEM-SOLVING STEPS

1. UNDERSTAND the problem. Some ways of doing this are:

Read and reread the problem.

Construct a drawing.

Assign a variable to an unknown in the problem.

The incubation period for a golden eagle is three times the incubation period for a hummingbird. If the total of their incubation periods is 60 days, find the incubation period for each bird. (*Source: Wildlife Fact File*, International Masters Publishers)

1. UNDERSTAND the problem. Then assign a variable. Let

$x =$ incubation period of a hummingbird

$3x =$ incubation period of a golden eagle

Section 11.5   Equations and Problem Solving *(continued)*

2. TRANSLATE the problem into an equation.

2. TRANSLATE.

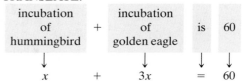

$$x \quad + \quad 3x \quad = \quad 60$$

3. SOLVE the equation.

3. SOLVE:

$$x + 3x = 60$$

$$4x = 60$$

$$\frac{\overset{1}{\cancel{4}}x}{\underset{1}{\cancel{4}}} = \frac{60}{4}$$

$$x = 15$$

4. INTERPRET the results. *Check* the proposed solution in the stated problem and *state* your conclusion.

4. INTERPRET the solution in the stated problem. The incubation period for the hummingbird is 15 days. The incubation period for the golden eagle is $3x = 3 \cdot 15 = 45$ days.

Since 15 days + 45 days = 60 days and 45 is 3(15), the solution checks.

State your conclusion: The incubation period for the hummingbird is 15 days. The incubation period for the golden eagle is 45 days.

Are you prepared for your final exam?

To prepare for your final exam, try the following study techniques.

- Review the material that you will be responsible for on your exam. Also check your notebook for any lecture notes that you highlighted.

- Review any formulas that you may need to memorize.

- Check to see if your instructor or math department will be conducting a final exam review.

- Check with your instructor to see whether there are final exams from previous semesters/quarters that are available to students for study.

- Use your previously taken tests as a practice final exam. To do so, rewrite the test questions in mixed order on blank sheets of paper. This will help you prepare for exam conditions.

- If you are unsure of a few topics, see your instructor or visit a learning lab for further assistance. Also, viewing the video segment of a troublesome section will help.

- If you need further exercises to work, try the chapter tests at the end of appropriate chapters.

Good luck! I hope you have enjoyed this textbook and your basic mathematics skills course.

Chapter 11 Review

(11.1) *Evaluate each expression when* $x = 5$, $y = 0$, *and* $z = -2$.

1. $\dfrac{2x}{z}$

2. $4x - 3$

3. $\dfrac{x + 7}{y}$

4. $\dfrac{y}{5x}$

5. $x^3 - 2z$

6. $\dfrac{7 + x}{3z}$

7. $(y + z)^2$

8. $\dfrac{100}{x} + \dfrac{y}{3}$

△ **9.** Find the volume of a storage cube whose sides measure 2 feet. Use $V = s^3$.

2 feet

△ **10.** Find the volume of a wooden crate in the shape of a cube 4 feet on each side. Use $V = s^3$.

4 feet

11. Lamar deposited his $5000 bonus into an account paying 6% annual interest. How much interest will he earn in 6 years? Use $I = prt$.

12. Jennifer Lewis borrowed $2000 from her grandmother and agreed to pay her 5% simple interest. How much interest will she owe after 3 years? Use $I = prt$.

Simplify each expression by combining like terms.

13. $3y + 7y - 15$

14. $2y - 10 - 8y$

15. $8a + a - 7 - 15a$

16. $y + 3 - 9y - 1$

17. $1.7x - 3.2 + 2.9x - 8.7$

18. $3.6x - 10.2 - 5.7x - 9.8$

Multiply.

19. $-2(x + 5)$

20. $-3(y + 8)$

Simplify.

21. $7x + 3(x - 4) + x$

22. $10 - 2(m - 3) - m$

23. $3(5a - 2) - 20a + 10$

24. $6y + 3 + 2(3y - 6)$

Find the area of each figure.

△ **25.**

(2x − 1) yards

3 yards | Rectangle

△ **26.**

5y meters

Square

(11.2)

27. Is 4 a solution of $5(2 - x) = -10$?

28. Is 0 a solution of $6y + 2 = 23 + 4y$?

Solve.

29. $z - 5 = -7$

30. $x + 1 = 8$

31. $x + \dfrac{7}{8} = \dfrac{3}{8}$

32. $y + \dfrac{4}{11} = -\dfrac{2}{11}$

33. $c - 5 = -13 + 7$

34. $7x + 5 - 6x = -20$

35. $n + 18 = 10 - (-2)$

36. $15 = 8x + 35 - 7x$

37. $m - 3.9 = -2.6$

38. $z - 4.6 = -2.2$

(11.3) *Solve.*

39. $-3y = -21$

40. $-8x = 72$

41. $-5n = -5$

42. $-3a = 15$

43. $\dfrac{2}{3}x = -\dfrac{8}{15}$

44. $-\dfrac{7}{8}y = 21$

45. $-1.2x = 144$

46. $-0.8y = -10.4$

47. $-5x = 100 - 120$

48. $18 - 30 = -4x$

49. $-7x + 3x = -50 - 2$

50. $-x + 8x = -38 - 4$

(11.4) *Solve.*

51. $3x - 4 = 11$

52. $6y + 1 = 73$

53. $14 - y = -3$

54. $7 - z = 0$

55. $-\dfrac{5}{9}x + 23 = -12$

56. $-\dfrac{4}{3}x - 11 = -55$

57. $6.8 + 4y = -2.2$

58. $-9.6 + 5y = -3.1$

59. $5(n - 3) = 7 + 3n$

60. $7(2 + x) = 4x - 1$

61. $2x + 7 = 6x - 1$

62. $5x - 18 = -4x + 36$

Write each sentence as an equation.

63. The difference of 20 and -8 is 28.

64. The product of 5 and the sum of 2 and -6 yields -20.

65. The quotient of −75 and the sum of 5 and 20 is equal to −3.

66. Nineteen subtracted from −2 amounts to −21.

(11.5)   *Write each phrase as an algebraic expression. Use x to represent "a number."*

67. Eleven added to twice a number

68. The product of −5 and a number decreased by 50

69. The quotient of 70 and the sum of a number and 6

70. Twice the difference of a number and 13

Write each sentence as an equation using x as the variable.

71. Twice a number minus 8 is 40.

72. Twelve subtracted from the quotient of a number and 2 is 10.

73. The difference of a number and 3 is the quotient of the number and 4.

74. The product of some number and 6 is equal to the sum of the number and 2.

Solve.

75. Five times a number subtracted from 40 is the same as three times the number. Find the number.

76. The product of a number and 3 is twice the difference of that number and 8. Find the number.

77. In an election the incumbent received 14,000 votes of the 18,500 votes cast. Of the remaining votes, the Democratic candidate received 272 more than the Independent candidate. Find how many votes the Democratic candidate received.

78. Rajiv Puri has twice as many cassette tapes as he has compact discs. Find the number of CDs if he has a total of 126 music recordings.

Chapter 11 Test

1. Evaluate: $\dfrac{3x - 5}{2y}$ when $x = 7$ and $y = -8$

2. Simplify $7x - 5 - 12x + 10$ by combining like terms.

3. Multiply: $-2(3y + 7)$

4. Simplify: $5(3z + 2) - z - 18$

△ **5.** Find the area.

3 meters

Rectangle | $(3x - 1)$ meters

Solve.

6. $x - 17 = -10$

7. $y + \dfrac{3}{4} = \dfrac{1}{4}$

8. $-4x = 48$

9. $-\dfrac{5}{8}x = -25$

10. $5x + 12 - 4x - 14 = 22$

11. _____

12. _____

13. _____

14. _____

15. _____

16. _____

17. _____

18. _____

19. _____

20. _____

21. _____

22. a. _____

b. _____

23. _____

24. _____

25. _____

11. $2 - c + 2c = 5$ **12.** $3x - 5 = -11$ **13.** $-4x + 7 = 15$

14. $3.6 - 2x = -5.4$ **15.** $3(4 + 2y) = 12$ **16.** $5x - 2 = x - 10$

17. $10y - 1 = 7y + 20$ **18.** $6 + 2(3n - 1) = 28$ **19.** $4(5x + 3) = 2(7x + 6)$

Solve.

△ **20.** A lawn is in the shape of a trapezoid with a height of 60 feet and bases of 70 feet and 130 feet. Find the area of the lawn. Use
$$A = \frac{1}{2} \cdot h \cdot (B + b).$$

△ **21.** If the height of a triangularly shaped jib sail is 12 feet and its base is 5 feet, find the area of the sail. Use
$$A = \frac{1}{2} \cdot b \cdot h.$$

22. Translate the following phrases into mathematical expressions. Use x to represent "a number."
   a. The product of a number and 17
   b. Twice a number subtracted from 20

23. The difference of three times a number and five times the same number is 4. Find the number.

24. In a championship basketball game, Paula Zimmerman made twice as many points as Maria Kaminsky. If the total number of points made by both women was 51, find how many free throws Paula made.

25. In a 10-kilometer race, there are 112 more men entered than women. Find the number of women runners if the total number of runners in the race is 600.

Name _____ Section _____ Date _____

Cumulative Review

1. Multiply: 0.0531 × 16

2. Given the rectangle shown:

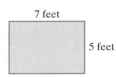

7 feet

5 feet

    a. Find the ratio of its width to its length.
    b. Find the ratio of its length to its perimeter.

3. 12% of what number is 0.6?

4. What percent of 12 is 9?

5. Convert 3 pounds to ounces.

6. Divide 18.08 ml by 16.

7. Identify each figure as a line, a ray, a line segment, or an angle.

a.

D
C

b.
E   F

c.
M
N
O

d. P
T

△ **8.** Find the diameter of the circle.

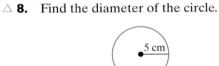

5 cm

△ **9.** Find the perimeter of the room shown below.

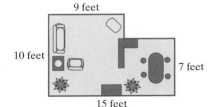

9 feet

10 feet

7 feet

15 feet

△ **10.** Find the area of the triangle.

8 centimeters

14 centimeters

Answers

1. _____

2. a. _____

   b. _____

3. _____

4. _____

5. _____

6. _____

7. a. _____

   b. _____

   c. _____

   d. _____

8. _____

9. _____

10. _____

11. _____

12. _____

13. a. _____

    b. _____

14. _____

15. _____

16. _____

17. _____

18. _____

19. _____

20. _____

21. _____

22. _____

23. _____

24. _____

25. _____

714

11. Simplify: $3\sqrt{81} - \sqrt{4}$

12. Mel is a 6-foot-tall park ranger who needs to know the height of a particular tree. He notices that when the shadow of the tree is 69 feet long, his own shadow is 9 feet long. Find the height of the tree.

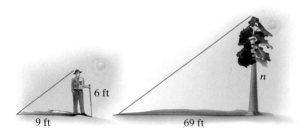

13. The following bar graph shows the number of endangered species in 2001. Use this graph to answer the questions.

How Many U.S. Animal Species Are Endangered?

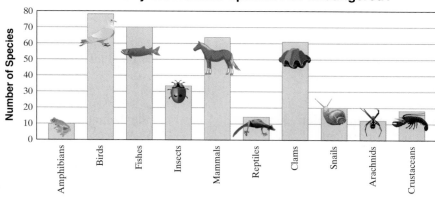

Source: U.S. Fish and Wildlife Service

  a. Approximate the number of endangered species that are reptiles.
  b. Which category has the most endangered species?

Insert $<$ or $>$ between each pair of numbers to make a true statement.

14. $-2.9 \quad 1$

15. $-9 \quad -11$

Find the absolute value.

16. $|-2|$

17. $|1.2|$

Add.

18. $-5 + (-1)$

19. $-\dfrac{3}{8} + \left(-\dfrac{1}{8}\right)$

Subtract.

20. $8 - 15$

21. $-4 - (-5)$

Multiply.

22. $-2(-5)$

23. $\left(-\dfrac{1}{2}\right)\left(-\dfrac{2}{3}\right)$

24. Simplify: $(-3) \cdot |-5| - (-2) + 4^2$

25. Solve: $3(2x - 6) + 6 = 0$

Addition Table and One Hundred Addition Facts

+	0	1	2	3	4	5	6	7	8	9
0	0	1	2	3	4	5	6	7	8	9
1	1	2	3	4	5	6	7	8	9	10
2	2	3	4	5	6	7	8	9	10	11
3	3	4	5	6	7	8	9	10	11	12
4	4	5	6	7	8	9	10	11	12	13
5	5	6	7	8	9	10	11	12	13	14
6	6	7	8	9	10	11	12	13	14	15
7	7	8	9	10	11	12	13	14	15	16
8	8	9	10	11	12	13	14	15	16	17
9	9	10	11	12	13	14	15	16	17	18

APPENDIX A: ONE HUNDRED ADDITION FACTS

Knowledge of the basic addition facts found above is an important prerequisite for a course in prealgebra. Study the table above and then perform the additions. Check your answers either by comparing them with those found in the back-of-the-book answer section or by using the table. Review any facts that you missed.

1.   1   **2.**   5   **3.**   2   **4.**   7   **5.**   3   **6.**   6   **7.**   4   **8.**   0   **9.**   9   **10.** 8
   +4   +6   +3   +8   +9   +1   +4   +6   +5   +2

11. 5   **12.** 3   **13.** 5   **14.** 1   **15.** 8   **16.** 6   **17.** 2   **18.** 3   **19.** 9   **20.** 5
   +7   +2   +5   +1   +1   +6   +9   +5   +9   +2

21.	6 +4	22.	0 +0	23.	1 +9	24.	3 +7	25.	9 +8	26.	0 +8	27.	4 +9	28.	3 +0	29.	7 +5	30.	8 +9

31. 9 +7   32. 2 +6   33. 4 +3   34. 8 +5   35. 3 +1   36. 0 +3   37. 7 +1   38. 3 +4   39. 8 +0   40. 6 +3

41. 2 +4   42. 0 +9   43. 8 +8   44. 5 +3   45. 3 +6   46. 6 +9   47. 4 +8   48. 0 +1   49. 2 +5   50. 6 +0

51. 2 +0   52. 4 +2   53. 8 +3   54. 7 +4   55. 1 +7   56. 4 +6   57. 0 +5   58. 9 +1   59. 8 +6   60. 5 +1

61. 6 +7   62. 4 +0   63. 1 +6   64. 4 +5   65. 0 +7   66. 5 +8   67. 7 +6   68. 7 +0   69. 4 +1   70. 5 +4

71. 0 +4   72. 1 +2   73. 7 +9   74. 3 +8   75. 7 +7   76. 9 +4   77. 1 +0   78. 4 +7   79. 2 +2   80. 1 +3

81. 2 +8   82. 5 +9   83. 6 +2   84. 9 +6   85. 5 +0   86. 8 +7   87. 7 +3   88. 0 +2   89. 9 +2   90. 3 +3

91. 9 +3   92. 1 +5   93. 2 +7   94. 6 +5   95. 7 +2   96. 1 +8   97. 6 +8   98. 8 +4   99. 9 +0   100. 2 +1

APPENDIX B

Multiplication Table and One Hundred Multiplication Facts

×	1	2	3	4	5	6	7	8	9
1	1	2	3	4	5	6	7	8	9
2	2	4	6	8	10	12	14	16	18
3	3	6	9	12	15	18	21	24	27
4	4	8	12	16	20	24	28	32	36
5	5	10	15	20	25	30	35	40	45
6	6	12	18	24	30	36	42	48	54
7	7	14	21	28	35	42	49	56	63
8	8	16	24	32	40	48	56	64	72
9	9	18	27	36	45	54	63	72	81

APPENDIX B: ONE HUNDRED MULTIPLICATION FACTS

Knowledge of the basic multiplication facts found above is an important prerequisite for a course in prealgebra. Study the table above and then perform the multiplications. Check your answers either by comparing them with those found in the back-of-the-book answer section or by using the table. Review any facts that you missed.

1. $\begin{array}{r} 1 \\ \times 1 \\ \hline \end{array}$ **2.** $\begin{array}{r} 5 \\ \times 7 \\ \hline \end{array}$ **3.** $\begin{array}{r} 7 \\ \times 8 \\ \hline \end{array}$ **4.** $\begin{array}{r} 3 \\ \times 3 \\ \hline \end{array}$ **5.** $\begin{array}{r} 8 \\ \times 4 \\ \hline \end{array}$ **6.** $\begin{array}{r} 9 \\ \times 5 \\ \hline \end{array}$ **7.** $\begin{array}{r} 4 \\ \times 7 \\ \hline \end{array}$ **8.** $\begin{array}{r} 7 \\ \times 1 \\ \hline \end{array}$ **9.** $\begin{array}{r} 2 \\ \times 2 \\ \hline \end{array}$ **10.** $\begin{array}{r} 0 \\ \times 5 \\ \hline \end{array}$

11. $\begin{array}{r} 9 \\ \times 7 \\ \hline \end{array}$ **12.** $\begin{array}{r} 8 \\ \times 8 \\ \hline \end{array}$ **13.** $\begin{array}{r} 3 \\ \times 2 \\ \hline \end{array}$ **14.** $\begin{array}{r} 6 \\ \times 0 \\ \hline \end{array}$ **15.** $\begin{array}{r} 5 \\ \times 6 \\ \hline \end{array}$ **16.** $\begin{array}{r} 2 \\ \times 5 \\ \hline \end{array}$ **17.** $\begin{array}{r} 4 \\ \times 6 \\ \hline \end{array}$ **18.** $\begin{array}{r} 0 \\ \times 7 \\ \hline \end{array}$ **19.** $\begin{array}{r} 6 \\ \times 3 \\ \hline \end{array}$ **20.** $\begin{array}{r} 8 \\ \times 9 \\ \hline \end{array}$

21. 5 ×8	22. 7 ×2	23. 4 ×8	24. 1 ×2	25. 9 ×6	26. 3 ×1	27. 8 ×7	28. 2 ×8	29. 6 ×9	30. 5 ×5
31. 2 ×1	32. 8 ×0	33. 4 ×9	34. 8 ×3	35. 6 ×2	36. 4 ×5	37. 9 ×4	38. 2 ×9	39. 3 ×4	40. 1 ×6
41. 8 ×6	42. 9 ×8	43. 1 ×8	44. 5 ×1	45. 9 ×0	46. 7 ×4	47. 9 ×3	48. 0 ×3	49. 3 ×5	50. 6 ×8
51. 5 ×9	52. 2 ×6	53. 1 ×0	54. 3 ×9	55. 9 ×9	56. 5 ×4	57. 0 ×6	58. 1 ×9	59. 5 ×0	60. 6 ×1
61. 9 ×2	62. 1 ×7	63. 1 ×3	64. 7 ×3	65. 6 ×6	66. 4 ×0	67. 7 ×9	68. 4 ×3	69. 7 ×5	70. 2 ×0
71. 6 ×7	72. 0 ×8	73. 8 ×5	74. 2 ×4	75. 0 ×1	76. 3 ×8	77. 9 ×1	78. 7 ×0	79. 5 ×3	80. 4 ×4
81. 1 ×5	82. 6 ×5	83. 3 ×0	84. 1 ×4	85. 3 ×7	86. 4 ×2	87. 0 ×2	88. 7 ×7	89. 8 ×2	90. 6 ×4
91. 0 ×0	92. 2 ×7	93. 4 ×1	94. 0 ×4	95. 2 ×3	96. 8 ×1	97. 3 ×6	98. 5 ×2	99. 0 ×9	100. 7 ×6

APPENDIX C

Review of Geometric Figures

Plane Figures Have Length and Width but No Thickness or Depth.		
Name	**Description**	**Figure**
Polygon	Union of three or more coplanar line segments that intersect with each other only at each end point, with each end point shared by two segments.	
Triangle	Polygon with three sides (sum of measures of three angles is 180°).	
Scalene Triangle	Triangle with no sides of equal length.	
Isosceles Triangle	Triangle with two sides of equal length.	
Equilateral Triangle	Triangle with all sides of equal length.	
Right Triangle	Triangle that contains a right angle.	leg hypotenuse leg
Quadrilateral	Polygon with four sides (sum of measures of four angles is 360°).	
Trapezoid	Quadrilateral with exactly one pair of opposite sides parallel.	base leg parallel sides leg base
Isosceles Trapezoid	Trapezoid with legs of equal length.	
Parallelogram	Quadrilateral with both pairs of opposite sides parallel.	
Rhombus	Parallelogram with all sides of equal length.	

(continued)

Name	Description	Figure
Rectangle	Parallelogram with four right angles.	
Square	Rectangle with all sides of equal length.	
Circle	All points in a plane the same distance from a fixed point called the **center**.	

Solid Figures Have Length, Width, and Height or Depth.		
Name	**Description**	**Figure**
Rectangular Solid	A solid with six sides, all of which are rectangles.	
Cube	A rectangular solid whose six sides are squares.	
Sphere	All points the same distance from a fixed point, called the **center**.	
Right Circular Cylinder	A cylinder having two circular bases that are perpendicular to its altitude.	
Right Circular Cone	A cone with a circular base that is perpendicular to its altitude.	

APPENDIX D

Table of Percents, Decimals, and Fraction Equivalents

Percent	Decimal	Fraction
1%	0.01	$\frac{1}{100}$
5%	0.05	$\frac{1}{20}$
10%	0.1	$\frac{1}{10}$
12.5% or $12\frac{1}{2}$%	0.125	$\frac{1}{8}$
$16.\overline{6}$% or $16\frac{2}{3}$%	$0.1\overline{6}$	$\frac{1}{6}$
20%	0.2	$\frac{1}{5}$
25%	0.25	$\frac{1}{4}$
30%	0.3	$\frac{3}{10}$
$33.\overline{3}$% or $33\frac{1}{3}$%	$0.\overline{3}$	$\frac{1}{3}$
37.5% or $37\frac{1}{2}$%	0.375	$\frac{3}{8}$
40%	0.4	$\frac{2}{5}$
50%	0.5	$\frac{1}{2}$
60%	0.6	$\frac{3}{5}$
62.5% or $62\frac{1}{2}$%	0.625	$\frac{5}{8}$
$66.\overline{6}$% or $66\frac{2}{3}$%	$0.\overline{6}$	$\frac{2}{3}$
70%	0.7	$\frac{7}{10}$
75%	0.75	$\frac{3}{4}$
80%	0.8	$\frac{4}{5}$
$83.\overline{3}$% or $83\frac{1}{3}$%	$08.\overline{3}$	$\frac{5}{6}$
87.5% or $87\frac{1}{2}$%	0.875	$\frac{7}{8}$
90%	0.9	$\frac{9}{10}$
100%	1.0	1
110%	1.1	$1\frac{1}{10}$
125%	1.25	$1\frac{1}{4}$
$133.\overline{3}$% or $133\frac{1}{3}$%	$1.\overline{3}$	$1\frac{1}{3}$
150%	1.5	$1\frac{1}{2}$
$166.\overline{6}$% or $166\frac{2}{3}$%	$1.\overline{6}$	$1\frac{2}{3}$
175%	1.75	$1\frac{3}{4}$
200%	2.0	2

APPENDIX E

Table of Squares and Square Roots

n	n^2	$\sqrt{n}$	n	n^2	$\sqrt{n}$
1	1	1.000	51	2601	7.141
2	4	1.414	52	2704	7.211
3	9	1.732	53	2809	7.280
4	16	2.000	54	2916	7.348
5	25	2.236	55	3025	7.416
6	36	2.449	56	3136	7.483
7	49	2.646	57	3249	7.550
8	64	2.828	58	3364	7.616
9	81	3.000	59	3481	7.681
10	100	3.162	60	3600	7.746
11	121	3.317	61	3721	7.810
12	144	3.464	62	3844	7.874
13	169	3.606	63	3969	7.937
14	196	3.742	64	4096	8.000
15	225	3.873	65	4225	8.062
16	256	4.000	66	4356	8.124
17	289	4.123	67	4489	8.185
18	324	4.243	68	4624	8.246
19	361	4.359	69	4761	8.307
20	400	4.472	70	4900	8.367
21	441	4.583	71	5041	8.426
22	484	4.690	72	5184	8.485
23	529	4.796	73	5329	8.544
24	576	4.899	74	5476	8.602
25	625	5.000	75	5625	8.660
26	676	5.099	76	5776	8.718
27	729	5.196	77	5929	8.775
28	784	5.292	78	6084	8.832
29	841	5.385	79	6241	8.888
30	900	5.477	80	6400	8.944
31	961	5.568	81	6561	9.000
32	1024	5.657	82	6724	9.055
33	1089	5.745	83	6889	9.110
34	1156	5.831	84	7056	9.165
35	1225	5.916	85	7225	9.220
36	1296	6.000	86	7396	9.274
37	1369	6.083	87	7569	9.327
38	1444	6.164	88	7744	9.381
39	1521	6.245	89	7921	9.434
40	1600	6.325	90	8100	9.487
41	1681	6.403	91	8281	9.539
42	1764	6.481	92	8464	9.592
43	1849	6.557	93	8649	9.644
44	1936	6.633	94	8836	9.695
45	2025	6.708	95	9025	9.747
46	2116	6.782	96	9216	9.798
47	2209	6.856	97	9409	9.849
48	2304	6.928	98	9604	9.899
49	2401	7.000	99	9801	9.950
50	2500	7.071	100	10,000	10.000

APPENDIX F

Compound Interest Table

Compounded Annually

	5%	6%	7%	8%	9%	10%	11%	12%	13%	14%	15%	16%	17%	18%
1 year	1.05000	1.06000	1.07000	1.08000	1.09000	1.10000	1.11000	1.12000	1.13000	1.14000	1.15000	1.16000	1.17000	1.18000
5 years	1.27628	1.33823	1.40255	1.46933	1.53862	1.61051	1.68506	1.76234	1.84244	1.92541	2.01136	2.10034	2.19245	2.28776
10 years	1.62889	1.79085	1.96715	2.15892	2.36736	2.59374	2.83942	3.10585	3.39457	3.70722	4.04556	4.41144	4.80683	5.23384
15 years	2.07893	2.39656	2.75903	3.17217	3.64248	4.17725	4.78459	5.47357	6.25427	7.13794	8.13706	9.26552	10.53872	11.97375
20 years	2.65330	3.20714	3.86968	4.66096	5.60441	6.72750	8.06231	9.64629	11.52309	13.74349	16.36654	19.46076	23.10560	27.39303

Compounded Semiannually

	5%	6%	7%	8%	9%	10%	11%	12%	13%	14%	15%	16%	17%	18%
1 year	1.05063	1.06090	1.07123	1.08160	1.09203	1.10250	1.11303	1.12360	1.13423	1.14490	1.15563	1.16640	1.17723	1.18810
5 years	1.28008	1.34392	1.41060	1.48024	1.55297	1.62889	1.70814	1.79085	1.87714	1.96715	2.06103	2.15892	2.26098	2.36736
10 years	1.63862	1.80611	1.98979	2.19112	2.41171	2.65330	2.91776	3.20714	3.52365	3.86968	4.24785	4.66096	5.11205	5.60441
15 years	2.09757	2.42726	2.80679	3.24340	3.74532	4.32194	4.98395	5.74349	6.61437	7.61226	8.75496	10.06266	11.55825	13.26768
20 years	2.68506	3.26204	3.95926	4.80102	5.81636	7.03999	8.51331	10.28572	12.41607	14.97446	18.04424	21.72452	26.13302	31.40942

Compounded Quarterly

	5%	6%	7%	8%	9%	10%	11%	12%	13%	14%	15%	16%	17%	18%
1 year	1.05095	1.06136	1.07186	1.08243	1.09308	1.10381	1.11462	1.12551	1.13648	1.14752	1.15865	1.16986	1.18115	1.19252
5 years	1.28204	1.34686	1.41478	1.48595	1.56051	1.63862	1.72043	1.80611	1.89584	1.98979	2.08815	2.19112	2.29891	2.41171
10 years	1.64362	1.81402	2.00160	2.20804	2.43519	2.68506	2.95987	3.26204	3.59420	3.95926	4.36038	4.80102	5.28497	5.81636
15 years	2.10718	2.44322	2.83182	3.28103	3.80013	4.39979	5.09225	5.89160	6.81402	7.87809	9.10513	10.51963	12.14965	14.02741
20 years	2.70148	3.29066	4.00639	4.87544	5.93015	7.20957	8.76085	10.64089	12.91828	15.67574	19.01290	23.04980	27.93091	33.83010

Compounded Daily

	5%	6%	7%	8%	9%	10%	11%	12%	13%	14%	15%	16%	17%	18%
1 year	1.05127	1.06183	1.07250	1.08328	1.09416	1.10516	1.11626	1.12747	1.13880	1.15024	1.16180	1.17347	1.18526	1.19716
5 years	1.28400	1.34983	1.41902	1.49176	1.56823	1.64861	1.73311	1.82194	1.91532	2.01348	2.11667	2.22515	2.33918	2.45906
10 years	1.64866	1.82203	2.01362	2.22535	2.45933	2.71791	3.00367	3.31946	3.66845	4.05411	4.48031	4.95130	5.47178	6.04696
15 years	2.11689	2.45942	2.85736	3.31968	3.85678	4.48077	5.20569	6.04786	7.02625	8.16288	9.48335	11.01738	12.79950	14.86983
20 years	2.71810	3.31979	4.05466	4.95216	6.04831	7.38703	9.02202	11.01883	13.45751	16.43582	20.07316	24.51533	29.94039	36.56577

Answers to Selected Exercises

Chapter 1 THE WHOLE NUMBERS

CHAPTER 1 PRETEST
1. hundreds; 1.2A   **2.** twenty-three thousand, four hundred ninety; 1.2B   **3.** 87; 1.3A   **4.** 3717; 1.6B   **5.** 626; 1.4A   **6.** 32; 1.4B
7. 136 pages; 1.4C   **8.** 9050; 1.5A   **9.** 3100; 1.5B   **10.** $9 \cdot 3 + 9 \cdot 11$; 1.6A   **11.** 25 in.; 1.3B   **12.** 184 sq yd; 1.6C   **13.** 576 seats; 1.6D
14. 243; 1.7A   **15.** 446 R9; 1.7B   **16.** 39; 1.7D   **17.** $9880; 1.8A   **18.** 9^7; 1.9A   **19.** 2401; 1.9B   **20.** 39; 1.9C

EXERCISE SET 1.2
1. tens   **3.** thousands   **5.** hundred-thousands   **7.** millions   **9.** five thousand, four hundred twenty   **11.** twenty-six thousand, nine
hundred ninety   **13.** one million, six hundred twenty thousand   **15.** fifty-three million, five hundred twenty thousand, one hundred seventy
17. sixty-two thousand, nine hundred ninety-seven   **19.** one thousand, four hundred eighty-three   **21.** thirteen million, six hundred thousand
23. 6508   **25.** 29,900   **27.** 6,504,019   **29.** 3,000,014   **31.** 755   **33.** $1,400,000   **35.** 1244   **37.** 400 + 6   **39.** 5000 + 200 + 90
41. 60,000 + 2000 + 400 + 7   **43.** 30,000 + 600 + 80   **45.** 30,000,000 + 9,000,000 + 600,000 + 80,000   **47.** five thousand, five hundred
thirty-two   **49.** 5000 + 400 + 90 + 2   **51.** Mt. Washington   **53.** Labrador retriever; one hundred seventy-two thousand, eight hundred
forty-one   **55.** Golden retriever   **57.** 7632   **59.** answers may vary   **61.** Canton

CALCULATOR EXPLORATIONS
1. 134   **3.** 340   **5.** 2834

MENTAL MATH
1. 12   **3.** 9000   **5.** 1620

EXERCISE SET 1.3
1. 36   **3.** 92   **5.** 49   **7.** 5399   **9.** 117   **11.** 71   **13.** 117   **15.** 25   **17.** 62   **19.** 212   **21.** 94   **23.** 910   **25.** 8273
27. 11,926   **29.** 1884   **31.** 16,717   **33.** 1110   **35.** 8999   **37.** 35,901   **39.** 612,389   **41.** 29 in.   **43.** 25 ft   **45.** 24 in.   **47.** 8 yd
49. 6684 ft   **51.** 340 ft   **53.** 4967 passes   **55.** 28,000 people   **57.** 1975   **59.** California   **61.** 312 stores   **63.** 595 stores
65. answers may vary   **67.** 166,510,192

CALCULATOR EXPLORATIONS
1. 770   **3.** 109   **5.** 8978

MENTAL MATH
1. 7   **3.** 5   **5.** 0   **7.** 400   **9.** 500

EXERCISE SET 1.4
1. 44   **3.** 60   **5.** 265   **7.** 254   **9.** 545   **11.** 600   **13.** 25   **15.** 45   **17.** 146   **19.** 288   **21.** 168   **23.** 6   **25.** 447   **27.** 5723
29. 504   **31.** 89   **33.** 79   **35.** 39,914   **37.** 32,711   **39.** 5041   **41.** 31,213   **43.** 4   **45.** 20   **47.** 7   **49.** 264 pages
51. 173 points   **53.** 6065 ft   **55.** $175   **57.** 358 mi   **59.** $389   **61.** 3,044,452 people   **63.** 29,393 cocker spaniels   **65.** Jo, by 271 votes
67. 5920 sq ft   **69.** 331 restaurants   **71.** Atlanta Hartsfield International   **73.** 272   **75.** DaimlerChrysler AG and General Motors Corp.
77. $243,770,900   **79.** $3,611,502,200   **81.** 5269 − 2385 = 2884   **83.** answers may vary

EXERCISE SET 1.5
1. 630   **3.** 640   **5.** 790   **7.** 400   **9.** 1100   **11.** 43,000   **13.** 248,700   **15.** 36,000   **17.** 100,000   **19.** 60,000,000   **21.** 5280; 5300; 5000
23. 9440; 9400; 9000   **25.** 14,880; 14,900; 15,000   **27.** 356,000   **29.** 38,000   **31.** 281,000,000   **33.** $2,300,000   **35.** 130   **37.** 380
39. 5500   **41.** 300   **43.** 8500   **45.** correct   **47.** incorrect   **49.** correct   **51.** correct   **53.** $3100   **55.** 80 mi   **57.** 6000 ft
59. 1,400,000 people   **61.** 14,000,000 votes   **63.** 107,000 children   **65.** $600,000,000   **67.** $224,000,000   **69.** 8550   **71.** 1,549,999
73. 21,900 mi

CALCULATOR EXPLORATIONS
1. 3456   **3.** 15,322   **5.** 272,291

MENTAL MATH
1. 24   **3.** 0   **5.** 0   **7.** 87

Exercise Set 1.6

1. $4 \cdot 3 + 4 \cdot 9$ **3.** $2 \cdot 4 + 2 \cdot 6$ **5.** $10 \cdot 11 + 10 \cdot 7$ **7.** 252 **9.** 1872 **11.** 1662 **13.** 5310 **15.** 4172 **17.** 10,857 **19.** 11,326 **21.** 24,800 **23.** 0 **25.** 5900 **27.** 59,232 **29.** 142,506 **31.** 1,821,204 **33.** 456,135 **35.** 64,790 **37.** 240,000 **39.** 300,000 **41.** 63 sq m **43.** 390 sq ft **45.** 375 cal **47.** $1890 **49.** 192 cans **51.** 9900 sq ft **53.** 495,864 sq m **55.** 5828 pixels **57.** 1500 characters **59.** 1280 cal **61.** 71,343 mi **63.** 506 windows **65.** 21,700,000 qts **67.** $5,148,000,000 **69.** 50 students **71.** apple and orange **73.** 2, 9 **75.** answers may vary **77.** 1938 points

Calculator Explorations

1. 53 **3.** 62 **5.** 261 **7.** 0

Mental Math

1. 5 **3.** 9 **5.** 0 **7.** 9 **9.** 1 **11.** 5 **13.** undefined **15.** 7 **17.** 0 **19.** 8

Exercise Set 1.7

1. 12 **3.** 37 **5.** 338 **7.** 16 R 2 **9.** 563 R 1 **11.** 37 R 1 **13.** 265 R 1 **15.** 49 **17.** 13 **19.** 97 R 40 **21.** 206 **23.** 506 **25.** 202 R 7 **27.** 45 **29.** 98 R 100 **31.** 202 R 15 **33.** 202 **35.** 58 students **37.** $252,000 **39.** 415 bushels **41.** 88 bridges **43.** yes, she needs 176 ft; she has 9 ft left over **45.** 26 touchdowns **47.** 1760 yd **49.** 26 **51.** 498 **53.** 79 **55.** 16° **57.** $2,277,489,500 **59.** increase; answers may vary **61.** no; answers may vary

Integrated Review

1. 148 **2.** 6555 **3.** 1620 **4.** 562 **5.** 79 **6.** undefined **7.** 9 **8.** 1 **9.** 0 **10.** 0 **11.** 0 **12.** 3 **13.** 2433 **14.** 9826 **15.** 213 R 3 **16.** 79,317 **17.** 27 **18.** 9 **19.** 138 **20.** 276 **21.** 1099 R 2 **22.** 111 R 1 **23.** 663 R 6 **24.** 1076 R 60 **25.** 1024 **26.** 9899 **27.** 30,603 **28.** 47,500 **29.** 0 **30.** undefined **31.** 0 **32.** 0 **33.** 86 **34.** 22 **35.** 8630; 8600; 9000 **36.** 1550; 1600; 2000 **37.** 10,900; 10,900; 11,000 **38.** 432,200; 432,200; 432,000 **39.** perimeter: 20 ft; area: 25 sq ft **40.** perimeter: 42 in.; area: 98 sq in. **41.** 26 mi **42.** 26 m

Exercise Set 1.8

1. 49 **3.** 237 **5.** 42 **7.** 600 **9.** 9600 sq ft **11.** $15,500 **13.** 168 hr **15.** 70,000 cu ft **17.** 129 yr **19.** 312 billion bricks **21.** $40,000,000 **23.** $16 **25.** 55 cal **27.** 2,641,152 tickets **29.** 38,034,000 students **31.** 3987 mi **33.** 13 paychecks **35.** $239 **37.** $1045 **39.** $12 **41.** yes, by $4 **43.** 296 guest rooms **45.** Fujitsu Limited **47.** 56 patents **49.** Motorola; 49 more

Calculator Explorations

1. 729 **3.** 1024 **5.** 2048 **7.** 2526 **9.** 4295 **11.** 8

Exercise Set 1.9

1. 3^4 **3.** 7^8 **5.** 12^3 **7.** $6^2 \cdot 5^3$ **9.** $9^3 \cdot 8$ **11.** $3 \cdot 2^5$ **13.** $3 \cdot 2^2 \cdot 5^3$ **15.** 25 **17.** 125 **19.** 64 **21.** 1024 **23.** 7 **25.** 243 **27.** 256 **29.** 64 **31.** 81 **33.** 729 **35.** 100 **37.** 10,000 **39.** 10 **41.** 1920 **43.** 729 **45.** 21 **47.** 8 **49.** 29 **51.** 4 **53.** 17 **55.** 46 **57.** 28 **59.** 10 **61.** 7 **63.** 4 **65.** 14 **67.** 72 **69.** 2 **71.** 35 **73.** 4 **75.** undefined **77.** 52 **79.** 44 **81.** 12 **83.** 13 **85.** 400 sq mi **87.** 64 sq cm **89.** 10,000 sq m **91.** $(2 + 3) \cdot 6 - 2$ **93.** $24 \div (3 \cdot 2) + 2 \cdot 5$ **95.** 1400 ft **97.** 6,384,814

Chapter 1 Review

1. hundreds **2.** ten millions **3.** five thousand, four hundred eighty **4.** forty-six million, two hundred thousand, one hundred twenty **5.** $6000 + 200 + 70 + 9$ **6.** $400,000,000 + 3,000,000 + 200,000 + 20,000 + 5000$ **7.** 59,800 **8.** 6,304,000,000 **9.** 1,630,553 **10.** 2,968,528 **11.** 531,341 **12.** 76,704 **13.** 13 **14.** 17 **15.** 3 **16.** 10 **17.** 38 **18.** 68 **19.** 56 **20.** 40 **21.** 110 **22.** 120 **23.** 950 **24.** 1250 **25.** 1711 **26.** 9867 **27.** 8032 mi **28.** $197,699 **29.** 276 ft **30.** 66 km **31.** 33 **32.** 43 **33.** 14 **34.** 33 **35.** 362 **36.** 65 **37.** 304 **38.** 476 **39.** 2114 **40.** 321 **41.** 397 pages **42.** $25,626 **43.** May **44.** August **45.** July, August, September **46.** April, May **47.** 90 **48.** 50 **49.** 470 **50.** 500 **51.** 4800 **52.** 58,000 **53.** 50,000,000 **54.** 800,000 **55.** 7400 **56.** 4100 **57.** 68,000,000 **58.** 680,000 **59.** 42 **60.** 24 **61.** 0 **62.** 0 **63.** 1410 **64.** 2898 **65.** 800 **66.** 900 **67.** 3696 **68.** 1694 **69.** 0 **70.** 0 **71.** 16,994 **72.** 8954 **73.** 113,634 **74.** 44,763 **75.** 411,426 **76.** 636,314 **77.** 1500 **78.** 240,000 **79.** 1,040,000 **80.** 7,020,000 **81.** 24 g **82.** $4,897,341 **83.** 60 sq mi **84.** 500 sq cm **85.** 3 **86.** 4 **87.** 6 **88.** 5 **89.** 5 R 2 **90.** 4 R 2 **91.** undefined **92.** 0 **93.** 1 **94.** 10 **95.** undefined **96.** 0 **97.** 15 **98.** 19 R 7 **99.** 24 R 2 **100.** 56 **101.** 1 R 17 **102.** 35 R 15 **103.** 500 **104.** 21 R 6 **105.** 506 **106.** 16 **107.** 199 R 8 **108.** 200 **109.** 458 ft **110.** 51 **111.** 25 **112.** 15 **113.** 100 **114.** 4 **115.** 27 boxes **116.** $192 **117.** 7 billion tablets **118.** 75¢ **119.** 7^4 **120.** $6^2 \cdot 3^3$ **121.** $4 \cdot 2^3 \cdot 3^2$ **122.** $5^2 \cdot 7^3 \cdot 2^2$ **123.** 49 **124.** 64 **125.** 1125 **126.** 19,600 **127.** 13 **128.** 10 **129.** 3 **130.** 1 **131.** 32 **132.** 33 **133.** 49 sq m **134.** 9 sq in.

Chapter 1 Test

1. 141 **2.** 113 **3.** 14,880 **4.** 766 R 42 **5.** 200 **6.** 48 **7.** 98 **8.** 0 **9.** undefined **10.** 33 **11.** 21 **12.** 36 **13.** 52,000 **14.** 13,700 **15.** 1600 **16.** $17 **17.** $119 **18.** $126 **19.** 167 players **20.** 360 cal **21.** 20 cm, 25 sq cm **22.** 60 yd, 200 sq yd **23.** $233,630,478 **24.** $44,014,054 **25.** $171,034,409

Chapter 2   Multiplying and Dividing Fractions

Chapter 2 Pretest

1. $\frac{3}{8}$; 2.1B   **2. a.** $\frac{25}{7}$; 2.1D   **b.** $\frac{77}{8}$; 2.1D   **3. a.** 4; 2.1E   **b.** $11\frac{5}{6}$; 2.1E   **4.** 1, 2, 3, 4, 6, 9, 12, 18, 36; 2.2A   **5.** prime; 2.2B

6. composite; 2.2B   **7.** $2 \cdot 2 \cdot 2 \cdot 2 \cdot 5$ or $2^4 \cdot 5$; 2.2C   **8.** $\frac{7}{12}$; 2.3A   **9.** $\frac{1}{7}$; 2.3A   **10.** equal; 2.3B   **11.** not equal; 2.3B   **12.** $\frac{2}{3}$; 2.3C

13. $\frac{6}{35}$; 2.4A   **14.** $\frac{1}{21}$; 2.4A   **15.** 2; 2.4B   **16.** \$320; 2.4C   **17.** $\frac{19}{12}$; 2.5A   **18.** 6; 2.5B   **19.** $\frac{15}{8}$ or $1\frac{7}{8}$; 2.5C   **20.** 23 mi; 2.5 D

Mental Math

1. numerator = 1; denominator = 2   **3.** numerator = 10; denominator = 3   **5.** numerator = 3; denominator = 7

Exercise Set 2.1

1. 1   **3.** 0   **5.** 13   **7.** 0   **9.** 0   **11.** 1   **13.** undefined   **15.** 9   **17.** $\frac{1}{3}$ **19.** $\frac{4}{7}$ **21.** $\frac{7}{12}$ **23.** $\frac{3}{7}$ **25.** $\frac{4}{9}$

27. 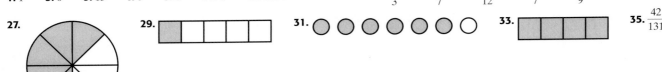   **29.**   **31.**   **33.**   **35.** $\frac{42}{131}$

37. $89; \frac{89}{131}$ **39.** $\frac{4}{10}$ **41.** $\frac{8}{43}$ **43. a.** $2\frac{3}{4}$ **b.** $\frac{11}{4}$ **45. a.** $3\frac{2}{3}$ **b.** $\frac{11}{3}$ **47. a.** $1\frac{1}{2}$ **b.** $\frac{3}{2}$ **49. a.** $1\frac{1}{3}$ **b.** $\frac{4}{3}$ **51. a.** $2\frac{5}{6}$ **b.** $\frac{17}{6}$

53. a. $1\frac{5}{9}$ **b.** $\frac{14}{9}$ **55.** $\frac{7}{3}$ **57.** $\frac{11}{3}$ **59.** $\frac{21}{8}$ **61.** $\frac{41}{15}$ **63.** $\frac{83}{7}$ **65.** $\frac{53}{8}$ **67.** $\frac{18}{5}$ **69.** $\frac{109}{24}$ **71.** $\frac{84}{13}$ **73.** $\frac{187}{20}$ **75.** $3\frac{2}{5}$

77. $3\frac{3}{13}$ **79.** $3\frac{2}{15}$ **81.** 33   **83.** 15   **85.** $4\frac{5}{8}$ **87.** $1\frac{1}{17}$ **89.** $10\frac{17}{23}$ **91.** $4\frac{2}{11}$ **93.** $66\frac{2}{3}$ **95.** 9   **97.** 125   **99.** 49   **101.** 24

103. $\frac{3043}{79}$ **105.** $\frac{5}{13}$ **107.** $\frac{55}{171}$ **109.** answers may vary

Exercise Set 2.2

1. 1, 2, 4, 8   **3.** 1, 5, 25   **5.** 1, 2, 4   **7.** 1, 2, 3, 6, 9, 18   **9.** 1, 7   **11.** 1, 2, 4, 5, 8, 10, 16, 20, 40, 80   **13.** 1, 2, 3, 4, 6, 12   **15.** 1, 2, 17, 34
17. prime   **19.** composite   **21.** composite   **23.** prime   **25.** composite   **27.** composite   **29.** prime   **31.** composite   **33.** $2^2 \cdot 3$
35. $3 \cdot 5$ **37.** $2^3 \cdot 5$ **39.** $2^2 \cdot 3^2$ **41.** $3 \cdot 13$ **43.** $2^4 \cdot 3$ **45.** $2 \cdot 3^3$ **47.** $2^2 \cdot 3 \cdot 5$ **49.** $2 \cdot 5 \cdot 11$ **51.** $2^3 \cdot 11$ **53.** 2^7 **55.** $2 \cdot 3 \cdot 5^2$
57. $2^2 \cdot 3 \cdot 5^2$ **59.** $2^4 \cdot 3 \cdot 5$ **61.** $3^3 \cdot 5 \cdot 7$ **63.** $2^2 \cdot 5^2 \cdot 7$ **65.** $2 \cdot 3^2 \cdot 7^2$ **67.** $7^2 \cdot 13$ **69.** 4300   **71.** 4,286,340   **73.** 55,300   **75.** 3500
77. 1000   **79.** $2^4 \cdot 3^2 \cdot 5^2$ **81.** $2^2 \cdot 3^5 \cdot 5 \cdot 7$ **83.** $3^4 \cdot 5^3 \cdot 13$ **85.** answers may vary

Exercise Set 2.3

1. $\frac{1}{4}$ **3.** $\frac{1}{5}$ **5.** $\frac{7}{8}$ **7.** $\frac{4}{5}$ **9.** $\frac{5}{6}$ **11.** $\frac{7}{8}$ **13.** $\frac{3}{7}$ **15.** $\frac{3}{5}$ **17.** $\frac{4}{7}$ **19.** $\frac{4}{5}$ **21.** $\frac{5}{8}$ **23.** $\frac{3}{10}$ **25.** $\frac{3}{2}$ or $1\frac{1}{2}$ **27.** $\frac{5}{8}$ **29.** $\frac{5}{14}$

31. $\frac{3}{14}$ **33.** $\frac{11}{17}$ **35.** $\frac{7}{8}$ **37.** $\frac{3}{14}$ **39.** $\frac{3}{5}$ **41.** equivalent   **43.** not equivalent   **45.** not equivalent   **47.** equivalent

49. equivalent   **51.** not equivalent   **53.** $\frac{1}{4}$ of a shift   **55.** $\frac{1}{2}$ mi   **57.** $\frac{29}{46}$ of individuals   **59. a.** $\frac{3}{10}$ **b.** 35 states   **c.** $\frac{7}{10}$ **61.** 364

63. 2322   **65.** 2520   **67.** false; $1 \cdot 1 = 1$ **69.** $\frac{3}{5}$ **71.** $\frac{9}{25}$ **73.** $\frac{1}{25}$ **75.** $\frac{3}{20}$

Integrated Review

1. $\frac{1}{4}$ **2.** $\frac{3}{6}$ **3.** $\frac{7}{4}$ **4.** $\frac{73}{85}$ **5.** 1   **6.** 17   **7.** 0   **8.** undefined   **9.** undefined   **10.** 1   **11.** 25   **12.** 0   **13.** $\frac{25}{8}$ **14.** $\frac{28}{5}$

15. $\frac{69}{7}$ **16.** $\frac{141}{7}$ **17.** $2\frac{6}{7}$ **18.** 5   **19.** $4\frac{7}{8}$ **20.** $8\frac{10}{11}$ **21.** 1, 5, 7, 35   **22.** 1, 2, 4, 5, 8, 10, 20, 40   **23.** 1, 2, 3, 4, 6, 8, 9, 12, 18, 24, 36, 72

24. 1, 13   **25.** $5 \cdot 13$ **26.** $2 \cdot 5 \cdot 7$ **27.** $2^2 \cdot 3^2 \cdot 7$ **28.** $3^2 \cdot 5 \cdot 7$ **29.** $3^2 \cdot 7^2$ **30.** $2 \cdot 11 \cdot 13$ **31.** $\frac{1}{7}$ **32.** $\frac{5}{6}$ **33.** $\frac{9}{19}$ **34.** $\frac{21}{55}$

35. $\frac{1}{2}$ **36.** $\frac{9}{10}$ **37.** $\frac{2}{5}$ **38.** $\frac{3}{8}$ **39.** $\frac{11}{14}$ **40.** $\frac{7}{11}$ **41.** not equivalent   **42.** equivalent   **43.** $\frac{1}{25}$ **44.** $\frac{146}{761}$

Exercise Set 2.4

1. $\frac{2}{15}$ **3.** $\frac{6}{35}$ **5.** $\frac{9}{80}$ **7.** $\frac{7}{12}$ **9.** $\frac{5}{28}$ **11.** $\frac{1}{15}$ **13.** $\frac{45}{32}$ or $1\frac{13}{32}$ **15.** $\frac{1}{70}$ **17.** 0 **19.** $\frac{18}{55}$ **21.** 0 **23.** $\frac{7}{2}$ or $3\frac{1}{2}$ **25.** $\frac{5}{16}$

27. $\frac{1}{110}$ **29.** $\frac{27}{80}$ **31.** $\frac{1}{8}$ **33.** $\frac{1}{56}$ **35.** $\frac{1}{56}$ **37.** $\frac{2}{105}$ **39.** 0 **41.** $\frac{1}{90}$ **43.** $\frac{3}{4}$ **45.** $\frac{5}{2}$ or $2\frac{1}{2}$ **47.** $\frac{1}{5}$ **49.** $\frac{5}{3}$ or $1\frac{2}{3}$ **51.** $\frac{2}{3}$

53. $\frac{77}{10}$ or $7\frac{7}{10}$ **55.** $\frac{836}{35}$ or $23\frac{31}{35}$ **57.** 6 **59.** $\frac{25}{2}$ or $12\frac{1}{2}$ **61.** 15 **63.** $\frac{45}{4}$ or $11\frac{1}{4}$ **65.** $\frac{49}{3}$ or $16\frac{1}{3}$ **67.** 0 **69.** $\frac{208}{7}$ or $29\frac{5}{7}$ **71.** $\frac{3}{2}$ or $1\frac{1}{2}$ in.

73. $\frac{55}{2}$ or $27\frac{1}{2}$ mi **75.** $\frac{17}{2}$ or $8\frac{1}{2}$ in. **77.** $\frac{8}{5}$ or $1\frac{3}{5}$ ft **79.** 600 cal **81.** \$1838 **83.** $\frac{39}{2}$ or $19\frac{1}{2}$ in. **85.** $\frac{1}{14}$ sq ft **87.** $\frac{7}{2}$ or $3\frac{1}{2}$ sq yd

89. $\frac{121}{4}$ or $30\frac{1}{4}$ sq in. **91.** 3840 mi **93.** 2400 mi **95.** 206 **97.** 56 R 12 **99.** 19,140,000 Americans **101.** answers may vary

Exercise Set 2.5

1. $\frac{7}{4}$ **3.** 11 **5.** $\frac{1}{15}$ **7.** $\frac{7}{12}$ **9.** $\frac{4}{5}$ **11.** $\frac{1}{6}$ **13.** $\frac{16}{9}$ or $1\frac{7}{9}$ **15.** $\frac{18}{35}$ **17.** $\frac{3}{4}$ **19.** $\frac{121}{60}$ or $2\frac{1}{60}$ **21.** $\frac{1}{100}$ **23.** $\frac{1}{3}$ **25.** $\frac{3}{4}$

27. $\frac{21}{20}$ or $1\frac{1}{20}$ **29.** $\frac{35}{36}$ **31.** $\frac{14}{37}$ **33.** $\frac{8}{45}$ **35.** 1 **37.** $\frac{11}{119}$ **39.** undefined **41.** $\frac{9}{5}$ or $1\frac{4}{5}$ **43.** 0 **45.** $\frac{7}{10}$ **47.** $\frac{1}{6}$

49. $\frac{40}{3}$ or $13\frac{1}{3}$ **51.** 5 **53.** $\frac{5}{28}$ **55.** $\frac{36}{35}$ or $1\frac{1}{35}$ **57.** 96 **59.** $\frac{35}{11}$ or $3\frac{2}{11}$ **61.** $\frac{26}{51}$ **63.** undefined **65.** $\frac{33}{50}$ **67.** 0 **69.** $\frac{35}{18}$ or $1\frac{17}{18}$

71. $\frac{19}{30}$ **73.** $\frac{5}{6}$ Tbsp **75.** $\frac{235}{34}$ or $6\frac{31}{34}$ gal **77.** \$5362 million **79.** 14 lb **81.** $\frac{625}{13}$ or $48\frac{1}{13}$ hr **83.** $4\frac{2}{3}$ m **85.** 201 **87.** 196
89. 1569 **91.** 640 **93.** 16 games

Chapter 2 Review

1. proper **2.** improper **3.** proper **4.** mixed number **5.** proper **6.** mixed number **7.** $\frac{2}{6}$ **8.** $\frac{4}{7}$ **9.** $\frac{7}{3}$ **10.** $\frac{9}{4}$ **11.** $\frac{11}{12}$

12. $\frac{23}{131}$ **13.** $3\frac{3}{4}$ **14.** 3 **15.** $45\frac{5}{6}$ **16.** $31\frac{1}{4}$ **17.** $\frac{6}{5}$ **18.** $\frac{26}{9}$ **19.** $\frac{47}{12}$ **20.** $\frac{95}{17}$ **21.** composite **22.** prime **23.** composite

24. composite **25.** 1, 2, 3, 6, 7, 14, 21, 42 **26.** 1, 2, 3, 5, 6, 10, 15, 30 **27.** $2^2 \cdot 17$ **28.** $2 \cdot 3^2 \cdot 5$ **29.** $5 \cdot 157$ **30.** $3 \cdot 5 \cdot 17$ **31.** $\frac{3}{7}$

32. $\frac{5}{9}$ **33.** $\frac{1}{3}$ **34.** $\frac{1}{2}$ **35.** $\frac{29}{32}$ **36.** $\frac{18}{23}$ **37.** $\frac{5}{3}$ or $1\frac{2}{3}$ **38.** $\frac{7}{5}$ or $1\frac{2}{5}$ **39.** 8 **40.** 6 **41.** $\frac{14}{15}$ **42.** $\frac{3}{5}$ **43.** $\frac{3}{10}$ **44.** $\frac{5}{14}$

45. $\frac{7}{12}$ **46.** $\frac{1}{4}$ **47.** 9 **48.** $\frac{1}{2}$ **49.** $\frac{35}{8}$ or $4\frac{3}{8}$ **50.** $\frac{5}{2}$ or $2\frac{1}{2}$ **51.** $\frac{5}{3}$ or $1\frac{2}{3}$ **52.** $\frac{49}{3}$ or $16\frac{1}{3}$ **53.** $\frac{13}{12}$ or $1\frac{1}{12}$ **54.** $\frac{15}{11}$ or $1\frac{4}{11}$ **55.** 10

56. $\frac{51}{4}$ or $12\frac{3}{4}$ **57.** $\frac{99}{4}$ or $24\frac{3}{4}$ **58.** $\frac{1}{6}$ **59.** 203 cal **60.** $\frac{40}{3}$ or $13\frac{1}{3}$ g **61.** $\frac{119}{80}$ or $1\frac{39}{80}$ sq in. **62.** $\frac{275}{8}$ or $34\frac{3}{8}$ sq m **63.** $\frac{1}{7}$ **64.** 8

65. $\frac{23}{14}$ **66.** $\frac{5}{17}$ **67.** 2 **68.** $\frac{15}{4}$ or $3\frac{3}{4}$ **69.** 9 **70.** $\frac{27}{2}$ or $13\frac{1}{2}$ **71.** $\frac{5}{6}$ **72.** $\frac{8}{3}$ or $2\frac{2}{3}$ **73.** $\frac{21}{4}$ or $5\frac{1}{4}$ **74.** $\frac{121}{46}$ or $2\frac{29}{46}$ **75.** $\frac{7}{3}$ or $2\frac{1}{3}$

76. $\frac{32}{5}$ or $6\frac{2}{5}$ **77.** \$200 **78.** $\frac{21}{20}$ or $1\frac{1}{20}$ mi

Chapter 2 Test

1. $\frac{4}{3}$ or $1\frac{1}{3}$ **2.** $\frac{4}{3}$ or $1\frac{1}{3}$ **3.** $\frac{1}{4}$ **4.** $\frac{16}{45}$ **5.** 16 **6.** $\frac{9}{2}$ or $4\frac{1}{2}$ **7.** $\frac{4}{11}$ **8.** 9 **9.** $\frac{64}{3}$ or $21\frac{1}{3}$ **10.** $\frac{45}{2}$ or $22\frac{1}{2}$ **11.** $\frac{18}{5}$ or $3\frac{3}{5}$

12. $\frac{20}{3}$ or $6\frac{2}{3}$ **13.** $\frac{23}{3}$ **14.** $\frac{39}{11}$ **15.** $4\frac{3}{5}$ **16.** $18\frac{3}{4}$ **17.** $\frac{34}{27}$ or $1\frac{7}{27}$ sq mi **18.** 24 mi **19.** $\frac{4}{35}$ **20.** $\frac{3}{5}$ **21.** 8800 sq yd
22. $2^3 \cdot 5 \cdot 7$ **23.** $2^2 \cdot 3 \cdot 7$ **24.** \$90 per share

Cumulative Review

1. ten-thousands; Sec. 1.2, Ex. 1 **2.** 805; Sec. 1.2, Ex. 8 **3.** 184,046; Sec. 1.3, Ex. 2 **4.** 13 in.; Sec. 1.3, Ex. 5 **5.** \$120,171; Sec. 1.3, Ex. 7
6. 7321; Sec. 1.4, Ex. 2 **7. a.** R **b.** 3; Sec. 1.4, Ex. 7 **8.** 570; Sec. 1.5, Ex. 1 **9.** 1800; Sec 1.5, Ex. 5 **10. a.** 6 **b.** 0 **c.** 45
d. 0; Sec. 1.6, Ex. 1 **11. a.** $3 \cdot 4 + 3 \cdot 5$ **b.** $10 \cdot 6 + 10 \cdot 8$ **c.** $2 \cdot 7 + 2 \cdot 3$; Sec. 1.6, Ex. 2 **12. a.** 0 **b.** 0 **c.** 0
d. undefined; Sec. 1.7, Ex. 3 **13.** 208; Sec. 1.7, Ex. 5 **14.** 7 boxes; Sec. 1.7, Ex. 11 **15.** 40 ft; Sec. 1.8, Ex. 5 **16.** 4^3; Sec. 1.9, Ex. 1
17. $6^3 \cdot 8^5$; Sec. 1.9, Ex. 4 **18.** 7; Sec. 1.9, Ex. 9 **19.** $\frac{3}{4}$; Sec. 2.1, Ex. 4 **20. a.** $\frac{38}{9}$ **b.** $\frac{19}{11}$; Sec. 2.1, Ex. 8 **21.** 1, 2, 4, 5, 10, 20; Sec. 2.2, Ex. 1
22. $\frac{7}{11}$; Sec. 2.3, Ex. 2 **23.** $\frac{35}{12}$ or $2\frac{11}{12}$; Sec. 2.4, Ex. 7 **24.** $\frac{3}{1}$ or 3; Sec. 2.5, Ex. 3 **25.** $\frac{5}{12}$; Sec. 2.5, Ex. 6

Chapter 3   ADDING AND SUBTRACTING FRACTIONS

CHAPTER 3 PRETEST

1. $\frac{8}{9}$; 3.1A   **2.** 1; 3.1A   **3.** $\frac{11}{12}$; 3.3A   **4.** $\frac{5}{12}$; 3.3A   **5.** $9\frac{7}{36}$; 3.4A   **6.** $\frac{5}{17}$; 3.1B   **7.** $\frac{1}{9}$; 3.1B   **8.** $\frac{29}{42}$; 3.3B   **9.** $\frac{2}{15}$; 3.3B

10. $5\frac{17}{30}$; 3.4B   **11.** 30; 3.2A, B   **12.** 360; 3.2A, B   **13.** $\frac{14}{35}$; 3.2C   **14.** $\frac{39}{27}$; 3.2C   **15.** $<$; 3.5A   **16.** $\frac{27}{125}$; 3.5B   **17.** $\frac{7}{18}$; 3.5D

18. $\frac{2}{5}$; 3.5D   **19.** $\frac{7}{8}$ lb; 3.3C   **20.** $5\frac{3}{8}$ ft; 3.4C   **21.** $6\frac{7}{8}$ c; 3.6A

MENTAL MATH

1. unlike   **3.** like   **5.** like   **7.** unlike

EXERCISE SET 3.1

1. $\frac{3}{7}$ **3.** $\frac{1}{5}$ **5.** $\frac{2}{3}$ **7.** $\frac{7}{20}$ **9.** $\frac{1}{2}$ **11.** $\frac{13}{11} = 1\frac{2}{11}$ **13.** $\frac{7}{13}$ **15.** $\frac{2}{3}$ **17.** $\frac{6}{11}$ **19.** $\frac{3}{5}$ **21.** 1   **23.** $\frac{3}{4}$ **25.** $\frac{5}{6}$

27. $\frac{4}{5}$ **29.** $\frac{19}{33}$ **31.** 1 in.   **33.** 2 m   **35.** $1\frac{1}{2}$ hr   **37.** $\frac{7}{10}$ of a mi   **39.** $\frac{12}{25}$ **41.** $\frac{9}{20}$ **43.** $\frac{21}{50}$ **45.** $2 \cdot 5$ **47.** 2^3 **49.** $5 \cdot 11$

51. $\frac{5}{8}$ **53.** $\frac{8}{11}$ **55.** $\frac{29}{50}$ **57.** answers may vary

EXERCISE SET 3.2

1. 12   **3.** 45   **5.** 36   **7.** 72   **9.** 126   **11.** 75   **13.** 24   **15.** 42   **17.** 150   **19.** 68   **21.** 588   **23.** 900   **25.** 357

27. 363   **29.** 216   **31.** 60   **33.** $\frac{20}{35}$ **35.** $\frac{14}{21}$ **37.** $\frac{10}{25}$ **39.** $\frac{15}{30}$ **41.** $\frac{30}{21}$ **43.** $\frac{21}{28}$ **45.** $\frac{30}{45}$ **47.** $\frac{36}{81}$ **49.** $\frac{12}{9}$

51. $\frac{18}{10}$ **53.** $\frac{5}{10} = \frac{1}{2}$ **55.** $\frac{2}{5}$ **57.** $\frac{8}{18} = \frac{4}{9}$ **59.** $\frac{9}{9} = 1$ **61.** $\frac{814}{3630}$ **63.** answers may vary

EXERCISE SET 3.3

1. $\frac{5}{6}$ **3.** $\frac{5}{6}$ **5.** $\frac{8}{33}$ **7.** $\frac{9}{14}$ **9.** $\frac{3}{5}$ **11.** $\frac{19}{36}$ **13.** $\frac{53}{60}$ **15.** $\frac{1}{6}$ **17.** $\frac{193}{280}$ **19.** $\frac{98}{143}$ **21.** $\frac{75}{56} = 1\frac{19}{56}$ **23.** $\frac{16}{11} = 1\frac{5}{11}$ **25.** $\frac{13}{27}$

27. $\frac{89}{60} = 1\frac{29}{60}$ **29.** $\frac{11}{16}$ **31.** $\frac{17}{42}$ **33.** $\frac{33}{56}$ **35.** $\frac{4}{33}$ **37.** $\frac{1}{35}$ **39.** $\frac{11}{36}$ **41.** $\frac{1}{20}$ **43.** $\frac{1}{84}$ **45.** $\frac{9}{1000}$ **47.** $\frac{1}{104}$ **49.** $\frac{34}{15} = 2\frac{4}{15}$ cm

51. $\frac{17}{10} = 1\frac{7}{10}$ m   **53. a.** $\frac{1}{4}$ **b.** 125 sheets   **55.** $\frac{49}{100}$ of students   **57.** $\frac{77}{100}$ of Americans   **59.** 5   **61.** $\frac{16}{29}$ **63.** $\frac{19}{3} = 6\frac{1}{3}$ **65.** $\frac{49}{44} = 1\frac{5}{44}$

67. answers may vary   **69.** $\frac{7}{19}$ **71.** 7,000,000 sq mi   **73.** $\frac{1}{60}$

INTEGRATED REVIEW

1. 30   **2.** 21   **3.** 14   **4.** 25   **5.** 100   **6.** 90   **7.** $\frac{9}{24}$ **8.** $\frac{28}{36}$ **9.** $\frac{10}{40}$ **10.** $\frac{12}{30}$ **11.** $\frac{55}{75}$ **12.** $\frac{40}{48}$ **13.** $\frac{1}{2}$ **14.** $\frac{2}{5}$

15. $\frac{13}{15}$ **16.** $\frac{7}{12}$ **17.** $\frac{3}{4}$ **18.** $\frac{2}{15}$ **19.** $\frac{17}{45}$ **20.** $\frac{19}{50}$ **21.** $\frac{37}{40}$ **22.** $\frac{11}{36}$ **23.** 0   **24.** $\frac{1}{17}$ **25.** $\frac{5}{33}$ **26.** $\frac{1}{42}$ **27.** $\frac{5}{18}$

28. $\frac{5}{13}$ **29.** $\frac{11}{18}$ **30.** $\frac{37}{50}$ **31.** $\frac{47}{30} = 1\frac{17}{30}$ **32.** $\frac{7}{30}$ **33.** $\frac{3}{5}$ **34.** $\frac{27}{20} = 1\frac{7}{20}$ **35.** $\frac{279}{350}$ **36.** $\frac{309}{350}$ **37.** $\frac{98}{5} = 19\frac{3}{5}$

38. $\frac{9}{250}$ **39.** $10\frac{1}{3}$ **40.** $1\frac{29}{64}$ **41.** $\frac{49}{54}$ **42.** $\frac{83}{48} = 1\frac{35}{48}$ **43.** $3\frac{87}{101}$ **44.** $2\frac{1}{72}$ **45.** $\frac{106}{135}$ **46.** $\frac{67}{224}$

EXERCISE SET 3.4

1. $6\frac{4}{5}$ **3.** $13\frac{11}{14}$ **5.** $17\frac{7}{25}$ **7.** $7\frac{5}{24}$ **9.** $7\frac{5}{8}$ **11.** $20\frac{1}{15}$ **13.** $28\frac{7}{12}$ **15.** $56\frac{53}{270}$ **17.** $13\frac{13}{24}$ **19.** $47\frac{53}{84}$ **21.** $2\frac{3}{5}$ **23.** $7\frac{5}{14}$

25. $\frac{24}{25}$ **27.** $5\frac{11}{14}$ **29.** $2\frac{7}{15}$ **31.** $23\frac{31}{72}$ **33.** $1\frac{4}{5}$ **35.** $15\frac{7}{8}$ **37.** $2\frac{1}{48}$ **39.** $1\frac{13}{15}$ **41.** $3\frac{5}{9}$ **43.** $15\frac{3}{4}$ **45.** $17\frac{9}{11}$ **47.** $17\frac{11}{12}$

49. $2\frac{3}{8}$ hr   **51.** $10\frac{1}{4}$ hr   **53.** $7\frac{13}{20}$ in.   **55.** no, she will be $\frac{1}{12}$ of a ft short   **57.** $10\frac{5}{8}$ lb   **59.** $352\frac{1}{3}$ yd   **61.** $92\frac{99}{100}$ m   **63.** $1\frac{4}{5}$ min

65. 7 mi   **67.** $21\frac{5}{24}$ m   **69.** 8   **71.** 25   **73.** 81   **75.** 64   **77.** Supreme is heavier by $\frac{1}{8}$ lb   **79.** answers may vary

EXERCISE SET 3.5

1. $>$ **3.** $<$ **5.** $<$ **7.** $>$ **9.** $>$ **11.** $<$ **13.** $<$ **15.** $>$ **17.** $\frac{1}{16}$ **19.** $\frac{8}{125}$ **21.** $\frac{64}{343}$ **23.** $\frac{4}{81}$ **25.** $\frac{1}{6}$

27. $\frac{18}{125}$ **29.** $\frac{11}{15}$ **31.** $\frac{3}{35}$ **33.** $\frac{5}{9}$ **35.** $10\frac{4}{99}$ **37.** $\frac{1}{12}$ **39.** $\frac{9}{11}$ **41.** 0 **43.** 0 **45.** $\frac{2}{5}$ **47.** $\frac{2}{77}$ **49.** $\frac{17}{60}$ **51.** $\frac{5}{8}$ **53.** $\frac{1}{2}$
55. $\frac{29}{10} = 2\frac{9}{10}$ **57.** $\frac{27}{32}$ **59.** $\frac{1}{81}$ **61.** $\frac{9}{64}$ **63.** $\frac{3}{4}$ **65.** $\frac{1}{4}$ **67.** $\frac{5}{6}$ **69.** $\frac{5}{6}$ **71.** $\frac{2}{3}$ **73.** standard mail **75.** savings account
77. no; answers may vary

EXERCISE SET 3.6

1. $\frac{5}{6}$ c **3.** $12\frac{1}{2}$ in. **5.** $21\frac{1}{2}$ mi per gal **7.** $1\frac{1}{2}$ yr **9.** $9\frac{2}{5}$ in. **11.** no; $\frac{1}{4}$ yd **13.** 5 pieces **15.** $\frac{9}{8}$ or $1\frac{1}{8}$ in. **17.** $3\frac{3}{4}$ c **19.** $1\frac{1}{4}$ min
21. $11\frac{1}{4}$ sq in. **23.** 67 sheets **25.** yes; $1\frac{1}{3}$ ft left over **27.** $2\frac{15}{16}$ lb **29.** area: $\frac{9}{128}$ sq in; perimeter: $1\frac{1}{8}$ in. **31.** $4\frac{3}{4}$ ft **33.** $\frac{5}{26}$ ft
35. answers may vary **37.** $485\frac{1}{3}$ cu ft **39.** $26\frac{8}{9}$ ft

CHAPTER 3 REVIEW

1. $\frac{10}{11}$ **2.** $\frac{2}{3}$ **3.** $\frac{1}{6}$ **4.** $\frac{1}{5}$ **5.** $\frac{2}{3}$ **6.** $\frac{1}{7}$ **7.** $\frac{3}{5}$ **8.** $\frac{3}{5}$ **9.** 1 **10.** 1 **11.** $\frac{19}{25}$ **12.** $\frac{16}{21}$ **13.** $\frac{1}{2}$ hr **14.** $\frac{3}{8}$ of a gal
15. $\frac{3}{4}$ of his homework **16.** $\frac{3}{2}$ mi $= 1\frac{1}{2}$ mi **17.** 55 **18.** 60 **19.** 120 **20.** 84 **21.** 252 **22.** 72 **23.** $\frac{56}{64}$ **24.** $\frac{20}{30}$ **25.** $\frac{21}{33}$
26. $\frac{20}{26}$ **27.** $\frac{16}{60}$ **28.** $\frac{25}{60}$ **29.** $\frac{11}{18}$ **30.** $\frac{7}{26}$ **31.** $\frac{7}{12}$ **32.** $\frac{11}{12}$ **33.** $\frac{27}{55}$ **34.** $\frac{7}{15}$ **35.** $\frac{17}{36}$ **36.** $\frac{1}{6}$ **37.** $\frac{23}{18} = 1\frac{5}{18}$ **38.** $\frac{3}{14}$
39. $2\frac{1}{9}$ m **40.** $1\frac{1}{2}$ ft **41.** $\frac{1}{4}$ of a yd **42.** $\frac{7}{10}$ has been cleaned **43.** $45\frac{16}{21}$ **44.** 60 **45.** $32\frac{13}{22}$ **46.** $3\frac{19}{60}$ **47.** $111\frac{5}{18}$ **48.** $20\frac{7}{24}$
49. $5\frac{16}{35}$ **50.** $3\frac{4}{55}$ **51.** $6\frac{7}{20}$ lb **52.** $44\frac{1}{2}$ yd **53.** $7\frac{4}{5}$ in. **54.** $11\frac{1}{6}$ ft **55.** 5 ft **56.** $\frac{1}{40}$ oz **57.** < **58.** > **59.** <
60. > **61.** > **62.** > **63.** $\frac{9}{49}$ **64.** $\frac{64}{125}$ **65.** $\frac{9}{400}$ **66.** $\frac{9}{100}$ **67.** $\frac{81}{196}$ **68.** $\frac{1}{7}$ **69.** $\frac{11}{25}$ **70.** $\frac{1}{8}$ **71.** $\frac{1}{27}$ **72.** $3\frac{3}{5}$
73. $1\frac{17}{28}$ **74.** $\frac{5}{6}$ **75.** 21 **76.** $15\frac{5}{8}$ acres **77.** $2\frac{2}{15}$ ft **78.** 5 recipes **79.** $4\frac{1}{4}$ in. **80.** $\frac{7}{10}$ yd **81.** perimeter: $1\frac{6}{11}$ mi; area: $\frac{3}{22}$ sq mi
82. perimeter: $2\frac{1}{3}$ m; area: $\frac{5}{16}$ sq m

CHAPTER 3 TEST

1. 60 **2.** 72 **3.** < **4.** < **5.** $\frac{8}{9}$ **6.** $\frac{2}{5}$ **7.** $\frac{13}{10} = 1\frac{3}{10}$ **8.** $\frac{8}{21}$ **9.** $\frac{13}{24}$ **10.** $\frac{1}{7}$ **11.** $\frac{67}{60} = 1\frac{7}{60}$ **12.** $\frac{7}{50}$ **13.** $\frac{3}{2} = 1\frac{1}{2}$
14. $14\frac{1}{40}$ **15.** $30\frac{13}{45}$ **16.** $1\frac{7}{24}$ **17.** $16\frac{8}{11}$ **18.** $\frac{5}{3} = 1\frac{2}{3}$ **19.** $\frac{16}{81}$ **20.** $\frac{10}{3} = 3\frac{1}{3}$ **21.** $\frac{153}{200}$ **22.** $\frac{3}{8}$ **23.** $3\frac{3}{4}$ ft **24.** 2368
25. $\frac{5}{16}$ **26.** 125,000 backpacks **27.** $9\frac{1}{2}$ m **28.** perimeter: $3\frac{1}{3}$ ft; area: $\frac{2}{3}$ sq ft **29.** $1\frac{2}{3}$ in.

CUMULATIVE REVIEW

1. eighty-five; Sec. 1.2, Ex. 4 **2.** one hundred twenty-six; Sec. 1.2, Ex. 5 **3.** 159; Sec. 1.3, Ex. 1 **4.** 14; Sec. 1.4, Ex. 3
5. 278,000; Sec. 1.5, Ex. 2 **6.** 20,296; Sec. 1.6, Ex. 4 **7. a.** 8 **b.** 11 **c.** 1 **d.** 1 **e.** 10 **f.** 1; Sec. 1.7, Ex. 2 **8.** 1038 mi; Sec. 1.8, Ex. 1
9. 64; 1.9, Ex. 5 **10.** 32; Sec. 1.9, Ex. 7 **11.** improper fraction: $\frac{4}{3}$; mixed number: $1\frac{1}{3}$; Sec. 2.1, Ex. 6
12. improper fraction: $\frac{5}{2}$; mixed number: $2\frac{1}{2}$; Sec. 2.1, Ex. 7 **13.** 3, 11, 17; Sec. 2.2, Ex. 2 **14.** $2^2 \cdot 3^2 \cdot 5$; Sec. 2.2, Ex. 4
15. $\frac{36}{13}$ or $2\frac{10}{13}$; Sec. 2.3, Ex. 5 **16.** not equivalent; Sec. 2.3, Ex. 8 **17.** $\frac{10}{33}$; Sec. 2.4, Ex. 1 **18.** $\frac{1}{8}$; Sec. 2.4, Ex. 2 **19.** $\frac{11}{51}$; Sec. 2.5, Ex. 9
20. $\frac{51}{23}$ or $2\frac{5}{23}$; Sec. 2.5, Ex. 10 **21.** $\frac{5}{8}$; Sec. 3.1, Ex. 2 **22.** 24; Sec. 3.2, Ex. 1 **23.** $\frac{6}{6}$ or 1; Sec. 3.3, Ex. 4 **24.** $4\frac{5}{21}$; Sec. 3.4, Ex. 4
25. $\frac{6}{13}$; Sec. 3.5, Ex. 11

Chapter 4 DECIMALS

CHAPTER 4 PRETEST

1. 5.04; 4.1B **2.** $\frac{17}{25}$; 4.1C **3.** 0.081; 4.1D **4.** <; 4.2A **5.** 364.7; 4.2B **6.** 72.245; 4.3A **7.** 47.717; 4.3B **8.** 1.404; 4.4A
9. 0.02701; 4.4B **10.** $22\pi \approx 69.08$ in.; 4.4C **11.** 5000; 4.5A **12.** 4.5; 4.5A **13.** 0.01723; 4.5B **14.** 17.1; 4.6A **15.** 516.84; 4.6A
16. 4.48; 4.6B **17.** 1.14; 4.6B **18.** 0.555; 4.7A **19.** $\frac{1}{3}$, 0.37, $\frac{7}{18}$; 4.7B **20.** $24.70; 4.4D

MENTAL MATH

1. tens   **3.** tenths

EXERCISE SET 4.1

1. six and fifty-two hundredths   **3.** sixteen and twenty-three hundredths   **5.** two hundred five thousandths
7. one hundred sixty-seven and nine thousandths   **9.** two hundred and five thousandths   **11.** thirty-one and four hundredths
13. one and eight tenths   **15.** 6.5   **17.** 9.08   **19.** 5.625   **21.** 0.0064   **23.** 32.52   **25.** 15.8   **27.** $\frac{3}{10}$ **29.** $\frac{27}{100}$ **31.** $\frac{4}{5}$ **33.** $\frac{3}{20}$

35. $5\frac{47}{100}$ **37.** $\frac{6}{125}$ **39.** $7\frac{1}{125}$ **41.** $15\frac{401}{500}$ **43.** $\frac{601}{2000}$ **45.** $487\frac{8}{25}$ **47.** 0.6   **49.** 0.45   **51.** 3.7   **53.** 0.268   **55.** 0.09
57. 4.026   **59.** 0.028   **61.** 56.3   **63.** 47,260   **65.** 47,000   **67.** answers may vary
69. twenty-six million, eight hundred forty-nine thousand, five hundred seventy-six hundred-billionths   **71.** 17.268

EXERCISE SET 4.2

1. <   **3.** >   **5.** <   **7.** =   **9.** <   **11.** >   **13.** 0.006, 0.0061, 0.06   **15.** 0.03, 0.042, 0.36   **17.** 1.01, 1.09, 1.1, 1.16
19. 20.905, 21.001, 21.03, 21.12   **21.** 0.6   **23.** 0.23   **25.** 0.594   **27.** 98,210   **29.** 12.3   **31.** 17.67   **33.** 0.5   **35.** 0.130   **37.** 3830
39. $0.07   **41.** $42,650   **43.** $27   **45.** $0.20   **47.** 0.26499; 0.25786   **49.** 40,000 people   **51.** 2.40 hr   **53.** 24.623 hr   **55.** 2.8 min
57. 5766   **59.** 71   **61.** 243   **63.** 228.040; Parker Bohn III
65. 228.040, 226.130, 225.490, 225.370, 222.980, 222.830, 222.008, 221.000, 219.702, 218.208   **67.** answers may vary   **69.** answers may vary

CALCULATOR EXPLORATION

1. 328.742   **3.** 5.2414   **5.** 865.392

MENTAL MATH

1. 0.5   **3.** 1.26   **5.** 8.9   **7.** 0.6

EXERCISE SET 4.3

1. 3.5   **3.** 6.83   **5.** 0.094   **7.** 622.012   **9.** 583.09   **11.** 465.56   **13.** 115.123   **15.** 27.0578   **17.** 56.432   **19.** 6.5   **21.** 15.3
23. 598.23   **25.** 1.83   **27.** 861.6   **29.** 376.89   **31.** 876.6   **33.** 194.4   **35.** 2.9988   **37.** 16.3   **39.** $454.71   **41.** $0.14   **43.** $7.52
45. 13.3 lb   **47.** 15.81 in.   **49.** 167.607 mph   **51.** 240.8 in.   **53.** 67.44 ft   **55.** $0.075   **57.** 715.05 hr   **59.** Switzerland   **61.** 8.1 lb
63.   **65.** 138   **67.** 960   **69.** $\frac{1}{125}$ **71.** $\frac{5}{12}$ **73.** answers may vary   **75.** 4.59 m

Country	Pounds of Chocolate per Person
Switzerland	22.0
Norway	16.0
Germany	15.8
United Kingdom	14.5
Belgium	13.9

EXERCISE SET 4.4

1. 0.12   **3.** 0.6   **5.** 1.3   **7.** 22.26   **9.** 43.274   **11.** 8.23854   **13.** 11.2746   **15.** 84.97593   **17.** 65   **19.** 0.65   **21.** 709.3
23. 0.0006   **25.** 9100   **27.** 0.03762   **29.** 5,500,000,000   **31.** 36,400,000   **33.** 49,800,000   **35.** $8\pi \approx 25.12$ m   **37.** $10\pi \approx 31.4$ cm
39. $18.2\pi \approx 57.148$ yd   **41.** 24.8 g   **43.** $4462.50   **45.** 64.9605 in.   **47.** $555.20   **49.** 26   **51.** 36   **53.** 8   **55.** 9
57. 3,831,600 mi   **59.** $250\pi \approx 785$ ft   **61.** answers may vary

EXERCISE SET 4.5

1. 0.094   **3.** 300   **5.** 5.8   **7.** 6.6   **9.** 0.413   **11.** 7   **13.** 4.8   **15.** 2100   **17.** 30   **19.** 7000   **21.** 9.8   **23.** 9.6   **25.** 45
27. 200   **29.** 23.87   **31.** 110   **33.** 0.54982   **35.** 0.0129   **37.** 8.7   **39.** 24 mo   **41.** $4417.46   **43.** 202.1 lb   **45.** 5.1 m
47. 11.4 boxes   **49.** 112.8 mph   **51.** 20.19 points   **53.** 345.22   **55.** 1001.0   **57.** 20   **59.** 18   **61.** 85   **63.** 18.125 million, or
18,125,000, CDs   **65.** 45.2 cm   **67.** answers may vary

INTEGRATED REVIEW

1. 2.57   **2.** 4.05   **3.** 8.9   **4.** 3.5   **5.** 0.16   **6.** 0.24   **7.** 0.27   **8.** 0.52   **9.** 4.8   **10.** 6.09   **11.** 75.56   **12.** 289.12   **13.** 25.026
14. 44.125   **15.** 8.6   **16.** 5.4   **17.** 280   **18.** 1600   **19.** 224.938   **20.** 145.079   **21.** 0.56   **22.** 0.63   **23.** 27.6092   **24.** 145.6312
25. 5.4   **26.** 17.74   **27.** 414.44   **28.** 1295.03   **29.** 34   **30.** 28   **31.** 116.81   **32.** 18.79   **33.** 156.2   **34.** 1.562   **35.** 25.62
36. 5.62   **37.** 45.1   **38.** 304.876   **39.** 114.66   **40.** 119.86   **41.** 0.000432   **42.** 0.000075   **43.** 0.0672   **44.** 0.0275

EXERCISE SET 4.6

1. 12 **3.** 2.8 **5.** 2898.66 **7.** 4.2 **9.** 149.8 **11.** 22.89 **13.** 36 in. **15.** 39 ft **17.** 43.96 m **19.** 47 gal **21.** about $12,000
23. about 53 mi **25.** $2300 million **27.** 400,000 people **29.** 0.16 **31.** 3 **33.** 0.28 **35.** 80.52 **37.** 5.5 **39.** 5.29 **41.** 7.6
43. 0.2025 **45.** 129 **47.** 114.2 **49.** 29.23 **51.** 72.6 **53.** 144.4 **55.** $\frac{5}{16}$ **57.** $\frac{3}{4}$ **59.** $\frac{1}{12}$ **61.** 43.388569
63. subtract; answers may vary

EXERCISE SET 4.7

1. 0.2 **3.** 0.5 **5.** 0.75 **7.** 0.08 **9.** 0.375 **11.** $0.91\overline{6}$ **13.** 0.425 **15.** 0.45 **17.** $0.\overline{3}$ **19.** 0.4375 **21.** $0.\overline{2}$ **23.** $1.\overline{6}$
25. 0.33 **27.** 0.44 **29.** 0.2 **31.** 1.7 **33.** 0.68 **35.** 0.4 **37.** 0.62 **39.** < **41.** > **43.** < **45.** < **47.** < **49.** >
51. < **53.** < **55.** 0.32, 0.34, 0.35 **57.** 0.49, 0.491, 0.498 **59.** 0.73, $\frac{3}{4}$, 0.78 **61.** 0.412, 0.453, $\frac{4}{7}$ **63.** 5.23, $\frac{42}{8}$, 5.34 **65.** $\frac{17}{8}$, 2.37, $\frac{12}{5}$
67. 25.65 sq in. **69.** 9.36 sq cm **71.** 0.248 sq yd **73.** 8 **75.** 72 **77.** $\frac{1}{81}$ **79.** $\frac{9}{25}$ **81.** $\frac{5}{2}$ **83.** 0.210 **85.** 7400 **87.** 0.625
89. answers may vary

CHAPTER 4 REVIEW

1. tenths **2.** hundred-thousandths **3.** twenty-three and forty-five hundredths **4.** three hundred forty-five hundred-thousandths
5. one hundred nine and twenty-three hundredths **6.** two hundred and thirty-two millionths **7.** 2.15 **8.** 503.102 **9.** 16,025.0014
10. $\frac{4}{25}$ **11.** $12\frac{23}{1000}$ **12.** $1\frac{9}{2000}$ **13.** $\frac{231}{100,000}$ **14.** $25\frac{1}{4}$ **15.** 0.9 **16.** 0.25 **17.** 0.045 **18.** 0.07 **19.** > **20.** = **21.** >
22. < **23.** 0.6 **24.** 0.94 **25.** 42.90 **26.** 16.349 **27.** $0.26 **28.** $12.46 **29.** $123.00 **30.** $3646.00 **31.** 13,500
32. $10\frac{3}{4}$ **33.** 9.5 **34.** 5.1 **35.** 1.7 **36.** 2.49 **37.** 7.28 **38.** 26 **39.** 320.312 **40.** 148.74236 **41.** 459.7 **42.** 100.278
43. 65.02 **44.** 189.98 **45.** $0.44 **46.** 22.2 in. **47.** 72 **48.** 9345 **49.** 78.246 **50.** 73,246.446 **51.** 14π m; 43.96 m **52.** 63.8 mi
53. 70 **54.** 0.21 **55.** 4900 **56.** 23.904 **57.** 8.059 **58.** 15.825 **59.** 0.0267 **60.** 9.3 **61.** 7.3 m **62.** 45 mo **63.** 18.2 **64.** 50
65. 99.05 **66.** 54.1 **67.** 35.5782 **68.** 0.3526 **69.** 32.7 **70.** 30.4 **71.** 8932 sq ft **72.** yes **73.** 16.94 **74.** 3.89 **75.** 0.1024
76. 3.6 **77.** 0.8 **78.** 0.923 **79.** 0.429 **80.** $0.21\overline{6}$ or 0.217 rounded **81.** 0.1125 or 0.113 rounded **82.** 51.057 **83.** = **84.** <
85. < **86.** < **87.** 0.832, 0.837, 0.839 **88.** 0.42, $\frac{3}{7}$, 0.43 **89.** $\frac{19}{12}$, 1.63, $\frac{18}{11}$ **90.** $\frac{3}{4}$, $\frac{6}{7}$, $\frac{8}{9}$ **91.** 6.9 sq ft **92.** 5.46 sq in.

CHAPTER 4 TEST

1. forty-five and ninety-two thousandths **2.** 3000.059 **3.** 34.9 **4.** 0.862 **5.** < **6.** < **7.** $\frac{69}{200}$ **8.** $24\frac{73}{100}$ **9.** 0.5 **10.** 0.941
11. 17.583 **12.** 11.4 **13.** 43.86 **14.** 56 **15.** 6.673 **16.** 12,690 **17.** 47.3 **18.** 1.21 **19.** 6.2 **20.** 2.31 sq mi **21.** 198.08 oz
22. 18π mi; 56.52 mi **23.** 12 CDs **24.** 54 mi

CUMULATIVE REVIEW

1. one hundred six million, fifty-two thousand, four hundred forty-seven; Sec. 1.2, Ex. 6 **2.** 445 baseball cards; Sec. 1.3, Ex. 8
3. 726; Sec. 1.4, Ex. 4 **4.** 2300; Sec. 1.5, Ex. 4 **5.** 15,540 thousand bytes; Sec. 1.6, Ex. 7 **6.** 401 R 2; Sec. 1.7, Ex. 8 **7.** 47; Sec. 1.9, Ex. 11
8. numerator: 3; denominator: 7; Sec. 2.1, Ex. 1 **9.** $\frac{1}{10}$; Sec. 2.3, Ex. 6 **10.** $\frac{15}{1}$ or 15; Sec. 2.4, Ex. 9 **11.** $\frac{63}{16}$; Sec. 2.5, Ex. 5
12. $\frac{15}{4}$ or $3\frac{3}{4}$; Sec. 2.4, Ex. 8 **13.** $\frac{3}{20}$; Sec. 2.5, Ex. 8 **14.** $\frac{7}{9}$; Sec. 3.1, Ex. 4 **15.** $\frac{1}{4}$; Sec. 3.1, Ex. 5 **16.** $\frac{15}{20}$; Sec. 3.2, Ex. 8
17. $\frac{13}{30}$; Sec. 3.3, Ex. 2 **18.** $4\frac{7}{40}$ lb; Sec. 3.4, Ex. 7 **19.** $\frac{1}{16}$; Sec. 3.5, Ex. 3 **20.** $\frac{3}{256}$; Sec. 3.5, Ex. 5 **21.** $\frac{43}{100}$; Sec. 4.1, Ex. 7
22. > ; Sec. 4.2, Ex. 1 **23.** 11.568; Sec. 4.3, Ex. 4 **24.** 2370.2; Sec. 4.4, Ex. 5 **25.** 768.05; Sec. 4.4, Ex.8

Chapter 5 RATIO AND PROPORTION

CHAPTER 5 PRETEST

1. $\frac{27}{8}$; 5.1A **2.** $\frac{4}{37}$; 5.1A **3.** $\frac{5}{12}$; 5.1B **4.** $\frac{4}{3}$; 5.1B **5.** $\frac{6}{19}$; 5.1B **6.** $\frac{4}{11}$; 5.1B **7.** $\frac{35 \text{ students}}{2 \text{ instructors}}$; 5.2A
8. $\frac{7 \text{ c of flour}}{2 \text{ cakes}}$; 5.2A **9.** $\frac{78.75 \text{ km}}{1 \text{ hr}}$ or 78.75 km/hr; 5.2B **10.** $\frac{3 \text{ cookies}}{1 \text{ child}}$ or 3 cookies/child; 5.2B **11.** 12 oz; 5.2C
12. 12 doughnuts; 5.2C **13.** $\frac{5}{80} = \frac{15}{240}$; 5.3A **14.** false; 5.3B **15.** true; 5.3B **16.** $n = 16$; 5.3C **17.** $n = 4.5$; 5.3C
18. $n = 133$; 5.3C **19.** $12,000; 5.4A **20.** $472.75; 5.4A

Exercise Set 5.1

1. $\dfrac{11}{14}$ **3.** $\dfrac{23}{10}$ **5.** $\dfrac{151}{201}$ **7.** $\dfrac{2.8}{7.6}$ **9.** $\dfrac{5}{7\frac{1}{2}}$ **11.** $\dfrac{3\frac{3}{4}}{1\frac{2}{3}}$ **13.** $\dfrac{2}{3}$ **15.** $\dfrac{77}{100}$ **17.** $\dfrac{463}{821}$ **19.** $\dfrac{3}{4}$ **21.** $\dfrac{5}{12}$ **23.** $\dfrac{8}{25}$ **25.** $\dfrac{12}{7}$

27. $\dfrac{16}{23}$ **29.** $\dfrac{2}{5}$ **31.** $\dfrac{5}{3}$ **33.** $\dfrac{17}{40}$ **35.** $\dfrac{5}{4}$ **37.** $\dfrac{3}{1}$ **39.** $\dfrac{15}{1}$ **41.** $\dfrac{1}{3}$ **43.** $\dfrac{5}{4}$ **45.** 2.3   **47.** 0.15   **49.** $\dfrac{2}{23}$ **51.** $\dfrac{10}{23}$

53. no, the shipment should not be refused   **55. a.** $\dfrac{17}{50}$ **b.** $\dfrac{17}{33}$ **c.** no; answers may vary

Exercise Set 5.2

1. $\dfrac{1 \text{ shrub}}{3 \text{ ft}}$ **3.** $\dfrac{3 \text{ returns}}{20 \text{ sales}}$ **5.** $\dfrac{2 \text{ phone lines}}{9 \text{ employees}}$ **7.** $\dfrac{9 \text{ gal}}{2 \text{ acres}}$ **9.** $\dfrac{3 \text{ flight attendants}}{100 \text{ passengers}}$ **11.** $\dfrac{71 \text{ cal}}{2 \text{ fl oz}}$ **13.** 71 riders/car   **15.** 110 cal/oz
17. 6 diapers/baby   **19.** $50,000/yr   **21.** ≈ 6.67 km/min   **23.** 7600 sq mi/county   **25.** 300 good/defective   **27.** 0.48 tons of dust and dirt/acre   **29.** $58,000/species   **31.** $46,600/house   **33. a.** 31.25 computer boards/hr   **b.** ≈ 33.3 computer boards/hr   **c.** Lamont
35. $11.50 per compact disc   **37.** $0.17 per banana   **39.** 8 oz: $0.149 per oz; 12 oz: $0.133 per oz; 12 oz
41. 16 oz: $0.106 per oz; dozen: $0.115 per egg; 16 oz   **43.** 12 oz: $0.191 per oz; 8 oz: $0.186 per oz; 8 oz
45. 100: $0.006 per napkin; 180: $0.005 per napkin; 180 napkins   **47.** 10.2   **49.** 4.44   **51.** 1.9
53. Miles driven: 257,352,347; Miles per gallon: 19.2, 22.3, 21.6   **55.** 1.5 steps/ft   **57.** 575 students/school   **59.** no; answers may vary

Integrated Review

1. $\dfrac{9}{10}$ **2.** $\dfrac{9}{25}$ **3.** $\dfrac{43}{50}$ **4.** $\dfrac{8}{23}$ **5.** $\dfrac{173}{139}$ **6.** $\dfrac{6}{7}$ **7.** $\dfrac{7}{26}$ **8.** $\dfrac{20}{33}$ **9.** $\dfrac{2}{3}$ **10.** $\dfrac{1}{8}$ **11.** $\dfrac{204}{161}$ **12.** $\dfrac{3553}{1127}$ **13.** $\dfrac{36}{761}$ **14.** $\dfrac{2}{3}$
15. $\dfrac{1 \text{ office}}{4 \text{ graduate assistants}}$ **16.** $\dfrac{2 \text{ lights}}{5 \text{ ft}}$ **17.** $\dfrac{2 \text{ senators}}{1 \text{ state}}$ **18.** $\dfrac{1 \text{ teacher}}{28 \text{ students}}$ **19.** $\dfrac{16}{25}$ **20.** $\dfrac{269}{25}$ **21.** 55 mi/hr   **22.** 140 ft/sec
23. 21 employees/fax line   **24.** 17 phone calls/teenager   **25.** 23 mi/gal   **26.** 16 teachers/computer   **27.** 6 books/student
28. 154 lb/adult   **29.** 8 lb: $0.27 per lb; 18 lb: $0.28 per lb; 8 lb   **30.** 100: $0.020 per plate; 500: $0.018 per plate; 500 paper plates
31. 3 packs: $0.80 per pack; 8 packs: $0.75 per pack; 8 packs   **32.** 4: $0.92 per battery; 10: $0.99 per battery; 4 batteries

Mental Math

1. true   **3.** false   **5.** true

Exercise Set 5.3

1. $\dfrac{10 \text{ diamonds}}{6 \text{ opals}} = \dfrac{5 \text{ diamonds}}{3 \text{ opals}}$ **3.** $\dfrac{3 \text{ printers}}{12 \text{ computers}} = \dfrac{1 \text{ printer}}{4 \text{ computers}}$ **5.** $\dfrac{6 \text{ eagles}}{58 \text{ sparrows}} = \dfrac{3 \text{ eagles}}{29 \text{ sparrows}}$ **7.** $\dfrac{2\frac{1}{4} \text{ c flour}}{24 \text{ cookies}} = \dfrac{6\frac{3}{4} \text{ c flour}}{72 \text{ cookies}}$
9. $\dfrac{22 \text{ vanilla wafers}}{1 \text{ c cookie crumbs}} = \dfrac{55 \text{ vanilla wafers}}{2.5 \text{ c cookie crumbs}}$ **11.** true   **13.** false   **15.** true   **17.** true   **19.** false   **21.** true   **23.** true   **25.** false

27. 3   **29.** 5   **31.** 4   **33.** 3.2   **35.** 19.2   **37.** $\dfrac{9}{20}$ **39.** 1   **41.** 12   **43.** $\dfrac{3}{4}$ **45.** 0.0025   **47.** 25   **49.** <   **51.** >

53. <   **55.** 0   **57.** 1400   **59.** 252.5   **61.** answers may vary

Exercise Set 5.4

1. 12 passes   **3.** 165 min   **5.** 180 students   **7.** 23 ft   **9.** 270 sq ft   **11.** 56 mi   **13.** 450 km   **15.** 24 oz   **17.** 16 bags   **19.** $162,000

21. 15 hits   **23.** 27 people   **25.** 86 wk   **27.** 6 people   **29.** 112 ft; 11-in. difference   **31.** $2\frac{2}{3}$ lb   **33.** 28 adults   **35.** $3 \cdot 5$ **37.** $2^2 \cdot 5$

39. $2^3 \cdot 5^2$ **41.** 2^5 **43.** $4\frac{2}{3}$ ft   **45.** answers may vary

Chapter 5 Review

1. $\dfrac{5}{4}$ **2.** $\dfrac{11}{13}$ **3.** $\dfrac{9}{40}$ **4.** $\dfrac{14}{5}$ **5.** $\dfrac{4}{15}$ **6.** $\dfrac{1}{2}$ **7.** $\dfrac{1}{2}$ **8.** $\dfrac{7}{150}$ **9.** $\dfrac{1 \text{ stillborn birth}}{125 \text{ live births}}$ **10.** $\dfrac{3 \text{ professors}}{10 \text{ assistants}}$ **11.** $\dfrac{5 \text{ pages}}{2 \text{ min}}$
12. $\dfrac{4 \text{ computers}}{3 \text{ hr}}$ **13.** 52 mi/hr   **14.** 15 ft/sec   **15.** $0.31/pear   **16.** $1.74/diskette   **17.** 65 km/hr   **18.** $1\frac{1}{3}$ gal/acre
19. $36.80/course   **20.** 13 bushels/tree   **21.** 8-oz size   **22.** 18-oz size   **23.** 1-gal size   **24.** 32-oz size   **25.** $\dfrac{20 \text{ men}}{14 \text{ women}} = \dfrac{10 \text{ men}}{7 \text{ women}}$
26. $\dfrac{50 \text{ tries}}{4 \text{ successes}} = \dfrac{25 \text{ tries}}{2 \text{ successes}}$ **27.** $\dfrac{16 \text{ sandwiches}}{8 \text{ players}} = \dfrac{2 \text{ sandwiches}}{1 \text{ player}}$ **28.** $\dfrac{12 \text{ tires}}{3 \text{ cars}} = \dfrac{4 \text{ tires}}{1 \text{ car}}$ **29.** no   **30.** yes   **31.** no   **32.** yes   **33.** 5
34. 15   **35.** 32.5   **36.** 5.625   **37.** 32   **38.** 13.5   **39.** 60   **40.** $7\frac{1}{5}$ **41.** 0.94   **42.** 0.36   **43.** 14   **44.** 35   **45.** 8 bags
46. 16 bags   **47.** no   **48.** 79 gal   **49.** $54.600   **50.** $1023.50   **51.** $40\frac{1}{2}$ ft   **52.** $8\frac{1}{4}$ in.

CHAPTER 5 TEST

1. $\frac{9}{13}$ **2.** $\frac{15}{2}$ **3.** $\frac{7 \text{ men}}{1 \text{ woman}}$ **4.** $\frac{3 \text{ in.}}{10 \text{ days}}$ **5.** 81.25 km/hr **6.** $\frac{2}{3}$ in./hr **7.** 28 students/teacher **8.** 8-oz size **9.** 16-oz size

10. true **11.** false **12.** 5 **13.** $4\frac{4}{11}$ **14.** $\frac{7}{3}$ **15.** 8 **16.** $49\frac{1}{2}$ ft **17.** $3\frac{3}{4}$ hr **18.** $53\frac{1}{3}$ g **19.** $144\frac{2}{3}$ cartons **20.** 4266 adults

CUMULATIVE REVIEW

1. **a.** 3; Sec. 1.4, Ex. 1 **b.** 5; Sec. 1.4, Ex. 1 **c.** 0; Sec. 1.4, Ex. 1 **d.** 7; Sec. 1.4, Ex. 1 **2.** 249.000; Sec. 1.5, Ex. 3 **3.** 200; Sec. 1.6, Ex. 3

4. $17.820; Sec. 1.8, Ex. 3 **5.** $3 \cdot 3 \cdot 5$ or $3^2 \cdot 5$; Sec. 2.2, Ex. 3 **6.** $\frac{3}{5}$; Sec. 2.3, Ex. 1 **7.** $\frac{6}{5}$; Sec. 2.4, Ex. 5 **8.** $\frac{2}{5}$; Sec. 2.4, Ex. 6

9. $\frac{5}{7}$; Sec. 3.1, Ex. 1 **10.** 2; Sec. 3.1, Ex. 3 **11.** 18; Sec. 3.2, Ex. 2 **12.** $\frac{7}{14}$; Sec. 3.2, Ex. 9 **13.** $\frac{8}{33}$; Sec. 3.3, Ex. 6 **14.** $\frac{1}{6}$ hr; Sec. 3.3, Ex. 9

15. $7\frac{17}{24}$; Sec. 3.4, Ex. 1 **16.** > ; Sec. 3.5, Ex. 1 **17.** one and three-tenths; Sec. 4.1, Ex. 1 **18.** 736.2; Sec. 4.2, Ex. 3 **19.** 25.454; Sec. 4.3, Ex. 1

20. 0.0849; Sec. 4.4, Ex. 2 **21.** 0.125; Sec. 4.5, Ex. 3 **22.** 3.7; Sec. 4.6, Ex. 5 **23.** $\frac{4}{9}, \frac{9}{20}$, 0.456; Sec. 4.7, Ex. 6 **24.** $\frac{2.6}{3.1}$; Sec. 5.1, Ex. 2

25. $\frac{1\frac{1}{2}}{7\frac{3}{4}}$; Sec. 5.1, Ex. 3

Chapter 6 PERCENT

CHAPTER 6 PRETEST

1. 48%; 6.1A **2.** 0.73; 6.1B **3.** 0.068; 6.1B **4.** 3%; 6.1C **5.** 210%; 6.1C **6.** $\frac{11}{50}$; 6.2A **7.** $\frac{3}{200}$; 6.2A **8.** 90%; 6.2B

9. 220%; 6.2B **10.** 0.035; 6.2C **11.** $4 = n \cdot 28$; 6.3A **12.** $\frac{70}{b} = \frac{18}{100}$; 6.4A **13.** 40; 6.3B, 6.4B **14.** 6.4; 6.3B, 6.4B

15. discount: $46; sale price: $184; 6.5A **16.** 21%; 6.5B **17.** sales tax: $2.52; total price: $44.52; 6.6A **18.** 4%; 6.6B **19.** $144; 6.7A
20. $103.33; 6.7C

EXERCISE SET 6.1

1. 81% **3.** 9% **5.** chocolate chip; 52% **7.** 75% **9.** 0.48 **11.** 0.06 **13.** 1 **15.** 0.613 **17.** 0.028 **19.** 0.006 **21.** 3
23. 0.3258 **25.** 0.67 **27.** 0.045 **29.** 0.212 **31.** 98% **33.** 310% **35.** 2900% **37.** 0.3% **39.** 22% **41.** 530% **43.** 5.6%
45. 33.28% **47.** 300% **49.** 70% **51.** 10% **53.** 9.3% **55.** 38% **57.** 0.25 **59.** 0.65 **61.** 0.9 **63.** computer engineers
65. 0.77 **67.** answers may vary

MENTAL MATH

1. 13% **3.** 87% **5.** 1%

EXERCISE SET 6.2

1. $\frac{3}{25}$ **3.** $\frac{1}{25}$ **5.** $\frac{9}{200}$ **7.** $1\frac{3}{4}$ **9.** $\frac{73}{100}$ **11.** $\frac{1}{8}$ **13.** $\frac{1}{16}$ **15.** $\frac{2}{25}$ **17.** $\frac{31}{300}$ **19.** $\frac{179}{800}$ **21.** 75% **23.** 70% **25.** 40%

27. 59% **29.** 34% **31.** $37\frac{1}{2}$% **33.** $31\frac{1}{4}$% **35.** 160% **37.** $66\frac{2}{3}$% **39.** 65% **41.** 250% **43.** 190% **45.** 63.64%

47. 26.67% **49.** 14.29% **51.** 91.67% **53.** 0.35, $\frac{7}{20}$; 20%, 0.2; 50%, $\frac{1}{2}$; 0.7, $\frac{7}{10}$; 37.5%, 0.375

55. 0.4, $\frac{2}{5}$; $23\frac{1}{2}$%, $\frac{47}{200}$; 80%, 0.8; 0.3333, $\frac{1}{3}$; 87.5%, 0.875; 0.075, $\frac{3}{40}$ **57.** $\frac{37}{250}$ **59.** 0.402 **61.** 28.2% **63.** 0.0825 **65.** 22% **67.** $n = 15$
69. $n = 10$ **71.** $n = 12$ **73.** 0.266; 26.6% **75.** 1.155; 115.5% **77.** greater **79.** answers may vary

MENTAL MATH

1. percent: 42; base: 50; amount: 21 **3.** percent: 125; base: 86; amount: 107.5

EXERCISE SET 6.3

1. $15\% \cdot 72 = n$ **3.** $30\% \cdot n = 80$ **5.** $n \cdot 90 = 20$ **7.** $1.9 = 40\% \cdot n$ **9.** $n = 9\% \cdot 43$ **11.** 3.5 **13.** 7.28 **15.** 600 **17.** 10

19. 110% **21.** 32% **23.** 1 **25.** 45 **27.** 500 **29.** 400% **31.** 25.2 **33.** 45% **35.** 35 **37.** $n = 30$ **39.** $n = 3\frac{7}{11}$

41. $\frac{17}{12} = \frac{n}{20}$ **43.** $\frac{8}{9} = \frac{14}{n}$ **45.** 686.625 **47.** 12,285

MENTAL MATH

1. amount: 12.6; base: 42; percent: 30   **3.** amount: 102; base: 510; percent: 20

EXERCISE SET 6.4

1. $\dfrac{a}{65} = \dfrac{32}{100}$ **3.** $\dfrac{75}{b} = \dfrac{40}{100}$ **5.** $\dfrac{70}{200} = \dfrac{p}{100}$ **7.** $\dfrac{2.3}{b} = \dfrac{58}{100}$ **9.** $\dfrac{a}{130} = \dfrac{19}{100}$ **11.** 5.5   **13.** 18.9   **15.** 400   **17.** 10   **19.** 125%

21. 28%   **23.** 29   **25.** 1.92   **27.** 1000   **29.** 210%   **31.** 55.18   **33.** 45%   **35.** 85   **37.** $\dfrac{7}{8}$ **39.** $3\dfrac{2}{15}$ **41.** 0.7   **43.** 2.19

45. 12,011.226 or 12,011.2 rounded   **47.** 7270.6

INTEGRATED REVIEW

1. 12%   **2.** 68%   **3.** 25%   **4.** 50%   **5.** 520%   **6.** 780%   **7.** 6%   **8.** 44%   **9.** 250%   **10.** 325%   **11.** 3%   **12.** 5%

13. 0.65   **14.** 0.31   **15.** 0.08   **16.** 0.07   **17.** 1.42   **18.** 5.38   **19.** 0.029   **20.** 0.066   **21.** $\dfrac{3}{100}$ **22.** $\dfrac{2}{25}$ **23.** $\dfrac{21}{400}$ **24.** $\dfrac{51}{400}$

25. $\dfrac{19}{50}$ **26.** $\dfrac{9}{20}$ **27.** $\dfrac{37}{300}$ **28.** $\dfrac{1}{6}$ **29.** 8.4   **30.** 100   **31.** 250   **32.** 120%   **33.** 28%   **34.** 76   **35.** 11   **36.** 130%

37. 86%   **38.** 37.8   **39.** 150   **40.** 62

EXERCISE SET 6.5

1. 1600 bolts   **3.** 30 hr   **5.** 15%   **7.** 295 components   **9.** 29.2%   **11.** 496 chairs; 6696 chairs   **13.** $136   **15.** $867.87; $20,153.87

17. 97,680 physician assistants   **19.** 10; 25%   **21.** 102; 120%   **23.** 2; 25%   **25.** 120; 75%   **27.** 44%   **29.** 21.5%   **31.** 12.6%

33. 149.4%   **35.** 587.5%   **37.** 4.56   **39.** 11.18   **41.** 58.54   **43.** 8.2%   **45.** answers may vary

EXERCISE SET 6.6

1. $7.50   **3.** $858.93   **5.** 9%   **7.** $130.20   **9.** $1917   **11.** $11,500   **13.** $112.35   **15.** 6%   **17.** $49,474.24   **19.** 14%

21. $1888.50   **23.** 3%   **25.** $6.80; $61.20   **27.** $48.25; $48.25   **29.** $75.25; $139.75   **31.** $3255; $18,445   **33.** $45; $255   **35.** 1200

37. 132   **39.** 16   **41.** $26,838.45   **43.** a 20% discount followed by an additional 40% off; answers may vary

CALCULATOR EXPLORATIONS

1. 1.56051   **3.** 8.06231   **5.** $634.50

EXERCISE SET 6.7

1. $32   **3.** $73.60   **5.** $750   **7.** $33.75   **9.** $700   **11.** $78,125   **13.** $5562.50   **15.** $12,580   **17.** $46,815.40   **19.** $2327.15

21. $58,163.60   **23.** $240.75   **25.** $938.66   **27.** $971.90   **29.** $260.31   **31.** $637.26   **33.** 32 yd   **35.** 35 m   **37.** answers may vary

39. answers may vary

CHAPTER 6 REVIEW

1. 37%   **2.** 77%   **3.** 0.83   **4.** 0.75   **5.** 0.735   **6.** 0.015   **7.** 1.25   **8.** 1.45   **9.** 0.005   **10.** 0.007   **11.** 2.00 or 2   **12.** 4.00 or 4

13. 0.2625   **14.** 0.8534   **15.** 260%   **16.** 5.5%   **17.** 35%   **18.** 102%   **19.** 72.5%   **20.** 25%   **21.** 7.6%   **22.** 8.5%   **23.** 75%

24. 65%   **25.** 400%   **26.** 900%   **27.** $\dfrac{1}{100}$ **28.** $\dfrac{1}{10}$ **29.** $\dfrac{1}{4}$ **30.** $\dfrac{17}{200}$ **31.** $\dfrac{51}{500}$ **32.** $\dfrac{1}{6}$ **33.** $\dfrac{1}{3}$ **34.** $1\dfrac{1}{10}$ **35.** 20%

36. 70%   **37.** $83\dfrac{1}{3}$%   **38.** 62.5%   **39.** $166\dfrac{2}{3}$%   **40.** 125%   **41.** 60%   **42.** 6.25%   **43.** 100,000   **44.** 8000   **45.** 23%

46. 114.5   **47.** 3000   **48.** 150%   **49.** 418   **50.** 300   **51.** 64.8   **52.** 180%   **53.** 110%   **54.** 165   **55.** 66%   **56.** 16%

57. 106.25%   **58.** 20.9%   **59.** $206,400   **60.** $13.23   **61.** $263.75   **62.** $1.15   **63.** $5000   **64.** $300.38

65. discount: $900; sale price: $2100   **66.** discount: $9; sale price: $81   **67.** $120   **68.** $1320   **69.** $30,104.64   **70.** $17,506.56

71. $80.61   **72.** $32,830.10

CHAPTER 6 TEST

1. 0.85   **2.** 5   **3.** 0.006   **4.** 5.6%   **5.** 610%   **6.** 35%   **7.** $\dfrac{6}{5}$ **8.** $\dfrac{77}{200}$ **9.** $\dfrac{1}{500}$ **10.** 55%   **11.** 37.5%   **12.** 175%   **13.** 33.6

14. 1250   **15.** 75%   **16.** 38.4 lb   **17.** $56,750   **18.** $358.43   **19.** 5%   **20.** discount: $18; sale price: $102   **21.** $395   **22.** 1%

23. $647.50   **24.** $2005.64   **25.** $427

CUMULATIVE REVIEW

1. 206 cases; 12 cans; yes; Sec. 1.8, Ex. 2   **2. a.** $4\dfrac{2}{7}$ **b.** $1\dfrac{1}{15}$ **c.** 14; Sec. 2.1, Ex. 8   **3.** $2 \cdot 2 \cdot 2 \cdot 3$ or $2^3 \cdot 3$; Sec. 2.2, Ex. 7   **4.** $\dfrac{10}{27}$; Sec. 2.3, Ex. 3

5. $\dfrac{23}{56}$; Sec. 2.4, Ex. 4   **6.** $\dfrac{8}{11}$; Sec. 2.5, Ex. 2   **7.** $\dfrac{4}{5}$ in.; Sec. 3.1, Ex. 6   **8.** 60; Sec. 3.2, Ex. 4   **9.** $\dfrac{2}{3}$; Sec. 3.3, Ex. 1   **10.** $3\dfrac{5}{14}$; Sec. 3.4, Ex. 5

11. $\dfrac{7}{16}$; Sec. 3.5, Ex. 6   **12.** $\dfrac{2}{33}$; Sec. 3.5, Ex. 8   **13.** 0.8; Sec. 4.1, Ex. 12   **14.** 8.7; Sec. 4.1, Ex. 13   **15.** $1.03; Sec. 4.2, Ex. 5

16. 829.6561; Sec. 4.3, Ex. 2   **17.** 18.408; Sec. 4.4, Ex. 1   **18.** 0.0786; Sec. 4.5, Ex. 7   **19.** 0.012; Sec. 4.5, Ex. 8   **20.** 1.69; Sec. 4.6, Ex. 6

21. 0.25; Sec. 4.7, Ex. 1   **22.** $\dfrac{5 \text{ nails}}{3 \text{ ft}}$; Sec. 5.2, Ex. 1   **23.** no; Sec. 5.3, Ex. 3   **24.** $17\dfrac{1}{2}$ mi; Sec. 5.4, Ex. 1   **25.** $n = 25\% \cdot 0.008$; Sec. 6.3, Ex. 3

Chapter 7   MEASUREMENT

PRETEST

1. 72 in.; 7.1A   **2.** 8 ft 2 in.; 7.1B   **3.** 14,784 ft; 7.1A   **4.** 980 cm; 7.1D   **5.** 112 oz; 7.2A   **6.** 2.9 g; 7.2C   **7.** 600 g; 7.2C
8. $4\frac{1}{2}$ gal; 7.3A   **9.** 6.28 kl; 7.3C   **10.** 113°F; 7.4A   **11.** 25°C; 7.4B   **12.** 11,670 ft-lb; 7.5A   **13.** 13 ft 5 in.; 7.1C
14. 4 yd 2 ft; 7.1C   **15.** 48.6 mm; 7.1E   **16.** 1 ton 825 lb; 7.2B   **17.** 20.991 g or 20,991 mg; 7.2D   **18.** 66 qt 1 pt; 7.3B   **19.** 9 c; 7.3B
20. $0.325; 7.3D

CALCULATOR EXPLORATIONS

1. ≈22.96 ft   **3.** ≈21.59 cm   **5.** ≈3.1 mi

MENTAL MATH

1. 1 ft   **3.** 2 ft   **5.** 1 yd   **7.** no   **9.** yes   **11.** no

EXERCISE SET 7.1

1. 5 ft   **3.** 36 ft   **5.** 8 mi   **7.** $8\frac{1}{2}$ ft   **9.** $3\frac{1}{3}$ yd   **11.** 33,792 ft   **13.** 13 yd 1 ft   **15.** 3 ft 5 in.   **17.** 1 mi 4720 ft   **19.** 62 in.
21. 17 ft   **23.** 84 in.   **25.** 12 ft 3 in.   **27.** 22 yd 1 ft   **29.** 8 ft 5 in.   **31.** 5 ft 6 in.   **33.** 3 ft 4 in.   **35.** 50 yd 2 ft   **37.** 10 ft 6 in.
39. 13 ft 11 in.   **41.** 15 ft 9 in.   **43.** 3 ft 1 in.   **45.** 86 ft 6 in.   **47.** $105\frac{1}{3}$ yd   **49.** 4000 cm   **51.** 4.0 cm   **53.** 0.3 km   **55.** 1.4 m
57. 15 m   **59.** 83 mm   **61.** 0.201 dm   **63.** 40 mm   **65.** 8.94 m   **67.** 2.94 m or 2940 mm   **69.** 1.29 cm or 12.9 mm
71. 12.640 km or 12,640 m   **73.** 54.9 m   **75.** 1.55 km   **77.** 9.12 m   **79.** 26.7 mm   **81.** 41.25 m or 4125 cm   **83.** 3.35 m
85. 6.009 km or 6009 m   **87.** 15 tiles   **89.** 21%   **91.** 13%   **93.** 25%   **95.** answers may vary   **97.** 6.575 m

CALCULATOR EXPLORATIONS

1. ≈425.25 g   **3.** ≈15.4 lb   **5.** ≈0.175 oz

MENTAL MATH

1. 1 lb   **3.** 2000 lb   **5.** 16 oz   **7.** 1 ton   **9.** no   **11.** yes   **13.** no

EXERCISE SET 7.2

1. 32 oz   **3.** 10,000 lb   **5.** 6 tons   **7.** $3\frac{3}{4}$ lb   **9.** $1\frac{3}{4}$ tons   **11.** 260 oz   **13.** 9800 lb   **15.** 76 oz   **17.** 1.5 tons   **19.** 53 lb 10 oz
21. 9 tons 390 lb   **23.** 3 tons 175 lb   **25.** 8 lb 11 oz   **27.** 31 lb 2 oz   **29.** 1 ton 700 lb   **31.** 5 lb 8 oz   **33.** 35 lb 14 oz   **35.** 130 lb
37. 211 lb   **39.** 5 lb 1 oz   **41.** 0.5 kg   **43.** 4000 mg   **45.** 25,000 g   **47.** 0.048 g   **49.** 0.0063 kg   **51.** 15,140 mg   **53.** 4010 g
55. 13.5 mg   **57.** 5.815 g or 5815 mg   **59.** 1850 mg or 1.850 g   **61.** 1360 g or 1.360 kg   **63.** 13.52 kg   **65.** 2.125 kg   **67.** 8.064 kg
69. 30 mg   **71.** 250 mg   **73.** 144 mg   **75.** 6.12 kg   **77.** 850 g or 0.85 kg   **79.** 2.38 kg   **81.** 0.25   **83.** 0.16   **85.** 0.875
87. answers may vary

CALCULATOR EXPLORATIONS

1. ≈4.73 L   **3.** ≈4.62 gal   **5.** ≈3.785 L

MENTAL MATH

1. 1 pt   **3.** 1 gal   **5.** 1 qt   **7.** 1 c   **9.** 2 c   **11.** 4 qt   **13.** no   **15.** no

EXERCISE SET 7.3

1. 4 c   **3.** 16 pt   **5.** $2\frac{1}{2}$ gal   **7.** 5 pt   **9.** 8 c   **11.** $3\frac{3}{4}$ qt   **13.** 768 fl oz   **15.** 9 c   **17.** 22 pt   **19.** 10 gal 1 qt   **21.** 4 c 4 fl oz
23. 1 gal 1 qt   **25.** 2 gal 3 qt 1 pt   **27.** 2 qt 1 c   **29.** 17 gal   **31.** 4 gal 3 qt   **33.** 48 fl oz   **35.** 2 qt   **37.** yes   **39.** 2 qt 1 c   **41.** 9 qt
43. 5000 ml   **45.** 4.5 L   **47.** 0.41 kl   **49.** 0.064 L   **51.** 160 L   **53.** 3600 ml   **55.** 0.00016 kl   **57.** 22.5 L   **59.** 4.5 L or 4500 ml
61. 8410 ml or 8.410 L   **63.** 10,600 ml or 10.6 L   **65.** 3840 ml   **67.** 162.4 L   **69.** 1.59 L   **71.** 18.954 L   **73.** $0.316   **75.** 474 ml
77. $\frac{7}{10}$ **79.** $\frac{3}{100}$ **81.** $\frac{3}{500}$ **83.** answers may vary

INTEGRATED REVIEW

1. 3 ft   **2.** 2 mi   **3.** $6\frac{2}{3}$ yd   **4.** 18 ft   **5.** 11,088 ft   **6.** 38.4 in.   **7.** 3000 cm   **8.** 2.4 cm   **9.** 2 m   **10.** 18 m   **11.** 72 mm
12. 0.6 km   **13.** 10,000 lb   **14.** 5.5 tons   **15.** 136 oz   **16.** 2.5 lb   **17.** 3.5 lb   **18.** 80 oz   **19.** 28,000 g   **20.** 1.4 g   **21.** 0.0056 kg
22. 6000 g   **23.** 0.67 g   **24.** 0.0036 kg   **25.** 12 pt   **26.** 2.5 qt   **27.** 3.5 gal   **28.** 8.5 pt   **29.** 7 c   **30.** 6.5 gal   **31.** 7000 ml
32. 0.35 kl   **33.** 0.047 L   **34.** 970 L   **35.** 126 L   **36.** 0.075 L

MENTAL MATH

1. yes   **3.** no   **5.** no   **7.** yes

EXERCISE SET 7.4

1. 5°C   **3.** 40°C   **5.** 140°F   **7.** 239°F   **9.** 16.7°C   **11.** 61.2°C   **13.** 197.6°F   **15.** 61.3°F   **17.** 50°C   **19.** 80.6°F   **21.** 21.1°C
23. 37.9°C   **25.** 244.4°F   **27.** 260°C   **29.** 462.2°C   **31.** 62 m   **33.** 15 ft   **35.** 10.4 m   **37.** 4988°C

MENTAL MATH

1. 30 ft-lb   **3.** 60 ft-lb   **5.** 90 cal   **7.** 5 cal

EXERCISE SET 7.5

1. 1140 ft-lb   **3.** 3696 ft-lb   **5.** 425,000 ft-lb   **7.** 23,340 ft-lb   **9.** 778,000 ft-lb   **11.** 15,560,000 ft-lb   **13.** 26,553,140 ft-lb
15. 10,283 BTU   **17.** 805 cal   **19.** 750 cal   **21.** 1440 cal   **23.** 2.6 hr   **25.** 17.5 mi   **27.** $\frac{4}{5}$ **29.** $\frac{3}{5}$ **31.** $\frac{9}{10}$ **33.** 92.925 ft-lb
35. 256 lb

CHAPTER 7 REVIEW

1. 9 ft   **2.** 24 yd   **3.** 13,200 ft   **4.** 75 in.   **5.** 17 yd 1 ft   **6.** 3 ft 10 in.   **7.** 4200 cm   **8.** 820 mm   **9.** 0.01218 m   **10.** 0.00231 km
11. 21 yd 1 ft   **12.** 7 ft 5 in.   **13.** 41 ft 3 in.   **14.** 3 ft 8 in.   **15.** 9.5 cm or 95 mm   **16.** 5.26 m or 526 cm   **17.** 9117 m or 9.117 km
18. 1.1 m or 1100 mm   **19.** 169 yd 2 ft   **20.** 126 ft 8 in.   **21.** 108.5 km   **22.** 0.24 sq m   **23.** 4.125 lb   **24.** 4600 lb   **25.** 3 lb 4 oz
26. 4 tons 200 lb   **27.** 1.4 g   **28.** 40,000 g   **29.** 21 dag   **30.** 0.0003 dg   **31.** 3 lb 9 oz   **32.** 10 tons 800 lb   **33.** 2 tons 750 lb
34. 33 lb 8 oz   **35.** 4.9 g or 4900 mg   **36.** 9 kg or 9000 g   **37.** 8.1 g or 8100 mg   **38.** 50.4 kg   **39.** 4 lb 4 oz   **40.** 9 tons 1075 lb
41. 7.85 kg   **42.** 1.1625 kg   **43.** 8 qt   **44.** 5 c   **45.** 27 qt   **46.** 17 c   **47.** 4 qt 1 pt   **48.** 3 gal 3 qt   **49.** 3800 ml   **50.** 0.042 dl
51. 1.4 kl   **52.** 3060 cl   **53.** 1 gal 1 qt   **54.** 7 gal 1 qt   **55.** 736 ml or 0.736 L   **56.** 15.5 L or 15,500 ml   **57.** 2 gal 3 qt   **58.** 6 fl oz
59. 10.88 L   **60.** yes   **61.** 473°F   **62.** 320°F   **63.** 107.6°F   **64.** 186.8°F   **65.** 34°C   **66.** 11°C   **67.** 5.2°C   **68.** 26.7°C
69. 1.7°C   **70.** 329°F   **71.** 67.2 ft-lb   **72.** 136.5 ft-lb   **73.** 108,000 ft-lb   **74.** 9,336,000 ft-lb   **75.** 2600 BTU   **76.** 1125 cal
77. 15,120 cal   **78.** 2 hr 20 min

CHAPTER 7 TEST

1. 23 ft 4 in.   **2.** 10 qt   **3.** 1.875 lb   **4.** 5600 lb   **5.** $4\frac{3}{4}$ gal   **6.** 0.04 g   **7.** 2400 g   **8.** 36 mm   **9.** 0.43 g   **10.** 830 ml
11. 1 gal 2 qt   **12.** 3 lb 13 oz   **13.** 8 ft 3 in.   **14.** 2 gal 3 qt   **15.** 66 mm or 6.6 cm   **16.** 2.256 km or 2256 m   **17.** 28.9°C   **18.** 54.7°F
19. 5.6 m   **20.** 4 gal 3 qt   **21.** 105.8°F   **22.** 91.4 m   **23.** 16 ft 6 in.   **24.** 679 ft-lb   **25.** 20,228,000 ft-lb   **26.** 900 cal
27. 493 ft 6 in.   **28.** 150.368 m   **29.** 53.9°C

CUMULATIVE REVIEW

1. 2010; Sec. 1.3, Ex. 4   **2.** $2 \cdot 2 \cdot 2 \cdot 2 \cdot 5$ or $2^4 \cdot 5$; Sec. 2.2, Ex. 5   **3.** 33; Sec. 3.2, Ex. 7   **4.** $5\frac{1}{15}$; Sec. 3.4, Ex. 2   **5.** $\frac{1}{8}$; Sec. 4.1, Ex. 9
6. $105\frac{83}{1000}$; Sec. 4.1, Ex. 11   **7.** <; Sec. 4.2, Ex. 2   **8.** 67.69; Sec. 4.3, Ex. 6   **9.** 4.21; Sec. 4.4, Ex. 7   **10.** 0.0092; Sec. 4.4, Ex. 9
11. 7.53; Sec. 4.5, Ex. 2   **12.** 200 mi; Sec. 4.6, Ex. 4   **13.** $0.\overline{6}$; Sec. 4.7, Ex. 3   **14.** $\frac{12}{17}$; Sec. 5.1, Ex. 1   **15.** 24.5 mi/gal; Sec. 5.2, Ex. 5
16. $n = \frac{35}{6}$ or $5\frac{5}{6}$; Sec. 5.3, Ex. 6   **17.** 7 bags; Sec. 5.4, Ex. 3   **18.** 0.23; Sec. 6.1, Ex. 3   **19.** 8.33%; Sec. 6.2, Ex. 9   **20.** 14; Sec. 6.3, Ex. 7
21. $\frac{75}{30} = \frac{p}{100}$; Sec. 6.4, Ex. 5   **22.** 18%; Sec. 6.5, Ex. 5   **23.** discount: $16.25; sale price: $48.75; Sec. 6.6, Ex. 5   **24.** $120; Sec. 6.7, Ex. 1
25. $4\frac{1}{2}$ tons; Sec. 7.2, Ex. 1

Chapter 8 GEOMETRY

CHAPTER 8 PRETEST

1. obtuse; 8.1B   **2.** acute; 8.1B   **3.** 36°; 8.1C   **4.** $x = 120°$, $y = 60°$, $z = 120°$; 8.1D   **5.** 70°; 8.2A   **6.** 24.4 in.; 8.2B   **7.** 20 m; 8.3A
8. 56.52 cm; 8.3B   **9.** 15 sq ft; 8.4A   **10.** 510 cu in.; 8.5A   **11.** $\frac{2048}{3}\pi$ cu cm; 8.5A   **12.** 6; 8.6A   **13.** $\frac{9}{7}$; 8.6A   **14.** 8.544 yd; 8.6C
15. $\frac{14}{5}$; 8.7C

Exercise Set 8.1

1. line; line yz or $\overleftrightarrow{yz}$ **3.** line segment; line segment LM or $\overline{LM}$ **5.** line segment; line segment PQ or $\overline{PQ}$ **7.** ray; ray UW or $\overrightarrow{UW}$ **9.** 15° **11.** 50° **13.** 65° **15.** 95° **17.** 90° **19.** 0°; 90° **21.** straight **23.** right **25.** obtuse **27.** right **29.** 73° **31.** 163° **33.** 42° **35.** 55° **37.** $\angle MNP$ and $\angle RNO$; $\angle PNQ$ and $\angle QNR$ **39.** $\angle SPT$ and $\angle TPQ$; $\angle SPR$ and $\angle RPQ$; $\angle SPT$ and $\angle SPR$; $\angle TPQ$ and $\angle QPR$ **41.** 32° **43.** 75° **45.** $\angle x = 35°$; $\angle y = 145°$; $\angle z = 145°$ **47.** $\angle x = 77°$; $\angle y = 103°$; $\angle z = 77°$ **49.** $\angle x = 100°$; $\angle y = 80°$; $\angle z = 100°$ **51.** $\angle x = 134°$; $\angle y = 46°$; $\angle z = 134°$ **53.** $\frac{9}{8}$ or $1\frac{1}{8}$ **55.** $\frac{7}{32}$ **57.** $\frac{5}{6}$ **59.** $\frac{4}{3}$ or $1\frac{1}{3}$ **61.** 54.8° **63.** $\angle a = 60°$; $\angle b = 50°$; $\angle c = 110°$; $\angle d = 70°$; $\angle e = 120°$

Exercise Set 8.2

1. equilateral **3.** scalene **5.** isosceles **7.** 25° **9.** 13° **11.** 40° **13.** diameter **15.** rectangle **17.** parallelogram **19.** hypotenuse **21.** true **23.** true **25.** false **27.** 14 m **29.** 14.5 cm **31.** pentagon **33.** hexagon **35.** cylinder **37.** rectangular solid **39.** cone **41.** 14.8 in. **43.** 13 mi **45.** cube **47.** rectangular solid **49.** sphere **51.** pyramid **53.** 114.4 **55.** 108 **57.** 72,368 mi **59.** answers may vary

Exercise Set 8.3

1. 64 ft **3.** 36 cm **5.** 21 in. **7.** 120 cm **9.** 48 ft **11.** 66 in. **13.** 21 ft **15.** 60 ft **17.** 346 yd **19.** 22 ft **21.** $66 **23.** 36 in. **25.** 28 in. **27.** $24.08 **29.** 96 m **31.** 66 ft **33.** 128 mi **35.** 17π cm; 53.38 cm **37.** 16π mi; 50.24 mi **39.** 26π m; 81.64 m **41.** 31.43 ft **43.** 12,560 ft **45.** 23 **47.** 1 **49.** 10 **51.** 216 **53.** perimeter **55.** area **57.** area **59.** perimeter **61. a.** 62.8 m; 125.6 m **b.** yes **63.** $44 + 10\pi \approx 75.4$ m

Exercise Set 8.4

1. 7 sq m **3.** $9\frac{3}{4}$ sq yd **5.** 15 sq yd **7.** 2.25π sq in. ≈ 7.065 sq in. **9.** 36.75 sq ft **11.** 28 sq m **13.** 22 sq yd **15.** $36\frac{3}{4}$ sq ft **17.** $22\frac{1}{2}$ sq in. **19.** 25 sq cm **21.** 86 sq mi **23.** 24 sq cm **25.** 36π sq in. ≈ 113.1 sq in. **27.** 168 sq ft **29.** 113,625 sq ft **31.** 128 sq in.; $\frac{8}{9}$ sq ft **33.** 510 sq in. **35.** 168 sq ft **37.** 9200 sq ft **39.** 381 sq ft **41.** 14π in. ≈ 43.96 in. **43.** 25 ft **45.** $12\frac{3}{4}$ ft **47.** 12-in. pizza **49.** $1\frac{1}{3}$ sq ft; 192 sq in. **51.** 4605 ft

Exercise Set 8.5

1. 72 cu in. **3.** 512 cu cm **5.** $12\frac{4}{7}$ cu yd **7.** $523\frac{17}{21}$ cu in. **9.** 75 cu cm **11.** $2\frac{10}{27}$ cu in. **13.** 8.4 cu ft **15.** $10\frac{5}{6}$ cu in. **17.** 960 cu cm **19.** $\frac{1372}{3}\pi$ cu in. or $\left(457\frac{1}{3}\right)\pi$ cu in. **21.** $7\frac{1}{2}$ cu ft **23.** $12\frac{4}{7}$ cu cm **25.** 36π cu in. ≈ 113.04 cu in. **27.** 25 **29.** 9 **31.** 5 **33.** 20 **35.** 2,583,283 cu m **37.** 2,583,669 cu m **39.** 26,696.5 cu ft **41.** answers may vary

Integrated Review

1. 153°; 63° **2.** $\angle x = 75°$; $\angle y = 105°$; $\angle z = 75°$ **3.** $\angle x = 128°$; $\angle y = 52°$; $\angle z = 128°$ **4.** $\angle x = 52°$ **5.** 20 m; 25 sq m **6.** 12 ft; 6 sq ft **7.** 6π cm ≈ 18.84 cm; 9π sq cm ≈ 28.26 sq cm **8.** 32 mi; 44 sq mi **9.** 62 ft; 238 sq ft **10.** 64 cu in. **11.** 30.6 cu ft **12.** 400 cu cm **13.** $\frac{2048}{3}\pi$ cu mi $\approx 2145\frac{11}{21}$ cu mi

Calculator Explorations

1. 32 **3.** 3.873 **5.** 9.849

Exercise Set 8.6

1. 2 **3.** 25 **5.** $\frac{1}{9}$ **7.** $\frac{12}{8} = \frac{3}{2}$ **9.** 16 **11.** $\frac{3}{2}$ **13.** 1.732 **15.** 3.873 **17.** 3.742 **19.** 6.856 **21.** 2.828 **23.** 5.099 **25.** 8.426 **27.** 2.646 **29.** 13 in. **31.** 6.633 cm **33.** 5 **35.** 8 **37.** 17.205 **39.** 16.125 **41.** 12 **43.** 44.822 **45.** 42.426 **47.** 1.732 **49.** 141.42 yd **51.** 25.0 ft **53.** 340 ft **55.** $n = 4$ **57.** $n = 45$ **59.** $n = 6$ **61.** 6, 7 **63.** 10, 11 **65.** answers may vary

Exercise Set 8.7

1. congruent **3.** congruent **5.** $\frac{2}{1}$ **7.** $\frac{3}{2}$ **9.** 4.5 **11.** 6 **13.** 5 **15.** 13.5 **17.** 17.5 **19.** 8 **21.** 21.25 **23.** 500 ft **25.** 60 ft **27.** 14.4 ft **29.** $3\frac{8}{9}$ in.; no **31.** 17.5 **33.** 81 **35.** 8.4 **37.** answers may vary

Chapter 8 Review

1. right **2.** straight **3.** acute **4.** obtuse **5.** 65° **6.** 75° **7.** 108° **8.** 89° **9.** 58° **10.** 98° **11.** 90° **12.** 25° **13.** 133° and 47° **14.** 43° and 47°; 58° and 32° **15.** $\angle x = 100°$; $\angle y = 80°$; $\angle z = 80°$ **16.** $\angle x = 155°$; $\angle y = 155°$; $\angle z = 25°$ **17.** $\angle x = 53°$; $\angle y = 53°$; $\angle z = 127°$ **18.** $\angle x = 42°$; $\angle y = 42°$; $\angle z = 138°$ **19.** 103° **20.** 60° **21.** 60° **22.** 65° **23.** 4.2 m

24. 7 ft   **25.** 9.5 m   **26.** 15.2 cm   **27.** cube   **28.** cylinder   **29.** pyramid   **30.** rectangular solid   **31.** 18 in.   **32.** 2.35 m
33. pentagon   **34.** hexagon   **35.** 88 m   **36.** 30 cm   **37.** 36 m   **38.** 90 ft   **39.** 32 ft   **40.** 440 ft   **41.** 5.338 in.   **42.** 31.4 yd
43. 240 sq ft   **44.** 140 sq m   **45.** 600 sq cm   **46.** 189 sq yd   **47.** 49π sq ft $\approx$ 153.86 sq ft   **48.** 4π sq in. $\approx$ 12.56 sq in.   **49.** 119 sq in.
50. 1248 sq cm   **51.** 144 sq m   **52.** 432 sq ft   **53.** 130 sq ft   **54.** $15\frac{5}{8}$ cu in.   **55.** 84 cu ft   **56.** $62,857\frac{1}{7}$ cu cm   **57.** $346\frac{1}{2}$ cu in.
58. $2\frac{2}{3}$ cu ft   **59.** 307.72 cu in.   **60.** $7\frac{1}{2}$ cu ft   **61.** 0.5π cu ft   **62.** 1260 cu ft   **63.** 28.728 cu ft   **64.** 8   **65.** 12   **66.** 6   **67.** 1
68. $\frac{2}{5}$ **69.** $\frac{1}{10}$ **70.** 13   **71.** 29   **72.** 10.7   **73.** 93   **74.** 86.6   **75.** 28.28 cm   **76.** 88.2 ft   **77.** $37\frac{1}{2}$ **78.** $13\frac{1}{3}$ **79.** $17\frac{2}{5}$
80. $6\frac{1}{2}$ **81.** 33 ft   **82.** $x = \frac{5}{6}$ in.; $y = 2\frac{1}{6}$ in.

CHAPTER 8 TEST

1. 12°   **2.** 56°   **3.** 50°   **4.** $\angle x = 118°$; $\angle y = 62°$; $\angle z = 118°$ **5.** $\angle x = 73°$; $\angle y = 73°$; $\angle z = 73°$ **6.** 6.2 m   **7.** 10 in.   **8.** 26°
9. circumference $= 18\pi \approx 56.52$ in.; area $= 81\pi \approx 254.34$ sq in.   **10.** perimeter $= 24.6$ yd; area $= 37.1$ sq yd
11. perimeter $= 68$ in.; area $= 185$ sq in.   **12.** $62\frac{6}{7}$ cu in.   **13.** 30 cu ft   **14.** 7   **15.** 12.530   **16.** $\frac{8}{10} = \frac{4}{5}$ **17.** 16 in.   **18.** 18 cu ft
19. 62 ft   **20.** 5.66 cm   **21.** 198.08 oz   **22.** 7.5   **23.** 69 ft

CUMULATIVE REVIEW

1. nineteen and five thousand twenty-three ten-thousandths; Sec. 4.1, Ex. 3   **2.** 736.24; Sec. 4.2, Ex. 4   **3.** 47.06; Sec. 4.3, Ex. 3
4. 76.8; Sec. 4.4, Ex. 4   **5.** 76,300; Sec. 4.4, Ex. 6   **6.** 38.6; Sec. 4.5, Ex. 1   **7.** 6.35; Sec. 4.6, Ex. 7   **8.** $>$; Sec. 4.7, Ex. 4
9. $\frac{26}{31}$; Sec. 5.1, Ex. 5   **10.** yes; Sec. 5.3, Ex. 2   **11.** 17%; Sec. 6.1, Ex. 1   **12.** $\frac{19}{1000}$; Sec. 6.2, Ex. 2   **13.** $\frac{5}{4}$ or $1\frac{1}{4}$; Sec. 6.2, Ex. 3
14. 255; Sec. 6.3, Ex. 8   **15.** 52; Sec. 6.4, Ex. 9   **16.** 775 freshmen; Sec. 6.5, Ex. 1   **17.** $1710; Sec. 6.6, Ex. 3   **18.** 96 in.; Sec. 7.1, Ex. 1
19. 3200 g; Sec. 7.2, Ex. 7   **20.** 3 gal 3 qt; Sec. 7.3, Ex. 3   **21.** 84.2°F; Sec. 7.4, Ex. 2   **22.** 50°; Sec. 8.2, Ex. 1   **23.** 28 in.; Sec. 8.3, Ex. 1
24. $\frac{2}{5}$; Sec. 8.6, Ex. 3   **25.** $\frac{12}{17}$; Sec 8.7, Ex. 2

Chapter 9   STATISTICS AND PROBABILITY

PRETEST

1. April; 9.1D   **2.** July; 9.1D   **3.** 700; 9.1D   **4.**   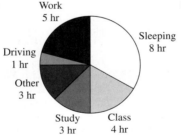 ; 9.2B   **5.** ||; 2; 9.3B   **6.** ||||; 4; 9.3B   **7.** ||||; 4; 9.3B

8. ⊪⊪ |; 6; 9.3B   **9.**   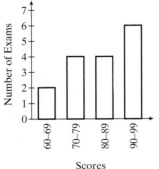 ; 9.3B   **10.** 2.65; 9.4A   **11.** 64; 9.4B   **12.** $\frac{1}{6}$; 9.5B   **13.** $\frac{1}{2}$; 9.5B   **14.** $\frac{1}{3}$; 9.5B

EXERCISE SET 9.1

1. 2000   **3.** 4000 cars   **5.** 1996, 1997, 2001   **7.** 1995, 1998   **9.** 22.5 oz   **11.** 1997, 1999, 2001   **13.** 3 oz per week
15. consumption of chicken is increasing   **17.** April   **19.** 19 deaths   **21.** February, March, April, May, June
23. Tokyo; 34.5 million or 34,500,000   **25.** New York City; 21.4 million or 21,400,000   **27.** 16 million or 16,000,000

29.

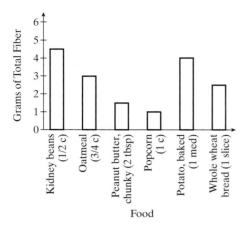

Fiber Content of Selected Foods

31.

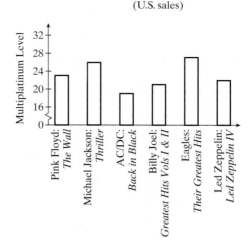

Best-Selling Albums of All Time
(U.S. sales)

33. 3.4 goals   **35.** 1998   **37.** decrease   **39.** 3.6   **41.** 6.2   **43.** 25%   **45.** 34%   **47.** 83°F   **49.** Sunday; 68°F   **51.** Tuesday; 13°F
53. bar graph; answers may vary

EXERCISE SET 9.2

1. parent or guardian's home   **3.** $\frac{9}{35}$ **5.** $\frac{9}{16}$ **7.** Asia   **9.** 37%   **11.** 17,100,000 sq mi   **13.** 2,850,000 sq mi   **15.** 55%
17. nonfiction   **19.** 31,400 books   **21.** 27,632 books   **23.** 25,120 books   **25.**   **27.** $2^2 \times 5$ **29.** $2^3 \times 5$

31. 5×17 **33.** answers may vary   **35.** 129,600,002 sq km   **37.** 55,542,858 sq km   **39.** 583 respondents   **43.** 2128 respondents

EXERCISE SET 9.3

1. 15 adults   **3.** 61 adults   **5.** 24 adults   **7.** 12 adults   **9.** $\frac{9}{100}$ **11.** 45–54   **13.** 17 million householders   **15.** 45 million householders
17. answers may vary   **19.** |; 1   **21.** ⊞|||; 8   **23.** ⊞|; 6   **25.** ⊞|; 6   **27.** ||; 2   **29.**   **31.** 90

33. 20   **35.** 27   **37.** 1996   **39.** 1999

INTEGRATED REVIEW

1. 69 lb   **2.** 78 lb   **3.** 1985   **4.** 1995 and 2000   **5.** Oroville Dam; 755 ft   **6.** New Bullards Bar Dam; 635 ft   **7.** 15 ft   **8.** 4 dams
9. Thursday and Saturday; 100°F   **10.** Monday; 82°F   **11.** Sunday, Monday, and Tuesday   **12.** Wednesday, Thursday, Friday, and Saturday

13. 70 qt containers   **14.** 52 qt containers   **15.** 2 qt containers   **16.** 6 qt containers   **17.** ||; 2   **18.** |; 1   **19.** |||; 3   **20.** ||||| |; 6
21. |||||; 5   **22.**

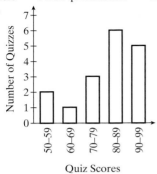

MENTAL MATH

1. 4   **3.** 3

EXERCISE SET 9.4

1. mean: 29; median: 28; no mode   **3.** mean: 8.1; median: 8.2; mode: 8.2   **5.** mean: 0.6; median: 0.6; mode: 0.2 and 0.6
7. mean: 370.9; median: 313.5; no mode   **9.** 1416 ft   **11.** 1332 ft   **13.** answers may vary   **15.** 2.79   **17.** 3.46   **19.** 6.8   **21.** 6.9

23. 85.5   **25.** 73   **27.** 70 and 71   **29.** 9 rates   **31.** $\frac{1}{3}$ **33.** $\frac{3}{5}$ **35.** $\frac{11}{15}$ **37.** 35, 35, 37, 43

MENTAL MATH

1. $\frac{1}{2}$ **3.** $\frac{1}{2}$

EXERCISE SET 9.5

1.

Outcomes

a — 1 a, 1
    2 a, 2
    3 a, 3
e — 1 e, 1
    2 e, 2
    3 e, 3
i — 1 i, 1
    2 i, 2
    3 i, 3
o — 1 o, 1
    2 o, 2
    3 o, 3
u — 1 u, 1
    2 u, 2
    3 u, 3

15 outcomes

3.

Outcomes

Red   Red
Blue   Blue
Yellow Yellow

3 outcomes

5.

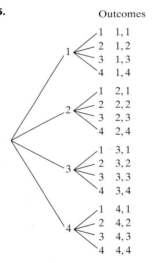

Outcomes

1 — 1, 1
    1, 2
    1, 3
    1, 4
2 — 2, 1
    2, 2
    2, 3
    2, 4
3 — 3, 1
    3, 2
    3, 3
    3, 4
4 — 4, 1
    4, 2
    4, 3
    4, 4

16 outcomes

7.

Outcomes

Red — 1 Red, 1
      2 Red, 2
      3 Red, 3
      4 Red, 4
Blue — 1 Blue, 1
       2 Blue, 2
       3 Blue, 3
       4 Blue, 4
Yellow — 1 Yellow, 1
         2 Yellow, 2
         3 Yellow, 3
         4 Yellow, 4

12 outcomes

9.

Outcomes

H — 1 H, 1
    2 H, 2
    3 H, 3
    4 H, 4
T — 1 T, 1
    2 T, 2
    3 T, 3
    4 T, 4

8 outcomes

11. $\frac{1}{6}$ **13.** $\frac{1}{3}$ **15.** $\frac{1}{2}$ **17.** $\frac{1}{3}$ **19.** $\frac{2}{3}$ **21.** $\frac{1}{7}$

23. $\frac{2}{7}$ **25.** $\frac{19}{100}$ **27.** $\frac{1}{20}$ **29.** $\frac{5}{6}$ **31.** $\frac{1}{6}$ **33.** $\frac{20}{3}$ or $6\frac{2}{3}$ **35.** $\frac{1}{52}$

37. $\frac{1}{13}$ **39.** $\frac{1}{4}$ **41.** $\frac{1}{12}$ **43.** 0   **45.** answers may vary

CHAPTER 9 REVIEW

1. 4,000,000 **2.** 1,750,000 **3.** South **4.** Northeast **5.** Midwest, South, and West **6.** Northeast **7.** 7.5% **8.** 2000
9. 1980, 1990, 2000 **10.** answers may vary **11.** $2,100,000 **12.** $1,200,000 **13.** 1991 and 1992 **14.** 1999 and 2000

15. 1990, 1991, 1992, 1993, 1994 **16.** mortgage payment **17.** utilities **18.** $1225 **19.** $700 **20.** $\frac{39}{160}$ **21.** $\frac{7}{40}$ **22.** $\frac{5}{7}$ **23.** 20 states

24. 11 states **25.** 1 state **26.** 29 states **27.** 4 employees **28.** 1 employee **29.** 9 employees **30.** 18 employees **31.** ⊪⊪; 5
32. ⦀; 3 **33.** ⦀⦀; 4 **34.**

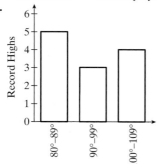

35. mean: 17.8; median: 14; no mode **36.** mean: 55.2; median: 60; no mode

37. mean: $24,500; median: $20,000; mode: $20,000 **38.** mean: 447.3; median: 420; mode: 400 **39.** 3.25 **40.** 2.57

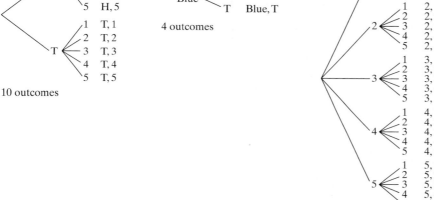

41.
Outcomes

H — 1 H, 1
2 H, 2
3 H, 3
4 H, 4
5 H, 5

T — 1 T, 1
2 T, 2
3 T, 3
4 T, 4
5 T, 5

10 outcomes

42.
Outcomes

Red — H Red, H
T Red, T

Blue — H Blue, H
T Blue, T

4 outcomes

43.
Outcomes

1 — 1 1,1
2 1,2
3 1,3
4 1,4
5 1,5

2 — 1 2,1
2 2,2
3 2,3
4 2,4
5 2,5

3 — 1 3,1
2 3,2
3 3,3
4 3,4
5 3,5

4 — 1 4,1
2 4,2
3 4,3
4 4,4
5 4,5

5 — 1 5,1
2 5,2
3 5,3
4 5,4
5 5,5

25 outcomes

44.
Outcomes

Red — Red Red, Red
Blue Red, Blue

Blue — Red Blue, Red
Blue Blue, Blue

4 outcomes

45.
Outcomes

1 — Red 1, Red
Blue 1, Blue

2 — Red 2, Red
Blue 2, Blue

3 — Red 3, Red
Blue 3, Blue

4 — Red 4, Red
Blue 4, Blue

5 — Red 5, Red
Blue 5, Blue

10 outcomes

46. $\frac{1}{6}$ **47.** $\frac{1}{6}$ **48.** $\frac{1}{5}$ **49.** $\frac{1}{5}$ **50.** $\frac{3}{5}$ **51.** $\frac{2}{5}$

CHAPTER 9 TEST

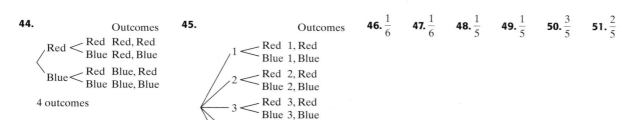

1. $225 **2.** 3rd week; $350 **3.** $1100 **4.** Ford F-Series **5.** Ford Focus **6.** Ford Focus, Ford Taurus, Honda Civic

7.

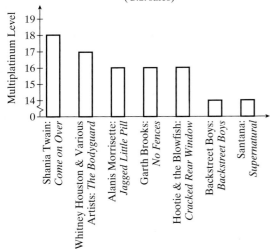

Best-Selling Albums of the 1990s
(U.S. sales)

8. 1.5%   **9.** 1990, 1991, 2000   **10.** 1994–1995, 1995–1996, 1998–1999, 1999–2000

11. $\frac{17}{40}$ **12.** $\frac{31}{22}$ **13.** 40,920,000 people   **14.** 21,120,000 people   **15.** 9 students   **16.** 11 students

17.

Class Interval (Scores)	Tally	Class Frequency (Number of Students)
40–49	|	1
50–59	|||	3
60–69	||||	4
70–79	⊞	5
80–89	⊞ |||	8
90–99	||||	4

18.

19. mean: 38.4; median: 42; no mode

20. mean: 12.625; median: 21.5; mode: 12 and 16   **21.** 3.07

22.

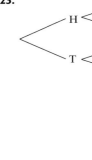

23.

24. $\frac{1}{10}$ **25.** $\frac{1}{5}$

CUMULATIVE REVIEW

1. 28; Sec. 1.9, Ex. 10   **2.** $\frac{5}{18}$; Sec. 2.3, Ex. 4   **3.** $8\frac{3}{10}$; Sec. 3.4, Ex. 3   **4.** 8.4 sq ft; Sec. 4.7, Ex. 7   **5.** $\frac{2}{3}$; Sec. 5.1, Ex. 4

6. $\frac{\$180}{1\ \text{week}}$; Sec. 5.2, Ex. 2   **7.** $\frac{45\ \text{mi}}{2\ \text{gal}}$; Sec. 5.2, Ex. 3   **8.** yes; Sec. 5.3, Ex. 4   **9.** 22.4 cc; Sec. 5.4, Ex. 2   **10.** 0.046; Sec. 6.1, Ex. 4

11. 1.9; Sec. 6.1, Ex. 5   **12.** $\frac{2}{5}$; Sec. 6.2, Ex. 1   **13.** $\frac{1}{3}$; Sec. 6.2, Ex. 4   **14.** $5 = n \cdot 20$; Sec. 6.3, Ex. 1

15. sales tax: $6.41; total price: $91.91; Sec. 6.6, Ex. 1   **16.** $2400; Sec. 6.7, Ex. 3   **17.** $2\frac{1}{3}$ yd; Sec. 7.1, Ex. 2   **18.** 4 lb 11 oz; Sec. 7.2, Ex. 5

19. 0.0235 g; Sec. 7.2, Ex. 8   **20.** 3.21 L; Sec. 7.3, Ex. 7   **21.** 59°F; Sec. 7.4, Ex. 1   **22.** 42°; Sec. 8.1, Ex. 4   **23.** $\frac{1}{6}$; Sec. 8.6, Ex. 2   **24.** 14, 77;

Sec. 9.4, Ex. 5   **25.** $\frac{1}{4}$; Sec. 9.5, Ex.3

Chapter 10 SIGNED NUMBERS

CHAPTER 10 PRETEST

1.

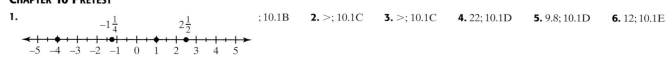

; 10.1B   **2.** >; 10.1C   **3.** >; 10.1C   **4.** 22; 10.1D   **5.** 9.8; 10.1D   **6.** 12; 10.1E

7. 19; 10.2A   **8.** −24; 10.3A   **9.** −15; 10.3A   **10.** 5; 10.3B   **11.** 39; 10.3B   **12.** 54; 10.4A   **13.** 120; 10.4A   **14.** 81; 10.4A
15. 46; 10.4B   **16.** −1.2; 10.4B   **17.** −19; 10.5A   **18.** −28; 10.5A   **19.** −$8; 10.3C   **20.** −72; 10.4A

EXERCISE SET 10.1

1. −1445   **3.** +14,410   **5.** −15   **7.** −317   **9.** −20,786 thousand   **11.** cooler   **13.** −91

15.

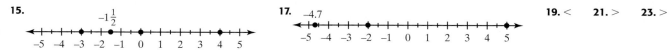

17.

19. <   **21.** >   **23.** >

25. >   **27.** <   **29.** <   **31.** <   **33.** >   **35.** 5   **37.** 8   **39.** 0   **41.** 5   **43.** 8.1   **45.** $\frac{9}{10}$ **47.** $\frac{3}{8}$ **49.** 7.6   **51.** −5
53. 4   **55.** −23.6   **57.** $\frac{9}{16}$ **59.** 0.7   **61.** $-\frac{17}{18}$ **63.** 7   **65.** −20   **67.** −3   **69.** 8
71. 14   **73.** 29   **75.** 13   **77.** 35   **79.** 360   **81.** true   **83.** true   **85.** answers may vary

CALCULATOR EXPLORATIONS

1. −159   **3.** 44   **5.** −894.855

MENTAL MATH

1. 5   **3.** −35

EXERCISE SET 10.2

1.

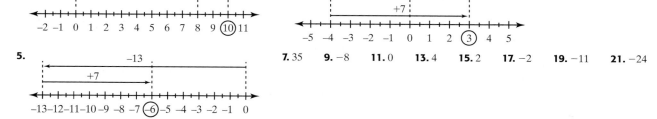

3.

5.

7. 35   **9.** −8   **11.** 0   **13.** 4   **15.** 2   **17.** −2   **19.** −11   **21.** −24

23. −57   **25.** −223   **27.** 0   **29.** 7   **31.** −3   **33.** −9   **35.** 30   **37.** 20   **39.** 51   **41.** −8.5   **43.** 4.6   **45.** $-\frac{5}{6}$ **47.** $-\frac{7}{10}$
49. −7   **51.** −24.6   **53.** 16   **55.** 13   **57.** 2°C   **59.** −165 + (−16) = −181; 181 ft below the surface   **61.** Team 1, 7; Team 2, 6; winning
team, Team 1   **63.** −19°F   **65.** −7679 m   **67.** 44   **69.** 0   **71.** 28   **73.** true   **75.** false   **77.** answers may vary

MENTAL MATH

1. 0   **3.** 0

EXERCISE SET 10.3

1. 0   **3.** 5   **5.** −5   **7.** 14   **9.** 3   **11.** −18   **13.** −14   **15.** −3.3   **17.** 4.02   **19.** $\frac{2}{5}$ **21.** $-\frac{3}{10}$ **23.** $\frac{5}{6}$ **25.** −38   **27.** −17
29. 13   **31.** 2   **33.** 0   **35.** −1   **37.** −27   **39.** 40   **41.** 262°F   **43.** −$16   **45.** 3919 ft   **47.** 652 ft   **49.** 144 ft   **51.** 382°C
53. Saturn; 39°C   **55.** 0   **57.** 8   **59.** 1058   **61.** −4   **63.** 0   **65.** −12   **67.** false   **69.** answers may vary   **71.** answers may vary

INTEGRATED REVIEW

1. +29,028   **2.** −35,840   **3.** −26,322   **4.** +28,250   **5.**

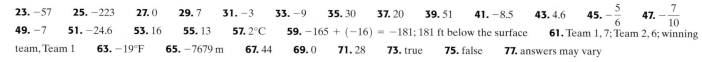

6. >   **7.** <   **8.** <

9. > **10.** > **11.** > **12.** > **13.** < **14.** 1 **15.** 1 **16.** 0 **17.** −4 **18.** 8.6 **19.** 100.3 **20.** $\frac{3}{4}$ **21.** $-\frac{3}{4}$ **22.** −6

23. 3 **24.** −89.1 **25.** $\frac{2}{9}$ **26.** 5 **27.** −20.8 **28.** $\frac{2}{9}$ **29.** −2 **30.** $\frac{3}{4}$ **31.** −4 **32.** −9.6 **33.** −6 **34.** 12 **35.** 84

36. −3.44 **37.** $-\frac{11}{10} = -1\frac{1}{10}$ **38.** 6 **39.** 1 **40.** −19 **41.** 12 **42.** −4 **43.** 189°F

EXERCISE SET 10.4

1. 6 **3.** −36 **5.** −3.12 **7.** 0 **9.** $\frac{6}{35}$ **11.** −48 **13.** −8 **15.** 80 **17.** 0 **19.** 4 **21.** −27 **23.** 25 **25.** −125 **27.** 0

29. $-\frac{21}{32}$ **31.** 1 **33.** 43.4 **35.** −8 **37.** −4 **39.** −5 **41.** 8 **43.** 0 **45.** −13 **47.** −26 **49.** $\frac{7}{2}$ **51.** −5 **53.** −6

55. 3 **57.** −300 **59.** $-\frac{4}{5}$ **61.** $3 \cdot (-4) = -12$ yd **63.** $5 \cdot (-20) = -100$ ft **65.** −1.75 **67.** −$472 million **69. a.** −8 condors;

b. −1 condor per year **c.** 157 condors; **d.** 11 condors per year **71.** 225 **73.** 109 **75.** 8 **77.** false **79.** true
81. a. −2232 Land Rovers; **b.** −15,624 Land Rovers; **c.** 13,756 Land Rovers **83.** answers may vary

CALCULATOR EXPLORATIONS

1. 4.8 **3.** −258

EXERCISE SET 10.5

1. 3 **3.** −1 **5.** −7 **7.** −14 **9.** $-\frac{2}{9}$ **11.** −8 **13.** 7 **15.** −1 **17.** 4 **19.** −3 **21.** −55 **23.** 24 **25.** 16 **27.** 15

29. −3 **31.** −3 **33.** 4 **35.** 16 **37.** −27 **39.** 34 **41.** 0.65 **43.** −59 **45.** −7 **47.** $\frac{5}{36}$ **49.** −11 **51.** 36 **53.** −117

55. 30 **57.** −3 **59.** −2.21 **61.** 4050 **63.** 45 **65.** 32 in. **67.** 30 ft **69.** $2 \cdot (7 - 5) \cdot 3 = 12$ **71.** $-6 \cdot (10 - 4) = -36$
73. 20,736 **75.** answers may vary

CHAPTER 10 REVIEW

1. −1435 **2.** +7562 **3.** **4.**

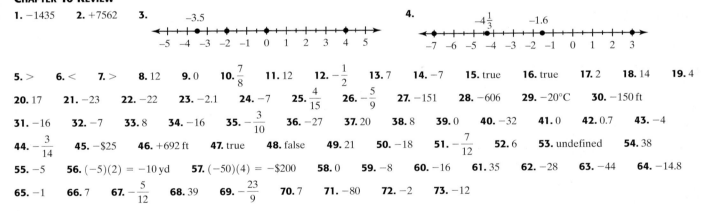

5. > **6.** < **7.** > **8.** 12 **9.** 0 **10.** $\frac{7}{8}$ **11.** 12 **12.** $-\frac{1}{2}$ **13.** 7 **14.** −7 **15.** true **16.** true **17.** 2 **18.** 14 **19.** 4

20. 17 **21.** −23 **22.** −22 **23.** −2.1 **24.** −7 **25.** $\frac{4}{15}$ **26.** $-\frac{5}{9}$ **27.** −151 **28.** −606 **29.** −20°C **30.** −150 ft

31. −16 **32.** −7 **33.** 8 **34.** −16 **35.** $-\frac{3}{10}$ **36.** −27 **37.** 20 **38.** 8 **39.** 0 **40.** −32 **41.** 0 **42.** 0.7 **43.** −4

44. $-\frac{3}{14}$ **45.** −$25 **46.** +692 ft **47.** true **48.** false **49.** 21 **50.** −18 **51.** $-\frac{7}{12}$ **52.** 6 **53.** undefined **54.** 38

55. −5 **56.** $(-5)(2) = -10$ yd **57.** $(-50)(4) = -$200 **58.** 0 **59.** −8 **60.** −16 **61.** 35 **62.** −28 **63.** −44 **64.** −14.8

65. −1 **66.** 7 **67.** $-\frac{5}{12}$ **68.** 39 **69.** $-\frac{23}{9}$ **70.** 7 **71.** −80 **72.** −2 **73.** −12

CHAPTER 10 TEST

1. 3 **2.** −6 **3.** −10 **4.** 4 **5.** $-\frac{1}{2}$ **6.** 12 **7.** 65 **8.** 5 **9.** 12 **10.** −6 **11.** 50 **12.** −2 **13.** −11 **14.** −46

15. −117 **16.** 3456 **17.** $-\frac{14}{81}$ **18.** −213 **19.** −1 **20.** −2 **21.** 2 **22.** −5 **23.** 34,335 ft **24.** $14,893 + (-147) = +14,746$;

14,746 ft **25.** +41

CUMULATIVE REVIEW

1. 78,875; Sec. 1.6, Ex. 5 **2.** $\frac{4}{5}$; Sec. 2.5, Ex. 7 **3.** $\frac{17}{21}$; Sec. 3.3, Ex. 3 **4.** 48.26; Sec. 4.1, Ex. 5 **5.** 6.095; Sec. 4.1, Ex. 6

6. 3.432; Sec. 4.3, Ex. 5 **7.** 5.6881; Sec. 4.6, Ex. 8 **8.** $\frac{4500 \text{ dollars}}{1 \text{ month}}$ or 4500 dollars/month; Sec. 5.2, Ex. 4 **9.** $n = 2$; Sec. 5.3, Ex. 5

10. 46%; Sec. 6.1, Ex. 2 **11.** 45%; Sec. 6.2, Ex. 6 **12.** 150%; Sec. 6.2, Ex. 8 **13.** 200; Sec. 6.3, Ex. 10 **14.** $\frac{101}{200} = \frac{p}{100}$; Sec. 6.4, Ex. 2

15. $160; Sec. 6.7, Ex. 2 **16.** 9 ft 1 in.; Sec. 7.1, Ex. 5 **17.** 37°C; Sec. 7.4, Ex. 5 **18.** 73°; Sec. 8.1, Ex. 5 **19.** 60°; Sec. 8.2, Ex. 2

20. 5.1 sq mi; Sec. 8.4, Ex. 2 **21.** $113\frac{1}{7}$ cu in.; Sec. 8.5, Ex. 2 **22.** 57; Sec. 9.4, Ex. 3 **23.** $\frac{1}{3}$; Sec. 9.5, Ex. 4 **24.** −23; Sec. 10.2, Ex. 3

25. 82; Sec. 10.5, Ex. 6

Chapter 11 INTRODUCTION TO ALGEBRA

CHAPTER 11 PRETEST

1. -9; 11.1A **2.** $-10x$; 11.1B **3.** $7x - 4y + 20$; 11.1B **4.** $-12m + 6$; 11.1C **5.** no; 11.2A **6.** yes; 11.2A **7.** $y = 2$; 11.2B
8. $x = -8$; 11.2B **9.** $z = -8$; 11.3A **10.** $a = 35$; 11.3A **11.** $x = -2$; 11.3A **12.** $x = 7$; 11.3A **13.** $b = 2$; 11.4A **14.** $n = 3$; 11.4A
15. $y = 1$; 11.4B **16.** $w = 2$; 11.4B **17.** $3 \cdot 12 = 36$; 11.4C **18.** $2(4 + 6) = 20$; 11.4C **19.** $\dfrac{10}{3x} = 40$; 11.5B **20.** 15; 11.5C

MENTAL MATH

1. unlike **3.** like **5.** unlike **7.** like

EXERCISE SET 11.1

1. -3 **3.** 8 **5.** 4 **7.** -3 **9.** 1 **11.** 133 **13.** 20 **15.** -4 **17.** $-\dfrac{4}{3}$ **19.** -4 **21.** 2000 sq ft **23.** 64 ft **25.** $360

27. 78.5 sq ft **29.** 23°F **31.** 288 cu in. **33.** $8x$ **35.** $-4n$ **37.** $-2c$ **39.** $-4x$ **41.** $13a - 8$ **43.** $-0.9x + 11.2$
45. $2x - 7$ **47.** $-5x + 4y - 5$ **49.** $30x$ **51.** $-22y$ **53.** $2y + 4$ **55.** $5a - 40$ **57.** $-12x - 28$ **59.** $2x - 9$ **61.** $27n - 20$
63. $7w + 15$ **65.** $-x - 12$ **67.** $16y^2$ sq cm **69.** -3 **71.** 8 **73.** 0 **75.** 45,996.2 sq in. **77.** $4824q + 12,274$
79. answers may vary **81.** $(20x + 16)$ sq mi

MENTAL MATH

1. 2 **3.** -1

EXERCISE SET 11.2

1. yes **3.** no **5.** yes **7.** yes **9.** 18 **11.** -8 **13.** 9 **15.** -16 **17.** 3 **19.** $\dfrac{1}{8}$ **21.** 6 **23.** 8 **25.** 5.3 **27.** -1

29. -20.1 **31.** 2 **33.** 0 **35.** -28 **37.** answers may vary **39.** 1 **41.** 1 **43.** 1 **45.** 162,964 **47.** 2199 yd
49. $14,722,244,000

EXERCISE SET 11.3

1. 4 **3.** -4 **5.** 0 **7.** -17 **9.** 50 **11.** 25 **13.** -30 **15.** $\dfrac{1}{3}$ **17.** $-\dfrac{2}{3}$ **19.** $-\dfrac{15}{64}$ **21.** $\dfrac{2}{3}$ **23.** -4 **25.** 8 **27.** -0.1

29. 2 **31.** -8 **33.** -3 **35.** 1 **37.** 0 **39.** -2 **41.** 25 **43.** 1 **45.** -10 **47.** answers may vary **49.** -3648 **51.** 6.5 hr
53. 58.8 mph

INTEGRATED REVIEW

1. 4 **2.** -3 **3.** 1 **4.** -4 **5.** $8x$ **6.** $-4y$ **7.** $-2a - 2$ **8.** $-2x + 3y - 7$ **9.** $-8x$ **10.** $5y + 10$ **11.** $3x + 12$
12. $11x + 16$ **13.** $(12x - 6)$ sq m **14.** $25y^2$ sq in. **15.** 13 **16.** -9 **17.** 5 **18.** 0 **19.** -4 **20.** 25 **21.** -1 **22.** -3

23. 6 **24.** 8 **25.** $-\dfrac{9}{11}$ **26.** $\dfrac{7}{10}$

CALCULATOR EXPLORATIONS

1. yes **3.** no **5.** yes

EXERCISE SET 11.4

1. 3 **3.** 4 **5.** -4 **7.** -3 **9.** -12 **11.** 100 **13.** -3.9 **15.** -4 **17.** 5 **19.** 1 **21.** -4 **23.** 2 **25.** 5 **27.** -3
29. 2 **31.** -2 **33.** 3 **35.** -1 **37.** 4 **39.** -4 **41.** 3 **43.** -1 **45.** 64 **47.** -45 **49.** 5 **51.** -5 **53.** 8 **55.** 0

57. -22 **59.** 4 **61.** 1 **63.** -30 **65.** $-42 + 16 = -26$ **67.** $-5(-29) = 145$ **69.** $3(-14 - 2) = -48$ **71.** $\dfrac{100}{2(50)} = 1$

73. 50 million returns **75.** 38 million returns **77.** 123.26°F **79.** -9.4°F

EXERCISE SET 11.5

1. $x + 5$ **3.** $x + 8$ **5.** $20 - x$ **7.** $512x$ **9.** $\dfrac{x}{2}$ **11.** $(17 + x) + 5x$ **13.** $5x$ **15.** $11 - x$ **17.** $50 - 8x$ **19.** $-5 + x = -7$

21. $3x = 27$ **23.** $-20 - x = 104$ **25.** 8 **27.** 9 **29.** 5 **31.** 12 **33.** 8 **35.** 5 **37.** 16 U.S. titles **39.** 80 books
41. truck: 35 mph; car: 70 mph **43.** $225 **45.** 68 points **47.** 590 **49.** 1000 **51.** 3000 **53.** answers may vary

CHAPTER 11 REVIEW

1. -5 **2.** 17 **3.** undefined **4.** 0 **5.** 129 **6.** -2 **7.** 4 **8.** 20 **9.** 8 cu ft **10.** 64 cu ft **11.** $1800 **12.** $300
13. $10y - 15$ **14.** $-6y - 10$ **15.** $-6a - 7$ **16.** $-8y + 2$ **17.** $4.6x - 11.9$ **18.** $-2.1x - 20$ **19.** $-2x - 10$ **20.** $-3y - 24$
21. $11x - 12$ **22.** $-3m + 16$ **23.** $-5a + 4$ **24.** $12y - 9$ **25.** $(6x - 3)$ sq yd **26.** $25y^2$ sq m **27.** yes **28.** no **29.** -2
30. 7 **31.** $-\dfrac{1}{2}$ **32.** $-\dfrac{6}{11}$ **33.** -1 **34.** -25 **35.** -6 **36.** -20 **37.** 1.3 **38.** 2.4 **39.** 7 **40.** -9 **41.** 1 **42.** -5

43. $-\dfrac{4}{5}$ **44.** -24 **45.** -120 **46.** 13   **47.** 4   **48.** 3   **49.** 13   **50.** -6 **51.** 5   **52.** 12   **53.** 17   **54.** 7   **55.** 63   **56.** 33

57. -2.25 **58.** 1.3   **59.** 11   **60.** -5 **61.** 2   **62.** 6   **63.** $20 - (-8) = 28$ **64.** $5[2 + (-6)] = -20$ **65.** $\dfrac{-75}{5 + 20} = -3$

66. $-2 - 19 = -21$ **67.** $2x + 11$ **68.** $-5x - 50$ **69.** $\dfrac{70}{x + 6}$ **70.** $2(x - 13)$ **71.** $2x - 8 = 40$ **72.** $\dfrac{x}{2} - 12 = 10$

73. $x - 3 = \dfrac{x}{4}$ **74.** $6x = x + 2$ **75.** 5   **76.** -16 **77.** 2386 votes   **78.** 42 CDs

CHAPTER 11 TEST

1. -1 **2.** $-5x + 5$ **3.** $-6y - 14$ **4.** $14z - 8$ **5.** $(9x - 3)$ sq m   **6.** 7   **7.** $-\dfrac{1}{2}$ **8.** -12 **9.** 40   **10.** 24   **11.** 3   **12.** -2
13. -2 **14.** 4.5   **15.** 0   **16.** -2 **17.** 7   **18.** 4   **19.** 0   **20.** 6000 sq ft   **21.** 30 sq ft   **22. a.** $17x$ **b.** $20 - 2x$ **23.** -2
24. 34 points   **25.** 244 women

CUMULATIVE REVIEW

1. 0.8496; Sec. 4.4, Ex. 3   **2. a.** $\dfrac{5}{7}$ **b.** $\dfrac{7}{24}$; Sec. 5.1, Ex. 7   **3.** 5; Sec. 6.3, Ex. 9   **4.** 75%; Sec. 6.3, Ex. 11   **5.** 48 oz; Sec. 7.2, Ex. 2

6. 1.13 ml; Sec. 7.3, Ex. 10   **7. a.** line   **b.** line segment   **c.** angle   **d.** ray; Sec. 8.1, Ex. 1   **8.** 10 cm; Sec. 8.2, Ex. 3   **9.** 50 ft; Sec. 8.3, Ex. 6
10. 56 sq cm; Sec. 8.4, Ex. 1   **11.** 25; Sec. 8.6, Ex. 5   **12.** 46 ft; Sec. 8.7, Ex. 4   **13. a.** 15 reptile species   **b.** birds; Sec. 9.1, Ex. 2
14. <; Sec. 10.1, Ex. 6   **15.** >; Sec. 10.1, Ex. 5   **16.** 2; Sec. 10.1, Ex. 10   **17.** 1.2; Sec. 10.1, Ex. 14   **18.** -6; Sec. 10.2, Ex. 4
19. $-\dfrac{1}{2}$; Sec. 10.2, Ex. 6   **20.** -7; Sec. 10.3, Ex. 6   **21.** 1; Sec. 10.3, Ex. 7   **22.** 10; Sec. 10.4, Ex. 2   **23.** $\dfrac{1}{3}$; Sec. 10.4, Ex. 4

24. 3; Sec. 10.5, Ex. 7   **25.** 2; Sec. 11.4, Ex. 7

SOLUTIONS TO SELECTED EXERCISES

Chapter 1

Exercise Set 1.2

1. The place value of the 5 in 352 is tens.

5. The place value of the 5 in 62,500,000 is hundred-thousands.

9. 5420 is written as five thousand, four hundred twenty.

13. 1,620,000 is written as one million, six hundred twenty thousand.

17. 62,997 is written as sixty-two thousand, nine hundred ninety-seven.

21. 13,600,000 is written as thirteen million, six hundred thousand.

25. Twenty-nine thousand, nine hundred in standard form is 29,900.

29. Three million, fourteen in standard form is 3,000,014.

33. One million, four hundred thousand dollars in standard form is $1,400,000.

37. $406 = 400 + 6$

41. $62,407 = 60,000 + 2000 + 400 + 7$

45. $39,680,000 = 30,000,000 + 9,000,000 + 600,000 + 80,000$

49. $5492 = 5000 + 400 + 90 + 2$

53. Shih Tzu with thirty seven thousand, five hundred ninety nine.

57. The largest number is achieved when the largest number available is used for each place value when reading from left to right. Thus, the largest number possible is 7632.

61. The town is Canton.

Exercise Set 1.3

1.
$$\begin{array}{r} 14 \\ + 22 \\ \hline 36 \end{array}$$

5.
$$\begin{array}{r} 12 \\ 13 \\ + 24 \\ \hline 49 \end{array}$$

9.
$$\begin{array}{r} \overset{1}{5}3 \\ + 64 \\ \hline 117 \end{array}$$

13.
$$\begin{array}{r} \overset{11}{3}8 \\ + 79 \\ \hline 117 \end{array}$$

17.
$$\begin{array}{r} \overset{2}{6} \\ 21 \\ 14 \\ 9 \\ + 12 \\ \hline 62 \end{array}$$

21.
$$\begin{array}{r} \overset{1}{6}2 \\ 18 \\ + 14 \\ \hline 94 \end{array}$$

25.
$$\begin{array}{r} \overset{111}{7}542 \\ 49 \\ + 682 \\ \hline 8273 \end{array}$$

29.
$$\begin{array}{r} \overset{1\ 2}{6}27 \\ 628 \\ + 629 \\ \hline 1884 \end{array}$$

33.
$$\begin{array}{r} \overset{111}{5}07 \\ 593 \\ + 10 \\ \hline 1110 \end{array}$$

37.
$$\begin{array}{r} \overset{1122}{4}9 \\ 628 \\ 5762 \\ + 29,462 \\ \hline 35,901 \end{array}$$

41. $8 + 3 + 5 + 7 + 5 + 1 = 8 + 1 + 3 + 7 + 5 + 5$
$$= 9 + 10 + 10 = 29$$
The perimeter is 29 inches.

45. Opposite sides of a rectangle have the same lengths, $4 + 8 + 4 + 8 = 12 + 12 = 24$
The perimeter is 24 inches.

49.
$$\begin{array}{r} 3560 \\ + 3124 \\ \hline 6684 \end{array}$$
Mt. Mitchell is 6684 feet above sea level.

53.
$$\begin{array}{r} 2305 \\ + 2662 \\ \hline 4967 \end{array}$$
Dan Marino completed 4967 passes.

57.
$$\begin{array}{r} 1940 \\ + 35 \\ \hline 1975 \end{array}$$
Marion Jones was born in 1975.

61. Texas, Florida, and California have the most Target stores
$$\begin{array}{r} \overset{2\,1}{1}53 \\ 68 \\ + 91 \\ \hline 312 \end{array}$$
These three states have the most Target stores, with a total of 312 stores.

65. answers may vary

Exercise Set 1.4

1.
$$\begin{array}{r} 67 \\ - 23 \\ \hline 44 \end{array}$$
Check:
$$\begin{array}{r} 44 \\ + 23 \\ \hline 67 \end{array}$$

5.
$$\begin{array}{r} 389 \\ - 124 \\ \hline 265 \end{array}$$
Check:
$$\begin{array}{r} 265 \\ + 124 \\ \hline 389 \end{array}$$

9.
$$\begin{array}{r} 998 \\ - 453 \\ \hline 545 \end{array}$$
Check:
$$\begin{array}{r} 545 \\ + 453 \\ \hline 998 \end{array}$$

13.
$$\begin{array}{r} 62 \\ - 37 \\ \hline 25 \end{array}$$
Check:
$$\begin{array}{r} \overset{1}{2}5 \\ + 37 \\ \hline 62 \end{array}$$

17.
$$\begin{array}{r} 938 \\ - 792 \\ \hline 146 \end{array}$$
Check:
$$\begin{array}{r} \overset{1}{1}46 \\ + 792 \\ \hline 938 \end{array}$$

21.
$$\begin{array}{r} 600 \\ - 432 \\ \hline 168 \end{array}$$
Check:
$$\begin{array}{r} \overset{1\,1}{1}68 \\ + 432 \\ \hline 600 \end{array}$$

25.
$$\begin{array}{r} 923 \\ - 476 \\ \hline 447 \end{array}$$
Check:
$$\begin{array}{r} \overset{1\,1}{4}47 \\ + 476 \\ \hline 923 \end{array}$$

29.
$$\begin{array}{r} 533 \\ - 29 \\ \hline 504 \end{array}$$
Check:
$$\begin{array}{r} \overset{1}{5}04 \\ + 29 \\ \hline 533 \end{array}$$

33.
$$\begin{array}{r} 1983 \\ - 1904 \\ \hline 79 \end{array}$$
Check:
$$\begin{array}{r} \overset{1}{7}9 \\ + 1904 \\ \hline 1983 \end{array}$$

37.
$$\begin{array}{r} 50,000 \\ - 17,289 \\ \hline 32,711 \end{array}$$
Check:
$$\begin{array}{r} \overset{1\,1}{3}2,711 \\ + 17,289 \\ \hline 50,000 \end{array}$$

41.
$$\begin{array}{r} 51,111 \\ - 19,898 \\ \hline 31,213 \end{array}$$
Check:
$$\begin{array}{r} \overset{1\,1}{3}1,213 \\ + 19,898 \\ \hline 51,111 \end{array}$$

45.
$$\begin{array}{r} 41 \\ - 21 \\ \hline 20 \end{array}$$
Check:
$$\begin{array}{r} 20 \\ + 21 \\ \hline 41 \end{array}$$

49.
$$\begin{array}{r} 503 \\ -\,239 \\ \hline 264 \end{array}$$
Dyllis has 264 more pages to read.

53.
$$\begin{array}{r} 20{,}320 \\ -\,14{,}255 \\ \hline 6065 \end{array}$$
Mt. McKinley is 6065 feet higher than Long's Peak.

57.
$$\begin{array}{r} 645 \\ -\,287 \\ \hline 358 \end{array}$$
The distance between Hays and Denver is 358 miles.

61.
$$\begin{array}{r} 15{,}982{,}378 \\ -\,12{,}937{,}926 \\ \hline 3{,}044{,}452 \end{array}$$
The increase was 3,044,452 people.

65. The total number of votes cast for Jo was:
$$\begin{array}{r} {\scriptstyle 1\,2\,1} \\ 276 \\ 362 \\ 201 \\ +\,179 \\ \hline 1018 \end{array}$$
The total number of votes cast for Trudy was:
$$\begin{array}{r} {\scriptstyle 2\,1} \\ 295 \\ 122 \\ 312 \\ +\,182 \\ \hline 911 \end{array} \qquad \begin{array}{r} 1018 \\ -\,911 \\ \hline 107 \end{array}$$
Since more votes were cast for Jo than for Trudy, Jo won the election by 107 votes.

69.
$$\begin{array}{r} 2817 \\ -\,2486 \\ \hline 331 \end{array}$$
There were 331 new restaurants added during 2000.

73.
$$\begin{array}{r} 909 \\ -\,637 \\ \hline 272 \end{array}$$
Chicago O'Hare has 272 more airplane movements than Phoenix Sky Harbor International Airport.

77.
$$\begin{array}{r} 826{,}194{,}200 \\ -\,582{,}423{,}300 \\ \hline 243{,}770{,}900 \end{array}$$
DaimlerChrysler spent $243,770,900 more on ads than General Motors.

81.
$$\begin{array}{r} 5269 \\ -\,2385 \\ \hline 2884 \end{array}$$

Exercise Set 1.5

1. To round 632 to the nearest ten, observe that the digit in the ones place is 2. Since this digit is less than 5, we do not add 1 to the digit in tens place. The number 632 rounded to the nearest ten is 630.

5. To round 792 to the nearest ten, observe that the digit in the ones place is 2. Since this digit is less than 5, we do not add 1 to the digit in the tens place. The number 792 rounded to the nearest ten is 790.

9. To round 1096 to the nearest ten, observe that the digit in the ones place is 6. Since this digit is at least 5, we need to add 1 to the digit in the tens place. The number 1096 rounded to the nearest ten is 1100.

13. To round 248,695 to the nearest hundred, observe, that the digit in the tens is 9. Since this digit is at least 5, we need to add 1 to the digit in the hundreds place. The number 248,695 rounded to the nearest hundred is 248,700.

17. To round 99,995 to the nearest ten, observe that the digit in the ones place is 5. Since this digit is at least 5, we need to add 1 to the digit in the tens place. The number 99,995 rounded to the nearest ten is 100,000.

		Ten	Hundred	Thousand
21.	5281	5280	5300	5000
25.	14,876	14,880	14,900	15,000

29. To round 38,387 to the nearest thousand, observe that the digit in the hundreds place is 3. Since this digit is less than 5 we do not add 1 to the digit in the thousands place. The number 38,387 rounded to the nearest thousand is 38,000.

33. To round $2,264,403 to the nearest hundred-thousand, observe that the digit in the ten-thousands place is 6. Since this digit is at least 5, we need to add 1 to the digit in the hundred-thousands place. The number $2,264,403 rounded to the nearest hundred-thousand is $2,300,000.

37.
$$\begin{array}{rll} 649 & \text{rounds to} & 650 \\ -\,272 & \text{rounds to} & -\,270 \\ \hline & & 380 \end{array}$$
The estimated difference is 380.

41.
$$\begin{array}{rll} 1774 & \text{rounds to} & 1800 \\ -\,1492 & \text{rounds to} & -\,1500 \\ \hline & & 300 \end{array}$$
The estimated difference is 300.

45. $362 + 419$ is approximately $360 + 420 = 780$. The answer of 781 is correct.

49. $7806 + 5150$ is approximately $7800 + 5200 = 13{,}000$. The answer of 12,956 is correct.

53.
$$\begin{array}{rll} 799 & \text{rounds to} & 800 \\ 1299 & \text{rounds to} & 1300 \\ +\,999 & \text{rounds to} & +\,1000 \\ \hline & & 3100 \end{array}$$
The total cost is approximately $3100.

57.
$$\begin{array}{rll} 20{,}320 & \text{rounds to} & 20{,}000 \\ -\,14{,}410 & \text{rounds to} & -\,14{,}000 \\ \hline & & 6{,}000 \end{array}$$
The difference in elevation is approximately 6000 feet.

61.
$$\begin{array}{rll} 41{,}126{,}333 & \text{rounds to} & 41{,}000{,}000 \\ -\,27{,}174{,}898 & \text{rounds to} & -\,27{,}000{,}000 \\ \hline & & 14{,}000{,}000 \end{array}$$
Johnson won the election by approximately 14,000,000 votes.

65. 582,423,300 rounds to 600,000,000.
General Motors Corp spent approximately $600,000,000 on television advertising.

69. The smallest possible number that rounds to 8600 is 8550.

Exercise Set 1.6

1. $4(3 + 9) = 4 \cdot 3 + 4 \cdot 9$

5. $10(11 + 7) = 10 \cdot 11 + 10 \cdot 7$

9.
$$\begin{array}{r} 624 \\ \times\ 3 \\ \hline 1872 \end{array}$$

13.
$$\begin{array}{r} 1062 \\ \times\ 5 \\ \hline 5310 \end{array}$$

17.
$$\begin{array}{r} 231 \\ \times\ 47 \\ \hline 1617 \\ 9240 \\ \hline 10{,}857 \end{array}$$

21.
$$\begin{array}{r} 620 \\ \times\ 40 \\ \hline 0 \\ 24{,}800 \\ \hline 24{,}800 \end{array}$$

25. $(590)(1)(10) = 5900$

29.
$$\begin{array}{r} 609 \\ \times\ 234 \\ \hline 2436 \\ 18{,}270 \\ 121{,}800 \\ \hline 142{,}506 \end{array}$$

33.
$$\begin{array}{r} 1941 \\ \times\ 235 \\ \hline 9705 \\ 58{,}230 \\ 388{,}200 \\ \hline 456{,}135 \end{array}$$

37.
$$\begin{array}{ll} 576 & \text{rounds to} \quad 600 \\ \times\ 354 & \text{rounds to} \quad \times\ 400 \\ & \hline \quad\quad\quad 240{,}000 \end{array}$$

576×354 is approximately 240,000.

41. Area = length · width
   = (9 meters)(7 meters)
   = 63 square meters
The area is 63 square meters.

45.
$$\begin{array}{r} 125 \\ \times\ 3 \\ \hline 375 \end{array}$$
There are 375 calories in 3 tablespoons of olive oil.

49.
$$\begin{array}{r} 12 \\ \times\ 8 \\ \hline 96 \end{array}$$
$2 \times 96 = 192$
There are 192 cans in a case.

53.
$$\begin{array}{r} 776 \\ \times\ 639 \\ \hline 6984 \\ 23{,}280 \\ 465{,}600 \\ \hline 495{,}864 \end{array}$$
The floor area is 495,864 square meters.

57.
$$\begin{array}{r} 60 \\ \times\ 25 \\ \hline 300 \\ 1200 \\ \hline 1500 \end{array}$$
There are 1500 characters in 25 lines.

61.
$$\begin{array}{r} 7927 \\ \times\ 9 \\ \hline 71{,}343 \end{array}$$
Saturn has a diameter of 71,343 miles.

65.
$$\begin{array}{r} 700{,}000 \\ \times\ 31 \\ \hline 700{,}000 \\ 21{,}000{,}000 \\ \hline 21{,}700{,}000 \end{array}$$
21,700,000 quarts of milk would be used in March.

69. $5 \times 10 = 50$
50 students chose grapes as their favorite fruit.

73. The result of multiplying 3 by the digit in the first blank is a number ending in 6. Only 2 works. The result of multiplying the digit in the second blank by 42 is 378, so the digit in the second blank is 9,
$$\begin{array}{r} 42 \\ \times\ 93 \\ \hline 126 \\ 3780 \\ \hline 3906 \end{array}$$

77. $61 \times 3 = 183$ and $640 \times 2 = 1280$
$$\begin{array}{r} 183 \\ 1280 \\ +\ 475 \\ \hline 1938 \end{array}$$
Kobe Bryant scored 1938 points in the 2000–2001 season.

Exercise Set 1.7

1.
$$\begin{array}{r} 12 \\ 9\overline{)108} \\ -9 \\ \hline 18 \\ -18 \\ \hline 0 \end{array}$$
Check: $12 \cdot 9 = 108$

5.
$$\begin{array}{r} 338 \\ 3\overline{)1014} \\ -9 \\ \hline 11 \\ -9 \\ \hline 24 \\ -24 \\ \hline 0 \end{array}$$
Check: $338 \cdot 3 = 1014$

9.
$$\begin{array}{r} 563 \ \text{R } 1 \\ 2\overline{)1127} \\ -10 \\ \hline 12 \\ -12 \\ \hline 07 \\ -6 \\ \hline 1 \end{array}$$
Check:
$563 \cdot 2 + 1 = 1127$

13.
$$\begin{array}{r} 265 \ \text{R } 1 \\ 8\overline{)2121} \\ -16 \\ \hline 52 \\ -48 \\ \hline 41 \\ -40 \\ \hline 1 \end{array}$$
Check:
$265 \cdot 8 + 1 = 2121$

17.
$$\begin{array}{r} 13 \\ 55\overline{)715} \\ -55 \\ \hline 165 \\ -165 \\ \hline 0 \end{array}$$
Check: $13 \cdot 55 = 715$

21.
$$\begin{array}{r} 206 \\ 18\overline{)3708} \\ -36 \\ \hline 10 \\ -0 \\ \hline 108 \\ -108 \\ \hline 0 \end{array}$$
Check: $206 \cdot 18 = 3708$

25.
$$\begin{array}{r} 202 \ \text{R } 7 \\ 46\overline{)9299} \\ -92 \\ \hline 09 \\ -0 \\ \hline 99 \\ -92 \\ \hline 7 \end{array}$$
Check: $202 \cdot 46 + 7 = 9299$

29.
$$
\begin{array}{r}
98 \quad \text{R } 100 \\
103\overline{)10194} \\
\underline{-927} \\
924 \\
\underline{-824} \\
100
\end{array}
$$
Check: $98 \cdot 103 + 100 = 10{,}194$

33.
$$
\begin{array}{r}
202 \\
223\overline{)45046} \\
\underline{-446} \\
44 \\
\underline{-0} \\
446 \\
\underline{-446} \\
0
\end{array}
$$
Check: $202 \cdot 223 = 45{,}046$

37.
$$
\begin{array}{r}
252000 \\
21\overline{)5292000} \\
\underline{-42} \\
109 \\
\underline{-105} \\
42 \\
\underline{-42} \\
00 \\
\underline{-0} \\
00 \\
\underline{-0} \\
00 \\
\underline{-0} \\
0
\end{array}
$$
Each person receives $252,000.

41.
$$
\begin{array}{r}
88 \quad \text{R } 1 \\
3\overline{)265} \\
\underline{-24} \\
25 \\
\underline{-24} \\
1
\end{array}
$$
There are 88 bridges in 265 miles.

45.
$$
\begin{array}{r}
26 \\
6\overline{)156} \\
\underline{-12} \\
36 \\
\underline{-36} \\
0
\end{array}
$$
He scored 26 touchdowns during 2000.

49. There are six numbers.
$$
\begin{array}{r}
14 \\
22 \\
45 \\
18 \\
30 \\
\underline{+27} \\
156
\end{array}
\qquad
\begin{array}{r}
26 \\
6\overline{)156} \\
\underline{-12} \\
36 \\
\underline{-36} \\
0
\end{array}
$$
Average $= \dfrac{156}{6} = 26$

53. There are five numbers.
$$
\begin{array}{r}
86 \\
79 \\
81 \\
69 \\
\underline{+80} \\
395
\end{array}
\qquad
\begin{array}{r}
79 \\
5\overline{)395} \\
\underline{-35} \\
45 \\
\underline{-45} \\
0
\end{array}
$$
Average $= \dfrac{395}{5} = 79$

57. Add the two largest amounts, then divide by 2.
$$
\begin{array}{r}
2{,}883{,}215{,}000 \\
+\ 1{,}671{,}764{,}000 \\
\hline
4{,}554{,}979{,}000
\end{array}
$$
$$
\begin{array}{r}
2277489500 \\
2\overline{)4554979000} \\
\underline{-4} \\
05 \\
\underline{-4} \\
15 \\
\underline{-14} \\
14 \\
\underline{-14} \\
09 \\
\underline{-8} \\
17 \\
\underline{-16} \\
19 \\
\underline{-18} \\
10 \\
\underline{-10} \\
00 \\
\underline{-0} \\
00 \\
\underline{-0} \\
0
\end{array}
$$
The average amount spent by the top two companies is $2,277,489,500.

61. No, because all the numbers are greater than 86.

Exercise Set 1.8

1.
41	increased by	8	is	some number
↓	↓	↓	↓	↓
41	+	8	=	some number

$$
\begin{array}{r}
41 \\
\underline{+\ 8} \\
49
\end{array}
$$
The number is 49.

5.
The total of	35	and	7	is	some number
	↓	↓	↓	↓	↓
	35	+	7	=	some number

$$
\begin{array}{r}
35 \\
\underline{+\ 7} \\
42
\end{array}
$$
The number is 42.

9.
Area	is	length	times	width
↓	↓	↓	↓	↓
Area	=	120	×	80

$$
\begin{array}{r}
120 \\
\underline{\times\ 80} \\
9600
\end{array}
$$
The area is 9600 square feet.

13.
Hours per week	is	hours per day	times	days per week
↓	↓	↓	↓	↓
Hours per week	=	24	×	7

$$
\begin{array}{r}
24 \\
\underline{\times\ 7} \\
168
\end{array}
$$
There are 168 hours in a week.

17.

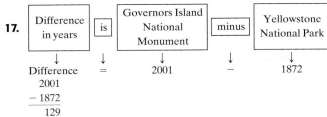

$$\begin{array}{ccc} \text{Difference} & = & 2001 & - & 1872 \\ 2001 & & & & \\ -1872 & & & & \\ \hline 129 & & & & \end{array}$$

Yellowstone is 129 years older than Governors Island.

21.

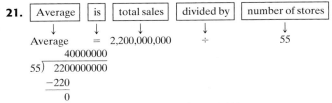

Average $= 2,200,000,000 \div 55$

$$\begin{array}{r} 40000000 \\ 55\overline{)2200000000} \\ \underline{-220} \\ 0 \end{array}$$

The average sales by each store is $40,000,000.

25.

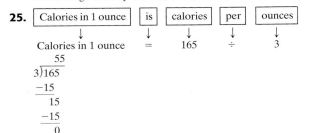

Calories in 1 ounce $= 165 \div 3$

$$\begin{array}{r} 55 \\ 3\overline{)165} \\ \underline{-15} \\ 15 \\ \underline{-15} \\ 0 \end{array}$$

There are 55 calories in 1 ounce of canned tuna.

29.

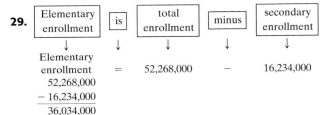

Elementary enrollment $= 52,268,000 - 16,234,000$

$$\begin{array}{r} 52,268,000 \\ -16,234,000 \\ \hline 36,034,000 \end{array}$$

There will be 36,034,000 students enrolled in elementary schools in 2008.

33.

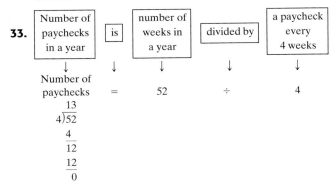

Number of paychecks $= 52 \div 4$

$$\begin{array}{r} 13 \\ 4\overline{)52} \\ \underline{4} \\ 12 \\ \underline{12} \\ 0 \end{array}$$

Marcel will get 13 paychecks in a year.

37.

Amount $= \$950 + \$300 - \$205$

$$\begin{array}{cc} 950 & 1250 \\ +\ 300 \quad \text{and} & -\ 205 \\ \hline 1250 & 1045 \end{array}$$

Amount in the account is $1045.

41. a. Add up the costs of the items:
$6(3) + 4(3) + 4(1) = 18 + 12 + 4 = 34$.

Order a) costs $34.

b. Add up the costs of the items:
$4(4) + 4(2) + 2(1) + 4(1) = 16 + 8 + 2 + 4 = 30$

Order b) costs $30.

Order b) costs less by $34 - \$30 = \$4$.

45. Fujitsu Limited received the least patents in 2000.

49. Motorola received more patents by 49.

$$\begin{array}{r} 1196 \\ -1147 \\ \hline 49 \end{array}$$

Exercise Set 1.9

1. $3 \cdot 3 \cdot 3 \cdot 3 = 3^4$

5. $12 \cdot 12 \cdot 12 = 12^3$

9. $9 \cdot 9 \cdot 9 \cdot 8 = (9 \cdot 9 \cdot 9)8 = 9^3 \cdot 8$

13. $3 \cdot 2 \cdot 2 \cdot 5 \cdot 5 \cdot 5 = 3(2 \cdot 2)(5 \cdot 5 \cdot 5) = 3 \cdot 2^2 \cdot 5^3$

17. $5^3 = 5 \cdot 5 \cdot 5 = 125$

21. $2^{10} = 2 \cdot 2 \cdot 2 \cdot 2 \cdot 2 \cdot 2 \cdot 2 \cdot 2 \cdot 2 \cdot 2 = 1024$

25. $3^5 = 3 \cdot 3 \cdot 3 \cdot 3 \cdot 3 = 243$

29. $4^3 = 4 \cdot 4 \cdot 4 = 64$

33. $9^3 = 9 \cdot 9 \cdot 9 = 729$

37. $10^4 = 10 \cdot 10 \cdot 10 \cdot 10 = 10,000$

41. $1920^1 = 1920$

45. $15 + 3 \cdot 2 = 15 + 6$
$= 21$

49. $5 \cdot 9 - 16 = 45 - 16$
$= 29$

53. $14 + \dfrac{24}{8} = 14 + 3$
$= 17$

57. $0 \div 6 + 4 \cdot 7 = 0 + 4 \cdot 7$
$= 0 + 28$
$= 28$

61. $(6 + 8) \div 2 = 14 \div 2$
$= 7$

65. $(3 + 5^2) \div 2 = (3 + 25) \div 2$
$= 28 \div 2$
$= 14$

69. $\dfrac{18 + 6}{2^4 - 4} = \dfrac{24}{16 - 4}$
$= \dfrac{24}{12}$
$= 2$

73. $\dfrac{7(9 - 6) + 3}{3^2 - 3} = \dfrac{7 \cdot 3 + 3}{9 - 3}$
$= \dfrac{21 + 3}{6}$
$= \dfrac{24}{6}$
$= 4$

77. $3^4 - [35 - (12 - 6)] = 3^4 - [35 - 6]$
$= 81 - 29$
$= 52$

81. $8 \cdot [4 + (6-1) \cdot 2] - 50 \cdot 2 = 8 \cdot [4 + 5 \cdot 2] - 50 \cdot 2$
$$= 8 \cdot [4 + 10] - 100$$
$$= 8 \cdot 14 - 100$$
$$= 112 - 100$$
$$= 12$$

85. Area of a square $= (\text{side})^2$
$$= (20 \text{ miles})^2$$
$$= 400 \text{ square miles}$$

89. Area of base $= (\text{side})^2$
$$= (100 \text{ meters})^2$$
$$= 10{,}000 \text{ square meters}$$

93. $24 \div (3 \cdot 2) + 2 \cdot 5 = 24 \div 6 + 2 \cdot 5$
$$= 4 + 10$$
$$= 14$$

97. $(7 + 2^4)^5 - (3^5 - 2^4)^2 = (7 + 16)^5 - (3^5 - 2^4)^2$
$$= 23^5 - (243 - 16)^2$$
$$= 23^5 - 227^2$$
$$= 6{,}434{,}343 - 51{,}529$$
$$= 6{,}384{,}814$$

Chapter 1 Test

1.
$$\begin{array}{r} \overset{1}{5}9 \\ + 82 \\ \hline 141 \end{array}$$

5. $2^3 \cdot 5^2 = 2 \cdot 2 \cdot 2 \cdot 5 \cdot 5 = 200$

9. $62 \div 0$ is defined

13. To round 52,369 to the nearest thousand observe that the digit in the hundreds place is 3. Since this digit is less than 5, we do not add 1 to the digit in the thousands place. The number 52,369 rounded to the nearest thousand is 52,000.

17.

Total cost	is	17	times	7

Total cost $=$ 17 $\times$ 7
$$\begin{array}{r} 17 \\ \times 7 \\ \hline 119 \end{array}$$
The total cost of the tickets is $119.

21. Perimeter $= 4(\text{side}) = 4(5 \text{ centimeters}) = 20 \text{ centimeters}$
Area $= (\text{side})^2 = (5 \text{ centimeters})^2 = 25 \text{ square centimeters}$

25.
$$\begin{array}{r} 431{,}065{,}444 \\ - 260{,}031{,}035 \\ \hline 171{,}034{,}409 \end{array}$$
The difference is $171,034,409.

Chapter 2

Exercise Set 2.1

1. 1   **5.** 13   **9.** 0   **13.** undefined

17. 1 out of 3 equal parts is shaded: $\dfrac{1}{3}$

21. 7 out of 12 equal parts are shaded: $\dfrac{7}{12}$

25. 4 out of 9 equal parts are shaded: $\dfrac{4}{9}$

29.    **33.**

37. $131 - 42 = 89$ non-freshmen
non-freshmen $\rightarrow \dfrac{89}{}$
students $\rightarrow \dfrac{}{131}$
$\dfrac{89}{131}$ of the students are not freshmen.

41. born in Virginia $\rightarrow \dfrac{8}{}$
U.S. presidents $\rightarrow \dfrac{}{43}$
$\dfrac{8}{43}$ of U.S. presidents were born in Virginia.

45. Each part is $\dfrac{1}{3}$, and there are 11 parts shaded, or 3 wholes and 2 more parts.
a. $3\dfrac{2}{3}$ **b.** $\dfrac{11}{3}$

49. Each part is $\dfrac{1}{3}$, and there are 4 parts shaded, or 1 whole and 1 more part.
a. $1\dfrac{1}{3}$ **b.** $\dfrac{4}{3}$

53. Each part is $\dfrac{1}{9}$, and there are 14 parts shaded, or 1 whole and 5 more parts.
a. $1\dfrac{5}{9}$ **b.** $\dfrac{14}{9}$

57. $3\dfrac{2}{3} = \dfrac{3 \cdot 3 + 2}{3} = \dfrac{11}{3}$

61. $2\dfrac{11}{15} = \dfrac{15 \cdot 2 + 11}{15} = \dfrac{41}{15}$

65. $6\dfrac{5}{8} = \dfrac{8 \cdot 6 + 5}{8} = \dfrac{53}{8}$

69. $4\dfrac{13}{24} = \dfrac{24 \cdot 4 + 13}{24} = \dfrac{109}{24}$

73. $9\dfrac{7}{20} = \dfrac{20 \cdot 9 + 7}{20} = \dfrac{187}{20}$

77.
$$\begin{array}{r} 3 \text{R } 3 \\ 13\overline{)42} \\ -39 \\ \hline 3 \end{array}$$
$$\dfrac{42}{13} = 3\dfrac{3}{13}$$

81.
$$\begin{array}{r} 33 \\ 6\overline{)198} \\ -18 \\ \hline 18 \\ -18 \\ \hline 0 \end{array}$$
$$\dfrac{198}{6} = 33$$

85.
$$\begin{array}{r} 4 \text{R } 5 \\ 8\overline{)37} \\ -32 \\ \hline 5 \end{array}$$
$$\dfrac{37}{8} = 4\dfrac{5}{8}$$

89.
$$\begin{array}{r} 10 \quad \text{R } 17 \\ 23{\overline{)}}247 \\ -23 \\ \hline 17 \\ -0 \\ \hline 17 \end{array}$$

$$\frac{247}{23} = 10\frac{17}{23}$$

93.
$$\begin{array}{r} 66 \quad \text{R } 2 \\ 3{\overline{)}}200 \\ -18 \\ \hline 20 \\ -18 \\ \hline 2 \end{array}$$

$$\frac{200}{3} = 66\frac{2}{3}$$

97. $5^3 = 5 \cdot 5 \cdot 5 = 125$

101. $2^3 \cdot 3 = 8 \cdot 3 = 24$

105. Total number of

Centers $= 5 + 3 + 2 + 1 + 1 + 1$
$= 13.$

Fraction is

$\dfrac{5}{13}.$

109. answers may vary

Exercise Set 2.2

1. $1 \cdot 8 = 8$
$2 \cdot 4 = 8$
The factors of 8 are 1, 2, 4, and 8.

5. $1 \cdot 4 = 4$
$2 \cdot 2 = 4$
The factors of 4 are 1, 2, and 4.

9. $1 \cdot 7 = 7$
The factors of 7 are 1 and 7.

13. $1 \cdot 12 = 12$
$2 \cdot 6 = 12$
$3 \cdot 4 = 12$
The factors of 12 are 1, 2, 3, 4, 6, and 12.

17. Prime, since its only factors are 1 and 7.

21. Composite, since its factors are 1, 2, 5, and 10.

25. Composite, since its factors are 1, 2, 3, and 6.

29. Prime, since its only factors are 1 and 31.

33.
$$\begin{array}{r} 3 \\ 2{\overline{)}}\ 6 \\ 2{\overline{)}}\ 12 \end{array}$$

$12 = 2^2 \cdot 3$

37.
$$\begin{array}{r} 5 \\ 2{\overline{)}}\ 10 \\ 2{\overline{)}}\ 20 \\ 2{\overline{)}}\ 40 \end{array}$$

41.
$$\begin{array}{r} 13 \\ 3{\overline{)}}\ 39 \end{array}$$

$39 = 3 \cdot 13$

45.
$$\begin{array}{r} 3 \\ 3{\overline{)}}\ 9 \\ 3{\overline{)}}\ 27 \\ 2{\overline{)}}\ 54 \end{array}$$

$54 = 2 \cdot 3^3$

49.
$$\begin{array}{r} 11 \\ 5{\overline{)}}\ 55 \\ 2{\overline{)}}\ 110 \end{array}$$

$110 = 2 \cdot 5 \cdot 11$

53.
$$\begin{array}{r} 2 \\ 2{\overline{)}}\ 4 \\ 2{\overline{)}}\ 8 \\ 2{\overline{)}}\ 16 \\ 2{\overline{)}}\ 32 \\ 2{\overline{)}}\ 64 \\ 2{\overline{)}}\ 128 \end{array}$$

$128 = 2^7$

57.
$$\begin{array}{r} 5 \\ 5{\overline{)}}\ 25 \\ 3{\overline{)}}\ 75 \\ 2{\overline{)}}\ 150 \\ 2{\overline{)}}\ 300 \end{array}$$

$300 = 2^2 \cdot 3 \cdot 5^2$

61.
$$\begin{array}{r} 7 \\ 5{\overline{)}}\ 35 \\ 3{\overline{)}}\ 105 \\ 3{\overline{)}}\ 315 \\ 3{\overline{)}}\ 945 \end{array}$$

$945 = 3^3 \cdot 5 \cdot 7$

65.
$$\begin{array}{r} 7 \\ 7{\overline{)}}\ 49 \\ 3{\overline{)}}\ 147 \\ 3{\overline{)}}\ 441 \\ 2{\overline{)}}\ 882 \end{array}$$

$882 = 2 \cdot 3^2 \cdot 7^2$

69. To round 4267 to the nearest hundred, observe that the digit in the tens place is 6. Since this digit is at least 5, we need to add 1 to the digit in the hundreds place. The number 4267 rounded to the nearest hundred is 4300.

73. To round 55,342 to the nearest hundred, observe that the digit in the tens place is 4. Since this digit is less than 5, we do not add 1 to the digit in the hundreds place. The number 55,342 rounded to the nearest hundred is 55,300.

77. To round 1247 to the nearest thousand, observe that the digit in the hundreds place is 2. Since this digit is less than 5, we do not add 1 to the digit in the thousands place. The number 1247 rounded to the nearest thousand is 1000.

81.
$$\begin{array}{r} 7 \\ 5{\overline{)}}\ 35 \\ 3{\overline{)}}\ 135 \\ 3{\overline{)}}\ 315 \\ 3{\overline{)}}\ 945 \\ 3{\overline{)}}\ 2835 \\ 3{\overline{)}}\ 8505 \\ 2{\overline{)}}\ 17010 \\ 2{\overline{)}}\ 34020 \end{array}$$

$34{,}020 = 2^2 \cdot 3^5 \cdot 5 \cdot 7$

85. answers may vary

Exercise Set 2.3

1. $\dfrac{3}{12} = \dfrac{3}{2 \cdot 2 \cdot 3} = \dfrac{1}{2 \cdot 2} = \dfrac{1}{4}$

5. $\dfrac{14}{16} = \dfrac{2 \cdot 7}{2 \cdot 2 \cdot 2 \cdot 2} = \dfrac{7}{2 \cdot 2 \cdot 2} = \dfrac{7}{8}$

9. $\dfrac{35}{42} = \dfrac{5 \cdot 7}{2 \cdot 3 \cdot 7} = \dfrac{5}{2 \cdot 3} = \dfrac{5}{6}$

13. $\dfrac{21}{49} = \dfrac{3 \cdot 7}{7 \cdot 7} = \dfrac{3}{7}$

17. $\dfrac{36}{63} = \dfrac{2 \cdot 2 \cdot 3 \cdot 3}{3 \cdot 3 \cdot 7} = \dfrac{2 \cdot 2}{7} = \dfrac{4}{7}$

21. $\dfrac{25}{40} = \dfrac{5 \cdot 5}{2 \cdot 2 \cdot 2 \cdot 5} = \dfrac{5}{2 \cdot 2 \cdot 2} = \dfrac{5}{8}$

25. $\dfrac{36}{24} = \dfrac{2 \cdot 2 \cdot 3 \cdot 3}{2 \cdot 2 \cdot 2 \cdot 3} = \dfrac{3}{2} \text{ or } 1\dfrac{1}{2}$

29. $\dfrac{70}{196} = \dfrac{2 \cdot 5 \cdot 7}{2 \cdot 2 \cdot 7 \cdot 7} = \dfrac{5}{2 \cdot 7} = \dfrac{5}{14}$

33. $\dfrac{55}{85} = \dfrac{5 \cdot 11}{5 \cdot 17} = \dfrac{11}{17}$

37. $\dfrac{75}{350} = \dfrac{3 \cdot 5 \cdot 5}{2 \cdot 5 \cdot 5 \cdot 7} = \dfrac{3}{2 \cdot 7} = \dfrac{3}{14}$

41. Equivalent, since the cross products are equal:
$10 \cdot 9 = 90$ and $15 \cdot 6 = 90$.

45. Not equivalent, since the cross products are not equal:
$10 \cdot 15 = 150$ and $13 \cdot 12 = 156$.

49. Equivalent, since the cross products are equal:
$4 \cdot 15 = 60$ and $10 \cdot 6 = 60$.

53. $\dfrac{2\text{ hours}}{8\text{ hours}} = \dfrac{2}{2 \cdot 2 \cdot 2} = \dfrac{1}{2 \cdot 2} = \dfrac{1}{4}$

2 hours represents $\dfrac{1}{4}$ of a work shift.

57. $\dfrac{261}{414} = \dfrac{3 \cdot 3 \cdot 29}{3 \cdot 3 \cdot 46} = \dfrac{29}{46}$

$\dfrac{29}{46}$ of individuals who have flown in space were Americans.

61.
$\begin{array}{r} 91 \\ \times\, 4 \\ \hline 364 \end{array}$

65.
$\begin{array}{r} 72 \\ \times\, 35 \\ \hline 360 \\ 2160 \\ \hline 2520 \end{array}$

69. $\dfrac{372}{620} = \dfrac{2 \cdot 2 \cdot 3 \cdot 31}{2 \cdot 2 \cdot 5 \cdot 31} = \dfrac{3}{5}$

73. $3 + 1 = 4$

$\dfrac{4}{100} = \dfrac{2 \cdot 2}{2 \cdot 2 \cdot 5 \cdot 5} = \dfrac{1}{5 \cdot 5} = \dfrac{1}{25}$

$\dfrac{1}{25}$ of donors have type AB (either Rh-positive or Rh-negative).

Exercise Set 2.4

1. $\dfrac{1}{3} \cdot \dfrac{2}{5} = \dfrac{1 \cdot 2}{3 \cdot 5} = \dfrac{2}{15}$

5. $\dfrac{3}{10} \cdot \dfrac{3}{8} = \dfrac{3 \cdot 3}{10 \cdot 8} = \dfrac{9}{80}$

9. $\dfrac{2}{7} \cdot \dfrac{5}{8} = \dfrac{2 \cdot 5}{7 \cdot 8} = \dfrac{\overset{1}{\cancel{2}} \cdot 5}{7 \cdot \underset{4}{\cancel{2}}} = \dfrac{5}{28}$

13. $\dfrac{5}{8} \cdot \dfrac{9}{4} = \dfrac{5 \cdot 9}{8 \cdot 4} = \dfrac{45}{32}$ or $1\dfrac{13}{32}$

17. $0 \cdot \dfrac{8}{9} = 0$

21. $\dfrac{19}{37} \cdot 0 = 0$

25. $\dfrac{14}{21} \cdot \dfrac{15}{16} \cdot \dfrac{1}{2} = \dfrac{14 \cdot 15 \cdot 1}{21 \cdot 16 \cdot 2}$

$= \dfrac{\overset{1}{\cancel{2}} \cdot \overset{1}{\cancel{7}} \cdot \overset{1}{\cancel{3}} \cdot 5}{\underset{1}{\cancel{3}} \cdot \underset{1}{\cancel{7}} \cdot \underset{1}{\cancel{2}} \cdot 8 \cdot 2}$

$= \dfrac{5}{8 \cdot 2} = \dfrac{5}{16}$

29. $\dfrac{3}{8} \cdot \dfrac{9}{10} = \dfrac{3 \cdot 9}{8 \cdot 10} = \dfrac{27}{80}$

33. $\dfrac{7}{72} \cdot \dfrac{9}{49} = \dfrac{7 \cdot 9}{72 \cdot 49}$

$= \dfrac{\overset{1}{\cancel{7}} \cdot \overset{1}{\cancel{9}}}{\underset{1}{\cancel{9}} \cdot 8 \cdot \underset{1}{\cancel{7}} \cdot 7}$

$= \dfrac{1}{8 \cdot 7}$

$= \dfrac{1}{56}$

37. $\dfrac{1}{3} \cdot \dfrac{2}{7} \cdot \dfrac{1}{5} = \dfrac{1 \cdot 2 \cdot 1}{3 \cdot 7 \cdot 5} = \dfrac{2}{105}$

41. $\dfrac{3}{14} \cdot \dfrac{6}{25} \cdot \dfrac{5}{27} \cdot \dfrac{7}{6} = \dfrac{3 \cdot 6 \cdot 5 \cdot 7}{14 \cdot 25 \cdot 27 \cdot 6}$

$= \dfrac{\overset{1}{\cancel{3}} \cdot \overset{1}{\cancel{6}} \cdot \overset{1}{\cancel{5}} \cdot \overset{1}{\cancel{7}}}{\underset{1}{\cancel{7}} \cdot 2 \cdot 5 \cdot \underset{1}{\cancel{5}} \cdot \underset{1}{\cancel{3}} \cdot 9 \cdot \underset{1}{\cancel{6}}} = \dfrac{1}{2 \cdot 5 \cdot 9} = \dfrac{1}{90}$

45. $\dfrac{5}{8} \cdot 4 = \dfrac{5}{8} \cdot \dfrac{4}{1} = \dfrac{5 \cdot \overset{1}{\cancel{4}}}{\underset{1}{\cancel{4}} \cdot 2 \cdot 1} = \dfrac{5}{2}$ or $2\dfrac{1}{2}$

49. $\dfrac{2}{5} \cdot 4\dfrac{1}{6} = \dfrac{2}{5} \cdot \dfrac{25}{6} = \dfrac{2 \cdot 25}{5 \cdot 6}$

$= \dfrac{\overset{1}{\cancel{2}} \cdot \overset{1}{\cancel{5}} \cdot 5}{\underset{1}{\cancel{5}} \cdot \underset{1}{\cancel{2}} \cdot 3} = \dfrac{5}{3}$ or $1\dfrac{2}{3}$

53. $2\dfrac{1}{5} \cdot 3\dfrac{1}{2} = \dfrac{11}{5} \cdot \dfrac{7}{2} = \dfrac{11 \cdot 7}{5 \cdot 2}$

$= \dfrac{77}{10}$ or $7\dfrac{7}{10}$

57. $\dfrac{3}{4} \cdot 16 = \dfrac{3}{4} \cdot \dfrac{16}{1} = \dfrac{3 \cdot 16}{4 \cdot 1} = \dfrac{3 \cdot \overset{1}{\cancel{4}} \cdot 4}{\underset{1}{\cancel{4}} \cdot 1} = \dfrac{3 \cdot 4}{1} = \dfrac{12}{1} = 12$

61. $1\dfrac{1}{5} \cdot 12\dfrac{1}{2} = \dfrac{6}{5} \cdot \dfrac{25}{2} = \dfrac{6 \cdot 25}{5 \cdot 2} = \dfrac{\overset{1}{\cancel{2}} \cdot 3 \cdot \overset{1}{\cancel{5}} \cdot 5}{\underset{1}{\cancel{5}} \cdot \underset{1}{\cancel{2}}} = \dfrac{3 \cdot 5}{1} = \dfrac{15}{1} = 15$

65. $3\dfrac{1}{2} \cdot 1\dfrac{3}{4} \cdot 2\dfrac{2}{3} = \dfrac{7}{2} \cdot \dfrac{7}{4} \cdot \dfrac{8}{3} = \dfrac{7 \cdot 7 \cdot 8}{2 \cdot 4 \cdot 3} = \dfrac{7 \cdot 7 \cdot \overset{1}{\cancel{2}} \cdot \overset{1}{\cancel{4}}}{\underset{1}{\cancel{2}} \cdot \underset{1}{\cancel{4}} \cdot 3} = \dfrac{49}{3}$ or $16\dfrac{1}{3}$

69. $12\dfrac{4}{5} \cdot 6\dfrac{7}{8} \cdot \dfrac{26}{77} = \dfrac{64}{5} \cdot \dfrac{55}{8} \cdot \dfrac{26}{77}$

$= \dfrac{64 \cdot 55 \cdot 26}{5 \cdot 8 \cdot 77} = \dfrac{8 \cdot \overset{1}{\cancel{8}} \cdot \overset{1}{\cancel{5}} \cdot \overset{1}{\cancel{11}} \cdot 26}{\underset{1}{\cancel{5}} \cdot \underset{1}{\cancel{8}} \cdot 7 \cdot \underset{1}{\cancel{11}}}$

$= \dfrac{8 \cdot 26}{7} = \dfrac{208}{7}$ or $29\dfrac{5}{7}$

73. $5\dfrac{1}{2} \cdot 5 = \dfrac{11}{2} \cdot \dfrac{5}{1} = \dfrac{11 \cdot 5}{2 \cdot 1} = \dfrac{55}{2}$ or $27\dfrac{1}{2}$

She drives $27\dfrac{1}{2}$ miles every week.

77. $4 \cdot \dfrac{2}{5} = \dfrac{4}{1} \cdot \dfrac{2}{5} = \dfrac{4 \cdot 2}{1 \cdot 5} = \dfrac{8}{5}$ or $1\dfrac{3}{5}$

The part of the post underground is $1\dfrac{3}{5}$ foot.

81. $\dfrac{2}{3} \cdot 2757 = \dfrac{2}{3} \cdot \dfrac{2757}{1}$

$= \dfrac{2 \cdot 2757}{3 \cdot 1}$

$= \dfrac{2 \cdot \overset{1}{\cancel{3}} \cdot 919}{\underset{1}{\cancel{3}} \cdot 1}$

$= \dfrac{2 \cdot 919}{1}$

$= \dfrac{1838}{1}$

$= 1838$

The cruise sale price is \$1838.

85. $\dfrac{5}{14} \cdot \dfrac{1}{5} = \dfrac{\overset{1}{\cancel{5}} \cdot 1}{14 \cdot \underset{1}{\cancel{5}}} = \dfrac{1}{14}$

The area is $\dfrac{1}{14}$ square foot.

89. $5\frac{1}{2} \cdot 5\frac{1}{2} = \frac{11}{2} \cdot \frac{11}{2}$

$= \frac{11 \cdot 11}{2 \cdot 2}$

$= \frac{121}{4}$ or $30\frac{1}{4}$

The area of the chip mole is $30\frac{1}{4}$ square inches.

93. $\frac{1}{5} \cdot 12{,}000 = \frac{1}{5} \cdot \frac{12{,}000}{1}$

$= \frac{1 \cdot 12{,}000}{5 \cdot 1}$

$= \frac{1 \cdot \overset{1}{\cancel{5}} \cdot 2400}{\underset{1}{\cancel{5}} \cdot 1}$

$= \frac{2400}{1}$

$= 2400$

The Rodriguez family drove 2400 miles on family business.

97.
$$
\begin{array}{r}
56 \quad \text{R } 12 \\
23\overline{)1300} \\
-115 \\
\hline
150 \\
-138 \\
\hline
12
\end{array}
$$

Exercise Set 2.5

1. The reciprocal of $\frac{4}{7}$ is $\frac{7}{4}$.

5. The reciprocal of 15 or $\frac{15}{1}$ is $\frac{1}{15}$.

9. $\frac{2}{3} \div \frac{5}{6} = \frac{2}{3} \cdot \frac{6}{5}$

$= \frac{2 \cdot 6}{3 \cdot 5}$

$= \frac{2 \cdot \overset{1}{\cancel{3}} \cdot 2}{\underset{1}{\cancel{3}} \cdot 5}$

$= \frac{4}{5}$

13. $\frac{8}{9} \div \frac{1}{2} = \frac{8}{9} \cdot \frac{2}{1} = \frac{8 \cdot 2}{9 \cdot 1} = \frac{16}{9}$ or $1\frac{7}{9}$

17. $\frac{3}{5} \div \frac{4}{5} = \frac{3}{5} \cdot \frac{5}{4} = \frac{3 \cdot 5}{5 \cdot 4} = \frac{3}{4}$

21. $\frac{1}{10} \div \frac{10}{1} = \frac{1}{10} \cdot \frac{1}{10} = \frac{1 \cdot 1}{10 \cdot 10} = \frac{1}{100}$

25. $\frac{3}{7} \div \frac{4}{7} = \frac{3}{7} \cdot \frac{7}{4} = \frac{3 \cdot 7}{7 \cdot 4} = \frac{3}{4}$

29. $\frac{7}{45} \div \frac{4}{25} = \frac{7}{45} \cdot \frac{25}{4}$

$= \frac{7 \cdot 25}{45 \cdot 4}$

$= \frac{7 \cdot 5 \cdot \overset{1}{\cancel{5}}}{\underset{1}{\cancel{5}} \cdot 9 \cdot 4}$

$= \frac{7 \cdot 5}{9 \cdot 4}$

$= \frac{35}{36}$

33. $\frac{3}{25} \div \frac{27}{40} = \frac{3}{25} \cdot \frac{40}{27} = \frac{3 \cdot 40}{25 \cdot 27} = \frac{\overset{1}{\cancel{3}} \cdot \overset{1}{\cancel{5}} \cdot 8}{5 \cdot \underset{1}{\cancel{5}} \cdot \underset{1}{\cancel{3}} \cdot 9} = \frac{8}{5 \cdot 9} = \frac{8}{45}$

37. $\frac{11}{85} \div \frac{7}{5} = \frac{11}{85} \cdot \frac{5}{7} = \frac{11 \cdot 5}{85 \cdot 7} = \frac{11 \cdot \overset{1}{\cancel{5}}}{\underset{1}{\cancel{5}} \cdot 17 \cdot 7} = \frac{11}{17 \cdot 7} = \frac{11}{119}$

41. $\frac{27}{100} \div \frac{3}{20} = \frac{27}{100} \cdot \frac{20}{3} = \frac{27 \cdot 20}{100 \cdot 3} = \frac{9 \cdot \overset{1}{\cancel{3}} \cdot \overset{1}{\cancel{20}}}{5 \cdot \underset{1}{\cancel{20}} \cdot \underset{1}{\cancel{3}}} = \frac{9}{5}$ or $1\frac{4}{5}$

45. $\frac{25}{126} \div \frac{125}{441} = \frac{25}{126} \cdot \frac{441}{125} = \frac{25 \cdot 441}{126 \cdot 125}$

$= \frac{\overset{1}{\cancel{5}} \cdot \overset{1}{\cancel{5}} \cdot \overset{1}{\cancel{3}} \cdot \overset{1}{\cancel{3}} \cdot \overset{1}{\cancel{7}} \cdot 7}{\underset{1}{\cancel{3}} \cdot \underset{1}{\cancel{3}} \cdot 2 \cdot \underset{1}{\cancel{7}} \cdot \underset{1}{\cancel{5}} \cdot \underset{1}{\cancel{5}} \cdot 5} = \frac{7}{2 \cdot 5} = \frac{7}{10}$

49. $8 \div \frac{3}{5} = \frac{8}{1} \cdot \frac{5}{3} = \frac{8 \cdot 5}{1 \cdot 3} = \frac{40}{3}$ or $13\frac{1}{3}$

53. $\frac{5}{12} \div 2\frac{1}{3} = \frac{5}{12} \div \frac{7}{3} = \frac{5}{12} \cdot \frac{3}{7} = \frac{5 \cdot 3}{12 \cdot 7}$

$= \frac{5 \cdot \overset{1}{\cancel{3}}}{\underset{1}{\cancel{3}} \cdot 4 \cdot 7} = \frac{5}{4 \cdot 7} = \frac{5}{28}$

61. $1\frac{4}{9} \div 2\frac{5}{6} = \frac{13}{9} \div \frac{17}{6} = \frac{13}{9} \cdot \frac{6}{17} = \frac{13 \cdot 6}{9 \cdot 17}$

$= \frac{13 \cdot 2 \cdot \overset{1}{\cancel{3}}}{\underset{1}{\cancel{3}} \cdot 3 \cdot 17} = \frac{13 \cdot 2}{3 \cdot 17} = \frac{26}{51}$

65. $\frac{33}{50} \div 1 = \frac{33}{55} \div \frac{1}{1} = \frac{33}{50} \cdot \frac{1}{1} = \frac{33 \cdot 1}{50 \cdot 1} = \frac{33}{50}$

69. $1 \div \frac{18}{35} = \frac{1}{1} \div \frac{18}{35} = \frac{1}{1} \cdot \frac{35}{18} = \frac{1 \cdot 35}{1 \cdot 18} = \frac{35}{18}$

73. $3\frac{1}{3} \div 4 = \frac{10}{3} \div \frac{4}{1} = \frac{10}{3} \cdot \frac{1}{4} = \frac{10 \cdot 1}{3 \cdot 4}$

$= \frac{\overset{1}{\cancel{2}} \cdot 5}{3 \cdot 2 \cdot \underset{1}{\cancel{2}}} = \frac{5}{3 \cdot 2} = \frac{5}{6}$

Sharon takes $\frac{5}{6}$ tablespoon of medicine.

77. $\frac{7660}{1} \cdot \frac{7}{10} = \frac{7660 \cdot 7}{10} = \frac{\overset{1}{\cancel{10}} \cdot 766 \cdot 7}{\underset{1}{\cancel{10}}}$

$= \frac{766 \cdot 7}{1} = \frac{5362}{1} = 5362$

The amount received for R-rated movies is \$5362 million.

81. $\frac{125}{1} \div 2\frac{3}{5} = \frac{125}{1} \div \frac{13}{5} = \frac{125}{1} \cdot \frac{5}{13} = \frac{125 \cdot 5}{1 \cdot 13}$

$= \frac{625}{13}$ or $48\frac{1}{13}$

It takes $48\frac{1}{13}$ hours to complete the order.

85.
$$
\begin{array}{r}
\overset{2\ 2}{27} \\
76 \\
+\ 98 \\
\hline
201
\end{array}
$$

89.
$$
\begin{array}{r}
2000 \\
-\ 431 \\
\hline
1569
\end{array}
$$

93. $4 \div \frac{1}{4} = \frac{4}{1} \cdot \frac{4}{1} = \frac{4 \cdot 4}{1 \cdot 1} = \frac{16}{1} = 16$

The Ravens played 16 games.

Chapter 2 Test

1. $\frac{4}{4} + \frac{3}{4} = \frac{4}{4} \cdot \frac{4}{3} = \frac{\overset{1}{\cancel{4}} \cdot 4}{\underset{1}{\cancel{4}} \cdot 3} = \frac{4}{3}$ or $1\frac{1}{3}$

5. $8 \div \dfrac{1}{2} = \dfrac{8}{1} \div \dfrac{1}{2} = \dfrac{8}{1} \cdot \dfrac{2}{1} = \dfrac{8 \cdot 2}{1 \cdot 1} = \dfrac{16}{1} = 16$

9. $\dfrac{16}{3} \div \dfrac{3}{12} = \dfrac{16}{3} \cdot \dfrac{12}{3}$

$= \dfrac{16 \cdot 12}{3 \cdot 3}$

$= \dfrac{16 \cdot \overset{1}{\cancel{3}} \cdot 4}{\underset{1}{\cancel{3}} \cdot 3}$

$= \dfrac{64}{3} \text{ or } 12\dfrac{1}{3}$

13. $7\dfrac{2}{3} = \dfrac{3 \cdot 7 + 2}{3} = \dfrac{23}{3}$

17. $1\dfrac{8}{9} \cdot \dfrac{2}{3} = \dfrac{17}{9} \cdot \dfrac{2}{3} = \dfrac{17 \cdot 2}{9 \cdot 3} = \dfrac{34}{27} \text{ or } 1\dfrac{7}{27}$

The area of the figure 1 $1\dfrac{7}{27}$ square miles.

21. Area $=$ length $\cdot$ width

$= (100 + 10 + 10) \cdot \left(53\dfrac{1}{3} + 10 + 10\right)$

$= 120 \cdot 73\dfrac{1}{3}$

$= \dfrac{120}{1} \cdot \dfrac{220}{3}$

$= \dfrac{120 \cdot 220}{1 \cdot 3}$

$= \dfrac{\overset{1}{\cancel{3}} \cdot 40 \cdot 220}{1 \cdot \underset{1}{\cancel{3}}}$

$= \dfrac{40 \cdot 220}{1}$

$= \dfrac{8800}{1}$

$= 8800$

8800 square yards of turf are needed.

Chapter 3

Exercise Set 3.1

1. $\dfrac{1}{7} + \dfrac{2}{7} = \dfrac{1 + 2}{7} = \dfrac{3}{7}$

5. $\dfrac{2}{9} + \dfrac{4}{9} = \dfrac{2 + 4}{9} = \dfrac{6}{9} = \dfrac{2 \cdot 3}{3 \cdot 3} = \dfrac{2}{3}$

9. $\dfrac{3}{14} + \dfrac{4}{14} = \dfrac{3 + 4}{14} = \dfrac{7}{14} = \dfrac{1 \cdot 7}{2 \cdot 7} = \dfrac{1}{2}$

13. $\dfrac{4}{13} + \dfrac{2}{13} + \dfrac{1}{13} = \dfrac{4 + 2 + 1}{13} = \dfrac{7}{13}$

17. $\dfrac{10}{11} - \dfrac{4}{11} = \dfrac{10 - 4}{11} = \dfrac{6}{11}$

21. $\dfrac{7}{4} - \dfrac{3}{4} = \dfrac{7 - 3}{4} = \dfrac{4}{4} = \dfrac{1 \cdot 4}{1 \cdot 4} = 1$

25. $\dfrac{25}{12} - \dfrac{15}{12} = \dfrac{25 - 15}{12} = \dfrac{10}{12} = \dfrac{5 \cdot 2}{6 \cdot 2} = \dfrac{5}{6}$

29. $\dfrac{27}{33} - \dfrac{8}{33} = \dfrac{27 - 8}{33} = \dfrac{19}{33}$

33. The perimeter is the distance around.
A rectangle has 2 sets of equal sides.
Add the lengths of the sides.

$\dfrac{5}{12} + \dfrac{7}{12} + \dfrac{5}{12} + \dfrac{7}{12} = \dfrac{5 + 7 + 5 + 7}{12}$

$= \dfrac{24}{12}$

$= \dfrac{2 \cdot 12}{1 \cdot 12}$

$= 2$

The perimeter is 2 meters.

37. Find the remaining amount of track to be impected. Subtract $\dfrac{5}{20}$ of a mile form $\dfrac{19}{20}$ of a mile.

$\dfrac{19}{20} - \dfrac{5}{20} = \dfrac{19 - 5}{20} = \dfrac{14}{20} = \dfrac{7 \cdot 2}{10 \cdot 2} = \dfrac{7}{10}$

She needs to inspect $\dfrac{7}{10}$ of a mile more.

41. Add $\dfrac{23}{100}$ to $\dfrac{22}{100}$

$\dfrac{23}{100} + \dfrac{22}{100} = \dfrac{45}{100} = \dfrac{9 \cdot 5}{20 \cdot 5} = \dfrac{9}{20}$

$\dfrac{9}{20}$ of adults spend less than \$100 per online book order.

45. $10 = 2 \cdot 5$

49. $55 = 5 \cdot 11$

53. $\dfrac{4}{11} + \dfrac{5}{11} - \dfrac{3}{11} + \dfrac{2}{11} = \dfrac{4 + 5 - 3 + 2}{11} = \dfrac{9 - 3 + 2}{11}$

$= \dfrac{6 + 2}{11} = \dfrac{8}{11}$

57. answers may vary

Exercise Set 3.2

1. Multiples of 3: $\quad$ 3, 6, 9, ⑫, 15, . . .
Multiples of 4: $\quad$ 4, 8, ⑫, 16, . . .
LCM: 12

5. Multiples of 12: $\quad$ 12, 24, ㊱, 48, 60, 72, . . .
Multiples of 18: $\quad$ 18, ㊱, 54, . . .
LCM: 36

9. $18 = ②\cdot ③\cdot 3$
$21 = 3 \cdot ⑦$
The LCM is
$2 \cdot 3 \cdot 3 \cdot 7 = 126.$

13. $8 = ②\cdot 2 \cdot 2$
$24 = 2 \cdot 2 \cdot 2 \cdot ③$
The LCM is
$2 \cdot 2 \cdot 2 \cdot 3 = 24.$

17. $25 = ⑤\cdot 5$
$15 = ③\cdot 5$
$6 = ②\cdot 3$
The LCM is $2 \cdot 3 \cdot 5 \cdot 5 = 150.$

21. $84 = 2 \cdot 2 \cdot 3 \cdot 7$
$294 = 2 \cdot 3 \cdot 7 \cdot 7$
The LCM is $2 \cdot 2 \cdot 3 \cdot 7 \cdot 7 = 588.$

25. $3 = ③$
$21 = 3 \cdot ⑦$
$51 = 3 \cdot ⑰$
The LCM is $3 \cdot 7 \cdot 17 = 357.$

29. $8 = \boxed{2 \cdot 2 \cdot 2}$
$6 = 2 \cdot 3$
$27 = \boxed{3 \cdot 3 \cdot 3}$
The LCM is $2 \cdot 2 \cdot 2 \cdot 3 \cdot 3 \cdot 3 = 216$.

33. $\dfrac{4}{7} = \dfrac{4 \cdot 5}{7 \cdot 5} = \dfrac{20}{35}$

37. $\dfrac{2}{5} = \dfrac{2 \cdot 5}{5 \cdot 5} = \dfrac{10}{25}$

41. $\dfrac{10}{7} = \dfrac{10 \cdot 3}{7 \cdot 3} = \dfrac{30}{21}$

45. $\dfrac{2}{3} = \dfrac{2 \cdot 15}{3 \cdot 15} = \dfrac{30}{45}$

49. $\dfrac{4}{3} = \dfrac{4 \cdot 3}{3 \cdot 3} = \dfrac{12}{9}$

53. $\dfrac{7}{10} - \dfrac{2}{10} = \dfrac{7 - 2}{10} = \dfrac{5}{10} = \dfrac{1 \cdot 5}{2 \cdot 5} = \dfrac{1}{2}$

57. $\dfrac{23}{18} - \dfrac{15}{18} = \dfrac{23 - 15}{18} = \dfrac{8}{18} = \dfrac{4 \cdot 2}{9 \cdot 2} = \dfrac{4}{9}$

61. $\dfrac{37}{165} = \dfrac{37 \cdot 22}{165 \cdot 22} = \dfrac{814}{3630}$

63. answers may vary

Exercise Set 3.3

1. **Step 1.** The LCD for the denominators 3 and 6 is 6.

Step 2. $\dfrac{2}{3} = \dfrac{2 \cdot 2}{3 \cdot 2} = \dfrac{4}{6}, \dfrac{1}{6}$ already has a denominator of 6.

Step 3. $\dfrac{2}{3} + \dfrac{1}{6} = \dfrac{4}{6} + \dfrac{1}{6} = \dfrac{5}{6}$

Step 4. $\dfrac{5}{6}$ is in simplest form.

5. **Step 1.** The LCD for the denominators 11 and 33 is 33.

Step 2. $\dfrac{2}{11} = \dfrac{2 \cdot 3}{11 \cdot 33} = \dfrac{6}{33}, \dfrac{2}{3}$ already has a denominator of 33.

Step 3. $\dfrac{2}{11} + \dfrac{2}{33} = \dfrac{6}{33} + \dfrac{2}{33} = \dfrac{8}{33}$

Step 4. $\dfrac{8}{33}$ is in simplest form.

9. **Step 1.** The LCD for the denominators 35 and 7 is 35.

Step 2. $\dfrac{11}{35}$ already has a denominator of $35, \dfrac{2}{7} = \dfrac{2 \cdot 5}{7 \cdot 5} = \dfrac{10}{35}$

Step 3. $\dfrac{11}{35} + \dfrac{2}{7} = \dfrac{11}{35} + \dfrac{10}{35} = \dfrac{21}{35}$

Step 4. $\dfrac{21}{35} = \dfrac{3 \cdot 7}{5 \cdot 7} = \dfrac{3}{5}$

13. **Step 1.** The LCD for the denominators 15 and 12 is 60.

Step 2. $\dfrac{7}{15} = \dfrac{7 \cdot 4}{15 \cdot 4} = \dfrac{28}{60}, \dfrac{5}{12} = \dfrac{5 \cdot 5}{12 \cdot 5} = \dfrac{25}{60}$

Step 3. $\dfrac{7}{15} + \dfrac{5}{12} = \dfrac{28}{60} + \dfrac{25}{60} = \dfrac{53}{60}$

Step 4. $\dfrac{53}{60}$ is in simplest form.

17. **Step 1.** The LCD for the denominators 40 and 14 is 280.

Step 2. $\dfrac{19}{40} = \dfrac{19 \cdot 7}{40 \cdot 7} = \dfrac{133}{280}, \dfrac{3}{14} = \dfrac{3 \cdot 20}{14 \cdot 20} = \dfrac{60}{280}$

Step 3. $\dfrac{19}{40} + \dfrac{3}{14} = \dfrac{133}{280} + \dfrac{60}{280} = \dfrac{193}{280}$

Step 4. $\dfrac{193}{280}$ is in simplest form.

21. **Step 1.** The LCD for the denominators 7, 8, and 2 is 56.

Step 2. $\dfrac{5}{7} = \dfrac{5 \cdot 8}{7 \cdot 8} = \dfrac{40}{56}$,

$\dfrac{1}{8} = \dfrac{1 \cdot 7}{8 \cdot 7} = \dfrac{7}{56}$,

$\dfrac{1}{2} = \dfrac{1 \cdot 28}{2 \cdot 28} = \dfrac{28}{56}$

Step 3. $\dfrac{5}{7} + \dfrac{1}{8} + \dfrac{1}{2} = \dfrac{40}{56} + \dfrac{7}{56} + \dfrac{28}{56} = \dfrac{75}{56}$

Step 4. $\dfrac{75}{56}$ or $1\dfrac{19}{56}$ is in simplest form.

25. **Step 1.** The LCD for the denominators 3, 9, and 27 is 27.

Step 2. $\dfrac{1}{3} = \dfrac{1 \cdot 9}{3 \cdot 9} = \dfrac{9}{27}, \dfrac{1}{9} = \dfrac{1 \cdot 3}{9 \cdot 3} = \dfrac{3}{27}, \dfrac{1}{27}$ already has a denominator of 27.

Step 3. $\dfrac{1}{3} + \dfrac{1}{9} + \dfrac{1}{27} = \dfrac{9}{27} + \dfrac{3}{27} + \dfrac{1}{27} = \dfrac{13}{27}$

Step 4. $\dfrac{13}{27}$ is in simplest form.

29. **Step 1.** The LCD for the denominators 8 and 16 is 16.

Step 2. $\dfrac{7}{8} = \dfrac{7 \cdot 2}{8 \cdot 2} = \dfrac{14}{16}, \dfrac{3}{16}$ already has a denominator of 16.

Step 3. $\dfrac{7}{8} - \dfrac{3}{16} = \dfrac{14}{16} - \dfrac{3}{16} = \dfrac{11}{16}$

Step 4. $\dfrac{11}{16}$ is in simplest form.

33. **Step 1.** The LCD for the denominators 7 and 8 is 56.

Step 2. $\dfrac{5}{7} = \dfrac{5 \cdot 8}{7 \cdot 8} = \dfrac{40}{56}, \dfrac{1}{8} = \dfrac{1 \cdot 7}{8 \cdot 7} = \dfrac{7}{56}$

Step 3. $\dfrac{5}{7} - \dfrac{1}{8} = \dfrac{40}{56} - \dfrac{7}{56} = \dfrac{33}{56}$

Step 4. $\dfrac{33}{56}$ is in simplest form.

37. **Step 1.** The LCD for the denominators 35 and 7 is 35.

Step 2. $\dfrac{11}{35}$ already has a denominator of 35,

$\dfrac{2}{7} = \dfrac{2 \cdot 5}{7 \cdot 5} = \dfrac{10}{35}$

Step 3. $\dfrac{11}{35} - \dfrac{2}{7} = \dfrac{11}{35} - \dfrac{10}{35} = \dfrac{1}{35}$

Step 4. $\dfrac{1}{35}$ is in simplest form.

41. **Step 1.** The LCD for the denominators 15 and 12 is 60.

Step 2. $\dfrac{7}{15} = \dfrac{7 \cdot 4}{15 \cdot 4} = \dfrac{28}{60}, \dfrac{5}{12} = \dfrac{5 \cdot 5}{12 \cdot 5} = \dfrac{25}{60}$

Step 3. $\dfrac{7}{15} - \dfrac{5}{12} = \dfrac{28}{60} - \dfrac{25}{60} = \dfrac{3}{60}$

Step 4. $\dfrac{3}{60} = \dfrac{1 \cdot 3}{20 \cdot 3} = \dfrac{1}{20}$

45. **Step 1.** The LCD for the denominators and 1000 is 1000.

Step 2. $\dfrac{1}{100} = \dfrac{1 \cdot 10}{100 \cdot 10} = \dfrac{10}{1000}, \dfrac{1}{1000}$ already has a denominator of 1000

Step 3. $\dfrac{1}{100} - \dfrac{1}{1000} = \dfrac{10}{1000} - \dfrac{1}{1000} = \dfrac{9}{1000}$

Step 4. $\dfrac{9}{1000}$ is in simplest form.

49. Add the lengths of the 4 sides.

Step 1. The LCD for the denominators $3, 5, 3,$ and 5 is 15.

Step 2. $\dfrac{1}{3} = \dfrac{1 \cdot 5}{3 \cdot 5} = \dfrac{5}{15},$

$\dfrac{4}{5} = \dfrac{4 \cdot 3}{5 \cdot 3} = \dfrac{12}{15},$

$\dfrac{1}{3} = \dfrac{1 \cdot 5}{3 \cdot 5} = \dfrac{5}{15},$

$\dfrac{4}{5} = \dfrac{4 \cdot 3}{5 \cdot 3} = \dfrac{12}{15},$

Step 3. $\dfrac{1}{3} + \dfrac{4}{5} + \dfrac{1}{3} + \dfrac{4}{5} = \dfrac{5}{15} + \dfrac{12}{15} + \dfrac{5}{15} + \dfrac{12}{15}$

$= \dfrac{34}{15}$

Step 4. $\dfrac{34}{15}$ or $2\dfrac{4}{15}$ is in simplest form.

The perimeter is $\dfrac{34}{15}$ or $2\dfrac{4}{15}$ centimeters.

53. a. $\dfrac{5}{8} \cdot \dfrac{2}{5} = \dfrac{5 \cdot 2}{8 \cdot 5} = \dfrac{\overset{1}{\cancel{5}} \cdot \overset{1}{\cancel{2}}}{4 \cdot \cancel{2} \cdot \cancel{5}} = \dfrac{1}{4}$

$\dfrac{1}{4}$ of the ream is in the printer.

Amount remaining is $1 - \dfrac{1}{4} = \dfrac{4}{4} - \dfrac{1}{4}$

$= \dfrac{4-1}{4} = \dfrac{3}{4}.$

b. $500 \cdot \dfrac{3}{4} = \dfrac{500}{1} \cdot \dfrac{3}{4} = \dfrac{500 \cdot 3}{1 \cdot 4} = \dfrac{\overset{1}{\cancel{4}} \cdot 125 \cdot 3}{\underset{1}{\cancel{4}}}$

$= \dfrac{125 \cdot 3}{1} = \dfrac{375}{1} = 375.$

375 sheet of paper remain in the ream.

57. Add the fractions for 1 or 2 times per week and 3 times per week.

Step 1. The LCD for the denominators 50 and 100 is 100.

Step 2. $\dfrac{23}{50} = \dfrac{23 \cdot 2}{50 \cdot 2} = \dfrac{46}{100}, \dfrac{31}{100}$ already has a denominator of 100.

Step 3. $\dfrac{23}{50} + \dfrac{31}{100} = \dfrac{46}{100} + \dfrac{31}{100} = \dfrac{77}{100}$

Step 4. $\dfrac{77}{100}$ is in simplest form.

The fraction of Americans that eat pasta 1, 2, or 3 times per week is $\dfrac{77}{100}$.

61. $4 + 7\dfrac{1}{4} = \dfrac{4}{1} + \dfrac{29}{4} = \dfrac{4}{1} \cdot \dfrac{4}{29} = \dfrac{4 \cdot 4}{1 \cdot 29} = \dfrac{16}{29}$

65. **Step 1.** The LCD for the denominators 55 and 1760 is 1760.

Step 2. $\dfrac{30}{55} = \dfrac{30 \cdot 32}{55 \cdot 32} = \dfrac{960}{1760}, \dfrac{1000}{1760}$ already has a denominator of 1760.

Step 3. $\dfrac{30}{55} + \dfrac{1000}{1760} = \dfrac{960}{1760} + \dfrac{1000}{1760} = \dfrac{1960}{1760}$

Step 4. $\dfrac{1960}{1760} = \dfrac{49 \cdot 40}{44 \cdot 40} = \dfrac{49}{44}$ or $1\dfrac{5}{44}$

69. Find the sum of the fractions for Asia and Europe. The LCD is already 57.

$\dfrac{17}{57} + \dfrac{4}{57} = \dfrac{21}{57} = \dfrac{3 \cdot 7}{3 \cdot 19}$

$= \dfrac{7}{19}$

The fraction of the world's land area that is accounted for by Asia and Europe is $\dfrac{7}{19}$.

73. $\dfrac{2}{3} - \dfrac{1}{4} - \dfrac{2}{5} = \dfrac{2 \cdot 20}{3 \cdot 20} - \dfrac{1 \cdot 15}{4 \cdot 15} - \dfrac{2 \cdot 12}{5 \cdot 12}$

$= \dfrac{40}{60} - \dfrac{15}{60} - \dfrac{24}{60} = \dfrac{40 - 15 - 24}{60} = \dfrac{25 - 24}{60} = \dfrac{1}{60}$

Exercise Set 3.4

1.
$$\begin{aligned} 4\tfrac{7}{10}& \\ +\,2\tfrac{1}{10}& \\ \hline 6\tfrac{8}{10} &= 6\tfrac{4}{5} \end{aligned}$$

5.
$$\begin{aligned} 9\tfrac{1}{5} &= \quad 9\tfrac{5}{25} \\ +\,8\tfrac{2}{25} &= +\,8\tfrac{2}{25} \\ \hline & \quad 17\tfrac{7}{25} \end{aligned}$$

9.
$$\begin{aligned} 3\tfrac{1}{2} &= \quad 3\tfrac{4}{8} \\ +\,4\tfrac{1}{8} &= +\,4\tfrac{1}{8} \\ \hline & \quad 7\tfrac{5}{8} \end{aligned}$$

13.
$$\begin{aligned} 15\tfrac{1}{6} &= \quad 15\tfrac{2}{12} \\ +\,13\tfrac{5}{12} &= +\,13\tfrac{5}{12} \\ \hline & \quad 28\tfrac{7}{12} \end{aligned}$$

17.
$$\begin{aligned} 3\tfrac{5}{8} &= \quad 3\tfrac{15}{24} \\ 2\tfrac{1}{6} &= \quad 2\tfrac{4}{24} \\ +\,7\tfrac{3}{4} &= +\,7\tfrac{18}{24} \\ \hline & \quad 12\tfrac{37}{24} = 13\tfrac{13}{24} \end{aligned}$$

21.
$$4\frac{7}{10}$$
$$-2\frac{1}{10}$$
$$2\frac{6}{10} = 2\frac{3}{5}$$

25.
$$9\frac{1}{5} = \quad 9\frac{5}{25} = \quad 8\frac{30}{25}$$
$$-8\frac{6}{25} = -8\frac{6}{25} = -8\frac{6}{25}$$
$$\frac{24}{25}$$

29.
$$5\frac{2}{3} = \quad 5\frac{10}{15}$$
$$-3\frac{1}{5} = -3\frac{8}{15}$$
$$2\frac{7}{15}$$

33.
$$10 = \quad 9\frac{5}{5}$$
$$-8\frac{1}{5} = -8\frac{1}{5}$$
$$1\frac{4}{5}$$

37.
$$7\frac{3}{16} = \quad 7\frac{27}{144}$$
$$-5\frac{3}{18} = -5\frac{24}{144}$$
$$2\frac{3}{144} = 2\frac{1}{48}$$

41.
$$6 = \quad 5\frac{9}{9}$$
$$-2\frac{4}{9} = -2\frac{4}{9}$$
$$3\frac{5}{9}$$

45.
$$29\frac{9}{11}$$
$$-12$$
$$17\frac{9}{11}$$

49. The phrase "How much longer" tells us to subtract. Subtract $3\frac{1}{2}$ hours from $5\frac{7}{8}$ hours.

$$5\frac{7}{8} = \quad 5\frac{7}{8}$$
$$-3\frac{1}{2} = -3\frac{4}{8}$$
$$2\frac{3}{8}$$

It takes him $2\frac{3}{8}$ hours longer to prepare the business return.

53. The phrase "how much more" tells us to subtract. Subtract $3\frac{3}{5}$ inches from $11\frac{11}{4}$ inches.

$$11\frac{1}{4} = \quad 11\frac{5}{20} = \quad 10\frac{25}{20}$$
$$-3\frac{3}{5} = -3\frac{12}{20} = -3\frac{12}{20}$$
$$7\frac{13}{20}$$

Tucson gets $7\frac{13}{20}$ inches more, on average, than Yuma.

57. Subtract 50 pounds from $60\frac{5}{8}$ pounds.

$$60\frac{5}{8}$$
$$-50$$
$$10\frac{5}{8}$$

The traveler will have to pay charges on $10\frac{5}{8}$ pounds.

61. The phrase "overall height" tells us to add. Add the two heights.

$$46\frac{1}{20} = \quad 46\frac{5}{100}$$
$$+46\frac{47}{50} = +46\frac{94}{100}$$
$$92\frac{99}{100}$$

The overall height is $92\frac{99}{100}$ meters.

65. Find the distance around. Add the lengths of the three sides.

$$2\frac{1}{3}$$
$$2\frac{1}{3}$$
$$+2\frac{1}{3}$$
$$6\frac{3}{3} = 7$$

The perimeter is 7 miles.

69. $2^3 = 2 \cdot 2 \cdot 2 = 8$

73. $3^4 = 3 \cdot 3 \cdot 3 \cdot 3 = 81$

77. First, find the total weight of the Supreme box by adding the two weights. Then find the total weight of the Deluxe box by adding the two weights. Finally, subtract the lighter box from the heavier box to see how much more the heavier one weighs.

$$2\frac{1}{4} = \quad 2\frac{1}{4}$$
$$+3\frac{1}{2} = +3\frac{2}{4}$$
$$5\frac{3}{4}$$

The Supreme box weighs $5\frac{3}{4}$ pounds.

$$1\frac{3}{8} = 1\frac{3}{8}$$
$$+4\frac{1}{4} = +4\frac{2}{8}$$
$$\overline{5\frac{5}{8}}$$

The Deluxe box weighs $5\frac{5}{8}$ pounds.

$$5\frac{3}{4} = 5\frac{6}{8}$$
$$-5\frac{5}{8} = +5\frac{5}{8}$$
$$\overline{\frac{1}{8}}$$

The Supreme box is heavier by $\frac{1}{8}$ pound.

Exercise Set 3.5

1. Since $7 > 6$, then $\frac{7}{9} > \frac{6}{9}$.

5. The LCD is 42.

$\frac{9}{42}$ has a denominator of 42 $\frac{5}{21} = \frac{5 \cdot 2}{21 \cdot 2} = \frac{10}{42}$

Since $9 < 10$, then $\frac{9}{42} < \frac{10}{42}$ or $\frac{9}{42} < \frac{5}{21}$

9. The LCD is 12.

$\frac{3}{4} = \frac{3 \cdot 3}{4 \cdot 3} = \frac{9}{12}, \frac{2}{3} = \frac{2 \cdot 4}{3 \cdot 4} = \frac{8}{12}$

Since $9 > 8$, then $\frac{9}{12} > \frac{8}{12}$ so $\frac{3}{4} > \frac{2}{3}$

13. The LCD is 100.

$\frac{27}{100}$ has a denominator of 100 $\frac{7}{25} = \frac{7 \cdot 4}{25 \cdot 4} = \frac{28}{100}$

Since $27 < 28$, then $\frac{27}{100} < \frac{28}{100}$, so $\frac{27}{100} < \frac{7}{25}$

17. $\left(\frac{1}{2}\right)^4 = \frac{1}{2} \cdot \frac{1}{2} \cdot \frac{1}{2} \cdot \frac{1}{2} = \frac{1}{16}$

21. $\left(\frac{4}{7}\right)^3 = \frac{4}{7} \cdot \frac{4}{7} \cdot \frac{4}{7} = \frac{64}{343}$

25. $\left(\frac{3}{4}\right)^2 \cdot \left(\frac{2}{3}\right)^3 = \left(\frac{3}{4} \cdot \frac{3}{4}\right) \cdot \left(\frac{2}{3} \cdot \frac{2}{3} \cdot \frac{2}{3}\right)$

$$= \frac{3 \cdot 3 \cdot 2 \cdot 2 \cdot 2}{4 \cdot 4 \cdot 3 \cdot 3 \cdot 3}$$

$$= \frac{1}{6}$$

29. $\frac{2}{15} + \frac{3}{5} = \frac{2}{15} + \frac{3 \cdot 3}{5 \cdot 3} = \frac{2}{15} + \frac{9}{15} = \frac{2+9}{15} = \frac{11}{15}$

33. $1 - \frac{4}{9} = \frac{9}{9} - \frac{4}{9} = \frac{9-4}{9} = \frac{5}{9}$

37. $\frac{5}{6} - \frac{3}{4} = \frac{5 \cdot 2}{6 \cdot 2} - \frac{3 \cdot 3}{4 \cdot 3} = \frac{10}{12} - \frac{9}{12} = \frac{10-9}{12} = \frac{1}{12}$

41. $0 \cdot \frac{9}{10} = 0$

45. $\frac{20}{35} \cdot \frac{7}{10} = \frac{20 \cdot 7}{35 \cdot 10} = \frac{2 \cdot 10 \cdot 7}{5 \cdot 7 \cdot 10} = \frac{2}{5}$

49. $\frac{1}{5} + \frac{1}{3} \cdot \frac{1}{4} = \frac{1}{5} + \frac{1}{12} = \frac{12}{60} + \frac{5}{60} = \frac{17}{60}$

53. $\frac{1}{5} \cdot \left(2\frac{5}{6} - \frac{1}{3}\right) = \frac{1}{5} \cdot \left(\frac{17}{6} - \frac{1}{3}\right) = \frac{1}{5} \cdot \left(\frac{17}{6} - \frac{2}{6}\right) = \frac{1}{5} \cdot \left(\frac{15}{6}\right)$

$$= \frac{1 \cdot 15}{5 \cdot 6} = \frac{3 \cdot 5}{5 \cdot 2 \cdot 3} = \frac{1}{2}$$

57. $\left(\frac{3}{4}\right)^2 \div \left(\frac{3}{4} - \frac{1}{12}\right) = \frac{9}{16} \div \left(\frac{3}{4} - \frac{1}{12}\right) = \frac{9}{16} \div \left(\frac{9}{12} - \frac{1}{12}\right)$

$$= \frac{9}{16} \div \frac{8}{12} = \frac{9}{16} \div \frac{12}{8} = \frac{9 \cdot 12}{16 \cdot 8} = \frac{9 \cdot 3 \cdot 4}{16 \cdot 2 \cdot 4}$$

$$= \frac{9 \cdot 3}{16 \cdot 2} = \frac{27}{32}$$

61. $\left(\frac{3}{4} + \frac{1}{8}\right)^2 - \left(\frac{1}{2} + \frac{1}{8}\right) = \left(\frac{6}{8} + \frac{1}{8}\right)^2 - \left(\frac{4}{8} + \frac{1}{8}\right)$

$$= \left(\frac{7}{8}\right)^2 - \left(\frac{5}{8}\right) = \frac{49}{64} - \frac{5}{8} = \frac{49}{64} - \frac{40}{64} = \frac{9}{64}$$

65. $\left(\frac{1}{3} + \frac{1}{4} + \frac{1}{6}\right) + 3 = \left(\frac{4}{12} + \frac{3}{12} + \frac{2}{12}\right) + 3$

$$= \frac{9}{12} + 3$$

$$= \frac{9}{12} \cdot \frac{1}{3}$$

$$= \frac{3 \cdot 3 \cdot 1}{2 \cdot 2 \cdot 3 \cdot 3}$$

$$= \frac{1}{4}$$

69. $\frac{20}{24} = \frac{2 \cdot 2 \cdot 5}{2 \cdot 2 \cdot 2 \cdot 3} = \frac{5}{2 \cdot 3} = \frac{5}{6}$

73. $\frac{11}{67} = \frac{11 \cdot 2}{67 \cdot 2} = \frac{22}{134}$.

Since $\frac{75}{134} > \frac{22}{134}$, the greater portion of mail was standard mail.

77. $\frac{2^3}{3} = \frac{2 \cdot 2 \cdot 2}{3} = \frac{8}{3}$ whereas

$\left(\frac{2}{3}\right)^3 = \left(\frac{2}{3}\right)\left(\frac{2}{3}\right)\left(\frac{2}{3}\right) = \frac{8}{27}$. Thus, $\frac{8}{3} > \frac{8}{27}$.

Exercise Set 3.6

1.

Total cheese	is	cheddar cheese	plus	jalapeno cheese
↓	↓	↓	↓	↓
Total	=	$\frac{1}{3}$	+	$\frac{1}{2}$

$$= \frac{1}{3} + \frac{1}{2} = \frac{1 \cdot 2}{3 \cdot 2} + \frac{1 \cdot 3}{2 \cdot 3} = \frac{2}{6} + \frac{3}{6} = \frac{5}{6}$$

The total cheese in a nacho recipe is $\frac{5}{6}$ cup.

5.

Miles per gallon	is	miles	divided by	gallons
↓	↓	↓	↓	↓

$$\text{mpg} = 290\frac{1}{4} \div 13\frac{1}{2}$$

$$= 290\frac{1}{4} \div 13\frac{1}{2}$$

$$= \frac{1161}{4} \div \frac{27}{2} = \frac{1161}{4} \cdot \frac{2}{27} = \frac{1161 \cdot 2}{2 \cdot 2 \cdot 27} = \frac{2 \cdot 3 \cdot 3 \cdot 3 \cdot 43}{2 \cdot 2 \cdot 3 \cdot 3 \cdot 3}$$

$$= \frac{43}{2} \text{ or } 21\frac{1}{2}.$$

They got $21\frac{1}{2}$ miles per gallon in their vehicle.

9.

Difference in gauge	is	Spain gauge	minus	U.S.A. gauge
↓	↓	↓	↓	↓

$$\text{Difference} = 65\frac{9}{10} - 56\frac{1}{2}$$

$$
\begin{array}{cc}
65\frac{9}{10} & 65\frac{9}{10} \\
-56\frac{1}{2} & -56\frac{5}{10} \\
\hline
& 9\frac{4}{10} = 9\frac{2}{5}.
\end{array}
$$

The difference in standard gauges is $9\frac{2}{5}$ inches.

13.

Number of pieces	is	total pipe	cut up into	smaller pieces
↓	↓	↓	↓	↓

$$\text{Number} = 10 \div \frac{9}{5}$$

$$= 10 \div \frac{9}{5} = \frac{10}{1} \cdot \frac{5}{9} = \frac{50}{9} = 5\frac{5}{9}.$$

The plumber gets 5 full pieces of pipe with a small piece $\left(\frac{5}{9}\right)$ left over.

17.

Total flour	is	amount of flour	times	size of recipe
↓	↓	↓	↓	↓

$$\text{Total} = 2\frac{1}{2} \times 1\frac{1}{2}$$

$$= 2\frac{1}{2} \cdot 1\frac{1}{2} = \frac{5}{2} \cdot \frac{3}{2} = \frac{5 \cdot 3}{2 \cdot 2} = \frac{15}{4} = 3\frac{3}{4}.$$

Use $3\frac{3}{4}$ cups of flour to make $1\frac{1}{2}$ recipes of cookies.

21.

Area of photograph	is	length	times	width
↓	↓	↓	↓	↓

$$\text{Area} = 4\frac{1}{2} \times 2\frac{1}{2}$$

$$= 4\frac{1}{2} \cdot 2\frac{1}{2} = \frac{9}{2} \cdot \frac{5}{2} = \frac{9 \cdot 5}{2 \cdot 2} = \frac{45}{4} \text{ or } 11\frac{1}{4}.$$

The area of the photograph is $11\frac{1}{4}$ square inches.

25.

Number of pieces	is	total pipe	cut up into	smaller pieces
↓	↓	↓	↓	↓
Number	=	10	÷	$\frac{3}{4}$

$$= 10 \div \frac{3}{4} = \frac{10}{1} \cdot \frac{4}{3} = \frac{10 \cdot 4}{1 \cdot 3} = \frac{40}{3} = 13\frac{1}{3}.$$

Since he wants 12 pieces but can get $13\frac{1}{3}$ pieces, so he has enough tubing. The amount left over is $13\frac{1}{3} - 12 = 1\frac{1}{3}$ pieces and the *length* of the left over part is $1\frac{1}{3} \cdot \frac{3}{4} = \frac{4}{3} \cdot \frac{3}{4} = \frac{4 \cdot 3}{3 \cdot 4} = \frac{12}{12} = 1$. He has I foot of tubing left over.

29.

Perimeter	is	two	times	length	plus	two	times	width
↓	↓	↓	↓	↓	↓	↓	↓	↓
Perimeter	=	2	×	l	+	2	×	w

$$= 2 \cdot \frac{3}{16} + 2 \cdot \frac{3}{8}$$

$$= \frac{2 \cdot 3}{16} + \frac{2 \cdot 3}{8}$$

$$= \frac{6}{16} + \frac{6}{8} = \frac{6}{16} + \frac{12}{16} = \frac{18}{16} = \frac{9}{8} = 1\frac{1}{8}.$$

The perimeter is $1\frac{1}{8}$ inches.

Area	is	length	times	width
↓	↓	↓	↓	↓
Area	=	l	×	w

$$= \frac{3}{16} \cdot \frac{3}{8} = \frac{3 \cdot 3}{16 \cdot 8} = \frac{9}{128}$$

The area is $\frac{9}{128}$ square inch.

33.

Width of each strip	is	width	divided by	number of strips
↓	↓	↓	↓	↓
width	=	$2\frac{1}{2}$	÷	13

$$= 2\frac{1}{2} \div 13$$

$$= \frac{5}{2} \div \frac{13}{1} = \frac{5}{2} \cdot \frac{1}{13} = \frac{5 \cdot 1}{2 \cdot 13} = \frac{5}{26}.$$

The width of each strip is $\frac{5}{26}$ foot.

37.

Volume	is	length	times	width	times	height
↓	↓	↓	↓	↓	↓	↓
Volume	=	$13\frac{1}{3}$	×	$8\frac{3}{4}$	×	$4\frac{4}{25}$

$$= 13\frac{1}{3} \cdot 8\frac{3}{4} \cdot 4\frac{4}{25}$$

$$= \frac{40}{3} \cdot \frac{35}{4} \cdot \frac{104}{25} = \frac{40 \cdot 35 \cdot 104}{3 \cdot 4 \cdot 25} = \frac{4 \cdot 2 \cdot 5 \cdot 5 \cdot 7 \cdot 104}{3 \cdot 4 \cdot 5 \cdot 5}$$

$$= \frac{2 \cdot 7 \cdot 104}{3} = \frac{1456}{3} = 485\frac{1}{3}.$$

The volume of the suitcase is $485\frac{1}{3}$ cubic feet.

Chapter 3 Test

1. $4 = \boxed{2 \cdot 2}$
$15 = \boxed{3} \cdot \boxed{5}$
The LCM is $2 \cdot 2 \cdot 3 \cdot 5 = 60$.

5. $\dfrac{7}{9} + \dfrac{1}{9} = \dfrac{7 + 1}{9} = \dfrac{8}{9}$

9. **Step 1.** The LCD for the denominators 8 and 3 is 24.

Step 2. $\dfrac{7}{8} = \dfrac{7 \cdot 3}{8 \cdot 3} = \dfrac{21}{24}, \dfrac{1}{3} = \dfrac{1 \cdot 8}{3 \cdot 8} = \dfrac{8}{24}$

Step 3. $\dfrac{7}{8} - \dfrac{1}{3} = \dfrac{21}{24} - \dfrac{8}{24} = \dfrac{13}{24}$

Step 4. $\dfrac{13}{24}$ is in simplest form.

13. **Step 1.** The LCD for the denominators 12, 8, and 24 is 24.

Step 2. $\dfrac{11}{12} = \dfrac{11 \cdot 2}{12 \cdot 2} = \dfrac{22}{24}, \dfrac{3}{8} = \dfrac{3 \cdot 3}{8 \cdot 3} = \dfrac{9}{24}, \dfrac{5}{24}$
already has a denominator of 24.

Step 3. $\dfrac{11}{12} + \dfrac{3}{8} + \dfrac{5}{24} = \dfrac{22}{24} + \dfrac{9}{24} + \dfrac{5}{24} = \dfrac{36}{24}$

Step 4. $\dfrac{36}{24} = \dfrac{3 \cdot 12}{2 \cdot 12} = \dfrac{3}{2}$ or $1\dfrac{1}{2}$

17. $\begin{aligned} 19 &= 18\dfrac{11}{11} \\ -2\dfrac{3}{11} &= -2\dfrac{3}{11} \\ \hline &\quad 16\dfrac{8}{11} \end{aligned}$

21. $\left(\dfrac{4}{5}\right)^2 + \left(\dfrac{1}{8}\right)^3 = \dfrac{16}{25} + \dfrac{1}{8}$

$= \dfrac{16 \cdot 8}{25 \cdot 8} + \dfrac{1 \cdot 25}{8 \cdot 25} = \dfrac{128}{200} + \dfrac{25}{200} = \dfrac{153}{200}$

25. $\dfrac{1}{8} + \dfrac{3}{16} = \dfrac{1 \cdot 2}{8 \cdot 2} + \dfrac{3}{16} = \dfrac{2}{16} + \dfrac{3}{16} = \dfrac{5}{16}$

29. Perimeter is sum of 5 sides.

$P = \dfrac{2}{15} + \dfrac{8}{15} + \dfrac{1}{3} + \dfrac{6}{15} + \dfrac{4}{15}$

$= \dfrac{2}{15} + \dfrac{8}{15} + \dfrac{5}{15} + \dfrac{6}{15} + \dfrac{4}{15}$

$= \dfrac{2 + 8 + 5 + 6 + 4}{15} = \dfrac{25}{15} = \dfrac{5 \cdot 5}{5 \cdot 3} = \dfrac{5}{3} = 1\dfrac{2}{3}$

The perimeter is $1\dfrac{2}{3}$ inches

Chapter 4

Exercise Set 4.1

1. 6.52 in words is six and fifty-two hundredths.

5. 0.205 in words is two hundred five thousandths.

9. 200.005 in words is two hundred and five thousandths.

13. 1.8 in words is one and eight tenths.

17. Nine and eight hundredths is 9.08.

21. Sixty-four ten-thousandths is 0.0064.

25. Fifteen and eight tenths is 15.8.

29. $0.27 = \dfrac{27}{100}$

33. $0.15 = \dfrac{15}{100} = \dfrac{3}{20}$

37. $0.048 = \dfrac{48}{1000} = \dfrac{6}{125}$

41. $15.802 = 15\dfrac{802}{1000} = 15\dfrac{401}{500}$

45. $487.32 = 487\dfrac{32}{100} = 487\dfrac{8}{25}$

49. $\dfrac{45}{100} = 0.45$

53. $\dfrac{268}{1000} = 0.268$

57. $\dfrac{4026}{1000} = 4.026$

61. $\dfrac{563}{10} = 56.3$

65. To round 47,261 to the nearest thousand, observe that the digit in the hundreds place is 2. Since the digit is less than 5, we do not add 1 to the digit in the thousands place. The number 47,261 rounded to the nearest thousand is 47,000.

69. 0.00026849576 in words is twenty-six million, eight hundred forty-nine, five hundred seventy-six hundred-billionths.

Exercise Set 4.2

1. $\begin{array}{cc} 0.15 & 0.16 \\ \uparrow & \uparrow \\ 5 < & 6 \text{ so} \end{array}$
$0.15 < 0.16$

5. $\begin{array}{cc} 0.098 & 0.1 \\ \uparrow & \uparrow \\ 0 < & 1 \text{ so} \end{array}$
$0.098 < 0.1$

9. $\begin{array}{cc} 167.908 & 167.980 \\ \uparrow & \uparrow \\ 0 < & 8 \text{ so} \end{array}$
$167.908 < 167.980$

13. Smallest to largest:
0.006, 0.0061, 0.06

17. Smallest to largest:
1.01, 1.09, 1.1, 1.16

21. To round 0.57 to the nearest tenth, observe that the digit in the hundredths place is 7. Since this digit is at least 5, we need to add 1 to the digit in the tenths place. The number 0.57 rounded to the nearest tenth is 0.6.

25. To round 0.5942 to the nearest thousandth, observe that the digit in the ten-thousandths place is 2. Since this digit is less than 5, we do not add to the digit in the thousandths place. The number 0.5942 rounded to the nearest thousandth is 0.594.

29. To round 12.342 to the nearest tenth, observe that the digit in the hundredths place is 4. Since this digit is less than 5, we do not add 1 to digit in the tenths place. The number 12.342 rounded to the nearest tenth is 12.3.

33. To round 0.501 to the nearest tenth, observe that the digit in the hundredths place is 0. Since this digit is less than 5, we do not add 1 to the digit in the tenths place. The number 0.501 rounded to the nearest tenth is 0.5.

37. To round 3829.34 to the nearest ten, observe that the digit in the ones place is 9. Since this digit is at least 5, we need to add 1 to the digit in the tens place. The number 3829.34 rounded to the nearest ten is 3830.

41. To round 42,650.14 to the nearest one, observe that the digit in the tenths is 1. Since this digit is less than 5, we do not add 1 to the digit in the ones place. The number 42,650.14 rounded to the nearest one is 42,650. The amount is $42,650.

45. To round 0.1992 to the nearest hundredth, observe that the digit in the thousandths place is 9. Since this digit is at least 5, we need to add 1 to the digit in the hundredths place. The number 0.1992 rounded to the nearest hundredth is 0.2. The amount is $0.20.

49. To round 39,867 to the nearest thousand, observe that the digit in the hundreds place is 8. Since this digit is at least 5, we need to add 1 to the digit in the thousands place. The number 39,867 rounded to the nearest thousand is 40,000.

53. To round 24.6229 to the nearest thousandth, observe that the digit in the ten-thousandths place is 9. Since this digit is at least 5, we need to add 1 to the digit in the thousandths place. The number 24.6229 rounded to the nearest thousandth is 24.623. The lenght is 24.623 hours.

57.
$$\begin{array}{r} 3452 \\ + 2314 \\ \hline 5766 \end{array}$$

61.
$$\begin{array}{r} 482 \\ - 239 \\ \hline 243 \end{array}$$
Check:
$$\begin{array}{r} \overset{1}{243} \\ + 239 \\ \hline 482 \end{array}$$

65. Compare each place value and list from greatest to least.

$$\left\{ \begin{array}{l} 228.040, 226.130, 225.490, 225.370, \\ 222.980, 222.830, 222.008, 221.000, \\ 219.702, 218.208 \end{array} \right.$$

Exercise Set 4.3

1.
$$\begin{array}{r} 1.3 \\ + 2.2 \\ \hline 3.5 \end{array}$$

5.
$$\begin{array}{r} 0.003 \\ + 0.091 \\ \hline 0.094 \end{array}$$

9.
$$\begin{array}{r} \overset{1}{490.000} \\ + 93.09 \\ \hline 583.09 \end{array}$$

13.
$$\begin{array}{r} \overset{1\ \ 11}{100.009} \\ 6.080 \\ + 9.034 \\ \hline 115.123 \end{array}$$

17.
$$\begin{array}{r} \overset{1\ \ 1}{45.023} \\ 3.006 \\ + 8.403 \\ \hline 56.432 \end{array}$$

21.
$$\begin{array}{r} 18.0 \\ - 2.7 \\ \hline 15.3 \end{array}$$
Check:
$$\begin{array}{r} \overset{1}{15.3} \\ + 2.7 \\ \hline 18.0 \text{ or } 18 \end{array}$$

25.
$$\begin{array}{r} 5.90 \\ - 4.07 \\ \hline 1.83 \end{array}$$
Check:
$$\begin{array}{r} \overset{1}{1.83} \\ + 4.07 \\ \hline 5.90 \text{ or } 5.9 \end{array}$$

29.
$$\begin{array}{r} 500.34 \\ - 123.45 \\ \hline 376.89 \end{array}$$
Check:
$$\begin{array}{r} \overset{111\ 1}{376.89} \\ + 123.45 \\ \hline 500.34 \end{array}$$

33.
$$\begin{array}{r} 200.0 \\ - 5.6 \\ \hline 194.4 \end{array}$$
Check:
$$\begin{array}{r} \overset{111}{194.4} \\ + 5.6 \\ \hline 200.0 \text{ or } 200 \end{array}$$

37.
$$\begin{array}{r} 23.0 \\ - 6.7 \\ \hline 16.3 \end{array}$$
Check:
$$\begin{array}{r} \overset{11}{16.3} \\ + 6.7 \\ \hline 23.0 \text{ or } 23 \end{array}$$

41. The phrase "By how much did the price change" tells us to subtract. Subtract 1.339 from 1.479.

$$\begin{array}{r} 1.479 \\ - 1.339 \\ \hline 0.140 \end{array}$$
Check:
$$\begin{array}{r} 1.339 \\ + 0.140 \\ \hline 1.479 \end{array}$$
The price changed by $0.14.

45. The phrase "How much more" tells us to subtract. Subtract 136.8 from 150.1.

$$\begin{array}{r} 150.1 \\ - 136.8 \\ \hline 13.3 \end{array}$$
Check:
$$\begin{array}{r} \overset{1}{136.8} \\ + 13.3 \\ \hline 150.1 \end{array}$$
The difference in consumption is 13.3 pounds per person.

49. To find the new record we must add the old record to the increase.

$$\begin{array}{r} 153.601 \\ + 14.006 \\ \hline 167.606 \end{array}$$
The new record is 167.606 mph.

53. Find the sum of the lenghts of the three sides.

$$\begin{array}{r} \overset{1\ 1}{12.40} \\ 29.34 \\ + 25.70 \\ \hline 67.44 \end{array}$$
The architect needs 67.44 feet of border material.

57. Add the durations of all four missions.

$$\begin{array}{r} \overset{2\ 12\ 11}{330.583} \\ 94.567 \\ 147.000 \\ + 142.900 \\ \hline 715.050 \end{array}$$
James A. Lovell has spent 715.05 hours in spaceflight.

61. Subtract 13.9 from 22.

$$\begin{array}{r} 22.0 \\ + 13.9 \\ \hline 8.1 \end{array}$$
Check:
$$\begin{array}{r} \overset{11}{8.1} \\ + 13.9 \\ \hline 22.0 \text{ or } 22 \end{array}$$
The difference in consumption is 8.1 pounds.

65.
$$\begin{array}{r} 46 \\ \times 3 \\ \hline 138 \end{array}$$

69. $\left(\dfrac{1}{5}\right)^3 = \dfrac{1}{5} \cdot \dfrac{1}{5} \cdot \dfrac{1}{5} = \dfrac{1}{125}$

73. answers may vary

Exercise Set 4.4

1.
$$\begin{array}{r} 0.2 \\ \times 0.6 \\ \hline 12 \\ 00 \\ \hline 0.12 \end{array}$$

5.
$$\begin{array}{r} 0.26 \\ \times 5 \\ \hline 1.30 \text{ or } 1.3 \end{array}$$

9.
$$\begin{array}{r} 5.62 \\ \times 7.7 \\ \hline 3934 \\ 39340 \\ \hline 43.274 \end{array}$$

13.
$$\begin{array}{r} 490.2 \\ \times 0.023 \\ \hline 14706 \\ 98040 \\ 000 \\ 0000 \\ \hline 11.2746 \end{array}$$

17. To find 6.5×10, note that 10 has 1 zero. Therefore, we move the decimal point of 6.5 to the right 1 place. The product is 65.

21. To find 7.093×100, note that 100 has 2 zeros. Therefore, we move the decimal point of 7.093 to the right 2 places. The product is 709.3.

25. To find 9.1×1000, note that 1000 has 3 zeros. Therefore, we move the decimal point of 9.1 to the right 3 places. The product is 9100.

29. 5.5 billion $= 5.5 \times 1,000,000,000$
$= 5,500,000,000$
They can make 5,500,000,000 chocolate bars.

33. 49.8 million $= 49.8 \times 1,000,000$
$= 49,800,000$
49,800,000 people have riden the Blue Streak roller coaster.

37. Circumference $= \pi \cdot$ diameter
$C = \pi \cdot 10 = 10\pi$
$C \approx 10(3.14) = 31.4$
The circumference is 10π centimeters, which is approximately 31.4 centimeters.

41. Multiply the number of ounces by the number of grams of fat in 1 ounce to get the total amount of fat.
$$\begin{array}{r} 6.2 \\ \times 4 \\ \hline 24.8 \end{array}$$
There are 24.8 grams of fat in a 4-ounce serving of cream cheese.

45. Multiply 39.37 by 1.65
$$\begin{array}{r} 39.37 \\ \times 1.65 \\ \hline 19685 \\ 236220 \\ 393700 \\ \hline 64.9605 \end{array}$$
She is about 64.9605 inches tall.

49.
$$\begin{array}{r} 26 \\ 5)\overline{130} \\ -10 \\ \hline 30 \\ -30 \\ \hline 0 \end{array}$$

53.
$$\begin{array}{r} 8 \\ 365)\overline{2920} \\ -2920 \\ \hline 0 \end{array}$$

57. $(20.6)(1.86)(100,000) = 3,831,600$ miles

61. answers may vary

Exercise Set 4.5

1.
$$\begin{array}{r} 0.094 \\ 5)\overline{0.470} \\ -45 \\ \hline 20 \\ -20 \\ \hline 0 \end{array}$$

5. $0.82)\overline{4.756}$ becomes
$$\begin{array}{r} 5.8 \\ 82.)\overline{475.6} \\ -410 \\ \hline 65\ 6 \\ -65\ 6 \\ \hline 0 \end{array}$$

9.
$$\begin{array}{r} 0.413 \\ 15)\overline{6.195} \\ -60 \\ \hline 19 \\ -15 \\ \hline 45 \\ -45 \\ \hline 0 \end{array}$$

13. $0.27)\overline{1.296}$ becomes
$$\begin{array}{r} 4.8 \\ 27.)\overline{129.6} \\ -108 \\ \hline 21\ 6 \\ -21\ 6 \\ \hline 0 \end{array}$$

17. $0.6)\overline{18}$ becomes
$$\begin{array}{r} 30 \\ 6.)\overline{180} \\ -18 \\ \hline 00 \\ -0 \\ \hline 0 \end{array}$$

21. $7.2)\overline{70.56}$ becomes
$$\begin{array}{r} 9.8 \\ 72.)\overline{705.6} \\ -648 \\ \hline 57\ 6 \\ -57\ 6 \\ \hline 0 \end{array}$$

25. $0.027)\overline{1.215}$ becomes
$$\begin{array}{r} 45 \\ 27.)\overline{1215.} \\ -108 \\ \hline 135 \\ -135 \\ \hline 0 \end{array}$$

29. $0.23)\overline{0.549}$ becomes
$$\begin{array}{r} 23.869 \approx 23.87 \\ 23.)\overline{549.000} \\ -46 \\ \hline 89 \\ -69 \\ \hline 20\ 0 \\ -18\ 4 \\ \hline 1\ 60 \\ -1\ 38 \\ \hline 220 \\ -207 \\ \hline 13 \end{array}$$

33. To find $54.982 + 100$, note that 100 has 2 zeros. Therefore, we move the decimal point of 54.982 to the left 2 places. The quotient in 0.54982.

37. To find $87 + 10$, note that 10 has 1 zero. Therefore, we move the decimal point of 87 to the left 1 place. The quotient in 8.7.

41. There are 52 weeks per year and 40 hours per week. Therefore, there are $52 \times 40 = 2080$ hours per year.

$$
\begin{array}{r}
4417.462 \approx 4417.46 \\
2080\overline{)9{,}188.321.000} \\
-8\,320 \\
\hline
868\,3 \\
-832\,0 \\
\hline
36\,32 \\
-20\,80 \\
\hline
15\,521 \\
-14\,560 \\
\hline
9\,610 \\
-8\,320 \\
\hline
1\,2900 \\
-1\,2480 \\
\hline
4200 \\
-4160 \\
\hline
40
\end{array}
$$

His hourly wage was $4417.46.

45. $39.37\overline{)200}$ becomes

$$
\begin{array}{r}
5.08 \approx 5.1 \\
3937.\overline{)20{,}000.00} \\
-19\,685 \\
\hline
315\,0 \\
-0 \\
\hline
315\,00 \\
-314\,96 \\
\hline
4
\end{array}
$$

There are 5.1 meters in 200 inches.

49.
$$
\begin{array}{r}
112.80 = 112.8 \\
24\overline{)2707.44} \\
-24 \\
\hline
30 \\
-24 \\
\hline
67 \\
-48 \\
\hline
19\,4 \\
-19\,2 \\
\hline
24 \\
-24 \\
\hline
0
\end{array}
$$

Their average speed was 112.8 mph.

53. To round 345.219 to the nearest hundredth, observe that the digit in the thousandths place is 9. Since this digit is at least 5, we need to add 1 to the digit in the hundredths place. The number 345.219 rounded to the nearest hundredth is 345.22.

57. $2 + 3 \cdot 6 = 2 + 18 = 20$

61. $(86 + 78 + 91 + 85) \div 4 = 340 \div 4$

$$
\begin{array}{r}
85 \\
4\overline{)340} \\
-32 \\
\hline
20 \\
-20 \\
\hline
0
\end{array}
$$

The average is 85.

65.
$$
\begin{array}{r}
45.2 \\
4.\overline{)180.8} \\
-16 \\
\hline
20 \\
-20 \\
\hline
0\,8 \\
-8 \\
\hline
0
\end{array}
$$

The length of a side is 45.2 centimeters.

Exercise Set 4.6

1. $2.1 + 5.8 + 4.1 = 12.0$
   $2 + 6 + 4 = 12$
   Yes, the result and estimate agree.

5.
$$
\begin{array}{rr}
6 & 6 \\
\times 483.11 & \times 500 \\
\hline
2898.66 & 3000
\end{array}
$$
   3000 is close to 2898.66, so the answer is reasonable.

9.
$$
\begin{array}{rr}
69.2 & 70 \\
32.1 & 30 \\
+ 48.5 & + 50 \\
\hline
149.8 & 150
\end{array}
$$
   149.8 is close to 150, so the answer is reasonable.

13. $2(12.2) + 2(5.9) \approx 2(12) + 2(6)$
   $= 24 + 120 = 36$ inches

17. $3.14(7)(2) = 43.96$ meters

21. 198.79 is approximately 200
   $(200)(12)(5) = \$12{,}000$

25. Estimate
$$
\begin{array}{r}
600.7 \text{ million} \\
460.9 \text{ million} \\
431.1 \text{ million} \\
399.8 \text{ million} \\
+ 356.8 \text{ million} \\
\hline
2249.3 \text{ million}
\end{array}
$$
   The total estimate is $2249.3 million.

29. $(0.4)^2 = (0.4)(0.4) = 0.16$

33. $1.4(2 - 1.8) = 1.4(0.2) = 0.28$

37. $7.8 - 4.83 + 2.1 = 7.8 - 2.3$
   $= 5.5$

41. $(3.1 + 0.7)(2.9 - 0.9) = (3.8)(2.0)$
   $= 7.6$

45. $\dfrac{7 + 0.74}{0.06} = \dfrac{7.74}{0.06} = 129$

49. $(0.3)^2 + 2(14.6 - 0.03) = 0.09 + 2(14.6 - 0.03)$
   $= 0.09 + 2(14.57) = 0.09 + 29.14 = 29.23$

53. $6 \div 0.1 + 8.9 \times 10 - 4.6 = 60 + 89 - 4.6 = 149 - 4.6$
   $= 144.4$

57. $\dfrac{36}{36} + \dfrac{30}{35} = \dfrac{36}{56} \cdot \dfrac{35}{30}$
   $= \dfrac{2 \cdot 2 \cdot 3 \cdot 3 \cdot 5 \cdot 7}{2 \cdot 2 \cdot 2 \cdot 7 \cdot 2 \cdot 3 \cdot 5}$
   $= \dfrac{3}{4}$

61. $1.96(7.852 - 3.147)^2 = 1.96(4.705)^2$
   $= 1.96(22.137025)$
   $= 43.388569$
   Estimate: $2(8 - 3)^2 = 2(5)^2 = 2(25) = 50$

Exercise Set 4.7

1. $\frac{1}{5} = 0.2$

$$5\overline{)1.0}$$
$$0.2$$

5. $\frac{3}{4} = 0.75$

$$4\overline{)3.00}$$
$$0.75$$
$$\underline{-2\,8}$$
$$20$$
$$\underline{-20}$$
$$0$$

9. $\frac{3}{8} = 0.375$

$$8\overline{)3.000}$$
$$0.375$$
$$\underline{-24}$$
$$60$$
$$\underline{-56}$$
$$40$$
$$\underline{-40}$$
$$0$$

13. $\frac{17}{40} = 0.425$

$$40\overline{)17.000}$$
$$0.425$$
$$\underline{-16\,0}$$
$$1\,00$$
$$\underline{-80}$$
$$200$$
$$\underline{-200}$$
$$0$$

17. $\frac{1}{3} = 0.\overline{3}$

$$3\overline{)1.00}$$
$$0.33\ldots$$
$$\underline{-9}$$
$$10$$
$$\underline{-9}$$
$$1$$

21. $\frac{2}{9} = 0.\overline{2}$

$$9\overline{)2.00}$$
$$0.22\ldots$$
$$\underline{-1\,8}$$
$$20$$
$$\underline{-18}$$
$$2$$

25. $0.\overline{3} = 0.333\ldots \approx 0.33$

29. $0.\overline{2} = 0.222\ldots \approx 0.2$

33.

$$25\overline{)17.00}$$
$$0.68$$
$$\underline{-15\,0}$$
$$2\,00$$
$$\underline{-2\,00}$$
$$0$$

$\frac{17}{25} = 0.68$

37.

$$91\overline{)56.000}$$
$$0.615 \approx 0.62$$
$$\underline{-54\,6}$$
$$1\,40$$
$$\underline{-91}$$
$$490$$
$$\underline{-455}$$
$$35$$

$\frac{56}{91} \approx 0.62$

41. $2 > 1$, so $0.823 > 0.813$

45. $\frac{2}{3} = \frac{4}{6}$, so $\frac{2}{3} < \frac{5}{6}$

49. $\frac{4}{7} \approx 0.5714$

$0.5714 > 0.14$, so $\frac{4}{7} > 0.14$

53. $\frac{456}{64} = 7.125$

$7.123 < 7.125$ so $7.123 < \frac{456}{64}$

57. $0.49 = 0.490$

$0.49, 0.491, 0.498$

61. $\frac{4}{7} \approx 0.571$

$0.412, 0.453, \frac{4}{7}$

65. $\frac{12}{5} = 24$

$\frac{17}{8} = 2.125$

$\frac{17}{8}, 2.37, \frac{12}{5}$

69. Area $= \frac{1}{2} \times$ base $\times$ base height

$= \frac{1}{2} \times 5.2 \times 3.6$

$= 0.5 \times 5.2 \times 3.6$

$= 9.36$

Area is 9.36 square centimeters.

73. $2^3 = (2)(2)(2) = 8$

77. $\left(\frac{1}{3}\right)^4 = \frac{1}{3} \cdot \frac{1}{3} \cdot \frac{1}{3} \cdot \frac{1}{3} = \frac{1}{81}$

81. $\left(\frac{2}{5}\right)\left(\frac{5}{2}\right)^2 = \frac{2}{5} \cdot \frac{5}{2} \cdot \frac{5}{2} = \frac{5}{2}$

85.

	Estimate
2249	2200
1557	1600
1426	1400
1135	1100
$-\,1118$	$+\,1100$
	7400

The estimate is 7400 stations.

89. answers may vary

Chapter 4 Test

1. 45.092 in words is forty-five and ninety-two thousandths

5. $25.0909 < 25.9090$

9. $\frac{13}{26} = 0.5$

13.

$$\begin{array}{r} 10.2 \\ \times\,4.3 \\ \hline 306 \\ 4080 \\ \hline 43.86 \end{array}$$

17. $\frac{473}{10} = 47.3$

21. $A = 123.8(80) = 9904$

9904 square feet of lawn

$9904 \times 0.02 = 198.08$

She needs 198.08 ounces of insecticide.

Chapter 5

Exercise Set 5.1

1. 11 to 14 is $\frac{11}{14}$

5. 151 to 201 is $\frac{151}{201}$

9. 5 to $7\frac{1}{2}$ is $\dfrac{5}{7\frac{1}{2}}$

13. $\dfrac{16}{24} = \dfrac{2 \cdot 8}{3 \cdot 8} = \dfrac{2}{3}$

17. $\dfrac{4.63}{8.21} = \dfrac{4.63 \times 100}{8.21 \times 100} = \dfrac{463}{821}$

21. $\dfrac{10 \text{ hours}}{24 \text{ hours}} = \dfrac{5 \cdot 2}{12 \cdot 2} = \dfrac{5}{12}$

25. $\dfrac{24 \text{ days}}{14 \text{ days}} = \dfrac{12 \cdot 2}{7 \cdot 2} = \dfrac{12}{7}$

29. $\dfrac{8 \text{ inches}}{20 \text{ inches}} = \dfrac{2 \cdot 4}{5 \cdot 4} = \dfrac{2}{5}$

33. perimeter $= 8 + 15 + 17 = 40$ feet

$\dfrac{\text{hypotenuse}}{\text{perimeter}} = \dfrac{14 \text{ feet}}{40 \text{ feet}} = \dfrac{17}{40}$

37. $6000 - 4500 = 1500$ married

$\dfrac{4500 \text{ single}}{1500 \text{ married}} = \dfrac{3 \text{ single}}{1 \text{ married}}$

41. $\dfrac{19}{57} = \dfrac{19}{3 \cdot 19} = \dfrac{1}{3}$

45.
$$\begin{array}{r} 2.3 \\ 9\overline{)20.7} \\ 18 \\ \hline 2\,7 \\ 2\,7 \\ \hline 0 \end{array}$$

49. $\dfrac{4 \text{ states}}{46 \text{ states}} = \dfrac{2 \cdot 2}{2 \cdot 23} = \dfrac{2}{23}$

53. $\dfrac{3 \text{ bruised}}{33 \text{ total}} = \dfrac{1 \cdot 3}{11 \cdot 3} = \dfrac{1}{11}$

$\dfrac{1}{11} < \dfrac{1}{10}$

No, the shipment should not be refused.

Exercise Set 5.2

1. $\dfrac{5 \text{ shrubs}}{15 \text{ feet}} = \dfrac{1 \text{ shrub}}{3 \text{ feet}}$

5. $\dfrac{8 \text{ phone lines}}{36 \text{ employees}} = \dfrac{2 \text{ phone lines}}{9 \text{ employees}}$

9. $\dfrac{6 \text{ flight attendants}}{200 \text{ passengers}} = \dfrac{3 \text{ flight attendants}}{100 \text{ passengers}}$

13. $\dfrac{375 \text{ riders}}{5 \text{ subway cars}} = \dfrac{75 \text{ riders}}{1 \text{ subway car}}$
$= 75$ riders/subway car

17. $\dfrac{144 \text{ diapers}}{24 \text{ babies}} = \dfrac{6 \text{ diapers}}{1 \text{ baby}}$
$= 6$ diapers/baby

21. $\dfrac{600 \text{ kilometers}}{90 \text{ minutes}} = 6\frac{2}{3}$ kilometers/minute

25. $\dfrac{12{,}000 \text{ good}}{40 \text{ defective}} = 300$ good/defective

29. $\dfrac{\$29{,}000{,}000}{500 \text{ species}} = \dfrac{\$58{,}000}{1 \text{ species}}$ or \$58,000/species

33. a. $\dfrac{250 \text{ boards}}{8 \text{ hours}} = 32.25$ computer boards/hour

b. $\dfrac{400 \text{ boards}}{12 \text{ hours}} \approx 33.3$ computer boards/hour

c. $33.3 > 31.25$ so Lamont can assemble computer boards faster.

37. $\dfrac{\$1.19}{7 \text{ bananas}} = \dfrac{\$0.17}{1 \text{ banana}}$ or \$0.17 per banana

41. $\dfrac{\$1.69}{16 \text{ ounces}} \approx 0.106$ per ounce

$\dfrac{\$0.69}{6 \text{ ounces}} \approx \$0.115$ per ounce

The 16-ounce size is the better buy.

45. $\dfrac{\$0.59}{100 \text{ napkins}} = \$0.006$ per napkin

$\dfrac{\$0.93}{180 \text{ napkins}} \approx \$.005$ per napkin

The package of 180 napkins is the better buy.

49.
$$\begin{array}{r} 3.7 \\ \times\ 1.2 \\ \hline 74 \\ 370 \\ \hline 4.44 \end{array}$$

53. Fill in the column labeled Miles Driven by subtracting Beginning Odometer Reading from Ending Odometer Reading. Fill in the column labeled Miles per Gallon by dividing Miles driven by Gallons of Gas Used, and round to the nearest tenth.

Miles Driven	Miles Per Gallon
257	19.2
352	22.3
347	21.6

57. $\dfrac{45{,}000{,}000 \text{ students}}{78{,}300 \text{ schools}} \approx 575$ students/school

Exercise Set 5.3

1. $\dfrac{10 \text{ diamonds}}{6 \text{ opals}} = \dfrac{5 \text{ diamonds}}{3 \text{ opals}}$

5. $\dfrac{6 \text{ eagles}}{58 \text{ sparrows}} = \dfrac{3 \text{ eagles}}{29 \text{ sparrows}}$

9. $\dfrac{22 \text{ vanilla wafers}}{1 \text{ cup cookie crumbs}} = \dfrac{55 \text{ vanilla wafers}}{2.5 \text{ cups cookie crumbs}}$

13. $\dfrac{8}{6} = \dfrac{9}{7}$
$8 \cdot 7 = 9 \cdot 6$
$56 = 54$
false

17. $\dfrac{5}{8} = \dfrac{625}{1000}$
$5 \cdot 1000 = 8 \cdot 625$
$5000 = 5000$
true

21. $\dfrac{4.2}{8.4} = \dfrac{5}{10}$
$(4.2)(10) = 5(8.4)$
$42 = 42$
true

25.
$$\frac{2\frac{2}{5}}{\frac{2}{3}} = \frac{\frac{10}{9}}{\frac{1}{4}}$$

$$2\frac{2}{5}\cdot\frac{1}{4} = \frac{10}{9}\cdot\frac{2}{3}$$

$$\frac{12}{5}\cdot\frac{1}{4} = \frac{10}{9}\cdot\frac{2}{3}$$

$$\frac{3}{5} = \frac{20}{27}$$

false

29. $\frac{30}{10} = \frac{15}{n}$

$30n = 150$

$n = 5$

33. $\frac{n}{6} = \frac{8}{15}$

$15n = 48$

$n = 3.2$

37. $\frac{\frac{1}{3}}{\frac{3}{8}} = \frac{\frac{2}{5}}{x}$

$$\frac{1}{3}x = \frac{3}{8}\cdot\frac{2}{5}$$

$$\frac{1}{3}x = \frac{3}{20}$$

$$x = \frac{3}{20}\cdot\frac{3}{1}$$

$$x = \frac{9}{20}$$

41. $\frac{\frac{2}{3}}{\frac{6}{9}} = \frac{12}{n}$

$$\frac{2}{3}n = 12\cdot\frac{6}{9}$$

$$\frac{2}{3}n = 8$$

$$n = 8\cdot\frac{3}{2}$$

$$n = 12$$

49. 8.01   8.1

↑   ↑

0 < 1 so

8.01 < 8.1

53. $5\frac{1}{3} = \frac{16}{3}$

$6\frac{2}{3} = \frac{20}{3}$

$16 < 20$, so $5\frac{1}{3} < 6\frac{2}{3}$

57. $\frac{n}{1150} = \frac{588}{483}$

$(588)(1150) = 483n$

$676{,}200 = 483n$

$1400 = n$

61. answers may vary

Exercise Set 5.4

1. Let n = number of completed passes

$\frac{4\text{ completed}}{9\text{ attempted}} = \frac{n\text{ completed}}{27\text{ attempted}}$

$27\cdot 4 = 9n$

$108 = 9n$

$12 = n$

12 passes were completed.

5. Let n = number of student accepted

$\frac{2\text{ accepts}}{7\text{ applicants}} = \frac{n\text{ accepts}}{630\text{ applicants}}$

$1620 = 7n$

$180 = n$

180 students were accepted.

9. Let n = amount of floor space

$\frac{9\text{ square feet}}{1\text{ student}} = \frac{n\text{ square feet}}{30\text{ students}}$

$270 = n$

270 square feet is required.

13. Let n = distance from Milan to Rome

$\frac{1\text{ centimeter}}{30\text{ kilometers}} = \frac{15\text{ centimeters}}{n\text{ kilometers}}$

$n = 30\cdot 15$

$n = 450$

It is 450 km from Milan to Rome.

17. Let n = number of bags of fertilizer

Area of lawn = $260 \times 180 = 46{,}800$ square feet

$\frac{1\text{ bag}}{3000\text{ square feet}} = \frac{n}{46{,}800\text{ square feet}}$

$46{,}800 = 3000n$

$15.6 = n$

16 bags are needed.

21. Let n = number of hits expected

$\frac{3\text{ hits}}{8\text{ at bats}} = \frac{n\text{ hits}}{40\text{ at bats}}$

$3\cdot 40 = 8n$

$120 = 8n$

$15 = n$

He would be expected to make 15 hits.

25. Let n = number of weeks

$\frac{5\text{ boxes}}{3\text{ weeks}} = \frac{144\text{ boxes}}{n\text{ weeks}}$

$5n = 432$

$n = 86.4$

The envelopes will last 86 weeks.

29. Let n = estimated height of Statue of Liberty

$\frac{\text{statue}}{\text{student}}$ $\frac{42\text{ feet}}{2\text{ feet}} = \frac{n\text{ feet}}{5\frac{1}{3}\text{ feet}}$

$$2n = 42\cdot 5\frac{1}{3}$$

$$2n = \frac{42}{1}\cdot\frac{16}{3}$$

$$2n = 224$$

$$n = 224 + 2$$

$$n = 112$$

$$112 - 111\frac{1}{12} = \frac{11}{12}$$

The estimate is 112 feet. The difference is 11 inches.

33. Let n = number of restaurant workers

$\frac{1\text{ worker}}{3\text{ adults}} = \frac{n\text{ workers}}{84\text{ adults}}$

$3n = 1\cdot 84$

$3n = 84$

$n = 28$

There are 28 restaurant workers.

37. $\begin{array}{r}5\\2\overline{)10}\\2\overline{)20}\end{array}$

$20 = 2^2\cdot 5$

41.
$$2\overline{)\,4}$$
$$2\overline{)\,8}$$
$$2\overline{)16}$$
$$2\overline{)32}$$
$$32 = 2^5$$

45. answers may vary

Chapter 5 Test

1. $\dfrac{4500 \text{ trees}}{6500 \text{ trees}} = \dfrac{9 \cdot 500}{13 \cdot 500} = \dfrac{9}{13}$

5. $\dfrac{650 \text{ kilometers}}{8 \text{ hours}} = 81.25 \text{ kilometers per hour}$

$$
\begin{array}{r}
81.25 \\
8\overline{)650.00} \\
-64 \\
\hline
10 \\
-8 \\
\hline
2\,0 \\
-1\,6 \\
\hline
40 \\
-40 \\
\hline
0
\end{array}
$$

9. $\dfrac{\$1.49}{16 \text{ ounces}} \approx \dfrac{\$0.09}{1 \text{ ounce}}$

$\dfrac{\$2.39}{24 \text{ ounces}} \approx \dfrac{\$0.10}{1 \text{ ounce}}$

Therefore, the 16-ounce size is the better buy.

13. $\dfrac{8}{n} = \dfrac{11}{6}$

$11 \cdot n = 8 \cdot 6$

$11n = 48$

$n = 48 \div 11$

$n = 4\dfrac{4}{11}$

17. Let n = length of time

$\dfrac{3}{80} = \dfrac{n}{100}$

$80n = 300$

$n = \dfrac{15}{4}$

$= 3\dfrac{3}{4}$

It will take $3\dfrac{3}{4}$ hours.

Chapter 6

Exercise Set 6.1

1. $\dfrac{81}{100} = 81\%$

5. The largest section of the circle graph is chocolate chip. Therefore, chocolate chip was the most preferred cookie.

$\dfrac{52}{100} = 52\%$

9. $48\% = 48.0\% = 0.48$

13. $100\% = 1.00\% = 1.00 = 1$

17. $2.8\% = 0.028$

21. $300\% = 3$

25. $67\% = 0.67$

29. $21.2\% = 0.212$

33. $3.1 = 3.10 = 310\%$

37. $0.003 = 0.00\,3 = 0.3\%$

41. $5.3 = 5.30 = 530\%$

45. $0.3328 = 33.28\%$

49. $0.7 = 0.70 = 70\%$

53. $0.093 = 9.3\%$

57. $\dfrac{1}{4} = \dfrac{1 \cdot 25}{4 \cdot 25} = \dfrac{25}{100} = 0.25$

61. $\dfrac{9}{10} = 0.9$

65. $77\% = 0.77$

Exercise Set 6.2

1. $12\% = \dfrac{12}{100} = \dfrac{4 \cdot 3}{4 \cdot 25} = \dfrac{3}{25}$

5. $4.5\% = \dfrac{4.5}{100} = \dfrac{45}{1000} = \dfrac{5 \cdot 9}{5 \cdot 200} = \dfrac{9}{200}$

9. $73\% = \dfrac{73}{100}$

13. $6.25\% = \dfrac{6.25}{100} = \dfrac{625}{10,000} = \dfrac{625 \cdot 1}{625 \cdot 16} = \dfrac{1}{16}$

17. $10\dfrac{1}{3}\% = \dfrac{10\frac{1}{3}}{100} = \dfrac{\frac{31}{3}}{100} = \dfrac{31}{3} \div 100 = \dfrac{31}{3} \cdot \dfrac{1}{100} = \dfrac{31}{300}$

21. $\dfrac{3}{4} = \dfrac{3}{4} \cdot 100\% = \dfrac{300}{4}\% = 75\%$

25. $\dfrac{2}{5} = \dfrac{2}{5} \cdot 100\% = \dfrac{200}{5}\% = 40\%$

29. $\dfrac{17}{50} = \dfrac{17}{50} \cdot 100\% = \dfrac{1700}{50}\% = 34\%$

33. $\dfrac{5}{16} = \dfrac{5}{16} \cdot 100\% = \dfrac{500}{16}\% = 31\dfrac{1}{4}\%$

37. $\dfrac{2}{3} = \dfrac{2}{3} \cdot 100\% = \dfrac{200}{3}\% = 66\dfrac{2}{3}\%$

41. $2\dfrac{1}{2} = \dfrac{5}{2} \cdot 100\% = \dfrac{500}{2}\% = 250\%$

45. $\dfrac{7}{11} = \dfrac{7}{11} \cdot 100\% = \dfrac{700}{11}\% \approx 63.64\%$

$$
\begin{array}{r}
63.636 \approx 63.64 \\
11\overline{)700.000} \\
-66 \\
\hline
40 \\
-33 \\
\hline
7\,0 \\
-6\,6 \\
\hline
40 \\
-33 \\
\hline
70 \\
-66 \\
\hline
4
\end{array}
$$

49. $\dfrac{1}{7} = \dfrac{1}{7} \cdot 100\% = \dfrac{100}{7} \approx 14.29\%$

$$\begin{array}{r} 14.285 \approx 14.29 \\ 7\overline{)100.000} \\ \underline{-7} \\ 30 \\ \underline{-28} \\ 2\,0 \\ \underline{-1\,4} \\ 60 \\ \underline{-56} \\ 40 \\ \underline{-35} \\ 5 \end{array}$$

53.

Percent	Decimal	Fraction
35%	0.35	$\frac{7}{20}$
20%	0.2	$\frac{1}{3}$
50%	0.5	$\frac{1}{2}$
70%	0.7	$\frac{1}{2}$
37.5%	0.375	$\frac{3}{8}$

57. $14.8\% = 0.148 = \dfrac{148}{1000} = \dfrac{4 \cdot 37}{4 \cdot 250} = \dfrac{37}{250}$

61. $\dfrac{141}{500} = 0.282 = 28.2\%$

$$\begin{array}{r} 0.282 \\ 500\overline{)141.000} \\ \underline{-100\,0} \\ 41\,00 \\ \underline{-40\,00} \\ 1\,000 \\ \underline{-1\,000} \\ 0 \end{array}$$

65. $\dfrac{11}{50} = \dfrac{11 \cdot 2}{50 \cdot 2} = \dfrac{22}{100} = 22\%$

69. $8 \cdot n = 80$

$\dfrac{8 \cdot n}{8} = \dfrac{80}{8}$

$n = 10$

73. $\dfrac{21}{79} \approx 0.266 = 26.6\%$

77. A fraction written as a percent is greater than 100% when the numerator is *greater* than the denominator.

81. answers may vary

Exercise Set 6.3

1. $15\% \cdot 72 = n$

5. $n \cdot 90 = 20$

9. $n = 9\% \cdot 43$

13. $n = 14\% \cdot 52$
$n = 0.14 \cdot 52$
$n = 7.28$

17. $1.2 = 12\% \cdot n$
$1.2 = 0.12n$
$\dfrac{1.2}{0.12} = n$
$10 = n$

21. $16 = n \cdot 50$
$\dfrac{16}{50} = n$
$0.32 = n$
$32\% = n$

25. $125\% \cdot 36 = n$
$1.25 \cdot 36 = n$
$45 = n$

29. $126 = n \cdot 31.5$
$\dfrac{126}{31.5} = n$
$4 = n$
$n = 4 = 400\%$

33. $n \cdot 150 = 67.5$
$n = \dfrac{67.5}{150}$
$n = 0.45$
$n = 45\%$

37. $\dfrac{27}{n} = \dfrac{9}{10}$
$9 \cdot n = 27 \cdot 10$
$n = \dfrac{270}{9}$
$n = 30$

41. $\dfrac{17}{12} = \dfrac{n}{20}$

45. $1.5\% \cdot 45,775 = n$
$0.015 \cdot 45,775 = n$
$686.625 = n$

Exercise Set 6.4

1. $\dfrac{a}{65} = \dfrac{32}{100}$

5. $\dfrac{70}{200} = \dfrac{p}{100}$

9. $\dfrac{a}{130} = \dfrac{19}{100}$

13. $\dfrac{a}{105} = \dfrac{18}{100}$
$\dfrac{a}{105} = \dfrac{9}{50}$
$a \cdot 50 = 105 \cdot 9$
$a \cdot 50 = 945$
$a = \dfrac{945}{50}$
$a = 18.9$
Therefore, 18% of 105 is 18.9.

17. $\dfrac{7.8}{b} = \dfrac{78}{100}$
$\dfrac{7.8}{b} = \dfrac{39}{50}$
$7.8 \cdot 50 = b \cdot 39$
$390 = b \cdot 39$
$\dfrac{390}{39} = b$
$10 = b$
Therefore, 78% of 10 is 7.8.

21. $\dfrac{14}{50} = \dfrac{p}{100}$
$\dfrac{7}{25} = \dfrac{p}{100}$
$7 \cdot 100 = 25 \cdot p$
$700 = 25 \cdot p$
$\dfrac{700}{25} = p$
$28 = p$
Therefore, 28% of 50 is 14.

25. $\dfrac{a}{80} = \dfrac{2.4}{100}$
$a \cdot 100 = 80 \cdot 2.4$
$a \cdot 100 = 192$
$a = \dfrac{192}{100}$
$a = 1.92$
Therefore, 2.4% of 80 is 1.92.

29. $\dfrac{348.6}{166} = \dfrac{p}{100}$
$348.6 \cdot 100 = 166 \cdot p$
$34,860 = 166 \cdot p$
$\dfrac{34,860}{166} = p$
$210 = p$
Therefore, 210% of 166 is 348.6.

33. $\dfrac{3.6}{8} = \dfrac{p}{100}$
$3.6 \cdot 100 = 8 \cdot p$
$360 = 8 \cdot p$
$\dfrac{360}{8} = p$
$45 = p$
Therefore, 45% of 8 is 3.6.

37. $\dfrac{11}{16} + \dfrac{3}{16} = \dfrac{11 + 3}{16} = \dfrac{14}{16} = \dfrac{7 \cdot 2}{8 \cdot 2} = \dfrac{7}{8}$

41. $\quad\overset{1}{0.41}$
$\underline{+\;0.29}$
$\quad0.70 \text{ or } 0.7$

45. $\dfrac{a}{53,862} = \dfrac{22.3}{100}$
$\quad a \cdot 100 = 53,862 \cdot 22.3$
$\quad a \cdot 100 = 1,201.122.6$
$\qquad a = \dfrac{1,201.122.6}{100}$
$\qquad a = 12,011.226$

Therefore, 22.3% of 53,862 is 12,011.226.

Exercise Set 6.5

1. 1.5% of what number is 24?
$1.5\% \cdot n = 24$
$0.015 \cdot n = 24$
$\qquad n = \dfrac{24}{0.015}$
$\qquad n = 1600$
1600 bolts were inspected.

5. Let n = percent spent on food
$\quad\$300 = n \cdot \$2000$
$\quad\dfrac{300}{2000} = \dfrac{n \cdot 2000}{2000}$
$\quad0.15 = n$
$0.15 \cdot 100\% = n$
$\quad15\% = n$

She spends 15% of her monthly income on food.

9. Let n = percent of calories from fat.
The number of fat calories is what percent of total calories?
$35 = n \cdot 120$
$35 = 120n$
$n = \dfrac{35}{120} \approx 0.292 = 29.2\%$

29.2% of the food's total calories is from fat.

13. 20% of $170 is what number?
$20\% \cdot \$170 = n$
$0.2 \cdot \$170 = n$
$\quad\$34.00 = n$
new bill $= \$170 - \$34$
$\qquad = \$136$

Their new bill is $136.

17. 48% of 66,000 is what number?
$48\% \cdot 66,000 = n$
$0.48 \cdot 66,000 = n$
$\quad31,680 = n$

Predicted number $= 66,000 + 31,680 = 97,680.$

The predicted number of physician assistants is 97,680.

	Original Amount	New Amount	Amount of Increase	Percent Increase
21.	85	187	102	120%

	Original Amount	New Amount	Amount of Decrease	Percent Decrease
25.	160	40	120	75%

29. percent increase $= \dfrac{23.7 - 19.5}{19.5}$
$= \dfrac{4.2}{19.5}$
≈ 0.215
$= 21.5\%$

33. percent increase $= \dfrac{434 - 174}{174}$
$= \dfrac{260}{174}$
≈ 1.494
$= 149.4\%$

37. $\quad0.12$
$\underline{\times38}$
$\quad\;\;96$
$\underline{\;360\;}$
$\quad4.56$

41. $\quad78.00$
$\underline{-19.46}$
$\quad58.54$

Exercise Set 6.6

1. tax $= 5\% \cdot \$150.000$
$= 0.05 \cdot \$150.000$
$= \$7.50$
The sales tax is $7.50.

5. $\quad\$54 = r \cdot \$600$
$\quad\dfrac{\$54}{\$600} = r$
$\quad0.09 = r$
$\quad9\% = r$
The sales tax rate is 9%.

9. tax $= 6.5\% \cdot \$1800$
$= 0.065 \cdot \$1800$
$= \$117$
total $= \$1800 + \$117 = \$1917$
The total price is $1917.

13. total purchase $= \$90 + \$15 = \$105$
$\quad$ tax $= 7\% \cdot \$105$
$= 0.07 \cdot \$105$
$= \$7.35$
$\quad$ total $= \$105 + \$7.35 = \$112.35$
The total price is $112.35.

17. commission $= 4\% \cdot \$1.236.856$
$= 0.04 \cdot \$1.236.856$
$= \$49,474.24$
Her commission was $49,474.24.

21. commission $= 1.5\% \cdot \$125,900$
$= 0.015 \cdot \$125,900$
$= \$1888.50$
His commission is $1888.50.

	Original Price	Discount Rate	Amount of Discount	Sale Price
25.	$68.00	10%	$6.80	$61.20
29.	$215.00	35%	$75.25	$139.75

33. discount $= 15\% \cdot \$300$
$= 0.15 \cdot \$300$
$= \$45$
sale price $= \$300 - \$45$
$= \$255$

37. $400 \cdot 0.03 \cdot 11 = 12 \cdot 11 = 132$

41.
$$\begin{aligned} \text{tax} &= 7.5\% \cdot \$24{,}966 \\ &= 0.075 \cdot \$24{,}966 \\ &= \$1872.45 \end{aligned}$$
$$\begin{aligned} \text{total price} &= \$24{,}966 + \$1872.45 \\ &= \$26{,}838.45 \end{aligned}$$
The total price is $26,838.45.

Exercise Set 6.7

1.
$$\begin{aligned} \text{simple interest} &= \text{principal} \cdot \text{rate} \cdot \text{time} \\ &= (\$200)(8\%)(2) \\ &= (\$200)(0.08)(2) \\ &= \$32 \end{aligned}$$

5.
$$\begin{aligned} \text{simple interest} &= \text{principal} \cdot \text{rate} \cdot \text{time} \\ &= (\$5000)(10\%)\left(1\frac{1}{2}\right) \\ &= (\$5000)(0.10)(1.5) \\ &= \$750 \end{aligned}$$

9.
$$\begin{aligned} \text{simple interest} &= \text{principal} \cdot \text{rate} \cdot \text{time} \\ &= (\$2500)(16\%)\left(\frac{21}{12}\right) \\ &= (\$2500)(0.16)(1.75) \\ &= \$700 \end{aligned}$$

13.
$$\begin{aligned} \text{simple interest} &= \text{principal} \cdot \text{rate} \cdot \text{time} \\ &= \$5000(9\%)\left(\frac{15}{12}\right) \\ &= \$5000(0.09)(1.25) \\ &= \$562.50 \end{aligned}$$
$$\text{Total} = \$5000 + \$562.50 = \$5562.50$$

17.
$$\begin{aligned} \text{Total amount} &= \text{original principal} \cdot \text{compound interest factor} \\ &= \$6150(7.61226) \\ &= \$46{,}815.399 \end{aligned}$$
The total amount is $46,815.40.

21.
$$\begin{aligned} \text{Total amount} &= \text{original principal} \cdot \text{compound interest factor} \\ &= \$10{,}000(5.81636) \\ &= \$58{,}163.60 \end{aligned}$$
The total amount is $58,163.60.

25.
$$\begin{aligned} \text{Total amount} &= \text{original principal} \cdot \text{compound interest factor} \\ &= \$2000(1.46933) \\ &= \$2938.66 \end{aligned}$$
$$\begin{aligned} \text{Compound interest} &= \text{total amount} - \text{original principal} \\ &= \$2938.66 - \$2000 \\ &= \$938.66 \end{aligned}$$

29.
$$\begin{aligned} \text{monthly payment} &= \frac{\text{principal} + \text{simple interest}}{\text{number of payments}} \\ &= \frac{\$1500 + \$61.88}{6} \\ &= \frac{\$1561.88}{6} \\ &\approx \$260.31 \end{aligned}$$
The monthly payment is $260.31.

33.
$$\begin{aligned} \text{perimeter} &= 10 + 6 + 10 + 6 \\ &= 32 \end{aligned}$$
The perimeter is 32 yards.

37. answers may vary

Chapter 6 Test

1. $85\% = 0.85$

5. $6.1 = 610\%$

9. $0.2\% = \dfrac{0.2}{100} = \dfrac{2}{1000} = \dfrac{1}{500}$

13.
$$\begin{aligned} n &= 42\% \cdot 80 \\ n &= 0.42 \cdot 80 \\ n &= 33.6 \end{aligned}$$
Therefore, 42% of 80 is 33.6.

17. 20% of what is $11.350?
$$\begin{aligned} 0.20n &= \$11.350 \\ n &= \frac{\$11.350}{0.20} \\ n &= \$56.750 \end{aligned}$$
The value is $56.750.

21.
$$\begin{aligned} \text{commission} &= 4\% \cdot \$9875 \\ &= 0.04 \cdot \$9875 \\ &= \$395 \end{aligned}$$
His commission is $395.

25.
$$\begin{aligned} \text{simple interest} &= \text{principal} \cdot \text{rate} \cdot \text{time} \\ &= (\$400)(13.5\%)\left(\frac{6}{12}\right) \\ &= (\$400)(0.135)(0.5) \\ &= \$27.00 \end{aligned}$$
$$\begin{aligned} \text{Total amount due the bank} &= \$400 + \$27 \\ &= \$427 \end{aligned}$$

Chapter 7

Exercise Set 7.1

1. $60 \text{ inches} = 60 \text{ inches} \cdot \dfrac{1 \text{ foot}}{12 \text{ inches}} = 5 \text{ feet}$

5. $42{,}240 \text{ feet} = 42{,}240 \text{ feet} \cdot \dfrac{1 \text{ mile}}{5280 \text{ feet}} = 5 \text{ miles}$

9. $10 \text{ feet} = 10 \text{ feet} \cdot \dfrac{1 \text{ yard}}{3 \text{ feet}} = 3\frac{1}{3} \text{ yards}$

13. $40 \text{ feet} \div 3 = 13 \text{ with remainder } 1$
$40 \text{ feet} = 13 \text{ yards } 1 \text{ foot}$

17. $10{,}000 \text{ feet} \div 5280 = 1 \text{ with remainder } 4720$
$10{,}000 = 1 \text{ mile } 4720 \text{ feet}$

21. $5 \text{ yards} = 5 \text{ yards} \cdot \dfrac{3 \text{ feet}}{1 \text{ yard}} = 15 \text{ feet}$
$5 \text{ yards } 2 \text{ feet} = 15 \text{ feet} + 2 \text{ feet} = 17 \text{ feet}$

25.
$$\begin{array}{r} 5 \text{ feet } 8 \text{ inches} \\ + 6 \text{ feet } 7 \text{ inches} \\ \hline \end{array}$$
$11 \text{ feet } 15 \text{ inches} = 11 \text{ feet} + 1 \text{ foot } 3 \text{ inches}$
$= 12 \text{ feet } 3 \text{ inches}$

29.
$$\begin{array}{r} 24 \text{ feet } 8 \text{ inches} \\ - 16 \text{ feet } 3 \text{ inches} \\ \hline 8 \text{ feet } 5 \text{ inches} \end{array}$$

33.
$$\begin{array}{r} 3 \text{ feet } 4 \text{ inches} \\ 2\overline{)6 \text{ feet } 8 \text{ inches}} \\ -6 \text{ feet} \\ \hline 0 \qquad 8 \text{ inches} \\ - \qquad 8 \text{ inches} \\ \hline 0 \end{array}$$

37. 6 feet 10 inches
+ 3 feet 8 inches
9 feet 18 inches = 9 feet + 1 foot 6 inches
= 10 feet 6 inches
The bamboo is 10 feet 6 inches tall.

41. 1 foot 9 inches
× 9
9 feet 81 inches = 9 feet + 6 feet 9 inches
= 15 feet 9 inches
9 stacks would extend 15 feet 9 inches from the wall.

45. $P = 2L + 2W$
= 2(24 feet 9 inches) + 2(18 feet 6 inches)
= 48 feet 18 inches + 36 feet 12 inches
= 84 feet 30 inches
= 84 feet + 2 feet 6 inches
= 86 feet 6 inches
She must buy 86 feet 6 inches of fencing material.

49. 40 meters to centimeters   *Move the decimal 2 places to the right.*
40 meters = 4000 centimeters

53. 300 meters to kilometers   *Move the decimal 3 places to the left.*
300 meters = 0.300 kilometer

57. 1500 centimeters to meters   *Move the decimal 2 places to the left.*
1500 centimeters = 15 meters

61. 20.1 millimeters to decimeters   *Move the decimal 2 places to the left.*
20.1 millimeters = 0.201 decimeter

65. 8.6 meters
+ 0.34 meters
8.94 meters

69. 24.8 millimeters − 1.19 centimeters
= 2.48 centimeters − 1.19 centimeters
= 1.29 centimeters or 12.9 millimeters

73. 18.3 meters × 3 = 54.9 meters

77. 3.4 meters + 5.8 meters − 8 centimeters
= 3.4 meters + 5.8 meters − 0.08 meter
= 9.12 meters
The tied ropes are 9.12 meters long.

81. 25(1 meter + 65 centimeters)
= 25(100 centimeters + 65 centimeters)
= 25(165 centimeters)
= 4125 centimeters or 41.25 meters
4125 centimeters of 41.25 meters of wood must be ordered.

85. 5.988 kilometers + 21 meters
= 5988 meters + 21 meters
= 6009 meters or 6.009 kilometers
The elevation is 6009 meters or 6.009 kilometers.

89. 0.21 = 21%

93. $\frac{1}{4}$ = 0.25 = 25%

97. A square has 4 equal sides.

$$4)\overline{26.300} = 6.575$$

−24
2 3
−2 0
30
−28
20
−20
0

Each side will be 6.575 meters long.

Exercise Set 7.2

1. 2 pounds = 2 pounds $\cdot \frac{16 \text{ ounces}}{1 \text{ pound}}$ = 32 ounces

5. 12,000 pounds = 12,000 pounds $\cdot \frac{1 \text{ ton}}{2000 \text{ pounds}}$
= 6 tons

9. 3500 pounds = 3500 pounds $\cdot \frac{1 \text{ ton}}{2000 \text{ pounds}}$
= $\frac{7}{4}$ tons
= $1\frac{3}{4}$ tons

13. 4.9 tons = 4.9 tons $\cdot \frac{2000 \text{ pounds}}{1 \text{ ton}}$
= 9800 pounds

17. 2950 pounds = 2950 pounds $\cdot \frac{1 \text{ ton}}{2000 \text{ pounds}}$
= $\frac{2950 \text{ tons}}{2000}$
= 1.475 tons
≈ 1.5 tons

21. 6 tons 1540 pounds
+ 2 tons 850 pounds
8 tons 2390 pounds = 8 tons + 1 ton 390 pounds
= 9 tons 390 pounds

25. 12 pounds 4 ounces   11 pounds 20 ounces
− 3 pounds 9 ounces   − 3 pounds 9 ounces
8 pounds 11 ounces

29. 5)6 tons 1500 pounds   1 ton 700 pounds
− 5 tons
1 ton = 2000 pounds
3500 pounds
− 3500 pounds
0

33. 64 pounds 8 ounces   63 pounds 24 ounces
− 28 pounds 10 ounces   − 28 pounds 10 ounces
35 pounds 14 ounces
Her zucchini was 35 pounds 14 ounces below the record.

37. 55 pounds 4 ounces   54 pounds 20 ounces
− 2 pounds 8 ounces   − 2 pounds 8 ounces
52 pounds 12 ounces

$$\begin{array}{r} 52 \text{ pounds } 12 \text{ ounces} \\ \times\, 4 \\ \hline \end{array}$$
$208 \text{ pounds } 48 \text{ ounces} = 208 \text{ pounds} + 3 \text{ pounds}$
$= 211 \text{ pounds}$

41. 500 grams to kilograms   *Move the decimal 3 places to the left.*

500 grams = 0.5 kilograms

45. 25 kilograms to grams   *Move the decimal 3 places to the right.*

25 kilograms = 25,000 grams

49. 6.3 grams to kilograms   *Move the decimal 3 places to the left.*

6.3 grams = 0.0063 kilograms

53. 4.01 kilograms to grams   *Move the decimal 3 places to the right.*

4.01 kilograms = 4010 grams

57. 205 mg = 0.205 g or 5.61 g = 5610 mg

$$\begin{array}{ll} 0.205 \text{ g} & 205 \text{ mg} \\ \underline{5.610 \text{ g}} & \underline{5610 \text{ mg}} \\ 5.815 \text{ g} & 5815 \text{ mg} \end{array}$$

61. 1.61 kg = 1610 g or 250 g = 0.250 kg

$$\begin{array}{ll} 1610 \text{ g} & 1.610 \text{ kg} \\ \underline{-\,250 \text{ g}} & \underline{-\,0.250 \text{ kg}} \\ 1360 \text{ g} & 1.360 \text{ kg} \end{array}$$

65. 17 kilograms ÷ 8 = 2.125 kilograms

69. 0.09 grams − 60 milligrams
= 90 milligrams − 60 milligrams = 30 milligrams
The extra-strength tablet contains 30 milligrams more medication.

73. 3 · 16 · 3 milligrams = 144 milligrams
3 cartons contain 144 milligrams of preservatives.

77. 0.3 kilograms + 0.15 kilograms + 400 grams
= 0.3 kilograms + 0.15 kilograms + 0.4 kilograms
= 0.85 kilograms or 850 grams
The package weighs 0.85 kilograms or 850 grams.

81. $\dfrac{1}{4} = \dfrac{1 \cdot 25}{4 \cdot 25} = \dfrac{25}{100} = 0.25$

85. $\dfrac{7}{8} = \dfrac{7 \cdot 125}{8 \cdot 125} = \dfrac{875}{1000} = 0.875$

Exercise Set 7.3

1. $32 \text{ fluid ounces} = 32 \text{ fluid ounces} \cdot \dfrac{1 \text{ cup}}{8 \text{ fluid ounces}}$
$= 4 \text{ cups}$

5. $10 \text{ quarts} = 10 \text{ quarts} \cdot \dfrac{1 \text{ gallon}}{4 \text{ quarts}} = 2\dfrac{1}{2} \text{ gallons}$

9. $2 \text{ quarts} = 2 \text{ quarts} \cdot \dfrac{4 \text{ cups}}{1 \text{ quart}} = 8 \text{ cups}$

13. 6 gallons
$= 6 \text{ gallons} \cdot \dfrac{4 \text{ quarts}}{1 \text{ gallon}} \cdot \dfrac{2 \text{ pints}}{1 \text{ quart}} \cdot \dfrac{2 \text{ cups}}{1 \text{ pint}} \cdot \dfrac{8 \text{ fluid ounces}}{1 \text{ cup}}$
$= 768 \text{ fluid ounces}$

17. $2\dfrac{3}{4} \text{ gallons} = \dfrac{11}{4} \text{ gallons} \cdot \dfrac{4 \text{ quarts}}{1 \text{ gallon}} \cdot \dfrac{2 \text{ pints}}{1 \text{ quart}} = 22 \text{ pints}$

21.
$$\begin{array}{r} 1 \text{ c } 5 \text{ fl oz} \\ +\,2 \text{ c } 7 \text{ fl oz} \\ \hline \end{array}$$
$3 \text{ c } 12 \text{ fl oz} = 3 \text{ c} + 1 \text{ c } 4 \text{ fl oz} = 4 \text{ c } 4 \text{ fl oz}$

25.
$$\begin{array}{ll} 3 \text{ gal } 1 \text{ qt} & 2 \text{ gal } 4 \text{ qt } 2 \text{ pt} \\ \underline{-\,\qquad 1 \text{ qt } 1 \text{ pt}} & \underline{-\,\qquad 1 \text{ qt } 1 \text{ pt}} \\ & 2 \text{ gal } 3 \text{ qt } 1 \text{ pt} \end{array}$$

29.
$$\begin{array}{r} 8 \text{ gal } 2 \text{ qt} \\ \times\, 2 \\ \hline \end{array}$$
$16 \text{ gal } 4 \text{ qt} = 17 \text{ gal}$

33. $1\dfrac{1}{2} \text{ quarts} = \dfrac{3}{2} \text{ quarts} \cdot \dfrac{2 \text{ pints}}{1 \text{ quart}} \cdot \dfrac{2 \text{ cups}}{1 \text{ pint}} \cdot \dfrac{8 \text{ fluid ounces}}{1 \text{ cup}}$
$= 48 \text{ fluid ounces}$

37.
$$\begin{array}{r} 5 \text{ pints } 1 \text{ cup} \\ +\,2 \text{ pints } 1 \text{ cup} \\ \hline 7 \text{ pints } 2 \text{ cups} \end{array}$$
Since 8 pints = 1 gallon, the fruit punch can be poured into the container.

41. 12 fluid ounces · 24 = 288 fluid ounces
$= 288 \text{ fluid ounces} \cdot \dfrac{1 \text{ cup}}{8 \text{ fluid ounces}} \cdot \dfrac{1 \text{ pint}}{2 \text{ cups}} \cdot \dfrac{1 \text{ quart}}{2 \text{ pints}}$
$= 9 \text{ quarts}$

45. 4500 ml to liters   *Move the decimal 3 places to the left.*

4500 ml = 4.5 liters

49. 64 ml to liters   *Move the decimal 3 places to the left.*

64 ml = 0.064 liters

53. 3.6 L to milliliters   *Move the decimal 3 places to the right.*

3.6 L = 3600 milliliters

57. 2.9 L + 19.6 L = 22.5 L

61. 8.6 L = 8600 ml or 190 ml = 0.190 L

$$\begin{array}{ll} 8600 \text{ ml} & 8.60 \text{ L} \\ \underline{-\,190 \text{ ml}} & \underline{-\,0.19 \text{ L}} \\ 8410 \text{ ml} & 8.41 \text{ L} \end{array}$$

65. 480 ml × 8 = 3840 ml

69. 2 L − 410 ml = 2 L − 0.41 L = 1.59 L
1.59 L remains in the bottle.

73. \$14.00 ÷ 44.3 L ≈ \$0.316
The cost is about \$0.316 per liter.

77. $0.7 = \dfrac{7}{10}$

81. $0.006 = \dfrac{6}{1000} = \dfrac{3}{500}$

Exercise Set 7.4

1. $C = \dfrac{5}{9} \cdot (41 - 32)$
$= \dfrac{5}{9} \cdot (9)$
$= 5$
$41°\text{F} = 5°\text{C}$

5. $F = \dfrac{9}{5} \cdot (60) + 32$

$= 9(12) + 32$

$= 108 + 32$

$= 140$

$60°C = 140°F$

9. $C = \dfrac{5}{9} \cdot (62 - 32)$

$= \dfrac{5}{9} \cdot (30)$

$= \dfrac{150}{9}$

≈ 16.7

$62°F \approx 16.7°C$

13. $F = 1.8(92) + 32$

$= 165.6 + 32$

$= 197.6$

$92°C = 197.6°F$

17. $C = \dfrac{5}{9} \cdot (122 - 32)$

$= \dfrac{5}{9} \cdot (90)$

$= 50$

$122°F = 50°C$

21. $C = \dfrac{5}{9} \cdot (70 - 32)$

$= \dfrac{5}{9} \cdot (38)$

$= \dfrac{190}{9}$

≈ 21.1

$70°F \approx 21.1°C$

25. $F = 1.8(118) + 32$

$= 212.4 + 32$

$= 244.4$

$188°C = 244.4°F$

29. $C = \dfrac{5}{9} \cdot (864 - 32)$

$= \dfrac{5}{9} \cdot (832)$

$= \dfrac{4160}{9}$

≈ 462.2

$864°F \approx 462.2°C$

33. $P = 3 \text{ feet} + 3 \text{ feet} + 3 \text{ feet} + 3 \text{ fett} + 3 \text{ feet}$

$= 15 \text{ feet}$

37. $C = \dfrac{5}{9} \cdot (9010 - 32)$

$= \dfrac{5}{9} \cdot (8978)$

≈ 4988

$9010°F$ is about $4988°C$

Exercise Set 7.5

1. energy $= 3 \text{ pounds} \cdot 380 \text{ feet}$

$= 1140 \text{ foot-pounds}$

5. energy $= \dfrac{2.5 \text{ tons}}{1} \cdot \dfrac{2000 \text{ pounds}}{1 \text{ ton}} \cdot 85 \text{ feet}$

$= 425,000 \text{ foot-pounds}$

9. $1000 \text{ BTU} = 1000 \text{ BTU} \cdot \dfrac{778 \text{ foot-pounds}}{1 \text{ BTU}}$

$= 778,000 \text{ foot-pounds}$

13. $34,130 \text{ BTU} = 34,130 \text{ BTU} \cdot \dfrac{778 \text{ foot-pounds}}{1 \text{ BTU}}$

$= 26,553,140 \text{ foot-pounds}$

17. total hours $= 7 \text{ hours}$

calories $= 7 \cdot 115 = 805 \text{ calories}$

21. total hours $= \dfrac{1}{3} \cdot 6 = 2 \text{ hours}$

calories $= 2 \cdot 720 = 1440 \text{ calories}$

25. miles $= \dfrac{3500}{200} = 17.5 \text{ miles}$

29. $\dfrac{27}{45} = \dfrac{3 \cdot 3 \cdot 3}{3 \cdot 3 \cdot 5} = \dfrac{3}{5}$

33. energy $= 123.9 \text{ pounds} \cdot \dfrac{9 \text{ inches}}{1} \cdot \dfrac{1 \text{ foot}}{12 \text{ inches}}$

$= 92.925 \text{ foot-pounds}$

Chapter 7 Test

1. $\begin{array}{r} 23 \\ 12\overline{)280} \\ -24 \\ \hline 40 \\ -36 \\ \hline 4 \end{array}$

$280 \text{ inches} = 23 \text{ feet } 4 \text{ inches}$

5. $38 \text{ pints} = 38 \text{ pints} \cdot \dfrac{1 \text{ gallon}}{8 \text{ pints}} = 4\dfrac{3}{4} \text{ gallons}$

9. $4.3 \text{ decigrams} = 0.43 \text{ grams}$

13. $\begin{array}{r} 2 \text{ feet } 9 \text{ inches} \\ \times 3 \\ \hline 6 \text{ feet } 27 \text{ inches} \end{array} = 6 \text{ feet} + 2 \text{ feet } 3 \text{ inches}$

$= 8 \text{ feet } 3 \text{ inches}$

17. $C = \dfrac{5}{9} \cdot (84 - 32)$

$= \dfrac{5}{9} \cdot (52)$

$= \dfrac{260}{9}$

$= 28.\overline{8}$

≈ 28.9

$84°F \approx 28.9°C$

21. $F = 1.8(41) + 32 = 73.8 + 32 = 105.8$

Her fever is $105.8°F$.

25. $26,000 \text{ BTU} = 26,000 \text{ BTU} \cdot \dfrac{778 \text{ foot-pounds}}{1 \text{ BTU}}$

$= 20,228,000 \text{ foot-pounds}$

29. $C = \dfrac{5}{9} \cdot (129 - 32)$

$\quad = \dfrac{5}{9} \cdot (97)$

$\quad = 53.\overline{8}$

$\quad \approx 53.9$

$\quad 129°F \approx 53.9°C$

Chapter 8

Exercise Set 8.1

1. The figure extends indefinitely in two directions. It is line yz, or $\overleftrightarrow{yz}$.

5. The figure has two endpoints. It is line segment PQ, or $\overline{PQ}$.

9. $\angle ABC = 15°$

13. $\angle DBA = 50° + 15°$

$\quad\quad\quad = 65°$

17. $90°$

21. $\angle S$ is a straight angle.

25. $\angle Q$ is an obtuse angle. It measures between $89°$ and $180°$.

29. The complement of an angle that measures $17°$ is an angle that measures $90° - 17° = 73°$.

33. The complement of a $45°$ angle is an angle that measures $90° - 45° = 45°$.

37. $\angle MNP$ and $\angle RNO$; are complementary angles since $60° + 30° = 90°$. Also, $\angle PNQ$ and $\angle QNR$ are complementary angles since $52° + 38° = 90°$.

41. $\angle x = 120° - 88°$

$\quad\quad = 32°$

45. Since $\angle x$ and the $35°$ angle are vertical angles, $\angle x = 35°$, $\angle x$ and $\angle y$ are adjacent angles, so $\angle y = 180° - 35° = 45°$, $\angle y$ and $\angle z$ are vertical angles, so $\angle z = 145°$.

49. $\angle x$ and the $80°$ angle are adjacent angles, so $\angle x = 180° - 80° = 100°$, $\angle y$ and the $80°$ angle are alternate interior angles, so $\angle y = 80°$, $\angle x$ and $\angle z$ are corresponding angles, so $\angle z = 100°$.

53. $\dfrac{7}{8} + \dfrac{1}{4} = \dfrac{7}{8} + \dfrac{2}{8} = \dfrac{9}{8}$ or $1\dfrac{1}{8}$

57. $3\dfrac{1}{3} - 2\dfrac{1}{2} = \dfrac{10}{3} - \dfrac{5}{2} = \dfrac{20}{6} - \dfrac{15}{6} = \dfrac{5}{6}$

61. The supplement of a $125.2°$ angle is $180° - 125.2° = 54.8°$.

Exercise Set 8.2

1. The triangle is an equilateral triangle because all three sides are the same length.

5. The triangle is an isosceles triangle since two sides are the same length.

9. $\angle x = 180° - 95° - 72° = 13°$

13. diameter   **17.** parallelogram

21. true   **25.** false

29. $r = \dfrac{d}{2}$

$\quad = \dfrac{29}{2}$

$\quad = 14.5$ centimeters

33. hexagon   **37.** rectangular solid

41. $d = 2 \cdot r$

$\quad = 2 \cdot 7.4$

$\quad = 14.8$ inches

45. cube   **49.** sphere

53. $4(28.6) = 114.4$

57. $d = 2 \cdot r$

$\quad = 2 \cdot 36.184$

$\quad = 72.368$ miles

Exercise Set 8.3

1. $P = 2l + 2w$

$\quad = 2(17\text{ feet}) + 2(15\text{ feet})$

$\quad = 34\text{ feet} + 30\text{ feet}$

$\quad = 64\text{ feet}$

5. $P = a + b + c$

$\quad = 5\text{ inches} + 7\text{ inches} + 9\text{ inches}$

$\quad = 21\text{ inches}$

9. $P = 10\text{ feet} + 8\text{ feet} + 8\text{ feet} + 15\text{ feet} + 7\text{ feet}$

$\quad = 48\text{ feet}$

13. $P = 5\text{ feet} + 3\text{ feet} + 2\text{ feet} + 7\text{ feet} + 4\text{ feet}$

$\quad = 21\text{ feet}$

17. $P = 2l + 2w$

$\quad = 2(120\text{ yards} + 2(53\text{ yards})$

$\quad = 240\text{ yards} + 106\text{ yards}$

$\quad = 346\text{ yards}$

21. $\text{cost} = 3(22)$

$\quad\quad\quad = \$66$

25. $P = 4s$

$\quad = 4(7\text{ inches})$

$\quad = 28\text{ inches}$

29. The missing lengths are:

$28\text{ meters} - 20\text{ meters} = 8\text{ meters}$

$20\text{ meters} - 17\text{ meters} = 3\text{ meters}$

$P = 8\text{ meters} + 3\text{ meters} + 20\text{ meters} + 20\text{ meters}$

$\quad + 28\text{ meters} + 17\text{ meters}$

$\quad = 96\text{ meters}$

33. The missing lengths are:

$12\text{ miles} + 10\text{ miles} = 22\text{ miles}$

$8\text{ miles} + 34\text{ miles} = 42\text{ miles}$

$P = 12\text{ miles} + 34\text{ miles} + 10\text{ miles} + 8\text{ miles}$

$\quad + 22\text{ miles} + 42\text{ miles}$

$\quad = 128\text{ miles}$

37. $C = 2\pi r$

$\quad = 2\pi \cdot 8\text{ miles}$

$\quad = 16\pi\text{ miles}$

$\quad \approx 16 \cdot 3.14\text{ miles}$

$\quad = 50.24\text{ miles}$

41. $C = 2\pi r$

$\quad = 2\pi \cdot 5\text{ feet}$

$\quad = 10\pi\text{ feet}$

$\quad \approx 10 \cdot \dfrac{22}{7}\text{ feet}$

$\quad = \dfrac{220}{7}\text{ feet}$

$\quad \approx 31.43\text{ feet}$

45. $5 + 6 \cdot 3 = 5 + 18 = 23$

49. $(18 + 8) - (12 + 4) = 26 - 16 = 0$

53. perimeter   **57.** area

61. a. Circumference of the smaller circle:

$C = 2\pi r$

$= 2\pi \cdot 10$ meters

$= 20\pi$ meters

$\approx 20 \cdot 3.14$ meters

$= 62.8$ meters,

Circumference of the larger circle:

$C = 2\pi r$

$= 2\pi \cdot 20$ meters

$= 40\pi$ meters

$\approx 40 \cdot 3.14$ meters

$= 125.6$ meters

b. Yes, the circumference is doubled.

Exercise Set 8.4

1. $A = l \cdot w$

$= (3.5 \text{ meters})(2 \text{ meters})$

$= 7$ square meters

5. $A = \frac{1}{2} \cdot b \cdot h$

$= \frac{1}{2}(6 \text{ yards})(5 \text{ yards})$

$= \frac{1}{2} \cdot 6 \cdot 5$ square yards

$= 15$ square yards

9. $A = b \cdot h$

$= (7 \text{ feet})(5.25 \text{ feet})$

$= 36.75$ square feet

13. $A = \frac{1}{2} \cdot (b + B) \cdot h$

$= \frac{1}{2} \cdot (4 \text{ yards} + 7 \text{ yards}) \cdot (4 \text{ yards})$

$= \frac{1}{2} \cdot 11 \cdot 4$ square yards

$= 22$ square yards

17. $A = b \cdot h$

$= (5 \text{ inches})\left(4\frac{1}{2} \text{ inches}\right)$

$= 5 \cdot \frac{9}{2}$ square inches

$= \frac{45}{2}$ or $22\frac{1}{2}$ square inches

21.

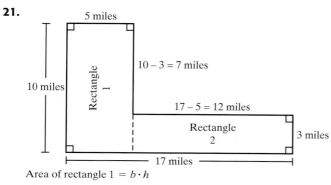

Area of rectangle 1 $= b \cdot h$

$= (5 \text{ miles})(10 \text{ miles})$

$= 50$ square miles

Area of rectangle 2 $= b \cdot h$

$= (12 \text{ miles})(3 \text{ miles})$

$= 36$ square miles

Total area $= 50$ square miles $+ 36$ square miles

$= 86$ square miles

25. $A = \pi \cdot r^2$

$= \pi (6 \text{ inches})^2$

$= 36\pi$ square inches

$\approx 36 \cdot \frac{22}{7}$ square inches

$= \frac{792}{7}$ square inches

≈ 113.1 square inches

29. $A = l \cdot w$

$= (505 \text{ feet})(225 \text{ feet})$

$= 113,625$ square feet

33. $A = l \cdot w$

$= (20 \text{ inches})\left(25\frac{1}{2} \text{ inches}\right)$

$= 20 \cdot \frac{51}{2}$ square inches

$= \frac{1020}{2}$ or 510 square inches

37. $A = \frac{1}{2} \cdot (b + B) \cdot b$

$= \frac{1}{2} \cdot (90 \text{ feet} + 140 \text{ feet}) \cdot (80 \text{ feet})$

$= \frac{1}{2} \cdot 230 \cdot 80$ square feet

$= 9200$ square feet

41. $C = \pi d$

$= \pi \cdot (14 \text{ inches})$

$= 14\pi$ inches

$= \approx 14 \cdot 3.14$ inches

$= 43.96$ inches

45. Perimeter $= 6 \cdot s$

$= 6 \cdot \left(2\frac{1}{8} \text{ feet}\right)$

$= \left(6 \cdot \frac{17}{8}\right)$ feet

$= \frac{102}{8}$ or $12\frac{3}{4}$ feet

49. $8 \text{ inches} = \frac{8}{12} \text{ foot} = \frac{2}{3} \text{ foot}$

$A = l \cdot w$

$= (2 \text{ feet}) \cdot \left(\frac{2}{3} \text{ foot}\right)$

$= \frac{4}{3}$ or $1\frac{1}{3}$ square feet

$2 \text{ feet} = 24 \text{ inches}$

$A = l \cdot w$

$= (24 \text{ inches}) \cdot (8 \text{ inches})$

$= 192$ square inches

Exercise Set 8.5

1. $V = l \cdot w \cdot h$

$= (6 \text{ inches}) \cdot (4 \text{ inches}) \cdot (3 \text{ inches})$

$= 72$ cubic inches

5. $V = \frac{1}{3} \cdot \pi \cdot r^2 \cdot h$

$\qquad = \frac{1}{3} \cdot \pi \cdot (2 \text{ yards})^2 \cdot (3 \text{ yards})$

$\qquad = 4\pi \text{ cubic yards}$

$\qquad \approx 4 \cdot \frac{22}{7} \text{ cubic yards}$

$\qquad = \frac{88}{7} \text{ or } 12\frac{4}{7} \text{ cubic yards}$

9. $V = \frac{1}{3} \cdot s^2 \cdot h$

$\qquad = \frac{1}{3} \cdot (5 \text{ centimeters})^2 \cdot (9 \text{ centimeters})$

$\qquad = 75 \text{ cubic centimeters}$

13. $V = l \cdot w \cdot h$

$\qquad = (2 \text{ feet}) \cdot (1.4 \text{ feet}) \cdot (3 \text{ feet})$

$\qquad = 8.4 \text{ cubic feet}$

17. $V = \frac{1}{3} \cdot s^2 \cdot h$

$\qquad = \frac{1}{3} \cdot (12 \text{ centimeters})^2 \cdot (20 \text{ centimeters})$

$\qquad = \frac{2880}{3} \text{ cubic centimeters}$

$\qquad = 960 \text{ cubic centimeters}$

21. $V = l \cdot w \cdot h$

$\qquad = (2 \text{ feet}) \cdot \left(2\frac{1}{2} \text{ feet}\right) \cdot \left(1\frac{1}{2} \text{ feet}\right)$

$\qquad = 2 \cdot \frac{5}{2} \cdot \frac{3}{2} \text{ cubic feet}$

$\qquad = \frac{30}{4} \text{ or } 7\frac{1}{2} \text{ cubic feet}$

25. $r = \frac{d}{2} = \frac{6 \text{ inches}}{2} = 3 \text{ inches}$

$\qquad V = \frac{4}{3} \cdot \pi \cdot r^3 = \frac{4}{3} \cdot \pi \cdot (3 \text{ inches})^3$

$\qquad = \frac{4}{3} \cdot \pi \cdot 27 \text{ cubic inches}$

$\qquad = 36\pi \text{ cubic inches}$

$\qquad \approx 36 \cdot 3.14 \text{ cubic inches}$

$\qquad = 113.04 \text{ cubic inches}$

29. $3^2 = 3 \cdot 3 = 9$

33. $4^2 + 2^2 = 16 \cdot 4 = 20$

37. $V = \frac{1}{3} \cdot 5^2 \cdot h$

$\qquad = \frac{1}{3} \cdot (344 \text{ meters})^2 \cdot (65.5 \text{ meters})$

$\qquad \approx 2{,}583.669 \text{ cubic meters.}$

41. No, answers may vary

Exercise Set 8.6

1. $\sqrt{4} = 2$ because $2^2 = 4$

5. $\sqrt{\frac{1}{81}} = \frac{1}{9}$ because $\frac{1}{9} \cdot \frac{1}{9} = \frac{1}{81}$

9. $\sqrt{256} = 16$ because $16^2 = 256$

13. $\sqrt{3} \approx 1.732$

17. $\sqrt{14} \approx 3.742$

21. $\sqrt{8} \approx 2.828$

25. $\sqrt{71} \approx 8.426$

29. $\quad a^2 + b^2 = c^2$

$\qquad 5^2 + 12^2 = c^2$

$\qquad 25 + 144 = c^2$

$\qquad\qquad 169 = c^2$

$\qquad\qquad\quad c = \sqrt{169}$

$\qquad\qquad\quad c = 13 \text{ inches}$

33.

$\quad a^2 + b^2 = c^2$

$\quad 3^2 + 4^2 = c^2$

$\quad 9 + 16 = c^2$

$\qquad 25 = c^2$

$\qquad\quad c = \sqrt{125}$

$\qquad\quad c = 5$

37.

$\quad a^2 + b^2 = c^2$

$\quad 10^2 + 14^2 = c^2$

$\quad 100 + 196 = c^2$

$\qquad\quad 296 = c^2$

$\qquad\qquad c = \sqrt{296}$

$\qquad\qquad c \approx 17.205$

41.

$\quad a^2 + b^2 = c^2$

$\quad 5^2 + b^2 = 13^2$

$\quad 25 + b^2 = 169$

$\qquad\quad b^2 = 144$

$\qquad\quad b = 12$

45.

$\quad a^2 + b^2 = c^2$

$\quad 30^2 + 30^2 = c^2$

$\quad 900 + 900 = c^2$

$\qquad\quad 1800 = c^2$

$\qquad\qquad c = \sqrt{1800}$

$\qquad\qquad c \approx 42.426$

49. $\qquad a^2 + b^2 = c^2$

$\qquad 100^2 + 100^2 = c^2$

$\quad 10{,}000 + 10{,}000 = c^2$

$\qquad\qquad 20{,}000 = c^2$

$\qquad\qquad\quad c = \sqrt{20{,}000}$

$\qquad\qquad\quad c \approx 141.42$

Approximately 141.42 yards.

53.
$$a^2 + b^2 = c^2$$
$$300^2 + 160^2 = c^2$$
$$90{,}000 + 25{,}000 = c^2$$
$$115{,}600 = c^2$$
$$c = \sqrt{15{,}600}$$
$$c = 340$$

340 feet

57.
$$\frac{9}{11} = \frac{n}{55}$$
$$11 \cdot n = 55 \cdot 9$$
$$11n = 495$$
$$\frac{11n}{11} = \frac{495}{11}$$
$$n = 45$$

61. $\sqrt{38}$ is between 6 and 7.
$\sqrt{38} \approx 6.164$

65. answers may vary

Exercise Set. 8.7

1. The triangles are congruent by Side-Side-Side.

5. Since the triangles are similar, we can compare any of the corresponding sides to find the ratio.
$$\frac{22}{11} = \frac{2}{1}$$

9.
$$\frac{3}{n} = \frac{6}{9}$$
$$6n = 27$$
$$n = 4.5$$

13.
$$\frac{n}{3.75} = \frac{12}{9}$$
$$9n = 45$$
$$n = 5$$

17.
$$\frac{n}{3.25} = \frac{17.5}{3.25}$$
$$3.25n = 56.875$$
$$n = 17.5$$

21.
$$\frac{34}{n} = \frac{16}{10}$$
$$16n = 340$$
$$n = 21.25$$

25.
$$\frac{5}{4} = \frac{n}{48}$$
$$4n = 240$$
$$n = 60$$
The height of the tree is 60 feet.

29.
$$\frac{7}{9} = \frac{w}{5}$$
$$9w = 35$$
$$w = \frac{35}{9} \approx 3\frac{8}{9}$$
The width is $3\frac{8}{9}$ inches. The print will not fit on an index card.

33. Average $= \dfrac{76 + 79 + 88}{3}$
$$= \frac{243}{3} = 81$$

37. answers may vary

Chapter 8 Test

1. The complement of a 78° angle is $90° - 78° = 12°$.

5. $\angle x$ and the 73° angle are vertical angles, so $\angle x = 73°$, $\angle y$ and the 73° angle are corresponding angles, so $\angle y = 73°$, $\angle z$ and $\angle y$ are vertical angles, so $\angle z = 73°$.

9. $C = 2 \cdot \pi \cdot r$
$$= 2 \cdot \pi \cdot (9 \text{ inches})$$
$$= 18\pi \text{ inches}$$
$$\approx 18 \cdot 3.14 \text{ inches}$$
$$= 56.62 \text{ inches}$$
$A = \pi \cdot r^2$
$$= \pi \cdot (9 \text{ inches})^2$$
$$= 81\pi \text{ square inches}$$
$$\approx 81 \cdot 3.14 \text{ square inches}$$
$$= 254.34 \text{ square inches}$$

13. $V = l \cdot w \cdot h$
$$= (3 \text{ feet}) \cdot (5 \text{ feet}) \cdot (2 \text{ feet})$$
$$= 30 \text{ cubic feet}$$

17. $P = 4 \cdot s$
$$= 4 \cdot (4 \text{ inches})$$
$$= 16 \text{ inches}$$

21. $A = l \cdot w$
$$= (123.8 \text{ feet})(80 \text{ feet})$$
$$= 9904 \text{ square feet}$$
$$\frac{0.02 \text{ ounces}}{1 \text{ square foot}} = \frac{x}{9904 \text{ square feet}}$$
$$x = 198.08 \text{ ounces}$$
198.08 ounces of insecticide are required.

Chapter 9

Exercise Set 9.1

1. The year 2000 has the most cars, so the greatest number of automobiles was manufactured in 2000.

5. Compare each year with the previous year to see that the production of automobiles decreased from the previous year in 1996, 1997, and 2001.

9. The year 1997 has seven chickens and each chicken represents 3 ounces of chicken, so approximately $7 \cdot 3.5 = 22.5$ ounces of chicken were consumed per week in 1997.

13. There is one more chicken for 2001 than for 1995, so there was an increase of approximately $1 \cdot 3 = 3$ ounces per week.

17. The tallest bar corresponds to the month of April, so April has the most tornado-related deaths.

21. Look for bars that extend above the horizontal line at 5. The months of February, March, April, May, and June have over 5 tornado-related deaths.

25. The two U.S. cities are New York City and Los Angeles, New York City is the largest, with an estimated population of 21.4 million or 21,400,000.

29.
Fiber Content of Selected Foods

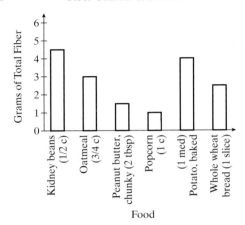

33. 3.4 goals

37. decrease

41. $(0.10)(62) = 6.2$

45. $\dfrac{17}{50} = \dfrac{17 \cdot 2}{50 \cdot 2} = \dfrac{34}{100} = 0.34 = 34\%$

49. 68°F on Sunday

53. answers may vary

Exercise Set 9.2

1. The largest sector, or 320 students, corresponds to where most college students live. Most college students live at parent or guardian's home.

5. $\dfrac{\text{students living in campus housing}}{\text{students living at home}} = \dfrac{180}{320} = \dfrac{9}{16}$

9. $30\% + 7\% = 37\%$

13. 5% of 57,000,000
$= (0.05)(57,000,000)$
$= 2,850,000$ square miles

17. The second-largest sector is 25% which corresponds to "Nonfiction." Nonfiction is the second-largest category of books.

21. amount $= 22\% \cdot 125,600 = 27,632$ books

25.
Europe
5%

Asia
30%

United States
65%

29.

$$
\begin{array}{r}
5 \\
2\overline{)10} \\
2\overline{)20} \quad 2\overline{)40}
\end{array}
$$

$40 = 2^3 \cdot 5$

33. answers may vary

37. Indian Ocean: $21\% \cdot 264,489,800$
$= 55,542,858$ square kilometers

41. $18.7\% + 59.8\% = 78.5\%$
Then, $78.5\% \cdot 2711$
$= 0.785 \cdot 2711 \times 2128$ users

Exercise Set 9.3

1. The height of the bar representing 100–149 miles, is 15. Therefore, 15 adults drive 100–149 miles per week.

5. There are two bars that fit this description, namely 100–149 miles and 150–199 miles. Therefore, $15 + 9 = 24$ adults drive 100–199 miles per week.

9. $\dfrac{\text{number of adults who drive } 100-199 \text{ miles per week}}{\text{total number of adults}}$
$= \dfrac{9}{100}$

13. The height of the bar representing 55–64 years old is 17. Therefore, 17 million householders are 55–64 years old.

17. answers may vary

21.

Class Intervals (Scores)	Tally	Class Frequency (Number of Games)
90–99	ⅢⅢ ‖‖	8

25.

Class Intervals (Account Balances)	Tally	Class Frequency (Number of People)
$200–$299	ⅢⅢ ‖	6

29.

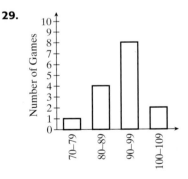

Golf Scores

33. $\dfrac{12 + 28 + 20}{3} = \dfrac{60}{3} = 20$

37. The greatest difference in the heights of the bars occurs in 1996. Thus, 1996 is the desired year.

Exercise Set 9.4

1. Mean:
$$\dfrac{21 + 28 + 16 + 42 + 38}{5} = \dfrac{145}{5} = 29$$
Median: Write the numbers in order.
16, 21, 28, 38, 48
The middle number is 28.
Mode: There is no mode, since each number occurs once.

5. Mean:
$$\frac{0.2 + 0.3 + 0.5 + 0.6 + 0.9 + 0.2 + 0.7 + 1.1}{9} = \frac{5.1}{9} \approx 0.6$$
0.2, 0.3, 0.5, 0.6, 0.9, 0.2, 0.7, 1.1
The middle number is 0.6.
Mode: Since 0.2 and 0.6 occur twice, there are two modes, 0.2 and 0.6.

9. Mean:
$$\frac{1483 + 1483 + 1450 + 1381 + 1283}{5} = \frac{7080}{5} = 1416 \text{ feet}$$

13. answers may vary

17. $\text{gpa} = \dfrac{4 \cdot 3 + 4 \cdot 3 + 3 \cdot 4 + 3 \cdot 1 + 3 \cdot 2}{3 + 3 + 4 + 1 + 2} = \dfrac{45}{13} \approx 3.46$

21. Mode: 6.9 since this score appears twice.

25. Mean:
$$\frac{\text{sum of 15 scores}}{15} = \frac{1095}{15} = 73$$

29. There are 9 scores (rates) lower than the mean. They are 66, 68, 71, 64, 71, 70, 65, 70, and 72.

33. $\dfrac{18}{30} = \dfrac{2 \cdot 3 \cdot 3}{5 \cdot 2 \cdot 3} = \dfrac{3}{5}$

37. Use the numbers 35, 35, 37, and 43.

Exercise Set 9.5

1.

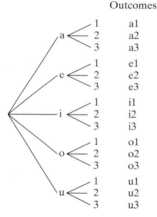

Outcomes

There are 15 outcomes.

5.

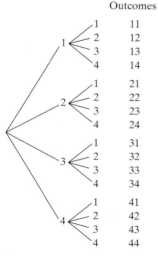

Outcomes

There are 16 outcomes.

9.

Outcomes

$$
\begin{array}{ccl}
 & 1 & \text{H1} \\
 & 2 & \text{H2} \\
\text{H} & 3 & \text{H3} \\
 & 4 & \text{H4} \\
 & 1 & \text{T1} \\
\text{T} & 2 & \text{T2} \\
 & 3 & \text{T3} \\
 & 4 & \text{T4} \\
\end{array}
$$

There are 8 outcomes.

13. probability $= \dfrac{2}{6} = \dfrac{1}{3}$

17. probability $= \dfrac{1}{3}$

21. probability $= \dfrac{1}{7}$

25. probability $= \dfrac{38}{200} = \dfrac{19}{100}$

29. $\dfrac{1}{2} + \dfrac{1}{3} = \dfrac{1}{2} \cdot \dfrac{3}{3} + \dfrac{1}{3} \cdot \dfrac{2}{2} = \dfrac{3}{6} + \dfrac{2}{6} = \dfrac{5}{6}$

33. $5 \div \dfrac{3}{4} = \dfrac{5}{1} \cdot \dfrac{4}{3} = \dfrac{5 \cdot 4}{1 \cdot 3} = \dfrac{20}{3}$ or $6\dfrac{2}{3}$

37. There are four Kings.
probability $= \dfrac{4}{52} = \dfrac{1}{13}$

41.

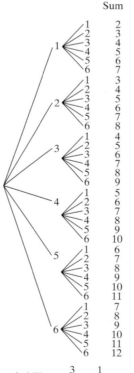

Sum

probability $= \dfrac{3}{36} = \dfrac{1}{12}$

Chapter 9 Test

1. The second week has $4\dfrac{1}{2}$ bills and each bill represents \$50.
So, $4.5 \cdot \$50 = \$225$ was collected during the second week.

5. Ford Focus had the least sales.

9. 1990, 1991, and 2000.

13. Services = 31% of 132,000,000
= (0.31)(132,000,000)
= 40,920,000

17.

Class Interval	Tally	Class Frequency
40–49	|	1
50–59	|||	3
60–69	||||	4
70–79	⊞	5
80–89	⊞ |||	8
90–99	||||	4

21.

Grade	Point Value	Credit Hours	$\left(\dfrac{\text{Point}}{\text{Value}}\right) \cdot \left(\dfrac{\text{Credit}}{\text{Hours}}\right)$
A	4	3	12
B	3	3	9
C	2	3	6
B	3	4	12
A	4	1	4
	Totals	14	43

Grade point average $= \dfrac{43}{14} \approx 3.07$

25. probability $= \dfrac{2}{10} = \dfrac{1}{5}$

Chapter 10

Exercise Set 10.1

1. If 0 represents ground level, then 1445 feet underground is -1445

5. If 0 represents the line of scrimmage, a loss of 15 yards is -15.

9. If 0 represents \$0, a loss of \$20,786 thousand is $-20,786$ thousand.

13. If 0 represents 0%, a loss of 91% is -91.

17.

21. $4 > 0$

25. $0 > -3$

29. $-4.6 < -27$

33. $\dfrac{1}{4} > -\dfrac{8}{11}$

37. $|-8| = 8$, because -8 is 8 units from 0.

41. $|-5| = 5$, because -5 is 5 units from 0.

45. $\left|\dfrac{9}{10}\right| = \dfrac{9}{10}$ because $\dfrac{9}{10}$ is $\dfrac{9}{10}$ units from 0.

49. $|7.6| = 7.6$ because 7.6 is 7.6 units from 0.

53. The opposite of -4 is $-(-4) = 4$.

57. The opposite of $-\dfrac{9}{16}$ is $-\left(-\dfrac{9}{16}\right) = \dfrac{9}{16}$.

61. The opposite of $\dfrac{17}{18}$ is $-\dfrac{17}{18}$.

65. $-|20| = {}^-20$

The opposite of the absolute value of 20 is the opposite of 20.

69. $-(-8) = 8$
The opposite of negative 8 is 8.

73. $-(-29) = 29$
The opposite of negative 29 is 29.

77.
$$\begin{array}{r} 15 \\ + 20 \\ \hline 35 \end{array}$$

81. True. Consider the values on a number line.

85. answers may vary

Exercise Set 10.2

1. $8 + 2 = 10$

5. $-13 + 7 = -6$

9. $|-6| + |-2| = 8$
Common sign is negative, so -8.

13. $|6| - |-2| = 4$
$6 > 2$, so answer is $+4$.

17. $|-5| - |3| = 2$
$5 > 3$ so answer is -2.

21. $|-12| + |-12| = 24$
The common sign is negative, so -24.

25. $|-123| + |-100| = 223$
The common sign is negative, so -223.

29. $|12| - |-5| = 7$
$12 > 5$, so answer is $+7$.

33. $|-12| - |3| = 9$
$12 > 3$, so answer is -9.

37. $|57| - |-37| = 20$
$57 > 37$, so answer is $+20$.

41. $|-6.3| + |-2.2| = 8.5$
Common sign is negative, so -8.5.

45. $\left|-\dfrac{2}{3}\right| + \left|-\dfrac{1}{6}\right| = \dfrac{4}{6} + \dfrac{1}{6} = \dfrac{5}{6}$

Common sign is negative, so $-\dfrac{5}{6}$.

49. $-4 + 2 + (-5) = -2 + (-5) = -7$

53. $12 + (-4) + (-4) + 12 = 8 + (-4) + 12$
$= 4 + 12$
$= 16$

57. $-10° + 12° = 2°$
A rise of 12° from $-10°$ Celsius puts the 11 P.M. temperature at 2°C.

61. Team 1: $(-2) + (-13) + 20 + 2$
$$= -15 + 20 + 2$$
$$= 5 + 2$$
$$= 7$$

Team 2: $5 + 11 + (-7) + (-3)$
$$= 16 + (-7) + (-3)$$
$$= 9 + (-3)$$
$$= 6$$

Team 1.7; Team 2.6; Winning team, Team 1

65. $-10,924 + 3245 = -7679$
The depth of the Pacific's Aleutian Trench is $-7679 meters$.

69. $52 - 52 = 0$

73. True. Add any two negative numbers on a number line to verify.

77. answers may vary

Exercise Set 10.3

1. $-5 - (-5) = -5 + 5 = 0$

5. $3 - 8 = 3 + (-8) = -5$

9. $-5 - (-8) = -5 + 8 = 3$

13. $2 - 16 = 2 + (-16) = -14$

17. $3.62 - (-0.4) = 3.62 + 0.4 = 4.02$

21. $\frac{2}{5} - \frac{7}{10} = \frac{2}{5} + \left(-\frac{7}{10}\right) = \frac{4}{10} + \left(-\frac{7}{10}\right) = -\frac{3}{10}$

25. $-20 - 18 = -20 + (-18) = -38$

29. $2 - (-11) = 2 + 11 = 13$

33. $12 - 5 - 7 = 12 + (-5) + (-7)$
$$= 7 + (-7)$$
$$= 0$$

37. $-10 + (-5) - 12 = -10 + (-5) + (-12)$
$$= -15 + (-12)$$
$$= -27$$

41. To find out how many degrees warmer, subtract the coldest temperature from the warmest temperature.
$136 - (-126) = 136 + 126 = 262$
$136°F$ is $262°F$ warmer than $-126°F$.

45. To find the difference in elevation, subtract the two depths.
$28,374 - 24,455 = 3919$ feet

49. $144 - 0 = 144$ feet

53. $-176 - (-215) = -176 + 215 = 39$
Saturn has the warmest average temperature because $-176 > -215$. It $39°C$ warner than Uranus.

57. $1 \cdot 8 = 8$

61. $|-3| - |-7| = 3 - 7 = 3 + (-7) = -4$

65. $|-17| - |-29| = 17 - 29 = 17 + (-29) = -12$

69. answers may vary

Exercise Set 10.4

1. $-2(-3) = 6$

5. $(2.6)(-1.2) = -3.12$

9. $-\frac{3}{5}\left(-\frac{2}{7}\right) = \frac{(-3) \cdot (-2)}{5 \cdot 7} = \frac{6}{35}$

13. $-1(-2)(-4) = 2(-4) = -8$

17. $10(-5)(0) = -50(0) = 0$

21. $(-3)^3 = (-3)(-3)(-3) = 9(-3) = -27$

25. $(-5)^3 = (-5)(-5)(-5) = 25(-5) = -125$

29. $\frac{3}{4}\left(-\frac{7}{8}\right) = \frac{3 \cdot (-7)}{4 \cdot 8} = -\frac{21}{32}$

33. $-1(2)(7)(-3.1) = -2(7)(-3.1)$
$$= -14(-3.1)$$
$$= 43.4$$

37. $-24 + 6 = -4$

41. $\frac{-88}{-11} = 8$

45. $\frac{39}{-3} = -13$

49. $-\frac{7}{12} \div \left(-\frac{1}{6}\right) = -\frac{7}{12} \cdot \left(-\frac{6}{1}\right) = \frac{-7 \cdot (-6)}{12 \cdot 1} = \frac{7}{2}$

53. $240 \div (-40) = -6$

57. $\frac{-120}{0.4} = -300$

61. $(-4)(3) = -12$ yards

65. $\frac{0 + (-5) + (-1) + (-1)}{4} = \frac{-7}{4} = -1.75$
Her average score per round was -1.75.

69. a. $27 - 35 = -8$
There was a change of -8 condors.

b. 1979 to 1987 is 8 years.
$$\frac{-8}{8} = -1$$
There was a change of -1 condor per year.

c. $184 - 27 = 157$
There was a change of 157 condors.

d. $\frac{157}{14} \approx 11.$
There was a change of 11 condors per year.

73. $90 + 12^2 - 5^3 = 90 + 12 \cdot 12 - 5 \cdot 5 \cdot 5$
$$= 90 + 144 - 125$$
$$= 234 - 125$$
$$= 109$$

77. False; The product of two negative numbers is always a positive number.

81. a. $27,148 - 29,380 = -2232.$
There was a change of -2232 Land Rovers.

b. $7(-2232) = -15,624$
The change will be $-15,624$ Land Rovers.

c. For the year 2006:
$7(-2232) + 29,380$
$$= -15,624 + 29,380 = 13,756 \text{ Land Rovers.}$$

Exercise Set 10.5

1. $-1(-2) + 1 = 2 + 1 = 3$

5. $9 - 12 - 4 = 9 + (-12) + (-4)$
$$= -3 + (-4)$$
$$= -7$$

9. $\frac{4}{9}\left(\frac{2}{10} - \frac{7}{10}\right) = \frac{4}{9}\left(-\frac{5}{10}\right) = \frac{4}{9}\left(-\frac{1}{2}\right) = -\frac{2}{9}$

13. $25 \div (-5) + 12 = -5 + 12 = 7$

17. $\dfrac{24}{10 + (-4)} = \dfrac{24}{6} = 4$

21. $(-19) - 12(3) = (-19) - 36$
$= (-19) + (-36)$
$= -55$

25. $[8 + (-4)]^2 = [4]^2 = 16$

29. $(3 - 12) \div 3 = (-9) \div 3 = -3$

33. $(5 - 9)^2 \div (4 - 2)^2 = (-4)^2 \div (2)^2$
$= 16 \div 4$
$= 4$

37. $(-12 - 20) \div 16 - 25 = (-32) \div 16 - 25$
$= -2 - 25$
$= -27$

41. $(0.2 - 0.7)(0.6 - 1.9) = (-0.5)(-1.3) = 0.65$

45. $(-36 \div 6) - (4 \div 4) = -6 - 1 = -7$

49. $(-5)^2 - 6^2 = 25 - 36 = -11$

53. $2(8 - 10)^2 - 5(1 - 6)^2 = 2(-2)^2 - 5(-5)^2$
$= 2(-2)(-2) - 5(-5)(-5)$
$= 2(-2)(-2) - 5(-5)(-5)$
$= (-4)(-2) + 25(-5)$
$= 8 + (-125)$
$= -117$

57. $\dfrac{(-7)(-3) - (4)(3)}{3[7 \div (3 - 10)]} = \dfrac{(-7)(-3) + (-4)(3)}{3[7 \div (3 + (-10))]}$
$= \dfrac{21 + (-12)}{3[7 \div (-7)]}$
$= \dfrac{9}{3(-1)}$
$= \dfrac{9}{-3}$
$= -3$

61. $45 \cdot 90 = 4050$

65. $P = 4(8 \text{ in.}) = 32 \text{ inches}$

69. $2 \cdot (7 - 5) \cdot 3 = 2 \cdot (2) \cdot 3 = 4 \cdot 3 = 12$

73. $(-12)^4 = (-12)(-12)(-12)(-12)$
$= 144(-12)(-12)$
$= -1728(-12)$
$= 20{,}736$

Chapter 10 Test

1. $-5 + 8 = 3$

5. $-\dfrac{3}{11} + \left(-\dfrac{5}{22}\right) = -\dfrac{6}{22} + \left(-\dfrac{5}{22}\right) = -\dfrac{11}{22} = -\dfrac{1}{2}$

9. $|-25| + (-13) = 25 + (-13) = 12$

13. $(-8) + 9 \div (-3) = -8 + (-3) = -11$

17. $\left(\dfrac{5}{9} - \dfrac{7}{9}\right)^2 + \left(-\dfrac{2}{9}\right) = \left(-\dfrac{2}{9}\right)^2 + \left(-\dfrac{2}{9}\right)$
$= \dfrac{4}{81} + \left(-\dfrac{2}{9}\right)$
$= \dfrac{4}{81} + \left(-\dfrac{18}{81}\right)$
$= -\dfrac{14}{81}$

21. $\dfrac{-3(-2) + 12}{-1(-4 - 5)} = \dfrac{-3(-2) + 12}{-1[-4 + (-5)]}$
$= \dfrac{6 + 12}{-1(-9)}$
$= \dfrac{18}{9}$
$= 2$

25. $237 - 157 - 17 + 38 = 80 - 77 + 38$
$= 3 + 38$
$= +41$
Her balance is $+41$ dollars.

Chapter 11

Exercise Set 11.1

1. $3 + 2z = 3 + 2(-3)$
$= 3 - 6$
$= -3$

5. $z - x + y = -3 - (-2) + 5$
$= -3 + 2 + 5$
$= 4$

9. $8 - (y - x) = 8 - [5 - (-2)]$
$= 8 - (5 + 2)$
$= 8 - 7$
$= 1$

13. $\dfrac{6xy}{z} = \dfrac{6 \cdot (-2) \cdot 5}{-3}$
$= \dfrac{-60}{-3}$
$= 20$

17. $\dfrac{x + 2y}{2z} = \dfrac{-2 + 2 \cdot 5}{2 \cdot (-3)}$
$= \dfrac{-2 + 10}{-6}$
$= \dfrac{8}{-6}$
$= -\dfrac{4}{3}$

21. $A = lw$
$= (50 \text{ feet}) \cdot (40 \text{ feet})$
$= 2000 \text{ square feet}$
The area is 2000 square feet.

25. $I = prt$
$= 3000 \cdot 0.06 \cdot 2$
$= 360$
It will earn \$360 in interest.

29. $F = \dfrac{9}{5}C + 32$
$= \dfrac{9}{5} \cdot (-5) + 32$
$= \dfrac{9 \cdot (-5)}{5 \cdot 1} + 32$
$= \dfrac{-45}{5} + 32$
$= -9 + 32$
$= 23$
The temperature is $23°$F.

33. $3x + 5x = (3 + 5)x$
$= 8x$

37. $4c + c - 7c = (4 + 1 - 7)c$
$$= -2c$$

41. $4a + 3a + 6a - 8 = (4 + 3 + 6)a - 8$
$$= 13a - 8$$

45. $3x + 7 - x - 14 = 3x - x + 7 - 14$
$$= 2x - 7$$

49. $6(5x) = (6 \cdot 5)x = 30x$

53. $2(y + 2) = 2 \cdot y + 2 \cdot 2 = 2y + 4$

57. $-4(3x + 7) = -4 \cdot 3x + (-4) \cdot 7$
$$= -12x - 28$$

61. $4(6n - 5) + 3n = 4 \cdot 6n + 4 \cdot (-5) + 3n$
$$= 24n - 20 + 3n$$
$$= 27n - 20$$

65. $-2(3x + 1) + 5(x - 2)$
$$= -2 \cdot 3x + (-2) \cdot 1 + 5 \cdot x + 5 \cdot (-2)$$
$$= -6x - 2 + 5x - 10$$
$$= -6x + 5x - 2 - 10$$
$$= -x - 12$$

69. $-13 + 10 = -3$

73. $-4 + 4 = 0$

77. $9684q - 686 - 4860q + 12.960$
$$= (9685 - 4860)q + (12,960 - 686)$$
$$= 4824q + 12,274$$

81. Add the areas of the two rectangles.
Area = (length) $\cdot$ (width)
$$\begin{pmatrix} \text{Area of the} \\ \text{rectangle} \\ \text{on the left} \end{pmatrix} + \begin{pmatrix} \text{Area of the} \\ \text{rectangle} \\ \text{on the right} \end{pmatrix}$$
$$= 7(2x + 1) + 3(2x + 3)$$
$$= 7(2x) + 7(1) + 3(2x) + 3(3)$$
$$= 14x + 7 + 6x + 9$$
$$= 20x + 16$$
The area is $(20x + 16)$ square miles.

Exercise Set 11.2

1. $x - 8 = 2$
$10 - 8 \overset{?}{=} 2$
$2 \overset{?}{=} 2$ True
Yes, 10 is a solution.

5. $7f - 8 = 64 - f$
$7(8) \overset{?}{=} 64 - 8$
$56 \overset{?}{=} 64 - 8$
$56 \overset{?}{=} 56$ True
Yes, 8 is a solution.

9. $a + 5 = 23$
$a + 5 - 5 = 23 - 5$
$a = 18$
Check: $a + 5 = 23$
$18 + 5 \overset{?}{=} 23$
$23 \overset{?}{=} 23$ True
The solution is 18.

13. $7 = y - 2$
$7 + 2 = y - 2 + 2$
$9 = y$
Check: $7 = y - 2$
$7 \overset{?}{=} 9 - 2$
$7 \overset{?}{=} 7$ True
The solution is 9.

17. $x + \dfrac{1}{2} = \dfrac{7}{2}$
$x + \dfrac{1}{2} - \dfrac{1}{2} = \dfrac{7}{2} - \dfrac{1}{2}$
$x = \dfrac{6}{2}$
$x = 3$
Check: $x + \dfrac{1}{2} = \dfrac{7}{2}$
$3 + \dfrac{1}{2} \overset{?}{=} \dfrac{7}{2}$
$\dfrac{6}{2} + \dfrac{1}{2} \overset{?}{=} \dfrac{7}{2}$
$\dfrac{7}{2} \overset{?}{=} \dfrac{7}{2}$ True
The solution is 3.

21. $x - 3 = -1 + 4$
$x - 3 = 3$
$x - 3 + 3 = 3 + 3$
$x = 6$
Check: $x - 3 = -1 + 4$
$6 - 3 \overset{?}{=} -1 + 4$
$3 \overset{?}{=} 3$ True
The solution is 6.

25. $x - 0.6 = 4.7$
$x - 0.6 + 0.6 = 4.7 + 0.6$
$x = 5.3$
Check: $x - 0.6 = 4.7$
$5.3 - 0.6 \overset{?}{=} 4.7$
$4.7 \overset{?}{=} 4.7$ True
The solution is 5.3.

29. $y + 2.3 = -9.2 - 8.6$
$y + 2.3 = -17.8$
$y + 2.3 - 2.3 = -17.8 - 2.3$
$y = -20.1$
Check: $y + 2.3 = -9.2 - 8.6$
$-20.1 + 2.3 \overset{?}{=} -9.2 - 8.6$
$-17.8 \overset{?}{=} -17.8$ True
The solution is -20.1.

33. $2 - 2 = 5x - 4x$
$0 = x$
Check: $2 - 2 = 5x - 4x$
$2 - 2 \overset{?}{=} 5(0) - 4(0)$
$0 \overset{?}{=} 0 - 0$
$0 \overset{?}{=} 0$ True
The solution is 0.

37. answers may vary

41. $\dfrac{1}{3} \cdot 3 = \dfrac{3}{3} = 1$

45. $x - 76,862 = 86,102$
$x - 76,862 + 76,862 = 86,102 + 76,862$
$x = 162,964$

47. Expenses = $\$15,326,552,000 - \$604,308,000$
$$= \$14,722,244,000$$

Exercise Set 11.3

1. $5x = 20$
$\dfrac{5x}{5} = \dfrac{20}{5}$
$x = 4$

5. $0.4y = 0$

$$\frac{0.4y}{0.4} = \frac{0}{0.4}$$

$$y = 0$$

9. $-0.3x = -15$

$$\frac{-0.3x}{0.3} = \frac{-15}{-0.3}$$

$$x = 50$$

13. $\frac{1}{6}y = -5$

$$6 \cdot \frac{1}{6}y = 6 \cdot (-5)$$

$$y = -30$$

17. $-\frac{2}{9}z = \frac{4}{27}$

$$-\frac{9}{2} \cdot \left(-\frac{2}{9}z\right) = -\frac{9}{2} \cdot \frac{4}{27}$$

$$z = -\frac{9 \cdot 4}{2 \cdot 27}$$

$$z = -\frac{2}{3}$$

21. $-\frac{3}{5}x = -\frac{6}{15}$

$$-\frac{5}{3} \cdot \left(-\frac{3}{5}x\right) = -\frac{5}{3} \cdot \left(-\frac{6}{15}\right)$$

$$x = \frac{-5 \cdot (-6)}{3 \cdot 15}$$

$$x = \frac{2}{3}$$

25. $16 = 10t - 8t$

$16 = 2t$

$$\frac{16}{2} = \frac{2t}{2}$$

$8 = t$

29. $4 - 10 = -3z$

$-6 = -3z$

$$\frac{-6}{-3} = \frac{-3z}{-3}$$

$2 = z$

33. $-36 = 9u + 3u$

$-36 = 12u$

$$\frac{-36}{12} = \frac{12u}{12}$$

$-3 = u$

37. $5 - 5 = 2x + 7x$

$0 = 9x$

$$\frac{0}{9} = \frac{9x}{9}$$

$0 = x$

41. $3x + 10 = 3(5) + 10$

$= 15 + 10$

$= 25$

45. $\dfrac{3x + 5}{x - 7} = \dfrac{3(5) + 5}{5 - 7}$

$$= \frac{15 + 5}{-2}$$

$$= \frac{20}{-2}$$

$$= -10$$

49. $-0.025x = 91.2$

$$\frac{-0.025x}{-0.025} = \frac{91.2}{-0.025}$$

$$x = -3648$$

53. $d = rt$

$294 = r \cdot 5$

$$\frac{294}{5} = \frac{r \cdot 5}{5}$$

$58.8 = r$

The driver should maintain a speed of 58.8 miler per hour.

Exercise Set 11.4

1. $2x - 6 = 0$

$2x - 6 + 6 = 0 + 6$

$2x = 6$

$$\frac{2x}{2} = \frac{6}{2}$$

$x = 3$

5. $6 - n = 10$

$6 - n - 6 = 10 - 6$

$-n = 4$

$$\frac{-n}{-1} = \frac{4}{-1}$$

$n = -4$

9. $3x - 7 = 4x + 5$

$3x - 7 + 7 = 4x + 5 + 7$

$3x = 4x + 12$

$3x - 4x = 4x + 12 - 4x$

$-x = 12$

$$\frac{-x}{-1} = \frac{12}{-1}$$

$x = -12$

13. $1.7 = 2y + 9.5$

$1.7 - 9.5 = 2y + 9.5 - 9.5$

$-7.8 = 2y$

$$\frac{-7.8}{2} = \frac{2y}{2}$$

$-3.9 = y$

17. $8 - t = 3$

$8 - t - 8 = 3 - 8$

$-t = -5$

$$\frac{-t}{-1} = \frac{-5}{-1}$$

$t = 5$

21. $2n + 8 = 0$

$2n + 8 - 8 = 0 - 8$

$2n = -8$

$$\frac{2n}{2} = \frac{-8}{2}$$

$n = -4$

25. $3r + 4 = 19$

$3r + 4 - 4 = 19 - 4$

$3r = 15$

$$\frac{3r}{3} = \frac{15}{3}$$

$r = 5$

29. $2 = 3z - 4$

$2 + 4 = 3z - 4 + 4$

$6 = 3z$

$$\frac{6}{3} = \frac{3z}{3}$$

$2 = z$

33.
$$-7c + 1 = -20$$
$$-7c + 1 - 1 = -20 - 1$$
$$-7c = -21$$
$$\frac{-7c}{-7} = \frac{-21}{-7}$$
$$c = 3$$

37.
$$4x + 3 = 2x + 11$$
$$4x + 3 - 3 = 2x + 11 - 3$$
$$4x = 2x + 8$$
$$4x - 2x = 2x + 8 - 2x$$
$$2x = 8$$
$$\frac{2x}{2} = \frac{8}{2}$$
$$x = 4$$

41.
$$-8n + 1 = -6n - 5$$
$$-8n + 1 - 1 = -6n - 5 - 1$$
$$-8n = -6n - 6$$
$$-8n + 6n = -6n - 6 + 6n$$
$$-2n = -6$$
$$\frac{-2n}{-2} = \frac{-6}{-2}$$
$$n = 3$$

45.
$$\frac{3}{8}x + 14 = \frac{5}{8}x - 2$$
$$\frac{3}{8}x + 14 - 14 = \frac{5}{8}x - 2 - 14$$
$$\frac{3}{8}x = \frac{5}{8}x - 16$$
$$\frac{3}{8}x - \frac{5}{8}x = \frac{5}{8}x - 16 - \frac{5}{8}x$$
$$-\frac{2}{8}x = -16$$
$$-\frac{8}{2} \cdot \left(-\frac{2}{8}x\right) = -\frac{8}{2} \cdot (-16)$$
$$x = 64$$

49.
$$3(x - 1) = 12$$
$$3x - 3 = 12$$
$$3x - 3 + 3 = 12 + 3$$
$$3x = 15$$
$$\frac{3x}{3} = \frac{15}{3}$$
$$x = 5$$

53.
$$35 = 17 + 3(x - 2)$$
$$35 = 17 + 3x - 6$$
$$35 = 11 + 3x$$
$$35 - 11 = 11 + 3x - 11$$
$$24 = 3x$$
$$\frac{24}{3} = \frac{3x}{3}$$
$$8 = x$$

57.
$$2t - 1 = 3(t + 7)$$
$$2t - 1 = 3t + 21$$
$$2t - 1 + 1 = 3t + 21 + 1$$
$$2t = 3t + 22$$
$$2t - 3t = 3t + 22 - 3t$$
$$-1t = 22$$
$$\frac{-t}{-1} = \frac{22}{-1}$$
$$t = -22$$

61.
$$10 + 5(z - 2) = 4z + 1$$
$$10 + 5z - 10 = 4z + 1$$
$$5z = 4z + 1$$
$$5z - 4z = 4z + 1 - 4z$$
$$z = 1$$

65.

the sum of −42 and 16	is	−26

$$-42 + 16 = -26$$
$$-42 + 16 = -26$$

69.

3 times	the difference of −14 and 2	amounts to	−48

$$3 \cdot \qquad (-14 - 2) \qquad = \qquad -48$$
$$3(-14 - 2) = -48$$

73. The height of the bar representing 2002 is approximately 50. Therefore, the number of returns filed electronically is 50 million.

77.
$$C = \frac{5}{9}(F - 32)$$
$$50.7 = \frac{5}{9}(F - 32)$$
$$50.7 = \frac{5}{9}F - \frac{160}{9}$$
$$50.7 + \frac{160}{9} = \frac{5}{9}F - \frac{160}{9} + \frac{160}{9}$$
$$\frac{456.3}{9} + \frac{160}{9} = \frac{5}{9}F$$
$$\frac{616.3}{9} = \frac{5}{9}F$$
$$\frac{9}{5} \cdot \frac{616.3}{9} = \frac{9}{5} \cdot \frac{5}{9}F$$
$$123.26 = F$$
The temperature is 123.26°F.

Exercise Set 11.5

1. $x + 5$

5. $20 - x$

9. $\dfrac{x}{2}$

13. $5x$

17. $50 - 8x$

21. $3x = 27$

25.
$$3x + 9 = 33$$
$$3x = 24$$
$$x = 8$$

29.
$$8 - x = \frac{15}{5}$$
$$8 - x = 3$$
$$-x = -5$$
$$x = 5$$

33.
$$3(x - 5) = \frac{108}{12}$$
$$3x - 15 = 9$$
$$3x = 24$$
$$x = 8$$

37. U.S. titles is x. International editions is $x + 86$.
$$x + (x + 86) = 118$$
$$2x + 86 = 118$$
$$2x = 32$$
$$x = 16$$
$$x + 86 = 102$$
Hearst Magazines published 16 U.S. titles.

41. Truck's speed is t. Car's speed is $2t$.
Combined speed is 105.
$$t + 2t = 105$$
$$3t = 105$$
$$t = 35$$
Truck's speed is 35 mph, and car's speed is $2 \cdot 35 = 70$ mph.

45. Purdue scored x points. Notre Dame scored $x + 2$ points.
$$x + (x + 2) = 134$$
$$2x + 2 = 134$$
$$2x = 132$$
$$x = 66$$
$$x + 2 = 68$$
Notre Dame scored 68 points.

49. 1026 rounded to the nearest hundred is 1000.

53. answers may vary

Chapter 11 Test

1.
$$\frac{3x - 5}{2y} = \frac{3(7) - 5}{2(-8)}$$
$$= \frac{21 - 5}{-16}$$
$$= \frac{16}{-16}$$
$$= -1$$

5. $A = lw$
$$= 3(3x - 1)$$
$$= 3 \cdot 3x + 3 \cdot (-1)$$
$$= 9x - 3$$
The area is $(9x - 3)$ square meters.

9.
$$-\frac{5}{8}x = -25$$
$$-\frac{8}{5} \cdot \left(-\frac{5}{8}x\right) = -\frac{8}{5} \cdot (-25)$$
$$x = 40$$

13.
$$-4x + 7 = 15$$
$$-4x + 7 - 7 = 15 - 7$$
$$-4x = 8$$
$$\frac{-4x}{-4} = \frac{8}{-4}$$
$$x = -2$$

17.
$$10y - 1 = 7y + 20$$
$$10y - 1 + 1 = 7y + 20 + 1$$
$$10y = 7y + 21$$
$$10y - 7y = 7y + 21 - 7y$$
$$3y = 21$$
$$\frac{3y}{3} = \frac{21}{3}$$
$$y = 7$$

21. $A = \frac{1}{2}bh$
$$= \frac{1}{2} \cdot 5 \cdot 12$$
$$= 30$$
The area is 30 square feet.

25. If there are x women entered, then there are $x + 112$ men entered.
$$x + (x + 112) = 600$$
$$2x + 112 = 600$$
$$2x + 112 - 112 = 600 - 112$$
$$2x = 488$$
$$x = 244$$
244 women entered the race.

SUBJECT INDEX

solving using addition and multiplication properties, 687–88, 704
solving with addition property of equality, 673–76, 687–88, 704
solving with multiplication property of equality, 704
steps for solving, 689
writing percent problems as, 353–54
writing sentences as, 689–90, 696
Equation-solving process, modeling with addition and subtraction, 672
Equilateral triangle, 480
Equivalencies
apothecaries' measures, 440
for dry capacity, 440
between units of capacity, 431
between units of length, 408, 409
Equivalent equations, 674
Equivalent fractions, 121, 123–24, 150, 213, 173–74
Estimation, problem solving by, 41–42
Evaluating the expression, 81–82, 703
Evaluating the expression for the variable, 664
Events, probability of, 586–87, 595
Expanded form, whole numbers written in, 11
Expenses, business, 646
Experiment, 585, 595
Exponential expressions, evaluating, 81–82
Exponential notation, using, 81
Exponents, 91, 214
defined, 81
evaluating expressions with, 199
evaluating with calculator, 84
and order of operations, 81–84
Expressions
algebraic, 663
simplifying those containing decimals, 268–69

F

Factorization, 115
Factors, 47
finding, 115
grouping of, 48
order of, 48
Factor trees, 118
Fahrenheit, Gabriel, 443
Fahrenheit degrees, converting to Celsius degrees, 444–45, 457
False proportions, 311, 326
False statements, as no solution of equation, 673
Fear of failure, 152
Feet, 405
converting to inches, 455
Ferris, George Washington Gale, 255
FG%. See Field goal percentage
FGA. See Field goals attempted
Field goal percentage, 352
Field goals attempted, 352
Financial information, 661
Flight time, calculating, 180
Floppy diskettes, storage capacity of, 418
Focus on business and career
Consumer Price Index, 330

in-demand occupations, 240
net income and net loss, 646
product packaging, 54
surveys, 556
Focus on history
development of units of measure, 426
inductive reasoning, 206
magic squares, 638
multicultural fraction use, 236
Pythagorean Theorem, 538
Focus on mathematical connections
modeling adding and subtracting integers, 628
modeling equation solving with addition and subtraction, 672
modeling fractions, 140
modeling multiplication with decimal numbers, 256
modeling subtraction of whole numbers, 32
stem-and-leaf displays, 570
Focus on real life
body dimensions, 320
Focus on the real world
computer storage capacity, 418
house affordability analysis, 340
judging distances, 92
misleading graphs, 562
mortgages, 374
road sign costs, 506
Foot, 405, 426
Foot-pound (ft-lb), converting BTU to, 449, 457
Ford, Henry, 335
Ford Motor Company, 335
Formulas. See also Symbols
area, 497–500, 535
circumference, 490–91
perimeter, 487–89, 535
volume, 507, 536
Fractional notation, 297
Fraction bar, 57, 91
simplifying expressions with, 645
Fractions, 101, 149
comparing, 148, 197–98
comparing decimals and, 276–77
converting, 347
decimals written as, 229–30
dividing, 141–42, 151
division of, and mixed numbers or whole numbers, 143
equivalent, 121, 123–24, 150, 173–74, 213
evaluating those raised to powers, 198
improper, 105, 149
like, 163
modeling, 140
multicultural use of, 236
multiplication of, and mixed numbers or whole numbers, 132–33
multiplying, 131–32, 151
percent written as, 335
problem solving by dividing, 143–44
problem solving steps with, 215
problem solving with, 203–5
proper, 105, 149
reciprocals of, 141
reviewing operations on, 199

simplest form of, 150
simplifying of, to 0 or 1, 103
solving area problems containing, 277
unit, 405, 406
unlike, 163
using, 212
writing as decimals, 230, 275–76, 285
writing as percents, 345–46
writing decimals as, 347
writing from real-life data, 105
writing in simplest form, 121, 122, 123
writing in simplest form and problem solving, 124
writing percent as, 345, 391
writing rates as, 303
writing ratios as, 297
writing to represent shaded areas of figures, 103–4
Freezing points, 443
French Academy of Sciences, 426
French Revolution, and metric system, 426
Frequency distribution table, histograms constructed from, 572

G

Gallons, 405, 431
Geometric figures, area formulas of, 497
Geometry, meaning of word, 467
Geometry icon, 4
Gigabytes, 418
Golf, 609
Grade point average (GPA), 579–80
Grams, 422, 423, 433, 456
converting to kilograms, 456
milligrams converted to, 423
Grand Canyon, 101
Graphing, signed numbers, 612
Graphs
bar, 551–54, 593
circle, 594
histogram, 571, 594
line, 554–55, 593
misleading, 562
pictograph, 551, 593
Greater number, 612
Greater than symbol, 197, 612
Great Pyramids, 467, 513
Greeks (ancient), magic squares understood by, 638
Grouping of factors, 48
Grouping symbols, 643
operations performed within, 82, 83
order of operations within, 91
performing operations on decimals within, 268

H

Hard drive, storage capacity of, 418
Health sciences, proportions used in, 317
Hebrews, units of measure used by, 426
Heights
comparing, 243–44
finding in meters, 411
Help, 4, 5. See also Study skills
Henry I (king of England), 426
Hershey, Milton S., 225
Hershey, Pennsylvania, 225

PHOTO CREDITS

TABLE OF CONTENTS   Chapter 1: Gerard Lacz/Peter Arnold, Inc.; Chapter 2: Gilda Schiff/Photo Researchers, Inc.; Chapter 3: Robert Frerck/Woodfin Camp & Associates; Chapter 4: Chuck Pefley/StockBoston; Chapter 5: Jane E. Cunningham Visuals Unlimited; Chapter 6: CORBIS; Chapter 7: Getty Images, Inc./PhotoDisc, Inc.; Chapter 8: Bill Gallery/StockBoston; Chapter 9: P.Lloyd/Weatherstock; Chapter 10: © John Zich/AFP/CORBIS; Chapter 11: Donovan Reese/Getty Images, Inc.

CHAPTER 1   Chapter Opener: Gerard Lacz/Peter Arnold, Inc.; **p. 3** Sepp Seitz; **p. 25** Tom Roberts/Getty Images, Inc.; **p. 35** PhotoDisc; **p. 36** Gordon E. Smith/Photo Researchers, Inc., SuperStock, Inc.; **p. 37** The Schiller Group, Ltd., Jerry L. Ferrara/Photo Researchers, Inc., PhotoDisc; **p. 44** Bruce Curtis/Peter Arnold, Inc.; **p. 54** Hyatt Corporation; **p. 71** Joseph R. Conner; **p. 76** Tony Freeman/PhotoEdit, Morton Beebe-S.F./CORBIS; **p. 77** Miro Vintoniv/StockBoston, Bree Underhill/Getty Images, Inc.; **p. 87** Jean-Claude LeJeune/StockBoston

CHAPTER 2   Chapter Opener: Gilda Schiff/Photo Researchers, Inc.; **p. 114** Margaret Miller/Photo Researchers, Inc.; **p. 126** Shuttle Mission Imagery/Getty Images, Inc., Kevin Horan/StockBoston; **p. 128** American Red Cross; **p. 133** Kevin Horan/Getty Images, Inc.; **p. 147** Kathryn M. Anderson

CHAPTER 3   Chapter Opener: Robert Frerck/Woodfin Camp & Associates; **p. 183** Richard Hutchings/PhotoEdit; **p. 193** Robert Harbison; **p. 194** Rev. Ronald Royer/Photo Researchers, Inc.; **p. 204** Edith Haun/StockBoston; **p. 208** Billy E. Barnes/StockBoston; **p. 209** Lehtikuva Oy/Woodfin Camp & Associates

CHAPTER 4   Chapter Opener: Chuck Pefley/StockBoston; **p. 228** AP/Wide World Photos; **p. 238** Jim Bourg/Getty Images, Inc.; **p. 263** Carl Wolinsky/StockBoston, Michael Newman/PhotoEdit; **p. 279** MarathonFoto, Kathryn M. Anderson

CHAPTER 5   Chapter Opener: Jane E. Cunningham/Visuals Unlimited; **p. 301** Bill Lai/The Image Works, SuperStock, Inc.; **p. 306** Grace Davies/Omni-Photo Communications, Inc., British Airways; **p. 308** Photo-Verlag Gyger/Switzerland Tourism; **p. 317** Michael Newman/PhotoEdit; **p. 318** PhotoDisc; **p. 319** StockBoston; **p. 322** PhotoDisc; **p. 332** Koichi Kamoshida/Getty Images, Inc.

CHAPTER 6   Chapter Opener: CORBIS; **p. 337** C.Borland/Getty Images, Inc./PhotoDisc, Inc.; **p. 377** PhotoDisc

CHAPTER 7   Chapter Opener: Getty Images, Inc./PhotoDisc, Inc.; **p. 409** PhotoDisc; **p. 422** PhotoDisc

CHAPTER 8   Chapter Opener: Bill Gallery/StockBoston; **p. 478** Brian K. Diggs/AP/Wide World Photos, Will & Deni McIntyre/Photo Researchers, Inc.; **p. 495** Tom Bean/DRK Photo; **p. 505** UPI/CORBIS; **p. 514** Bill Gallery/StockBoston; **p. 520** PhotoDisc; **p. 523** PhotoDisc; **p. 546** PhotoDisc

CHAPTER 9   Chapter Opener: P.Lloyd/Weatherstock; **p. 551** PhotoDisc; **p. 558** PhotoDisc; **p. 573** PhotoDisc; **p. 574** PhotoDisc; **p. 577** PhotoDisc; **p. 597** PhotoDisc; **p. 599** PhotoDisc; **p. 603** PhotoDisc; **p. 605** PhotoDisc

CHAPTER 10   Chapter Opener: © John Zich/AFP/CORBIS; **p. 627** PhotoDisc; **p. 631** PhotoDisc; **p. 638** The Bridgeman Art Library International Ltd.

CHAPTER 11   Chapter Opener: Donovan Reese/Getty Images, Inc.; **p. 671** Gary Gray/DRK Photo, Patti Murray/Earth Scenes; **p. 694** PhotoDisc